Selenium
Chemistry, Analysis, Function and Effects

Food and Nutritional Components in Focus

Series Editor:
Professor Victor R. Preedy, *School of Medicine, King's College London, UK*

Titles in the Series:
1: Vitamin A and Carotenoids: Chemistry, Analysis, Function and Effects
2: Caffeine: Chemistry, Analysis, Function and Effects
3: Dietary Sugars: Chemistry, Analysis, Function and Effects
4: B Vitamins and Folate: Chemistry, Analysis, Function and Effects
5: Isoflavones: Chemistry, Analysis, Function and Effects
6: Fluorine: Chemistry, Analysis, Function and Effects
7: Betaine: Chemistry, Analysis, Function and Effects
8: Imidazole Dipeptides: Chemistry, Analysis, Function and Effects
9: Selenium: Chemistry, Analysis, Function and Effects

How to obtain future titles on publication:
A standing order plan is available for this series. A standing order will bring delivery of each new volume immediately on publication.

For further information please contact:
Book Sales Department, Royal Society of Chemistry, Thomas Graham House, Science Park, Milton Road, Cambridge, CB4 0WF, UK
Telephone: +44 (0)1223 420066, Fax: +44 (0)1223 420247
Email: booksales@rsc.org
Visit our website at www.rsc.org/books

Selenium
Chemistry, Analysis, Function and Effects

Edited by

Victor R. Preedy
School of Medicine, King's College London, UK
Email: editor@publicationeditor.org.uk

ROYAL SOCIETY
OF CHEMISTRY

THE QUEEN'S AWARDS
FOR ENTERPRISE:
INTERNATIONAL TRADE
2013

Food and Nutritional Components in Focus No. 9

Print ISBN: 978-1-84973-891-0
PDF eISBN: 978-1-78262-221-5
ISSN: 2045-1695

A catalogue record for this book is available from the British Library

Published by The Royal Society of Chemistry,
Thomas Graham House, Science Park, Milton Road,
Cambridge CB4 0WF, UK

Registered Charity Number 207890

For further information see our web site at www.rsc.org

Preface

Recently, there have been major advances in our understanding of the chemistry and function of nutritional components. This has been enhanced by rapid developments in analytical techniques and instrumentation. Chemists, food scientists and nutritionists are, however, separated by divergent skills, and professional disciplines. Hitherto this transdisciplinary divide has been difficult to bridge.

The series *Food and Nutritional Components in Focus* aims to cover in a single volume the chemistry, analysis, function and effects of single components in the diet or its food matrix. Its aim is to link scientific disciplines so that information becomes more meaningful and applicable to health in general.

The series *Food and Nutritional Components in Focus* covers the latest knowledge base and has a structured format with major subsections covering

- Chemistry
- Analysis
- Function and Effects

In some books the section on Chemistry is also linked with Biochemistry and there are also additional sections such as Food Components in Context.

Each chapter has a novel cohort of features namely by containing:

- *Summary Points*
- *Key Facts* (areas of focus explained for the lay person)
- *Definitions of Words and Terms*

The series covers numerous classes of dietary components including, for example, minerals, vitamins, food additives, and so on. The chapters are written by national or international experts, specialists and leaders in the field.

Food and Nutritional Components in Focus No. 9
Selenium: Chemistry, Analysis, Function and Effects
Edited by Victor R. Preedy

Published by the Royal Society of Chemistry, www.rsc.org

In *Selenium* we have the following coverage: Section 1 Chemistry and Biochemistry has chapters on chemistry, biochemistry, molecular biology, selenoproteins, organoselenium derivatives, selenium dioxide, selenite ions, selenium species and fish. In Section 2 Analysis there are chapters on analytical strategies, techniques, selenoproteins, selenocysteine, selenomethionine, foodstuffs, 2D size-exclusion reversed-phase HPLC-ICP-MS, solid-phase extraction, atomic absorption spectrometry, selenometabolites, animals, plants, bioavailability, *in vivo* and *in vitro* testing, dietary supplements, Fourier transform infrared spectra and assessment by questionnaires. Section 3 Function and Effects covers functional foods, parenteral nutrition, biofortification, fruit, vegetables, legumes, semiarid conditions, saliva, selenohomolanthionine, medicine, nutrition, selenium cocktails, lycopene, cancers, skin cancer, cancer metastasis, cancer-prevention trials, autoimmune thyroiditis, brain, neuronal cells, vascular function, heart, Kashin–Beck disease, placenta, mitochondria, oxidative stress and microgels.

Selenium is specifically designed for chemists, analytical scientists, forensic scientists, food scientists, dieticians, nutritionists, food scientists and research academics. The series is also designed to be suitable for lecturers and teachers in food and nutritional sciences. Additionally and importantly, the series is also designed for college or university libraries as a reference guide.

Professor Victor R. Preedy
King's College London

Contents

Chemistry and Biochemistry

Food and Nutritional Components in Focus No. 9
Selenium: Chemistry, Analysis, Function and Effects
Edited by Victor R. Preedy

Published by the Royal Society of Chemistry, www.rsc.org

Analysis

Function and Effects

Section I

Chemistry and Biochemistry

CHAPTER 1

The Chemistry of Selenium

DANIEL PERRONE*[a], MARIANA MONTEIRO[b], AND JULIANA CÔRTES NUNES[c]

[a]Laboratório de Bioquímica Nutricional e de Alimentos, Departamento de Bioquímica, Instituto de Química, Universidade Federal do Rio de Janeiro, Av. Athos da Silveira Ramos 149, CT, Bloco A, sala 528A, 21941-909, Rio de Janeiro, Brazil; [b]Departamento de Nutrição Básica e Experimental, Instituto de Nutrição Josué de Castro, Universidade Federal do Rio de Janeiro, Av. Carlos Chagas Filho, 373, CCS, bloco J, 2° andar, sala 16, 21941-902, Rio de Janeiro, Brazil; [c]Departamento de Tecnologia de Alimentos, Escola de Nutrição, Universidade Federal do Estado do Rio de Janeiro, Av. Pasteur, 296, 3° andar, 22290-240, Rio de Janeiro, Brazil
*E-mail: danielperrone@iq.ufrj.br

1.1 Chemical Properties of Selenium

Selenium was discovered in 1818 by Jöns Jacob Berzelius, a Swedish chemist who also isolated and described the elements cerium (1803), zirconium (1824), silicon (1824) and thorium (1828). The name selenium derives from the Greek word for "moon", *selene*, in opposition to the Latin word for "earth", *tellus*. Selenium is a nonmetal with symbol Se and atomic number 34, located in the forth period of the periodic table in the chalcogen group, along with oxygen, sulfur and tellurium. Its atomic and physical properties are listed in Table 1.1.

In nature, selenium is found in ore minerals, partially substituting sulfur. Selenium can exist as multiple allotropes (Butterman and Brown, 2004)

Food and Nutritional Components in Focus No. 9
Selenium: Chemistry, Analysis, Function and Effects
Edited by Victor R. Preedy

Published by the Royal Society of Chemistry, www.rsc.org

Table 1.1 Atomic and physical properties of selenium.

Atomic number	34
Atomic weight	78.963 Da
Electronic configuration	[Ar] $3d^{10}$ $4s^2$ $4p^4$
Atomic radius	103 pm
Covalent radius	116 pm
van der Waals radius	190 pm
Common oxidation states	−2, 0, +4, +6
M–M bond energy	44 kcal mol^{-1}
M–H bond energy	67 kcal mol^{-1}
Ionization potential	1st: 941.0 kJ mol^{-1} 2nd: 2045.0 kJ mol^{-1} 3rd: 2973.7 kJ mol^{-1}
Electron affinity	−4.21 eV
Electronegativity (Pauling's scale)	2.55
Melting point	494 K, 221 °C
Boiling point	958 K, 685 °C

that is, it can exist in essentially different molecular forms with varying physical properties. Both amorphous (without crystalline shape) and crystalline varieties of selenium occur. Amorphous selenium is best known as the "vitreous" and the finely divided brick red forms, which are frequently described as two distinct allotropes. These forms are, however, identical at the microscopic level and their distinct appearance is only due to different routes of production. The crystalline allotropes include several monoclinic varieties, red to brown in color, as well as the so-called "metallic" gray or black selenium. The most thermodynamically stable allotrope is the hexagonal, crystalline form to which both major monoclinic forms convert at temperatures above 110 °C and to which amorphous selenium converts spontaneously at 70–120 °C.

Elemental selenium is relatively nontoxic and is considered to be an essential trace element. Commonly, it can be reduced to the −2 (selenide, Se^{-2}) oxidation state or oxidized to the +4 (selenite, SeO_3^{2-}) or +6 (selenate, SeO_4^{2-}) oxidation states. Hydrogen selenide (H_2Se) and other selenium compounds are extremely toxic (LD_{50} 6 ppm for guinea pigs), and resemble arsenic in their physiological reactions. Selenium in the −2 oxidation state is the dominant form in organoselenium chemistry. Selenide is an important metabolite in animals and some micro-organisms, since it is the precursor of selenocysteine at the active centers of many selenoenzymes (Rayman, 2004). Going down in the chalcogen group, bond strength becomes increasingly weaker (341 kJ mol^{-1}, 272 kJ mol^{-1} and 234 kJ mol^{-1} for the C–O, C–S and C–Se bonds, respectively) and longer (141 pm, 181 pm and 198 pm for the C–O, C–S and C–Se bonds, respectively). Selenium compounds are more nucleophilic and acidic than the corresponding sulfur compounds.

The selenium standard atomic mass is 78.963 Da and shows six naturally occurring isotopes, five of which are stable: ^{74}Se, ^{76}Se, ^{77}Se, ^{78}Se, ^{80}Se (Rosman and Taylor, 1998; Jörg *et al.*, 2010). The other naturally occurring isotope,

Table 1.2 Selenium naturally occurring isotopes data. (Number in parentheses represents the uncertainties of the last digits as reported in the literature.)

Isotope	Atomic mass (Da)	Representative isotopic composition (atom %)	Nuclear spin (I)
^{74}Se	73.9224746 (16)	0.89 (4)	0
^{76}Se	75.9192120 (16)	9.37 (29)	0
^{77}Se	76.9199125 (16)	7.63 (16)	½
^{78}Se	77.9173076 (16)	23.77 (28)	0
^{80}Se	79.9165196 (19)	49.61 (41)	0
^{82}Se	81.9166978 (23)	8.73 (22)	0

^{82}Se, is considered to be stable for practical reasons since it has a very long half-life (approximately 10^{20} years), yielding ^{82}Kr *via* a double beta decay. Data regarding selenium naturally occurring isotopes are listed in Table 1.2. Although many studies have used repletion of tissue selenium or glutathione peroxidase activity as criteria for determination of bioavailability, most of the selenium stable isotopes have been employed for the study of its role as an essential nutrient. These isotopes may be produced as highly enriched materials at an affordable cost (Young *et al.*, 1982) and may be analyzed by mass spectrometry or neutron activation analysis. More recently, stable ^{77}Se and ^{82}Se were employed to estimate selenium bioavailability to men (Fox *et al.*, 2004). This study was notable for the use of foods intrinsically labeled with ^{77}Se stable isotope, as well as for the double-label method (^{82}Se as selenate) that allowed each individual to serve as their own control.

So far, twenty-four unstable selenium isotopes have been described (Audi *et al.*, 2003), showing half-lives ranging from approximately 20 ms to 295 000 years (Table 1.3). Selenium radioactive isotopes are mainly used for the production of medical and industrial bromine radioisotopes. Since some selenium radioactive isotopes show a very long half-life, their study is of great relevance for dose calculations performed in the frame of geological disposal of long-lives radioactive waste.

The selenium valence electronic configuration ([Ar] $3d^{10}$ $4s^2$ $4p^4$) is equivalent to that of sulfur ([Ne] $3s^2$ $3p^4$), which explains their similar atomic and physicochemical properties such as atomic sizes (both in covalent and ionic states), bond energies, ionization potentials and electronegativities. Despite these similarities, selenium compounds differ from sulfur compounds in two important chemical features, which are pivotal to differentiate their role in biological systems: their redox behavior and their acidity. Selenium compounds tend to be metabolized to more reduced states while sulfur compounds show an opposite behavior, tending to be metabolized to more oxidized states. Moreover, although their oxyacids show similar acid strengths (pK_a 2.6 and pK_a 1.9 for species with Se^{+4} and S^{+4}; pK_a 3.0 for both species with Se^{+6} and S^{+6}), H_2Se is much more acidic (pK_a 3.8) than H_2S (pK_a 7.0). This difference in hydrides acidity is reflected in the dissociation behaviors of the selenohydryl group (−SeH) of selenocysteine (pK_a 5.24) and the sulfhydryl group (−SH) of cysteine (pK_a 8.25). Consequently, selenocysteine and other

Table 1.3 Selenium radioactive isotopes data.

Isotope	Atomic mass (Da)	Half-life[a]	Mode of decay[b]
^{65}Se	64.96466	<50 × 10^{-3} s	β^+ to ^{65}As
^{66}Se	65.95521	33 × 10^{-3} s	β^+ to ^{66}As
^{67}Se	66.95009	133 × 10^{-3} s	β^+ to ^{67}As
^{68}Se	67.94180	35.5 s	β^+ to ^{68}As
^{69}Se	68.93956	27.4 s	β^+ to ^{69}As
^{70}Se	69.93339	41.1 min	β^+ to ^{70}As
^{71}Se	70.93224	4.74 min	β^+ to ^{71}As
^{72}Se	71.92711	8.40 days	EC to ^{72}As
^{73}Se	72.92676	7.15 h	β^+ to ^{73}As
^{75}Se	74.92252	119.779 days	EC to ^{75}As
^{79}Se	78.91850	295 000 years	β^+ to ^{79}As
^{81}Se	80.91799	18.45 min	β^- to ^{81}Br
^{83}Se	82.91912	22.3 min	β^- to ^{83}Br
^{84}Se	83.91845	3.1 min	β^- to ^{84}Br
^{85}Se	84.92225	31.7 s	β^- to ^{85}Br
^{86}Se	85.92427	15.3 s	β^- to ^{86}Br
^{87}Se	86.92852	5.5 s	β^- to ^{87}Br
^{88}Se	87.93142	1.53 s	β^- to ^{88}Br
^{89}Se	88.93645	410 × 10^{-3} s	β^- to ^{89}Br
^{90}Se	89.93996	~300 × 10^{-3} s	ND
^{91}Se	90.94596	270 × 10^{-3} s	β^- to ^{91}Br
^{92}Se	91.94992	~100 × 10^{-3} s	ND
^{93}Se	92.95629	~50 × 10^{-3} s	ND
^{94}Se	93.96049	~20 × 10^{-3} s	ND

[a]Units: s = seconds; min = minutes; h = hours; d = days; y = years.
[b]β^+ = beta plus decay (positron emission); β^- = beta minus decay (electron emission); EC = electron capture; ND = not determined.

selenols are mainly dissociated at physiological pH, while cysteine and other thiols are mainly nondissociated or protonated under the same conditions (Jacob *et al.*, 2003).

1.2 Organoselenium Compounds

Selenium occurs in different chemical forms, which determine its bioavailability, function and toxicity to human health. While inorganic selenium can be found in different minerals in the form of selenites and selenates, main organic forms are selenoaminoacids, selenopeptides and selenoproteins (Maseko *et al.*, 2013). Thus, the physiology or the biochemistry of organic selenium is mediated basically by its incorporation into selenoproteins. Indeed, the majority of selenium in plants is found in selenoaminoacids, such as selenomethionine and methylselenocysteine (Dumont *et al.*, 2006; Figure 1.1).

Organoselenium compounds can be defined as chemical compounds that contain both selenium and carbon. Other elements, such as halogens, oxygen, sulfur, or nitrogen are also frequently present in their structure.

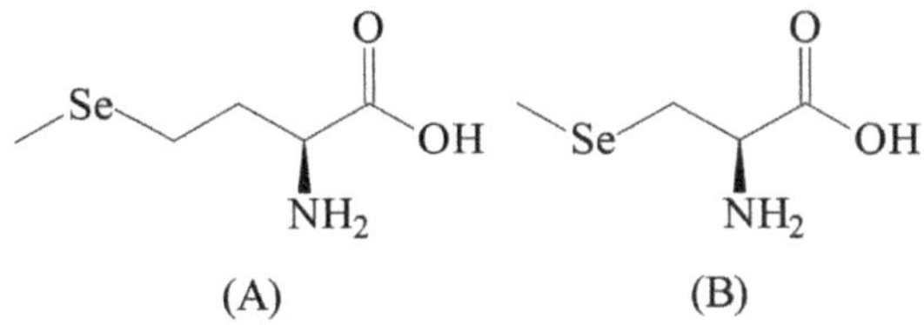

Figure 1.1 The chemical structures of selenomethionine (A) and methylselenocysteine (B).

In general, organoselenium compounds have greater bioavailability than that of inorganic selenium and also are usually found to be less toxic than the inorganic forms (Amoako *et al.*, 2009). In addition to their low toxicity, organoselenium compounds play important roles in biochemical processes such as antioxidants, anticancer and antiviral agent (Mugesh *et al.*, 2001; Soriano-Garcia, 2004). For this reason, in the last few years, a number of novel synthetic organoselenium compounds have been synthesized. Experimental evidence suggests that modification of organoselenium compounds can have a profound effect on their chemistry and consequently on their biological activity (Plano *et al.*, 2010).

The main classes of organoselenium compounds include selenols, diselenides, selenoxides, selenones, selenium acids, selenides, selenium halides and selenaheterocyclic compounds (Table 1.4). As mentioned before, the chemical properties of organoselenium compounds resemble their sulfur analogs and may be classified similarly. Therefore, selenols, selenides, and selenoxides correspond to thiols, sulfides, and sulfoxides. Despite their structural similarities, sulfur and selenium compounds are often different with respect to their stability, chemical properties and route of synthesis. Owing to the stronger metallic properties of selenium as compared to those of sulfur, the C—Se bond is more easily broken by halogens and oxidizing agents than is the C—S bond. For instance, Briviba *et al.* (1996) demonstrated that organoselenium compounds protected against peroxynitrite and that this protection was more efficient than that exerted by the corresponding sulfur analogs. This selectivity is attributed to a more easily oxidation of selenium in comparison to sulfur.

Selenols are one of the main classes of organoselenium compounds, and the most important selenol is the amino acid selenocysteine. Various enzymes found in mammals such as glutathione peroxidase, iodothyronine deiodinase and thioredoxin reductase contain selenols at their active sites (Pappas *et al.*, 2008). Selenols are strong nucleophiles that react with alkyl halides and epoxides yielding selenides (Back, 2011). They are also strong reducing agents that can convert N—O and N—N compounds into amines, dihalides into alkenes, and are also readily oxidized to diselenides by exposure to oxygen. Diselenides, such as diphenyl diselenide, have received significant attention, particularly with the discovery that these compounds possess a mode of antioxidant action similar to that of glutathione peroxidase

Table 1.4 The general formula and chemical structures of the major organoselenium compounds.

Class of compound	General formula	Chemical structure (example)
Selenols	R–SeH	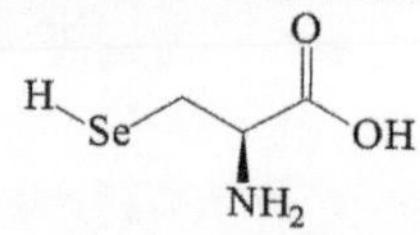 Selenocysteine
Selenides	R–Se–R	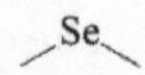 Dimethylselenide
Diselenides	R–Se–Se–R	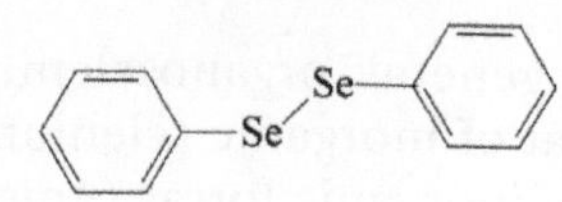 Diphenyldiselenide
Selenoxides	R–Se(O)–R	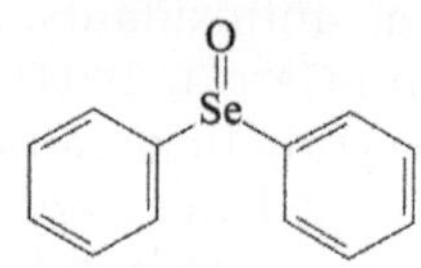 Phenylselenoxide
Selenones	R–Se(O)$_2$R	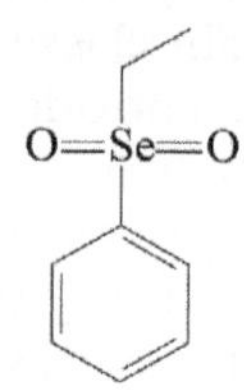 Ethylselenoyl benzene
Seleniumacids		
Seleninic acid	R–Se–(O)OH	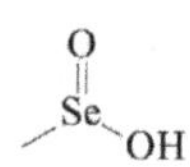 Methyl seleninic acid
Selenenic acid	RSe–OH	
Selenonic acid	RSe(O)$_3$H	
Selenocyanates	R–Se–CN	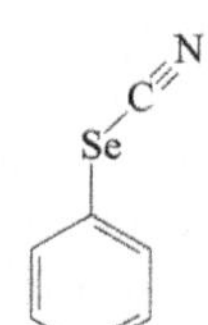 Phenylselenocyanate
Selenenyl halides	R–Se–Cl R–Se–Br	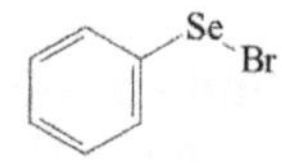 Phenylselenyl bromide
Selenaheterocyclic compounds		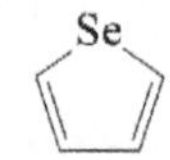 Selenophene

(Rosa *et al.*, 2007). Diselenides are usually prepared by either of two types of reaction: one in which the diselenide group is directly introduced into the molecule, or another in which a selenium-containing compound is oxidized or reduced to diselenide (Campbell *et al.*, 1952). More recently, Freudendahl *et al.* (2009) reported the synthesis of diselenides by reduction of aldehydes or imines in the presence of carbon monoxide. Diselenides can be also further reduced to selenols, oxidized to selenenic, seleninic, or selenonic acids. Finally, halogenation of diselenides leads to selenenyl halides.

Selenenic acids are usually unstable and disproportionate into mixtures of diselenides and seleninic acids. Seleninic acids are typically more stable, and the corresponding aryl derivatives serve as useful oxidizing agents toward a variety of organic functional groups (Back, 2011).

When selenium forms covalent bonds with both a hydrocarbon substituent and a carbonitrile group, selenocyanates are obtained. Selenocyanates have been synthesized by reaction of substrates with potassium selenocyanate, a product commercially available on large scale. The most usual reactions involving this reagent are the nucleophilic substitutions of halogen atoms or other leaving groups as tosylates, the reaction with diazonium salts and the addition on carbon double or triple bond (Jean-Claude, 2011). Selenocyanates can also be synthesized by the reaction of diselenides with trialkylsilylcyanide or metallic cyanides (Jean-Claude, 2011). Among the synthetic selenium derivatives, selenocyanates are also an important class that acts as bioactive compounds, showing antioxidative, antimutagenic and anticarcinogenic properties (Das *et al.*, 2007).

Selenaheterocyclic compounds demonstrates potential effects against various diseases acting as antioxidants, enzyme inhibitors, anti-inflammatory and anti-infective agents (Mugesh *et al.*, 2001). These compounds are key intermediates employed in the synthesis of alkenes, alkynes and selenium-free heterocycles (Młochowski *et al.*, 2007). The structures of selenaheterocyclic compounds are closely related to those of analogous sulfur compounds, but their properties are often distinct. In most selenaheterocyclic rings selenium is bonded to carbon or to another heteroatom, such as Se, O, N or S. Some selenaheterocyclic systems containing Se—Te, Se—P, Se—B, Se—Zn or Se—Sn bonds have also been reported (Młochowski *et al.*, 2007).

Selenides are the most prevalent organoselenium compounds. They may be obtained by synthetic methods or occur as byproducts of other reactions. Selenides are generally stable compounds that can be alkylated to afford selenonium salts or oxidized to either the corresponding selenoxides or selenones.

1.3 Selenium in Soil and Foods

The content of selenium in foods varies according to geographical origin both within and between countries (Navarro-Alarcon and Cabrera-Vique, 2008). Selenium levels in plants are directly affected by selenium content in the soil. In animal products, these levels are directly related to their diet (Navarro-Alarcon and Cabrera-Vique, 2008; Rayman *et al.*, 2008).

The distribution of selenium is uneven over the Earth's surface. Seleniferous and selenium deficient geo-ecosystems can be formed within limited geographic zones (Tan *et al.*, 2002). Several areas, such as the Enshi province (China), the great plains of the USA and Canada and portions of Ireland, Colombia and Venezuela are known seleniferous areas and have high soil concentrations of selenium (>5 ppm) (Tan *et al.*, 2002; Temmerman *et al.*, 2014; Zhang *et al.*, 2014). However, areas of low selenium (<0.05 ppm) or selenium-deficient soil are more common than areas of selenium-rich soil and according to the World Health Organization includes over 40 countries, such as China (Keshan area), Denmark, Finland, New Zealand and Russia (eastern and central Siberia) (Li *et al.*, 2007).

Selenium levels in soil are not the only factor that determines the contents of selenium in foods. Different physicochemical conditions of the soils such as pH and redox potential are also important (Rayman, 2004; Spadoni *et al.*, 2007; Zhao *et al.*, 2007). The two major forms of selenium in soil are selenate and selenite. While selenite is the major species in-well aerated and neutral to alkaline soils, selenite is found in higher levels in well-drained mineral soils with pH from acid to neutral (Rayman, 2008). These characteristics are important because transport of selenite is both lower and slower than that of selenite (Spadoni *et al.*, 2007; Zhu *et al.*, 2009).

In soils, selenite is less bioavailable than selenate because it is strongly absorbed by iron oxides and/or hydroxides (Spadoni *et al.*, 2007). In acid soils selenium is usually bound as a basic ferric selenite of extremely low solubility and it is essentially unavailable to plants. In alkaline soils selenium may be oxidized to selenate ions and become water soluble and readily available to plants (Rayman, 2008). However, as it is not easy to increase the pH of sandy soils extensively, fertilizers enriched with selenium could be an option to increase crop selenium concentrations (Temmerman *et al.*, 2014).

A further important factor is that plant species differ in their ability to assimilate and accumulate selenium and can be divided into three groups: nonaccumulators, secondary selenium accumulators and selenium accumulators (Broadley *et al.*, 2006; Rayman, 2008). The only selenium-accumulator plant that is used as a food source is *Bertholletia excelsa*, a tree that produces Brazil nuts, considered the richest selenium food source (800 to 8300 μg 100 g^{-1}) (Thomson *et al.*, 2008). *Brassica* (Indian mustard, rapeseed, broccoli, and cabbage), *Allium* (garlic, onions, leeks and wild leeks) and some mushrooms species of commercial importance can be described as secondary selenium accumulators (Broadley *et al.*, 2006; Navarro-Alarcon and Cabrera-Vique, 2008). Cereal crops such as wheat, oats, rye and barley are nonaccumulators (Rayman, 2008) and therefore are not sources of selenium.

According to the literature, the typical selenium contents in foods varies as follows: organ meats, meats, eggs and seafood, 12.5 to 311.5 μg per 100 g; dairy products, less than 0.1 to 30 μg per 100 g; cereals, 1.0 to 55.0 μg per 100 g, whereas selenium in bread is generally found at levels of 6.0 to 16.0 μg per 100 g; fruits and vegetables, 0.1 to 2.2 μg per 100 g (Dumont *et al.*, 2006; Navarro-Alarcon and Cabrera-Vique, 2008; Rayman, 2008). The main forms of selenium in plants and animal foods are selenomethionine, selenocysteine, methylselenocysteine and selenate. Selenomethionine is predominantly in cereals, methylselenocysteine in Brazil nuts and *Brassica* and *Allium* species and selenomethionine and selenate in animal foods (Dumont *et al.*, 2006; Navarro-Alarcon and Cabrera-Vique, 2008; Rayman *et al.*, 2008).

Food processing such as cooking reduces the selenium contents of most foods (Dumont *et al.*, 2006). For example, vegetables such as asparagus and mushrooms can lose 40% during boiling for some minutes and other studies have estimated that 50% of the selenium content is lost from vegetables and dairy products during cooking especially in low pH (WHO *et al.*, 1996; Dumont *et al.*, 2006; Navarro-Alarcon and Cabrera-Vique, 2008). However, other researchers even reported that processes such as cooking, aeration or lyophilization significantly increases selenium content in all food (Navarro-Alarcon and Cabrera-Vique, 2008).

Summary Points

- Selenium is a chalcogen element with atomic number 34.
- Selenium's most stable allotrope is the hexagonal and crystalline form.
- Three main oxidation states of selenium are found: selenide (−2), selenite (+4) and selenate (+6).
- Selenium's atomic mass is 78.963 and shows six stable and twenty-four radioactive isotopes.
- Sulfur and selenium show equivalent valence configuration but in biological systems selenium tends to be reduced while sulfur tends to oxidize.
- While selenols are dissociated at physiological pH, thiols are mainly nondissociated.
- Organoselenium compounds such as selenoaminoacids are of major biological relevance due to their role in active centers of selenoenzymes.
- Selenols are strong nucleophiles that yield selenides when reacted with alkyl halides and epoxides.
- The content of selenium in plants depends on soil features (selenium levels, pH and redox potential) and species.
- Plant species may be classified as nonaccumulators, secondary selenium accumulators and selenium accumulators.

Keys Facts of Selenoproteins

1. Selenoproteins are proteins that specifically incorporate selenocysteine into their amino acid chain.
2. The biological functions of selenium are exerted by selenoproteins.
3. Although 30 selenoproteins have been identified in mammalian systems, it is expected that up to 100 may exist.
4. Selenoproteins exist in archaea, bacteria and eukaryotes with selenocysteine as a key catalytic group.

5. Approximately 60% of selenium in plasma is incorporated in selenoprotein P.
6. The glutathione peroxidades enzymes were the first class of selenoproteins to be characterized and represent the major class of functionally important selenoproteins. They catabolize hydroperoxides and thereby have antioxidant functions.
7. Other selenoproteins include thioredoxin reductases that reduce thioredoxin and thereby regulate cellular protein thiol redox status.
8. Three iodothyronine deiodinases have been described. Expression of these selenoenzymes plays a role in embryonic development and in regulation of metabolic rate.

Definitions and Explanations of Key Terms

Allotropy. Chemical property of some elements, especially those found naturally as solids, to be found in nature in two or more structurally different forms in the same physical state.

Beta decay. Physical phenomenon observed when the neutron to proton ratio is either too great (β^- decay; electron emission) or too small (β^+ decay; positron emission).

Bioavailability. Fraction of a nutrient or non-nutrient present in food that is digested, absorbed, metabolized, transported and exerts its biological role.

Electron capture. Radioactive decay in which an electron is absorbed by the nucleus that simultaneously emits an electron neutrino. Observed in atom nuclei with an abundance of protons but not enough energy to undergo a positron emission.

Ionization energy. Corresponds to the minimum energy (usually in kJ mol^{-1}) needed to remove the outermost electron of an atom or molecule in the ground state.

Lyophilization. Dehydration process, also known as freeze drying, in which water is removed from food and other products by sublimation in a vacuum chamber.

Nuclear spin. Intrinsic quantum mechanical property that represents the total angular momentum of a nucleus.

Nucleophilicity. Ability of an electron-rich chemical species to donate two electrons forming a covalent bond with a positive nuclear charge of an electron-poor species, known as an electrophile.

Oxidation state. Number assigned to an element that represents the number of electrons that are gained, lost or shared when a chemical bond is formed with another element.

Radioisotopes. Isotopes that have an unstable nucleus and therefore emit particles and energy to reach a more stable balance between protons and neutrons.

Selenoaminoacids. Amino acids such as selenocysteine and selenomethionine that contain selenium instead of sulfur in their structure.

List of Abbreviations

LD_{50} Median lethal dose
pK_a Acid dissociation constant

References

Amoako, P. O., Uden, P. C. and Tyson, J. F., 2009. Speciation of selenium dietary supplements, formation of *S*-(methylseleno)cysteine and other selenium compounds. *Analytica Chimica Acta*. 652: 315–323.

Audi, G., Bersillon, O., Blachot, J. and Wapstra, A. H., 2003. The NUBASE evaluation of nuclear and decay properties. *Nuclear Physics A*. 729: 3–128.

Back, T. G., 2011. Selenium: organoselenium chemistry. In: *Encyclopedia of Inorganic and Bioinorganic Chemistry*.

Briviba, K., Roussyn, I., Sharov, V. S. and Sies, H., 1996. Attenuation of oxidation and nitration reactions of peroxynitrite by selenomethionine, selenocystine and ebselen. *Biochemical Journal*. 319: 13–15.

Broadley, M. R., White, P. J. and Bryson, R. J., 2006. Biofortification of UK food crops with selenium. *Proceedings of the Nutrition Society*. 65: 169–181.

Butterman, W. C. and Brown, R. D., 2004. *Mineral Commodity Profiles – Selenium*. U.S. USGS Numbered Series, 20 p.

Campbell, T. W., Walker, H. G. and Coppinger, G. M., 1952. Some aspects of the organic chemistry of selenium. *Chemical Reviews*. 50: 279–349.

Das, R. K., Banerjee, S. and Bhattacharya, S., 2007. Amelioration of benzo(*a*)pyreneinduced lung carcinogenesis in strain a mice by diphenylmethyl selenocyanate. *Experimental and Toxicologic Pathology*. 58: 351–360.

Dumont, E., Vanhaecke, F. and Cornelis, R., 2006. Selenium speciation from food source to metabolites: a critical review. *Analytical and Bioanalytical Chemistry*. 385: 1304–1323.

Fox, T. E., Van den Heuvel, E. G. H. M., Atherton, C., Dainty, J. R., Lewis, D. J., Langford, N. J., Crews, H. M., Luten, J. B., Lorentzen, M., Sieling, F. W., Aken-Schneyder, P., Hoek, M., Kotterman, M. J. J., Dael, P. and Fairweather-Tait, S. J., 2004. Bioavailability of selenium from fish, yeast and selenate: a comparative study in humans using stable isotopes. *European Journal of Clinical Nutrition*. 58: 343–349.

Freudendahl, D. M., Shahzad, S. A. and Wirth, T., 2009. Recent Advances in organoselenium chemistry. *European Journal of Organic Chemistry*. 1649–1664.

Jacob, C., Giles, G. I., Giles, N. M. and Sies, H., 2003. Sulfur and selenium: the role of oxidation state in protein structure and function. *Angewandte Chemie International Edition*. 42: 4742–4758.

Jean-Claude, G., 2011. Organic selenocyanates: synthesis, characterization and uses in chemistry and biology. *Current Organic Chemistry*. 15: 1670–1687.

Jörg, G., Buhnemann, R., Hollas, S., Kivel, N., Kossert, K., Winckel, S. V. and Gostomski, C. L., 2010. Preparation of radiochemically pure ^{79}Se and highly precise determination of its half-life. *Applied Radiation and Isotopes*. 68: 2339–2351.

Li, N., Gao, Z. D., Luo, D. G., Tang, X., Chen, D. F. and Hu, Y. H., 2007. Selenium level in the environment and the population of Zhoukoudian area, Beijing, China. *Science of the Total Environment*. 381: 105–111.

Maseko, T., Callahan, D. L., Dunshea, F. R., Doronila, A., Kolev, S. D. and Ng, K., 2013. Chemical characterization and speciation of organic selenium in cultivated selenium-enriched *Agaricus bisporus*. *Food Chemistry*. 141: 3681–3687.

Młochowski, J., Kloc, K., Lisiak, R., Potaczek, P. and Wójtowicz, H., 2007. Developments in the chemistry of selenaheterocyclic compounds of practical importance in synthesis and medicinal biology. *ARKIVOC*. 6: 14–46.

Mugesh, G., du Mont, W.-W. and Sies, H., 2001. Chemistry of biologically important synthetic organoselenium compounds. *Chemical Reviews*. 101: 2125–2179.

Navarro-Alarcon, M. and Cabrera-Vique, C., 2008. Selenium in food and the human body: a review. *Science of the Total Environment*. 400: 115–141.

Pappas, A. C., Zoidis, E., Surai, P. F. and Zervas, G., 2008. Selenoproteins and maternal nutrition. *Comparative Biochemistry and Physiology, Part B: Biochemistry & Molecular Biology*. 151: 361–372.

Plano, D., Baquedano, Y., Ibáñez, E., Jiménez, I., Palop, J. A., Spallholz, J. E. and Sanmartín, C., 2010. Antioxidant-prooxidant properties of a new organoselenium compound library. *Molecules*. 15: 7292–7312.

Rayman, M. P., 2004. Food-chain selenium and human health: emphasis on intake. *British Journal of Nutrition*. 92: 557–573.

Rayman, M. P., 2008. The use of high-selenium yeast to raise selenium status: how does it measure up? *British Journal of Nutrition*. 100: 254–268.

Rayman, M. P., Infante, H. G. and Sargent, M., 2008. Food-chain selenium and human health: spotlight on speciation. *British Journal of Nutrition*. 100: 238–253.

Rosa, R. M., Moura, D. J., Romano e Silva, A. C., Saffi, J. and Pegas-Henriques, J. A., 2007. Antioxidant activity of diphenyl diselenide prevents the genotoxicity of several mutagens in Chinese hamster V79 cells. *Mutation Research*. 63: 44–54.

Rosman, K. J. R. and Taylor, P. D. P., 1998. Isotopic composition of the elements 1997. *Pure and Applied Chemistry*. 70: 217–235.

Soriano-Garcia, M., 2004. Organoselenium compounds as potential therapeutic and chemo preventive agents: a review. *Current Medicinal Chemistry*. 11: 1657–1669.

Spadoni, M., Voltaggio, M., Carcea, M., Coni, E., Raggi, A. and Cubadda, F., 2007. Bioaccessible selenium in Italian agricultural soils: comparison of the biogeochemical approach with a regression model based on geochemical and pedoclimatic variables. *Science of the Total Environment*. 376: 160–177.

Tan, J. A., Zhu, W. Y., Wang, W. Y., Li, R. B., Hou, S. F., Wang, D. C. and Yang, L. S., 2002. Selenium in soil and endemic diseases in China. *Science of the Total Environment*. 284: 227–235.

Temmerman, L. D., Waegeneers, N., Thiry, C., Laing, G. D., Tack, F. and Ruttens, A., 2014. Selenium content of Belgian cultivated soils and its uptake by field crops and vegetables. *Science of the Total Environment*. 468: 77–82.

Thomson, C. D., Chisholm, A., McLachlan, S. K. and Campbell, J. M., 2008. Brazil nuts: an effective way to improve selenium status. *American Journal of Clinical Nutrition*. 87: 379–384.

World Health Organization, Food and Agriculture Organization and International Atomic Energy Agency Expert Group, 1996. *Trace Elements in Human Nutrition and Health*. WHO, Geneva.

Young, V. R., Nahapetian, A. and Janghorbani, M., 1982. Selenium bioavailability with reference to human nutrition. *American Journal of Clinical Nutrition*. 35: 1076–1088.

Zhang, H., Feng, X., Jiang, C., Li, Q., Liu, Y., Gu, C., Shang, L., Li, P. and Larssen, T., 2014. Understanding the paradox of selenium contamination in mercury mining areas: high soil content and low accumulation in rice. *Environmental Pollution*. 188: 27–36.

Zhao, F. J., Lopez-Bellido, F. J., Gray, C. W., Whalley, W. R., Clark, L. J. and McGrath, S. P., 2007. Effects of soil compaction and irrigation on the concentrations of selenium and arsenic in wheat grains. *Science of the Total Environment*. 372: 433–439.

Zhu, Y. G., Pilon-Smits, E. A. H., Zhao, F. J., Williams, P. N. and Meharg, A. A., 2009. Selenium in higher plants: understanding mechanisms for biofortification and phytoremediation. *Trends in Plants Science*. 14: 436–442.

CHAPTER 2

Biochemistry and Molecular Biology of Selenoproteins

NOELIA FRADEJAS[a], DOREEN BRAUN[a], SANDRA SEEHER[a], AND ULRICH SCHWEIZER*[a]

[a]Institut für Biochemie und Molekularbiologie, Rheinische Friedrich-Wilhelms-Universität Bonn, Nussallee 11, 53115 Bonn, Germany
*E-mail: uschweiz@uni-bonn.de

2.1 Selenoproteins in Mammals

In order to insert the amino acid selenocysteine (Sec; U) as the 21st proteinogenic amino acid into protein, a UGA "stop" codon is recoded to the rare amino acid (Hatfield and Gladyshev, 2002). Selenoproteins like glutathione peroxidases degrade hydrogen and lipid peroxides. For a long time, selenoproteins were therefore associated with protection from "oxidative stress". The identification of the second mammalian Sec containing enzyme, deiodinase type I, expanded the chemical functions of selenoproteins to deiodination of thyroid hormones. Mammalian genomes contain 24–25 genes encoding selenoproteins (Kryukov *et al.*, 2003).

2.1.1 Thioredoxin Reductases

Thioredoxin reductases (TXNRD) are oxidoreductases required for NADPH (nicotinamide adenine dinucleotide phosphate)-dependent reduction of endogenous cofactors like thioredoxin, glutathione or glutaredoxin 2

Food and Nutritional Components in Focus No. 9
Selenium: Chemistry, Analysis, Function and Effects
Edited by Victor R. Preedy

Published by the Royal Society of Chemistry, www.rsc.org

(Lee *et al.*, 2013b). These cofactors are essential for the regeneration of enzymes involved in the protection against oxidative stress (*e.g.* GPx) or the reduction of disulfide bridges in oxidized proteins to maintain an antioxidant environment within a cell. Three different genes encode the three different isoforms of TXNRD.

2.1.2 Glutathione Peroxidases

Glutathione peroxidases (GPx) not only maintain cellular redox homeostasis by protecting against reactive oxygen species (H_2O_2, hydroperoxides, lipidoxides and oxygen radicals), but also play a role in intracellular signaling and metabolism. For example, lack of GPx1 leads to increased H_2O_2 levels, which in turn enhance insulin signaling (Loh *et al.*, 2009). Deficiency of GPx4 leads to oxidative stress in testicular cells with the consequence of abnormal sperm development and male infertility (Ursini *et al.*, 1999).

Among eight glutathione peroxidases known in humans, only five of them are selenoenzymes. GPx5, GPx7 and GPx8 exhibit a cysteine residue instead of a Sec (Brigelius-Flohe and Maiorino, 2013).

GPx enzymes follow ping-pong kinetics (Figure 2.1): A first reaction oxidizes the selenolate to selenenic acid and the enzyme is regenerated by reduction through glutathione (Brigelius-Flohe and Maiorino, 2013).

2.1.3 Selenoprotein P

Selenoprotein P (SePP) is the major Sec-containing protein in plasma (Table 2.1) responsible for up to 60% of whole plasma selenium content. SePP is secreted by the liver as a protein that contains 10 Sec residues in its

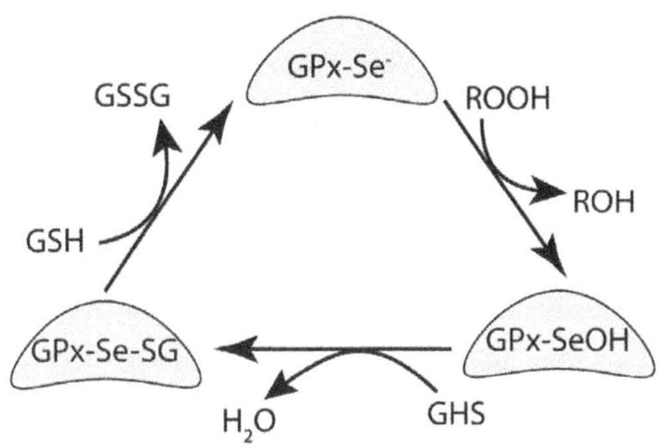

Figure 2.1 Ping-pong mechanism of GPx. GPx-mediated reduction of peroxides consists of a three-substrate reaction, which follows a ping-pong mechanism. First, the selenolate (GPx-Se-) is oxidized to selenenic acid (GPx-SeOH) to reduce the peroxide (ROOH). Then, the enzyme is regenerated by two molecules of glutathione (GSH). In the first step, GSH forms a selenenylsulfide (GPx-Se-SG) and a water molecule is released. In the second step, GSH regenerates the selenolate by formation of GSSG.

Table 2.1 Functions of selenoproteins and characteristics of selenoprotein knockout mouse models.

Selenoprotein	Expression	Function/characteristics	Animal model
Deiodinase type I (Dio1)	Plasma membrane of liver, kidney, thyroid	• Outer and inner ring deiodination of iodo-thyronines • Predicted as an iodine recycling enzyme	*Dio1*KO mouse model • No gross phenotype: no significant changes in thyroid axis • Increased renal iodine excretion
Deiodinase type II (Dio2)	Endoplasmatic reticulum of, *e.g.* brain, pituitary, muscle	• Outer ring deiodination of iodothyronines • Activation of T_4 to T_3 • Involved in thyroid hormone homeostasis by T_4 sensing in the pituitary	*Dio2*KO mouse model • Pituitary resistance to thyroxine • Impaired thermogenesis • Delayed cochlear development and deafness
Deiodinase type III (Dio3)	Plasma membrane of, *e.g.* brain, placenta, skin	• Inner ring deiodination of iodothyronines • Inactivation of T_4 to rT_3 and T_3 to $3,3'$-T_2 • Protects tissues (especially embryonic tissues) from excessive thyroid hormone levels	*Dio3*KO mouse model • Central hypothyroidism with low circulating T_3 and T_4 levels • Delayed development • Impaired auditory function with different pathogenesis to *Dio2*KO mice
Glutathione peroxidase 1 (GPx1)	Cytosol, mitochondria Enriched in liver, kidney, erythrocytes	• Identified as an enzyme that protects hemoglobin from oxidative damage • Reacts with H_2O_2 and low molecular mass hydroperoxides • Predicted as a Se-storage protein in liver	*GPx1*KO mouse model • Under normal conditions: no gross phenotype • Severe oxidative stress leads to lethality despite selenium supplementation • GPx1 enhance LPS-induced inflammatory response • Enhanced insulin sensitivity *GPx1*-transgenic mice • GPx1 overexpression leads to hyperglycemia, hyperinsulinemia, increased β-cell mass • Develop obesity and insulin resistance on high-fat diet

Glutathione peroxidase 2 (GPx2)	Cytosol Enriched in epithelium, *e.g.* gastrointestinal system and lung; liver	Suggested to protect against food-driven hydroperoxides	*GPx2*KO mouse model • GPx2-deficiency is partly compensated by GPx1 • GPx2 plays an anti-inflammatory role in colitis development
Glutathione peroxidase 3 (GPx3)	Secreted into plasma by the kidney; adipose tissue	Reduces H_2O_2 and peroxides in plasma	*GPx3*KO mouse model • No gross phenotype except involvement in inflammatory processes
Glutathione peroxidase 4 (GPx4)	Cytosolic GPx4 in liver, lung, kidney, brain; nuclear GPx4 in testis; mitochondrial GPx4 in sperm cells	• Differ in their N-terminal sequences • Reduces hydroperoxides of complex lipids involved in membrane formation • Involved in cell survival and proliferation	*GPx4*KO mouse model • Total and cytosolic GPx4: embryonic lethal due to increased apoptosis • Nuclear GPx4: no phenotype except delayed sperm chromatin condensation • Mitochondrial GPx4: infertility of male mice *GPx4*-transgenic mice • Cardiac dysfunction
Glutathione peroxidase 6 (GPx6)	Cytosol of the olfactory epithelium	• Sec-containing enzyme in humans, cysteine-containing enzyme in mice and rats • Detoxification of reactive oxygen species?	—
Selenoprotein 15 (Sep15)	Endoplasmatic reticulum	• Protein folding • Regulated by ER stress	*Sep15*KO mouse model • Development of lens cataracts due to improper protein folding
Selenoprotein H (SelH)	Nucleus	• H_2O_2 detoxification • Involved in mitochondrial cytochrome C production and GSH *de novo* synthesis	—

(*continued*)

Table 2.1 (*continued*)

Selenoprotein	Expression	Function/characteristics	Animal model
Selenoprotein I (SelI)	Membrane	• Unknown function	—
Selenoprotein K (SelK)	Endoplasmatic reticulum membrane	• Regulate Ca^{2+} influx of immune cells • Involved in protein palmitoylation	*SelK*KO mouse model • No gross phenotype except impaired immune cell function
Selenoprotein M (SelM)	Endoplasmatic reticulum	• Protein modification • Protects neurons from reactive oxygen species	—
Selenoprotein N (SelN, SEPN1)	Membrane of the endoplasmatic reticulum of, *e.g.* skeletal muscle, heart, lung, placenta	• Regulates the intracellular Ca^{2+} release by the ryanodine receptor • Responsible for Ca^{2+} homeostasis	*SelN*KO mouse model • Impaired motility during physical exercises • Impaired muscle regeneration
Selenoprotein O (SelO)	Mitochondria	• Unknown function	—
Selenoprotein P (SePP)	Plasma	• Secreted by the liver • Transport and storage of selenium	*SePP*KO mouse model • General selenium deficiency • Reduced selenoprotein expression • Neurological impairment with epilepsy and spastic movement • Male infertility
Selenoprotein R (SelR, MsrB1)	Cytosol	• Reduction of methionine-R-sulfoxides to methionine in proteins	*MsrB1*KO mouse model • No gross phenotype except some oxidative stress
Selenoprotein S (SelS, SEPS1)	Endoplasmatic reticulum membrane	• Involved in inflammatory processes • Upregulated by cytokines • Interacts with SelK for ERAD • Plays a role in diabetes mellitus	—

Selenoprotein T (SelT)	Endoplasmatic reticulum and golgi	• Redox regulation • Cell adhesion	—
Selenoprotein V (SelV)	Cytosol of spermatids	• Unknown function	—
Selenoprotein W (SelW)	Cytosol of skeletal muscles and other tissues	• Unknown function	—
Selenophosphate synthetase (SPS2, Sephs2)	Cytosol	• Synthesis of selenophosphate	—
Thioredoxin reductase 1 (TR1, TXNRD1)	Cytosol	• Reduction of oxidized cytosolic thioredoxin and glutaredoxin 2	*TR1*KO mouse model • Total *TR1*KO: embryonic lethal • Neuron-specific *TR1*KO: severe neurological injury • Cardiomyocyte-specific *TR1*KO: no phenotype
Thioredoxin reductase 2 (TR3, TXNRD2, SelZ)	Cytosol, mitochondria	• Reduction of proteins specific for the thioredoxin system	—
Thioredoxin reductase 3 (TR2, TXNRD3)	Mitochondria of testis	• Reduction of oxidized mitochondrial thioredoxin, glutathione and glutaredoxin 2	*TR3*KO mouse model • Total *TR3*KO: embryonic lethal • Neuron-specific *TR3*KO: no phenotype • Cardiomyocyte-specific *TR3*KO: lethal after birth because of heart failure

structure. As a selenium transport and storage protein it is very important for the maintenance of brain and testis selenium levels. During selenium deficiency whole-body and tissue selenium contents decrease except for the brain that keeps selenium concentration constant (Köhrle *et al.*, 2012). *Sepp*-deficient mice reveal irreversible neurological injury with epilepsy and paralysis due to reduced selenium content in brain (Hill *et al.*, 2004). $Sepp^{-/-}$ mice were thus the first animal model to demonstrate that the mammalian brain requires selenium, presumably as selenoproteins. This topic is expanded in Chapter 27 of this book, Selenoproteins in Brain.

2.1.4 Selenoprotein N

There are only three known human disorders related to deficiency of selenoproteins or its biosynthesis machinery. Mutations in selenoprotein N were identified in patients with either rigid spine muscular dystrophy (RSMD1), multiminicore disease (MmD), Mallory body-like desmin-related myopathy (MBDRM) or myopathy with congenital fiber-type disproportion (CFTD). These so-called *SEPN1*-related myopathies are characterized by muscle atrophy, muscle weakness, severe scoliosis and respiratory insufficiency (Moghadaszadeh *et al.*, 2001). *SePN*-deficient mice shed light on the physiological function of selenoprotein N in muscle regeneration, the maintenance of satellite cells (muscle progenitors) and the regulation of ryanodine receptor activity (Jurynec *et al.*, 2008; Castets *et al.*, 2011; Rederstorff *et al.*, 2011).

2.1.5 Iodothyronine Deiodinases

Three different deiodinases (deiodinase type I–III, Dio1–3) are responsible for activation and inactivation of thyroid hormones, respectively (Schomburg and Kohrle, 2008; Dentice and Salvatore, 2011). The thyroid gland mostly secrets the inactive prohormone T_4 (3,3′,5,5′-tetraiodothyronine, thyroxine) which is converted into the active T_3 (3,3′,5-triiodothyronine) by the action of Dio2 or inactivated to rT_3 (reverse 3,3′,5′-triiodothyronine) by Dio3 in the periphery. Thyroid hormone homeostasis is very important for embryogenesis, development and thermogenesis. For example, the action of Dio3 protects the embryo from excessive levels of maternal thyroid hormones during embryogenesis. Conversely, cochlear development depends on the precise concentration of activated T_3 mediated by the temporal expression pattern of Dio2 and Dio3 (Campos-Barros *et al.*, 2000; Ng *et al.*, 2009).

2.1.6 Selenoprotein R is a Methionine-Sulfoxide-Reductase

Selenoprotein R (SePR, MsrB1) bears R-methionine-sulfoxide-reductase activity (Kryukov *et al.*, 2002). Oxidation of methionine is not only a result of oxidative damage, but may play a role in reversible, regulatory protein modification (Lee *et al.*, 2013a).

2.1.7 Selenoproteins Involved in Protein Biosynthesis, Folding or Degradation

Selenoprotein S is associated with the endoplasmic reticulum-associated protein degradation (ERAD) pathway (Ye *et al.*, 2004), but it is not clear what role the Sec moiety plays in its function. Selenoprotein K clearly modifies immune function, possibly by palmitoylation of target proteins, although its mechanism is elusive. Selenoproteins Sep15 and SelM are located in the endoplasmic reticulum, but their function remains uncertain as little effects are found in respective knockout mouse models.

2.2 Selenoprotein Biosynthesis

2.2.1 Biosynthesis of Selenocysteine

2.2.1.1 Selenocysteine Biosynthesis Occurs on its tRNA

The selenocysteine tRNA ($tRNA^{[Ser]Sec}$) shows special features, which are responsible for substrate recognition and binding to the many different protein partners involved in selenocysteine biosynthesis. $tRNA^{[Ser]Sec}$ contains 90 nucleotides rather than the conventional 75 nucleotides of canonical tRNA molecules. It adopts a unique 9/4 fold with a 13-base pair acceptor-TΨC arm (where 9 and 4 reflect the number of base pairs in the acceptor and TΨC arms, respectively) and a long variable arm. A longer stem (6 base pairs) and a shorter loop (4 nucleotides) in the D-arm are characteristic of $tRNA^{[Ser]Sec}$ (Figure 2.2). Transfer $RNA^{[Ser]Sec}$ possesses some post-transcriptional modifications that contribute to correct folding and stabilization of the tertiary structure.

Selenocysteine biosynthesis comprises several steps. $tRNA^{[Ser]Sec}$ is first aminoacylated with serine (hence the Ser in brackets) by seryl-tRNA synthetase (SerRS). In Eukarya and Archaea, the seryl moiety of Ser-$tRNA^{[Ser]Sec}$ is phosphorylated by *O*-phosphoseryl-$tRNA^{Sec}$ kinase (PSTK) and finally converted to selenocysteine by the selenocysteine synthase (SepSecS) (Figure 2.3).

2.2.1.2 Seryl-tRNA Synthetase (SerRS)

SerRS catalyzes the transfer of serine to any cognate $tRNA^{Ser}$ to produce seryl-tRNA. However, SerRS can serylate $tRNA^{Ser}$ as well as $tRNA^{[Ser]Sec}$. The promiscuity of this enzyme can be explained because $tRNA^{[Ser]Sec}$ shares some identity elements of $tRNA^{Ser}$. The length of the variable arm and its orientation mimics perfectly the one of $tRNA^{Ser}$ (Wu and Gross, 1993; Itoh *et al.*, 2009). While many other tRNA synthetases read the anticodon, the major determinant for SerRS recognition is the long variable arm, which interacts with the N-terminal domain (NTD) of the enzyme. The catalytic activity of SerRS resides in its C-terminal domain (CTD). Another shared identity element is the discriminator base G73, the base preceding the CCA-end

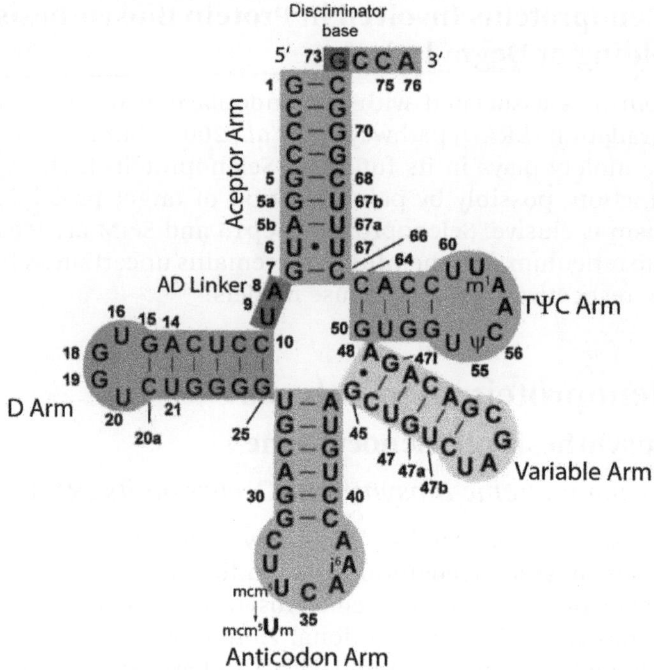

Figure 2.2 Cloverleaf structure of tRNA$^{[Ser]Sec}$. Transfer RNA$^{[Ser]Sec}$ is modified at several positions. Modifications within the anticodon loop enhance expression of selenoproteins. The discriminator base G^{73} and the long variable arm are shared with tRNA^{Ser}.

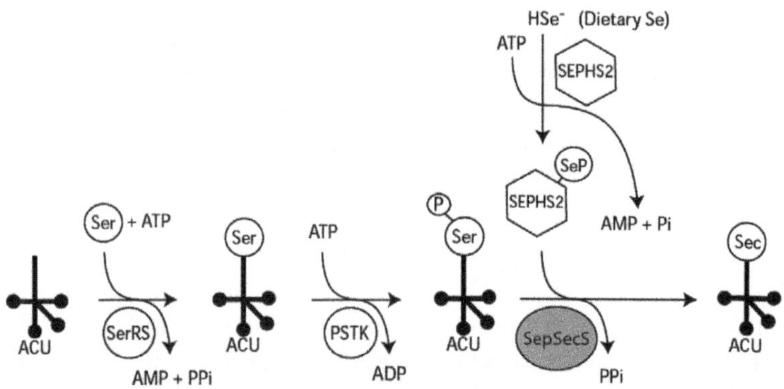

Figure 2.3 Synthesis of tRNA-RNA$^{[Ser]Sec}$. Sec synthesis occurs attached to tRNA$^{[Ser]Sec}$. The tRNA$^{[Ser]Sec}$ is first aminoacylated with serine by SerRS. The seryl moiety of Ser-tRNA$^{[Ser]Sec}$ is then phosphorylated by PSTK to phosphoseryl(Sep)-tRNA$^{[Ser]Sec}$. SepSecS catalyzes the elimination of phosphate from the phosphoseryl moiety of Sep-tRNA$^{[Ser]Sec}$ and subsequent addition of selenophosphate that hydrolyzes to form Sec-tRNA$^{[Ser]Sec}$. Selenophoshate is synthesized from selenite and ATP by SEPHS2 (SPS2), which is a selenoprotein itself.

(Wu and Gross, 1993; Itoh *et al.*, 2009). Since the same SerRS can charge either $tRNA^{Ser}$ or $tRNA^{[Ser]Sec}$, the selenocysteine biosynthesis machinery must strictly discriminate between both serylated tRNAs to avoid mistranslation of selenocysteine into proteins. Furthermore, the amount of $tRNA^{Ser}$ is much higher than $tRNA^{[Ser]Sec}$. Therefore, the selection of $tRNA^{[Ser]Sec}$ must be highly sensitive. In Archaea and Eukarya, PSTK is responsible for $tRNA^{[Ser]Sec}$ discrimination, since PSTK phosphorylates Ser-$tRNA^{[Ser]Sec}$ but not Ser-$tRNA^{Ser}$ (Chiba *et al.*, 2010).

2.2.1.3 *O-Phosphoseryl-tRNA^{Sec} Kinase (PSTK)*

PSTK is composed of an N-terminal domain, a C-terminal domain and a flexible linker, which connect both domains. The C-terminal domain interacts with the D-arm, whereas the N-terminal domain binds the acceptor arm and the CCA-terminus of Ser-$tRNA^{[Ser]Sec}$. The phosphotransferase activity resides in the N-terminal domain and the discriminator base is critical for the enzyme activity (Chiba *et al.*, 2010; Sherrer *et al.*, 2011). PSTK converts Ser-$tRNA^{[Ser]Sec}$ into serylphosphate(Sep)-$tRNA^{[Ser]Sec}$.

2.2.1.4 *Selenocysteine Synthetase (SepSecS)*

According to the human SepSecS crystal structure, SepSecS forms a stable tetramer. Each SepSecS monomer contains a pyridoxal-5-phosphate cofactor (PLP), which forms a Schiff base with the *N*ε-amino group of the conserved lysine 284. Two SepSecS monomers define a homodimer and in turn two homodimers bind into a tetramer. Two Sep-$tRNA^{[Ser]Sec}$ molecules are anchored to the SepSecS tetramer. Both tips of their acceptor arms interact with the same homodimer, called catalytic dimer. In this case, the most important interaction is between the discriminator base G73 in the $tRNA^{[Ser]Sec}$ and the highly conserved arginine 398 in the catalytic dimer. This interaction is critical for Sep-$tRNA^{[Ser]Sec}$ recognition (Palioura *et al.*, 2009). SepSecS catalyzes the conversion of the phosphoseryl moiety from Sep-$tRNA^{[Ser]Sec}$ into Sec-$tRNA^{[Ser]Sec}$ using selenophosphate as the selenium donor. The phosphoryl group from phosphoserine is required to properly position the ligand for catalysis. This explains why the obligate substrate for SepSecS is Sep-$tRNA^{[Ser]Sec}$ and not Ser-$tRNA^{[Ser]Sec}$. The reaction would start when Sep-$tRNA^{[Ser]Sec}$ is bound to SepSecS. Then, the amino group of phosphoserine would form the Schiff base with PLP. The next step is β-elimination of the phosphate group. The selenophosphate is then thought to undergo nucleophilic attack on the highly reactive dehydroalanine intermediate and subsequent release of phosphate. Sec-$tRNA^{[Ser]Sec}$ is then released and the active site of SepSecS would be regenerated by formation of a PLP Schiff base with Lys284 (Figure 2.4) (Palioura *et al.*, 2009).

Mutations in *SEPSECS* cause a severe early onset neurodegenerative disease, named progressive cerebellar cerebral atrophy (PCCA) first found in

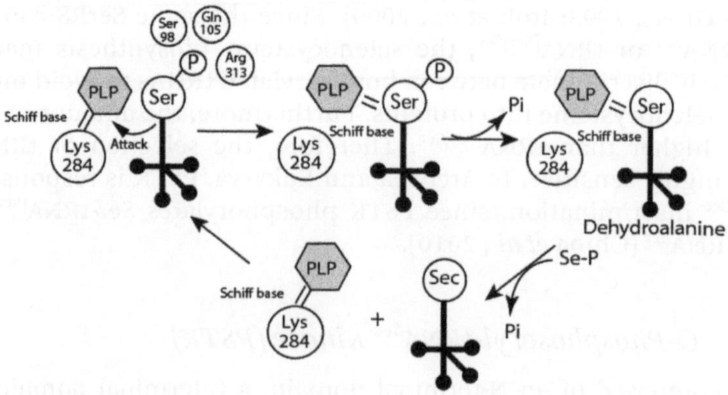

Figure 2.4 SepSecS reaction. First, Sep-tRNA[Ser]Sec is positioned into the active site of SepSecS in a manner that the amino group of Sep would attack the Schiff base formed between Lys[284] and PLP cofactor. The ligand is anchored through interactions between its phosphoryl group and the side chains of Ser[98], Gln[105] and Arg[313]. The formation of the new Schiff base between Sep-tRNA[Ser]Sec and PLP leads to β-elimination of phosphate and formation of dehydroalanine. Then, selenophosphate attacks dehydroalanine with subsequent release of phosphate. Finally, the restoration of the Schiff base between Lys[284] and PLP entails the release of free tRNA[Ser]Sec and the regeneration of the SepSecS active site.

Sephardic Jewish communities of Moroccan and Iraqi ancestry. Individuals affected by this disease suffer from severe spasticity, profound mental retardation and progressive microcephaly (Agamy *et al.*, 2010).

2.2.1.5 *Selenophosphate Synthetases*

Selenophosphate is the only selenium donor to SepSecS. Two selenophosphate synthetases exist in eukaryotes, SEPHS1 (SPS1) and SEPHS2 (SPS2). Selenophosphate is synthesized from selenite and ATP by SPS2, which is a selenoprotein itself. Lack of selenoprotein biosynthesis phenotype in SPS1 mutants and its presence in organisms that do not synthesize selenoproteins, indicates that this protein is not involved in selenoprotein biosynthesis (Lobanov *et al.*, 2008).

2.2.2 Translation of Selenoproteins

Translation of selenoproteins, in particular decoding of the in-frame UGA codon to selenocysteine (Sec) requires specific *trans-* and *cis*-acting factors. To date, it is known that the recoding event requires at least three *trans*-acting factors: Sec-tRNA[Ser]Sec, Sec-specific elongation factor (EFSec) and the Sec insertion sequence (SECIS) binding protein 2 (SECISBP2). Deletion or mutation of the SECIS element demonstrated its role as an essential *cis*-acting component. A second *cis*-acting factor, selenocysteine redefinition element (SRE), may only be needed for some of the selenoproteins.

2.2.2.1 The Hierarchy of Selenoproteins

An interesting observation in respect to selenoprotein synthesis was their hierarchical expression in case of selenium restriction (Schomburg and Schweizer, 2009). Essential housekeeping selenoproteins, like thioredoxin reductase 1 (Txnrd1) or glutathione peroxidase 4 (Gpx4), are more resistant to selenium limitation than nonessential, stress-related selenoproteins, such as glutathione peroxidase 1 (Gpx1) (Hill *et al.*, 1992; Lei *et al.*, 1995; Hadley and Sunde, 2001).

Less is known about the mechanisms by which dietary selenium affects selenoprotein synthesis. One mechanism that has been discussed is nonsense-mediated decay. During selenium shortage, the UGA-Sec codon in nonessential selenoproteins mRNAs (*e.g. Gpx1*) is recognized as a premature stop codon causing RNA degradation (Moriarty *et al.*, 1998; Weiss and Sunde, 1998). A second explanation involves different Sec incorporation efficiencies due to structural differences in SECIS elements resulting in altered ability to recruit the Sec incorporation machinery (Low *et al.*, 2000; Squires *et al.*, 2007; Budiman *et al.*, 2009). It is also known that selenium levels directly correlate with the degree of $tRNA^{[Ser]Sec}$ 2′-*O*-methylation (see below).

A recent study used ribosome profiling to examine the effect of dietary selenium levels on translational mechanisms controlling selenoprotein synthesis in mouse liver (Howard *et al.*, 2013). It was shown that dietary selenium levels control gene-specific selenoprotein expression primarily at the translational level by differential regulation of Sec incorporation efficiency. Nevertheless, effects on translation initiation and mRNA abundance were also observed. They further presented direct evidence for a dramatic increase in ribosome density downstream of UGA-Sec codons for a subset of selenoprotein mRNAs with increasing dietary selenium levels. Thus, in the presence of sufficient selenium, UGA read-through efficiency is unexpectedly high and outcompetes termination by release factors. Ribosome pausing near the UGA-Sec codon may be related to the decision whether to elongate with Sec or terminate at the UGA.

2.2.2.2 SECIS-Binding Protein 2 (SECISBP2)

UV-crosslinking with *Gpx4* SECIS was used to purify and clone SECISBP2 from rat testicular extracts (Copeland and Driscoll, 1999; Copeland *et al.*, 2000). Gel filtration chromatography showed SECISBP2 as being part of a large protein complex (~500 kDa). As SECISBP2 migrates at ~120 kDa in SDS-PAGE, this suggests the specific association of diverse factors (Copeland and Driscoll, 1999). Mutational analysis and sequence alignments displayed that SECISBP2 consists of three domains (Figure 2.5). An N-terminal domain with unknown function that is not necessary for Sec incorporation (rat aa 1–398) (Olieric *et al.*, 2009). A central Sec incorporation domain (SID or Lys-rich

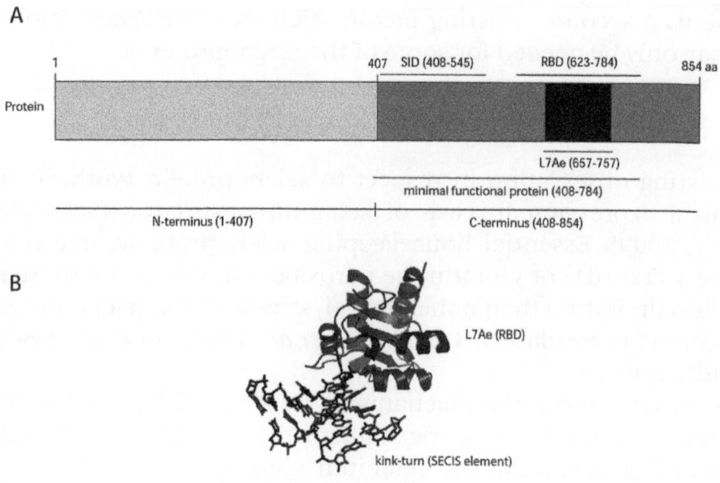

Figure 2.5 Structure of SECISBP2. (A) SBP2 is composed of three different domains: a N-terminal domain, which it is not necessary for Sec insertion; a central Sec incorporation domain (SID) and the L7Ae RNA binding domain in the C-terminus. (B) Representation of L7Ae motif interaction with the kink-turn of a SECIS element. Scaffold for the modeling was a crystal structure of the spliceosomal 15.5-kDa protein bound to an U4 snRNA fragment (PDB-ID:1E7K).

domain; rat aa 433–537) and a C-terminal RNA binding domain (RBD; rat aa 616–777) (Copeland and Driscoll, 1999; Copeland *et al.*, 2000; Copeland *et al.*, 2001; Allmang *et al.*, 2002; Lescure *et al.*, 2002; Bubenik and Driscoll, 2007; Caban *et al.*, 2007; Donovan *et al.*, 2008). The RBD consists of a L7Ae RNA binding motif. Proteins with this motif can interact with RNA kink-turns around their loop region and the α-helix.

Mutation of the conserved glycine residue (rat G669) in the L7Ae domain abolishes SECIS, but not ribosome binding (Copeland *et al.*, 2001). Only the SID and the RBD of SECISBP2 are required for full ribosome binding and Sec incorporation (Copeland *et al.*, 2001; Bubenik and Driscoll, 2007). A study showed that the interaction of the SID with the RBD is prevented when an $IILKE_{526-530}$ motif within the SID is mutated to penta-alanine (Donovan *et al.*, 2008; Takeuchi *et al.*, 2009). This mutation also abolishes Sec incorporation, but not SECIS binding. It was proposed that the RBD first has only weak contact to the SECIS element. RBD binding then triggers a conformational change and recruits the SID to stabilize the SECIS–RBD interaction. The stably formed SECISBP2–SECIS complex is then able to bind EFSec and Sec-$tRNA^{[Ser]Sec}$ (Figure 2.6) (Donovan *et al.*, 2008; Donovan and Copeland, 2010). The interaction of RBD with the ribosome could be disrupted by mutation of a specific region (rat aa $RFQDR_{647-651}$). Interestingly, Sec incorporation was also impaired, but not SECIS binding activity. This suggests that ribosome binding is an essential component of the Sec incorporation activity of SECISBP2 (Caban *et al.*, 2007).

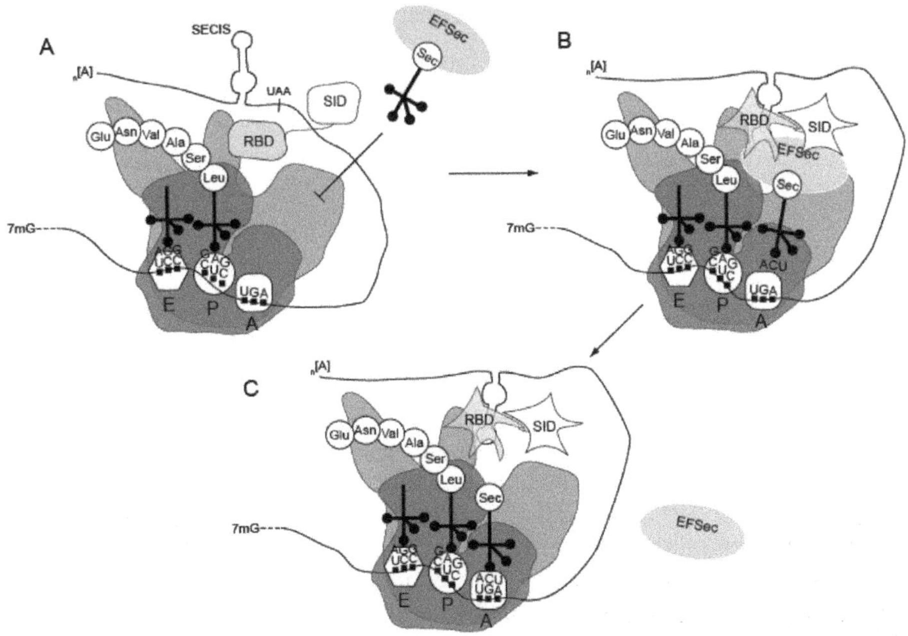

Figure 2.6 Translation of selenoproteins. (A) Sec incorporation requires an in-frame UGA Sec codon and a SECIS element in the 3′UTR of selenoprotein mRNA. Without SECISBP2, the EFSec-tRNA$^{[Ser]Sec}$ complex cannot enter the ribosomal A-site. (B) The RBD of SECISBP2 binds to the kink-turn motif of the SECIS element. Together with the SID it forms a stable RBD-SID-SECIS complex. As the RBD is also bound to the 28S rRNA of the ribosome, this directs the EFSec-tRNA$^{[Ser]Sec}$ complex to the ribosome. (C) EFSec releases the tRNA$^{[Ser]Sec}$ after GTP hydrolysis and dissociates from the ribosome. Sec can be incorporated into the growing peptide chain.

SECISBP2 binds the 28S rRNA of 80S ribosomes. Screening of the functional center of the 28S rRNA using selective 2′-hydroxyl acylation analyzed by primer extension (SHAPE) demonstrated that SECISBP2 specifically alters the reactivity of specific residues in Helix 89 and expansion segment 31. It was proposed that SECISBP2-induced conformational changes in this helix promote Sec incorporation by affecting Sec-tRNA$^{[Ser]Sec}$ accommodation and/or by stimulating the ribosome-dependent GTPase activation of EFSec (Figure 2.6) (Caban and Copeland, 2012).

Using a conditional *Secisbp2* knockout mouse model, a recent study showed that Secisbp2 depletion dramatically reduced, but not completely abolished, selenoprotein expression in liver. The results suggested that Secisbp2 is necessary for efficient selenoprotein expression *in vivo*, but there seems to exist a low level of Secisbp2-independent Sec incorporation. Secisbp2 appears to stabilize the mRNAs of some selenoproteins. Therefore, Secisbp2 might contribute to hierarchical selenoprotein expression in several ways (Seeher *et al.*, 2013).

2.2.2.3 Sec-Specific Eukaryotic Elongation Factor (EFSec)

Eukaryotic translation elongation factor EFSec is highly specific for Sec-tRNA$^{[Ser]Sec}$ but not for the Ser-tRNA$^{[Ser]Sec}$ precursor or any canonical aminoacyl-tRNAs (Fagegaltier *et al.*, 2000b; Tujebajeva *et al.*, 2000). EFSec was first identified by sequential homology searches of expressed sequence tag databases with the archaeon *Methanocaldococcus jannaschii* EFSec protein sequence (Fagegaltier *et al.*, 2000a; Tujebajeva *et al.*, 2000). The formation of a SECISBP2/SECIS/EFSec/GTP/Sec-tRNA$^{[Ser]Sec}$ complex is the central assumption for the proposed model for delivery of Sec-tRNA$^{[Ser]Sec}$ to the A-site of the ribosome.

2.2.2.4 SECp43

SECp43 was first identified from a degenerate PCR screen as a highly conserved RNA-binding protein, which consists of two RNA-binding domains and a polar/acidic C-terminus. Copurification of tRNA$^{[Ser]Sec}$ with SecP43 through a SECp43 antibody affinity column discovered its association with tRNA$^{[Ser]Sec}$ (Ding and Grabowski, 1999; Seeher *et al.*, 2012). Later, SECp43 was found in complex with SepSecS and SPS1 and its role as an accessory protein, which redistributes these proteins to the nucleus was proposed (Small-Howard *et al.*, 2006). The role of SECp43 in selenoprotein synthesis, however, remains to be clarified.

2.2.2.5 Selenocysteine Insertion Sequence (SECIS) Element

Sec incorporation into selenoproteins requires a selenocysteine insertion sequence (SECIS) element (Berry *et al.*, 1991). It was shown to act as a platform for SECISBP2 binding and recruitment of other factors of the translation machinery like EFSec (Donovan *et al.*, 2008). In contrast to bacterial selenoprotein mRNAs, where the SECIS is located directly downstream of the in-frame UGA-Sec codon, the eukaryotic SECIS is positioned in the 3′-UTR (Berry *et al.*, 1991). The SECIS element is a stem-loop structure with low sequence similarity among the selenoproteins. However, their secondary structures are highly conserved and contain consensus sequences that are indispensable for Sec incorporation (Lescure *et al.*, 1999; Castellano *et al.*, 2001; Kryukov *et al.*, 2003). The SECIS core is a putative kink-turn motif that allows the SECIS to act as a molecular switch by undergoing dynamic conformational changes in response to protein binding. SECISBP2 binds to the SECIS core and this binding is essential for Sec incorporation (Copeland and Driscoll, 1999; Copeland *et al.*, 2000). UGA recoding analysis of the 26 human SECIS elements in transfected cells and an *in vitro* translation system showed that the SECIS elements vary dramatically in their ability to direct recoding. It was suggested that intrinsic properties of the SECIS elements might lead to these differences (Latrèche *et al.*, 2009).

2.2.2.6 *Selenocysteine Redefinition Element (SRE)*

By bioinformatics, in 2005 and 2006 two independent groups identified a second *cis*-acting element, termed selenocysteine redefinition element – SRE (Howard *et al.*, 2005; Pedersen *et al.*, 2006). According to bacterial SECIS (bSECIS) elements (Zinoni *et al.*, 1990; Berg *et al.*, 1991), the SRE is located immediately downstream of UGA-Sec codon in a subset of selenoprotein mRNAs (*SEPN1*, *SPS2*, *SEPH*, *SEPO* and *SEPT*). The secondary structure of SREs appears to have little similarity to the bSECIS. The best studied is the *SEPN1* SRE. The sequence is highly conserved in all *SEPN1* genes identified, implies an important function for this element (Howard *et al.*, 2005). The potential stem-loop structure starts six nucleotides downstream of the UGA-Sec codon (Howard *et al.*, 2005, 2007). A role for the SRE was confirmed by identification of *SEPN1* mutations located in the SRE element and adjacent nucleotide (Moghadaszadeh *et al.*, 2001; Maiti *et al.*, 2009). SRE mutations lead to decreased UGA-Sec read-through and therefore to reduced selenoprotein N expression in the patients.

2.2.2.7 *tRNA$^{[Ser]Sec}$*

tRNA molecules normally contain different post-transcriptional modifications, which facilitate correct folding and stabilize the tertiary structure of tRNAs. As mentioned above, tRNA$^{[Ser]Sec}$ possesses several modified bases: pseudouridine (ψ^{55}) and 1-methyladenosine (m^1A^{58}) at positions 55 and 58, respectively. In the anticodon loop tRNA$^{[Ser]Sec}$ carries *N*6-isopentenyl-adenosine (i^6A^{37}). Within the anticodon resides a 5-methoxycarbonymethyl modification at the wobble position, mcm^5U^{34}, which is further methylated at the ribose to 5-methoxycarbonymethyl-2′-*O*-methyl-uridine (mcm^5Um^{34}) (Figure 2.2). Selenocysteine specific tRNA$^{[Ser]Sec}$ thus exists as two differentially modified subpopulations. The relative abundance of both isoforms is governed by selenium status, and high selenium availability increases methylation. Both isoforms appear to have a role in selenoprotein expression hierarchy. Thus, the expression of stress-related selenoproteins, such as GPx1 GPx3, SePR or SePT, correlates with the presence of mcm^5Um^{34} isoform and mcm^5U associates with essential selenoproteins expression, such GPx4 and TXNRD1. Interfering with isopentenylation of tRNA$^{[Ser]Sec}$ by overexpressing an A37G tRNA$^{[Ser]Sec}$ mutant in mice leads to reduced selenoprotein expression (Carlson *et al.*, 2005; Carlson *et al.*, 2007). Addition of i^6A on tRNA molecules is catalyzed by isopentenyl transferases, using dimethylallyl pyrophosphate (DMAPP) as a donor of the isopentenyl group. It was recently demonstrated that Trit1 is the enzyme responsible for isopentenylation of tRNA$^{[Ser]Sec}$ at adenine 37, since recombinant Trit1 was able to transfer a isopentenyl group from [^{14}C]DMAPP to tRNA$^{[Ser]Sec}$. Mutant A37G tRNA$^{[Ser]Sec}$ was not isopentenylated in the presence of the recombinant Trit1 protein. Knockdown of Trit1 in cells reduced selenoprotein expression (Fradejas *et al.*, 2013). The 2′-*O*-ribose methyltransferase that methylates tRNA$^{[Ser]Sec}$ at position 34 in response to selenium status has not yet been identified.

Summary Points

- Selenoproteins contain the rare amino acid selenocysteine (Sec).
- Selenocysteine is encoded by UGA codons.
- Selenocysteine acts as a nucleophile in selenoenzymes.
- Selenocysteine is synthesized attached to its tRNA.
- $tRNA^{[Ser]Sec}$ interacts with its own translation elongation factor EFSec.
- UGA stop → Sec recoding depends on the Sec insertion sequence (SECIS) element located in the 3′-UTR of selenoprotein mRNAs.
- Mutations in the SECISBP2 gene in humans lead to a multisystem disorder.
- Mutations in selenocysteine synthase lead to a neurodegenerative disorder in childhood.
- A hierarchy among selenoproteins distributes bioavailable selenium to a class of "housekeeping selenoproteins" at the expense of "stress-related selenoproteins".
- The mechanism of selenoprotein hierarchy is not completely understood.

Key Facts: Recoding of UGA

1. Recoding of the UGA codon as selenocysteine (Sec) is conserved among bacterial, archaeal, and eukaryotic genomes.
2. The Sec insertion machinery involves a SECIS (selenocysteine insertion sequence) element, a stem-loop structure in selenoprotein mRNAs, in all domains of life.
3. While the SECIS follows directly downstream of the UGA within the open reading frame of bacterial selenoproteins, eukaryotic SECIS elements are located in the 3′-untranslated region and therefore afar from the UGA codon.
4. The SECIS is recognized by the SECIS-binding protein 2 (SECISBP2).
5. SECISBP2 further interacts with EFSec, the Sec-specific translation factor.
6. In bacteria, the functions of EFSec and SECISBP2 are contained in a single protein SelB.

Key Facts: Mutations in SECISBP2

1. Mutations in the SECISBP2 gene have been identified in patients.
2. SECISBP2 mutations lead to a multisystem disorder with abnormalities in thyroid hormone levels (high T_4, low T_3, normal or slightly elevated TSH).
3. Short stature and bone maturation delay are clearly related to thyroid hormone insufficiency.
4. Other symptoms vary among patients, ranging from bilateral hearing loss, myopathy, azo-ospermia to photosensitivity.
5. Impaired selenoprotein expression appears to underlie these symptoms and is related to impaired SECIS binding (Dumitrescu *et al.*, 2005; Schoenmakers *et al.*, 2010).

Definitions

Deiodinases (Dio). Selenoenzymes that remove iodide atoms from iodothyronines (thyroid hormones and their metabolites). Removal of the 5′-iodine atom activates the prohormone T_4, while removal of a 5-iodine atom inactivates T_4 and T_3.

Elongation factor EFSec. Translation elongation factor specific for $tRNA^{[Ser]Sec}$ analogous to EF-1α.

Progressive cerebello cerebral atrophy (PCCA). A neurodegenerative disorder based on mutations in the *SEPSECS* gene encoding selenophosphate synthetase.

Protein palmitoylation. Membrane attachment of membrane associated proteins can be enhanced by acylation of serine close.

Reactive oxygen species (ROS). Molecules including peroxides, hydroxyl radicals, and superoxide may oxidatively damage proteins. Many selenoproteins are able to decompose or reduce ROS. While the traditional view is that selenoproteins thus protect cellular structures from oxidative damage, a more modern view stresses the signaling roles of ROS and oxidatively modified proteins. Selenoproteins may thus contribute to regulation of cellular physiology in addition to a "protective" role against oxidative damage.

Selenoproteins. Proteins containing the rare amino acid selenocysteine (Sec).

SECIS elements. Stem-loop structures in selenoprotein mRNAs that direct Sec insertion at UGA.

Selenocysteine redefinition elements (SRE). SECIS-like structures immediately downstream of UGA in eukaryotic selenoprotein mRNAs.

STOP codon. The codons UAA, UAG, and UGA interact with translational release factors, proteins that mediate chain termination during protein biosynthesis. The UGA codon can be recoded to mediate Sec incorporation in the presence of a cognate Sec-$tRNA^{[Ser]Sec}$. Chain elongation at stop codons is also called *nonsense-suppression.*

tRNA modifications. Occur at several positions in $tRNA^{[Ser]Sec}$, but only those in the anticodon loop, mcm^5Um^{34} and i^6A^{37} enhance selenoprotein expression.

List of Abbreviations

Dio	Deiodinase
EFSec	Translation elongation factor specific for $tRNA^{Sec}$
GPX	Glutathione peroxidase
RBD	RNA-binding domain
SID	Selenocysteine incorporation domain
Sec	Selenocysteine
SECIS	Selenocysteine insertion sequence
SepX	Selenoprotein X
SelX	Selenoprotein X (some instances) or *E. coli* gene in *Sel* operon
SRE	Selenocysteine redefinition element
Txnrd	Thioredoxin reductase

References

Agamy, O., Ben Zeev, B., Lev, D., Marcus, B., Fine, D., Su, D., Narkis, G., Ofir, R., Hoffmann, C., Leshinsky-Silver, E., *et al.*, 2010. Mutations disrupting selenocysteine formation cause progressive cerebello-cerebral atrophy. *American Journal of Human Genetics*. 87: 538–544.

Allmang, C., Carbon, P. and Krol, A., 2002. The SBP2 and 15.5 kD/Snu13p proteins share the same RNA binding domain: identification of SBP2 amino acids important to SECIS RNA binding. *RNA*. 8: 1308–1318.

Berg, B. L., Baron, C. and Stewart, V., 1991. Nitrate-inducible formate dehydrogenase in *Escherichia coli* K-12. II. Evidence that a mRNA stem-loop structure is essential for decoding opal (UGA) as selenocysteine. *Journal of Biological Chemistry*. 266: 22386–22391.

Berry, M. J., Banu, L., Chen, Y. Y., Mandel, S. J., Kieffer, J. D., Harney, J. W. and Larsen, P. R., 1991. Recognition of UGA as a selenocysteine codon in type I deiodinase requires sequences in the 3′ untranslated region. *Nature*. 353: 273–276.

Brigelius-Flohe, R. and Maiorino, M., 2013. Glutathione peroxidases. *Biochimica et Biophysica Acta*. 1830: 3289–3303.

Bubenik, J. L. and Driscoll, D. M., 2007. Altered RNA binding activity underlies abnormal thyroid hormone metabolism linked to a mutation in selenocysteine insertion sequence-binding protein 2. *Journal of Biological Chemistry*. 282: 34653–34662.

Budiman, M. E., Bubenik, J. L., Miniard, A. C., Middleton, L. M., Gerber, C. A., Cash, A. and Driscoll, D. M., 2009. Eukaryotic initiation factor 4a3 is a selenium-regulated RNA-binding protein that selectively inhibits selenocysteine incorporation. *Molecular Cell*. 35: 479–489.

Caban, K. and Copeland, P. R., 2012. Selenocysteine insertion sequence (SECIS)-binding protein 2 alters conformational dynamics of residues involved in tRNA accommodation in 80S ribosomes. *Journal of Biological Chemistry*. 287: 10664–10673.

Caban, K., Kinzy, S. A. and Copeland, P. R., 2007. The L7Ae RNA binding motif is a multifunctional domain required for the ribosome-dependent Sec incorporation activity of Sec insertion sequence binding protein 2. *Molecular and Cellular Biology*. 27: 6350–6360.

Campos-Barros, A., Amma, L. L., Faris, J. S., Shailam, R., Kelley, M. W. and Forrest, D., 2000. Type 2 iodothyronine deiodinase expression in the cochlea before the onset of hearing. *Proceedings of the National Academy of Sciences of the United States of America*. 97: 1287–1292.

Carlson, B. A., Moustafa, M. E., Sengupta, A., Schweizer, U., Shrimali, R., Rao, M., Zhong, N., Wang, S., Feigenbaum, L., Lee, B. J., *et al.*, 2007. Selective restoration of the selenoprotein population in a mouse hepatocyte selenoproteinless background with different mutant selenocysteine tRNAs lacking Um34. *Journal of Biological Chemistry*. 282: 32591–32602.

Carlson, B. A., Xu, X. M., Gladyshev, V. N. and Hatfield, D. L., 2005. Selective rescue of selenoprotein expression in mice lacking a highly specialized methyl group in selenocysteine tRNA. *Journal of Biological Chemistry*. 280: 5542–5548.

Castellano, S., Morozova, N., Morey, M., Berry, M. J., Serras, F., Corominas, M. and Guigo, R., 2001. In silico identification of novel selenoproteins in the *Drosophila melanogaster* genome. *EMBO Reports*. 2: 697–702.

Castets, P., Bertrand, A. T., Beuvin, M., Ferry, A., Le Grand, F., Castets, M., Chazot, G., Rederstorff, M., Krol, A., Lescure, A., *et al.*, 2011. Satellite cell loss and impaired muscle regeneration in selenoprotein N deficiency. *Human Molecular Genetics*. 20: 694–704.

Chiba, S., Itoh, Y., Sekine, S. and Yokoyama, S., 2010. Structural basis for the major role of *O*-phosphoseryl-tRNA kinase in the UGA-specific encoding of selenocysteine. *Molecular Cell*. 39: 410–420.

Copeland, P. R. and Driscoll, D. M., 1999. Purification, redox sensitivity, and RNA binding properties of SECIS-binding protein 2, a protein involved in selenoprotein biosynthesis. *Journal of Biological Chemistry*. 274: 25447–25454.

Copeland, P. R., Fletcher, J. E., Carlson, B. A., Hatfield, D. L. and Driscoll, D. M., 2000. A novel RNA binding protein, SBP2, is required for the translation of mammalian selenoprotein mRNAs. *EMBO Journal*. 19: 306–314.

Copeland, P. R., Stepanik, V. A. and Driscoll, D. M., 2001. Insight into mammalian selenocysteine insertion: domain structure and ribosome binding properties of Sec insertion sequence binding protein 2. *Molecular and Cellular Biology*. 21: 1491–1498.

Dentice, M. and Salvatore, D., 2011. Deiodinases: the balance of thyroid hormone: local impact of thyroid hormone inactivation. *Journal of Endocrinology*. 209: 273–282.

Ding, F. and Grabowski, P. J., 1999. Identification of a protein component of a mammalian tRNA(Sec) complex implicated in the decoding of UGA as selenocysteine. *RNA*. 5: 1561–1569.

Donovan, J., Caban, K., Ranaweera, R., Gonzalez-Flores, J. N. and Copeland, P. R., 2008. A novel protein domain induces high affinity selenocysteine insertion sequence binding and elongation factor recruitment. *Journal of Biological Chemistry*. 283: 35129–35139.

Donovan, J. and Copeland, P. R., 2010. The efficiency of selenocysteine incorporation is regulated by translation initiation factors. *Journal of Molecular Biology*. 400: 659–664.

Dumitrescu, A. M., Liao, X. H., Abdullah, M. S., Lado-Abeal, J., Majed, F. A., Moeller, L. C., Boran, G., Schomburg, L., Weiss, R. E. and Refetoff, S., 2005. Mutations in SECISBP2 result in abnormal thyroid hormone metabolism. *Nature Genetics*. 37: 1247–1252.

Fagegaltier, D., Hubert, N., Yamada, K., Mizutani, T., Carbon, P. and Krol, A., 2000a. Characterization of mSelB, a novel mammalian elongation factor for selenoprotein translation. *EMBO Journal*. 19: 4796–4805.

Fagegaltier, D., Lescure, A., Walczak, R., Carbon, P. and Krol, A., 2000b. Structural analysis of new local features in SECIS RNA hairpins. *Nucleic Acids Research*. 28: 2679–2689.

Fradejas, N., Carlson, B. A., Rijntjes, E., Becker, N. P., Tobe, R. and Schweizer, U., 2013. Mammalian Trit1 is a tRNA([Ser]Sec)-isopentenyl transferase required for full selenoprotein expression. *Biochemical Journal*. 450: 427–432.

Hadley, K. B. and Sunde, R. A., 2001. Selenium regulation of thioredoxin reductase activity and mRNA levels in rat liver. *Journal of Nutritional Biochemistry*. 12: 693–702.

Hatfield, D. L. and Gladyshev, V. N., 2002. How selenium has altered our understanding of the genetic code. *Molecular and Cellular Biology*. 22: 3565–3576.

Hill, K. E., Lyons, P. R. and Burk, R. F., 1992. Differential regulation of rat liver selenoprotein mRNAs in selenium deficiency. *Biochemical and Biophysical Research Communications*. 185: 260–263.

Hill, K. E., Zhou, J., McMahan, W. J., Motley, A. K. and Burk, R. F., 2004. Neurological dysfunction occurs in mice with targeted deletion of the selenoprotein P gene. *Journal of Nutrition*. 134: 157–161.

Howard, M. T., Aggarwal, G., Anderson, C. B., Khatri, S., Flanigan, K. M. and Atkins, J. F., 2005. Recoding elements located adjacent to a subset of eukaryal selenocysteine-specifying UGA codons. *EMBO Journal*. 24: 1596–1607.

Howard, M. T., Carlson, B. A., Anderson, C. B. and Hatfield, D. L., 2013. Translational redefinition of UGA codons is regulated by selenium availability. *Journal of Biological Chemistry*. 288: 19401–19413.

Howard, M. T., Moyle, M. W., Aggarwal, G., Carlson, B. A. and Anderson, C. B., 2007. A recoding element that stimulates decoding of UGA codons by Sec $tRNA^{[Ser]Sec}$. *RNA*. 13: 912–920.

Itoh, Y., Chiba, S., Sekine, S. and Yokoyama, S., 2009. Crystal structure of human selenocysteine tRNA. *Nucleic Acids Research*. 37: 6259–6268.

Jurynec, M. J., Xia, R., Mackrill, J. J., Gunther, D., Crawford, T., Flanigan, K. M., Abramson, J. J., Howard, M. T. and Grunwald, D. J., 2008. Selenoprotein N is required for ryanodine receptor calcium release channel activity in human and zebrafish muscle. *Proceedings of the National Academy of Sciences of the United States of America*. 105: 12485–12490.

Köhrle, J., Schweizer, U. and Schomburg, L., 2012. Selenium transport in mammals: selenoprotein P and its receptors. In: Hatfield, D. L., Berry, M. J. and Gladyshev, V. N. (ed.) *Selenium Its Molecular Biology and Role in Human Health*. Springer, New York, pp. 205–220.

Kryukov, G. V., Castellano, S., Novoselov, S. V., Lobanov, A. V., Zehtab, O., Guigo, R. and Gladyshev, V. N., 2003. Characterization of mammalian selenoproteomes. *Science*. 300: 1439–1443.

Kryukov, G. V., Kumar, R. A., Koc, A., Sun, Z. and Gladyshev, V. N., 2002. Selenoprotein R is a zinc-containing stereo-specific methionine sulfoxide reductase. *Proceedings of the National Academy of Sciences of the United States of America*. 99: 4245–4250.

Latrèche, L., Jean-Jean, O., Driscoll, D. M. and Chavatte, L., 2009. Novel structural determinants in human SECIS elements modulate the translational recoding of UGA as selenocysteine. *Nucleic Acids Research*. 37: 5868–5880.

Lee, B. C., Peterfi, Z., Hoffmann, F. W., Moore, R. E., Kaya, A., Avanesov, A., Tarrago, L., Zhou, Y., Weerapana, E., Fomenko, D. E., *et al.*, 2013a. MsrB1 and MICALs regulate actin assembly and macrophage function *via* reversible stereoselective methionine oxidation. *Molecular Cell*. 51: 397–404.

Lee, S., Kim, S. M. and Lee, R. T., 2013b. Thioredoxin and thioredoxin target proteins: from molecular mechanisms to functional significance. *Antioxidants & Redox Signaling*. 18: 1165–1207.

Lei, X. G., Evenson, J. K., Thompson, K. M. and Sunde, R. A., 1995. Glutathione peroxidase and phospholipid hydroperoxide glutathione peroxidase are differentially regulated in rats by dietary selenium. *Journal of Nutrition*. 125: 1438–1446.

Lescure, A., Fagegaltier, D., Carbon, P. and Krol, A., 2002. Protein factors mediating selenoprotein synthesis. *Current Protein & Peptide Science*. 3: 143–151.

Lescure, A., Gautheret, D., Carbon, P. and Krol, A., 1999. Novel selenoproteins identified in silico and in vivo by using a conserved RNA structural motif. *Journal of Biological Chemistry*. 274: 38147–38154.

Lobanov, A. V., Hatfield, D. L. and Gladyshev, V. N., 2008. Selenoproteinless animals: selenophosphate synthetase SPS1 functions in a pathway unrelated to selenocysteine biosynthesis. *Protein Science*. 17: 176–182.

Loh, K., Deng, H., Fukushima, A., Cai, X., Boivin, B., Galic, S., Bruce, C., Shields, B. J., Skiba, B., Ooms, L. M., *et al.*, 2009. Reactive oxygen species enhance insulin sensitivity. *Cell Metabolism*. 10: 260–272.

Low, S. C., Grundner-Culemann, E., Harney, J. W. and Berry, M. J., 2000. SECIS-SBP2 interactions dictate selenocysteine incorporation efficiency and selenoprotein hierarchy. *EMBO Journal*. 19: 6882–6890.

Maiti, B., Arbogast, S., Allamand, V., Moyle, M. W., Anderson, C. B., Richard, P., Guicheney, P., Ferreiro, A., Flanigan, K. M. and Howard, M. T., 2009. A mutation in the SEPN1 selenocysteine redefinition element (SRE) reduces selenocysteine incorporation and leads to SEPN1-related myopathy. *Human Mutation*. 30: 411–416.

Moghadaszadeh, B., Petit, N., Jaillard, C., Brockington, M., Roy, S. Q., Merlini, L., Romero, N., Estournet, B., Desguerre, I., Chaigne, D., *et al.*, 2001. Mutations in SEPN1 cause congenital muscular dystrophy with spinal rigidity and restrictive respiratory syndrome. *Nature Genetics*. 29: 17–18.

Moriarty, P. M., Reddy, C. C. and Maquat, L. E., 1998. Selenium deficiency reduces the abundance of mRNA for Se-dependent glutathione peroxidase 1 by a UGA-dependent mechanism likely to be nonsense codon-mediated decay of cytoplasmic mRNA. *Molecular and Cellular Biology*. 18: 2932–2939.

Ng, L., Hernandez, A., He, W., Ren, T., Srinivas, M., Ma, M., Galton, V. A., St Germain, D. L. and Forrest, D., 2009. A protective role for type 3 deiodinase, a thyroid hormone-inactivating enzyme, in cochlear development and auditory function. *Endocrinology*. 150: 1952–1960.

Olieric, V., Wolff, P., Takeuchi, A., Bec, G., Birck, C., Vitorino, M., Kieffer, B., Beniaminov, A., Cavigiolio, G., Theil, E., *et al.*, 2009. SECIS-binding protein 2, a key player in selenoprotein synthesis, is an intrinsically disordered protein. *Biochimie*. 91: 1003–1009.

Palioura, S., Sherrer, R. L., Steitz, T. A., Soll, D. and Simonovic, M., 2009. The human SepSecS-tRNASec complex reveals the mechanism of selenocysteine formation. *Science*. 325: 321–325.

Pedersen, J. S., Bejerano, G., Siepel, A., Rosenbloom, K., Lindblad-Toh, K., Lander, E. S., Kent, J., Miller, W. and Haussler, D., 2006. Identification and classification of conserved RNA secondary structures in the human genome. *PLoS Computational Biology*. 2: e33.

Rederstorff, M., Castets, P., Arbogast, S., Laine, J., Vassilopoulos, S., Beuvin, M., Dubourg, O., Vignaud, A., Ferry, A., Krol, A., *et al.*, 2011. Increased muscle stress-sensitivity induced by selenoprotein N inactivation in mouse: a mammalian model for SEPN1-related myopathy. *PLoS One*. 6: e23094.

Schoenmakers, E., Agostini, M., Mitchell, C., Schoenmakers, N., Papp, L., Rajanayagam, O., Padidela, R., Ceron-Gutierrez, L., Doffinger, R., Prevosto, C., *et al.*, 2010. Mutations in the selenocysteine insertion sequence-binding protein 2 gene lead to a multisystem selenoprotein deficiency disorder in humans. *Journal of Clinical Investigation*. 120: 4220–4235.

Schomburg, L. and Kohrle, J., 2008. On the importance of selenium and iodine metabolism for thyroid hormone biosynthesis and human health. *Molecular Nutrition & Food Research*. 52: 1235–1246.

Schomburg, L. and Schweizer, U., 2009. Hierarchical regulation of selenoprotein expression and sex-specific effects of selenium. *Biochimica et Biophysica Acta*. 1790: 1453–1462.

Seeher, S., Atassi, T., Mahdi, Y., Carlson, B. A., Braun, D., Wirth, E. K., Klein, M. O., Reix, N., Miniard, A. C., Schomburg, L., *et al.*, 2013. Secisbp2 is essential for embryonic development and enhances selenoprotein expression. *Antioxidants & Redox Signaling*. 21(6): 835–849.

Seeher, S., Mahdi, Y. and Schweizer, U., 2012. Post-transcriptional control of selenoprotein biosynthesis. *Current Protein & Peptide Science*. 13: 337–346.

Sherrer, R. L., Araiso, Y., Aldag, C., Ishitani, R., Ho, J. M., Soll, D. and Nureki, O., 2011. C-Terminal domain of archaeal *O*-phosphoseryl-tRNA kinase displays large-scale motion to bind the 7-bp D-stem of archaeal tRNA(Sec). *Nucleic Acids Research*. 39: 1034–1041.

Small-Howard, A., Morozova, N., Stoytcheva, Z., Forry, E. P., Mansell, J. B., Harney, J. W., Carlson, B. A., Xu, X. M., Hatfield, D. L. and Berry, M. J., 2006. Supramolecular complexes mediate selenocysteine incorporation in vivo. *Molecular and Cellular Biology*. 26: 2337–2346.

Squires, J. E., Stoytchev, I., Forry, E. P. and Berry, M. J., 2007. SBP2 binding affinity is a major determinant in differential selenoprotein mRNA translation and sensitivity to nonsense-mediated decay. *Molecular and Cellular Biology*. 27: 7848–7855.

Takeuchi, A., Schmitt, D., Chapple, C., Babaylova, E., Karpova, G., Guigo, R., Krol, A. and Allmang, C., 2009. A short motif in *Drosophila* SECIS binding protein 2 provides differential binding affinity to SECIS RNA hairpins. *Nucleic Acids Research*. 37: 2126–2141.

Tujebajeva, R. M., Copeland, P. R., Xu, X. M., Carlson, B. A., Harney, J. W., Driscoll, D. M., Hatfield, D. L. and Berry, M. J., 2000. Decoding apparatus for eukaryotic selenocysteine insertion. *EMBO Reports*. 1: 158–163.

Ursini, F., Heim, S., Kiess, M., Maiorino, M., Roveri, A., Wissing, J. and Flohé, L., 1999. Dual function of the selenoprotein PHGPx during sperm maturation. *Science*. 285: 1393–1396.

Weiss, S. L. and Sunde, R. A., 1998. *Cis*-acting elements are required for selenium regulation of glutathione peroxidase-1 mRNA levels. *RNA*. 4: 816–827.

Wu, X. Q. and Gross, H. J., 1993. The long extra arms of human tRNA((Ser) Sec) and tRNA(Ser) function as major identify elements for serylation in an orientation-dependent, but not sequence-specific manner. *Nucleic Acids Research*. 21: 5589–5594.

Ye, Y., Shibata, Y., Yun, C., Ron, D. and Rapoport, T. A., 2004. A membrane protein complex mediates retro-translocation from the ER lumen into the cytosol. *Nature*. 429: 841–847.

Zinoni, F., Heider, J. and Bock, A., 1990. Features of the formate dehydrogenase mRNA necessary for decoding of the UGA codon as selenocysteine. *Proceedings of the National Academy of Sciences of the United States of America*. 87: 4660–4664.

CHAPTER 3

Synthesis of Organoselenium Derivatives of Biological Relevance

ÓSCAR LÓPEZ*[a], PENÉLOPE MERINO-MONTIEL[b], AND JOSÉ G. FERNÁNDEZ-BOLAÑOS[a]

[a]Department of Organic Chemistry, Faculty of Chemistry, University of Seville, Profesor García González 1, E-41012, Seville, Spain; [b]Facultad de Ciencias Químicas, Benemérita Universidad Autónoma de Puebla, Ciudad Universitaria, Col. San Manuel, 72570, Puebla, Mexico
*E-mail: osc-lopez@us.es

3.1 Introduction

The metalloid selenium has been surrounded by controversy since its isolation in 1817 by the Swedish chemist Berzelius (Weekley and Harris, 2013). The first historical references reported livestock poisoning caused by the ingestion of this element coming from selenium-accumulating plants in seleniferous soils (Nogueira and Rocha, 2011), a process known as selenosis. Nevertheless, the discovery in the mid-20th century of selenium-containing proteins (*e.g.* glutathione peroxidase, GPx), which proved to act as natural defenses against oxidative stress provoked by reactive oxygen and nitrogen species (ROS, RNS), changed the conception of selenium, which started to be considered as an essential micronutrient (Iwaoka and Arai, 2013); up to now, more than 30 selenoproteins have been reported, and epidemiologic

Food and Nutritional Components in Focus No. 9
Selenium: Chemistry, Analysis, Function and Effects
Edited by Victor R. Preedy

Published by the Royal Society of Chemistry, www.rsc.org

studies revealed that selenium undergoes essential biological roles in living beings, like maintaining the redox status, or the thyroid hormone production (Tinggi, 2008). It has also been demonstrated that low concentrations of selenium in diet can also lead to the development of a series of diseases and disorders (Peng *et al.*, 2012), so a balanced diet must be fulfilled.

The interest in synthetic organoselenium chemistry has increased incessantly since the 1970s, on the one hand, due to the vast arsenal of applications of organoselenium compounds in modern organic synthesis (Freudendahl *et al.*, 2009), and on the other hand, due to the discovery of new encouraging biological properties exhibited by such compounds (Tiekink, 2012). As a result, a plethora of new derivatives have been conceived as antioxidants to fight oxidative stress (Bhabak and Mugesh, 2010), and also as antimicrobial (Victoria *et al.*, 2012), anti-inflammatory (Uzma *et al.*, 2011) or anticancer (Suzuki *et al.*, 2010) agents.

3.2 Isoselenocyanates and Related Selenoxo Compounds

Isoselenocyanates (R–N=C=Se) are highly versatile organoselenium building-blocks (Heimgartner *et al.*, 2008), as their intrinsic electrophilicity allows reactions with different nucleophiles, leading to the preparation of a plethora of structural families.

López *et al.* (2009) developed a practical procedure for the preparation of aryl and alkyl isoselenocyanates; it is based on the triphosgene-promoted dehydration of formamides and subsequent *in situ* coupling of the transient isocyanide with elemental black selenium (Scheme 3.1). Triphosgene (bis(trichloromethyl)carbonate) was claimed to be a safe substitute of hazardous phosgene.

Following the same synthetic approach, Sharma *et al.* (2008) accomplished the preparation of a series of phenylalkyl isoselenocyanates (up to 6 alkyl carbons) and evaluated *in vitro* their antiproliferative activity against a panel of 6 human cancer cell lines (breast, glioblastoma, prostate, fibrosarcoma, colon and melanoma); their IC_{50} values ranged from 10 to 49 μM. Phenylalkyl isoselenocyanates turned out to be stronger anticancer agents than their sulfur isosters (IC_{50} from 12 to >100 μM), and, generally, the activity was more pronounced upon increasing the alkyl chain length. Moreover, *in vivo* experiments in mice (melanoma xenograft model) showed that isothiocyanates required a 3-fold increase in concentration to exhibit the same potency as their selenium counterparts.

Coupling of phenyl isoselenocyanate with *O*-unprotected β-D-gluco(manno) pyranosylamine afforded the corresponding glycopyranosyl selenoureas **8, 9** in good yields (72 and 69%, respectively). Promising antioxidant activity was found for **8** (López *et al.*, 2009); such compound was slowly transformed into *trans*-fused bicyclic isourea **10** upon exposure to light, with selenium red being released (Scheme 3.1). This reaction was remarkably accelerated upon addition of H_2O_2, affording the formation of **10** in excellent yield (84%) in less than 1 h. It was suggested that the formation of **10** takes place *via* a non-detected carbodiimide with subsequent cyclization through the hydroxyl group on C-2.

Treatment of aryl isoselenocyanates (phenyl, *p*-tolyl, *o*-tolyl, 2-isopropyl-6-methylphenyl, and α-naphthyl) with refluxing ethanol, afforded the corresponding selenocarbamates **4** (Merino-Montiel *et al.*, 2013) upon nucleophilic addition of ethanol on the heterocumulene moiety (Scheme 3.1). Moreover, coupling reaction of phenyl and α-naphthyl isoselenocyanates with L-valine ethyl ester hydrochloride afforded the corresponding selenoureas as a new family of peptidomimetics; in the case of the phenyl isoselenocyanate, the corresponding selenourea **11** was not detected, as it underwent a spontaneous intramolecular cyclization involving Ar-NH and ester groups, to give the selenohydantoin **12** in a quantitative yield. Regarding α-naphthyl isoselenocyanate, a mixture of selenourea **13** and selenohydantoin **14** was isolated (53 and 25% yield, respectively); probably the higher steric hindrance involving the naphthyl residue decreased the rate of the cyclization (Scheme 3.1).

A new family of selenoureas (**5–7**) was reported by Fernández-Bolaños *et al.* (2012), by coupling of the corresponding aryl isoselenocyanates with dopamine hydrochloride; these compounds combine two antioxidant motifs: the selenoureido tether, and the catechol fragment, present in numerous natural polyphenols. These compounds were evaluated *in vitro* as scavengers of Fe^{3+}, free radicals, and also as inhibitors of lipid peroxidation. Results suggested a higher antioxidant potency than natural hydroxytyrosol 2-(3′,4′-dihydroxyphenyl)ethanol, an abundant natural polyphenol in olive tree. Furthermore, *ex vivo* experiments (inhibition of rat liver microsomal lipid peroxidation) suggested a similar bioactivity to that exhibited by α-tocopherol.

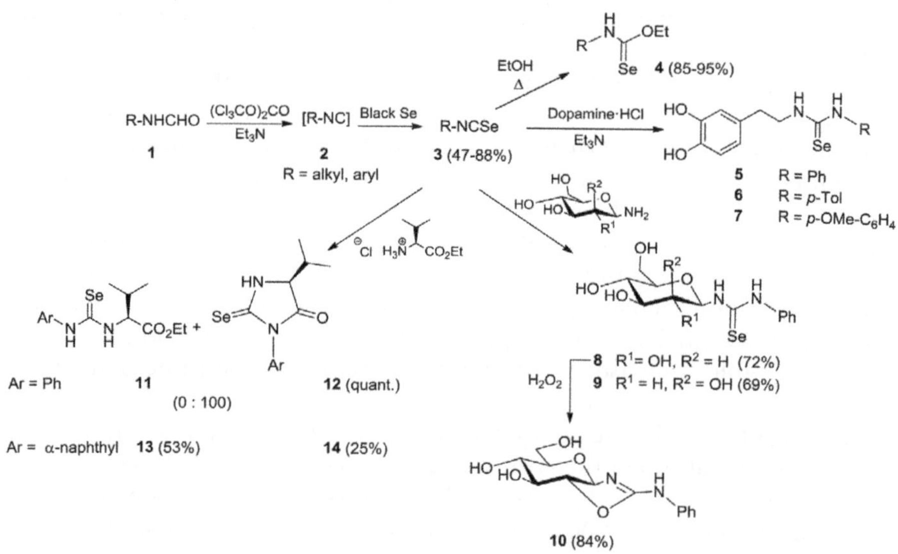

Scheme 3.1 Synthesis of alkyl and aryl-isoselenocyanates and their conversion into selenocarbamates, selenoureas and selenohydantoins.

López *et al.* (2009) also reported the preparation of sugar isoselenocyanates, derived from *O*-protected β-D-glucopyranosylamine (Scheme 3.2) and D-glucosamine, following a similar procedure to the one depicted in Scheme 3.1. Sugar formamides, the key intermediates in the synthesis, were obtained by treatment of the corresponding sugar-derived amine hydrohalides with AFA. Scheme 3.2 depicts the synthesis of *O*-acetylated β-D-glucopyranosyl isoselenocyanate **17** and its conversion into alkyl/aryl selenoureas **18** (including selenoureido-linked pseudodisaccharides) (López *et al.*, 2009), selenocarbamates **19** and peptidomimetics derived from L-aminoacids **20** (Merino-Montiel *et al.*, 2013).

Aryl selenocarbamates **4**, selenourea **13**, selenohydantoins **12**, **14**, sugar-derived selenocarbamates **19**, and peptidyl-selenoureas **20** were subjected to a structure–activity study concerning their antioxidant properties (Merino-Montiel *et al.*, 2013): free radical (DPPH), H_2O_2 and NO scavenging activities, lipid peroxidation inhibition and glutathione peroxidase mimicry. Relevant results are depicted in Table 3.1.

As indicated in Table 3.1, a remarkable difference in the antioxidant behavior was found for the selected compounds; thus, selenoureas, which turned out to be excellent lipophilic antioxidants, exhibited a much more potent antiradical (EC_{50} 19–40 μM) and GPx-like (half-life times 2.0–2.9 min) activities than related selenocarbamates (EC_{50} 43–323 μM; $t_{1/2}$ 100.0–108.0 min) and selenohydantoins (EC_{50} 147–379 μM; $t_{1/2}$ 180 min). It was claimed

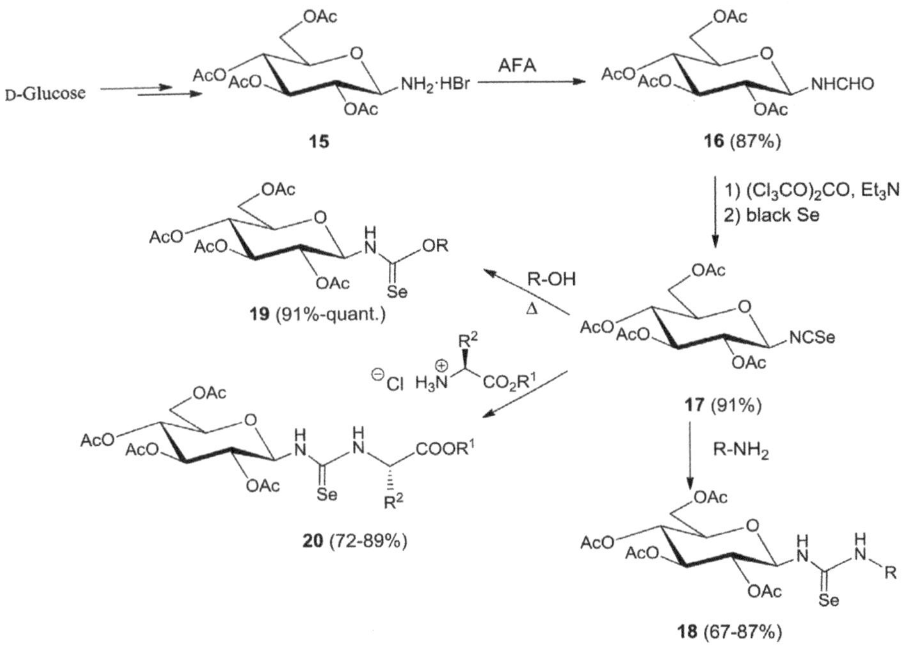

Scheme 3.2 Synthesis of carbohydrate-derived isoselenocyanates and their conversion into selenocarbamates and selenoureas.

that probably the major electronic density located on the selenium atom accounted for this improved activity.

Some selenosemicarbazones have also received attention because of their antiparasitic and antiproliferative activities. Derivatives **23** can be accessed by a condensation reaction between aromatic aldehydes/ketones and selenosemicarbazide (Pizzo *et al.*, 2012; Zaharia *et al.*, 2013), as depicted in Scheme 3.3. Some of the reported selenosemicarbazones behaved as potent inhibitors of cruzipain (Pizzo *et al.*, 2012), a pivotal protease in the life cycle of

Table 3.1 Selected antioxidant activities of selenocarbamates, selenoureas and selenohydantoins.

Compound	DPPH (EC_{50}, μM)	GPx-like activity ($t_{1/2}$, min)[a]
4a (R=Ph)	148	—[b]
4b (R=α-naphthyl)	43	108.0
19a (R=*n*-Pr)	323	100.0
12	147	—[b]
14	379	180.0
13	19	2.0
20a (Gly)	25	2.9
20b (Phe)	40	12.9

[a]A 5% molar ratio was used for each of the organoselenium derivatives.
[b]Not tested.

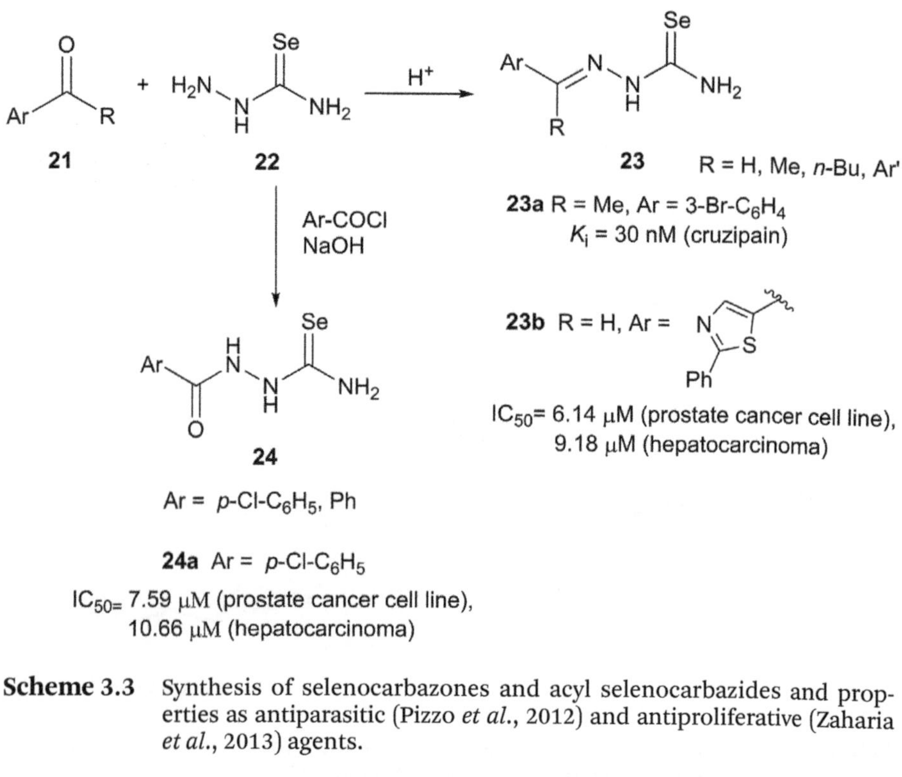

Scheme 3.3 Synthesis of selenocarbazones and acyl selenocarbazides and properties as antiparasitic (Pizzo *et al.*, 2012) and antiproliferative (Zaharia *et al.*, 2013) agents.

Tripanosomas cruzi, and therefore, a pharmaceutical target in the treatment of Chagas disease. Compound **23a** is remarkable, with K_i in the nanomolar range (30 nM); furthermore, these kinds of compounds showed improved activity when compared with benznidazole, an approved drug, and also, a better selectivity (Pizzo *et al.*, 2012). Moreover, when tested at low micromolar concentrations, they exhibited activity against amastigote intracellular infection.

Some of the selenosemicarbazones prepared by Zaharia *et al.* (2013) showed good antiproliferative activity against two cancer cell lines (prostatic and hepatocarcinome); the best results were found for derivative **23b**, with IC_{50} of 6.14 and 9.18 μM, respectively. The same group accomplished the preparation of *N*-acyl selenosemicarbazides **24**, by reaction of selenosemicarbazide with acid chlorides (Scheme 3.3); they showed antiproliferative capacity.

3.3 Selenosugars

Replacement of the endocyclic oxygen atom by selenium in carbohydrates has allowed the development of antioxidant agents, and also novel families of glycosidase inhibitors.

In this context, Iwaoka *et al.* (2001) reported the preparation of the water-soluble D,L-*trans*-3,4-dihydroxyselenolane **26**, its oxidized form **28**, diselenane **30**, and diselenothreitol **31**. Such compounds were designed with the aim of controlling the redox state of proteins by regenerating their disulfide bonds and therefore, their native structure when such bonds are cleaved. Compounds **26** and **30** were obtained by nucleophilic opening of diepoxide **25** with freshly prepared NaHSe or Na_2Se_2 (Scheme 3.4).

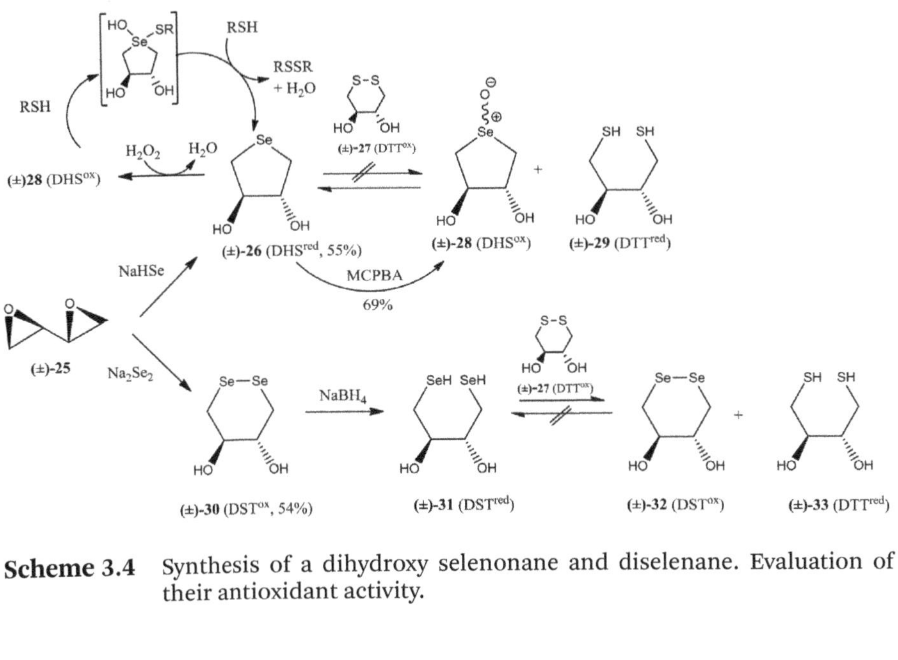

Scheme 3.4 Synthesis of a dihydroxy selenonane and diselenane. Evaluation of their antioxidant activity.

Moreover, oxidation of **26** with MCPBA, and reduction of diselenane **30** with $NaBH_4$ afforded derivatives **28** and **31**, respectively. Redox activity of these compounds was evaluated by considering reaction between selenoxide **28** and the reduced form of dithiothreitol **29**, and between **31** and dithiothreitol **27**, as depicted in Scheme 3.4; both reactions proceeded fast (within 10 s, and 5 min respectively), thus proving that the oxidizing power of selenoxide **28** is much stronger than that of DTT^{ox}, and that the reducing power of DST^{red} exceeds that of DTT^{red} (Iwaoka *et al.*, 2001).

The strong oxidation power of selenoxide **28** was used to study the oxidative regeneration pathways underwent by denatured bovine pancreatic ribonuclease A; oxidative folding took place by interaction of **28** with the corresponding cysteinyl residues (Iwaoka *et al.*, 2008).

Dihydroxyselenolane **26** was also investigated as a glutathione peroxidase mimic (Kumakura *et al.*, 2010), exhibiting good GPx-like catalytic activity for the reduction of H_2O_2 in the presence of thiol cofactors (Scheme 3.4).

The antioxidant properties of **26** were exploited in the treatment of indomethacin-induced gastric ulceration in mice, which was also accompanied by oxidative stress, leading to lipid peroxidation and oxidative degradation of proteins. This compound exhibited a better ulcer-healing profile than the control, omeprazole, and its activity can be attributed to its antioxidant properties, to its ability to induce mucin regeneration, and to the increase of prostaglandin synthesis by upregulating the cyclooxygenase (COX) system (Chakraborty *et al.*, 2012).

Other examples of water-soluble antioxidant selenosugars were reported by Storkey *et al.* (2011, 2012). The key step is a double nucleophilic displacement of freshly prepared NaHSe on a protected dimesylated polyol derivative. Thus, L-*gulo*-configured selenosugar **38** was efficiently prepared in a 5-step strategy starting from D-mannose (Scheme 3.5) in a 19% overall yield: regioselective protection of C-2–C-4 and C-6 hydroxyl groups using

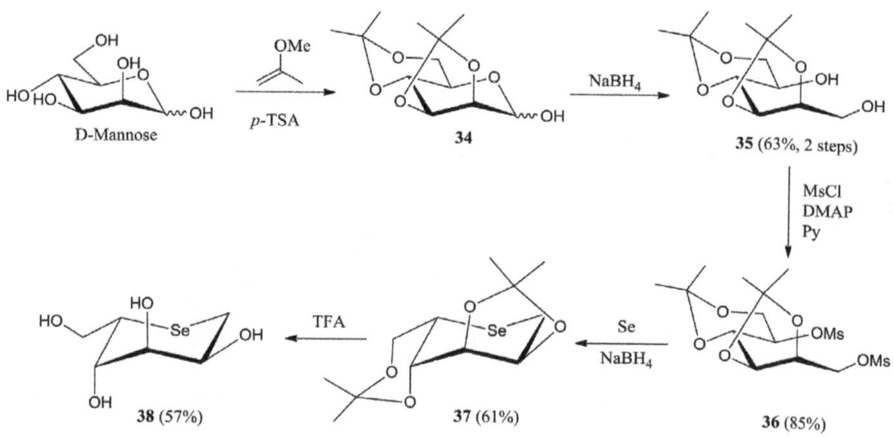

Scheme 3.5 Synthesis of 1,5-dideoxy-5-seleno-L-gulitol.

2-methoxypropene, reduction of the latent aldehyde group, di-*O*-mesylation, double nucleophilic displacement with NaHSe (generated *in situ* by reduction of elemental selenium with $NaBH_4$) and deprotection (Storkey *et al.*, 2011).

Selenosugar **43**, with D-*manno* configuration, was obtained also starting from D-mannose using a synthetic pathway that involved a double inversion on C-5 (Scheme 3.6); thus, diol **35** was treated with TBDMSCl to give **39** in a regioselective fashion. Subsequent Swern-type oxidation and reduction of the transient ketone **40** under Luche conditions ($NaBH_4$ in the presence of methanolic $CeCl_3$) allowed an inversion on C-5 to give **41**, with L-*gulo* configuration (Storkey *et al.*, 2011). Deprotection of the silylated primary alcohol with TBAF, and dimesylation of the transient diol derivative, afforded **42** in an 86% yield.

Double nucleophilic substitution exerted by freshly prepared NaHSe and deprotection of the acetal groups under acidic conditions afforded 5-selenopyranose **43**.

The same group accomplished the synthesis of 1,4-dideoxy-4-seleno-D-talitol **45** (Storkey *et al.*, 2012) from D-mannose, *via* a regioselective protection with dimethoxypropane of C-2/C-3 and C-5/C-6 hydroxyl groups (Scheme 3.6); such a synthetic pathway was again based on a double inversion involving C-4 (oxidation to ketone, Luche-type reduction and nucleophilic displacement exerted by NaHSe).

L-Gulo **38**, D-manno **43**, and D-talo **45** derivatives were assayed as potential scavengers of oxidant hypohalous acids (HOX; X = Cl, Br); such species are generated by NADPH oxidases-mediated oxidation of halides anions with H_2O_2. When high concentrations of hypohalous acids are generated in cells, reaction with pivotal biomolecules take place.

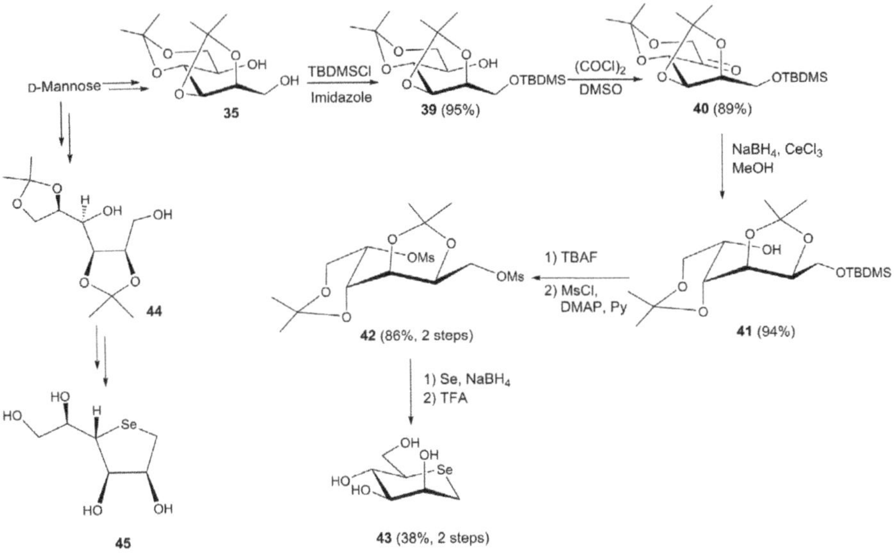

Scheme 3.6 Synthesis of selenosugars **43** and **45**.

Second-order constants for the scavenging activities of the above-mentioned selenosugars were measured, and compared with selenomethionine (SeMet), and glutathione (GSH) (Storkey *et al.*, 2012, and references therein). Data indicated that scavenging rates are remarkably higher for HOCl in comparison with HOBr for both, selenosugars ($8.0–10.0 \times 10^7\ M^{-1}\ s^{-1}$, $1.09–1.52 \times 10^7\ M^{-1}\ s^{-1}$, respectively) and reference compounds. Furthermore, no substantial differences were found among selenosugars, suggesting that stereochemistry and ring size do not exert any influences in the activity. Moreover, although HOCl scavenging rates of the selenosugars are slower than that of SeMet ($32 \times 10^7\ M^{-1}\ s^{-1}$), there are almost no differences from GSH ($10.9 \times 10^7\ M^{-1}\ s^{-1}$).

Selenosugar protection of bovine serum albumin (BSA) and human plasma against HOCl oxidation was also assayed, indicating that these compounds exerted a dose-dependent protection (Storkey *et al.*, 2012); provided data do not confirm whether the scavenging activity takes place either by removing HOCl or some of the transient intermediates.

Merino-Montiel *et al.* (2012) accomplished the preparation of the first examples of L-*fuco*-configured selenosugars (Scheme 3.7), and evaluated how the biological properties were affected by the electronic density on the selenium atom. Such compounds are seleno-isosters of isofucofagomine, an iminosugar exhibiting potent glycosidase inhibition.

O-Protected selenane **48** was obtained starting from L-arabinose in 9 steps; the key step is the double nucleophilic substitution by *in situ* generated Na_2Se

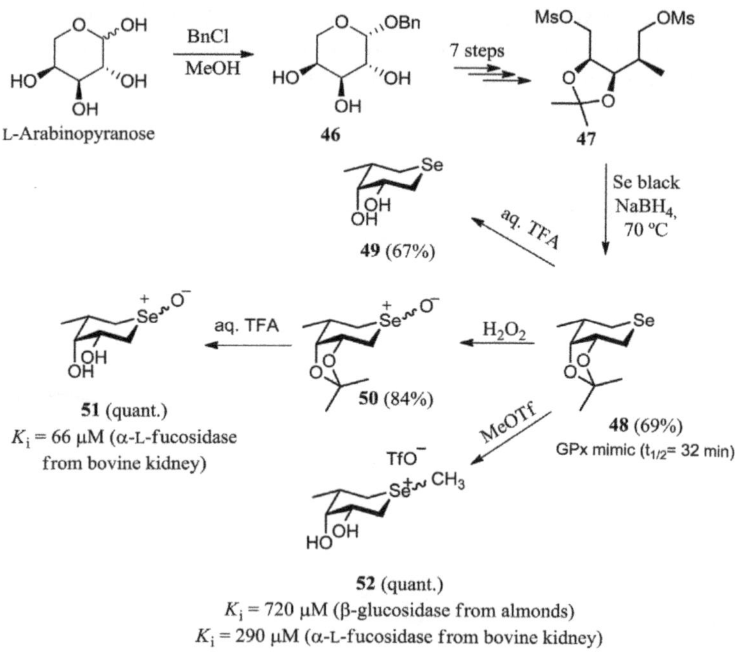

Scheme 3.7 Synthesis of L-*fuco*-selenosugars: selenane, selenoxide and selenonium derivatives.

on dimesylated derivative **47**. Nucleophilic attack of the selenium atom on methyl triflate, or H_2O_2-mediated oxidation afforded methyl selenonium **52** (spontaneous deprotection took place) and selenoxide **50**, respectively; removal of the acetal group under acidic conditions of selenane and selenoxide afforded *O*-unprotected derivatives **49** and **51**, respectively (Scheme 3.7). An intramolecular non-bonded interaction involving the OH on C-4 and an antibonding orbital of the selenium atom was detected for selenoxide **51**; remarkably, it has been reported that such interactions are responsible for the reactivity and bioactivity of numerous organoselenium derivatives (Bhabak and Mugesh, 2010).

O-Protected selenane **48** was evaluated as a GPx mimic; it was found that, when used at a 10% molar ratio, it enabled the removal of H_2O_2 in the presence of DTT as a thiol source ($t_{1/2}$ = 32 min).

Unprotected selenane **49**, selenoxide **51** and methyl selenonium **52** were assayed as glycosidase inhibitors (Scheme 3.7). The better inhibition profile was found against α-L-fucosidase by the selenonium and the selenoxide derivatives in comparison with parent selenane; this suggested that a certain positive charge density on selenium atom is required to improve the activity. Remarkably, a K_i of 66 μM was measured for selenoxide **51** (Merino-Montiel *et al.*, 2012), and although it is roughly 10 times higher than the value reported for isofucofagomine, it was the first example of a selenosugar inhibiting α-L-fucosidase.

Pinto's group has accomplished the synthesis of a vast arsenal of selenosugar selenonium salts as isosters of sulfonium derivatives isolated from *Salacia reticulata* and some other plants belonging to the *Salacia* genus; aqueous herbal extracts from *Salacia reticulata* have been widely used in the traditional Ayurvedic medicine for the treatment of type-2-diabetes. Such activity is due to the potent α-glucosidase inhibition exerted by these sulfonium compounds, like kotalanol **53**, ponkoranol **55**, salacinol **57** and their *O*-desulfonated species **54**, **56** and **58**, respectively (Figure 3.1).

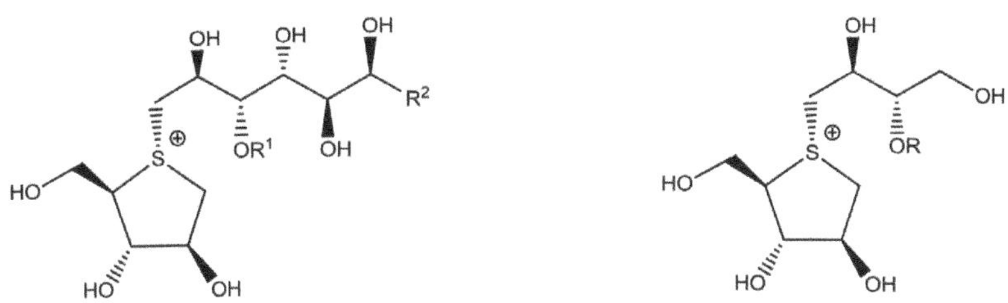

Kotalanol (**53**)	R^1= $-SO_3^-$, R^2 = $-CH_2OH$	Salacinol (**57**)	R= $-SO_3^-$
De-*O*-sulfonated kotalanol (**54**)	R^1= H, R^2 = $-CH_2OH$	De-*O*-sulfonated salacinol (**58**)	R = H
Ponkoranol (**55**)	R^1= $-SO_3^-$, R^2 = H		
De-*O*-sulfonated ponkoranol (**56**)	R^1 = R^2 = H		

Figure 3.1 Naturally occurring sulfonium derivatives isolated from *Salacia reticulate*.

In this context, an efficient synthesis of blintol **63**, the seleno-isoster of naturally occurring salacinol **57**, was reported by Liu and Pinto (2005) in 11 steps from L-xylose (Scheme 3.8). The key intermediate is PMB-protected 1,4-anhydro-4-seleno-D-arabinitol **61**; this compound can be accessed in good overall yield following a strategy based on the acetylation of L-xylose in the presence of boric acid in order to increase the furanose/pyranose ratio, and also on the use of an *n*-pentenyl glycoside, which can be removed without affecting the *O*-protecting groups. Coupling of **61** with the cyclic sulfate **62**, obtained in 9 steps from D-glucose, and subsequent deprotection with TFA furnished blintol in a 12% overall yield; the corresponding *cis*-epimer **64** on the selenium atom was also obtained as a minor compound.

Following similar synthetic pathways, Pinto and coworkers underwent structure–activity relationship studies with the aim of preparing more potent and selective glycosidase inhibitors based on polyhydroxyalkyl selenonium derivatives. Modifications on the pendant chain length and stereochemistry, together with the presence of absence of the sulfate group, led to the development of a series of analogs of the naturally occurring sulfonium compounds, with remarkable inhibitory profiles (Scheme 3.9).

Thus, coupling of *O*-PMB- or *O*-Bn-protected 1,4-anhydro-4-seleno-D-arabinitol motifs **61** or **65** with sulfates **66** (from D-arabinitol, Nasi *et al.*, 2007), **68** (with D-galacto and D-gluco configurations, Liu *et al.*, 2007), **71** (from D-perseitol, Mohan *et al.*, 2010), and with 6-*O*-tosyl-D-gluco and D-manno glycosides **74** (Eskandari *et al.*, 2011) furnished selenonium derivatives **67, 69, 70, 72, 73, 75** and **76** (Scheme 3.9), the selenium heteroanalogs of the products isolated from *Salacia* plants.

These compounds were evaluated *in vitro* as inhibitors of recombinant maltase glucoamylase (MGAM), the intestinal enzyme that catalyzes the hydrolysis of linear and branched oligosaccharides into glucose, such oligosaccharides

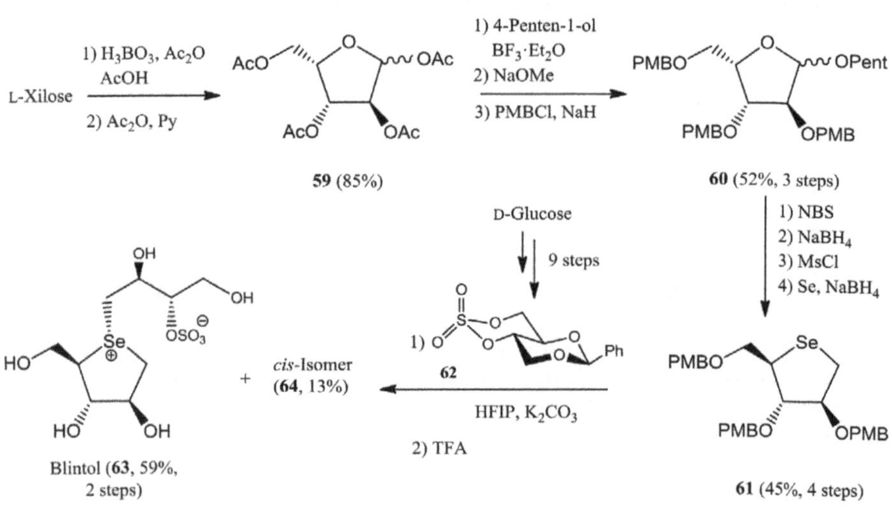

Scheme 3.8 Synthesis of blintol.

coming from previous partial hydrolysis of starch by α-amylase. MGAM exhibits two catalytic units (Jones *et al.*, 2011), named *N*-terminal (ntMGAM) and *C*-terminal (ctMGAM). Thus, selective inhibitors of one such subsite can provide more efficient drugs in the treatment of noninsulin-dependent diabetes, with reduced side-effects.

Data depicted in Figure 3.2 indicate a strong selectivity towards *C*-terminal subsite (18 nM) in comparison with *N*-terminal unit (490 nM) for blintol **63** (Ghavami *et al.*, 2001; Jones *et al.*, 2011); furthermore, for the latter catalytic site, there is a clear impairment in its inhibition when compared with natural sulfonium derivatives (30–190 nM). Compound **67**, a blintol analog bearing an extended chain with the same configuration at C-1′–C-3′ positions (Nasi *et al.*, 2007) showed an improvement of the inhibitory activity against ntMGAM (140 nM). Compounds **69** and **70** are diastereoisomers of **67**, where configurations at C-5′ or C-3′ and C-5′ are changed; inhibition constants (100 nM in both cases) suggest that elongation of the chain provokes again a positive effect compared with parent blintol **63**. Furthermore, configuration at C-3′ does not exert any influence on bioactivity, as both, **69** and **70** exhibit the same inhibition potency (Liu *et al.*, 2007).

Derivatives **72** and **73** are the selenoisosters of kotalanol **53** and de-*O*-sulfonated kotalanol **54**, respectively. For **72**, there is a roughly 2.4-fold increase in activity (80 nM *vs.* 190 nM) when compared with counterpart **53**, whereas there are almost no differences between both de-*O*-sulfonated species

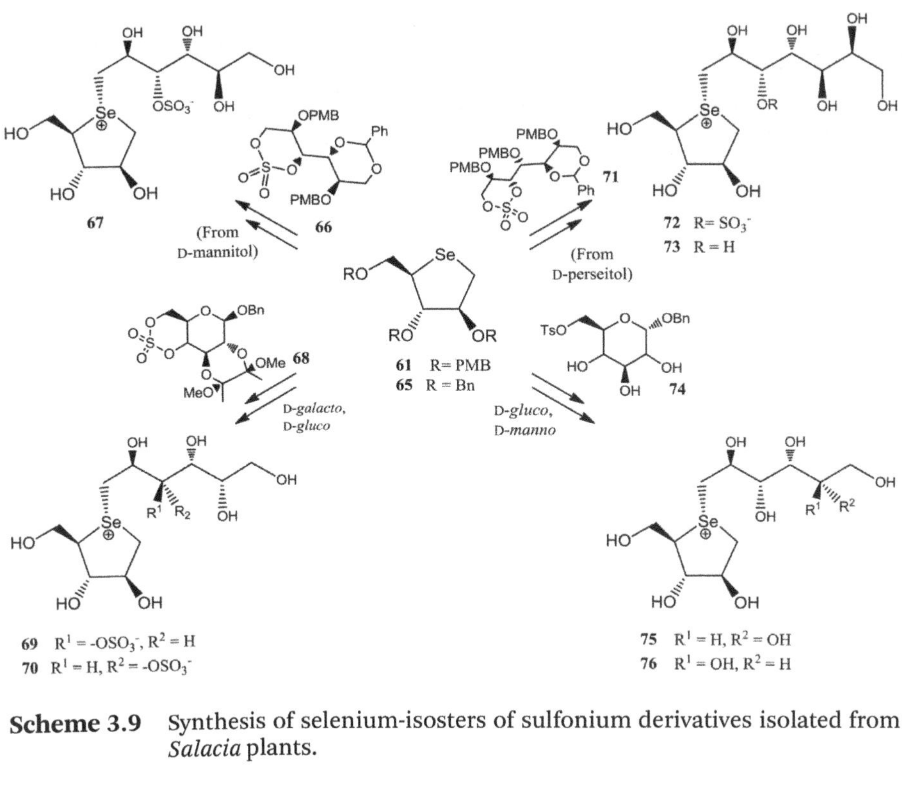

Scheme 3.9 Synthesis of selenium-isosters of sulfonium derivatives isolated from *Salacia* plants.

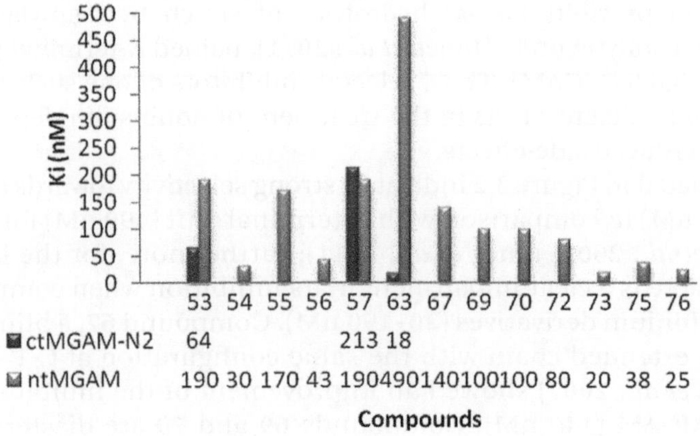

Figure 3.2 Inhibitory constants (K_i, nM) for selenonium derivatives compared with sulfonium counterparts **53–57** against ntMGAM. Data taken from Ghavami *et al.* (2001), Rossi *et al.* (2006), Liu *et al.* (2007), Nasi *et al.* (2007), Eskandari *et al.* (2010), Mohan *et al.* (2010), Eskandari *et al.* (2011) and Jones *et al.* (2011).

(20, 30 nM) (Mohan *et al.*, 2010). Furthermore, there is a remarkable increase in activity when desulfonation is considered, a situation already observed for kotalanol and ponkoranol (Eskandari *et al.*, 2010; Jones *et al.*, 2011).

Compounds **75** and **76** are the selenium analogs of de-*O*-sulfonated ponkoranol **55** and its C-5′ diastereoisomer, respectively; there is a slight improvement for the inhibition of ntMGAM when compared with their sulfur counterparts; furthermore, no inhibition was observed against ctMGAM, resulting in an excellent selectivity (Eskandari *et al.*, 2011).

Remarkably, all the depicted organoselenium derivatives were found to be much stronger inhibitors of ntMGAM than acarbose (K_i 62 000 nM), an approved pharmaceutical drug for the treatment of type 2 diabetes (Rossi *et al.*, 2006).

3.4 Ebselen and Derivatives

Up to date there is a vast number of selenium-containing heterocycles, but one that has attracted considerable attention is benzisoselenazol derivative **80** (Scheme 3.10), better known as ebselen.

Ebselen exhibits GPx-like activity (Bhabak and Mugesh, 2010) and also participates in biological redox processes; for instance, it was found to interact with thiol fragments of several proteins, which open a wide spectrum of possibilities for its pharmacological use as an antioxidant or anti-inflammatory agent (Parnham and Sies, 2013). Recently, it has been described that ebselen and ebselen oxide are effective antibiotic agents against *P. aeruginosa*, by inhibiting bacterial diguanylate cyclases (Lieberman *et al.*, 2014).

Ebselen inhibits melanin synthesis and melanosome transfer from melanocytes to keratinocytes, making it an excellent depigmenting agent (Kasraee *et al.*, 2012). Moreover, Singh *et al.* have reported that ebselen

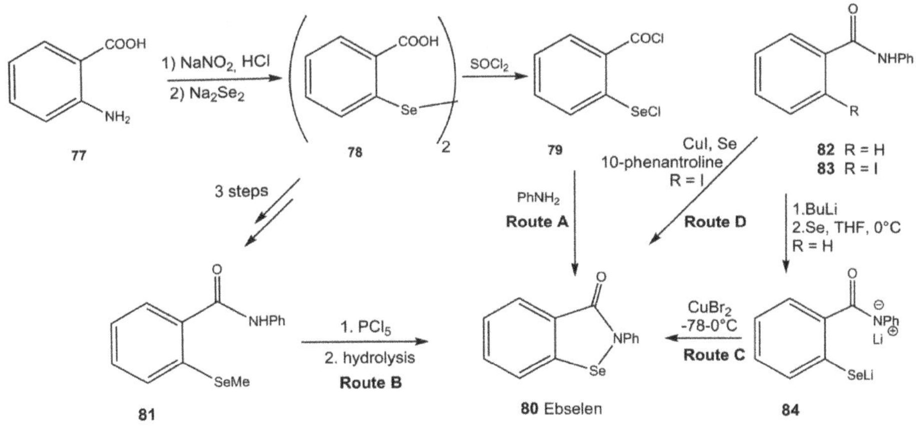

Scheme 3.10 Different synthetic pathways to prepare ebselen.

behaves as a safe lithium cation mimic by irreversibly inhibiting inositol monophosphatase (IC_{50} 1.5 μM); this might afford a potential use in the treatment of bipolar disorder (Singh *et al.*, 2013).

Although ebselen was firstly synthesized in the 1920s, there have been some latter approaches to improve its preparation. Scheme 3.10 depicts four of the most significant routes reported so far. Routes A and B are based on diselenide **78**, prepared from anthranilic acid upon diazotization and subsequent treatment with Na_2Se_2 (Welter *et al.*, 1984).

Engman and Hallberg (1989) reported a simpler procedure starting from benzanilide **82** (route C, Scheme 3.10), and based on an *ortho*-lithiation, followed by selenium insertion and $CuBr_2$-mediated cyclization. Recently Balkrishna *et al.* (2010) described an improved methodology starting from *o*-iodobenzanilide **83** (route D), and involving a novel aryl selenation reaction and Se–N bond formation catalyzed by Cu(I).

Nowadays, the most recurrent strategy for accessing ebselen derivatives is based on route A (Scheme 3.10), using the appropriate *o*-(chloroseleno) benzoyl chloride, with is coupled with different amines to furnish the benzisoselenazol-type ring. Following this approach, several ebselen analogs have been synthesized and their bioactivity tested (Scheme 3.11).

Compound **86a**, named ethaselen, and bearing two ebselen units (He *et al.*, 2012) proved to be a strong inhibitor of mammalian thioredoxin reductase, an enzyme considered as a pharmaceutical target in cancer treatment. Ethaselen and its methoxy analog **86b** exhibited anticancer activity against pancreatic carcinoma (IC_{50} 3.79 μM and 2.84 μM), glioma (IC_{50} 6.41 μM and 1.64 μM) and colon tumoral cells (IC_{50} 7.67 μM and 5.28 μM) (He *et al.*, 2012).

Moreover, derivatives **86c** and **86d**, holding dipeptide fragments Tyr–Val and Val–Ala, respectively were studied as GPx mimetics by scavening H_2O_2, *t*-butyl hydroperoxide (*t*BuOOH), and cumene hydroperoxide (Cum-OOH) using glutathione (GSH) as the thiol source; compound **86d** turned out to be a more potent antioxidant agent than ebselen for the three tested substrates (Satheeshkumar and Mugesh, 2011).

Ebselen derivatives **86e–h** (Scheme 3.11), recently synthesized by Li's group, incorporate tacrine (**86d** and **86f**) (Mao *et al.*, 2013) and donepezil (**86g** and **86h**) motifs (Luo *et al.*, 2013). These compounds, especially **86e** and **86h**, are good inhibitors of acetylcholinesterase (AChE) and butylcholinesterase (BuChE), and therefore are considered as potential drugs for the treatment of Alzheimer's disease (Figure 3.3).

Compounds **86e** and **86f** showed similar hydrogen peroxide scavenging activity (70% and 84% at 25 μM of concentration) as ebselen (90% at 25 μM of concentration).

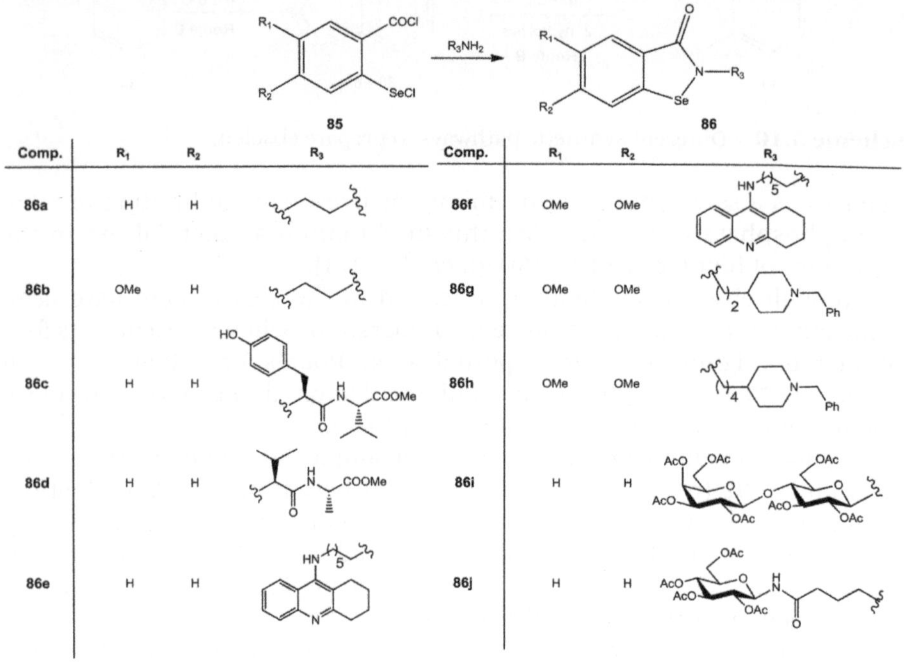

Scheme 3.11 Synthesis of ebselen analogs.

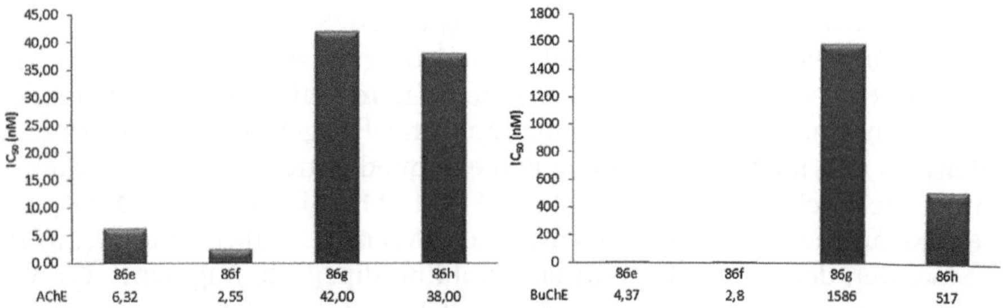

Figure 3.3 Inhibitory activity of ebelsen analogs on enzymes related to Alzheimer's disease. Inhibition of AChE and BuChE (IC_{50} in nM) by compounds **86e–h**.

Furthermore, **86i** and **86j**, bearing a sugar residue, showed better inhibition than ebselen of breast cancer proliferation (Bijian *et al.*, 2012) (IC_{50} 9.8, 18.9 and 42.3 μM, respectively); these ebselen analogs, unlike ebselen, also turned out to be good inhibitors of focal adhesion kinase.

3.5 Miscellaneous

Besides the aforementioned organoselenium derivatives, there are many other minor families with relevant biological importance.

Numerous diselenides have shown numerous biological activities, *e.g.* ROS-scavenging (Bhabak and Mugesh, 2010), neuroprotective (Pinton *et al.*, 2013) or antiparasitic (Plano *et al.*, 2012) agents.

For instance, (p-$MeOC_6H_4Se)_2$ has been proposed as a therapeutic drug for the treatment of Alzheimer's disease since it prevents β-amyloid-induced cell death by the inhibition of caspase 3 and 9; furthermore, *in vivo* assays showed that (p-$MeOC_6H_4Se)_2$ improves memory and learning performance in rodents (Pinton *et al.*, 2013). Diphenyl diselenide $(PhSe)_2$ exerts a protective action against the formation of arterial oxidized low-density lipoproteins, involved in the development of atherosclerosis (Straliotto *et al.*, 2013).

In this context Plano *et al.* (2012) have developed an efficient methodology for the synthesis of several diselenides *via* the formation of a selenocyanate and subsequent reduction with $NaBH_4$. Aromatic selenocyanates (*e.g.* **88**) can be obtained by treatment of malononitrile with SeO_2, followed by coupling with an aromatic amine (Scheme 3.12). Arylalkyl derivatives, like **91** can be accessed *via* nucleophilic reaction of KSeCN with arylalkyl halides (Plano *et al.*, 2012).

Diselenides **89** and **92** were tested against leishmaniasis in two stages of the parasite (promastigote and amastigote) and also against two cell lines of human leukemia (JurKat and THP-1); IC_{50} values (promastigote, amastigote, Jurkat and THP-1) for compound **89** (0.96, 0.65, 20.4, 15.3 μM) and for compound **92** (4.04, 0.38, 26.8, 29.1 μM) were compared with miltefosine (15.0, 2.84, 48.2, 18.5), a drug used in the treatment of leishmaniasis in children (Plano *et al.*, 2012); these values indicate a better antiparasitic and anticancer activity of such diselenides.

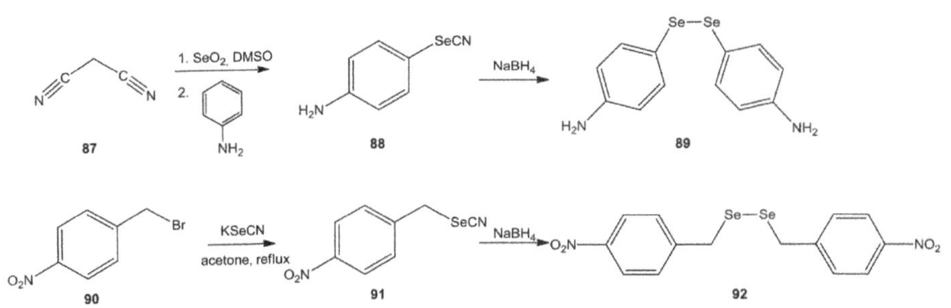

Scheme 3.12 Synthesis of diaryldiselenides.

Ibáñez *et al.* (2011) reported the preparation of symmetrically substituted alkylseleno-imidocarbamates **95** selenourea **93** their evaluation as antiproliferative agents. The synthesis of such compounds involved *Se*-alkylation with an alkyl chloride (methyl, benzyl), and subsequent acylation with an (heteroaryl)aryl acyl chloride (Scheme 3.13).

Many of these compounds exhibited GI_{50} values in the low micromolar concentrations, some of them, even in the low nanomolar range (*e.g.* **95a–f**) with better activity than some approved chemotherapeutic agents. *Se*–Me derivatives exhibited a better activity, as well as those bearing selenium instead of isosteric sulfur derivatives. Compound **95b**, with a good selectivity towards tumoral cells, also exhibited an excellent apoptotic profile.

Another family of compounds that have been targeted for the insertion of selenium atom is nucleosides, widely known compounds exhibiting antiviral and anticancer properties. Jeong *et al.* (2009) reported the preparation of 2′-deoxy-2′-fluoro-4′-selenoarabinofuranosyl-cytosine **102**; the key structural template is 4′-selenouridine **99**, already synthesized by the same group (Jeong *et al.*, 2008) starting from D-gulonic-γ-lactone (Scheme 3.14) *via* a double nucleophilic displacement by freshly prepared Na_2Se, oxidation to selenoxide **98** and a Pummerer-type base condensation as the main synthetic steps. C-2′ configuration was inverted *via* an anhydro-nucleoside to give partially protected *arabino*-configured derivative **100** (Scheme 3.14), and a fluorine atom

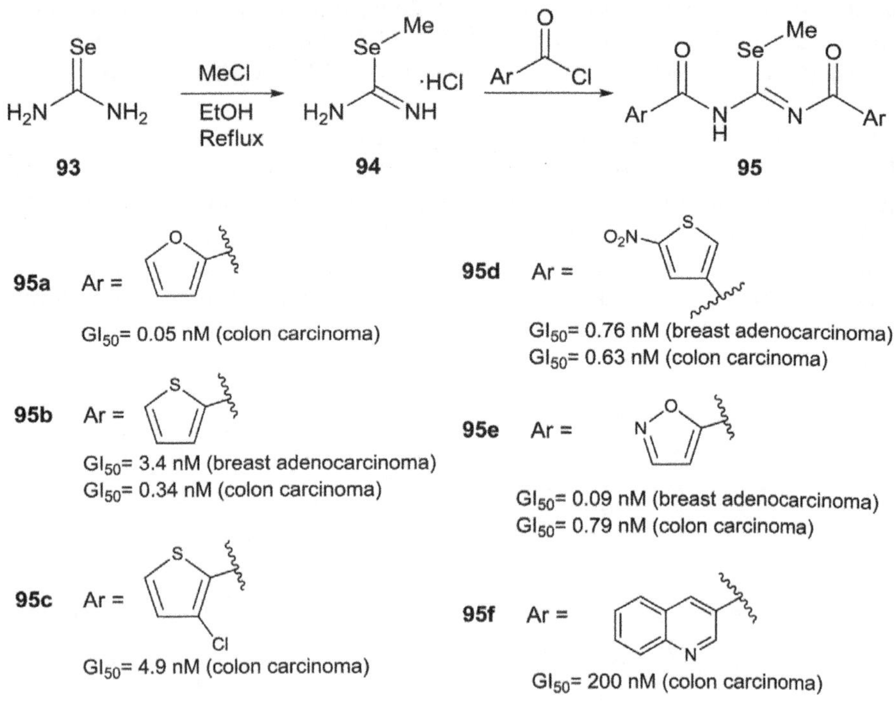

Scheme 3.13 Synthesis of symmetrically substituted methylseleno-imidocarbamates and antiproliferative activities.

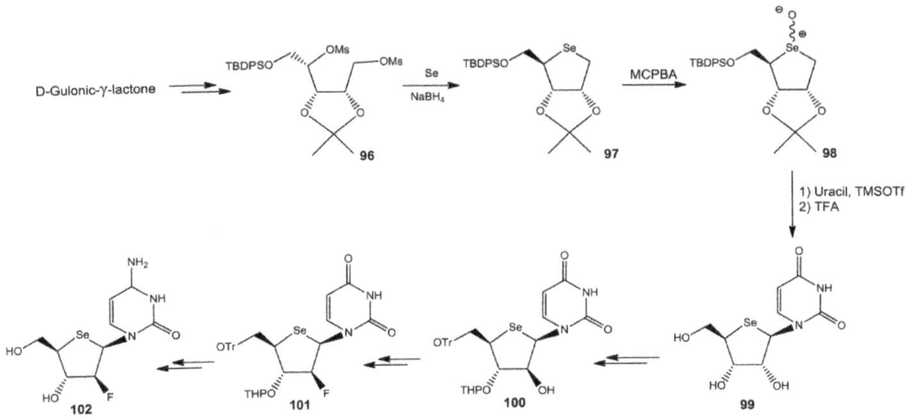

Scheme 3.14 Synthesis of 2′-F-4′-seleno-ara-C selenonucleoside.

was introduced on C-2'with retention of the configuration using DAST to give **101** as a minor compound in the crude reaction mixture. Final conversion of the uracil moiety into cytosine afforded **102**.

Selenonucleoside **102** was assayed as a potential anticancer agent against a panel of 6 human tumoral cells (colon, lung, stomach, breast, prostate, leukemia), with IC_{50} values in the low micromolar–submicromolar range (0.14–1.1 μM), displaying improved activity with regards to Ara-C, except for the leukemia cell line (Jeong *et al.*, 2009).

Thyroid has been found to possess the highest concentration of selenium in the human body, exerting a pivotal role in the correct functioning of this gland (Manna *et al.*, 2013), which is why numerous research groups have focused their attention on organoselenium compounds for the treatment of thyroid disorders like hypothyroidism. Muguesh's group has reported the preparation of selenium isosters of well-known drugs currently used in the treatment of this disease, bearing an imidazoline-2-thione or thiouracil motifs (methimazole, carbimazole, 6-*n*-propyl-2-thiouracil); such drugs act by inhibiting the iodination of tyrosine exerted by thyroid peroxidase (TPO) (Manna *et al.*, 2013).

Das *et al.* (2008) have recently prepared **106**, the seleno-isoster of carbimazole; substitution on both nitrogen atoms precludes the air-mediated oxidation into the corresponding diselenide *via* a zwitterionic structure, a situation found for the selenium-methimazole, leading to a reduced bioactivity. Compound **106** can be accessed using two synthetic approaches: lithiation of *N*-methylimidazole, followed by addition of elemental selenium, treatment with ethyl chloroformate and spontaneous migration of the ethoxy carbonyl group from Se to N. Alternatively, conversion of *N*-methylimidazole into the corresponding imidazolium salt, followed by *in situ* generation of an heterocyclic carbene and coupling with elemental selenium, also afforded **106** (Scheme 3.15). The same strategies were also employed in the synthesis of the sulfur counterpart carbimazole. Both isosters showed excellent activity in the inhibition of lactoperoxidase-catalyzed iodination of L-tyrosine, demonstrating that, on the one hand, the presence of the ethoxycarbonyl

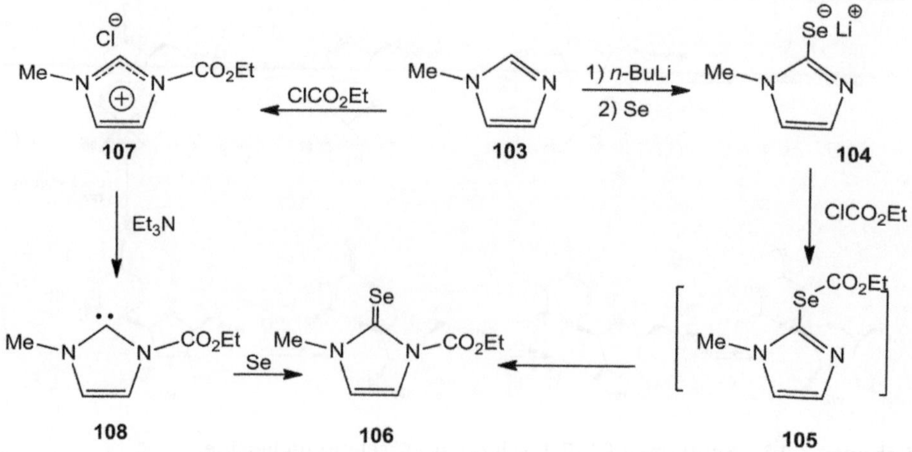

Scheme 3.15 Two synthetic approaches for the synthesis of seleno-isoster of carbimazole.

group prevents the formation of a nondesired disulfide or diselenide, and, on the other hand, such group can be cleaved to give reactive zwitterionic species which act as inhibitors of the tested enzyme.

Summary Points

- Numerous organoselenium derivatives are potent ROS and RNS scavengers.
- Numerous selenium-containing compounds are strong antiproliferative agents.
- Some selenosugars are strong glycosidase inhibitiors.
- Numerous organoselenium compounds act as glutathione peroxidase mimics.
- Isoselenocyanates are key synthetic intermediates in organic synthesis.

Key Facts

1. Oxidative stress produces damage to biomolecules and the development of diseases.
2. ROS and RNS are responsible for the oxidative stress.
3. Selenium-containing compounds are redox-active derivatives.
4. Organoselenium compounds can interact with proteins.

Definitions of Words and Terms

Oxidative stress: A cellular state where an imbalance between antioxidant agents and pro-oxidant species is produced.
ROS and RNS: Compounds usually obtained as intermediates in the cellular respiration cycle.

DPPH method: A widely used procedure to measure the antiradical activity of antioxidant agents.
Antitumoral xenograft model: A methodology for screening anticancer drugs, in which human tumoral cells are transplanted into experimental animals.
Selenoproteins: Proteins containing selenium in their catalytic site, where this atom exerts its role *via* redox-type reactions.
Glutathione peroxidase: A selenoprotein exerting defensive antioxidant activities.
Selenosugars: Compounds where a selenium atom replaces the endocyclic oxygen atom of a carbohydrate.
Ayurvedic medicine: The millenary medicine developed in India and Sri Lanka.
Caspases: Proteases involved in critical cellular processes, like apoptosis, necrosis and inflammation.
Glycosidases: Carbohydrate-processing enzymes involved in carbohydrate metabolism.

List of Abbreviations

AChE	Acetylcholinesterase
AFA	Acetic-formic anhydride
Ar	Aryl
Bn	Benzyl
BSA	Bovine serum albumin
BuChE	Butylcholinesterase
COX	Cyclooxynenase
DAST	Diethylaminosulfur trifluoride
DHS	Dihydroxyselenolane
DMAP	4,4-Dimethylaminopyridine
DMSO	Dimethylsulfoxide
DPPH	2,2-Diphenyl-1-picrylhydrazyl free radical
DST	Diselenothreitol
DTT	Dithiothreitol
GHS	Glutathione
$\mathbf{GI_{50}}$	Growth inhibition 50%
GPx	Glutathione peroxidase
HFIP	Hexafluoroisopropanol
$\mathbf{IC_{50}}$	Median inhibition concentration
K_i	Inhibition constant
NADPH	Nicotinamide adenine dinucleotide phosphate, reduced form
MCPBA	*meta*-Chloroperbenzoic acid
MGAM	Maltase glucoamylase
NBS	*N*-Bromosuccinimide
PMB	*para*-Methoxybenzyl
RNS	Reactive-nitrogen species
ROS	Reactive-oxygen species

TBDMS	*tert*-Butyldimethylsilyl
TBAF	Tetrabutylammonium fluoride
TFA	Trifluoroacetic acid
THP	Tetrahydropyranyl
TPO	Thyroid peroxidase
Tr	Trityl

Acknowledgements

We thank the Dirección General de Investigación of Spain and the Junta de Andalucía (grants numbers CTQ2008-02813, cofinanced with FEDER funds, P08-AGR-03751, and FQM134), and Mexican PROMEP (103.5/13/6823 BUAP-PTC-353) for financial support.

References

Balkrishna, S. J., Bhakuni, B. S., Chopra, D. and Kumar, S., 2010. Cu-catalized efficient synthetic methodology for ebselen and related Se–N heterocycles. *Organic Letters*. 12: 5394–5397.

Bhabak, K. P. and Mugesh, G., 2010. Functional mimics of glutathione peroxidase: bioinspired synthetic antioxidants. *Accounts of Chemical Research*. 43: 1408–1419.

Bijian, K., Zhang, Z., Xu, B., Jie, S., Chen, B., Wan, S., Wu, J., Jiang, T. and Alaoui-Jamali, M. A., 2012. Synthesis and biological activity of novel organoselenium derivatives targeting multiple kinases and capable of inhibiting cancer progression to metastases. *European Journal of Medicinal Chemistry*. 48: 143–152.

Chakraborty, S., Yadav, S. K., Subramanian, M., Priyadarsini, K. I., Iwaoka, M. and Chattopadhyay, S., 2012. DL-*trans*-3,4-Dihydroxy-1-selenolane (DHS_{red}) accelerates healing of indomethacin-induced stomach ulceration in mice. *Free Radical Research*. 46: 1378–1386.

Das, D., Roy, G. and Mugesh, G., 2008. Antithyroid drug carbimazole and its analogues: synthesis and inhibition of peroxidase-catalyzed iodination of L-tyrosine. *Journal of Medicinal Chemistry*. 51: 7313–7317.

Engman, L. and Hallberg, A., 1989. Expedient synthesis of ebselen and related compounds. *Journal of Organic Chemistry*. 54: 2964–2966.

Eskandari, R., Jones, K., Rose, D. R. and Pinto, M. B., 2011. The effect of heteroatom substitution of sulfur for selenium in glucosidase inhibitors on intestinal α-glucosidase activities. *Chemical Communication*. 47: 9134–9136.

Eskandari, R., Kuntz, D. A., Rose, D. R. and Pinto, B. M., 2010. Potent glucosidase inhibitors: De-*O*-sulfonated ponkoranol and its stereoisomer. *Organic Letters*. 12: 1632–1635.

Fernández-Bolaños Guzmán, J. M., Maya Castilla, I., López García, M. A., Fernández-Bolaños Guzmán, J., Rodríguez Gutiérrez, G. and Gómez Carretero, A., 2012. Synthesis, antioxidant and bioactive free radical scavenging activity of sulfur/selenium polyphenolic compounds, WO 2012164118 A1.

Freudendahl, D. M., Shahzad, S. A. and Wirth, T., 2009. Recent advances in organoselenium chemistry. *European Journal of Organic Chemistry*. 1649–1664.

Ghavami, A., Johnston, B. D. and Pinto, B. M., 2001. A new class of glycosidase inhibitor: synthesis of salacinol and its stereoisomers. *Journal of Organic Chemistry*. 66: 2312–2317.

He, J., Li, D., Xiong, K., Ge, Y., Jin, H., Zhang, G., Hong, M., Tian, Y., Yin, J. and Zeng, H., 2012. Inhibition of thioredoxin reductase by a novel series of bis-1,2-benzisoselenazol-3(2*H*)-ones: organoselenium compounds for cancer therapy. *Bioorganic & Medicinal Chemistry*. 20: 3816–3827.

Heimgartner, H., Zhou, Y., Atanassov, P. K. and Sommen, G. L., 2008. Isoselenocyanates as building blocks for selenium-containing heterocycles. *Phosphorus, Sulfur, and Silicon and the Related Elements*. 183: 840–855.

Ibáñez, E., Plano, D., Font, M., Calvo, A., Prior, C., Palop, J. A. and Sanmartín, C., 2011. Synthesis and antiproliferative activity of novel symmetrical alkylthio- and alkylseleno-imidocarbamates. *European Journal of Medicinal Chemistry*. 46: 265–274.

Iwaoka, M. and Arai, K., 2013. From sulfur to selenium. A new research arena in chemical biology and biological chemistry. *Current Chemical Biology*. 7: 2–24.

Iwaoka, M., Kumakura, F., Yoneda, M., Nakahara, T., Henmi, K., Aonuma, H., Nakatani, H. and Tomoda, S., 2008. Direct observation of conformational folding coupled with disulphide rearrangement by using a water-soluble selenoxide reagent-a case of oxidative regeneration of ribonuclease A under weakly basic conditions. *Journal of Biochemistry*. 144: 121–130.

Iwaoka, M., Takahashi, T. and Tomoda, S., 2001. Syntheses and structural characterization of water-soluble selenium reagents for the redox control of protein disulfide bonds. *Heteroatom Chemistry*. 12: 293–299.

Jeong, L. S., Tosh, D. K., Choi, W. J., Lee, S. K., Kang, Y.-J., Choi, S., Lee, J. H., Lee, H., Lee, H. W. and Kim, H. O., 2009. Discovery of a new template for anticancer agents: 2′-deoxy-2′-fluoro-4′-selenoarabinofuranosyl-cytosine (2′-F-4′-seleno-ara-C). *Journal of Medicinal Chemistry*. 52: 5303–5306.

Jeong, L. S., Tosh, D. K., Kim, H. O., Wang, T., Hou, X., Yun, H. S., Kwon, Y., Lee, S. K., Choi, J. and Zhao, L. X., 2008. First synthesis of 4′-selenonucleosides showing unusual southern conformation. *Organic Letters*. 10: 209–212.

Jones, K., Sim, L., Mohan, S., Kumarasamy, J., Liu, H., Avery, S., Naim, H. Y., Quezada-Calvillo, R., Nichols, B. L., Pinto, B. M. and Rose, D. R., 2011. Mapping the intestinal alpha-glucogenic enzyme specificities of starch digesting maltase-glucoamylase and sucrase-isomaltase. *Bioorganic & Medicinal Chemistry*. 19: 3929–3934.

Kasraee, B., Nikolic, D. S., Salomon, D., Carraux, P., Fontao, L., Piguet, V., Gholamhossein, R., Omrani, G. R., Sorg, O. and Saurat, J.-H., 2012. Ebselen is a new skin depigmenting agent that inhibits melanin biosynthesis and melanosomal transfer. *Experimental Dermatology*. 21: 19–24.

Kumakura, F., Mishra, B., Priyadarsini, K. I. and Iwaoka, M., 2010. A water-soluble cyclic selenide with enhanced glutathione peroxidase-like catalytic activities. *European Journal of Organic Chemistry*. 440–445.

Lieberman, O. J., Orr, M. W., Wang, Y. and Lee, V. T., 2014. High-throughput screening using the differential radial capillary action of ligand assay identifies ebselen as an inhibitor of diguanylate cyclases. *ACS Chemical Biology*. 9: 183–192.

López, Ó., Maza, S., Ulgar, V., Maya, I. and Fernández-Bolaños, J. G., 2009. Synthesis of sugar-derived isoselenocyanates, selenoureas, and selenazoles. *Tetrahedron*. 65: 2556–2566.

Liu, H. and Pinto, M. M., 2005. Efficient synthesis of the glucosidase inhibitor blintol, the selenium analogue of the naturally occurring glycosidase inhibitor salacinol. *Journal of Organic Chemistry*. 70: 753–755.

Liu, H., Nasi, R., Jayakanthan, K., Sim, L., Heipel, H., Rose, D. R. and Pinto, B. M., 2007. New synthetic routes to chain-extended selenium, sulfur, and nitrogen analogues of the naturally occurring glucosidase inhibitor salacinol and their inhibitory activities against recombinant human maltase glucoamylase. *Journal of Organic Chemistry*. 72: 6562–6572.

Luo, Z., Sheng, J., Sun, Y., Lu, C., Yan, J., Liu, A., Luo, H.-b., Huang, L. and Li, X., 2013. Synthesis and evaluation of multi-target-directed ligands against Alzheimer's disease based on the fusion of donepezil and ebselen. *Journal of Medicinal Chemistry*. 56: 9089–9099.

Manna, D., Roy, G. and Mugesh, G., 2013. Antithyroid drugs and their analogues: synthesis, structure and mechanism of action. *Accounts of Chemical research*. 46: 2706–2715.

Mao, F., Chen, J., Zhou, Q., Luo, Z., Huang, L. and Li, X., 2013. Novel tacrine–ebselen hybrids with improved cholinesterase inhibitory, hydrogen peroxide and peroxynitrite scavenging activity. *Bioorganic & Medicinal Chemistry Letters*. 23: 6737–6742.

Merino-Montiel, P., López, Ó. and Fernández-Bolaños, J. G., 2012. L-Isofucoselenofagomine and derivatives: dual activities as antioxidants and as glycosidase inhibitors. *Tetrahedron*. 68: 3591–3595.

Merino-Montiel, P., Maza, S., Martos, S., López, Ó., Maya, I. and Fernández-Bolaños, J. G., 2013. Synthesis and antioxidant activity of *O*-alkyl selenocarbamates, selenoureas and selenohydantoins. *European Journal of Pharmaceutical Sciences*. 48: 582–592.

Mohan, S., Jayakanthan, K., Nasi, R., Kuntz, D. A., Rose, D. A. and Pinto, B. M., 2010. Synthesis and biological evaluation of heteroanalogues of kotalanol and de-*O*-sulfonated kotalanol. *Organic Letters*. 12: 1088–1091.

Nasi, R., Sim, L., Rose, D. R. and Pinto, B. M., 2007. New chain-extended analogues of salacinol and blintol and their glycosidase inhibitory activities. Mapping the active-site requirements of human maltase glucoamylase. *Journal of Organic Chemistry*. 72: 180–186.

Nogueira, C. W. and Rocha, J. B., 2011. Toxicology and pharmacology of selenium: emphasis on synthetic organoselenium compounds. *Archives of Toxicology*. 85: 1313–1359.

Parnham, M. J. and Sies, H., 2013. The early research and development of ebselen. *Biochemical Pharmacology*. 86: 1248–1253.

Peng, X., Cui, H., Fang, J., Zuo, Z., Deng, J., Pan, K., Lai, W. and Zhou, Y., 2012. Low selenium diet alters cell cycle phase, apoptotic population and modifies oxidative stress markers of spleens in Broilers. *Biological Trace Element Research*. 148: 182–186.

Plano, D., Baquedano, Y., Moreno-Mateos, D., Font, M., Jiménez-Ruiz, A., Palop, J. A. and Sanmartín, C., 2012. Selenocyanates and diselenides: a new class of potent antileishmanial agents. *European Journal of Medicinal Chemistry*. 46: 3315–3323.

Pinton, S., Souza, A. C., Sari, M. H. M., Ramalho, R. M., Rodrigues, C. M. P. and Nogueira, C. W., 2013. *p,p'*-Methoxyl-diphenyl diselenide protects against amyloid-β induced cytotoxicity *in vitro* and improves memory deficits *in vivo*. *Behavioural Brain Research*. 247: 241–247.

Pizzo, C., Faral-Tello, P., Salinas, G., Flo, M., Robello, C., Wipf, P. and Mahler, S. G., 2012. Selenosemicarbazones as potent cruzipain inhibitors and their antiparasitic properties against *Trypanosoma cruzi*. *MedChemComm*. 3: 362–368.

Rossi, E. J., Sim, L., Kuntz, D. A., Hahn, D., Johnston, B. D., Ghavami, A., Szczepina, M. G., Kumar, N. S., Sterchi, E. E., Nichols, B. L., Pinto, B. M. and Rose, D. R., 2006. Inhibition of recombinant human maltase glucoamylase by salacinol and derivatives. *FEBS Journal*. 273: 2673–2683.

Satheeshkumar, K. and Mugesh, G., 2011. Synthesis and antioxidant activity of peptide-based ebselen analogues. *Chemistry – A European Journal*. 17: 4849–4857.

Sharma, A. K., Sharma, A., Desai, D., Madhunapantula, S. V., Huh, S. J., Robertson, G. P. and Amin, S., 2008. Synthesis and anticancer activity comparison of phenylalkyl isoselenocyanates with corresponding naturally occurring and synthetic isothiocyanates. *Journal of Medicinal Chemistry*. 51: 7820–7826.

Singh, N., Halliday, A. C., Thomas, J. M., Kuznetsova, O. V., Baldwin, R., Woon, E. C. Y., Aley, P. K., Antoniadou, I., Sharp, T., Vasudevan, S. R. and Churchill, G. C., 2013. A safe lithium mimetic for bipolar disorder. *Nature Communications*. 4: 1332.

Straliotto, M. R., Hort, M. A., Fiuza, B., Teixeira Rocha, J. B., Farina, M., Chiabrando, G. and Fabro de Bem, A., 2013. Diphenyl diselenide modulates oxLDL-induced cytotoxicity in macrophage by improving the redox signaling. *Biochimie*. 95: 1544–1551.

Tiekink, E. R. T., 2012. Therapeutic potential of selenium and tellurium compounds: opportunities yet unrealised. *Dalton Transactions*. 41: 6390–6395.

Tinggi, U., 2008. Selenium: its role as antioxidant in human health. *Environmental Health and Preventive Medicine*. 13: 102–108.

Storkey, C., Davies, M. J., White, J. M. and Schiesser, C. H., 2011. Synthesis and antioxidant capacity of 5-selenopyranose derivatives. *Chemical Communications*. 47: 9693–9695.

Storkey, C., Pattison, D. I., White, J. M., Schiesser, C. H. and Davies, M. J., 2012. Preventing protein oxidation with sugars: scavenging of hypohalous acids by 5-selenopyranose and 4-selenofuranose derivatives. *Chemical Research in Toxicology*. 25: 2589–2599.

Suzuki, M., Endo, M., Shinohara, F., Echigo, S. and Rikiishi, H., 2010. Differential apoptotic response of human cancer cells to organoselenium compounds. *Cancer Chemotherapy and Pharmacology*. 66: 475–484.

Uzma, N., Kumar, S. and Priyadarsini, K. I., 2011. Hepatoprotective, immunomodulatory, and anti-inflammatory activities of selenocystine in experimental liver injury of rats. *Biological Trace Element Research*. 142: 723–734.

Victoria, F. N., Radatz, C. S., Sachini, M., Jacob, R. G., Alves, D., Savegnago, L., Perin, G., Motta, A. S., Silva, W. P. and Lenardão, E. J., 2012. Further analysis of the antimicrobial activity of α-phenylseleno citronellal and α-phenylseleno citronellol. *Food Control*. 23: 95–99.

Weekley, C. M. and Harris, H. H., 2013. Which form is that? The importance of selenium speciation and metabolism in the prevention and treatment of disease. *Chemical Society Reviews*. 42: 8870–8894.

Welter, D., Christiaens, L. and Wirtz-Peitz, 1984. New benzisoselenazolones, process for producing the same and pharmaceutical preparations containing the same. EP 0044453A2.

Zaharia, V., Ignat, A., Ngameni, B., Kuete, V., Moungang, M. L., Fokunang, C. N., Vasilescu, M., Palibroda, N., Cristea, C., Silaghi-Dumitrescu, L. and Ngadjui, B. T., 2013. Heterocycles 23: synthesis, characterization and anticancer activity of new hydrazinoselenazole derivatives. *Medicinal Chemistry Research*. 22: 5670–5679.

CHAPTER 4

Chemistry and Activity of Selenium Dioxide and Selenite Ions

DANIEL L. MORRIS*[a]

[a]Department of Chemistry and Biochemistry, Rose-Hulman Institute of Technology, 5500 Wabash Avenue, Terre Haute, IN 47803-3920, USA
*E-mail: daniel.morris@rose-hulman.edu

4.1 Introduction

Selenium is an essential dietary trace element. Organic forms of Se, including selenomethionine (Se-Met), selenocysteine (Se-Cys) and Se-methylselenocysteine, are present in plants and animals to varying degrees, and are the predominant dietary forms (Whanger, 2002). Inorganic forms of Se, which include selenium dioxide (SeO_2), selenite ion (SeO_3^{2-}) and selenate ion (SeO_4^{2-}) are less abundant than organoselenium compounds as naturally occurring dietary sources of Se, but they are incorporated in the pathway of Se metabolism. Therefore, inorganic forms of Se (most notably SeO_3^{2-} and SeO_4^{2-}) are often used as nutritional supplements in animal feedstock and over-the-counter products for human consumption. Besides being active in important physiological metabolic pathways, the relevant chemical activity of SeO_2 and SeO_3^{2-} includes their behavior as both pro- and antioxidants, the ability to mediate the cytotoxic effects of toxic metals and the formation of coordination complexes with metal ions. These properties appear to

Food and Nutritional Components in Focus No. 9
Selenium: Chemistry, Analysis, Function and Effects
Edited by Victor R. Preedy

Published by the Royal Society of Chemistry, www.rsc.org

be associated with the oxidation state of Se in these inorganic sources. The selenite ion has been extensively studied because of its promise as a chemotherapeutic agent and its ability to increase the effectiveness of other chemotherapy regimens. This is directly related to its behavior as a pro-oxidant and its ability to form reactive oxygen species (ROS). The focus of this chapter is to address these various chemical behaviors of SeO_2 and SeO_3^{2-} in light of their applications as food and nutritional components.

4.2 Oxidation State and Speciation in Aqueous Solution

Selenium typically exists in one of three possible oxidation states (−2, +4 and +6) in naturally occurring compounds, and the oxidation state is largely responsible for the relevant chemistry and activity of Se species. Selenide ion (Se^{2-}, −2 oxidation state) is a pivotal intermediate in Se metabolism, and selenate ion (SeO_4^{2-}, +6 oxidation state) is an inorganic form of Se also used widely in nutritional supplements. SeO_2 and SeO_3^{2-} exhibit interesting chemical properties and reactivity as Se in both species exists in the intermediate +4 oxidation state. SeO_2 is an acid anhydride, and when dissolved in water it forms selenous or selenious acid (H_2SeO_3) according to eqn (4.1).

$$SeO_2 + H_2O \rightarrow H_2SeO_3 \quad (4.1)$$

H_2SeO_3 is a diprotic weak acid with $pK_{a1} = 2.6$ and $pK_{a2} = 8.3$. Stepwise loss of the two acidic protons produces the intermediate form hydrogen selenite or biselenite ion ($HSeO_3^-$) and the fully deprotonated selenite ion (SeO_3^{2-}). (See eqn (4.2) and (4.3).)

$$H_2SeO_3 \rightarrow HSeO_3^- + H^+ \quad (4.2)$$

$$HSeO_3^- \rightarrow SeO_3^{2-} + H^+ \quad (4.3)$$

The structures of SeO_2, H_2SeO_3, $HSeO_3^-$ and SeO_3^{2-} are presented in Figure 4.1.

Speciation in aqueous solutions of SeO_2 or SeO_3^{2-} is pH dependent with the predominant species being H_2SeO_3 at acidic pH, $HSeO_3^-$ at intermediate pH and SeO_3^{2-} at basic pH. Therefore, pH is expected to have a significant effect on the chemistry of these inorganic Se forms. Figure 4.2 shows the fraction of each Se species that exists at a given pH value.

Enzymatic reduction of SeO_3^{2-} to red, elemental Se^0 by intact yeast cells is optimum between pH 4.2 and 4.9 and is attributed to the ability of $HSeO_3^-$ (the dominant species in this pH range) to permeate the cell membrane. The optimum pH of SeO_3^{2-} reduction in cell-free medium is pH 7.0, which still reflects a high abundance of $HSeO_3^-$ (Falcone and Nickerson, 1963). This suggests that the ability of selenium compounds to transfer electrons is related to pH, and pro- and antioxidant behaviors may be determined or controlled by Se speciation (Cemeli *et al.*, 2003). However, aqueous solutions of

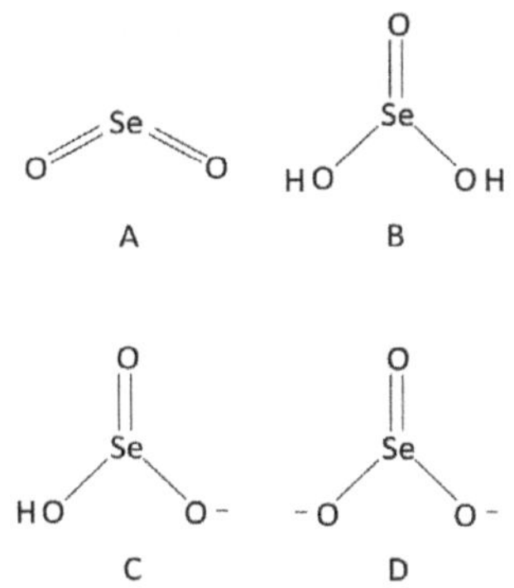

Figure 4.1 Structures of selenium dioxide (SeO_2), selenous acid (H_2SeO_3), biselenite ion ($HSeO_3^-$) and selenite ion (SeO_3^{2-}). SeO_2 (A) is expected to have "V-shaped" or "bent" geometry. H_2SeO_3 (B), $HSeO_3^-$ (C) and SeO_3^{2-} (D) should exhibit trigonal pyramidal geometry around the central Se atom. These geometries account for a nonbonding pair of electrons (not shown) on each Se atom. The Se atom exceeds the octet rule in each structure.

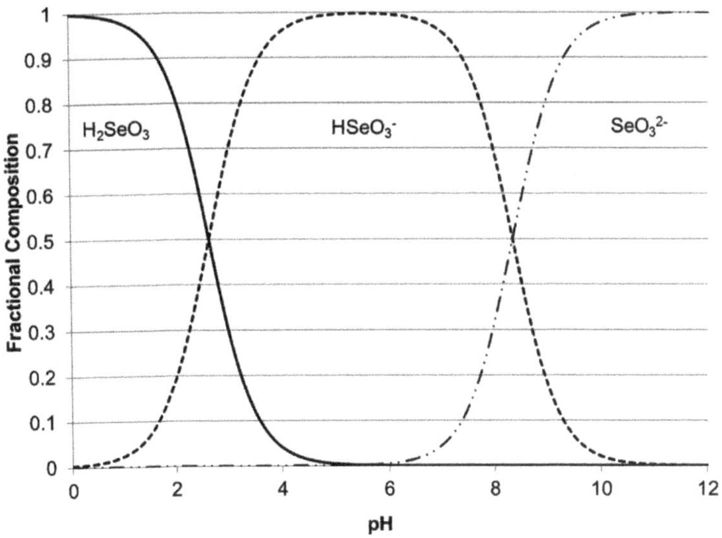

Figure 4.2 Fractions of H_2SeO_3, $HSeO_3^-$ and SeO_3^{2-} that exist in aqueous solution at various pH values. The fractions of H_2SeO_3 (solid line —), $HSeO_3^-$ (dashed line ----) and SeO_3^{2-} (alternating dashes and dots - ·· -) that exist over the pH range 0–12 were calculated based on $pK_{a1} = 2.6$ and $pK_{a2} = 8.3$. Aqueous solutions of SeO_2 should exist mainly as H_2SeO_3 at pH values less than pK_{a1}. At pH values between pK_{a1} and pK_{a2} $HSeO_3^-$ will be the predominant form. Fully deprotonated SeO_3^{2-} will be the most abundant species at pH values greater than pK_{a2}.

SeO_2 and SeO_3^{2-} at the same pH exhibit different pro- and antioxidant behaviors in some studies involving metal-ion coordination, suggesting that the metal-coordination complexes formed with SeO_2 and SeO_3^{2-} are not the same at a given pH. This behavior is addressed in Section 4.5.2.

4.3 SeO_3^{2-} as a Nutritional Supplement

Nutritional sources of Se include selenomethionine (Se-Met), selenocysteine (Se-Cys), SeO_4^{2-} and SeO_3^{2-}. Se-Met and Se-Cys are organic forms of Se that fall into the category of selenoamino acids. Se-Met is the main nutritional form of Se available in plants and crops grown in Se-adequate regions. It is also the major compound in Se-enriched yeast. Se-Met is converted to Se-Cys by animals that feed on these sources (Whanger, 2002). Se-Met residues are also incorporated into Se-containing proteins. Naturally occurring Se levels differ widely depending on geographic region, and in low-Se regions it is necessary to supplement animal feedstock. At low concentrations SeO_3^{2-} behaves as a micronutrient, and both SeO_3^{2-} and SeO_4^{2-} are inorganic forms of Se used widely as nutritional supplements. The advantage of using SeO_3^{2-} as a nutritional Se supplement is its ability to be absorbed readily by the body without accumulating in tissues and organs at effective doses. However, it can be reduced to a lower oxidation state by vitamin C and is not as effective as the naturally occurring organic form Se-Met (Schrauzer and Surai, 2009). These various Se sources (Se-Met, Se-Cys, SeO_4^{2-} and SeO_3^{2-}) are interrelated in the Se metabolic pathway, and several reviews address their metabolism in detail (Lu *et al.*, 1995; Suzuki and Ogra, 2002; Suzuki *et al.*, 2007; Schrauzer and Surai, 2009). A simplified schematic of the metabolic fate of the organic forms (Se-Met and Se-Cys) and inorganic forms (SeO_3^{2-} and SeO_4^{2-}) of Se is presented in Figure 4.3.

Selenite ion is taken up directly by red blood cells and reduced readily by glutathione (GSH) to produce selenodiglutathione (GSSeSG), GSSG (the oxidized form of GSH) and selenide (Se^{2-}). Selenide is a pivotal intermediate that can either be used in the synthesis of selenoproteins or undergo stepwise methylation reactions to form the excretory forms methylselenol (CH_3SeH), dimethyldiselenide ($(CH_3)_2Se$) and trimethylselenonium ion ($(CH_3)_3Se^+$).

4.4 Pro-Oxidant Behavior

4.4.1 Oxidative Activity of SeO_2

Much of the literature addressing the chemistry and activity of SeO_2 alone focuses on its use as a very effective catalyst for oxidizing organic compounds. It is used as a catalyst in conjunction with H_2O_2 and other peroxides to oxidize organic compounds such as anilines, pyrrolidines and olefins. In strongly acidic solutions aqueous SeO_2 (which exists in the form H_2SeO_3) is reported to be oxidized by H_2O_2 to produce selenate ion (SeO_4^{2-}) or selenic

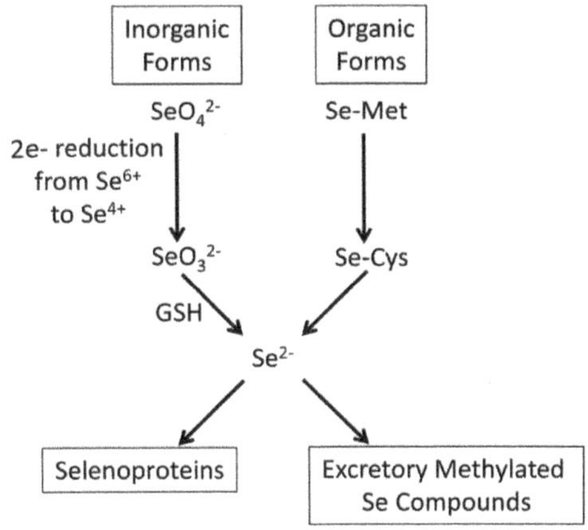

Figure 4.3 Metabolic pathway of inorganic and organic forms of Se to produce selenoproteins and methylated excretory forms. Inorganic Se can enter the metabolic pathway as SeO_4^{2-} (which is converted to SeO_3^{2-} by a 2 electron reduction) or as SeO_3^{2-} directly, and GSH reduces SeO_3^{2-} to Se^{2-}. Organic Se enters as the selenoamino acid Se-Met and is converted to Se-Cys and finally to Se^{2-}. Se^{2-} is the pivotal intermediate form of Se that produces selenoproteins and methylated forms that are excreted as waste.

acid (H_2SeO_4, with Se in a +6 oxidation state) according to eqn (4.4) (Hughes and Martin, 1955).

$$H_2SeO_3 + H_2O_2 \rightarrow H_2O + H_2SeO_4 \quad (4.4)$$

Peroxyselenous acid ($HOOSeO_2H$) appears to be an intermediate in the mechanism for this reaction, and formation of peroxyselenous acid appears to be a key component in the use of SeO_2 as a catalyst for oxidizing organic compounds.

4.4.2 ROS Formation and SeO_3^{2-}

Selenium is an essential dietary element, but high levels are toxic. In general, Se toxicity is associated with DNA damage (Biswas *et al.*, 2000) and oxidative stress induction (Wycherly *et al.*, 2004). High levels of inorganic Se species are also associated with acute toxicity related to DNA strand breaks (Combs and Gray, 1998). Selenite ion has been shown to react with reduced GSH to produce Se^0 and the reactive oxygen species superoxide radical ($O_2^{\cdot -}$) (Seko *et al.*, 1989). Dismutation of $O_2^{\cdot -}$ with superoxide dismutase produces H_2O_2, which can undergo Fenton or Fenton-like reactions with reduced forms of metals (*e.g.* Fe(II), Cr(III) and Cu(I)) to form the reactive oxygen species hydroxyl radical ($^{\cdot}OH$). Hydroxyl radical is considered to be the main culprit in oxidative DNA damage. High levels of SeO_3^{2-} and high levels of carcinogens increase the concentration of 8-hydroxy-2′-deoxyguanosine (8-OH-dG), an accepted oxidative DNA damage marker, suggesting that high dietary intake of SeO_3^{2-} may increase *in vivo* DNA oxidation (Wycherly *et al.*, 2004). Selenous acid (aqueous SeO_2) is also shown to exacerbate oxidative DNA damage in the presence of dichromate ($Cr_2O_7^{2-}$) in a manner similar to SeO_3^{2-} (Cemeli *et al.*, 2003). Selenite ions increase the levels of 8-OH-dG in primary rat hepatocyte cultures in a way that is related to formation of ROS rather than a decrease in 8-oxuguanine DNA glycosylase 1, which excises 8-OH-dG from DNA (Fujimoto *et al.*, 2009).

4.4.3 Chemopreventive and Chemotherapeutic Activity of SeO_3^{2-}

Intermediate SeO_3^{2-} concentrations exhibit chemopreventive activity, and SeO_3^{2-} shows great promise as a clinically relevant cancer therapy because of its low bioavailability (Reeves *et al.*, 2005). Its behavior as a chemopreventive agent is due to its ability to induce apoptosis of tumor cells. In general, Se-induced apoptosis in cancer cells is convoluted and depends on the chemical form of the Se compound and the types of cells involved, but a clear connection exists between cancer cell death and ROS production. Selenite ion is reported to induce apoptosis in cancer cells by producing the ROS superoxide radical ($O_2^{\cdot -}$) (Xiang *et al.*, 2009). ROS production is correlated strongly with oxidative DNA damage (including DNA strand breakage and site-specific modifications of DNA bases), and this oxidative damage also appears to be the basis of the use of SeO_3^{2-} as a chemotherapeutic substance. Selenite ions are among those that interact with DNA and RNA, and the interactions with DNA differ based on the anion/DNA(P) (P = phosphate) ratio. At low ratios, interactions with waters of hydration in grooves of the helix dominate, stabilizing the B-structure and DNA duplex. At higher ratios interactions with the base donor groups prevail. At very high anion ratios SeO_3^{2-} and other anions compete with the phosphate backbone for sodium ions, which destabilizes the helix and promotes A-type DNA formation (Tajmir-Riahi *et al.*, 1994). In RNA, the major interaction of SeO_3^{2-} is reported to be with backbone phosphate groups, while specific SeO_3^{2-}-base interactions play a lesser role (Nafisi *et al.*, 2011). These interactions are not without consequence because the ability of SeO_3^{2-} to form ROS (such as $O_2^{\cdot -}$) at or near DNA will produce generalized and site-specific oxidative DNA damage. Selenite ion has been shown to induce Top II-DNA cleavage complexes *in vitro*, which trigger ATM-dependent tumor cell apoptosis (Zhou *et al.*, 2003). This suggests that selenite's role in ATM-dependent damage response is associated with the generation of ROS, producing modified DNA bases (*e.g.* 8-OH-dG) and DNA strand breaks that trigger ATM activation pathways. This ultimately regulates G_2/M checkpoint response, which prevents damaged cells from entering mitosis (Qi *et al.*, 2010). Selenite also appears to activate early barriers of tumorigenesis by inducing ATM-dependent programmed cell death *via* redox regulation in noncancerous cells. This represents a novel mechanism of SeO_3^{2-} in suppressing tumorigenesis by limiting the proliferation of cells with damaged DNA (Wu *et al.*, 2010).

Brozmanova *et al.* have reviewed the use of SeO_2, SeO_3^{2-} and other Se compounds as chemotherapeutic agents for various cancers. Selenite has been shown to be an effective chemotherapeutic agent for human prostate cancer, mesothelomia, lung cancer and non-Hodgkin's lymphoma. Other cancers for which SeO_3^{2-} holds promise are acute myeloid leukemia and colorectal cancer. It is also shown to enhance the effectiveness of other chemotherapeutic regimens, including IR-radiation treatment. Selenium dioxide (in the form of selenous acid) is reported to increase the effectiveness of conventional cancer drugs, such as taxol and doxorubicin, and it aids in overcoming resistance of colorectal cancer cells to other chemotherapeutic agents (Brozmanova *et al.*, 2010). Addressing the specific roles of SeO_2 and SeO_3^{2-} as chemotherapeutic agents is beyond the scope of this chapter, but others have reviewed the mechanism and use of these and other selenium compounds in cancer prevention and cancer therapy (Letavayova *et al.*, 2006; Zhu and Jiang, 2008).

4.5 Interactions with Metal Ions

The interactions of SeO_2 and SeO_3^{2-} with metal ions is an important topic to consider given the roles of Se compounds and metal ions in essential biological processes. Some metal ions (such as Fe(II), Cu(II) and Zn) are endogenous to living organisms and critical in many physiological processes, including the proper functioning of enzymes. However, high levels of these metal ions (Fe(II) and Cu(II) in particular) are responsible for several diseases and clinical conditions often associated with ROS formation and subsequent oxidative damage. At intermediate and high concentrations SeO_3^{2-} is also associated with ROS formation and is considered toxic due to its pro-oxidant activity and its ability to induce oxidative stress (see Section 4.4.2). Other metal ions, including As(III), Hg(II), Cd(II) and Cr(III), are exogenous to living systems and toxic at relatively low levels. However, the interactions of some metal ions with SeO_2 and SeO_3^{2-} are known to mediate these toxic effects, and in combination exert an antagonistic effect on one another that decreases the toxicity of the mixture compared to that of the individual components.

4.5.1 Role of SeO_3^{2-} in Mediating Metal Toxicity

As(III) and Hg(II) are two highly toxic metal ions that are mediated by the presence of SeO_3^{2-} (Gailer, 2007). Compounds containing As–Se and Hg–Se bonds are reportedly formed in the bloodstream, establishing a link between the metabolism of As(III) and Hg(II) with Se metabolism that appears to be driven by GSH. While Se in the +4 oxidation state is reduced by GSH and Cys in mammalian metabolism, As(III) and Hg(II) (along with CH_3Hg^+ and Cd(II)) have a high binding affinity for thiol (SH) groups and manifest their toxicity by effectively blocking enzymes containing thiols as their active centers. The reaction involving As(III) and SeO_3^{2-} (in the forms

$As(OH)_3$ and $HSeO_3^-$ at neutral pH) proceeds by oxidation of GSH to GSSG and is reported to form an As=Se double bond (Gailer *et al.*, 2000). The reaction involving Hg(II) is similar to As(III) in that GSH gets oxidized to GSSG. The resulting compound is characterized as a Hg–Se core that has approximately 5 GS groups bonded to the surface through the Se atoms (Se–GS bonds). The species that forms *in vivo* appears to have the formula $(Hg{-}Se)_{100}(GS)_5$, which binds subsequently to selenoprotein P (Yoneda and Suzuki, 1997; Sasakura and Suzuki, 1998). With respect to decreasing As and Hg toxicity, these As–Se and Hg–Se bonds are apparently stronger than the As–S and Hg–S bonds that form with endogenous GSH. In contrast, Alp *et al.* saw no effect of SeO_3^{2-} in reducing the cytotoxic effects of As(III) on HEK293 human kidney cells, but in this case an increase in protein phosphorylation may be the defining mechanism behind the reduced cytotoxicity (Alp *et al.*, 2011).

4.5.2 SeO_2 and SeO_3^{2-} as Antioxidants

Thus far, the uses of SeO_2 and SeO_3^{2-} in oxidation reactions and reactions involving ROS formation have been addressed. However, one of the interesting aspects of SeO_2 and SeO_3^{2-} chemistry is their dual behavior as both pro- and antioxidants. Antioxidants are important in preventing numerous diseases and clinical conditions, including cancer, and selenium compounds serve this function (Battin and Brumaghim, 2009). Organic selenium compounds are reported to function as antioxidants in Fe(II) and Cu(I) mediated oxidative damage by coordinating these metal ions (Battin *et al.*, 2011), and this also appears to be the case for SeO_2 and SeO_3^{2-}. Selenite ion is reported to coordinate Cd(II) and Pb(II) ions in a 1 : 1 ratio, and it also appears to form a coordination complex with zinc ions. However, none of these metal ions appear to form complexes with SeO_4^{2-} (with Se in a +6 oxidation state) (Badiello *et al.*, 1996; Feroci *et al.*, 1997; Feroci *et al.*, 2005). In metal-mediated oxidative DNA damage studies where Fe(II) and H_2O_2 are combined to produce $^{\cdot}OH$ (Fenton reaction), Ramoutar and Brumaghim found that SeO_2 and SeO_3^{2-} exhibit both pro- and antioxidant behaviors in a manner dependent on both the concentrations of the selenium compounds and H_2O_2. Neither SeO_4^{2-} nor Se^{2-} appear to inhibit Fe(II)-mediated oxidative DNA damage, suggesting that Se oxidation state is a key component in the abilities of Se compounds to coordinate metal ions. Surprisingly, SeO_2 and SeO_3^{2-} exhibit different antioxidant behaviors in the Fe(II)-mediated reactions, suggesting that the two Se species are different despite being at the same pH. NMR spectra (^{77}Se) also suggest that the species are different with aqueous solutions of SeO_2 at pH 6 and 7 exhibiting singlets at δ 1317 and 1299 ppm, respectively, while SeO_3^{2-} solutions exhibit a singlet at δ 1274 ppm at both pH values (Ramoutar and Brumaghim, 2007). Hart *et al.* also demonstrated that SeO_2 and SeO_3^{2-} act as antioxidants and decrease production of the site-specific DNA damage marker 8-OH-dG in Fenton-type reactions involving Fe(II), Cr(III) and Cu(II). Their mechanism as antioxidants is also reported to involve forming coordination complexes with these metal ions, and the two Se species also appear to produce different complexes with a given metal despite being at the same pH. While metal-ion coordination appears to be responsible for the abilities of SeO_2 and SeO_3^{2-} to act as antioxidants, the fate of the metal–Se complexes is uncertain. Specifically, it is unclear whether the metal–Se complexes exist free in bulk solution or if they still interact with DNA, but it appears that the Se-coordinated metal ions are capable of reacting with H_2O_2 to produce ROS. The ability of Fe(II), Cr(III) and Cu(II) to produce ROS in the presence of SeO_2 and SeO_3^{2-} was investigated using DCF-DA as a fluorescent probe to detect ROS formation. The results indicate that the metal–Se complexes are capable of producing ROS in the presence of H_2O_2, suggesting that the complexes may react with H_2O_2 in a "sacrificial" manner that protects DNA from oxidative damage. It is also possible that ROS generation from the complexed metal ions produces generalized DNA damage (*e.g.* strand breaks) rather than site-specific base modifications (*e.g.* 8-OH-dG formation) (Hart *et al.*, 2013).

The behavior of SeO_2 and SeO_3^{2-} as antioxidants is not limited to their abilities to coordinate metal ions. Pre-exposure with SeO_3^{2-} lowered the Cu accumulation in the blue mussel *M. edulis*, and this was accompanied by a decrease in the levels of oxidative damage to DNA and protein thiols. It also served to prevent the loss of thioredoxin reductase activity that normally occurs during Cu exposure. Rather than coordinating Cu ions directly, however, it is suggested that SeO_3^{2-} supplementation induces other physiologically relevant antioxidant defenses involving GSH and GPx (Trevisan *et al.*, 2011). Selenite ion is also reported to be an antioxidant that functions by decomposing hydrogen peroxide to water *via* GPx (Fujimoto *et al.*, 2009).

4.6 Conclusion

Compared to naturally occurring organoselenium compounds, inorganic forms of selenium are limited in their abilities to meet nutritional needs fully. In addition, moderate to high levels of these inorganic forms are toxic. However, SeO_2 and SeO_3^{2-} exhibit some useful protective effects, and their usefulness as nutritional supplements cannot be dismissed. The fact that Se exists in a +4 oxidation state in these species appears to be intimately related to their dual role as pro- and antioxidants. The pro-oxidant behavior of SeO_3^{2-} makes it a promising chemotherapeutic agent, and the antioxidant behaviors of SeO_2 and SeO_3^{2-} establish their possible use as chemopreventive agents. A summary of roles and characteristics of SeO_2 and SeO_3^{2-} is presented in Figure 4.4.

The mechanisms by which SeO_2, SeO_3^{2-} and other selenium compounds fulfill these roles are not well understood, and further research is necessary to get a more complete understanding of their potential and utilize them to the fullest extent.

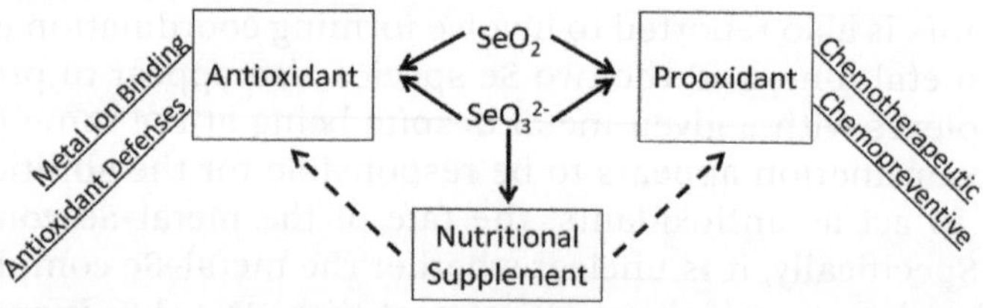

Figure 4.4 A summary of the various roles and characteristics of SeO_2 and SeO_3^{2-}. SeO_2 and SeO_3^{2-} exhibit dual behavior as both pro- and antioxidants. SeO_3^{2-} is an inorganic Se source employed commonly as a nutritional supplement, and it can have a direct effect as a pro- or antioxidant. Pro-oxidant behavior of SeO_2 and SeO_3^{2-} is a key component in their use as chemopreventive and chemotherapeutic agents. Antioxidant behavior is linked to their abilities to coordinate certain metal ions and trigger key antioxidant defenses.

Summary Points

- The oxidation state of Se in SeO_2 and SeO_3^{2-} is +4, which is intermediate between +6 for Se in selenate (SeO_4^{2-}) and −2 for selenide (Se^{2-}).
- Selenous or selenious acid (H_2SeO_3) is how SeO_2 will exist in aqueous solution.
- SeO_3^{2-} can be considered as the fully deprotonated form of H_2SeO_3. Se speciation should be identical for aqueous solutions of SeO_2 or SeO_3^{2-} at the same pH.
- SeO_3^{2-} (along with SeO_4^{2-}) is commonly used as a Se supplement in over-the-counter products, infant formulas, protein mixes and animal feeds.
- SeO_3^{2-} is able to mediate the toxic effects of heavy metals including As(III) and Hg(II). SeO_3^{2-} and these toxic metals exhibit an antagonistic effect on one another, making the toxic effects of the combined species less than that of the individual components.
- SeO_2, in combination with peroxides, is commonly used as a catalyst for oxidizing organic compounds.
- SeO_3^{2-} is capable of producing the ROS $O_2^{\cdot-}$ and is linked to production of oxidative DNA damage in the form of strand breaks and the accepted oxidative DNA damage marker 8-OH-dG.
- The ability of SeO_3^{2-} to form ROS and damage DNA is directly related to its use as a chemopreventive and chemotherapeutic agent.
- SeO_2 and SeO_3^{2-} act as pro- and antioxidants. They function as antioxidants in metal mediated oxidative DNA damage by coordinating metal ions. SeO_3^{2-} also functions as an antioxidant by inducing physiologically relevant antioxidant defenses.

Key Facts

Key facts of oxidative damage. Typical ROS that cause oxidative damage are listed, and oxidative damage to DNA (strand breaks and production of the modified base 8-OH-dG) is highlighted.

1. Oxidative damage is associated with at least 50 diseases and clinical conditions including atherosclerosis, rheumatoid arthritis, numerous cancers and aging.
2. Oxidative damage results from generation of reactive oxygen species (ROS) including superoxide radical ($O_2^{\cdot -}$), singlet oxygen ($^1O_2$), hydrogen peroxide (H_2O_2) and hydroxyl radical ($^{\cdot}OH$).
3. Antioxidants are species that minimize or prevent oxidative damage. They function by either suppressing or preventing ROS formation initially or by quenching ROS after they are formed.
4. Oxidative damage is a broad term that can be narrowed to a specific system, such as lipids and membranes, proteins, DNA and RNA.
5. Oxidative damage to DNA is an important area of study because it results in strand breaks and base modifications that interfere with normal cell metabolism.
6. In general, DNA nicking (single-strand breaks) is associated with ROS generated free in solution, while base modifications are indicative of more site-specific damage.
7. Despite the highly reactive nature of ROS, guanine is the most readily oxidized DNA base, and 8-hydroxy-2′-deoxyguanosine (8-OH-dG or 8-oxo-dG) is a widely accepted marker of oxidative DNA damage.
8. 8-OH-dG is excised from DNA as part of the natural repair mechanism for correcting oxidative damage and is excreted in urine. Increased urinary 8-OH-dG levels are correlated with oxidative stress.

Key facts regarding ROS formation from reactions of metal ions with H_2O_2 are presented. Site-specific ROS formation at metal–DNA binding sites is addressed along with antioxidants that function by coordinating metal ions.

1. Metal ions, such as Fe(II) and Cu(II), are essential to living organisms, and H_2O_2 is a normal product of aerobic cell metabolism.
2. Fe(II) reacts with H_2O_2 to produce the ROS hydroxyl radical ($^{\cdot}OH$) *via* the Fenton reaction. Other transition metals (including Cu(II) and Cr(III)) undergo Fenton-like reactions.
3. In the case of DNA, metal ions are capable of binding to the phosphate backbone and individual DNA bases.
4. ROS that are produced from these reactions are expected to react close to the site of formation due to their high reactivity. Oxidative damage is expected to be correlated with sites where metal ions bind.
5. Some antioxidants function by quenching ROS after they are formed. Species that show promise as antioxidants in metal-mediated ROS formation are expected to coordinate metal ions and prevent them from binding at critical sites on biomolecules or sequester them in a form that is unreactive with H_2O_2.
6. Metal-ion coordination appears to be the mechanism by which SeO_2, SeO_3^{2-} and organoselenium compounds function as antioxidants in metal-mediated ROS formation.

Definitions and Explanation of Key Terms

Apoptosis. Cell death that takes place in a programmed or controlled fashion. It is the direct opposite of necrosis, which describes fast and chaotic cell death.

ATM (Ataxia telangiectasia mutated) protein. A DNA-damage regulator that initiates DNA damage response.

Chemopreventive. An action or agent that prevents the development of cancer by inhibiting tumor development. Chemopreventive agents function by destroying tumors at an early stage.

Chemotherapeutic. Actions and therapies that are administered in order to destroy cancerous tumors after they are detected.

Cytotoxic. Toxic to cells. Agents that induce rapid or slow (programmed) cell death.

Dihydrodichlorofluorescein-diacetate (DCF-DA). DCF-DA is a compound commonly used to probe ROS formation. DCF-DA is nonfluorescent, but it is hydrolyzed under basic conditions to produce H_2DCF, which reacts with ROS (including $^{\cdot}OH$) to give fluorescent DCF.

Fenton reaction or "Fenton-like" reaction. The reaction of Fe(II) with H_2O_2 that produces $^{\cdot}OH$ ($Fe^{2+} + H_2O_2 \rightarrow Fe^{3+} + OH^- + {}^{\cdot}OH$). Fenton-like reactions involve other redox active metals that react with H_2O_2 to produce $^{\cdot}OH$ or other ROS that behave in a similar fashion.

G_2/M checkpoint. A critical step in the cell cycle that prevents cells from undergoing mitosis (duplication) before damaged DNA can be repaired.

Glutathione (GSH). A tripeptide composed of cysteine (Cys), glycine (Gly) and glutamic acid (Glu). It is responsible for taking available Se and reducing it to selenide (Se^{2-}), an intermediate form of Se in the metabolic pathway that leads to synthesis of selenoproteins and excretory Se compounds.

Glutathione peroxidase (GPx). A selenoenzyme responsible for preventing damage caused by oxidative stress. Its active site is the Se atom on a Se-Cys residue, and it functions by reducing hydroperoxides.

Metal-coordination complexes. A species formed between a metal ion and an anion or neutral species (*i.e.* ligand). The ligand possesses atoms with nonbonding electron pairs that are donated to form coordinate covalent bonds with the metal cation.

pK_a. The −log of a K_a value. The K_a value is the equilibrium constant for the loss of an acidic proton from a species (deprotonation). At pH = pK_a 50% of the acidic species is in the deprotonated form.

Reactive oxygen species (ROS). Highly reactive, oxygen-based species responsible for oxidative damage and aging. Examples include superoxide radical ($O_2^{\cdot -}$), singlet oxygen ($^1O_2$), hydrogen peroxide (H_2O_2) and hydroxyl radical ($^{\cdot}OH$).

Se-containing proteins. Proteins that contain Se-Met residues but have no specific biological functions.

Selenoproteins. Proteins containing Se-Cys residues that function as the active center in redox reactions of selenoenzymes.

Selenoprotein P. A selenoprotein found in plasma. Many believe that it plays a role in extracellular oxidant defense.

Thioredoxin reductase. A protein containing a Se-Cys residue that regulates intracellular redox reactions (including reduction of thioredoxin) and factors into repair mechanisms in DNA synthesis.

Topoisomerase II (Top II)-DNA-cleavage complexes. Top II is an enzyme that exhibits DNA-cleavage activity involved in processes such as DNA replication and transcription. Top II-DNA covalent complexes (cleavage complexes) are able to trigger cell apoptosis.

Tumorigenesis. The initial or early stages of tumor development.

List of Abbreviations

8-OH-dG	8-Hydroxy-2′-deoxyguanosine
ATM protein	Ataxia telangiectasia mutated protein
Cys	Cysteine
DCF-DA	Dihydrodichlorofluorescein-diacetate
DNA	Deoxyribonucleic acid
Glu	Glutamic acid
Gly	Glycine
GPx	Glutathione peroxidase
GSH	Glutathione
GSSeG	Selenodiglutathione
GSSG	Oxidized glutathione (GSH)
IR	Infrared
NMR	Nuclear magnetic resonance
RNA	Ribonucleic acid
ROS	Reactive oxygen species
Se-Cys	Selenocysteine
Se-Met	Selenomethionine
Top II	Topoisomerase II

References

Alp, O., Zhang, Y., Merino, E. J. and Caruso, J. A., 2011. Selenium effects on arsenic cytotoxicity and protein phosphorylation in human kidney cells using chip-based nanoLC-MS/MS. *Metallomics*. 3: 482–490.

Badiello, R., Feroci, G. and Fini, A., 1996. Interaction between trace elements: selenium and cadmium ions. *Journal of Trace Elements in Medicine and Biology*. 10: 156–162.

Battin, E. E. and Brumaghim, J. L., 2009. Antioxidant activity of sulfur and selenium: a review of reactive oxygen species scavenging, glutathione peroxidase, and metal-binding antioxidant mechanisms. *Cell Biochemistry and Biophysics*. 55: 1–23.

Battin, E. E., Zimmerman, M. T., Ramoutar, R. R., Quarles, C. E. and Brumaghim, J. L., 2011. Preventing metal-mediated oxidative DNA damage with selenium compounds. *Metallomics*. 3: 503–512.

Biswas, S., Talukder, G. and Sharma, A., 2000. Chromosome damage induced by selenium salts in human peripheral lymphocytes. *Toxicology in Vitro*. 14: 405–408.

Brozmanova, J., Manikova, D., Vlckova, V. and Chovanec, M., 2010. Selenium: a double-edged sword for defense and offence in cancer. *Archives of Toxicology*. 84: 919–938.

Cemeli, E., Carder, J., Anderson, D., Guillamet, E., Morillas, M. J., Creus, A. and Marcos, R., 2003. Antigenotoxic properties of selenium compounds on potassium dichromate and hydrogen peroxide. *Teratogenesis, Carcinogenesis, and Mutagenesis Supplement*. 2: 53–67.

Combs, G. F. and Gray, W. P., 1998. Chemopreventive agents: selenium. *Pharmacology and Therapeutics*. 79: 179–192.

Falcone, G. and Nickerson, W. J., 1963. Reduction of selenite by intact yeast cells and cell-free preparations. *Journal of Bacteriology*. 85: 754–762.

Feroci, G., Badiello, R. and Fini, A., 2005. Interactions between different selenium compounds and zinc, cadmium and mercury. *Journal of Trace Elements in Medicine and Biology*. 18: 227–234.

Feroci, G., Fini, A., Badiello, R. and Breccia, A., 1997. Interaction between selenium derivatives and heavy metal ions: Cu^{2+} and Pb^{2+}. *Microchemical Journal*. 57: 379–388.

Fujimoto, Y., Morinaga, K., Abe, M., Kitamura, T. and Sakuma, S., 2009. Selenite induces oxidative DNA damage in primary rat hepatocyte cultures. *Toxicology Letters*. 191: 341–346.

Gailer, J., 2007. Arsenic-selenium and mercury–selenium bonds in biology. *Coordination Chemistry Reviews*. 251: 234–254.

Gailer, J., George, G. N., Pickering, I. J., Prince, R. C., Ringwald, S. C., Pemberton, J. E., Glass, R. S., Younis, H. S., DeYoung, D. W. and Vasken, H., 2000. A metabolic link between arsenite and selenite: the seleno-bis(*S*-glutathionyl) arsinium ion. *Journal of the American Chemical Society*. 122: 4637–4639.

Hart, W. E., Marczak, S. P., Kneller, A. R., French, R. A. and Morris, D. L., 2013. The abilities of selenium dioxide and selenite ion to coordinate DNA-bound metal ions and decrease oxidative DNA damage. *Journal of Inorganic Biochemistry*. 125: 1–8.

Hughes, F. J. and Martin, D. S., 1955. Kinetics of the oxidation of selenious acid by hydrogen peroxide. *Journal of Physical Chemistry*. 59: 410–415.

Letavayova, L., Vickova, V. and Brozmanova, J., 2006. Selenium: from cancer prevention to DNA damage. *Toxicology*. 227: 1–14.

Lu, J., Jiang, C., Kaeck, M., Ganther, H., Vadhanavikit, S., IP, C. and Thompson, H., 1995. Dissociation of the genotoxic and growth inhibitory effects of selenium. *Biochemical Pharmacology*. 50: 213–219.

Nafisi, S., Manouchehri, F. and Montazeri, M., 2011. RNA adducts with Na_2SeO_4 and Na_2SeO_3-stability and structural features. *Journal of Molecular Structure*. 1006: 547–552.

Qi, Y., Schoene, N. W., Lartey, F. M. and Cheng, W.-H., 2010. Selenium compounds activate ATM-dependent DNA damage response *via* the mismatch repair protein hMLH1 in colorectal cancer cells. *Journal of Biological Chemistry*. 285: 33010–33017.

Ramoutar, R. R. and Brumaghim, J. L., 2007. Effects of inorganic selenium compounds on oxidative DNA damage. *Journal of Inorganic Biochemistry*. 101: 1028–1035.

Reeves, P. G., Leary, P. D., Gregoire, B. R., Finley, J. W., Lindlauf, J. E. and Johnson, L. K., 2005. Selenium bioavailability from buckwheat bran in rats fed a modified AIN-93G Torula yeast-based diet. *Journal of Nutrition*. 135: 2627–2633.

Sasakura, C. and Suzuki, K. T., 1998. Biological interaction between transition metals (Ag, Cd and Hg), selenide/sulfide and selenoprotein P. *Journal of Inorganic Biochemistry*. 71: 159–162.

Schrauzer, G. N. and Surai, P. F., 2009. Selenium in human and animal nutrition: resolved and unresolved issues. A partly historical treatise in commemoration of the fiftieth anniversary of the discovery of the biological essentiality of selenium, dedicated to the memory of Klaus Schwarz (1914–1978) on the occasion of the thirtieth anniversary of his death. *Critical Reviews in Biotechnology*. 29: 2–9.

Seko, Y., Saito, Y., Kitahara, J. and Imura, N., 1989. Active oxygen generation by the reaction of selenite with reduced glutathione *in vitro*. In: Wendel, A. (ed.) *Selenium in Biology and Medicine*. Springer, Berlin, Germany, pp. 70–73.

Suzuki, K. T., Kurasaki, K. and Suzuki, N., 2007. Selenocysteine β-lyase and methylselenol demethylase in the metabolism of Se-methylated selenocompounds into selenide. *Biochimica et Biophysica Acta*. 1770: 1053–1061.

Suzuki, K. T. and Ogra, Y., 2002. Metabolic pathway for selenium in the body: speciation by HPLC-ICP MS with enriched Se. *Food Additives and Contaminants*. 19: 974–983.

Tajmir-Riahi, H. A., Naoui, M., Diamantoglou, S. and Elmahmoudaoui, L., 1994. The effects of oxo-anions SeO_4^{2-}, SeO_3^{2-}, AsO_4^{3-}, AsO_3^{3-}, CrO_4^{2-}, CO_3^{2-}, HCO_3^-, SO_4^{2-}, SO_3^{2-}, PO_4^{3-}, PO_3^{3-}, NO_3^-, NO_2^-, ClO_4^- and CH_3COO^- on the solution structure of calf-thymus DNA. In: Collery, P., Poirier, L. A., Littlefield, N. A. and Etienne, J. C. (ed.) *Metal Ions in Biology and Medicine*. John Libbey Eurotext, Paris, France, pp. 53–57.

Trevisan, R., Mello, D. F., Fisher, A. S., Schuwerack, P. M., Dafre, A. L. and Moody, A. J., 2011. Selenium in water enhances antioxidant defenses and protects against copper-induced DNA damage in the blue mussel *Mytilus edulis*. *Aquatic Toxicology*. 101: 64–71.

Whanger, P. D., 2002. Selenocompounds in plants and animals and their biological significance. *Journal of the American College of Nutrition*. 21: 223–232.

Wu, M., Kang, M. M., Schoene, N. W. and Cheng, W.-H., 2010. Selenium compounds activate early barriers of tumorigenesis. *Journal of Biological Chemistry*. 285: 12055–12062.

Wycherly, B. J., Moak, M. A. and Christensen, M. J., 2004. High dietary intake of sodium selenite induces oxidative DNA damage in rat liver. *Nutrition and Cancer*. 48: 78–83.

Xiang, N., Zhao, R. and Zhong, W., 2009. Sodium selenite induces apoptosis by generation of superoxide *via* the mitochondrial-dependent pathway in human prostate cancer cells. *Cancer Chemotherapy and Pharmacology*. 63: 351–362.

Yoneda, S. and Suzuki, K. T., 1997. Detoxification of mercury by selenium by binding of equimolar Hg-Se complex to a specific plasma protein. *Toxicology and Applied Pharmacology*. 143: 274–280.

Zhou, N., Xiao, H., Li, T., Nur-E-Kamal, A. and Liu, L., 2003. DNA damage-mediated apoptosis induced by selenium compounds. *Journal of Biological Chemistry*. 278: 29532–29537.

Zhu, Z. and Jiang, W., 2008. Selenium in prevention of cancer: evidence and mechanism. *Biomedical Research on Trace Elements*. 19: 282–289.

CHAPTER 5

Chemistry and Biochemistry: Selenium Species in Fish

MAMORU HARATAKE*[a], KAORI KOGA[b], SAKURA YOSHIDA[b], TAKESHI FUCHIGAMI[b], AND MORIO NAKAYAMA*[b]

[a]Faculty of Pharmaceutical Sciences, Sojo University, 4-22-1 Ikeda, Nishi-ku, Kumamoto, Japan; [b]Graduate School of Biomedical Sciences, Nagasaki University, 1-14 Bunkyo-machi, Nagasaki, Japan
*E-mail: haratake@ph.sojo-u.ac.jp, morio@nagasaki-u.ac.jp

5.1 Background

Since the first report of biological importance of selenium by Schwarz and Foltz in 1957 (Schwarz and Foltz, 1957), research interests in studying its biochemical nature have increased and the importance of selenium as an essential micronutrient in many organisms has been well recognized (Rayman, 2012). Selenium occurs in proteins in the form of the 21st amino acid, seleno-L-cysteine (SeCys, Sec or U); eukaryotes have variable sets of SeCys-containing proteins (selenoproteomes), which are from zero selenoproteins in the higher plants and fungi to more than 30 in fish and algae (Lobanov *et al.*, 2009). The speciation analysis of selenium-containing materials in fish is important from the viewpoints of not only the marine biology of this trace element but also the nutritional and toxicological sciences. Selenium deficiency and toxic levels are close to each other, it is important to know its abundance and deficiency in foodstuffs and to estimate its appropriate balance. So far, no selenium-deficient diseases and related pathologies have

Food and Nutritional Components in Focus No. 9
Selenium: Chemistry, Analysis, Function and Effects
Edited by Victor R. Preedy

Published by the Royal Society of Chemistry, www.rsc.org

been reported in Japan, which is simply dependent on the diet. Fish, shellfish and other seafood materials are reported to be the major dietary sources of selenium for the Japanese population (~60% of daily intake) (Miyazaki *et al.*, 2004).

Nutritional selenium is thought to come from a variety of selenium compounds in dietary sources, mostly in organic forms such as seleno-L-methionine (SeMet), SeCys and their derivatives. While inorganic selenite (SeO_3^{2-}) is rare as a chemical form of the dietary source compounds, it is the most frequently used effective one for selenium swupplementation of medical treatments. Plant food sources of selenium contain mostly SeMet, but selenium-enriched yeast (a common selenium supplement form) is also found to contain significant amounts of selenite (B'Hymer and Caruso, 2000; Infante *et al.*, 2005).

The selenium content of food and beverages geographically varies both within and between countries and/or areas. The selenium contents of animal foodstuffs basically reflect the selenium content of their consumed diet (Barclay *et al.*, 1995), whereas the selenium content of plants is directly affected by the selenium concentrations in the soil in which they are grown. Fish can take up selenium from the water and/or by eating other marine species in the food chain or web. The accumulation of selenium in marine and freshwater animals from dietary sources (phyto- and zoo-planktons) is more important than that directly obtained from water (Dumont *et al.*, 2006). Selenium biology in humans, other animals and plants has been well covered in many books and reviews (Navarro-Alarcon and Cabrera-Vique, 2008). Selenium species in food products from terrestrial plants and animals are also described in other chapters of this book. In this chapter, we focus on the speciation analysis of fish-specific selenoproteins and low molecular mass organoselenium compounds from fish to be consumed.

5.2 Selenoproteins

Twenty-five kinds of selenoproteins are identified in humans based on a selenoproteome analysis (*cf.* 24 selenoproteins in mouse and rat) (Table 5.1) (Kryukov *et al.*, 2003), and their physiological roles and functions are being investigated (Chen and Berry, 2003; Cheng *et al.*, 2009; Whanger, 2009). Selenoproteins are ubiquitously expressed in all organs and tissues, such as glutathione peroxidase (GPx), thioredoxin reductase and iodothyronine deiodinase (Dio). The best-known selenoproteins are part of the GPx family (GPx1, GPx2, GPx3, GPx4 and GPx6) that can catalyze the reduction of certain peroxide species (R—OOH) to the corresponding alcohols (R—OH) at their active center SeCys residue. More specifically, phospholipid hydroperoxide GPx (GPx4) is the only antioxidative enzyme that can directly reduce phospholipid hydroperoxides generated in biological membranes, which plays a critical role in the developmental process and biological activity (Imai and Nakagawa, 2003; Yant *et al.*, 2003). It has been shown that selenoproteins can be a thousand times more effective in catalysis than their L-cysteine (Cys) homologs (Johansson *et al.*, 2005). Such a greater effectiveness of SeCys in catalysis is likely one of the major reasons that nature has developed selenium-dependent metabolic pathways and the specific machinery used for selenium insertion into proteins.

Selenoproteins are basically present in bacteria, archaea and eukaryota. However, some organisms do not use SeCys. For example, as yeast and higher

Table 5.1 Selenoproteins in fish and human.

Selenoprotein	Abbreviation	Fish	Human
15 kDa selenoprotein	Sep15	+	+
Fish 15 kDa selenoprotein-like protein	Fep15	+	−
Glutathione peroxidase 1a	GPx1	+	+
Glutathione peroxidase 1b	GPx1b	+	−
Glutathione peroxidase 2	GPx2	+	+
Glutathione peroxidase 3	GPx3	+	+
Glutathione peroxidase 3b	GPx3b	+	−
Glutathione peroxidase 4a	GPx4	+	+
Glutathione peroxidase 4b	GPx4b	+	−
Glutathione peroxidase 6	GPx6	−	+
Iodothyronine deiodinase 1	Dio1	+	+
Iodothyronine deiodinase 2	Dio2	+	+
Iodothyronine deiodinase 3a	Dio3	+	+
Iodothyronine deiodinase 3b	Dio3b	+	−
Methionine-*R*-sulfoxide reductase 1a	MsrB1	+	+
Methionine-*R*-sulfoxide reductase 1b	MsrB1b	+	−
Selenophosphate synthetase 2a	SPS2a	+	−
Selenophosphate synthetase 2b	SPS2b	−	+
Selenoprotein H	SelH	+	+
Selenoprotein I	SelI	+	+
Selenoprotein J	SelJ	+	−
Selenoprotein J2	SelJ2	+	−
Selenoprotein K	SelK	+	+
Selenoprotein L	SelL	+	−
Selenoprotein M	SelM	+	+
Selenoprotein N	SelN	+	+
Selenoprotein O	SelO	+	+
Selenoprotein O2	SelO2	+	−
Selenoprotein P	SelP	+	+
Selenoprotein Pb	SelPb	+	−
Selenoprotein S	SelS	+	+
Selenoprotein T1a	SelT1a	+	+
Selenoprotein T1b	SelT1b	+	−
Selenoprotein T2	SelT2	+	−
Selenoprotein U1a	SelU1	+	−
Selenoprotein U1b	SelU1b	+	−
Selenoprotein U1c	SelU1c	+	−
Selenoprotein V	SelV	−	+
Selenoprotein W1	SelW1	+	+
Selenoprotein W2a	SelW2a	+	−
Selenoprotein W2b	SelW2b	+	−
Selenoprotein W2c	SelW2c	+	−
Thioredoxin reductase 1	TR1	+	+
Thioredoxin reductase 3	TR3	+	+
Thioredoxin/glutathione reductase	TGR	−	+

plants lost the SeCys insertion machinery during evolution, they do not synthesize selenoproteins, and instead utilize Cys homologs of some selenoproteins (Fu *et al.*, 2002; Lobanov *et al.*, 2009). Selenium is essential to fish, and they can synthesize selenoproteins using the SeCys insertion machinery (Kryukov and Gladyshev, 2000; Tujebajeva *et al.*, 2000; Yeh *et al.*, 2009). The physiological functions of human selenoproteins have not yet been completely revealed. A study of selenoproteins using fish, *e.g.*, the zebrafish with 38 selenoproteins, could provide useful information, because the selenoproteome of fish contains most mammalian selenoproteins (Deniziak *et al.*, 2007). Several selenoproteins in fish, including GPx, have been identified and characterized. Fish GPx is purified in the liver of the Southern bluefin tuna (*Thunnus maccoyii*), which is present in a homotetramer with a native molecular mass of 85 kDa and a subunit molecular mass of approximately 24 kDa (Thompson *et al.*, 2006). The Southern bluefin tuna GPx is structurally similar to the classical GPx1 found in the liver and red blood cells, the gastrointestinal GPx2 and the plasma GPx3 of mammals.

Marine animals have more selenoproteins than terrestrial animals. The selenoproteomes of fish are greater in number than those of mammals; at 41 selenoproteins, the selenoproteomes of fish are among the largest known (Table 5.1) (Lobanov *et al.*, 2009; Mariotti *et al.*, 2012). The same core selenoprotein families are found in mammals and fish. In addition, fish have several species-specific selenoproteins (fish 15 kDa selenoprotein-like protein (Fep15), selenoprotein J (SelJ) and selenoprotein L (SelL)) that are missing in mammals, as well as several SeCys-containing copies of selenoproteins T, U and W (SelT1a, SelT1b and SelT2, SelU1, SelU1b and SelU1c, SelW2a, SelW2b and SelW2c), and two forms of the selenoprotein P (SelP and SelPb) (Lobanov *et al.*, 2009).

Fep15, absent in mammals, can be exclusively detected in fish and only in the SeCys-containing protein (Novoselov *et al.*, 2006). Fep15 is distantly related to members of the 15 kDa selenoprotein (Sep15) family. SeCys in Sep15 is present in the Cys–Gly SeCys motif that is a putative redox site, whereas Fep15 does not have Cys in the vicinity of SeCys. Moreover, the Fep15 sequences have no conserved Cys at all, and several Fep15s do not have any Cys. If SeCys has as an antioxidative activity like the GPxs, it is likely converted into selenenic acid (−SeOH) or selenenylsulfide (−Se−S−) is formed with a thiol of Cys in another protein or with a low molecular mass thiol such as glutathione. The biological function of this protein is not clarified yet.

SelJ was first discovered in the genome of the teleost fish *Tetraodon nigroviridis* (Jaillon *et al.*, 2004). Later, its preferential and homogeneous expression was found in the eye lens during the early stages of the zebrafish (*Danio rerio*) development (Castellano *et al.*, 2005). This selenoprotein has a restricted phylogenetic distribution and, in contrast to all known eukaryotic selenoproteins, is not involved in mammalian genomes, not even as a Cys homolog. In addition, SelJ is assumed to play a structural role, although most selenoproteins have enzymatic functions.

SelL was identified in diverse aquatic organisms, including fish, invertebrates, and marine bacteria (Shchedrina *et al.*, 2007). This selenoprotein contains two SeCys residues separated by two other residues to form a SeCys-X-X-SeCys motif (two SeCys separated by two X amino acid residues) similar to the catalytic Cys-X-X-Cys motif (two Cys separated by two X amino acid residues) in thioredoxin, which suggests a redox function of this selenoprotein.

Organisms living in aquatic habitats, such as fish, amphibians and some marine invertebrates, possess a particularly greater number of SeCys residues. For example, the number of SeCys residues in SelP is higher; 16–17 in fish, as opposed to 7–15 in mammals. Particularly, SelP of the sea urchin involves 28 SeCys residues (Lobanov *et al.*, 2008, 2009).

Several selenoproteins in fish have homologs in mammals in which Cys is present in place of SeCys (selenoprotein U (SelU1, SelU1b and SelU1c), a SelW-like protein radixin (Rdx) 12 and GPx6). In contrast, no fish could be found that had Cys orthologs of mammalian selenoproteins. Lower SeCys contents of SelP and unidirectional SeCys to Cys transitions in vertebrate selenoproteins suggest a trend toward the reduced utilization of selenium in mammals. Larger selenoproteomes in aquatic organisms may result from several marine-specific factors affecting the SeCys utilization, such as availability of selenium, gradients of temperature, pH, pressure, oxygen content, chemical environment, *etc.* (Lobanov *et al.*, 2007).

Freshwater fish also have selenoproteins. A water-soluble selenoprotein was isolated from muscle tissue of *Misgurnus anguillicaudatus* by Gong *et al.* (2012). The trypsin/flavorzyme combination was used to hydrolyze the selenoprotein for subsequent analysis by reverse-phase high-performance liquid chromatography (HPLC)-inductively coupled plasma-mass spectrometry (ICP-MS). The two major species found in the hydrolyzates were SeMet and seleno-L-cystine (SeCyst). Separately, when analyzing the enzyme-hydrolyzed samples by HPCL-electrospray ionization (ESI)-MS-MS, an unknown selenium-containing species with a similar polarity to SeCyst and fragments likely to be due SeMet-containing peptides were also detected. Selenoprotein in muscle tissue of *Misgurnus anguillicaudatus* may have a species-specific structure.

5.3 Low Molecular Mass Organoselenium Compounds

Food and beverages are generally the source of selenium for the general population and the selenium bioavailability is thought to mainly come from organoselenium compounds (generally more than 80%). A speciation analysis of selenium in various foodstuffs has been performed for past several decades (Figure 5.1). For example, SeMet, glutathione selenotrisulfide, selenite and selenate (SeO_4^{2-}) were detected in selenized yeast (Casiot *et al.*, 1999; Lindemann and Hintelmann, 2002; Encinar *et al.*, 2003; Dumont *et al.*, 2005). Selenium-enriched plant foods (*e.g.*, onion, garlic, shiitake mushroom) and

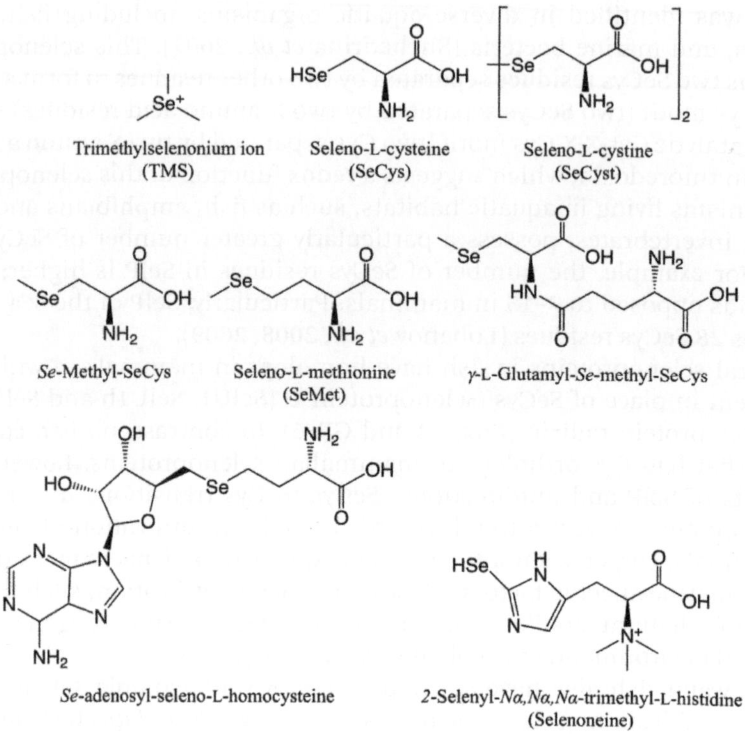

Figure 5.1 Chemical structures of low molecular mass organoselenium compounds in fish.

selenium-accumulating plants (*e.g.*, Indian mustard) contain the seleno-amino acids and their derivatives (Ogra *et al.*, 2004; Arnault and Auger, 2006; Dumont *et al.*, 2006; Freeman *et al.*, 2006). To date, the speciation analyses of biological samples have been mostly carried out using selenium-enriched plants. Several selenium compounds were identified from certain plant foods, while the selenium species in fish are hardly known, probably due to the extremely low selenium contents as compared with such selenium-enriched plant food.

Selenium is known to interact with several metal elements such as mercury. Ganther *et al.* showed that tuna contained enough selenium to modify the methyl mercury toxicity, that tuna diets reduced the methyl mercury toxicity in Japanese quail more than did diets based on plant sources of protein, and that tuna having a high content of mercury tended to accumulate selenium with mercury in a 1:1 molar ratio (Ganther *et al.*, 1972). Selenium was established as a naturally occurring antagonist in marine fish at levels capable of modifying the methyl mercury toxicity. Selenium has a protective effect regarding the toxicity of methyl mercury and forms an equimolar complex (Ganther, 1978). Cabanero *et al.* investigated the bioaccessibility of selenium and mercury in fish samples (tuna (*Thunnus* spp.), swordfish (*Aphanopus carbo*), and sardine (*Sardina pilchardus*)) by an *in vitro* gastrointestinal digestion method. The selenium and mercury bioaccessibility was found to be dependent on the type of fish analyzed. Simulated human gastric and intestinal digestion led to a high selenium bioaccessibility and low mercury bioaccessibility, and no modification during digestion of both species was found in all the fish samples. They concluded that the potential toxicity of fish cannot be independently evaluated by analyzing the total mercury or methyl mercury content, but also the selenium content that could significantly influence the mercury bioaccessible fraction (Cabanero *et al.*, 2004, 2007).

George *et al.* showed that mercury in fish (swordfish, *Xiphias gladius*) and fish digested with simulated gastric fluid was coordinated by a single thiolate donor, which resembled Cys. For the selenium, they found a mixture of organic forms that resembled SeMet and an aliphatic selenenylsulfide (–Se–S–) (George *et al.*, 2008). Furthermore, they detected a methyl mercury–Cys form in human hair samples taken from individuals with a high fish consumption (George *et al.*, 2010).

The water-soluble selenium species in commonly eaten fish were separated by size-exclusion chromatography with online detection by ICP-MS (Önning and Bergdahl, 1999; Önning, 2000). Cod (*Gadus morhua*), eel (*Anguilla anguilla*), rainbow trout (*Oncorhynchus mykiss*) and salmon (*Salmo salar*) mostly had the water-soluble species (up to nearly 90%) in a high molecular mass range (less than 10 kDa). Mackerel (*Scomber scombrus*), herring (*Clupea harengus*) and flat fish contained 0.26–0.50 μg selenium per g in the wet state of which 23–34% was water soluble. Half of the selenium species from mackerel and herring was separated in a low molecular mass range (less than 10 kDa), while all the flat fish, such as plaice (*Pleuronectes platessa*), turbot (*Psetta maxima*), flounder (*Platichthys flesus*) and dab (*Limanda limanda*) contained large amounts of low molecular mass organoselenium compounds (less than 2 kDa). SeMet and SeCyst were not detected in any of the fish tested in their study.

The selenium species in tuna and mussel were analyzed by enzymatic digestions and subsequent reverse-phase high-performance liquid chromatography in conjunction with ICP-MS. The total selenium contents were 1.3–4.6 μg selenium per g for tuna and 1.6–1.8 μg selenium per g for mussel. Their analytical method determined the organic (trimethylselenonium ion (TMS), SeCyst, SeMet and selenoethionine) and inorganic selenium species (selenite and selenate), but only TMS and SeMet were found in the samples. Unknown selenium species were also detected in tuna samples. The sum of the identified selenium species in the samples was only about 30% of the total selenium present in the enzymatic extract despite the fact that recoveries of the total hydrolyzed selenium were 93–102% (Quijano *et al.*, 2000).

Recently, Bjerregaard *et al.* showed that exposure to dietary selenium strongly reduces the retention of methyl mercury in freshwater fish (Bjerregaard *et al.*, 2011), but although this was very clearly established it may not be a universal phenomenon across all animal groups; for example, in mice, oral administration of selenium has little effect on the retention of methyl mercury (Glynn *et al.*, 1993; Glynn and Lind, 1995). It appears to be due to the species difference in the selenium metabolism.

Inorganic selenite and selenate were detected in dogfish (*Scyliorhinus canicula*) and swordfish (*Xiphias gladius*) muscle by microwave-assisted extraction (35% selenium recovery) and ion chromatography with online detection by ICP-MS. Unknown organoselenium compounds were also detected besides the inorganic selenium (Wang *et al.*, 2007).

The same analytical procedure was applied for the fractionation of selenium species in oyster (*Crassostrea gigas*) by aqueous extraction. The aqueous extract contained 35 ± 3% of the total selenium content (1.22 ± 0.03 μg selenium per g). The selenium species found in oyster tissues were TMS (9.8 ± 0.8%) and SeMet (46 ± 6%) (Moreno *et al.*, 2001). Moreno *et al.* also reported that SeMet within the range of 0.2–600 mg selenium per g was quantified in selected tissues, but SeCyst was not identified in tuna, trout, krill, oyster and mussel. TMS was quantified (0.1–0.3 μg selenium per g) in oyster, mussel and trout. Inorganic selenium as selenite was found in krill (Moreno *et al.*, 2004). In addition, several unidentified organoselenium compounds were detected.

Siwek *et al.* extracted approximately 24% of the total selenium of Antarctic krill (*Euphausia superba*) (2.4 mg selenium per g in dry state). They found that 80% of the extracted selenium was from organoselenium compounds with a molecular mass of 150–600 Da and the rest was bound to proteinous materials. A further analysis of the low molecular mass hydrolysates by HPLC-ICP-MS revealed the involvement of SeMet, SeCyst and its derivatives. The inorganic selenium species was negligible (Siwek *et al.*, 2005).

Huerta *et al.* compared four procedures (microwave digestion, methanol–HCl extraction, sodium dodecyl sulfate leaching, and enzymatic hydrolysis) for the extraction of selenium species from a cod muscle sample; enzymatic hydrolysis was the most effective (70% selenium recovery). Reverse-phase and size-exclusion HPLC analyses showed that most of the selenium in the sample is associated with proteins, and SeMet turns out to be the main organoselenium compound in the enzymatic hydrolysate. As SeMet does not appear after the methanol–HCl extraction, SeMet is not probably free but unspecifically incorporated into proteins in place of methionine (Huerta *et al.*, 2004).

Modern mass-spectrometric analyses, such as matrix-assisted laser desorption ionization time-of-flight mass spectrometry (MALDI TOF-MS) and ESI-MS are powerful techniques for the determination of the chemical structure of unknown selenium species according to its characteristic stable isotope pattern (Casiot *et al.*, 1999; Lindemann and Hintelmann, 2002; Encinar *et al.*, 2003; Dumont *et al.*, 2005). X-ray absorption near edge structure (XANES) spectroscopy is also useful for the detection of selenium species. This analytical technique is nondestructive and able to distinguish the chemical form of selenium (Pickering *et al.*, 1999; George *et al.*, 2008, 2010; Misra *et al.*, 2010). Misra *et al.* investigated the unique metabolism of selenite, selenate and SeMet in the isolated hepatocytes of rainbow trout (*Oncorhynchus mykiss*) by XANES spectroscopy besides other biochemical analyses;

the inorganic selenium compounds were metabolized into elemental selenium, while a major metabolite of SeMet was SeCyst.

To date, SeCys is thought to be a major selenium species in animal foodstuffs, since most of them involve SeCys that is incorporated into the selenoproteins, but organoselenium species other than selenoamino acids in fish are hardly ever observed. Although the selenium contents of seafood are higher than in terrestrial foodstuffs, little is known about the chemical forms of the selenium species in fish. Actually, not only proteinous selenium species like SeMet and SeCys derivatives, but also many low molecular mass nonproteinous ones were detected in fish samples. Many flat fish such as plaice contain low molecular mass selenium species other than SeMet and SeCys, but their chemical structures have not yet been identified. In most of the speciation studies, only using the limited standard materials could identify several selenium species, whereas unidentified peaks, of which their retention times were not consistent with those of the standard materials, still need to be evaluated.

An organoselenium compound, selenoneine was separated from the blood of bluefin tuna (*Thunnus orientalis*) (35 μg selenium per g blood) (Yamashita and Yamashita, 2010). The chemical structure of this compound was determined by ESI-MS and ^{1}H- and ^{13}C-nuclear magnetic resonance (NMR) techniques. It is a selenium analog of ergothioneine that was found in ergot and human blood. Selenoneine was also detected besides selenosugar (1β-methylseleno-*N*-acetyl-D-galactosamine) and TMS in the liver of sea turtles (*Eretmochelys imbricata* and *Chelonia mydas*) (43 μg selenium per g in wet state liver) (Anan *et al.*, 2011a). It should be noted that the selenium concentrations of both biological samples (several-ten micromolar) were originally much higher than the other fish samples described above (submicromolar or less).

Selenoneine appears to show a radical-scavenging activity, although the biological function of selenoneine is not fully yet known. The biological role of selenoneine in preventing toxicity against inorganic mercury and methyl mercury was studied by Anan *et al.* (2011b). An *in vitro* assay was performed by incubating liver homogenates of hawksbill turtle (as a source of selenoneine) with inorganic mercury and methyl mercury for 1 h at 37 °C in the absence and in the presence of exogenous stable isotope ^{82}Se labelled-selenite. Once incubated, the homogenates were analyzed by multimode gel permeation chromatography coupled to ICP-MS. The distribution of endogenous selenium and exogenous mercury in the liver cytosol showed that selenoneine seems to be not involved in the detoxification of mercury, whereas selenium(IV) reacted with inorganic mercury after being converted into selenide by GPx.

Yoshida *et al.* demonstrated that the niboshi (Figure 5.3(b)) contains a relatively high concentration of selenium (Table 5.2) (Yoshida *et al.*, 2012). Niboshi is a commonly used foodstuff that is processed from Japanese anchovy (*Engraulis japonicus*, Figure 5.2(a)), and niboshi extract (Figures 5.2(c) and 5.3 and Table 5.3) is used as a general base seasoning for a wide variety of Japanese cuisines, just like the *fond de veau* for the French cuisine. This extract contains 6 to 12% of the total selenium in the niboshi. Approximately 90% of selenium in the extracts was present as low molecular mass

Table 5.2 Selenium concentrations and its distribution in Japanese anchovy in lyophilized state and niboshi.

Item	Part	Selenium concentration (μg g^{-1})	Selenium distribution (%)
Japanese anchovy (lyophilized)	Head	1.27 ± 0.02	13.7
	Abdomen	2.91 ± 0.04	39.6
	Rest of all	0.89 ± 0.03	46.6
	Whole	1.14 ± 0.02	100.0
Niboshi	Head	1.30 ± 0.01	14.0
	Abdomen	3.29 ± 0.04	35.2
	Rest of all	0.93 ± 0.04	50.8
	Whole	1.20 ± 0.02	100.0

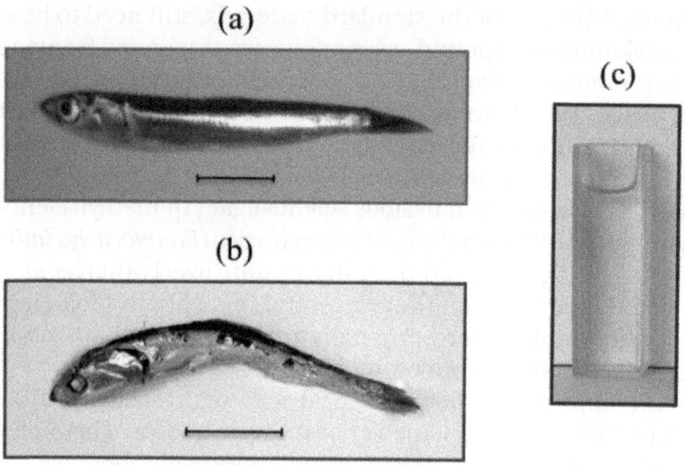

Figure 5.2 Japanese anchovy, niboshi and niboshi extract. (a) Japanese anchovy in raw, 1.14 ± 0.02 μg selenium per g in lyophilized state, (b) niboshi, 1.20 ± 0.02 μg selenium per g niboshi, (c) niboshi extract, 0.10 ± 0.02 μg selenium per g. Scale bars in (a) and (b): 1 cm.

organoselenium compounds (less than 5 kDa), including selenoamino acids and their derivatives. Selenium fractionation, by sequential separation with ion-exchange chromatography (Q-Sepharose in OH form and SP-Sepharose in H form columns) revealed that most of the selenium present in the niboshi has anionic and/or amphoteric characteristics.

The distribution and speciation of selenium in freshwater fish (brown trout (*Salmo trutta*), rainbow trout (*Oncorhynchus mykiss*), brook trout (*Salvelinus fontinalis*) and creole perch (*Percichthys trucha*)) from lakes in Argentina were investigated (Kristan *et al.*, 2013). Selenium concentrations in muscle range from 0.66 to 1.61 μg g^{-1}, while concentrations in the liver were from 4.46 to 73.71 μg g^{-1} on a dry-matter basis. Water-soluble selenium species were analyzed by ICP-MS subsequent to ion-exchange chromatographic separation;

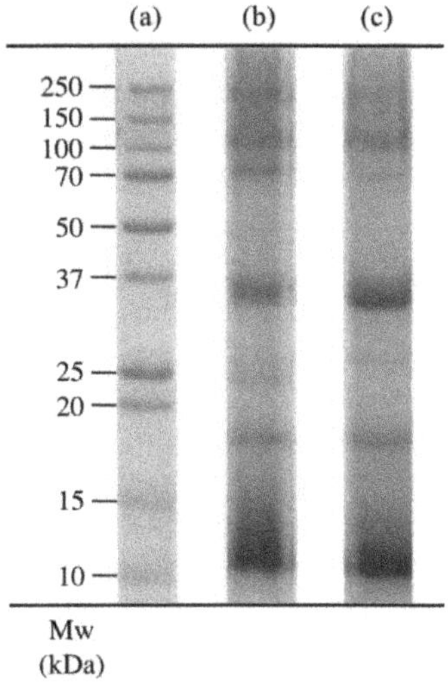

Figure 5.3 Sodium dodecyl sulfate-polyacrylamide gel electrophoresis of niboshi extract. A concentrated niboshi extract treated with or without 20 mM dithiothreitol was separated by 15% polyacrylamide gel in 0.1 M tris(hydroxymethyl)aminomethane, 0.05 M tricine, 0.1% sodium dodecyl sulfate solution at constant current 20 mA for 140 min. Gel was stained with Coomassie brilliant blue G-250.

Table 5.3 Analyses of components other than organoselenium in niboshi extract after lyophilization.[a]

Component	% by weight
Sodium	18.0
Magnesium	0.7
Phosphorus	2.7
Chlorine	14.2
Potassium	20.2
Calcium	≈0.0
Amino acids	11.2
Proteins	13.9
Inorganic selenium	≈0.0

[a]Sodium, magnesium, potassium and calcium were determined by atomic absorption spectrometry. The determination of the phosphorus content was based on vanadium (V) that is capable of bonding with the phosphate ion. The chlorine content was measured by the Mohr method. The amino acid contents were spectrofluorometrically determined after the reaction with *o*-phthalaldehyde and dithiothreitol to form a thio-substituted isoindole fluorophore. Protein content was directly measured by Lowry's method. Inorganic selenium was determined by fluorometry using 2,3-diaminonaphtharene.

47 to 55% of the total selenium in the muscle was extracted and the only selenium species identified was SeMet, which represented ~80% of the selenium species extracted. On the other hand, 61 to 76% of the total selenium in the liver was extracted and the contents of SeMet and SeCyst were only less than 17% of the selenium species extracted.

5.4 Conclusions

Selenoproteins occur in both eukaryotes and prokaryotes, but the selenoproteome is highly variable among organisms, and certain organisms do not utilize SeCys at all. Fish apparently retain, and in some cases, increase their selenoproteomes, whereas the selenoproteomes of some terrestrial organisms are reduced or completely lost. Many selenoproteins originate at the base of the eukaryotic domain and suggest that the sea environment that influences the selenium utilization plays a critical role in the fish selenoproteomes. Although recent advances in analytical techniques, especially mass spectrometry, significantly contribute to the speciation of low molecular mass organoselenium compounds in foodstuffs, selenium species in dietary seafood materials are still poorly known at present. Fish, including shellfish and other marine animals, are likely to have chemically diverse low molecular mass organoselenium species other than selenoamino acids and their derivatives, which may be responsible for their diets and/or fish-specific metabolic pathways. Consequently, fish appear to contain low-molecular-mass organoselenium compounds that have not yet been chemically identified.

Summary Points

- The speciation analysis of the fish-specific selenoproteins and low molecular mass organoselenium compounds from fish is described.
- Selenium is essential to fish as well as mammals.
- The selenoproteomes (sets of seleno-L-cysteine (SeCys)-containing proteins) of fish are greater in number than those of mammals (25 selenoproteins in humans); at 41 selenoproteins, the selenoproteomes of fish are among the largest known.
- Fish have several species-specific selenoproteins.
- Many low-mass organoselenium species are detected in fish samples other than seleno-L-methionine and SeCys derivatives.
- Although the selenium contents in fish are higher than in terrestrial foodstuffs, little is known about the chemical forms of such organoselenium species in fish.
- Fish appear to contain organoselenium compounds that have not yet been chemically identified.

Key Facts

Key Facts of Selenoprotein L (SelL)

1. SelL contains two U separated by two other residues to form a SeCys-X-X-SeCys motif.
2. SelL occurs in diverse aquatic organisms, including fish, invertebrates and marine bacteria.

3. Both eukaryotic and bacterial SelL genes use single SeCys insertion sequence (SECIS) elements for insertion of two SeCys.
4. SelL proteins possess a thioredoxin-like fold wherein the SeCys-X-X-SeCys motif corresponds to the catalytic Cys-X-X-Cys motif in thioredoxins.
5. A diselenide bond can be formed between the two Secs, establishing it as the first diselenide bond found in a natural protein.
6. *Danio rerio* SelL, transiently expressed in mammalian cells, incorporated two SeCys and localized to the cytosol.

Key Facts of Fish 15 kDa Selenoprotein-like Protein (Fep15)

1. Fep15 is distantly related to members of the 15 kDa selenoprotein (Sep15) family.
2. Fep15 is absent in mammals, can be detected only in fish and is present in these organisms only in the selenoprotein form.
3. Fep15 has only SeCys, in contrast with other members of the Sep15 family containing a putative active site composed of SeCys and Cys.
4. When transiently expressed in mammalian cells, Fep15 is targeted to the endoplasmic reticulum by its N-terminal signal peptide.
5. Phylogenetic analyses of Sep15 family members suggest that Fep15 evolved by gene duplication.

Definitions of Words and Terms

Antagonist. A chemical species (selenium species) that counteracts the effects and/or toxicity of another species (mercury species).

Bioaccessibility. The potential for a selenium compounds to be absorbed by an organism.

Bioavailability. The fraction of an administered selenium amount of selenium amount that are utilized to synthesize the selenoproteins.

Organoselenium compound. Organic compounds that involve covalently attached selenium atom(s).

Redox. Reduction and oxidation.

Selenium species. A chemical class of compounds having selenium atom(s).

Selenoamino acid. Seleno-L-cysteine and seleno-L-methionine that are substituted by selenium atoms in place of sulfur atoms of cysteine and methionine, respectively.

Selenoprotein. Proteins involving selenocysteine that are genetically incorporated in their polypeptide chains.

Selenoproteome. Variable sets of selenoproteins that are encoded by particular genes.

The 21st amino acid. Seleno-L-cysteine, three letter code "Sec", one letter code "U", It naturally occurs by UGA codon in all forms of life as a building block of selenoproteins.

List of Abbreviations

Cys	L-Cysteine
Dio	Iodothyronine deiodinase
ESI-MS	Electrospray ionization mass spectrometry
Fep15	Fish 15 kDa selenoprotein-like protein
GPx	Glutathione peroxidase
HPLC	High-performance liquid chromatography
ICP-MS	Inductively coupled plasma mass spectrometry
MALDI TOF-MS	Matrix-assisted laser desorption ionization time-of-flight mass spectrometry
Rdx	A SelW-like protein radixin
SeCys, Sec, U	Seleno-L-cysteine
SeCyst	Seleno-L-cystine
SeMet	Seleno-L-methionine
Sep15	15 kDa selenoprotein
TMS	Trimethylselenonium ion
XANES	X-ray absorption near edge structure

References

Anan, Y., Ishiwata, K., Suzuki, N., Tanabe, S. and Ogra, Y., 2011a. Speciation and identification of low molecular weight selenium compounds in the liver of sea turtles. *Journal of Analytical Atomic Spectrometry*. 26: 80–85.

Anan, Y., Tanabe, S. and Ogra, Y., 2011b. Comparison of selenoneine found in marine organisms with selenite in the interaction with mercury compounds *in vitro*. *Journal of Toxicological Sciences*. 36: 725–731.

Arnault, I. and Auger, J., 2006. Seleno-compounds in garlic and onion. *Journal of Chromatography A*. 1112: 23–30.

Barclay, M. N. I., MacPherson, A. and Dixon, J., 1995. Selenium content of a range of UK foods. *Journal of Food and Composition and Analysis*. 8: 307–318.

B'Hymer, C. and Caruso, J. A., 2000. Evaluation of yeast-based selenium food supplements using high-performance liquid chromatography and inductively coupled plasma mass spectrometry. *Journal of Analytical and Atomic Spectrometry*. 15: 1531–1539.

Bjerregaard, P., Fjordside, S., Hansen, M. G. and Petrova, M. B., 2011. Dietary selenium reduces retention of methyl mercury in freshwater fish. *Environmental Science and Technology*. 45: 9793–9798.

Cabanero, A. I., Madrid, Y. and Camara, C., 2004. Selenium and mercury bioaccessibility in fish samples: an *in vitro* digestion method. *Analytica Chimica Acta*. 526: 51–61.

Cabanero, A. I., Madrid, Y. and Camara, C., 2007. Mercury-selenium species ratio in representative fish samples and their bioaccessibility by an *in vitro* digestion method. *Biological Trace Element Research*. 119: 195–211.

Casiot, C., Vacchina, V., Chassaigne, H., Szpunar, J., Gautier, M. P. and Lobinski, R., 1999. An approach to the identification of selenium species in yeast extracts using pneumatically assited electrospray tandem mass spectrometry. *Analytical Communications*. 36: 77–80.

Castellano, S., Lobanov, A. V., Chapple, C., Novoselov, S. V., Albrecht, M., Hua, D., Lescure, A., Lengauer, T., Krol, A., Gladyshv, V. N. and Guigo, R., 2005. Diversity and functional plasticity of eukaryotic selenoproteins: identification and characterization of the SelJ family. *Proceedings of the National Academy of Sciences of the Unites States of America*. 102: 16188–16193.

Chen, J. and Berry, M. J., 2003. Selenium and selenoproteins in the brain and brain diseases. *Journal of Neurochemistry*. 86: 1–12.

Cheng, Q., Sandalova, T., Lindqvist, Y. and Arner, E. S. J., 2009. Crystal structure and catalysis of the selenoprotein thioredoxin reductase 1. *Journal of Biological Chemistry*. 284: 3998–4008.

Deniziak, M., Thisse, C., Rederstorff, M., Hindelang, C., Thisse, B. and Lescure, A., 2007. Loss of selenoprotein N function causes disruption of muscle architecture in the zebrafish embryo. *Experimental Cell Research*. 313: 156–167.

Dumont, E., Cremer, K. D., Hulle, M. V., Chery, C. C., Vanhaecke, F. and Cornelis, R., 2005. Identification of the major selenium compound, Se-methionine, in three yeast (*Saccharomyces cerevisiae*) dietary supplements by on-line narrowbore liquid chromatography-electrospray tandem mass spectrometry. *Journal of Chromatography A*. 1071: 191–196.

Dumont, E., Ogra, Y., Vanhaecke, F., Suzuki, K. T. and Cornelis, R., 2006. Liquid chromatography-mass spectrometry (LC-MS): a powerful combination for selenium speciation in garlic (*Allium sativum*). *Analytical and Bioanalytical Chemistry*. 384: 1196–1206.

Dumont, E., Vanhaecke, F. and Cornelis, R., 2006. Selenium speciation from food source to metabolites: a critical review. *Analytical and Bioanalytical Chemistry*. 385: 1304–1323.

Encinar, J. R., Ouerdane, L., Buchmann, W., Tortajada, J., Lobinski, R. and Szpunar, J., 2003. Identification of water-soluble selenium-containing proteins in selenized yeast by size-exclusion-reversed-phase HPLC/ICPMS followed by MALDI-TOF and electrospray Q-TOF mass spectrometry. *Analytical Chemistry*. 75: 3765–3774.

Freeman, J. L., Zhang, L. H., Marcus, M. A., Fakra, S., McGrath, S. P. and Pilon-Smits, E. A. H., 2006. Spatial imaging, speciation, and quantification of selenium in the hyperaccumulator plants *Astragalus bisulcatus* and *Stanleya pinnata*. *Plant Physiology*. 142: 124–134.

Fu, L. H., Wang, X. F., Eyal, Y., She, Y. M., Donald, L. J., Standing, K. G. and Ben-Hayyim, G., 2002. A selenoprotein in the plant kingdom. *Journal of Biological Chemistry*. 277: 25983–25991.

Ganther, H. E., Goudie, C., Sunde, M. L., Kopecky, M. J., Wagner, P., Oh, S. H. and Hoekstra, W. G., 1972. Selenium: relation to decreased toxicity of methylmercury added to diets containing tuna. *Science*. 175: 1122–1124.

Ganther, H. E., 1978. Modification of methylmercury toxicity and metabolism by selenium and vitamin E: possible mechanisms. *Environmental Health Perspectives*. 25: 71–76.

George, G. N., Singh, S. P., Prince, R. C. and Pickering, I. J., 2008. Chemical forms of mercury and selenium in fish following digestion with simulated gastric fluid. *Chemical Research in Toxicology*. 21: 2106–2110.

George, G. N., Singh, S. P., Myers, G. J., Watson, G. E. and Pickering, I. J., 2010. The chemical forms of mercury in human hair: a study using X-ray absorption spectroscopy. *Journal of Biological Inorganic Chemistry*. 15: 709–715.

Glynn, A. W., Ilback, N. G., Brabencova, D., Carlsson, L., Enqvist, E. C., Netzel, E. and Oskarsson, A., 1993. Influence of sodium selenite on ^{203}Hg absorption, distribution, and elimination in male mice exposed to methyl ^{203}Hg. *Biological Trace Element Research*. 39: 91–107.

Glynn, A. W. and Lind, Y., 1995. Effect of long-term sodium selenite supplementation on levels and distribution of mercury in blood, brain and kidneys of methyl mercury-exposed female mice. *Pharmacology and Toxicology*. 77: 41–47.

Gong, L. K., Xu, Q. B., Lee, C. and Zhang, H., 2012. Selenium speciation analysis of *Misgurnus anguillicaudatus* selenoprotein by HPLC-ICP-MS and HPLC-ESI-MS/MS. *European Food Research Technology*. 235: 169–176.

Huerta, V. D., Sanchez, M. L. F. and Sanz-Medel, A., 2004. Quantitative selenium speciation in cod muscle by isotope dilution ICP-MS with a reaction cell: comparison of different reported extraction procedures. *Journal of Analytical Atomic Spectroscopy*. 19: 644–648.

Imai, H. and Nakagawa, Y., 2003. Biological significance of phospholipid hydroperoxide glutathione peroxidase (PHGPx, GPx4) in mammalian cells. *Free Radical Biology and Medicine*. 34: 145–169.

Infante, H. G., Hearn, R. and Catterick, T., 2005. Current mass spectrometry strategies for selenium speciation in dietary sources of high-selenium. *Analytical and Bioanalytical Chemistry*. 382: 957–967.

Jaillon, O., Aury, J.-M., Brunet, F., Petit, J.-L., Stange-Thomann, N., Mauceli, E., Bouneau, L., Fischer, C., Ozouf-Costaz, C., Bernot, A., Nicaud, S., Jaffe, D., Fisher, S., Lutfalla, G., Dossat, C., Segurens, B., Dasilva, C., Salanoubat, M., Levy, M., Boudet, N., Castellano, S., Anthouard, V., Jubin, C., Castelli, V., Katinka, M., Vacherie, B., Biemont, C., Skalli, Z., Cattolico, L., Poulain, J., de Berardinis, V., Cruaud, C., Duprat, S., Brottier, P., Coutanceau, J. P., Gouzy, J., Parra, G., Lardier, G., Chapple, C., McKernan, K. J., McEwan, P., Bosak, S., Kellis, M., Volff, J. N., Guigo, R., Zody, M. C., Mesirov, J., Lindblad-Toh, K., Birren, B., Nusbaum, C., Kahn, D., Robinson-Rechavi, M., Laudet, V., Schachter, V., Quetier, F., Saurin, W., Scarpelli, C., Wincker, P., Lander, E. S., Weissenbach, J. and Crollius, H. R., 2004. Genome duplication in the teleost fish *Tetraodon nigroviridis* reveals the early vertebrate proto-karyotype. *Nature*. 431: 946–957.

Johansson, L., Gafvelin, G. and Arnér, E. S. J., 2005. Selenocysteine in proteins-properties and biotechnological use. *Biochimica et Biophysica Acta*. 1726: 1–13.

Kristan, U., Arribére, M. A. and Stibilj, V., 2013. Selenium species and their distribution in freshwater fish from Argentina. *Biological Trace Element Research*. 151: 240–246.

Kryukov, G. V. and Gladyshev, V. N., 2000. Selenium metabolism in zebrafish: multiplicity of selenoprotein genes and expression of a protein containing 17 selenocysteine residues. *Genes Cells*. 5: 1049–1060.

Kryukov, G. V., Castellano, S., Novoselov, S. V., Lobanov, A. V., Zehtab, O., Guigo, R. and Gladyshev, V. N., 2003. Characterization of mammalian selenoproteomes. *Science*. 300: 1439–1443.

Lindemann, T. and Hintelmann, H., 2002. Identification of selenium-containing glutathione S-conjugates in a yeast extract by two-dimensional liquid chromatography with inductively coupled plasma MS and nanoelectrospray MS/MS detection. *Analytical Chemistry*. 74: 4602–4610.

Lobanov, A. V., Fomenko, D. E., Zhang, Y., Sengupta, A., Hatfield, D. L. and Gladyshev, V. N., 2007. Evolutionary dynamics of eukaryotic selenoproteomes: large selenoproteomes may associate with aquatic life and small with terrestrial life. *Genome Biology*. 8: R198.

Lobanov, A. V., Hatfield, D. L. and Gladyshev, V. N., 2008. Reduced reliance on the trace element selenium during evolution of mammals. *Genome Biology*. 9: R62.

Lobanov, A. V., Hatfield, D. L. and Gladyshev, V. N., 2009. Eukaryotic selenoproteins and selenoproteomes. *Biochimica et Biophysica Acta*. 1790: 1424–1428.

Mariotti, M., Ridge, P. G., Zhang, Y., Lobanov, A. V., Pringle, T. H., Guigo, R., Hatfield, D. L. and Gladyshev, V. N., 2012. Composition and evolution of the vertebrate and mammalian selenoproteomes. *PLoS One*. 7: e33066.

Misra, S., Peak, D. and Niyogi, S., 2010. Application of XANES spectroscopy in understanding the metabolism of selenium in isolated rainbow trout hepatocytes: insights into selenium toxicity. *Metallomics*. 2: 710–717.

Miyazaki, Y., Koyama, H., Sasada, Y., Satoh, H., Nojiri, M. and Suzuki, S., 2004. Dietary habits and selenium intake of residents in mountain and coastal communities in Japan. *Journal of Nutritional Science and Vitaminology*. 50: 309–319.

Moreno, P., Quijano, M. A., Gutierrez, A. M., Perez-Conde, M. C. and Camara, C., 2001. Fractionation studies of selenium compounds from oysters, and their determination by high-performance liquid chromatography coupled to inductively coupled plasma mass spectrometry. *Journal of Analytical Spectrometry*. 16: 1044–1050.

Moreno, P., Quijano, M. A., Gutierrez, A. M., Perez-Conde, M. C. and Camara, C., 2004. Study of selenium species distribution in biological tissues by size exclusion and ion exchange chromatography inductively coupled plasma-mass spectrometry. *Analytica Chimica Acta*. 524: 315–327.

Navarro-Alarcon, M. and Cabrera-Vique, C., 2008. Selenium in food and the human body: a review. *Science of the Total Environment*. 400: 115–141.

Novoselov, S. V., Hua, D., Lobanov, A. V. and Gladyshev, V. N., 2006. Identification and characterization of Fep15, a new selenocysteine-containing member of the Sep15 protein family. *Biochemical Journal*. 394: 575–579.

Ogra, Y., Ishiwata, K., Encinar, J. R., Lobinski, R. and Suzuki, K. T., 2004. Speciation of selenium in selenium-enriched shiitake mushroom, *Lentinula edodes*. *Analytical and Bioanalytical Chemistry*. 379: 861–866.

Önning, G. and Bergdahl, I. A., 1999. Fractionation of soluble selenium compounds from fish using size-exclusion chromatography with on-line detection by inductively coupled plasma mass spectrometry. *Analyst*. 124: 1435–1438.

Önning, G., 2000. Separation of soluble selenium compounds in different fish species. *Food Chemistry*. 68: 133–139.

Pickering, I. J., George, G. N., Van Fleet-Stalder, V., Chasteen, T. G. and Prince, R. C., 1999. X-ray absorption spectroscopy of selenium-containing amino acids. *Journal of Biological Inorganic Chemistry*. 4: 791–794.

Quijano, M. A., Moreno, P., Gutierrez, A. M., Perez-Conde, M. C. and Camara, C., 2000. Selenium speciation in animal tissues after enzymatic digestion by high-performance liquid chromatography coupled to inductively coupled plasma mass spectrometry. *Journal of Mass Spectrometry*. 35: 878–884.

Rayman, M. P., 2012. Selenium and human health. *Lancet*. 379: 1256–1268.

Schwarz, K. and Foltz, C. M., 1957. Selenium as an integral part of factor 3 against dietary necrotic liver degeneration. *Journal of the American Chemical Society*. 79: 3292–3293.

Shchedrina, V. A., Novoselov, S. V., Malinouski, M. Y. and Gladyshev, V. N., 2007. Identification and characterization of a selenoprotein family containing a diselenide bond in a redox motif. *Proceedings of the National Academy of Sciences of the United States of America*. 104: 13919–13924.

Siwek, M., Galunsky, B. and Niemeyer, B., 2005. Isolation of selenium organic species from antarctic krill after enzymatic hydrolysis. *Analytical and Bioanalytical Chemistry*. 381: 737–741.

Thompson, J. L., Thomas, P. M. and Schuller, K. A., 2006. Purification and properties of a glutathione peroxidase from Southern bluefin tuna (*Thunnus maccoyii*) liver. *Comparative Biochemistry and Physiology, Part B: Biochemistry & Molecular Biology*. 144: 86–93.

Tujebajeva, R. M., Ransom, D. G., Harney, J. W. and Berry, M. J., 2000. Expression and characterization of nonmammalian selenoprotein P in the zebrafish, *Danio rerio*. *Genes Cells*. 5: 897–903.

Wang, R. Y., Hsu, Y. L., Chang, L. F. and Jiang, S. J., 2007. Speciation analysis of arsenic and selenium compounds in environmental and biological samples by ion chromatography-inductively coupled plasma dynamic reaction cell mass spectrometer. *Analytica Chimica Acta*. 590: 239–244.

Whanger, P. D., 2009. Selenoprotein expression and function-selenoprotein W. *Biochimica et Biophysica Acta*. 1790: 1448–1452.

Yamashita, Y. and Yamashita, M., 2010. Identification of a novel selenium-containing compound, selenoneine, as the predominant chemical form of organic selenium in the blood of bluefin tuna. *Journal of Biological Chemistry*. 285: 18134–18138.

Yant, L. J., Ran, Q., Rao, L., Remmen, H. V., Shibatani, T., Belter, J. G., Motta, L., Richardson, A. and Prolla, T. A., 2003. The selenoprotein GPX4 is essential for mouse development and protects from radiation and oxidative damage insults. *Free Radical Biology and Medicine*. 34: 496–502.

Yeh, S. P., Liu, K. F., Chiu, S. T., Jian, S. J., Cheng, W. and Liu, C. H., 2009. Identification and cloning of a selenium dependent glutathione peroxidase from giant freshwater prawn, *Macrobrachium rosenbergii*. *Fish and Shellfish Immunology*. 27: 181–191.

Yoshida, S., Haratake, M., Fuchigami, T. and Nakayama, M., 2012. Characterization of selenium species in extract from niboshi (a processed Japanese anchovy). *Chemical and Pharmaceutical Bulletin*. 60: 348–353.

Section II
Analysis

CHAPTER 6

Current Analytical Strategies and Techniques Related to Selenoproteins in Foods

MIGUEL ÁNGEL GARCÍA-SEVILLANO[a], MACARENA GONZÁLEZ-FERNÁNDEZ[a], TAMARA GARCÍA-BARRERA[a], AND JOSÉ LUIS GÓMEZ-ARIZA*[a]

[a]Department of Chemistry and Materials Science, Faculty of Experimental Sciences, University of Huelva, Av. Fuerzas Armadas s/n, 21007, Huelva, Spain
*E-mail: ariza@uhu.es

6.1 Introduction

Selenium (Se) is an essential trace element of fundamental importance for health due to its contribution in physiological functions by the bioactivity of selenium metabolites (mainly selenium amino acids) but also selenoproteins, since more than 30 mammalian selenoproteins or selenoenzymes are vital for the basic functions of life. Although the function of many metalloproteins has not been determined yet, their role in antioxidant defense, thyroid homeostasis and regulation of redox status can be mentioned. Many authors have suggested that Se is a potent protective agent for neurons and myocytes through selenoprotein expression in the brain, as well as for skeletal and cardiac muscles.

Food and Nutritional Components in Focus No. 9
Selenium: Chemistry, Analysis, Function and Effects
Edited by Victor R. Preedy

Published by the Royal Society of Chemistry, www.rsc.org

The main source of selenium in human beings is the diet; however, in several regions of the world the content of selenium in diet has been estimated to be insufficient for the correct functions of selenoproteins. The beneficial effects of selenium on human health are strongly dependent on the chemical forms and concentration of the element. Therefore, suitable analytical methods for selenium and selenoproteins speciation are necessary, considering that these species have to be analyzed in complex biological matrices, which requires well-designed extraction procedures for chemical species.

The speciation of heteroelement-containing proteins requires detection, identification and quantification of these biomolecules in biological samples, and particularly in food. Hyphenated techniques based on the coupling of chromatographic separation with inductively coupled plasma spectrometry (ICP-MS) and its combination with organic mass spectrometry (normally, ESI-QqQ-MS and MALDI-TOF-MS), besides isotopic dilution analysis (IDA), allow the characterization and quantification of selenium species and selenoproteins. Additionally, two-dimensional gel electrophoresis (2D-GE) can be used for this purpose.

6.1.1 Selenoproteins in Food

Selenium is found predominantly in organic forms in plants, food and feed, such as selenium amino acids, which incorporate the element specifically into selenoproteins as selenocysteine, using the active sites of glutathione peroxidases and iodothyronine deiodinases, or unspecifically in selenomethionine, at methionine position, since selenium can substitute sulfur in the protein due to their chemical similarity. These latter proteins with selenium are not considered as selenoproteins, and are called "selenium-containing proteins" to stress the lack of selenium-bonding specificity.

A wide array of selenoproteins present in food can be mentioned, which includes glutathione peroxidases (GPx1–GPx6), thioredoxin reductases (TrxR1–TrxR3), iodothyronine deiodinases (ID1–ID3), selenophosphate synthetase 2 (SPS2), 15 kDa Sel (Sel15), SelH, SelI, SelK, SelM, SelN, SelP, SelR, SelS, SelT, SelV, SelW, as well as the 15 kDa Sel (Fep15), SelJ and SelU found in fish (Table 6.1).

6.2 The Role of ICP-MS in Selenoproteins Speciation

ICP-MS is the detection technique more usually used at present for the analysis of selenium in food (Ruiz-Encinar *et al.*, 2003). This instrumental system has important benefits associated to high sensitivity and selectivity for multielemental analysis, detection limits in the range of parts per trillion, and capability for isotopic analysis. The potential of this technique in metalloproteins analysis has been highlighted in a recent review (Bettmer *et al.*, 2009). The main drawback of ICP-MS techniques is related to the presence of polyatomic and/or isobaric interferences from argon plasma and matrix

Table 6.1 Selenoproteins and their functions.

Selenoprotein	Abbreviation	Function
Cytosolic glutathione peroxidase	GPx1	Antioxidant defense
Gastrointestinal glutathione peroxidase	GPx2	Antioxidant defense
Extracellular glutathione peroxidase	GPx3	Maintenance of cellular redox status
Phospholipid hydroperoxide glutathione peroxidase	GPx4	Detoxification of lipid hydroperoxides
Epididymal glutathione peroxidase	GPx5	Antioxidant protection during spermiogenesis and sperm maturation
Olfactory glutathione peroxidase	GPx6	Antioxidant defense
Thioredoxin reductase Type I	TrxR1	Part of the thioredoxin system. Antioxidant defense, redox regulation, cell signaling
Thioredoxin reductase Type II	TrxR2	Part of the thioredoxin system. Antioxidant defense, redox regulation, cell signaling
Thioredoxin reductase Type III	TrxR3	Part of the thioredoxin system. Antioxidant defense, redox regulation, cell signaling
Iodothyronine deiodinase Type I	ID1	Conversion of T_4 to T_3 and T_4 to reverse T3
Iodothyronine deiodinase Type II	ID2	Conversion of T_4 to T_3
Iodothyronine deiodinase Type III	ID3	Conversion T_4 to reverse T_3
Selenophosphate synthetase	SPS2	Synthesis of selenophosphate
15 kDa selenoprotein	Sel15	Role in cell apoptosis and mediation of chemopreventive effects of Se
Selenoprotein H	SelH	Not fully known, possible upregulation of genes involved in glutathione synthesis
Selenoprotein I	SelI	Studies with *Escherichia coli* showed specific ethanolaminephosphotransferase activity
Selenoprotein J	SelJ	Structural role
Selenoprotein K	SelK	Possible antioxidant protection in cardiomyocytes
Selenoprotein M	SelM	Distantly related to Sel15. May be involved in cancer etiology
Selenoprotein N	SelN	It is linked with rigid spine syndrome
Selenoprotein P	SelP	Involved in Se transport, antioxidant defense
Selenoprotein R	SelR	Reduction of oxidized methionine residues in damaged proteins
Selenoprotein S	SelS	Cellular redox balance. Possible influence of inflammatory response
Selenoprotein T	SelT	Role in regulation of Ca^{2+} homeostasis and neuroendocrine secretion
Selenoprotein U	SelU	Unknown
Selenoprotein V	SelV	Possible role in redox regulation
Selenoprotein W	SelW	Antioxidant protection
Fish 15 kDa selenoprotein	Fep 15	Fish homolog of Sep15

effects. The use of a double-focusing high-resolution spectrometer is a good alternative to overcome spectral interferences in ICP-MS instrumentation, although quadrupole analyzers in combination with collision and dynamic reaction cells are more frequently used in routine analysis laboratories for this purpose. In this case an optimized flow of collision/reaction gas is introduced into the collision/reaction cell to reduce polyatomic interferences from plasma.

ICP-MS spectral interferences from isobaric elements and polyatomic species in the plasma arising from the sample matrix can be considerably removed during the chromatographic or electrophoretic separation required for speciation analysis. Unfortunately, there are spectral interferences that could not be eliminated by the chromatographic separation and it is necessary to use mathematical corrections, although it requires the monitorization of additional isotopes. In recent years, other analyzers, such as multicollector (MC), time-of-flight and triple quad (QqQ) present numerous applications in this field with the common objective of reducing and/or removing spectral interferences that are not possible with the use of collision/reaction cells (Goenaga-Infante *et al.*, 2002).

Samples can be introduced into the ICP-MS by direct nebulization or by direct hyphenation to separation systems that introduce the HPLC outflow in the nebulizer device, other separation systems such as capillary electrokinetic chromatography (CEC) and capillary electrophoresis (CE) can also be used for this purpose. Alternatively, laser ablation (LA) can introduce metal-containing biomolecules from living organism tissues or gel electrophoresis (GE) spots into ICP-MS.

6.3 Sample Preparation Procedures for Selenoproteins Speciation

Sample preparation could be considered a critical step in the analysis of selenoproteins, including sample matrix manipulations to make it suitable for analysis (Mesko *et al.*, 2011).

6.3.1 Conventional Sample Preparation Procedures for Selenoproteins Speciation

Sample preparation is crucial to isolate target analytes from the matrix. The main problems arise from analyte losses or contamination during sampling. In general, conventional sample preparation for selenoproteins speciation from biological matrices involves several steps:

(a) Cell disruption or lysis, or partial degradation of sample matrix for later extraction of the analyte using mechanical treatments in presence of liquid nitrogen (García-Sevillano *et al.*, 2012).
(b) Analyte extraction and solubilization assisted by different reagents, mainly suitable buffer solutions, *e.g.* Tris buffer or ammonium acetate (García-Sevillano *et al.*, 2012).

(c) Reduction of disulfide bonds to avoid the formation of protein disulfide bridges. For this purpose, free thiol-containing reducing agents such as dithiothreitol (DTT), dithioerythritol (DTE), reduced glutathione (GSH) and β-mercaptoethanol have been used. However, recently these reagents have been replaced for nonchelating mild-reducing compounds, such as tributylphosphine (TBP) or tris-carboxyethylphosphine (TCEP) (García-Sevillano *et al.*, 2012).
(d) Inhibition of proteases action to conserve selenoprotein molecule integrity, for this purpose proteases inhibitors without chelating properties, such as phenylmethylsulfonyl fluoride (PMSF), have been proposed (García-Sevillano *et al.*, 2012).

Optionally, when metal proteins are embedded into cell membranes the addition of detergents, such as SDS or CHAPS, is necessary, since these reagents release metal-binding proteins from the membrane preventing hydrophobic interactions (Mester and Sturgeon, 2003).

6.3.2 New Trends in Extraction Procedures for Selenoproteins from Food Matrices

6.3.2.1 Enzymatic Extraction

Recently, enzymes have been proposed to promote matrix degradation or selectively release the analytes present in food samples, *e.g.* selenoaminoacids (Montes-Bayón *et al.*, 2002; Mester and Sturgeon, 2003). Since enzymes can catalyze specific and recognized processes, no additional side effects in the sample are expected.

Examples of this treatment are: (a) extraction of selenium labile species from garlic using aqueous solution of enzyme-deactivating hydroxylamine hydrochloride (Larsen *et al.*, 2006); and (b) the joint use of cellulase, chitinase and glucanase to extract cell-wall-bound species completed by a proteolytic attack, in a sequential extraction procedure that also involves leaching with water and extraction with HCl to release the residual organic bound species, and final extraction with sulfite solution and CS_2 to complete the mass balance by the recovery of Se^0 and Se^{2-}, respectively (Mounicou *et al.*, 2009).

6.3.2.2 Protein Precipitation and Ultrafiltration

A good alternative in sample preparation for selenium speciation is to separate low from high molecular weight selenocompounds, using simple techniques based on size exclusion chromatography, protein precipitation or ultrafiltration with membranes of 10 kDa cut-off.

Jayasinghe and Caruso (2011) proposed the precipitation of proteins from crude protein mixture and enrichment of proteins of interest under nondenaturing conditions, using ammonium sulfate as precipitating agent. These authors have demonstrated the feasibility of fractional precipitation with ammonium sulfate for achieving matrix reduction and sample concentration simultaneously, in selenoproteomics studies. In addition, other authors proposed the use of trichloroacetic acid (TCA)–acetone or chloroform–methanol mixtures for this purpose, although it causes protein denaturation.

Ultrafiltration by offline use of membrane under centrifugal force allows separation of compounds according to the molecular mass. This approach is suitable for separation of salts and other low molecular weight solutes from high molecular weight species. Solutes with greater molecular weight or size are retained by the membrane and concentrated. Ultrafiltration offers a fast way to promote the separation with minimal denaturation of molecules compared to precipitation methods (Nuevo-Ordoñez *et al.*, 2010).

6.3.2.3 *Ultrasound Sonication*

The most common sonicators used in analytical applications are power bath and probe-type devices, which favor the penetration of the extractant into the sample, increasing the extraction efficiency. Thus, extraction with ultrasound probe and separation by 2D-GE were used for the characterization of Se-containing proteins in selenium-rich yeast. After centrifugation, the total Se content in the supernatant was measured by ICP-MS and water-soluble selenium-containing proteins were separated by 2D-GE for later characterization by consecutive use of nano-HPLC-ICP-MS and nano-high-performance liquid chromatography electrospray ionization mass spectrometry (nHPLC-ESI-MS). Another example is the extraction of selenium-rich yeast by ultrasonic probe with water containing DTT (antioxidant) and PMSF (protease activity inhibitor) in an ice bath. The efficiency of different extraction approaches were evaluated by total selenium analysis using SEC-ICP-MS. The extraction procedure preserves the protein integrity along the extraction and purification steps, for further separation by 2D-GE, mapping by nano-HPLC-ICP-MS and identification of selenium-containing peptides by nano-HPLC-ESI-MS (Tastet *et al.*, 2008).

6.3.3 Cleanup Procedures

The presence of different compounds in the matrix, such as hydrocarbons, polysaccharides, lipids, amino acids and proteins, limits the use of HPLC-ICP-MS in the speciation of selenium in biological samples, since it results in complex chromatograms and coelution of chemical components. For this reason, clean-up of samples is usually necessary.

Lyophilization, centrifugation, ultrafiltration, gel filtration, gel permeation, chromatography (GPC), solid-phase extraction (SPE) and column chromatography are some techniques used for cleanup and purification of extracts. For biological samples, the most common approaches for pretreatment prior to analysis deal with the use of C_{18} cartridges or ion-exchange resins. For example, anion-exchange resins (in the acetate form) have been used for removing inorganic selenium species from extracts of the metal hyperaccumulating plant *Brassica juncea* (Montes-Bayón *et al.*, 2002). On the other hand, online low-pressure size-exclusion gel filtration (Hitrap Desalting columns) allows a finer resolution of chemical species for subsequent separation steps, since it could act as a cleanup step for the elimination of high molecular mass biopolymers (polysaccharides, proteins) and other compounds that might be absorbed onto the chromatographic stationary phase or coeluted with selenoproteins during speciation analysis.

Reversed phases (C_{18} and C_4) have been used for concentration and purification of peptides, proteins and oligonucleotides after tryptic digestion in a buffer with high concentration of salts that assures enzyme stability. These phases are loaded at the end of 10 μL pipettes (ZipTip pipettes) that are used for peptide and protein desalting and concentration before the identification by mass spectrometry using MALDI-TOF/TOF-MS or DI-nESI-QqQ-TOF-MS (García-Sevillano *et al.*, 2012).

6.3.4 Stability of Selenoproteins during Storage

Suitable storage procedures to preserve the information contained in the species are a key requirement in the whole analytical process. Several drawbacks can be considered, especially species degradation, related to their chemical nature and external factors such as temperature that affects reaction rates and, consequently, species interconversion. In this way, selenoproteins, especially GPx and SelP, are readily degraded because of selenocysteine instability. The selenium atom in selenocysteine is more prone to oxidation than in selenomethionine (Mester and Sturgeon, 2003).

Sample stabilization can be performed by drying, especially freeze-drying for biological samples. Shock freezing under liquid nitrogen seems to be the safest technique to prevent species changes and can be performed at the sampling site (Mesko *et al.*, 2011).

Degradation of SeCys from the selenoproteins occurs during storage of sample at 4 °C as a function of time. Note that the selenium mass balance before and after storage shows the possibility of formation and losses of volatile species that are not observed in the chromatograms. However, storage at −20 °C prevents selenoproteins degradation for more than one year (Mester and Sturgeon, 2003).

Another factor affecting selenoproteins stability is oxidation; Palacios *et al.* (2006) indicate that protein is most affected by loss of selenium is GPx, whereas the SelP peak remains constant. The degradation was much less pronounced in the presence of a reducing agent such as DTT, which prevents the oxidation of SeCys present in GPx.

6.4 Current Analytical Strategies and Hyphenated Techniques for Selenoproteins Speciation in Foods

Different problems arise in the study of metallic species in biological systems since they are usually unknown and present a great variability of size and type of bond between metal and biomolecule. In addition, usually there are not standards available for these analyses. The need for the determination of individual chemical species of selenium in biological samples and in particular in food has driven the development of new analytical procedures designed for this purpose, which have deepened knowledge about selenium-species toxicity, mobility, and bioavailability.

The analytical speciation methods for selenoproteins identification and quantification include three main instrumental components: (i) a separation technique (*e.g.*, LC, CE, 2DE and 2D-PAGE) that contributes to the isolation of selenoproteins from the matrix and their time-resolved introduction into the detector—*selectivity component*; (ii) a high-sensitivity element detector, ICP-MS, for element quantification, mainly using isotopic dilution techniques—*sensitivity component*; and (iii) a molecule-specific detector, based on mass spectrometry: triple quadrupole (QqQ), ion-trap (IT), time-of-flight (TOF), or high resolution analyzers such as QqQ-TOF and Orbitrap—*structural component*. These basic components are integrated in different instrumental arrangements depending on the problem to be addressed.

6.4.1 Separation and Detections Techniques for Selenoproteins Based on Liquid Chromatography

Several modes of chromatographic separations can be used for species discrimination at the analytical, preparative, capillary or nanoscale, and usually two or three orthogonal systems can be offline combined to increase metal–biomolecules resolution, such as size-exclusion chromatography (SEC), reverse-phase (RP), ion-exchange chromatography (IEC), hydrophilic interaction liquid chromatography (HILIC) and affinity chromatography (AF). However, as noted before, the combination with organic mass spectrometry is mandatory to study unknown species in biochemical issues, using generally electrospray ionization (ESI) or matrix-assisted laser desorption ionization (MALDI) as ionization source for selenoproteins ionization. The use of ESI-MS is more suitable than MALDI-MS for tandem mass spectrometry and online couplings with separation techniques (HPLC, CE). However, MALDI-TOF-MS is recommended for matrices with low complexity, normally after 2D-GE separations, for spot characterization.

Size exclusion has been widely applied as first chromatographic dimension (SEC), followed by purification by other complementary liquid chromatographic devices in multidimensional schemes. This allows a first fractionation of selenocompounds by SEC followed by separation of collected fractions by other type of liquid chromatography, such as reverse-phase or ion-exchange chromatography. These multidimensional separation schemes has been applied to increase the resolution of the separation and remove matrix components that can influence selenocompounds characterization

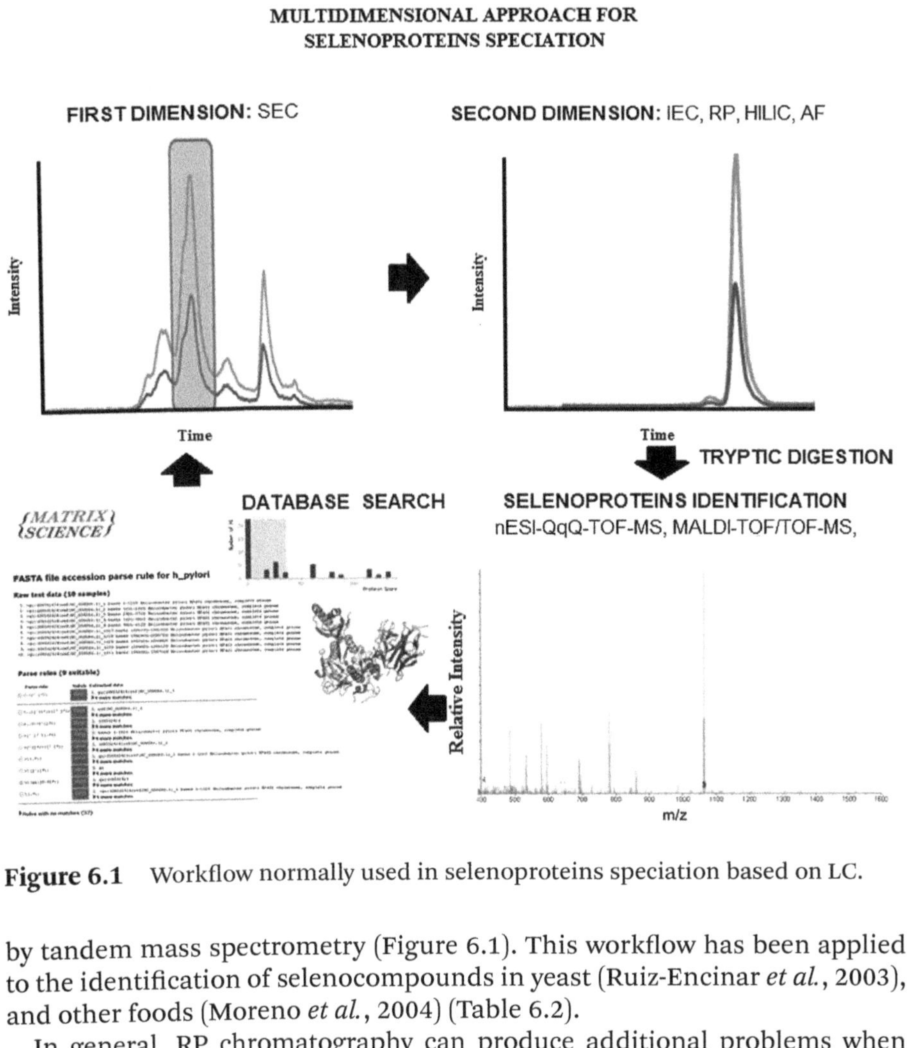

Figure 6.1 Workflow normally used in selenoproteins speciation based on LC.

by tandem mass spectrometry (Figure 6.1). This workflow has been applied to the identification of selenocompounds in yeast (Ruiz-Encinar *et al.*, 2003), and other foods (Moreno *et al.*, 2004) (Table 6.2).

In general, RP chromatography can produce additional problems when coupled to ICP-MS, mainly due to destabilization of plasma by organic solvents and gradient elution. To overcome this problem a response factor can be calculated on the base of organic phase gradient (Pröfrock and Prange, 2009). The use of micro- and nanoflow nebulizers with flow rates ranged from 200 nL min^{-1} to 4 μL min^{-1} allows the use of mobile phases with high organic content (Szpunar, 2005). Other alternatives are the introduction of oxygen to avoid clogging of sampling and skimmer cones by carbon deposition, and the use of inert Pt cones instead of the conventional ones made in Ni. Finally, recent advances in sample introduction, such as ultrasonic nebulizer/membrane desolvator ($U6000AT^{+}$, CETAC Technologies) provide enhanced desolvation of both aqueous and organic samples, reducing interferences and carbon deposition in the ICP-MS, and lowering the analyte detection limits.

Table 6.2 Examples of applications of different chromatographic liquid mechanisms for selenoproteins separation.

Species	Liquid chromatography mode	Column	Matrix	Detector	References
Selenoproteins	Ion-exchange chromatography	Mono Q 5/5 FPLC (50 × 5 mm)	Human serum	ICP-MS	Hinojosa-Reyes *et al.* (2003)
SelP, Se-albumin, p-GSH-Px and soluble compounds	Size-exclusion and affinity chromatography	Hitrap Desalting column Hitrap Heparin Sepharose, Hitrap blue-Sepharose	Human serum	ICP-MS	García-Sevillano *et al.* (2013)
Se-albumin and selenoproteins	Multiaffinity removal column	Kit provided by Agilent technology	Human serum	UV	Palacios *et al.* (2006)
SeMet, selenized, hemoglobin selenoprotein P and other soluble compounds	Size-exclusion chromatography	Superdex G-75 HR 10/30 (70–3 kDa)	Cow serum	ICP-MS	Palacios *et al.* (2005)
Selenoproteins	Size-exclusion chromatography	Superdex peptide 10/30 (14–0.18 kDa)	Onion leaves	UV/ICP-MS	Wrobel *et al.* (2004)
Selenoproteins	Size-exclusion chromatography	Superdex 75 HR 10/30 (70–3 kDa)	Plants (Indian mustard)	ICP-MS	Mounicou *et al.* (2004)
Selenoproteins	Size-exclusion chromatography Ion exchange chromatography	300 and 1 kDa (Biosep-SEC-2000) and 7 and 0.1 kDa (Superdex Peptide HR 10/30 Hamilton PRP-X200 column	Fish (tuna and trout), shellfish (krill, oyster and mussel), vegetal tissues (white clover and wheat flour) and selenized yeast	ICP-MS	Moreno *et al.* (2004)
Selenoproteins	Size-exclusion chromatography	Superdex G-75 HR 10/30	Cow blood	ICP-MS	Palacios *et al.* (2005)
Selenoproteins	2D size-exclusion ion-paring reversed phase chromatography	Superdex-200 column Hamilton PRP X-100 (150 mm × 4.6 mm, 5 μm) C8 Altima/Alltech (150 mm × 4.6 mm, 5 μm)	Whole milk	ICP-MS	Bierla *et al.* (2008)

6.4.2 Electrophoretic Techniques

Gel electrophoresis (GE) (native and SDS-PAGE) in a one- or two-dimensional format is a powerful tool for selenoprotein and Se-containing proteins separation providing better resolution than HPLC. Sodium dodecyl sulfate polyacrylamide gel electrophoresis (SDS-PAGE) is the most common separation method for this purpose. In this case, secondary protein structure is denaturized, and consequently its application is restricted to trace elements covalently bound to proteins, as is the case of selenoproteins, which do not lose the element during analysis. The Se-containing bands in 1DGE or spots in 2D GE can be traced by laser-ablation inductively coupled plasma-MS (LA-ICP-MS) (Fan *et al.*, 2002) (Figure 6.2). Additionally, matrix-assisted laser desorption/ionization mass spectrometry (MALDI-MS) or electrospray ionization mass spectrometry (ESI-MS) can be used for unknown selenoproteins identification.

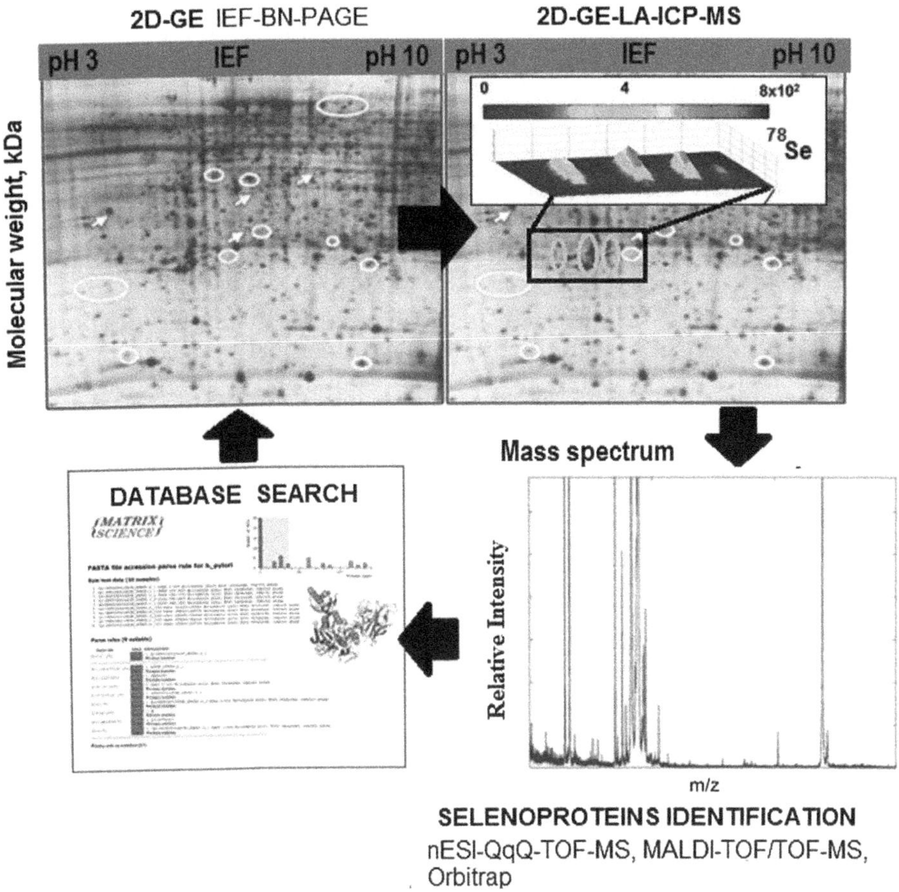

Figure 6.2 Workflow normally used in selenoproteins speciation based on 2D-GE.

6.4.3 Isotopic Dilution Analysis in Selenoproteins Quantification

Recently quantifications of selenoproteins in complex samples have been performed by isotope dilution analysis (IDA) using the multi-isotopic capability of ICP-MS. This approach has been used in many applications due to the methodological simplicity that only requires an isotopically enriched standard. In order to apply isotope dilution analysis in elemental speciation, a previous separation of the species is necessary prior to their analysis in the ICP-MS system.

For the application of IDA to selenoproteins speciation, new quantitation forms have been described: (i) postcolumn isotope dilution (species-unspecific mode, SUID), when the standard of a chemical species is not available. This procedure allows the accurate determination of elemental species even if the structure of the compounds is unknown and multiple isotopically enriched species are available for the evaluation and correction of species interconversion. However, this mode of spiking can only be applied when the ionization efficiency of the element is independent of the chemical form in which the element reaches the ion source; (ii) species-specific spiking mode (SSID), when standards for the species are available. This quantification procedure requires the use of a spike solution containing the species to be analyzed in an isotopically labeled form, similarly to the classical organic isotope dilution technique (Rodríguez-González *et al.*, 2005). In biological samples, several species occur at the same time, which can produce species degradation and interconversion that are corrected by the cited procedure (Rodríguez-González *et al.*, 2005). Despite the advantages of SSID mode its use is limited by the reduced number of selenium isotopically labeled species available.

When SUID is applied, the quantification of selenium-containing proteins and selenium-metabolites in the different chromatographic peaks is carried out by postcolumn species-unspecific isotopic dilution analysis as described by Sariego-Muñíz *et al.* (2001) for the analysis of selenium species in plasma and serum samples. The approach is based on the conversion of the intensity of selenium isotopes and polyatomic interferences in a mass-flow chromatogram for quantification. In addition, dead-time correction of ICP-MS detector has to be optimized using the procedure described by Vanhaecke *et al.* (1998), and mathematical treatments to correct polyatomic interferences usually present in plasma serum, such as BrH^+, as well as SeH^+. Mass bias corrections are applied using the $^{78}Se/^{74}Se$ and $^{80}Se/^{74}Se$ isotope ratios calculated (exponential mode) as previously described by Ruiz-Encinar *et al.* (2001). Finally, an online dilution equation has to be applied to each point of the chromatogram and the amount of selenium in each chromatographic peak can be calculated using suitable software.

The use of enriched stable isotopes as metabolic tracers with an ICP-MS detector has been extended rapidly in the last years, especially in human nutrition studies. Selenium isotopically enriched species has also been applied to discriminate between endogenous and supplemented selenium in lactating rats. Isotope pattern deconvolution (IPD) quantitative data of endogenous and exogenous essential (or toxic) elements has been used to discriminate the fate of endogenous and supplemented selenium and their catabolism species in rat serum and urine, by using ICP-MS equipped with a collision/reaction cell coupled to HPLC. In this case, two different enriched isotopes are necessary, a metabolic tracer (^{77}Se) and a quantitation tracer (^{74}Se) (González-Iglesias *et al.*, 2009).

6.4.4 Speciation Methods to Evaluate Selenium Status in Human

Some forms of dietary Se are beneficial for human health, reducing the risk for diseases such as cancer. Several chromatographic methods have been proposed for their study, mainly based on size exclusion chromatography (Palacios *et al.*, 2006) (SEC), anion-exchange chromatography (Hinojosa-Reyes *et al.*, 2003) (AEC) and affinity chromatography (Hinojosa-Reyes *et al.*, 2003) (AFC). SEC does not provide precise quantitative results for selenoproteins because the low chromatographic resolution that provokes overlapping between high-abundance selenium-containing proteins (Palacios *et al.*, 2006). On the other hand, AEC provides good analyte recovery, although chromatographic resolution is not acceptable either (Hinojosa-Reyes *et al.*, 2003). Finally, when AFC is used, the weakly retained eGPx is eluted beside nontarget components present in the matrix, which cause difficulties for the accurate quantification of the different selenium species that usually are quantified together (Hinojosa-Reyes *et al.*, 2003). Moreover, the high concentration of Cl^- and Br^- in biological fluids, such as plasma and serum (~3.5 g L^{-1} and ~3.5 mg L^{-1}, respectively) (Jitaru *et al.*, 2010) should be considered, which coelute with eGPx and Se metabolites with this procedure. To avoid this drawback AEC has been recently online coupled to AFC for the analysis of selenium-containing proteins alleviating the spectral interferences of Cl^- and Br^-, and making possible the use of ICP-ORS-MS (Jitaru *et al.*, 2008).

This analytical approach has been improved using species-unspecific isotope dilution (SUID)-ICP-ORS-qMS online coupled to 2D/SEC-AF-HPLC and 3D/SEC-AF-AEC-HPLC, which involve the use of five columns in one run with the aid of column-switching valves. Using this chromatographic arrangement, the spectral interference produced by bromide and chloride is removed and the total chromatographic runtime is less than 35 min. Consequently, a reliable speciation method for the analysis of eGPx, SeP, SeAlb and inorganic selenium compounds has been obtained for the first time (Figure 6.3) (García-Sevillano *et al.*, 2013).

6.5 Selected Cases Studies

6.5.1 Selenoproteins in Meat

The effect of organic selenium was investigated in chicken (*Gallus gallus*, broilers) studying changes in the expression of catalase (CAT) and phospholipid hydroperoxidase 4 (Se-GPx4) genes. This study showed that liver GPx4 mRNA levels could be down regulated by an excess of Se (Zoidis *et al.*, 2010).

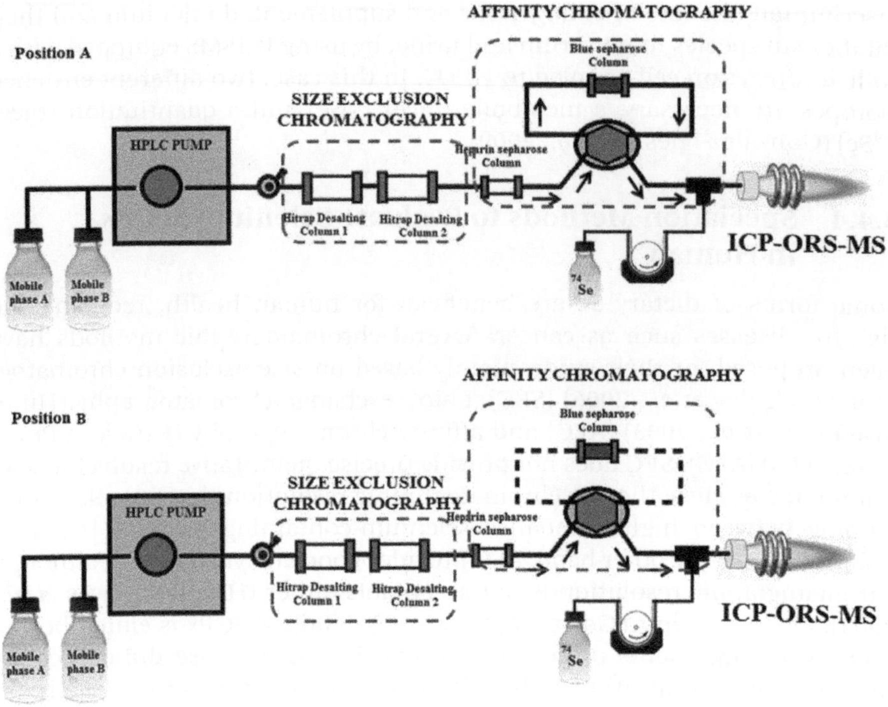

Figure 6.3 Chromatographic arrangement used for selenoproteins and selenometabolites speciation in human serum based on an online multidimensional approach using 2D/SEC-AF-SUID-ICP-ORS-qMS.

6.5.2 Selenoproteins in Seafood

Selenium is an essential element for fish, since it is a component of several antioxidant enzymes as glutathione peroxidase. Fish can accumulate selenium directly from the aquatic environment by the epidermis, but the main incorporation pathway is by feeding zoo- and phytoplankton through the food chain. Recent studies in carps (*Carassius auratus gibelio*), fed with inorganic and organic selenium, as selenite and selenomethionine, respectively, have shown the beneficial effect of selenium in animals grown, the increase of glutathione peroxides activity and the total selenium concentration accumulated into the muscle (Wang *et al.*, 2007).

In another study, Se species present in water-soluble fractions of proteins from lyophilized biological samples, such as fish (tuna and trout), shellfish (krill, oyster and mussel), vegetal tissues (white clover and wheat flour) and selenized yeast were evaluated. The water-soluble fractions obtained with and without 4% (w/v) sodium dodecyl sulfate were analyzed by SEC-ICP-MS and UV spectrophotometers (Moreno *et al.*, 2004). Two different columns with effective separation ranges between 300 and 1 kDa were used. The enzymatic digests of both water-soluble fractions and nonsoluble residues were analyzed by CEC-ICP-MS. The presence of high molecular weight selenoproteins (150–50 kDa) in the samples studied, except for fish and white clover with proteins <30 kDa, was demonstrated, and low molecular weight (<5 kDa) compounds were also found. The information obtained from the SEC-UV and SEC-ICP-MS chromatograms relating to fractions with high selenium to protein concentration ratios was very useful. These protein fractions could be suitable for further purification studies focused on amino acid sequence determination (Moreno *et al.*, 2004).

6.5.3 Selenoproteins in Milk

Bovine milk contains many nutrients necessary to meet the requirements of the newborn calf, including necessary trace elements such as selenium, zinc and copper, which are contained in proteins. In bovine milk GPx3 is the only identified selenoprotein so far, but many other protein-associated trace elements participate in the defense against oxidation and thereby maintain the quality of milk.

The coupling SEC-ICP-MS has been used to obtain the selenium-biomolecule profile in bovine whey and plasma, and consequently to study the effects of selenium supplementation on the distribution of selenium in these fluids (Hoac *et al.*, 2007). Fractionation of bovine whey by SEC-ICP-MS resulted in three main selenium fractions, associated to β-lactoglobulin, α-lactalbumin and selenomethionine or selenium compounds, equivalent to two-thirds of the selenium presents in the two protein fractions. Using this approach the selenium content and selenium profiles in milk and whey were compared (Hoac *et al.*, 2007). Using the same approach, bovine blood plasma was separated into six major selenium peaks with apparent molecular weights of >600, 195, 66, 11 and <2 kDa (two peaks). In addition, it was observed a 66 kDa peak and other less than 2 kDa peaks coeluting with bovine serum albumin and selenomethionine, respectively. A peak at 195 kDa can be related to selenoprotein P, but some SelP might also be in the void volume peak, and a peak at 11 kDa can not be related with any known selenium compound. The albumin fraction contributed approximately to 30% of the selenium in the sample.

In another experiment, the effect of selenium used as dietary supplement in yogurt has been evaluated. For this purpose, different concentrations of inorganic Se (ranging from 0.2 to 5000 μg per g) have been added to milk before the fermentation process. Biotransformation of inorganic Se into organic species has been carefully evaluated by IE, RP, or SEC, coupled to ICP-MS. The results showed that yogurt fermentation in the presence of up to 2 μg per g of Se(IV) produces a complete incorporation of this element into proteins, as has been demonstrated applying a dialysis procedure. Analysis by SEC-ICP-MS showed that most of the Se molecules have a molecular mass in the range of 30–70 kDa. Species determination after enzymatic hydrolysis has allowed the identification of Se-cystine using two different chromatographic systems (Alzate *et al.*, 2007).

6.5.4 Selenoproteins in Vegetables

The ability of several plants to accumulate and transform inorganic forms of selenium into bioactive organic compounds has important implications for human nutrition and health. Se-enriched *Allium* group vegetables such as garlic, onion and ramps have been the subject of several studies in the recent years. Recently, Goenaga *et al.* reported for the first time, the application of species-specific double-isotope dilution analysis by LC-ICP-MS to the accurate quantification of glutamylselenomethylselenocysteine in garlic sample extracts (Goenaga-Infante *et al.*, 2005).

The application of LA-ICP-MS proved to be a useful tool for the analysis of selenium-containing proteins in sunflower and allowed the identification of this element in a single protein spot after sample fractionation by gel electrophoresis. The study considers a method for the identification of selenium-containing proteins in an extract of sunflower leaves, which is based on the separation of the proteins by two-dimensional gel electrophoresis, followed by detection of selenium *via* laser-ablation inductively coupled plasma mass spectrometry (LA-ICP-MS). In this work, selenium was identified in 9.6% of the analyzed spots, indicating its random incorporation into the primary structure of the proteins (Oseas da Silva and Zezzi Arruda, 2012).

On the other hand, the high content of selenium in Brazil nut (*Bertholletia excelsa*) makes this seed a healthy food suitable as antifree-radical protector. Selenium was found to be distributed into the different nut protein fractions. The water-extracted fraction, which represented 17.7% of the nut protein, was the richest in selenium with 153 ppm. Analysis by HPLC-MS showed that selenium was linked by a covalent bond to two amino acids, selenomethionine and selenocystine. The selenomethionine represented a little less than 1% of the total amount of methionine. Extraction of proteins in three steps, water, salt solution, and NaOH solution, led to obtaining three protein families: albumin, globulin, and glutelin. The albumin family contained the most important part of selenium (153 ppm) and represented 17.7 g for 100 g of nut. The presence of glutelins was also remarkable, with 56 ppm of selenium and 3 g of proteins per 100 g of nut, similarly to globulins, with 63 ppm of selenium and 6 g of protein per 100 g of nut (Chunhieng *et al.*, 2004).

6.5.5 Selenoproteins in Food Supplements

6.5.5.1 Yeast

Yeast grown in selenium-rich media is the most popular supplement of organic Se for human and animals. Particular attention has been paid to sample preparation with regard to the fractionation of selenium into different classes of chemical species, the high-resolution fractionation of selenium from yeast water extracts by SEC and the characterization of water-soluble protein fraction by MALDI followed by TOF-MS and ESI-Q-TOF-MS. Selenium-specific ICP-MS detection allows molecular mass spectrometry to be focused on the selenium-containing fractions and is crucial for a successful analysis (Ruiz-Encinar *et al.*, 2003).

A proteomics approach based on 2D-GE followed by HPLC-electrospray Orbitrap MS/MS was developed to investigate the replacement and the degree of the Se/S substitution in methionine and cysteine in Se-rich yeast (Bierla *et al.*, 2013). In this work, capillary HPLC-inductively coupled plasma mass spectrometry (ICP-MS), employed in parallel to cap HPLC-ESI MS, indicated the virtual independence of the ESI-MS response of the peptide structure, and hence, the use of ESI-MS data to determine the SeCys/Cys and SeMet/Met substitution ratios. These ratios were determined in a large number of peptides resulting from the tryptic digestion of 19 Se-containing proteins, which were traced in the gel by LA-ICP MS imaging. The average Se/S substitution in methionine was 42.9 ± 3.5 and was protein dependent with ratios ranging from 5 to 160 for individual peptides. The substitution of sulfur in cysteine (14.1 ± 4.8%) in the cysteine-containing peptides was relatively similar (ratios from 9 to 23). Taking into account the cysteine/methionine average ratio (2:1) in the yeast protein fraction, the study concluded that 10–15% of selenium present in Se-enriched yeast is in the form of selenocysteine (Bierla *et al.*, 2013).

6.5.5.2 *Rice*

The accumulation of selenium (Se) in protein fractions of albumin, globulin, prolamin and glutelin extracted from Se-enriched brown rice was studied, and the molecular weight distribution of Se-containing proteins established. In this study, the amount of total Se (T-Se) and protein-bound Se (PBSe) in brown rice was significantly increased after germination with 10–60 μmol L^{-1} of sodium selenite. Except for prolamin, the amount of all the other three protein fractions decreased significantly with the increase of germination time. Low Se concentrations had promoting effects on degradation of albumin and globulin, while no significant effects were observed on prolamin and glutelin. The accumulation of T-Se and PB-Se were in the order albumin > glutelin > globulin > prolamin. SDS-PAGE analysis showed that Se was distributed in all the proteins with molecular weights from 13.6 to 121.4 kDa; however, 84.34% of the Se was present in proteins with molecular weights less than 36.3 kDa (Liu *et al.*, 2011).

Additionally, selenium distribution in Se-enriched rice and optimization of extraction for Se-containing proteins were studied by other authors (Fang *et al.*, 2010). This work was complemented by Se-availability studies in the Se-containing protein product simulating an *in vitro* gastrointestinal digestion. The results showed that Se was predominantly present as organic Se, whereas inorganic Se comprised only 2.85% of the total Se. The glutelin fraction contained the largest amount of Se, approximately 31.3% of the total Se in the rice gain. This rice protein product with high molecular weight Se-containing protein can readily be digested to low molecular weight peptides including selenomethionine (52.3% of the total Se in the protein extract) (Fang *et al.*, 2010).

6.5.5.3 Wheat

A method has been developed for a rapid and precise location of selenium-containing proteins in large 2D-GE of wheat. The sample was divided into four aliquots that were analyzed in parallel by 1D isoelectric focusing electrophoresis (IEF)-LA-ICP-MS, 1D-SDS PAGE-LA-ICP-MS, and, in duplicate, by 2D IEF-PAGE. On the basis of 1D electropherograms, areas supposed to contain the largest concentrations of Se were subjected to LA-ICP-MS imaging to locate precisely the position of Se-containing proteins that were then identified in parallel from 2D gel by ESI-Orbitrap-MS/MS. The method was applied to the identification and semiquantitative determination of selenium-storage proteins in wheat. MS evidence is presented for the Se–S substitution in plants not only in methionine but also in cysteine (Bianga *et al.*, 2013).

6.5.5.4 Mushroom

Using various chromatographic techniques (size exclusion, anion exchange, and cation exchange) combined with several detectors (neutron activation analysis and atomic fluorescence spectrometry); an attempt was made to characterize selenium compounds in some edible, selenium-accumulating mushrooms (*Albatrellus pes-caprae* and *Boletus edulis*). The mushrooms contained mostly low molecular weight (6 kDa) selenium compounds. After proteolysis, only a small fraction of the extractable selenium could be identified as selenite (3.0–9.2%, *Albatrellus pes-caprae*), selenocystine (minor, *Albatrellus pes-caprae*; 7.5%, *Boletus edulis*), or selenomethionine (1.0%, *Boletus edulis*), although most of the bulk has not been elucidated yet (Šlejkovec *et al.*, 2000).

In the same way, selenium species from Se-containing proteins in mushrooms (*Agaricus bisporus* and *Lentinula edodes*) were investigated with size-exclusion liquid chromatography coupled to UV and inductively coupled plasma mass spectrometry (ICP-MS). Different protein extraction protocols were investigated. Presumably, protein-bound selenoamino acids were released using enzymes (proteinase K, protease XIV and trypsin). The selenium speciation of the proteolytic extract of water-soluble proteins fraction was carried out by using RP-ion-pairing-(IP)-HPLC coupled online to ICP-MS for selenium specific detection. The presence of selenocystine, selenomethionine, methylselenocysteine and inorganic selenium were established in both samples using retention time standards and standard additions to the sample. This investigation was aimed to study the selenium-containing proteins fraction in two different mushroom species (Gergely *et al.*, 2006).

Summary Points

- Selenium and selenoproteins play a key role for basic functions of life ensuring human health.
- Current analytical strategies and techniques are necessary to characterize selenoproteins in foods.
- Hyphenated analytical techniques based on the coupling of chromatographic separation with inductively coupled plasma spectrometry (ICP-MS) followed by organic mass spectrometry are basic tools to identify and quantify selenium species and selenoproteins.
- Sample preparation is a crucial step for Se speciation in food, in which cell and tissues disaggregation, analyte extraction, protein disulfide bridges prevention and proteases inhibition are key points of these treatments.
- Specific sample-preparation procedures based on enzymatic extraction, protein precipitation, ultrafiltration and sonication have been recently proposed.
- Selenoproteins analysis based on these analytical techniques are crucial for different types of food, such as meat, seafood, milk, vegetables and food supplements (yeast, rice, wheat, mushroom, and others).

Key Facts of Selenium in Foods

1. Selenium is an essential element for human health involved in antioxidant defense, thyroid homeostasis and regulation of redox status, preventing cancer risk and thyroid disorders among others diseases.
2. Selenium is a potent protective agent for neurons as well as for skeletal and cardiac muscles.
3. The main source of selenium for human is diet. The chemical form of this element is crucial to assure its action, selenoproteins playing key roles in the basic functions of life.
4. Identification of selenoproteins and selenoaminoacids has to be performed in daily foods such as meat, seafood, milk, vegetables and food supplements.

Key Facts of Selenium Analysis in Foods

1. Selenium-containing proteins and selenoaminoacids analysis require instrumental coupling of chromatographic instruments with sensitive atomic devices (especially ICP-MS).
2. Molecular mass spectrometry allows identification of unknown selenium biomolecules.
3. Sample preparation, including selenium species preservation until analysis, species extraction from food matrices and sample extracts cleanup, is a very critical step in selenium speciation.

4. Quantification of selenoproteins in complex samples can be performed using isotope dilution analysis (IDA) using the multi-isotopic capability of ICP-MS.
5. Analysis of selenoproteins and selenoaminoacids in human serum (containing significant amounts of Br^- and Cl^-) to determine selenium status in this fluid can be performed by online combination of multichromatographic couplings, which involve up to five columns and ICP-MS detection.

Definitions of Words and Terms

Inductively coupled plasma mass spectrometry (ICP-MS). This is a type of mass spectrometry able to detect metals and several nonmetals at concentrations as low as one part in 10^{12} (part per trillion).

Laser ablation (LA). This is a form of direct introduction of elements from solid samples in the ICP-MS. LA can be used for analysis of single spots from 2D-gel electrophoresis, tissue samples from food, animals, and mineral or archeological samples, *etc.*

Speciation analysis. Speciation analysis is the identification and quantification of the different chemical forms of an element in a sample.

Isotopic dilution analysis (IDA). IDA is a quantification method of analysis based on the use of an isotopically enriched form of the analyte directly added to the sample and the measure isotopes ratio signals.

Mass spectrometry (MS). This is an analytical technique that produces spectra of the masses of atoms or molecules from a sample or material.

Two dimensional-gel electrophoresis (2D-GE). 2D-GE is a form of gel electrophoresis commonly used to analyze proteins using a combination of separation by isoelectric point followed by a second perpendicular separation in the gel plate by the molecular mass.

Matrix-assisted laser desorption/ionization (MALDI). MALDI this is a soft ionization technique used in mass spectrometry, allowing the analysis of large molecules. Ionization energy is transferred from laser source to molecules by an excess of matrix with UV absorption properties in which the analytes are embedded. MALDI produces few multiply charged ions.

Electrospray ionization (ESI). This is another soft ionization technique used in mass spectrometry producing multiply charged ions very suitable for the analysis of proteins and their associated polypeptide fragments.

List of Abbreviations

AF	Affinity chromatography
CE	Capillary electrophoresis
CEC	Capillary electrokinetic chromatography

CZE	Capillary zone electrophoresis
CHAPS	3-[(3-Cholamidopropyl)dimethylammonio]-1-propanesulfonate
CRMs	Certified reference materials
Cys	Cystine
DIMS	Direct infusion mass spectrometry
DTE	Dithioerythritol
DTT	Dithiothreitol
eGPx	Extracellular glutathione peroxidase
ESI	Electrospray ionization
GE	Gel electrophoresis
GPE	Gel permeation electrophoresis
GPx	Glutathione peroxidase
GSH	Reduced glutathione
HILIC	Hydrophilic interaction liquid chromatography
HPLC	High-performance liquid chromatography
ICP-MS	Inductively coupled plasma-mass spectrometry
ID	Iodothyronine deiodinase
IDA	Isotopic dilution analysis
IEC	Ionic exchange chromatography
IPD	Isotope pattern deconvolution
IT	Ion trap
LA	Laser ablation
LC-MS/MS	Liquid chromatography-tandem mass spectrometry
MALDI-TOF	Matrix-assisted laser desorption/ionization-time-of-flight
MC	Multicollector
ORS	Octopole reaction systems
PMSF	Phenylmethanesulfonyl fluoride
Q	Quadrupole
QTrap	Quadrupole trap
RP	Reverse phase
SDS-PAGE	Sodium dodecylsulfate-polyacrylamide gel electrophoresis
SeAlb	Selenoalbumin
SEC	Size-exclusion chromatography
SelP	Selenoprotein P
SeMet	Selenomethionine
SF	Sector field
SeCys	Selenocysteine
SPE	Solid-phase extraction
SSID	Species-specific isotopic dilution
SUID	Species-unspecific isotopic dilution
TBP	Tributylphosphine
TCA	Tricarboxylic acid cycle
TCEP	Tris(2-carboxyethyl)phosphine hydrochloride
TrxR	Thioredoxin reductase

References

Alzate, A., Cañas, B., Pérez-Munguía, S., Hernández-Mendoza, H., Pérez-Conde, C., Gutiérrez, A. M. and Cámara, C., 2007. *Journal of Agricultural and Food Chemistry*. 55: 9776.

Bettmer, J., Montes Bayón, M., Ruiz Encinar, J., Fernández Sánchez, M. L., Fernández de la Campa, M. R. and Sanz Medel, A., 2009. *Journal of Proteomics*. 72: 989.

Bianga, J., Govasmark, E. and Szpunar, J., 2013. *Analytical Chemistry*. 85: 2037.

Bierla, K., Szpunar, J. and Lobinski, R., 2008. *Analytica Chimica Acta*. 624: 195.

Bierla, K., Bianga, J., Ouerdane, L., Szpunar, J., Yiannikouris, A. and Lobinski, R., 2013. *Journal of Proteomics*. 87: 26.

Chunhieng, T., Petritis, K., Elfakir, C., Brochier, J., Goli, T. and Montet, D., 2004. *Journal of Agricultural and Food Chemistry*. 52: 4318.

Fan, T. W., Pruskowski, E. and Shuttewoth, S., 2002. *Journal of Analytical Atomic Spectrometry*. 17: 1621.

Fang, Y., Catron, B., Zhang, Y., Zhao, L., Caruso, J. A. and Hu, Q., 2010. *Journal of Agricultural and Food Chemistry*. 58: 9731.

García-Sevillano, M. A., González-Fernández, M., Jara-Biedma, R., García-Barrera, T., López-Barea, J., Pueyo, C. and Gómez-Ariza, J. L., 2012. *Analytical and Bioanalytical Chemistry*. 404: 1967.

García-Sevillano, M. A., García-Barrera, T. and Gómez-Ariza, J. L., 2013. *Journal of Chromatography A*. 1318: 171.

Gergely, V., Kubachka, K. M., Mounicou, S., Fodor, P. and Caruso, J. A., 2006. *Journal of Chromatography A*. 1101: 94.

Goenaga-Infante, H., Van Campenhout, K., Blustand, R. and Adams, F. C., 2002. *Journal of Analytical Atomic Spectrometry*. 17: 79.

Goenaga-Infante, H., O'Connor, G., Rayman, M., Wahlen, R., Spallholz, J. E., Hearn, J. and Catterick, T., 2005. *Journal of Analytical Atomic Spectrometry*. 20: 864.

González-Iglesias, H., Fernández-Sánchez, M. L., Rodríguez-Castrillón, J. A., García-Alonso, J. I., López-Sastre, J. and Sanz-Medel, A., 2009. *Journal of Analytical Atomic Spectrometry*. 24: 460.

Hinojosa-Reyes, L., Marchante-Gayón, J. M., García-Alonso, J. I. and Sanz-Medel, A., 2003. *Journal of Analytical Atomic Spectrometry*. 18: 1210.

Hoac, T., Lundh, T., Purup, S., Onning, G., Sejrsen, K. and Akesson, B., 2007. *Journal of Agricultural and Food Chemistry*. 55: 4237.

Jayasinghe, S. B. and Caruso, J. A., 2011. *International Journal of Mass Spectrometry*. 307: 16.

Jitaru, P., Prete, M., Cozzi, G., Turetta, C., Cairns, W., Seraglia, R., Traldi, P., Cecson, P. and Barbante, C., 2008. *Journal of Analytical Atomic Spectrometry*. 23: 402.

Jitaru, P., Cozzi, G., Seraglia, R., Traldi, P., Cescon, P. and Barbante, C., 2010. *Analytical Methods*. 2: 1382.

Larsen, E. H., Lobinski, R., Burger-Mayer, K., Hansen, M., Ruziuk, R., Mazurowska, L., *et al.*, 2006. *Analytical and Bioanalytical Chemistry*. 385: 1098.

Liu, K., Chen, F., Zhao, Y., Gu, Z. and Yang, H., 2011. *Food Chemistry*. 127: 1526.

Mesko, M. F., Hartwig, C. A., Bizzi, C. A., Pereira, J. S. F., Mello, P. A. and Flores, E. M. M., 2011. *International Journal of Mass Spectrometry*. 307: 123.

Mester, Z. and Sturgeon, R. (ed.), 2003. *Sample Preparation for Trace Element Analysis*. Elsevier, Amsterdam.

Montes-Bayón, M., Yanes, E. G., Leon, C. P., Jayasimhulu, K., Stalcup, A., Shann, J. and Caruso, J. A., 2002. *Analytical Chemistry*. 74: 107.

Moreno, P., Quijano, M. A., Gutiérrez, A. M., Pérez-Conde, M. C. and Cámara, C., 2004. *Analytica Chimica Acta*. 524: 315.

Mounicou, S., Dernovics, M., Bierla, K. and Szpunar, J., 2009. *Talanta*. 77: 1877.

Nuevo-Ordoñez, Y., Montes-Bayón, M., Blanco-Gonzalez, E. and Sanz-Medel, A., 2010. *Analytical Chemistry*. 82: 2387.

Oseas da Silva, M. A. and Zezzi Arruda, M. A., 2012. *Microchimica Acta*. 176: 131.

Palacios, O., Ruiz Encinar, J., Bertin, G. and Lobinski, R., 2005. *Analytical and Bioanalytical Chemistry*. 383: 516.

Palacios, O., Encinar, J. R., Schaumlöffel, D. and Lobinski, R., 2006. *Analytical and Bioanalytical Chemistry*. 384: 1276.

Pröfrock, D. and Prange, A., 2009. *Journal of Chromatography A*. 1216: 6706.

Rodríguez-González, P., Marchante-Gayón, J. M., García Alonso, J. I. and Sanz-Medel, A., 2005. *Spectrochimica Acta, Part B: Atomic Spectroscopy*. 60: 151.

Ruiz-Encinar, J., Śliwka-Kaszyńska, M., Połatajko, A., Vacchina, V. and Szpunar, J., 2003. *Analytica Chimica Acta*. 500: 171.

Ruiz-Encinar, J., García-Alonso, J. I., Sanz-Medel, A., Main, S. and Turner, P. J., 2001. *Journal of Analytical Atomic Spectrometry*. 16: 315.

Sariego-Muñíz, C., Marchante-Gayón, J. M., García-Alonso, J. I. and Sanz-Medel, A., 2001. *Journal of Analytical Atomic Spectrometry*. 16: 587.

Šlejkovec, Z., van Elteren, J. T., Woroniecka, U. D., Kroon, K. J., Falnoga, I. and Byrne, A. R., 2000. *Biological Trace Element Research*. 75: 139.

Szpunar, J., 2005. *Analyst*. 130: 442.

Tastet, L., Schaumloffel, D., Bouyssiere, B. and Lobinski, R., 2008. *Talanta*. 75: 1140.

Vanhaecke, F., Wannemacker, G., Moens, L., Dams, R., Latcoczy, C., Prohaska, T. and Stingerer, G., 1998. *Journal of Analytical Atomic Spectrometry*. 13: 567.

Wang, Y., Han, J., Li, W. and Xu, Z., 2007. *Animal Feed Science and Technology*. 134: 243.

Wrobel, K., Wrobel, K., Kannamukarath, S. S., Caruso, J. A., Wysocka, I. A., Bulska, E., Swiatek, J. and Wierzbicka, M., 2004. *Food Chemistry*. 86: 617.

Zoidis, E., Pappas, A. C., Georgiou, C. A., Komaitis, E. and Feggeros, K., 2010. *Comparative Biochemistry and Physiology, Part B: Biochemistry & Molecular Biology*. 155: 294.

CHAPTER 7

Focus on Determination of Selenocysteine and Selenomethionine in Foodstuffs of Animal Origin by 2D Size-Exclusion Reversed-Phase HPLC-ICP-MS

KATARZYNA BIERLA*[a], SIMON GODIN[a], RYSZARD LOBINSKI[a], AND JOANNA SZPUNAR[a]

[a]Laboratoire de Chimie Analytique Bio-inorganique et Environnement, CNRS UMR 5254, Hélioparc, 2, av. Angot, 64053 Pau, France
*E-mail: katarzynabierla@wp.pl

7.1 Introduction

Foodstuffs of animal origin such as meat, milk, fish, eggs, cheese and yogurt are popularly considered as indispensable to obtain appropriate amounts of nutrients. Dairy products are an important part of human diet from early childhood, whereas meat and eggs are consumed as a part of standard meals in developed countries.

Food and Nutritional Components in Focus No. 9
Selenium: Chemistry, Analysis, Function and Effects
Edited by Victor R. Preedy

Published by the Royal Society of Chemistry, www.rsc.org

Selenium is an essential element known to develop its biological activity *via* a selenocysteine residue in the catalytically active center of selenoproteins. Large variations observed in the selenium content of foods are determined by a combination of geologic and environmental factors and selenium supplementation of fertilizers (for edible plants) and animal feedstuffs (for edible animal tissues). Selected foodstuffs of animal origin containing high contents of selenium are listed in Table 7.1. The deficit of selenium in the diet in many countries spurred interest in supplementation of selenium in the form of either natural foods (*e.g.*, Brazil nuts), dedicated supplements, or Se-enriched products of animal origin: meat from animals (chicken, lamb, beef) fed on Se-rich diet, eggs or milk and its products. Several studies were carried out in order to develop efficient animal-feeding schemes to obtain selenium-enriched dairy products (Juniper *et al.*, 2006, 2008b; Petrera *et al.*, 2009; Calamari *et al.*, 2010), meat (Juniper *et al.*, 2008a,c, 2009; Vignola *et al.*, 2009) and eggs (Scheideler *et al.*, 2010; Pan *et al.*, 2011; Jlali *et al.*, 2013).

Also, the chemical forms of the element are absorbed and metabolized differently; although most dietary selenium is absorbed efficiently, the retention of organic forms is higher than that of inorganic ones (Fairweather-Tait *et al.*, 2010). As a result, the information on speciation of selenium is essential to understand the pathways of supplementation and to provide characteristics of the products for the consumer; it also helps to determine the commercial features of foodstuffs (*e.g.* oxidative stability, drip loss and color of meat). Whereas the determination of the total level of selenium in most samples can be considered as a routine task, the determination of selenium

Table 7.1 Selected foodstuffs of animal origin containing high contents of selenium.

Food	Micrograms (μg) per serving	Percent DV[a]
Tuna, yellowfin, cooked, dry heat, 3 ounces	92	131
Halibut, cooked, dry heat, 3 ounces	47	67
Sardines, canned in oil, drained solids with bone, 3 ounces	45	64
Ham, roasted, 3 ounces	42	60
Shrimp, canned, 3 ounces	40	57
Macaroni, enriched, cooked, 1 cup	37	53
Beef steak, bottom round, roasted, 3 ounces	33	47
Turkey, boneless, roasted, 3 ounces	31	44
Beef liver, pan fried, 3 ounces	28	40
Chicken, light meat, roasted, 3 ounces	22	31
Cottage cheese, 1% milkfat, 1 cup	20	29
Beef, ground, 25% fat, broiled, 3 ounces	18	26
Egg, hard-boiled, 1 large	15	21
Milk, 1% fat, 1 cup	8	11
Yogurt, plain, low fat, 1 cup	8	11

[a]DV = The U.S. Food and Drug Administration (FDA) approved daily value (70 mcg for adults and children aged 4 and older).

species turns out to be more difficult. The principal forms of "organic selenium" include two amino acids: selenomethionine and selenocysteine; the differentiation between them and inorganic selenium is a measure of the selenium bioaccessibility and quality of the supplement. They are incorporated in proteins, either in a nonspecific way (selenomethionine and selenocysteine replacing methionine and cysteine, respectively) or in a genetically coded way (selenocysteine only).

The quantitative determination of these selenoamino acids allowing the control of selenium balance was proposed as a means of assessment of selenium incorporation into animal proteins in foodstuffs. The optimized procedures were proposed for meat and offal tissues (Bierla *et al.*, 2008a), eggs (Lipiec *et al.*, 2010) and milk and milk products (Bierla *et al.*, 2008b) and are presented in this chapter.

7.2 Selenium in Foodstuffs of Animal Origin

7.2.1 Meat and Meat Products

Historians are unanimous in stating that man has been essentially carnivorous for millions of years; however, in modern times meat is less universally consumed due to development of eating habits specific to certain geographical regions, social environment (meat may not be accessible or available) or resulting from personal choice.

Beef, pork and poultry are an important source of selenium in developed countries (Hintze *et al.*, 2002). Selenium from these meats seems to be highly bioavailable for selenoprotein synthesis (Finley *et al.*, 2004) which raises interest in using such meat, after increasing the Se concentration by supplementation of animals, as a source of bioaccessible selenium. The concentration of selenium in beef varies dramatically among countries and regions (Zhang *et al.*, 2010); the values reported for 100 g of beef were 3.0–3.6, 2.2–8.3, 7.2–12.1 and 13.4–19.0 μg selenium in the UK, New Zealand, Australia and USA, respectively (McNaughton and Marks, 2002). Selenium contents in selected Spanish meat and meat products were also reported (Díaz-Alarcón *et al.*, 1996). The GSHPx activity is often used to assess selenium status in animals (Daun and Akesson, 2004a,b).

In recent years, much attention has been paid to develop meat and meat products with physiological functions to promote health and prevent the risk of diseases. The strategies to improve the functional value of meat and meat products can be realized by adding functional compounds including – among others – selenium in animal diets to improve animal production, carcass composition and fresh meat quality (Zhang *et al.*, 2010).

In Korea, selenium-enriched pork "Selen Pork" was produced by feeding yeast-bound selenium and sold as a functional food that can improve human health and nutrition. The meat contained approximately 10 times the selenium content of traditional pork and was leaner and juicier with a noticeably redder color (Fisinin *et al.*, 2009). In Korea, "Selen Chicken" has been developed as a premium chicken brand with a high content of selenium (Zhang *et al.*, 2010).

Attempts at characterization included speciation of water-soluble selenium compounds in two muscles from each of seven animal species (chicken, turkey, duck, ostrich, lamb, cattle and pig); soluble selenium compounds were separated by size-exclusion chromatography (SEC) and detected by inductively coupled plasma-mass spectrometry (ICP-MS) (Daun *et al.*, 2004). The water-soluble fraction of chicken and lamb muscles and lamb kidney, heart and liver (Bierla *et al.*, 2008a) and muscles of chicken, turkey, duck, ostrich, lamb, cattle and pig (Daun *et al.*, 2004). On the basis of peak identification by chromatographic mobility, glutathione peroxidase (GSHPx) and selenoprotein W were claimed to account for 70–100% of water-soluble selenium. The percentage of soluble selenium species in chicken tissues varied between 20 and 45% (Bierla *et al.*, 2008a), which was close to the values reported by Cabañero *et al.* (42%) (Cabañero *et al.*, 2005) but lower than the values of 65–76% reported by Daun *et al.* (2004). For lamb tissues, the water-extraction yield varied between 10 and 35% (Bierla *et al.*, 2008a) which was also lower than the 31–56% values reported by Daun *et al.* (Daun *et al.*, 2001, 2004; Daun and Akesson, 2004a,b). The method allows an insight into selenium speciation and comparison of different tissues from different animals but does not account for the insoluble selenium fraction and the identification of peaks is problematic.

7.2.2 Dairy Products

Among the food selenium sources, particular attention has been paid to milk due to its nutritional importance, especially in the diet of young children. Selenium concentration in milk is relatively low and some Se compounds can be removed during processing by hot-air treatment, further reducing the selenium concentration of commercial milks. Due to the concerns that the concentration of selenium in milk may not be sufficient, strategies of the Se supplementation increasing its uptake by dairy animals have been proposed resulting in its higher content in milk (Knowles *et al.*, 1999; Juniper *et al.*, 2006).

Most selenium (more than 70%) is contained in the (insoluble) casein fraction so a solubilization step is mandatory in procedures aimed at the selenium speciation analysis. The attempts to speciate selenium contained in intact proteins have concerned milk whey (Rivero Martino *et al.*, 2002; De La Flor St Remy *et al.*, 2004). A number of peaks were detected by SEC-ICP-MS, a clear increase in one of them is observed upon selenium supplementation in animal diet.

Some attention was paid to yogurt (Alzate *et al.*, 2007, 2008) and kefir (Alzate *et al.*, 2008) fermentation in the presence of inorganic selenium in the form of Se(IV) which results in the incorporation of this element into proteins, as has been demonstrated by applying a dialysis procedure. However, unlike Se(IV), the percentage of biotransformation observed for

Se(VI) was very low (Alzate *et al.*, 2007). Analysis by SEC-ICP-MS showed that most of the proteins had a molecular mass in the range of 30–70 kDa (Alzate *et al.*, 2007). Speciation studies of the enriched samples after enzymatic hydrolysis demonstrated the presence of $SeCys_2$ and MeSeCys being the main species generated in yogurt and SeMet in kefir; however, no quantitative data showing the species balance was given (Alzate *et al.*, 2007). Scanning electron microscopy (SEM) applied to the enriched samples showed segregated Se^0, at added concentrations higher than 5 μg g^{-1} (Alzate *et al.*, 2008).

7.2.3 Eggs

Selenium-enriched eggs were proposed as a delivery system of this trace mineral for humans. Currently, companies all over the world market selenium-enriched eggs including the UK, Ireland, Mexico, Columbia, Malaysia, Thailand, Australia, Turkey, Russia and the Ukraine (Fisinin *et al.*, 2008, 2009). However, only very scarce data exists on the enrichment factor of the commercial products. Increase of the whole-egg Se concentration was observed for quails fed Se from selenium-enriched kale sprout (Chinrasri *et al.*, 2013) and hens supplemented with both inorganic and organic selenium (the latter being more effective) (Payne *et al.*, 2005).

7.2.4 Fish

Fish is considered as one of major sources of selenium in the diet (*cf.* Table 7.1) and is often evoked in the context of its protective effects against methylmercury toxicity (Ralston and Raymond, 2010). The number of selenoproteins present in fish is higher than in mammals (30–37 groups in place of 25 described for humans) with some of them being species specific (Yoshida *et al.*, 2011). It was also reported that the amount of SeCys residues in selenoproteins is higher for aquatic organisms than for mammals (Yoshida *et al.*, 2011).

Selenium is important for pigment deposition and flesh color expression in salmonids; it was reported to enhance flesh texture and decrease its oil content (Silva de Lyons, 2007). However, no similar effects were observed for carp (Rodica Ani *et al.*, 2010).

Surprisingly, not much data is available on selenium speciation in fish. However, it was studied to some extent in the context of Se interaction with mercury species and possible detoxification effect of Se species on mercury (Cabañero *et al.*, 2005, 2007).

Selenium species were analyzed in water extracts of different fish and it was reported that in cod, salmon, eel and trout more than 76% of Se was present in high molecular weight compounds (>10 kDa) while in "flat" fish (*e.g.* plaice, flounder, turbot) the majority of Se compounds were in the low molecular weight (<2 kDa) fraction, but no SeMet or Cys was detected (Önning and Bergdahl, 1999; Önning, 2000). A simple enzymatic digestion (with no derivatization step) carried out on tuna and mussels revealed the presence of SeMet and trimethylselenonium ions only; they accounted for 30% of extracted selenium despite quantitative recovery of hydrolysis (Moreno *et al.*, 2004).

Díaz Huerta *et al.* (2004) tested different extraction methods obtaining the highest recovery (70%) for enzymatic hydrolysis (proteolysis) and finding SeMet as the most abundant and the only identified species present in extract, which suggests that Se in incorporated in Se-containing proteins. The morphology of the chromatographic profile is similar to those reported later for other protein-rich animal foodstuffs (Bierla *et al.*, 2008a,b; Lipiec *et al.*, 2010).

X-ray absorption near edge spectroscopy (XANES) was used to demonstrate the presence of selenomethionine, selenocysteine and selenocystine as the predominant forms of selenium in rainbow trout tissues; however, their relative proportion varied – in general, the organs primarily involved in selenium handling in fish (*e.g.*, liver, kidney) accumulated a higher percentage of selenocystine (Misra *et al.*, 2012). This finding is consistent with the observations concerning mammal farm animals (Bierla *et al.*, 2008a).

The most recent compound identified by ESI-MS and NMR (Yamashita and Yamashita, 2010) in fish (tuna) is selenoneine and till now its presence was reported in several other fish species but with the identification based only on comparison of elution time from SEC-ICP-MS (Yamashita *et al.*, 2011).

7.3 Determination of Selenomethionine and Selenocysteine by 2D HPLC-ICP-MS

An elegant approach to selenium speciation based on the simultaneous quantitative determination of selenomethionine (SeMet) and selenocysteine (SeCys) has been developed. The optimized procedures for different samples proposed elsewhere (Bierla *et al.*, 2008a,b; Lipiec *et al.*, 2010) were united and are presented here. Due to its quantitative character the method has the advantage of being able to account for the totality of selenium in the studied sample.

7.3.1 Analytical Procedure

The analytical protocol is summarized in Figure 7.1. It was based on the protein extraction with urea under reducing conditions (dithiothreitol), derivatization of SeCys and SeMet by carbamidomethylation with iodoacetamide (IAM) followed by quantitative proteolysis. The removal of fat and protein precipitation prior to derivatization was found to greatly improve the peak shape and resolution (Bierla *et al.*, 2008b). The mixture of the derivatized Se amino acids was purified by size-exclusion LC and analyzed by ion-paring reversed-phase HPLC-ICP-MS. The quantification of SeCys and SeMet was carried out by the method of standard additions. ^{77}Se-labeled SeMet was used to correct for the SeMet derivatization efficiency and recovery. The major difficulty in the validation of this method is the impossibility of spiking the analyte in exactly the same chemical form as it is present in the sample. The approach

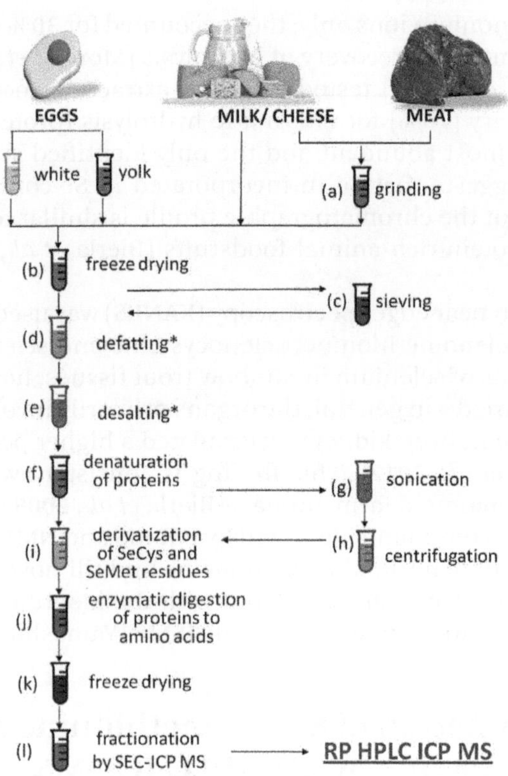

Figure 7.1 Outline of the analytical procedure for determination of selenoamino acids in foodstuffs of animal origin (* except egg white).

proposed by Bierla *et al.* involved the addition of SeCys-containing commercial protein: glutathione peroxidase. It allowed the verification of the protein behavior during the whole analytical process. An auxiliary proof of the validity of the procedure can be obtained by the determination of the mass balance of selenium where the sum of the determined species should correspond to the total selenium concentration (Bierla *et al.*, 2008a,b; Lipiec *et al.*, 2010).

The individual steps of the procedure were as follows (the numbers correspond to Figure 7.1):

(a) *Grinding.* All solid samples (*e.g.*, cheese, meat) were ground before freezing.
(b) *Freeze drying.* Frozen samples were dried by lyophilization.
(c) *Sieving.* In order to achieve homogeneous powder samples were sieved using a 1 mm sieve.
(d) *Defatting.* A sample (5 g) was shaken with 20 mL of cyclohexane for 10 min and centrifuged (10 min, 3000 rpm). The supernatant was removed and the procedure was repeated twice by shaking the residue with fresh portions of cyclohexane. Finally, the samples were dried in the oven at 37 °C.

(e) *Desalting (protein precipitation).* Dried defatted samples (for diary samples only) were dissolved in water and the proteins were precipitated by adding acetone to achieve a water–acetone ratio of 1 : 3. The solution was centrifuged, supernatant removed and the operation was repeated twice.

(f) *Denaturation.* 0.1 g of a freeze-dried sample was incubated with urea and sonicated in an ultrasonic bath until total solubilization of the sample. This step improved the digestion efficiency due to the use of a chaotropic reagent (urea) unfolding the proteins that are otherwise stabilized by the S—S bounds and therefore barely accessible to reagents.

(g) *Sonication* (used for meat only). Meat samples were sonicated using an ultrasonic probe in order to release the cell content to the denaturating solution. During sonication samples were kept on ice to avoid degradation of compounds.

(h) *Centrifugation* (used for meat only). As the meat matrix is much more complex total solubilization was not possible. For this reason samples were centrifuged and only the supernatant was used for further steps of the procedure.

(i) *Derivatization.* The solution was then incubated for 1 h at 25 °C in the dark with dithiothreitol (DTT) and 1 h with iodoacetamide (IAM) to break the Se—Se, S—S and Se—S bonds and to alkylate cysteine and selenocysteine residues. Then, a fresh aliquot of the DTT solution was added and the mixture was shaken for 1 h in order to destroy any excess of IAM. The amounts of DTT and IAM used were optimized for each kind of sample separately.

(j) *Enzymatic digestion (proteolysis).* Subsequently, an aliquot of protease solution (20 mg protease in 0.1 M TRIS buffer pH 7.5) was added and the sample was incubated for a minimum of 17 h at 37 °C. For meat samples an aliquot of protease/lipase solution (29 mg protease XIV and 16 mg lipase VII in 0.1 M TRIS buffer pH 7.5) was added after diluting the solution with 0.1 M Tris (until 1 M urea was reached), and the sample was incubated during 20 h at 37 °C.

(k) *Freeze drying.* After the proteolysis samples were frozen and freeze dried. The lyophilyzate was dissolved in 1 mL of water and centrifuged.

(l) *Fractionation.* The Superdex-200 column was calibrated with a mixture of thyroglobulin (669 kDa), ferritin (440 kDa), transferrin (81 kDa), bovine albumin (67 kDa), ovalbumin (43 kDa), myoglobin (17 kDa) and cobalamin (1.3 kDa) for fraction collection. Fractions were collected in the elution time corresponding to the elution time of small molecular weight compounds (Figure 7.2a). Collected fractions were frozen, freeze dried and resolubilized in buffer A used for RP HPLC separation.

The principle of the two-dimensional HPLC-ICP-MS determination of selenoamino acids is presented in Figure 7.2. The first, SEC step (Figure 7.2a) allows the verification of the completeness of the proteolytic digestion of

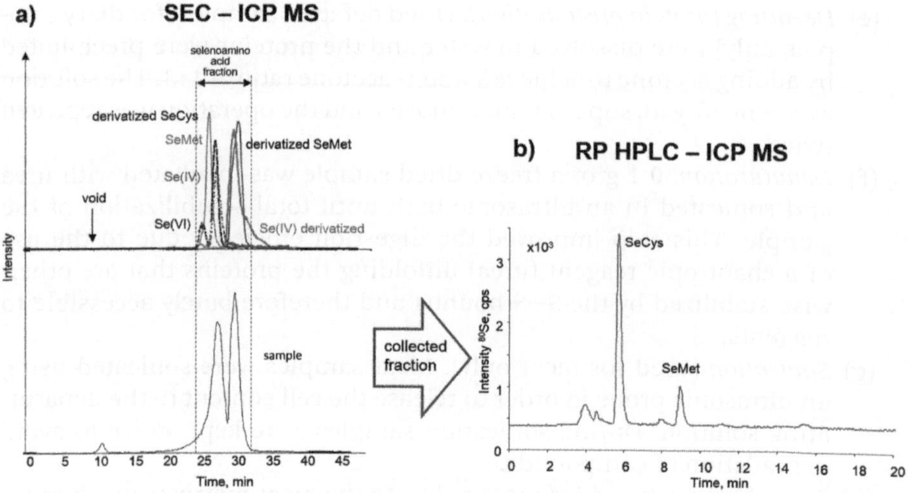

Figure 7.2 Principle of the 2D-HPLC-ICP-MS determination of selenoamino acids in foodstuffs; (a) SEC-ICP-MS (b) RP HPLC-ICP-MS; the chromatograms shown correspond to lamb kidney proteolysate. SEC conditions: Superdex Peptide GE Healthcare Amersham column (300 × 10 mm), isocratic elution with 50 mM ammonium acetate at pH = 7.5. RP-HPLC conditions: C8 Altima/Alltech column (150 × 4.6 mm, 5 μm), elution with A: 0.1% HFBA in water and B: 0.1% HFBA in MeOH using the following gradient: 0–5 min: 5% B, 5–30 min: 5–40% B, 30–40 min 40% B, 40–42 min: 40–45% B, 42–48 min 5% B.

the samples and – at the same time – purification of the extract prior to the quantitative RP HPLC-ICP-MS (Figure 7.2b) determination. The removal of the heavy matrix present in biological samples is a prerequisite of a successful long-term performance of reversed-phase chromatographic columns and high quality of the analytical data. Taking into account the principle of the method used (proteolytic digestion), and as demonstrated in Figure 7.2a the fraction collected is expected to contain a mixture of different forms of selenoamino acids. The method can be applied to any protein-rich foodstuff product.

The RP HPLC-ICP-MS chromatograms obtained for animal foodstuff matrices show the presence of up to five selenospecies. On the routine basis, the identification of these signals can be carried out by spiking with standards, as shown in Figure 7.3 for egg white and yolk. However, the experiments demonstrated that the selenomethionine is eluted in the form of two species. A more advanced study carried was out by ESI-MS/MS on the identification of the native and derivatized SeMet, derivatized SeCys and derivatized selenite. The spectra corresponding to these compounds are shown in Figure 7.4 and the detailed description of the molecular mass spectrometry procedure used was presented in detail by Lipiec *et al.* (2010).

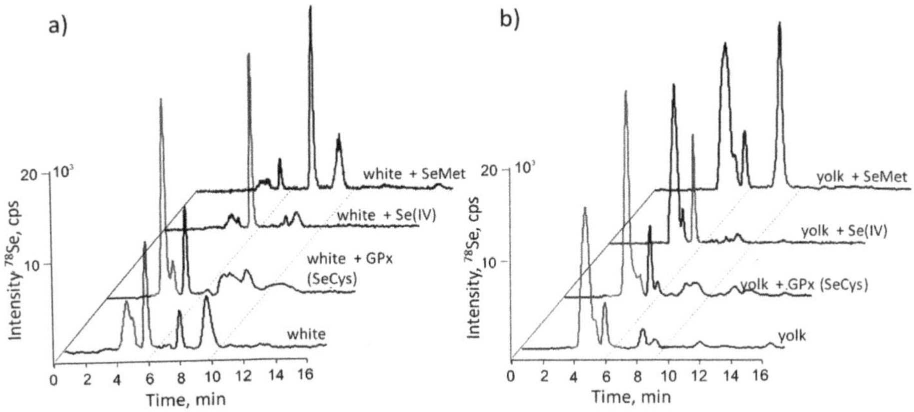

Figure 7.3 Identification of selenium species in RP HPLC-ICP-MS of egg proteolysate by spiking with standards: (a) white, (b) yolk (chromatographic conditions as in Figure 7.2).

7.3.2 Selenoamino Acid Distribution in Foodstuffs of Animal Origin

The analytical data obtained so far by the proposed method provide an insight into the incorporation of selenium into proteins of farm animals providing a variety of foodstuffs including meat and offal tissues (Bierla *et al.*, 2008a), eggs (Lipiec *et al.*, 2010), milk and cheese (Bierla *et al.*, 2008b). Significant differences were observed in selenoamino acid profiles of muscles with SeMet as a major species and liver and kidney where most of selenium was present as SeCys (Figure 7.6). Heart tissue contained similar quantities of both selenoamino acids (Bierla *et al.*, 2008a). These patterns are understandable from the point of view of selenium metabolism and its role in animal organisms; functional (SeCys-containing) proteins have an important role to play in liver and kidney, whereas storage (SeMet-containing) proteins can be found in muscles.

The presence of Se(IV) was detected in egg and milk samples only (Bierla *et al.*, 2008b; Lipiec *et al.*, 2010). The major form of selenium in yolk was SeCys, whereas SeMet dominated in white. Both yolk and white contained a small amount (*ca.* 10%) of selenite (Lipiec *et al.*, 2010). The supplementation led to the predominant incorporation of selenium in the white without discrimination between SeMet and SeCys; however, it did not lead to an increase in Se(IV) in the white but in the yolk (Lipiec *et al.*, 2010). No change was observed in the selenoamino acid pattern during milk processing as shown in Figure 7.5, proving the transfer of selenospecies from the raw material (milk) to the final commercial product (cheese) (Figure 7.5).

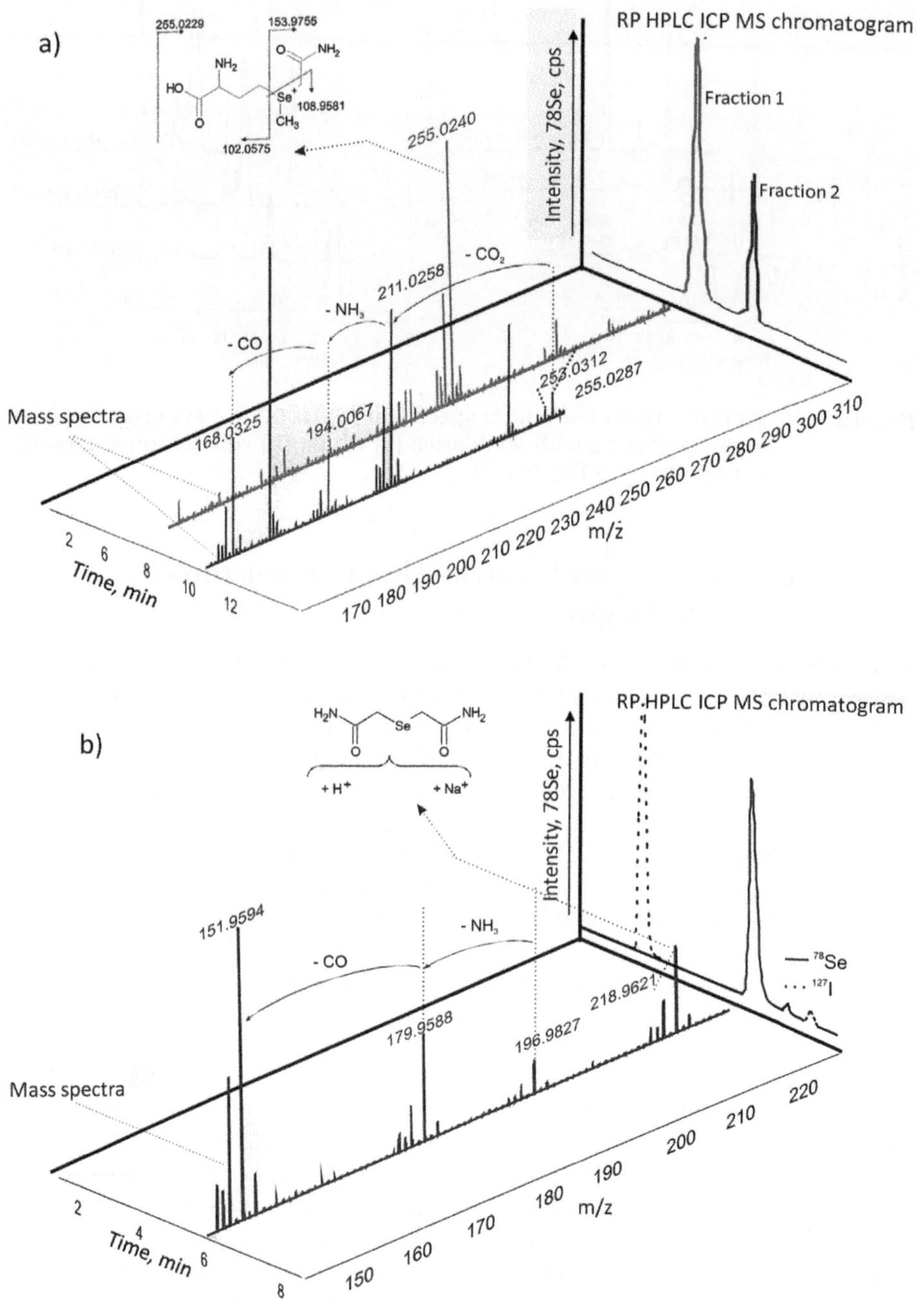

Figure 7.4 Identification of selenium species in RP HPLC-ICP-MS of egg-white proteolysate by ESI-MS; (a) derivatized SeMet, (b) derivatized selenite.

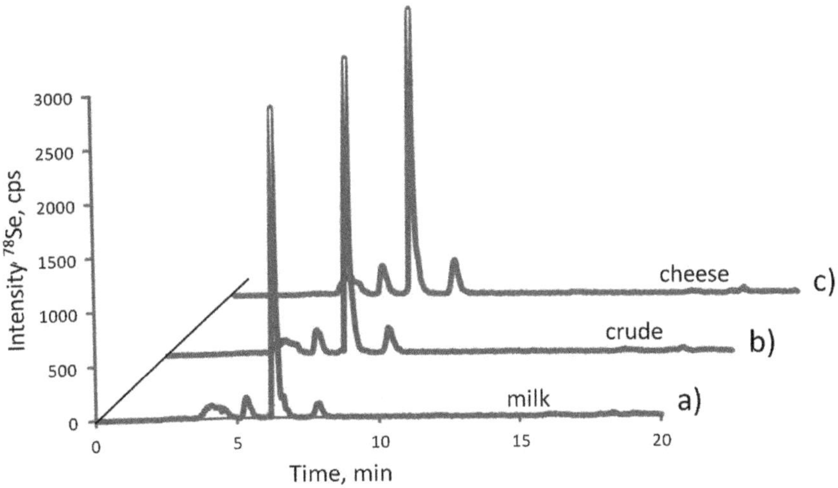

Figure 7.5 Evolution of selenium speciation during cheese production; RP HPLC-ICP-MS chromatograms of (a) milk, (b) crude and (c) cheese proteolysates (chromatographic conditions as in Figure 7.2).

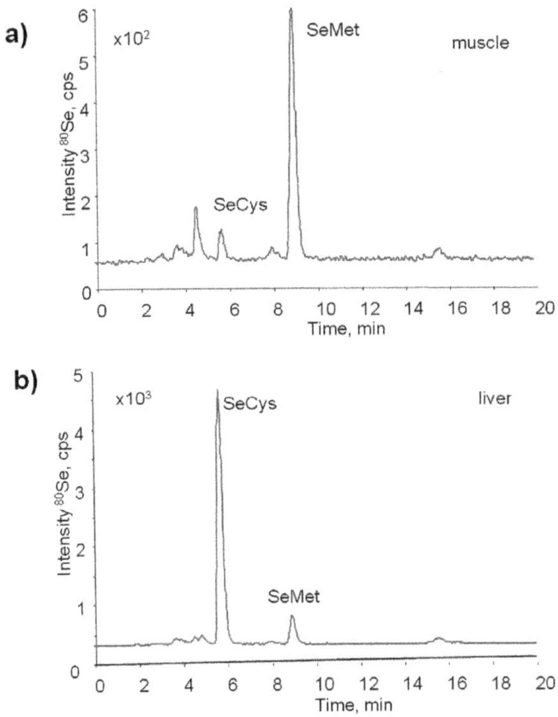

Figure 7.6 RP HPLC-ICP-MS chromatographic patterns of lamb (a) muscle and (b) liver (chromatographic conditions as in Figure 7.2).

7.4 Conclusions

The optimized sample-preparation method (defatting, the protein extraction with urea under reducing conditions (dithiothreitol), derivatization of SeCys by carbamidomethylation with iodoacetamide (IAM), and quantitative proteolysis) allowed a reliable simultaneous determination of SeMet and SeCys in foodstuff of animal origin. The development of quantitative isolation of selenoamino acid fraction by size-exclusion chromatography improved the separation efficiency, peak shape and reproducibility of ion-paring reversed-phase HPLC-ICP-MS opening up a way to a routine method for selenium speciation in a number of different samples. A number of tools contributing to the method validation were introduced. They included the use of ^{77}Se-labeled SeMet to control the SeMet derivatization efficiency and recovery, the use of spiking with a true selenoprotein for the validation of the selenocysteine determination and the mass balance of the Se aminoacids with regard to the total selenium concentration. The detection limits of developed methods are 20, 3 and 60 ng g^{-1} for SeCys and 20, 3 and 10 ng g^{-1} for SeMet for animal tissues, milk and eggs, respectively. These allow quantification of selenoamino acids in samples as usual level of them in animal tissues and eggs is in the low ppm range and around 50 ppb in dry milk.

Summary Points

- The chapter focuses on summarizing the status of selenium and state-of-the-art selenium analysis in food products.
- The procedures developed for quantitative determination of seleno-amino acids in food stuffs are presented.
- The crucial point of the methods is sample preparation prior to analysis.
- The sample-preparation step is strongly related to the protein/carbohydrates/fats/salts content of the sample.
- Two-dimensional chromatography must be applied to achieve satisfactory separation of the compounds of interest.

Key Facts of Selenoaminoacids Determination

1. Selenoaminoacids are selenomethionine and selenocysteine, they are analogs of methionine and cysteine in which the sulfur atom has been substituted by a selenium atom.
2. These are incorporated in the sequence of proteins.
3. Selenomethionine may substitute randomly a methionine in the sequence but has no actual biological activity.
4. Protein with only selenomethionine residues are called selenium-containing proteins.
5. Selenocysteine incorporation into proteins is genetically encoded.
6. Selenocysteine is sometimes referred as the 21st amino acid.
7. Selenocysteine has a specific biological activity through the selenol function that acts as an active site.

8. Only proteins in which a selenocysteine residue is included are called selenoproteins.
9. Selenocysteine is not stable as a free species and must be derivatized in order to be quantified.
10. The selenocysteine concentration is representative of the metabolically active amount of selenium.
11. Selenomethionine and selenocysteine cannot be determined if they are bound to other amino acid in a protein or a peptide.
12. ICP-MS coupled to liquid chromatography allows for a sensitive and specific detection of selenium species.
13. Elemental mass spectrometry is used for quantitative purpose.
14. Molecular mass spectrometry ensures compound identification, especially when standards are not available.

Definitions of Key Terms

Defatting. Removal of fat (interfering with the reactions carried out during the procedure) from the sample by dissolving it in organic solvent without any changes to compounds of interest.

Denaturation. Procedure of destroying the 3rd- and 4th-grade structure of proteins (unfolding) in order to make them more accessible for reaction.

Derivatization. Chemical reaction that aims to produce a compound (derivative), relatively stable in time, to allow its identification and quantification.

Sonication. Use of a cavitation phenomena (by applying ultrasound waves) in order to destroy cells and release their content into solution.

Proteolysis is a digestion procedure using enzymes that cut proteins non-specifically on a peptide bond. As a result, the mixture of individual amino acid is obtained.

Lyophilization (freeze drying) is a method of sample drying assuring preservation of chemical species present. It consists of decreasing pressure around a frozen sample that leads to removal of water in a sublimation process.

Size-exclusion chromatography. Chromatographic method in which separation is based on the size; the biggest compounds are eluting first.

Reversed-phase chromatography. Chromatographic method in which separation is based on the hydrophobic character of the compound.

Ion-pairing chromatography. Chromatographic method in which an ion-pairing agent is added to the mobile phase in order to create ion pairs. Stationary phase and molecules present in the sample are both involved in the formation of these ion pairs. The resulting modification of the chemical behavior of analytes and the stationary phase are the two main principles that dictate the retention obtained, which can be radically different from that of initial molecules, therefore allowing different separation to be obtained.

Two-dimensional chromatography (2D LC). Combining two different chromatographic separation mechanisms to achieve separation, purification of the sample which would not be possible with a single mechanism.

List of Abbreviations

DTT	Dithiothreitol
ESI-MS	Electrospray ionization mass spectrometry
HPLC	High-performance liquid chromatography
GSHPx	Glutathione peroxidase
IAM	Iodoacetoamide
ICP-MS	Inductively coupled plasma mass spectrometry
SEC	Size-exclusion chromatography
SeCys	Selenocysteine
$SeCys_2$	Selenocystine
SeMeCys	Selenomethyl selenocysteine
SeMet	Selenomethionine
RP	Reversed phase

References

Alzate, A. *et al.*, 2007. Evaluation of the inorganic selenium biotransformation in selenium-enriched yogurt by HPLC-ICP-MS. *Journal of Agricultural and Food Chemistry*. 55(24): 9776–9783.

Alzate, A. *et al.*, 2008. Comparison of biotransformation of inorganic selenium by *Lactobacillus* and *Saccharomyces* in lactic fermentation process of yogurt and kefir. *Journal of Agricultural and Food Chemistry*. 56(18): 8728–8736.

Bierla, K. *et al.*, 2008a. Determination of selenocysteine and selenomethionine in edible animal tissues by 2D size-exclusion reversed-phase HPLC-ICP-MS following carbamidomethylation and proteolytic extraction. *Analytical and Bioanalytical Chemistry*. 390(7): 1789–1798.

Bierla, K. *et al.*, 2008b. Specific determination of selenoaminoacids in whole milk by 2D size-exclusion-ion-paring reversed phase high-performance liquid chromatography-inductively coupled plasma mass spectrometry (HPLC-ICP-MS). *Analytica Chimica Acta*. 624(2): 195–202.

Cabañero, A. I. *et al.*, 2005. Quantification and speciation of mercury and selenium in fish samples of high consumption in Spain and Portugal. *Biological Trace Element Research*. 103(1): 17–35.

Cabañero, A. I. *et al.*, 2007. Mercury–selenium species ratio in representative fish samples and their bioaccessibility by an *in vitro* digestion method. *Biological Trace Element Research*. 119(3): 195–211.

Calamari, L. *et al.*, 2010. Effects of either sodium selenite or Se yeast (Sc CNCM I-3060) supplementation on selenium status and milk characteristics in dairy cows. *Livestock Science*. 128(1–3): 154–165.

Chinrasri, O. *et al.*, 2013. Productivity and selenium concentrations in egg and tissue of laying quails fed selenium from hydroponically produced selenium-enriched kale sprout (*Brassica oleracea* var. alboglabra L.). *Biological Trace Element Research*. 155(3): 381–386.

Daun, C. and Akesson, B., 2004a. Comparison of glutathione peroxidase activity, and of total and soluble selenium content in two muscles from chicken, turkey, duck, ostrich and lamb. *Food Chemistry*. 85: 295–303.

Daun, C. and Akesson, B., 2004b. Glutathione peroxidase activity, and content of total and soluble selenium in five bovine and porcine organs used in meat production. *Meat Science*. 66: 81–87.

Daun, C. *et al.*, 2001. Glutathione peroxidase activity, tissue and soluble selenium content in beef and pork in relation to meat ageing and pig RN phenotype. *Food Chemistry*. 73(3): 313–319.

Daun, C. *et al.*, 2004. Separation of soluble selenium compounds in muscle from seven animal species using size exclusion chromatography and inductively coupled plasma mass spectrometry. *Journal of Analytical Atomic Spectrometry*. 19(1): 129–134.

De La Flor St Remy, R. R. *et al.*, 2004. Multielemental distribution patterns in premature human milk whey and pre-term formula milk whey by size exclusion chromatography coupled to inductively coupled plasma mass spectrometry with octopole reaction cell. *Journal of Analytical Atomic Spectrometry*. 19: 1104–1110.

Díaz-Alarcón, J. P. *et al.*, 1996. Determination of selenium in meat products by hydride generation atomic absorption spectrometry—selenium levels in meat, organ meats, and sausages in Spain. *Journal of Agricultural and Food Chemistry*. 44: 1494–1497.

Díaz Huerta, V. *et al.*, 2004. Quantitative selenium speciation in cod muscle by isotope dilution ICP-MS with a reaction cell: comparison of different reported extraction procedures. *Journal of Analytical Atomic Spectrometry*. 19(5): 644–648.

Fairweather-Tait, S. J. *et al.*, 2010. Selenium bioavailability: current knowledge and future research requirements. *American Journal of Clinical Nutrition*. 91(5): 1484S–1491S.

Fisinin, V. I. *et al.*, 2008. Producing specialist poultry products to meet human nutrition requirements: selenium enriched eggs. *World's Poultry Science Journal*. 64(1): 85–97.

Fisinin, V. I. *et al.*, 2009. Producing selenium-enriched eggs and meat to improve the selenium status of the general population. *Critical Reviews in Biotechnology*. 29: 18–28.

Hintze, K. J. *et al.*, 2002. Selenium accumulation in beef: effect of dietary selenium and geographical area of animal origin. *Journal of Agriculture Food Chemistry*. 50(14): 3938–3942.

Jlali, M. *et al.*, 2013. Effect of 2-hydroxy-4-methylselenobutanoic acid as a dietary selenium supplement to improve the selenium concentration of table eggs. *Journal of Animal Science*. 91(4): 1745–1752.

Juniper, D. T. *et al.*, 2008a. Effect of dietary supplementation with selenium-enriched yeast or sodium selenite on selenium tissue distribution and meat quality in commercial-line turkeys. *Animal*. 5(11): 1751–1760.

Juniper, D. T. *et al.*, 2006. Selenium supplementation of lactating dairy cows: effect on selenium concentration in blood, milk, urine, and feces. *Journal of Dairy Science*. 89(9): 3544–3551.

Juniper, D. T. *et al.*, 2008b. Effect of dietary supplementation with selenium-enriched yeast or sodium selenite on selenium tissue distribution and meat quality in beef cattle. *Journal of Animal Science*. 86(11): 3100–3109.

Juniper, D. T. *et al.*, 2008c. Selenium persistency and speciation in the tissues of lambs following the withdrawal of dietary high-dose selenium-enriched yeast. *Animal*. 2(3): 375–380.

Juniper, D. T. *et al.*, 2009. Effect of high dose selenium enriched yeast diets on the distribution of total selenium and selenium species within lamb tissues. *Livestock Science*. 122(1): 63–67.

Keck and Finley, 2004. Cruciferous Vegetables: Cancer Protective Mechanisms of Glucosinolate Hydrolysis Products and Selenium. *Integrative Cancer Therapies*. 3(1): 5–12.

Knowles, S. O. *et al.*, 1999. Significance of amount and form of dietary selenium on blood, milk, and casein selenium concentrations in grazing cows. *Journal of Dairy Science*. 82(2): 429–437.

Lipiec, E. *et al.*, 2010. Determination of selenomethionine, selenocysteine, and inorganic selenium in eggs by HPLC-inductively coupled plasma mass spectrometry. *Analytical and Bioanalytical Chemistry*. 397(2): 731–741.

McNaughton, S. A. and Marks, G. C., 2002. Selenium content of Australian foods: a review of literature values. *Journal of Food Composition and Analysis*. 15: 169–182.

Misra, S. *et al.*, 2012. Tissue-specific accumulation and speciation of selenium in rainbow trout (*Oncorhynchus mykiss*) exposed to elevated dietary selenomethionine. *Comparative Biochemistry and Physiology, Part C: Toxicology & Pharmacology*. 155(4): 560–565.

Moreno, P. *et al.*, 2004. Study of selenium species distribution in biological tissues by size exclusion and ion exchange chromatography inductively coupled plasma-mass spectrometry. *Analytica Chimica Acta*. 524(1–2): 315–327.

Önning, G., 2000. Separation of soluble selenium compounds in different fish species. *Food Chemistry*. 68(2): 133–139.

Önning, G. and Bergdahl, I. A., 1999. Fractionation of soluble selenium compounds from fish using size-exclusion chromatography with on-line detection by inductively coupled plasma mass spectrometry. *Analyst*. 124(10): 1435–1438.

Pan, C. *et al.*, 2011. Effect of selenium-enriched probiotics on laying performance, egg quality, egg selenium content, and egg glutathione peroxidase activity. *Journal of Agricultural and Food Chemistry*. 59(21): 11424–11431.

Payne, R. L. *et al.*, 2005. Effect of inorganic versus organic selenium on hen production and egg selenium concentration. *Poultry Science*. 84(2): 232–237.

Petrera, F. *et al.*, 2009. Effect of either sodium selenite or Se-yeast supplementation on selenium status and milk characteristics in dairy goats. *Small Ruminant Research*. 82(2–3): 130–138.

Ralston, N. V. C. and Raymond, L. J., 2010. Dietary selenium's protective effects against methylmercury toxicity. *Toxicology*. 278(1): 112–123.

Rivero Martino, F. A. *et al.*, 2002. Multi-elemental fractionation in milk whey by size exclusion chromatography coupled on line to ICP-MS. *Journal of Analytical Atomic Spectrometry*. 17: 1271–1277.

Rodica Ani, A. *et al.*, 2010. Organic selenium (Sel-Plex) and its impact on the indices of growth, consumption and meat quality of carp (*Cyprinus carpio*), the Galitian variety. *Animal Science and Biotechnologies*. 43: 1–8.

Scheideler, S. E. *et al.*, 2010. Supplemental vitamin E and selenium effects on egg production, egg quality, and egg deposition of α-tocopherol and selenium. *Journal of Applied Poultry Research*. 19(4): 354–360.

Silva de Lyons, M. (2007). Organic selenium as a supplement for Atlantic Salmon: effects on meat quality, http://en.engormix.com/MA-aquaculture/articles/organic-selenium-supplement-atlantic-t747/p0.htm.

Vignola, G. *et al.*, 2009. Effects of selenium source and level of supplementation on the performance and meat quality of lambs. *Meat Science*. 81(4): 678–685.

Yamashita, Y. *et al.*, 2011. Selenoneine, total selenium, and total mercury content in the muscle of fishes. *Fisheries Science*. 77(4): 679–686.

Yamashita, Y. and Yamashita, M., 2010. Identification of a novel selenium-containing compound, selenoneine, as the predominant chemical form of organic selenium in the blood of bluefin tuna. *Journal of Biological Chemistry*. 285(24): 18134–18138.

Yoshida, S. *et al.*, 2011. Selenium in seafood materials. *Journal of Health Science*. 57(3): 215–224.

Zhang, W. *et al.*, 2010. Improving functional value of meat products. *Meat Science*. 86(1): 15–31.

CHAPTER 8

Some Aspects of Selenium Determination Employing Solid-Phase Extraction and Atomic Absorption Spectrometry

ANA LÚCIA SILVA FIGUEIREDO DE PAIVA[a], AYLA CAMPOS DO LAGO[b], GRAZIELLE CABRAL DE LIMA[c], CÉLIO WISNIEWSKI[d], CRISTIANA SCHMIDT DE MAGALHÃES[d], AND PEDRO ORIVAL LUCCAS*[c]

[a]Centro Superior de Ensino e Pesquisa de Machado/Fundação Educacional de Machado, Av Dr. Athaíde Pereira de Souza, 730, Machado, MG, Brazil, CEP 37750-000; [b]Departamento de Química, Universidade Federal de São Carlos, Rodovia Washington Luís, km 235, São Carlos, SP, Brazil, CEP 13565-905; [c]Instituto de Química, Universidade Federal de Alfenas, Rua Gabriel Monteiro da Silva, 714, Alfenas, MG, Brazil, CEP 37130-000; [d]Instituto de Ciências Exatas, Universidade Federal de Alfenas, Rua Gabriel Monteiro da Silva, 714, Alfenas, MG, Brazil, CEP 37130-000
*E-mail: pedro.luccas@unifal-mg.edu.br

Food and Nutritional Components in Focus No. 9
Selenium: Chemistry, Analysis, Function and Effects
Edited by Victor R. Preedy

Published by the Royal Society of Chemistry, www.rsc.org

8.1 Introduction

Selenium is a trace element that was first isolated in 1817 by the Swedish chemist, Jöns Jacob Berzelius. It was initially recognized for its toxicity, but has gained prominence due to its importance as a structural component of many enzymes required for homeostasis (Mehdi *et al.*, 2013).

This element is a semimetal of the chalcogenide group (6A), with atomic number 34 and molecular weight, 78.96 g mol^{-1}. At room temperature, it is solid and has six stable isotopes: ^{74}Se (natural fraction, 0.89%), ^{76}Se (9.37%), ^{77}Se (7.63%), ^{78}Se (23.77%), ^{80}Se (49.61%) and ^{82}Se (8.73%). It has four oxidation states: selenides (−2), elemental selenium (0), selenite (+4) and selenate (+6). Se is used as a semiconductor in solar cells, possessing the property of having electrons excited when light is focused on it. Monocyclic and vitreous Se, also called black Se, are their solid forms. The latter is used as a photoreceptor in xerographic photocopying process (Wake, 2004), see Table 8.1.

The oxyanions: selenite (SeO_3^{2-}) and selenate (SeO_4^{2-}) are the best known inorganic forms of the element (Table 8.1). When it is in organic form, it is present as selenoamine acids, the best known being selenocysteine, selenomethionine, selenoethionine and dimethylselenide compounds (Wake, 2004).

8.1.1 Se species and Concentrations in Different Samples

Se is found in the earth's crust, often associated with minerals containing sulfur. There are several Se species: selenides, elemental selenium, selenite and selenate. The concentration varies greatly from one geographical area to another, with the highest concentrations found in soils of volcanic origin. This element is found as HSe^- in air, metabolically produced by plants and, also, in the forms of elemental selenium, selenites and selenates. In urban areas, the concentration range is from 0.1 to 10 ng m^{-3} (WHO, 2011). Figure 8.1 shows the selenium biogeochemical cycle.

The Se level in the bloodstream is related to Se food intake. In the human body, there are about 10 mg of Se per 60 kg body weight. The total Se concentration found is 0.076 to 0.14 mg L^{-1} in the blood, in normal conditions. It is 0.05 to 0.15 mg L^{-1} in the serum/plasma and from 0.002 to 0.03 mg L^{-1} in the urine. The toxic concentration is 0.4 mg L^{-1}, in serum/plasma and urine, and it is lethal above the concentration of 1 mg L^{-1} in blood, and 2 mg L^{-1} in urine (Repetto and Repetto, 2005).

An important aspect that should be taken into consideration is the Se species present in each sample. In the bloodstream, there are the Se proteins. The serum Se protein glutathione peroxidase (GPx3) is an enzyme used as a biological marker. Its activity measurement can be correlated with the total Se level in blood. It has the maximal activity when the Se concentration in blood reaches values of about 100 ng mL^{-1}, which corresponds to the Se daily intake of 70 mg. Another serum marker is the P selenoprotein (SelP), which reaches its maximum expression when the Se level in the blood is about 124 ng mL^{-1}, which corresponds to the Se diet of 105 mg per day (Hurst *et al.*, 2010).

Table 8.1 Most commonly found selenium species.[a]

Name	Abbreviation	Structure
Selenite	Se(IV)	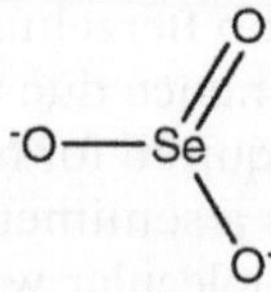
Selenate	Se(VI)	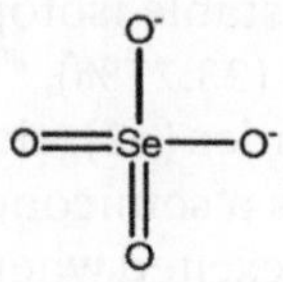
Selenomethionine	SeMet	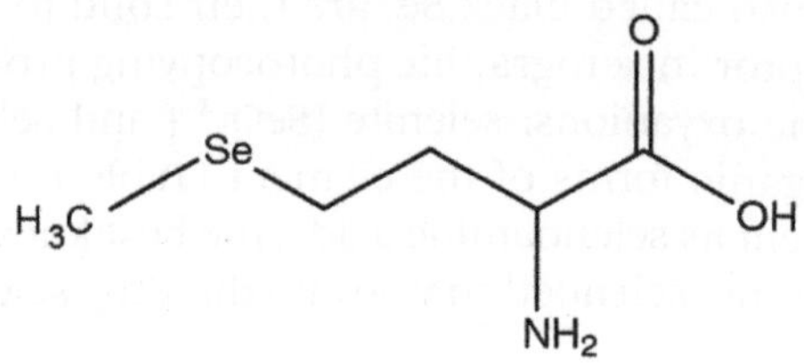
Selenocysteine	SeCys	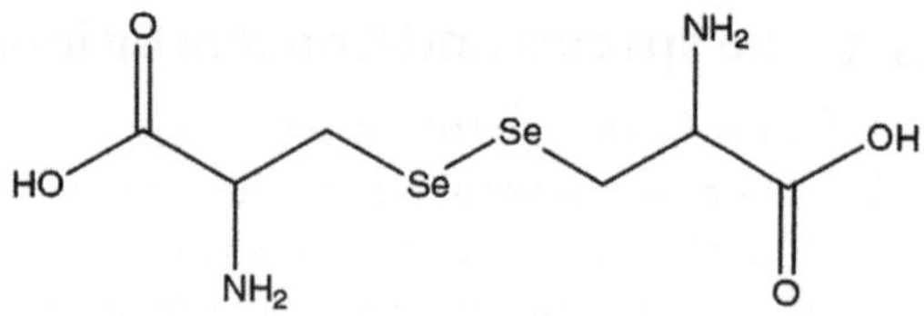
Selenoethionine	SeEt	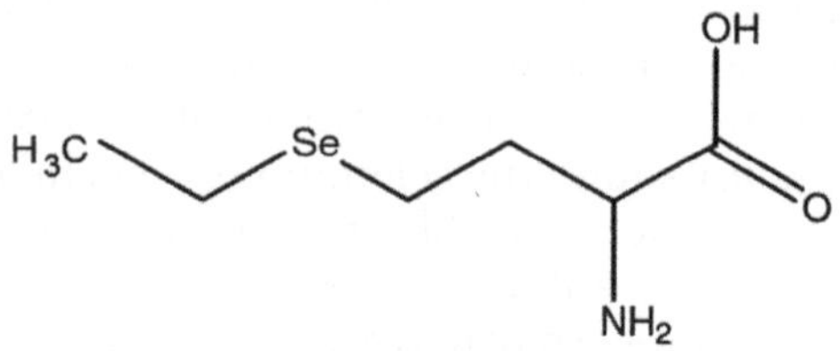
Dimethylselenide	DMSe	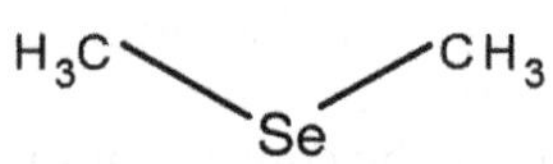

[a]Source: adapted from Wake (2004). Modern preconcentration methods for the selenium species determination in environmental water samples.

Se forms, found in urine, depend on the kind and amount of ingested Se. TMSe is the main urinary metabolite found in cases of dietary supplementation or toxic doses. Under the same conditions, DMSe is excreted *via* the respiratory route. In the case of ingestion below the toxic doses, the selenosugar Se-methyl-*N*-acetyl-galactosamine (MSeGalNAc) and minor amounts

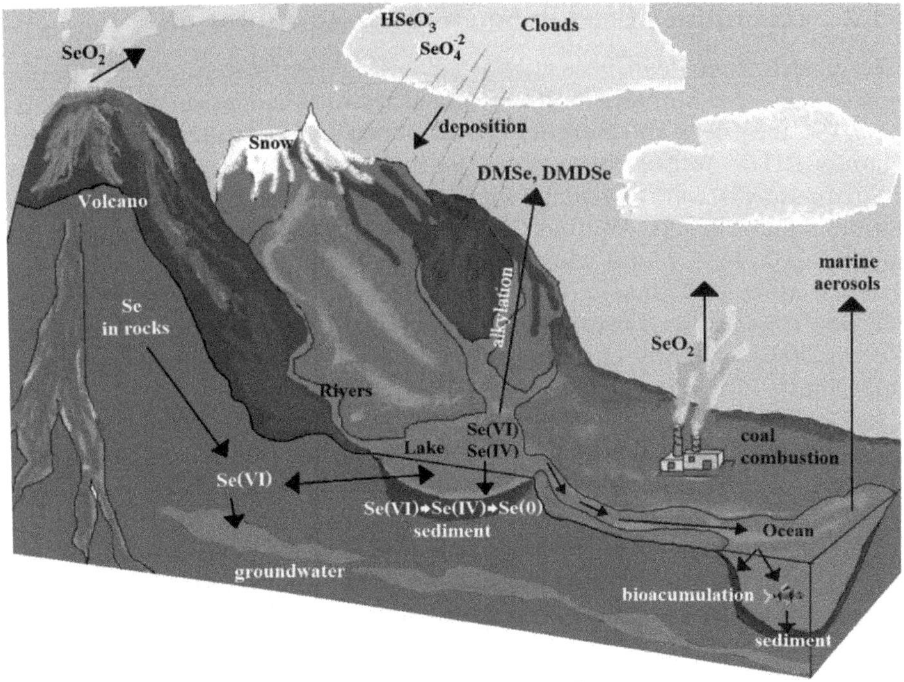

Figure 8.1 Selenium geochemical cycle.

of Se-methyl-*N*-acetyl-glucosamine (MSeGluNAc) and Se-methyl-amino-*N*-galactosamine (MSeGalNH2) are found in urine (Letsiou *et al.*, 2007).

The main Se species found in freshwater are selenite and selenate. Selenite and selenate sodium predominate in surface waters (Mehdi *et al.*, 2013). The total selenium maximum acceptable limit for human consumption is 50 mg L^{-1} (WHO, 2011).

The major Se intake source is dietary. The total Se amount varies greatly from one to another type of food, according to the soil conditions and geographical location. The concentration in bovine flesh is closely related to cattle feed.

The main Se species found in plants are SeMet, selenite, selenate and, in lower concentrations, SeCys, SeMCys and GGSeMCys (Roman *et al.*, 2014). The higher Se intakes come from cereals (0.1 to 10 mg kg^{-1}), as they tend to be ingested in large quantities. Garlic and onion plants may accumulate significant Se amounts, 133 mg kg^{-1} and 5 mg kg^{-1}, respectively. Mushrooms and broccoli, 248 mg kg^{-1} and 16 mg kg^{-1}. However, Brazil nuts represent the richest natural Se source, at about 19 mg kg^{-1} (USDA, 2012). Another great Se source is meat and seafood, as they are rich in proteins. The most common species are selenite, selenate, SeMet, SeCys in them. The total Se concentration is in the range of 0.3 to 0.5 mg kg^{-1}. In viscera, liver and kidney, the concentrations may exceed 1500 mg kg^{-1} (WHO, 2011).

8.1.2 Techniques Used in Selenium Determination

Several techniques can be used for selenium determination in different matrices, as an example, graphite furnace atomic absorption spectrometry (GF-AAS). Burguera and Burguera (2007), reviewed online settings for this technique. The authors have given the technical characteristics based on robustness, relatively low-cost analysis, and sensitivity for metal determination in μg L^{-1} levels. Another GF-AAS review for selenium determination applied to human blood samples was written by Sabé *et al.* (2000). The authors concluded that it is very sensitive. However, the pyrolysis temperature must be well established in order to avoid analyte loss. A modifier must be properly chosen, and the best results were obtained with palladium.

Inductively coupled plasma with mass spectrometry, atomic emission spectrometry with inductively coupled plasma, ultraviolet-visible spectrophotometry, X-ray fluorescence and voltammetry, have also been used in the selenium determination.

The analytical separation techniques are used with the main objective of promoting increased sensitivity and/or cleanup of the complex samples. After the sample undergoes separation, it is inserted at a specific detector. Therefore, separation, detection and the interface between them should be optimized.

There are several examples of hyphenated techniques for the selenium determination. Couplings of liquid-chromatography techniques with high-efficiency emission, atomic absorption or mass spectrometry and fluorescence detectors are frequently found. The coupling of high-performance liquid chromatography with mass spectrometry with inductively coupled plasma is quite common. An example is the work of Li *et al.* (2007), which determined organic and inorganic selenium compounds in urine samples from patients who ingested selenium. The limit of detection (LOD) varied from 0.05 to 1.48 μg L^{-1}.

Other hyphenated technique for selenium determination has been found by coupling capillary electrophoresis (CE) with direct detection in a UV-visible spectrometer.

The technique of hydride generation coupled to F-AAS, which was widely reported in the literature for selenium determination, resulted in high sensitivity in the range from 0.065 to 10.6 μg L^{-1}. In this technique, sodium borohydride in acidic media was used for the Se(IV) reduction in selenium hydride. This gas was atomized in a quartz tube and placed in the F-AAS equipment. As the purpose of the technique was the determination of total selenium, a step of reduction of Se(VI) to (IV) was necessary, since only selenite was converted into selenium hydride. For this reduction, HCl 5 mol L^{-1} at 95 °C for 20 min was used (Matusiewicz and Krawczyk, 2006).

If the procedure objective for carrying out is selenium speciation, the sample analysis is performed without the pretreatment described above. Therefore, Se(IV) is first determined. Next, the sample undergoes the above-mentioned treatment when the total selenium reading occurs, and the Se(VI) concentration is the difference.

Milk is an example of food containing high levels of organic compounds and its analysis usually involves digestion procedures that may be affected by losses and contamination. Aleixo and Nóbrega (2003), developed a procedure for the Se determination in bovine milk. The samples were diluted with a mixture of tertiary amine soluble in water, in order to improve the method repeatability. Using this strategy, the milk direct analysis was feasible using GF-AAS, without any digestion procedure, and palladium was used as Se chemical modifier. The pyrolysis and atomization temperatures were 1500 °C and 2400 °C, respectively. Se concentrations were found from 17 to 122 μg L^{-1}, in thirteen samples, being the highest concentration in a sample from Uruguay. The LOD varied depending on the type of milk. In samples of skim milk, the value was 1.2 μg L^{-1}, low-fat milk, 2.4 μg L^{-1} and in whole-milk, the LOD was 2.5 μg L^{-1}.

Se was analyzed in multivitamins by high-resolution continuum-source graphite furnace atomic absorption spectrometry (HR-CS-GFAAS) with slurry sampling. Cr, Cu, Fe, Mn, and Se concentrations were determined and compared to the amounts stated by the producers. The LOD for selenium slurry suspension was 0.06 μg g^{-1} (Krawczyk, 2014). Another study evaluated selenium levels in the plasma of Australian blood donors. Selenium levels were measured using GF-AAS with Zeeman background correction. The average selenium level in the donor's plasmas was 85.6 ± 0.5 μg L^{-1}, and the LOD 23.7 μg L^{-1}. Although the Se levels were found to be within the normal concentration range, 88.5% of the samples had it below 100 μg L^{-1}, a level that has been associated with suboptimal activity of the antioxidant enzyme glutathione peroxidase (McDonald *et al.*, 2013).

8.1.3 Solid-Phase Extraction (SPE) to Determine Se

The SPE technique is based on the extraction principle of an analyte retained on the solid-phase adsorbent for liquid or gaseous phase. The solid-phase extraction remedied the disadvantages offered by liquid–liquid extraction, such as excessive use of solvents, time-consuming extractions, and high costs.

The number of published papers using the SPE technique for selenium determination has tripled in the last decade (Figure 8.2). According to Wake (2004), SPE is among the most used preconcentration methods for selenium determination. The authors state that the main advantages of this technique are the high preconcentration factors, leading to lower LODs, easy automation and the possibility of online engagement. Table 8.2 presents some published works on SPE for selenium determination.

Sorption online separations are always carried out in columns. These are made by direct preconcentration flow coupling to specific detector equipment. Most papers employed online preconcentration to F-AAS. The first contribution of this coupling type occurred in 1983. This coupling is relatively simple, once the flame technique works with streams (Olsen *et al.*, 1983).

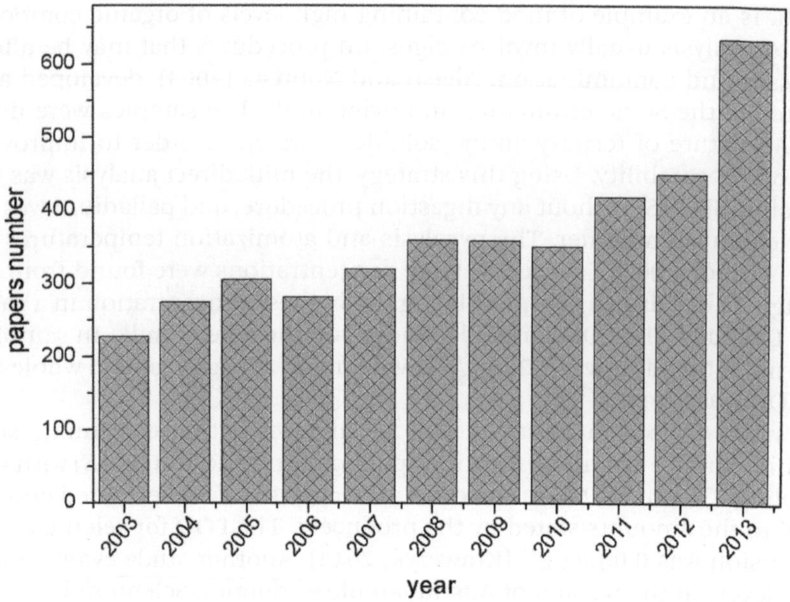

Figure 8.2 Number of published papers using the SPE technique for selenium determination in the last decade. Source: http://sciencedirect.com, key words: solid-phase extraction and selenium.

8.1.3.1 Flow Preconcentration Coupled with GF-AAS

The coupling of flow preconcentration with a graphite furnace is more complex. This takes place in a discontinuous, parallel and synchronized way with the atomization process. Another challenge is the low sample volume that must be used, being limited to a maximum of 100 μL. However, better LODs are achieved by this technique, when it is compared to F-AAS.

The first online preconcentration with GF-AAS work was published in 1990. Two peristaltic pumps automatically controlled by a computer, Tygon and PTFE tubes, a five-channel injector and a filled C_{18} column were used. Five stages were necessary from preconcentration until complete elution into the graphite tube. However, a manual step was used to insert the elution capillary into the graphite tube, before the preconcentration system starts (Fang *et al.*, 1990).

Finally, three published articles have demonstrated a preconcentration system, directly coupled to a GF-AAS detector. Queiroz *et al.* (2002), published a study, for copper determination, where the inline term was used to refer to this way of coupling. Our group published a similar study (Paiva *et al.*, 2010), where we used a minicolumn containing resin ion exchange, LCSAX, in-line coupling to GF-AAS for the Se determination. The method was suitable for Se determination/speciation in water samples. Some advantages of the inline coupling are: reduced use of reagents, increasing sampling frequency, low risk of sample contamination and loss of analyte.

Table 8.2 Published works using SPE and selenium detection by GFAAS.[a]

Species	Sorbent	Technique	LOD ($\mu g\ L^{-1}$)	Sample	References
Se(IV, VI) and total Se	Activated carbon	Spectrophotometry	0.012	Natural water	
Se(IV)	Nano-TiO_2	HG-AFS	0.0006	Waste water and metallurgical samples	Deng *et al.* (2012)
Total Se	Fiber coating	HG-HS-SPME-ICP-OES	0.8	Yeast	Tyburska *et al.* (2011)
Se(IV, VI)	Cellulose fiber with $La(OH)_3$	HG-AFS	0.009	Reference material	Chen *et al.* (2009)
Se(0)	Imprinted polymer (4-VP and VI)	HG-FAAS	0.053	Food sample	de Lima *et al.* (2013)
Se(IV)	Co-precipitated with $La(OH)_3$	HG-AFS	0.005	Surface waters	
Se(IV, VI)	Polypropylene fiber membrane	HF-LPME-GF-AAS	0.005 and 0.006	Environmental waters and soil	Ghasemi *et al.* (2010)
Se(IV, VI)	Activated alumina	GF-AAS	0.8 and 49.0	Tap and ground water	
Se(IV, VI)	(CTAB)-modified alkyl silica	ICP-OES	0.10	Natural water	Xiong *et al.* (2008)
Se(IV, VI) and selenocyanate	Anion exchange	HG-ICP-DRC-MS	0.00015 0.00027 and 0.00019	Sea and rain water	
Se(IV, VI)	Nano TiO_2	Ion chromatography-conductivity detection	800 and 400	Water and drink water	Xu *et al.* (2012)
Se(IV, VI)	Dowex 1X2	GF-AAS	0.0056	Ground water	
Se(IV, VI)	Dowex 1X8	GF-AAS	0.01	Drinking water	
Se(IV)	Amberlite CG-400 modified	GF-AAS	0.05	Natural water	
Se(IV, VI)	SAX	GF-AAS	0.031 and 0.049		Paiva *et al.* (2010)
Se(IV, VI)	SAX	GC-MS	0.0014 and 0.0016	Natural water	
Se(IV, VI)	Silica gel modified	HG-AAS	Se(IV): 0.8	Sea water	
Se(IV, VI)	Chitosan modified with DCCTS	ICP-OES	Se(VI): 0.012	Environmental water	Dai *et al.* (2011)

[a]LOD: limit of detection; HG: hydride generation, AFS: atomic fluorescence spectrometry; HS-SPME: headspace solid-phase microextraction; F-AAS: flame atomic absorption spectrometer, HF-LPME: hollow-fiber liquid-phase microextraction; GF: graphite furnace; CTAB: cetyltrimethylammonium bromide; ICP-OES: inductively coupled plasma-optical emission spectrometry, HG-ICP-DRC-MS: hydride generation-inductively coupled plasma-dynamic reaction cell-mass spectrometry; DCCTS: diethylene triamine; GC-MS: gas chromatography-mass spectrometry; 4VP: 4 vinylpyridine and VI: 1-vinylimidazole; SAX: strong anion exchanger.

The interface in this type of coupling occurs between the arm of the atomic absorption spectrometer autosampler and a polyethylene tube of FIA (flow injection analysis) system. The FIA system also be used to inject the sample into graphite furnace (Yan *et al.*, 1999).

SPE coupling to other detectors is also common. This coupling has been carried out by Xiong *et al.* (2008), employing modified silica with cetyltrimethylammonium bromide to ICP-OES. Se(VI) was determined after adsorption into columns. Total Se was determined after complete Se(IV) oxidation to Se(VI) with $KMnO_4$, and Se(IV) by the difference between the total Se and (VI). The selenite LOD for the raw water sample was 0.10 $\mu g\ L^{-1}$.

Yu *et al.* (2004), have achieved the LOD of 0.007 $\mu g\ L^{-1}$ to determine Se inorganic species in water samples. The SPE to ICP-MS coupling was carried out using a nonpolar C_{18} column. The selenite complex pyrrole ammonium dithiocarbamate was added to the sample. It was preconcentrated, and the effluent containing the selenate was conveyed for a reading. The total Se was determined using a sample aliquot, which did not undergo a preconcentration step. The selenite determination was performed by the difference in concentration between the total Se and Se(VI).

The principal separation mechanisms in SPE are: adsorption, partition, size exclusion and ion exchange. The latter is used for separation/preconcentration of analytes in their ionic form and usually the adsorbent possesses cationic or anionic groups attached to their surface.

Summary Points

- This chapter reports on the techniques for the determination of selenium and adsorbents used in SPE.
- Selenium acts in the human body as an antioxidant.
- The technique of hydride generation coupled to F-AAS has high sensitivity.
- The coupling of online preconcentration at the GF-AAS is more complex, but has a lower detection limit than the F-AAS.

Key Facts: Analytical Aspects in Selenium Determination

1. Selenium and its species have been found in volcanic soils.
2. Selenium is present in different matrices (blood, urine, plasma, serum), at different concentrations.
3. GPx3 and SelP are protein and enzymatic biological markers, respectively.
4. The greater sources of selenium intake are cereals, garlic, onions, Para's nuts, meat.
5. Selenium determination by GF-AAS is viable due to low analysis cost, sensibility, and direct selenium concentration determination in blood.

6. Milk samples can be directly introduced in GF-AAS after a simple treatment with a tertiary amine.
7. Selenium analysis can be performed by HR-CS-GFAAS.
8. SPE is the most used preconcentration technique for selenium determination, showing a high preconcentration factor, low limit of detection and hyphenation to FIA systems.
9. The preconcentration in flux coupled to GF-AAS is the most complex, however, it shows the inherent advantages to SPE and FIA systems, such as high reading frequency, lower reagents and samples consuming.
10. Selenium speciation is possible when using an ionic-change resin LCSAX, and coupling with GF-AAS.

Definitions of Words and Terms

Graphite furnace atomic absorption spectrometry (GF-AAS). This is a technique used to qualitatively determine metals, based on radiation absorption by excited atoms in a carbon furnace-type atomizer.

Inductively coupled plasma. This is a multielemental technique based on radiation emission by flame excited atoms. The analyte of interest is dispersed in a highly ionized plasma, which helps to maintain the high flame temperature.

Ultraviolet-visible spectrophotometry. This measures the electromagnetic radiation emitted by samples in the visible and ultraviolet regions.

X-ray fluorescence. This is a technique to qualitatively and quantitatively analyse elements in multielemental samples. It is based on sample compounds ionization by the photoelectric effect, using an X-ray source. When the excited atoms return to the fundamental level, they release energy emitting photons.

Voltammetry. This is an electrochemical technique for analytical determination, which measures the redox potential in an electrochemical cell when a potential is applied to the sample.

Liquid chromatography. This allows the compound separation of a mixture through a migration of them between the mobile and stationary phases.

Mass spectrometry. This is a technique that allows the ionized atom determination based on the mass/charge ratio.

Capillary electrophoresis (CE). This allows the separation of two solutes, according to their electrophoretic mobility differences, under the influence of an electric field.

Flame atomic absorption spectrometry (F-AAS). This allows metal determination by measuring the electromagnetic energy absorbed, when the analyte is atomized in a flame.

High-resolution continuum source (HC-CS). This is a type of background correction that uses a deuterium arc lamp as a continuum source. The primary source and continuum radiations pass alternatively through the atomizer, and after this reaching the detector.

Background. This is also called the background signal. It is the sum of all matrix interference contributions, sample injection system or anything else that may affect the analytical signal.

Zeeman correction. This is a background system based on applying a magnetic field that allows the atomic line splitting.

Solid-phase extraction (SPE). The analyte, present in an aqueous matrix, is extracted from it after passing through a column filled with sorbent material. Subsequently, a selective eluent passes through the column, extracting the analyte.

Flow-injection analysis (FIA). This is based on inserting the sample into a flux, properly a carrier, that transports it to the detector to measure the analytical signal.

List of Abbreviations

AFS	Atomic fluorescence spectrometry
C_{18}	Octadecyl column
CE	Capillary electrophoresis
CTAB	Cetyltrimethylammonium bromide
DCCTS	Diethylene triamine
DMSe	Dimethylselenide
F-AAS	Flame atomic absorption spectrometer
FIA	Flow injection analysis
GC-MS	Gas chromatography-mass spectrometry
GF-AAS	Graphite furnace atomic absorption spectrometry
G	Gram
GGSeMCys	γ-Glutamyl-Se-methylselenocysteine
GPX3	Glutathione peroxidase
HF-LPME	Hollow-fiber liquid-phase microextraction
HG	Hydride generation
HG-ICP-DRC-MS	Hydride generation-inductively coupled plasma-dynamic reaction cell-mass spectrometry
HR-CS-GFAAS	High-resolution continuum-source graphite furnace atomic absorption spectrometry
HS-SPME	Headspace solid-phase microextraction
ICP-MS	Inductively coupled plasma mass spectrometry
ICP-OES	Inductively coupled plasma-optical emission spectrometry
LCSAX	Liquid column strong anion exchanger
LOD	Limit of detection
MSe	Methylselenide
MSeGalNAc	SeMethyl-*N*-acetyl-galactosamine
MSeGalNH2	SeMethyl-amino-*N*-galactosamine
PTFE	Polytetrafluoroethylene

SAX	Strong anion exchanger
SeCys	Selenocysteine
SeEt	Selenoethionine
SelP	P selenoprotein
SeMCys	Se-methyl-selenocysteine
SeMet	Selenomethionine
SPE	Solid-phase extraction
TMSe	Trimethylselenide
USDA	United States Department of Agriculture
VI	1-Vinylimidazole
WHO	World Health Organization
4VP	4-Vinylpyridine

References

Aleixo, P. C. and Nóbrega, J. A., 2003. Direct determination of iron and selenium in bovine milk by graphite furnace atomic absorption spectrometry. *Food Chemistry*. 83: 457–462.

Burguera, M. and Burguera, L. J., 2007. On-line electrothermal atomic absorption spectrometry configurations: recent developments and trends. *Spectrochimica Acta, Part B: Atomic Spectroscopy*. 62: 884–896.

Chen, M., Yang, T. and Wang, J., 2009. Precipitate coating on cellulose fibre as sorption medium for selenium preconcentration and speciation with hydride generation atomic fluorescence spectrometry. *Analytica Chimica Acta*. 631: 74–79.

Dai, J., Ren, F. L., Tao, C. Y. and Bai, Y., 2011. Synthesis of cross-linked chitosan and application to adsorption and speciation of Se (VI) and Se (IV) in environmental water samples by inductively coupled plasma optical emission spectrometry. *International Journal Molecule Science*. 12: 4009–4020.

Deng, D., Zhou, J., Ai, X., Yang, L., Hou, X. and Zheng, C., 2012. Ultrasensitive determination of selenium by atomic fluorescence spectrometry using nano-TiO_2 pre-concentration and in situ hydride generation. *Journal of Analytical Atomic Spectrometry*. 27: 270–275.

Fang, Z., Sperling, M. and Welz, B., 1990. Flow injection on-line sorbent extraction preconcentration for graphite furnace atomic absorption spectrometry. *Journal of Analytical Atomic Spectrometry*. 5: 639–646.

Hurst, R., Armah, C. N., Dainty, J. R., Hart, D. J., Teucher, B., Goldson, A. J., Broadley, M. R., Motley, A. K. and Fairweather-Tait, S. T., 2010. Establishing optimal selenium status: results of a randomized, double-blind, placebo-controlled trial. *American Journal Clinical Nutrition*. 91: 923–931.

Krawczyk, M., 2014. Determination of macro and trace elements in multivitamin dietary supplements by high-resolution continuum source graphite furnace atomic absorption spectrometry with slurry sampling. *Journal of Pharmaceutical and Biomedical Analysis*. 88: 377–384.

Letsiou, S., Nischwitz, V., Traar, P., Francesconi, K. A. and Pergantis, S. A., 2007. Determination of selenosugars in crude human urine using high-performance liquid chromatography/atmospheric pressure chemical ionization tandem mass spectrometry. *Mass Spectrometry*. 21: 343–351.

Li, Yu-Feng, Chen, C., Li, B., Wang, Q., Wang, J., Gao, Y., Zhao, Y. and Chai, Z., 2007. Simultaneous speciation of selenium and mercury in human urine samples from long-term mercury-exposed populations with supplementation of selenium-enriched yeast by HPLC-ICP-MS. *Journal of Analytical Atomic Spectrometry*. 22: 925–930.

de Lima, G. C., do Lago, A. C., Chaves, A. A., Fadini, P. S. and Luccas, P. O., 2013. Determination of selenium using atomically imprinted polymer (AIP) and hydride generation atomic absorption spectrometry. *Analytica Chimica Acta*. 768: 35–40.

Matusiewicz, H. and krawczyk, M., 2006. On-line hyphenation of hydride generation with in situ trapping flame atomic absorption spectrometry for arsenic and selenium determination. *Analytical Science*. 22: 249–253.

McDonald, C., Colebourne, K., Faddy, H. M., Flower, R. and Fraser, J. F., 2013. Plasma selenium status in a group of Australian blood donors and fresh blood components. *Journal of Trace Elements in Medicine and Biology*. 27: 352–354.

Mehdi, Y., Hornick, J.-L., Istasse, L. and Dusfrasne, I., 2013. Selenium in the environment, metabolism and involvement in body functions. *Molecules*. 18: 3292–3311.

Olsen, S., Pessenda, L. C. R., Ruzicka, J. and Hansen, E. H., 1983. *Analyst*. 108: 905–917.

Paiva, A. L. S., Costa, F. F. A., Lima, G. F., Wisniewski, C., Tarley, C. R. T. and Luccas, P. O., 2010. Preconcentration and speciation of Se with flow injection analysis hyphenated in line to ETAAS. *Journal Flow Injection Analysis*. 27: 140–145.

Queiroz, Z. F., Rocha, F. R. P., Knapp, G. and Krug, F. J., 2002. Flow system with in-line separation/preconcentration coupled to graphite furnace atomic absorption spectrometry with W-Rh permanent modifier for copper determination in seawater. *Analytica Chimica Acta*. 463: 275–282.

Repetto, M. R. and Repetto, M., 2005. *Tabla de Concentraciones de xenobioticos en fluidos biologicos humanos como referencia para el diagnostico toxicologico*. Instituto Nacional de Toxicologia y Ciencias Forenses. Area de Toxicologia de la Universidad de Sevilla, España.

Roman, M., Jitaru, P. and Barbante, C., 2014. Selenium biochemistry and its role for human health. *Metallomics*. 6: 25–54.

Sabé, R., Rubio, R. and García-Beltrán, L. G., 2000. Determination of selenium in human blood specimens by electrothermal atomic absorption. *Analytica Chimica Acta*. 419: 121–135.

Tyburska, A., Jankowski, K. and Rodzik, A., 2011. Determination of arsenic and selenium by hydride generation and headspace solid phase microextraction coupled with optical emission spectrometry. *Spectrochimica Acta Part B*. 66: 517–521.

United States Department of Agriculture (USDA), 2012. National nutrient database for standard reference. Avaiable at: http://ndb.nal.usda.gov/. Accessed 20 January 2014.

Wake, B. D., 2004. Modern preconcentration methods for the determination of selenium species in environmental water samples. *Trends in Analytical Chemistry*. 23: 491–500.

World Health Organization (WHO), 2011. Guidelines for drinking-water. Available at: http://www.who.int/water_sanitation_health/dwq/chemicals/selenium/en/. Accessed 16 January 2014.

Xiong, C., He, M. and Hu, B., 2008. On-line separation and preconcentration of inorganic arsenic and selenium species in natural water samples with CTAB-modified alkyl silica microcolumn and determination by inductively coupled plasma-optical emission spectrometry. *Talanta*. 76: 772–779.

Xu, S., Zheng, M., Zhang, X., Zhang, J. and Lee, Y., 2012. Nano TiO_2-based preconcentration for the speciation analysis of inorganic selenium by using ion chromatography with conductivity detection. *Microchemical Journal*. 101: 70–74.

Yan, X., Sperling, M. and Welz, B., 1999. On-line coupling of flow injection microcolumn separation and preconcentration to electrothermal atomic absorption spectrometry for determination of (ultra)trace selenite and selenate in water. *Analytical Chemistry*. 71: 4353–4436.

Yu, C., Cai, Q., Guo, Z.-X., Yanga, Z. and Khoo, S. B., 2004. Simultaneous speciation of inorganic selenium and tellurium by inductively coupled plasma mass spectrometry following selective solid-phase extraction separation. *Journal of Analytical Atomic Spectrometry*. 19: 413–419.

CHAPTER 9

Identification of Selenometabolites in Animals and Plants by Mass Spectrometry

YASUMITSU OGRA*[a]

[a]Department of Toxicology and Environmental Health, Graduate School of Pharmaceutical Sciences, Chiba University, 1-8-1 Inohana, Chuo, Chiba 260-8675, Japan
*E-mail: ogra@chiba-u.jp

9.1 Analytical Techniques for the Detection and Identification of Selenometabolites

9.1.1 Inductively Coupled Argon Plasma Mass Spectrometry and Speciation

The inductively coupled plasma (ICP) is a high-energy ionization source, *i.e.*, the plasma normally operates at 5000–10 000 K at atmospheric pressure. Thus, ICP is a hard ionization source for the generation of ions of elemental atoms, and the generated ions are sampled by a mass spectrometer. Single-quadrupole (MS), triple-quadrupole (MS-MS), time-of-flight (TOF-MS), and sector-field (SF-MS) mass spectrometers have been employed with ICP. ICP equipped with a single-quadrupole mass spectrometer (hereinafter,

Food and Nutritional Components in Focus No. 9
Selenium: Chemistry, Analysis, Function and Effects
Edited by Victor R. Preedy

Published by the Royal Society of Chemistry, www.rsc.org

ICP-MS) is the most widely used among the ICP-mass spectrometers. ICP-MS is a powerful tool for elemental detection because of its multielement detection capability, low detection limit, ability to measure isotope ratio, and robustness to sample matrix. Hence, ICP-MS is used not only for Se detection but also Se speciation analysis.

The term "speciation" of an element (metal or metalloid) is defined as the distribution of an element among defined chemical species in a system, whereas speciation analysis is defined as the analytical activities undertaken to identify and/or measure the quantities of one or more chemical species in a sample (Templeton *et al.*, 2000). Se is transformed into Se-containing compounds having carbon–Se covalent bond(s), *i.e.*, selenometabolites, in the metabolic pathway. To reveal the entire metabolic pathway of Se, selenometabolites have to be separated by various separation methods and then detected with a Se-specific detector. As mentioned above, ICP-MS is the instrument of choice for the detection of Se. As the separation technique for selenometabolites, the most frequently used is HPLC because it has the inherent advantages of good separation and easy connection to an ICP-MS without the need for a special interface. Thus, ICP-MS can be used as the detector of HPLC. A schematic diagram is depicted in Figure 9.1A. As

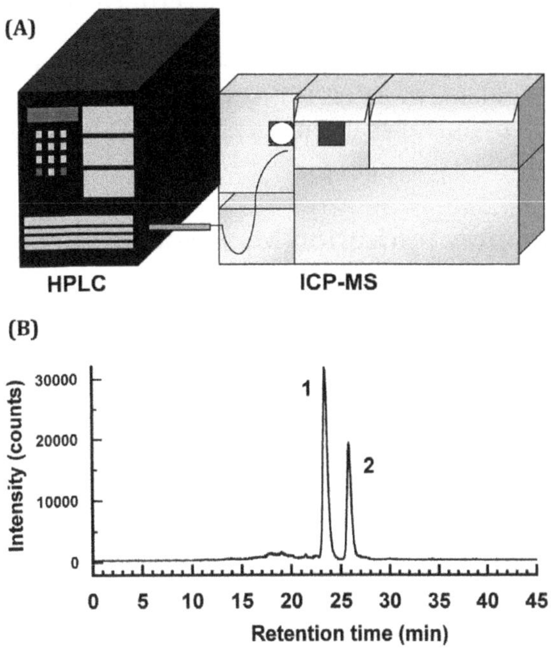

Figure 9.1 Schematic diagram of HPLC-ICP-MS (A) and typical elution profile of Se in urine sampled from Se-administered rat, obtained by multimode gel filtration HPLC (B). The urine sample was applied onto GS-320HQ (7.6 mm i.d. × 300 mm with a guard column measuring 7.6 mm i.d. × 75 mm). (1) MeSeGalNAc, (2) TMSe.

most selenometabolites are hydrophilic and nonvolatile, they are separable by HPLC. Therefore, HPLC-ICP-MS is the primary technique for the speciation analysis of selenometabolites. A typical chromatogram of Se in the urine sample of Se-fed rat is shown in Figure 9.1B. Two major selenometabolites are detected. Under the analytical conditions employed, no pretreatment of the urine sample is required prior to injection into the HPLC-ICP-MS equipped with a multimode/size-exclusion HPLC column. As will be mentioned later, the early and late eluting peaks of Se correspond to the major urinary selenosugar (*Se*-methylseleno-*N*-acetyl-galactosamine, MeGalNAc) and trimethylselenonium (TMSe), respectively. Speciation analysis gives us both qualitative and quantitative information simultaneously with high sensitivity and robustness to the sample matrix.

For volatile selenometabolites, such as dimethylselenide (DMSe), the use of gas chromatography coupled with ICP-MS for separation has been reported (Ghasemi *et al.*, 2011). Capillary electrophoresis is another separation technique for Se speciation (Lin *et al.*, 2006).

Although HPLC-ICP-MS is a reliable tool for Se speciation in biological samples, it has one crucial disadvantage, namely, as ICP-MS provides only elemental information of selenometabolites, the identification by HPLC-ICP-MS is possible only when the retention times of the eluted unknown selenometabolites can be matched with those of certified or authentic Se species (standard compounds). In other words, if the authentic Se compound of an unknown selenometabolite is not available, the unknown selenometabolite would not be identifiable by HPLC-ICP-MS. To overcome this disadvantage in the identification of unknown selenometabolites, other mass spectrometric techniques are used.

9.1.2 Electrospray Ionization Mass Spectrometry

As an alternative to HPLC-ICP-MS, electrospray ionization (ESI) or atmospheric-pressure chemical ionization (APCI)-mass spectrometry is used. Indeed, it was reported that unknown selenometabolites in human urine (Gammelgaard *et al.*, 2003); rat urine (Ogra *et al.*, 2002b); rat liver extract and pig liver extract (Lu and Pergantis, 2009); and Se-accumulating plants, such as *Astragalus praleongus* (Kotrebai *et al.*, 2000), *Brassica juncea* (Montes-Bayón *et al.*, 2002), and garlic (McSheehy *et al.*, 2000), were identified by mass spectrometry. As ESI and APCI are softer ionization sources than ICP, they can provide molecular information (Chassaigne *et al.*, 2000). Thus, those mass spectrometric techniques are often called molecular mass spectrometry or organic mass spectrometry, in contrast to ICP-MS, which is called elemental/atomic mass spectrometry or inorganic mass spectrometry. In addition, tandem mass spectrometry (MS-MS) and multistage cascade of mass spectrometry (MS^n) enable structure elucidation. The fact that Se has a characteristic cluster of signals that reflect its isotopic pattern (^{74}Se, 0.89%; ^{76}Se, 9.36%; ^{77}Se, 7.63%; ^{78}Se, 23.8%; ^{80}Se, 49.6%; ^{82}Se, 8.73%) has facilitated the scanning of Se-containing species by mass spectrometry. Thus, Se is the preferred target of organic mass spectrometry. Organic mass spectrometric techniques, such as ESI-MS-MS and ESI-MS^n, are currently available techniques for the identification of unknown selenometabolites. However, organic mass spectrometry has two weak points compared with ICP-MS. First, the detection limit of organic mass spectrometry for selenocompounds is inferior to that of ICP-MS. Recently, a Fourier transform orbital ion-trap mass spectrometer was used for the identification of selenometabolites (Bierla *et al.*, 2012). The instrument offers multistage fragmentation, accurate mass determination, and a large intrascan dynamic range with high sensitivity comparable to that of ICP-MS, although it still has to gain widespread popularity for the analysis of selenometabolites. Secondly, organic mass spectrometry is severely affected by the sample matrix.

9.1.3 Other Techniques

Matrix-assisted laser desorption ionization time-of-flight mass spectrometry (MALDI-TOF-MS) is another technique for the identification of selenocompounds (Ruiz Encinar *et al.*, 2004). As MALDI is a weaker ionization source than ESI, MALDI-TOF-MS is preferably used for the identification of macromolecules, such as selenoproteins and selenopeptides, rather than low molecular weight selenocompounds (Beilstein *et al.*, 1996; Ruiz Encinar *et al.*, 2003).

X-ray absorption fine structure (XAFS) is also used to analyze selenometabolites. XAFS is divided into two energy regions: the edge region and the beyond the edge region. Those two regions are referred to as X-ray absorption near edge spectroscopy (XANES) and extended X-ray absorption fine structure region (EXAFS), respectively. XANES provides the oxidation state and the coordination number of an element, whereas EXAFS provides structural information of the neighboring atoms of an element, such as the number of ligand atoms and the distance between the element and each ligand atom. XAFS is used for the speciation of selenometabolites in Se-hyperaccumulating plants (Lee *et al.*, 2001; Pickering *et al.*, 2003). The identification of selenometabolites by XAFS is accomplished by comparing the oxidation state and the coordination number of an element with those of a standard. However, at present, the application of XAFS to selenometabolites seems to be limited.

It is sometimes difficult to distinguish enantiomers and diastereomers by organic mass spectrometry and XAFS. As NMR can rigorously define the structure and/or configuration of a chemical species, it is the most valuable technique for identification. However, NMR requires more stringent sample purification methods and a larger amount of sample than MS. Thus, this technique seems to be not always applicable to the identification of trace selenometabolites in animal and plant samples. In studies conducted hitherto, the absolute configurations of selenosugars and selenoamino acids were determined by NMR (Kobayashi *et al.*, 2002; Block *et al.*, 2004; Ogra *et al.*, 2007).

9.2 Identification of Selenometabolites

9.2.1 Selenosugars

Se-Methylseleno-*N*-acetyl-galactosamine (MeSeGalNAc) was first identified as the major urinary selenometabolite in rat fed nutritional and low-toxicity levels of Se (Ogra *et al.*, 2002a). As shown in Figure 9.1, TMSe was detected in urine sampled from rat fed Se in an amount that exceeded the low-toxicity level. The mass spectra of the fragment ions of MeSeGalNac obtained by ESI-MS-MS are indicated in Figure 9.2.

To obtain structural information on MeSeGalNAc, HPLC-ESI-MS-MS was performed. Three precursor ions, *m*/*z* 298, 300, and 302, representing Se isotopes, ^{78}Se, ^{80}Se, and ^{82}Se, respectively, were extracted with the first quadrupole MS and then introduced into a collision cell to obtain their fragment ions. Thereafter, the fragment ions were detected with the second quadrupole MS.

At low collision energies, all parent ions produced the same major product ion at *m*/*z* 204, suggesting that an Se-containing moiety was cleaved off from each of the parent ions (Figure 9.2A–C). Because the difference in *m*/*z* between

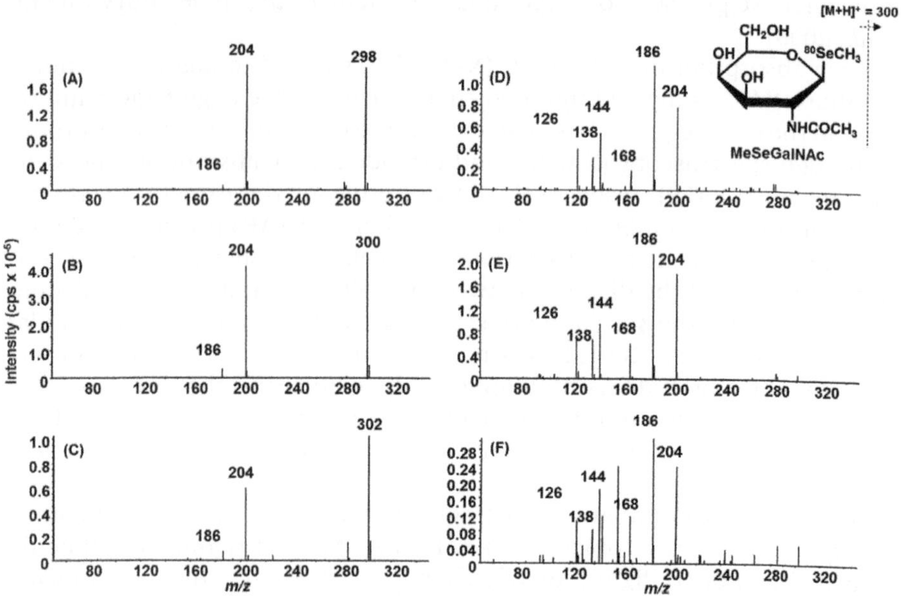

Figure 9.2 Collision-induced dissociation mass spectra (ESI-MS-MS) of the major urinary selenometabolite (MeSeGalNAc) in the urine of Se-administered rat. The dissociation of each Se-containing molecular ion, *i.e.*, 298 (^{78}Se, panels A and D), 300 (^{80}Se, panels B and E), or 302 (^{82}Se panels C and F), was induced in the collision cell with 10 (A–C) and 20 eV (D–F) collision energies and then, each fragment ion was detected with the second mass spectrometer.

the precursor ion and the fragment ion was 96, the common fragment ion was assumed to be a result of the removal of CH_3SeH from the precursor ion, and the presence of a methylselenyl group (CH_3Se-) in MeSeGalNAc was suggested. Another minor product ion appeared at *m*/*z* 186 from all the precursor ions (Figure 9.2A–C), which was simply assigned to the fragment ion that appeared after the loss of H_2O from the major product ion at *m*/*z* 204.

At high collision energies, common fragment ions originating in the precursor ion appeared at *m*/*z* 186, 144, 138, and 126. Three of the four fragment ions, *i.e.*, *m*/*z* 186, 144, and 126, were assigned to $[204\text{-}H_2O]^+$, $[204\text{-}CH_3COOH]^+$, and $[144\text{-}H_2O]^+$, respectively (Figure 9.2D–F). Consequently, the major urinary selenometabolite could be deconvoluted as shown in Figure 9.2.

In addition to MeSeGalNAc, *Se*-methylseleno-*N*-acetyl-glucosamine (MeSeGlcNAc) and *Se*-methylseleno-galactosamine (MeSeGalNH$_2$) were also identified in human urine as trace and minor selenometabolites, respectively (Gammelgaard *et al.*, 2003). Selenosugars are biologically inert; they are less toxic and less bioavailable than naturally occurring inorganic Se salts and selenoamino acids (Juresa *et al.*, 2007). However, it is still unclear why and how selenosugars are biosynthesized in animals. The task of elucidating the biological significance of selenosugars remains. In particular, determining why GalNAc is chosen as the sugar moiety of the major selenosugar would offer an important clue to understand the entire Se metabolic pathway.

9.2.2 Unique Selenoamino Acids

9.2.2.1 Se-Methylselenocysteine and Its Derivatives

Se-Methylselenocysteine (MeSeCys) and γ-glutamyl-*Se*-methylselenocysteine (γGluMeSeCys) are found in plants that have a high capacity for sulfur and Se assimilation, including garlic, onion, chive, leek, broccoli, and Indian mustard. When the Se-accumulating plants are exposed to inorganic Se salts, such as selenite and selenate, inorganic Se salts assimilated by the plants are mainly converted into selenoamino acids. The detection of these selenoamino acids by organic mass spectrometry has been reported (B'Hymer and Caruso, 2006; Pedrero and Madrid, 2009), and their structures are shown in Figure 9.3. As the minor derivative of MeSeCys, *Se*-propylselenocysteine has been reported. The fact that Se is not an essential element in plants indicates

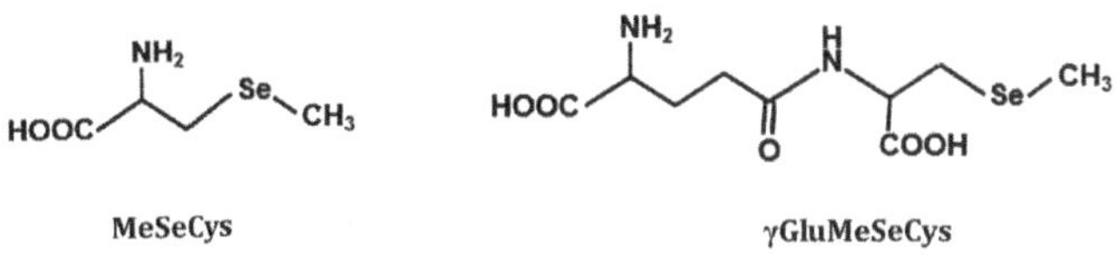

Figure 9.3 Structures of *Se*-methylselenocysteine (MeSeCys) and γ-glutamyl-*Se*-methylselenocysteine (γGluMeSeCys).

that no selenoprotein/selenoenzyme genes are coded in the plant genome. Selenoproteins/selenoenzymes in animals contain Se as selenocysteine (SeCys) in their primary structures. SeCys is required in the active center of selenoenzymes because of its high reactivity. Se-Accumulating plants may reduce the reactivity, *i.e.*, the toxicity of SeCys by masking the selenyl group (−SeH) with a methyl group.

9.2.2.2 Selenomethionine Derivatives

Although SeCys can be well discriminated from cysteine, selenomethionine (SeMet) is incorporated into proteins in an *in vitro* translation system using wheat germ extract (WGE) without discrimination from methionine (Met), and the fluorescence of green fluorescent protein (GFP) and the activity of JNK stimulatory phosphatase-1 (JSP-1) are not affected by the substitution of Met with SeMet by *in vitro* translation (Ogra *et al.*, 2008). Met is not only used for protein synthesis but also metabolized into coenzymes, such as methylmethionine (MeMet) and *S*-adenosylmethionine (AdoMet, SAM), which act as a methyl group donor.

Because it can be speculated that SeMet is metabolized into the above-mentioned coenzymes in higher eukaryotes, *i.e.*, plants and animals, the biological transformation of SeMet into SeMet derivatives in a model system of higher eukaryotes, *i.e.*, WGE, was evaluated by identifying Se-containing molecules with HPLC-ICP-MS and HPLC-ESI-MS-MS (Ogra *et al.*, 2009). Three Se-containing peaks tentatively named WGE1, WGE2, and WGE3 were detected by HPLC-ICP-MS (Figure 9.4).

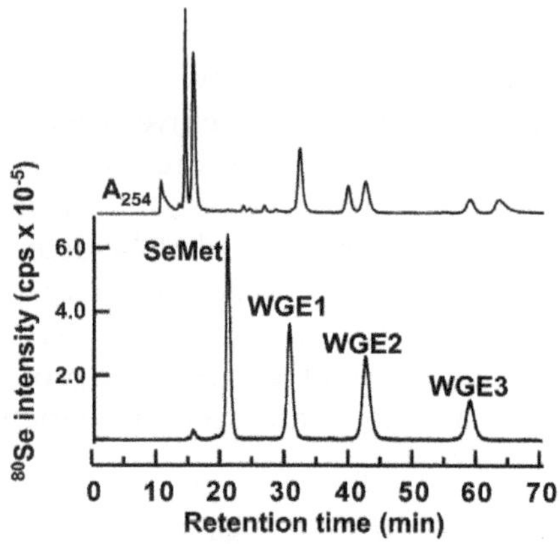

Figure 9.4 Elution profile of Se in the reaction mixture of wheat germ extract for the *in vitro* metabolism of SeMet. A 20 μL aliquot of the reaction mixture was subjected to HPLC equipped with a multimode/size-exclusion column (GS-320HQ) and coupled to a photodiode array detector (PDA) and ICP-MS to measure the absorbance at 254 nm (upper) and the Se distribution at *m*/*z* 80 (lower), respectively. The column was eluted with 10 mmol L^{-1} ammonium acetate, pH 6.5, at the flow rate of 0.6 mL min^{-1}.

HPLC-ESI-MS analysis revealed that WGE1 had $[M]^+$ at *m/z* 212, and its fragment ions indicated that WGE1 is *Se*-methylselenomethionine (MeSeMet) (Figure 9.5). WGE2 and WGE3 exhibited absorption bands at 254 nm and molecular ions at *m/z* 433 and 447, respectively. Their fragment ions revealed that WGE2 and WGE3 are *Se*-adenosylselenohomocysteine (AdoSeHcy) and *Se*-adenosylselenomethionine (AdoSeMet), respectively. Their structures were confirmed by the absorbance measurements of WGE2 and WGE3 at

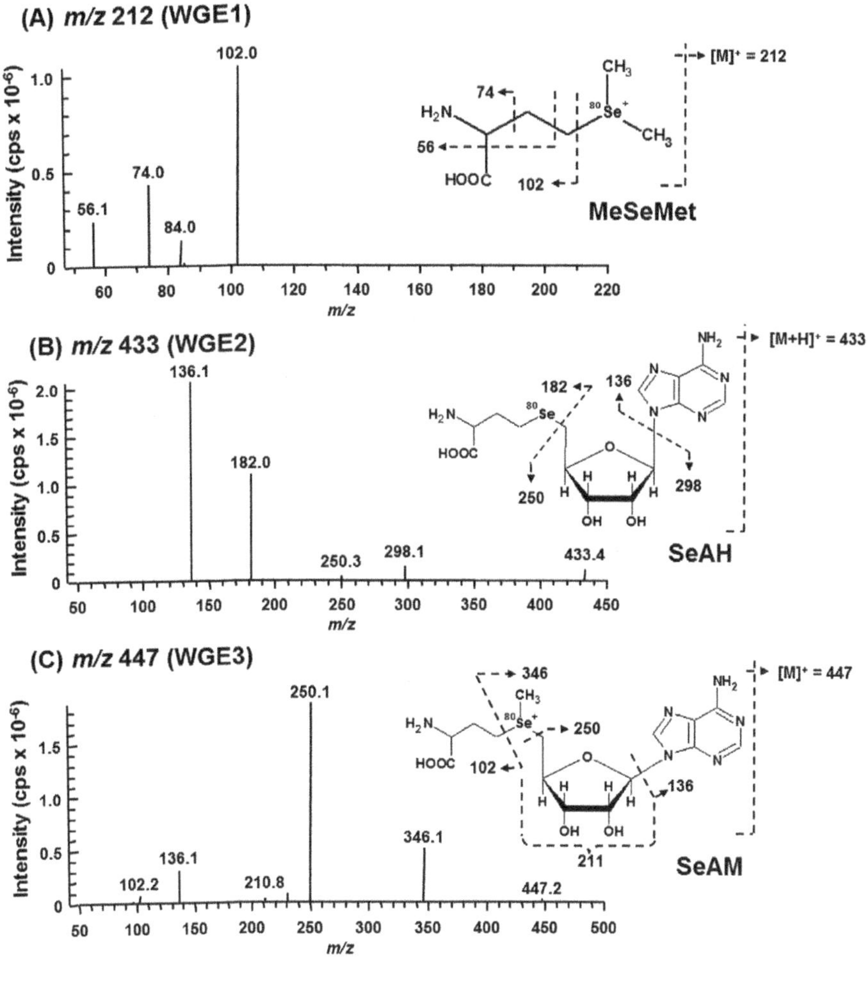

Figure 9.5 Collision-induced dissociation mass spectra (ESI-MS-MS) of Se-containing positive molecular ions of selenometabolites (MeSeMet, SeAH, and SeAM) in the reaction mixture of WGE and SeMet. The dissociation of each Se-containing molecular ion of WGE1 (A, MeSeMet), WGE2 (B, SeAH), and WGE3 (C, SeAM) at *m/z* 212, 433, and 447 was induced with 25 eV collision energy, and fragment ions were detected with the second mass spectrometer. The assignments of the fragments obtained are shown on the right of each MS-MS spectrum.

254 nm. In addition, a trace amount of AdoSeMet was suggested to exist also in rabbit reticulocyte lysate, a mammalian *in vitro* translation system. These results imply that SeMet and Met share a common metabolic pathway, namely, SeMet is not only incorporated into proteins in place of Met but also metabolized into Met/SeMe-containing coenzymes. Although it is reported that there are no differences in enzyme activity between Met- and SeMet-containing proteins (Ogra *et al.*, 2008), there is no solid evidence indicating that SeMet-containing coenzymes have the same activity as Met-containing ones.

9.2.2.3 *Selenohomolanthionine and Its Derivatives*

Selenohomolanthionine (SeHLan) is the common name of 4,4′-selenobis[2-aminobutanoic acid] or *Se*-(3-amino-3-carboxypropyl)-homocysteine. This unique selenoamino acid was first identified in Se-fortified Japanese pungent radish (*Raphanus sativus* L. cv. "Yukibijin"). The mass spectra of SeHLan are shown in Figure 9.6. The biological, pharmacological, and toxicological effects of SeHLan are mentioned in Chapter 20. A derivative of SeHLan, 2,3-dihydroxy-propionyl-SeHLan (2,3-DHP-SeHLan), was also identified by ESI-quadrupole time-of-flight tandem mass spectrometry (ESI-Q-TOF MS-MS) and ESI-hybrid linear ion trap/Orbitrap MS^n (Far *et al.*, 2010).

9.2.2.4 *Selenoneine and Its Derivatives*

Selenoneine (2-selenyl-*N*α,*N*α,*N*α-trimethyl-L-histidine) was first identified in bluefin tuna and other fish species (Yamashita and Yamashita, 2010). It was also reported that sea turtles accumulate Se in their livers in spite of the low hepatic Hg concentration, and that the majority of Se in the liver exists as selenoneine (Anan *et al.*, 2011b). ESI-MS-MS has also contributed significantly to the identification of selenoneine (Figure 9.7).

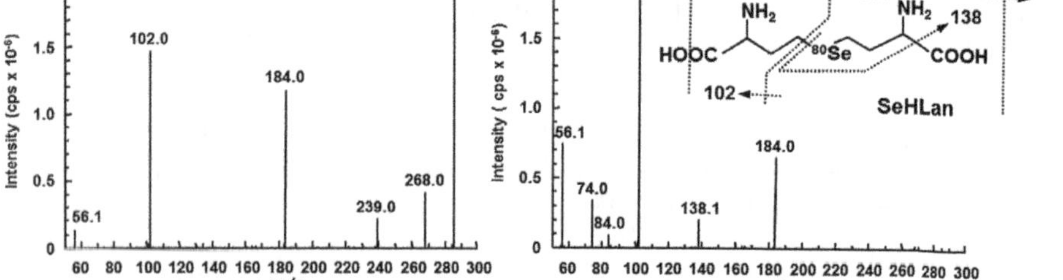

Figure 9.6 Collision-induced dissociation mass spectra (ESI-MS-MS) of Se-containing positive molecular ions of a selenometabolite (SeHLan) in Se-fortified Japanese pungent radish. The precursor ion at *m*/*z* 285 was induced with 20 (A) and 30 (B) eV collision energies, and fragment ions were detected with the second mass spectrometer. The assignment of the fragments obtained is shown.

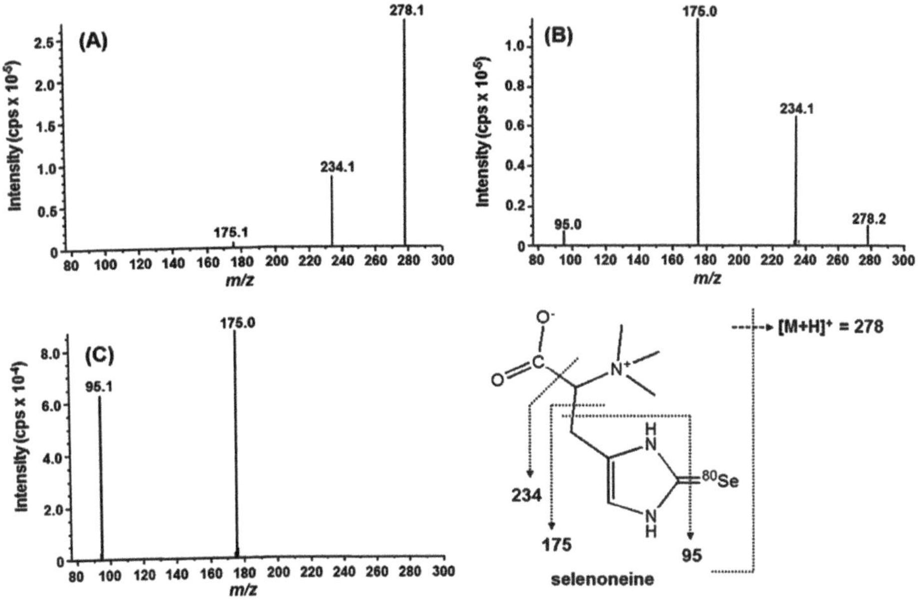

Figure 9.7 Collision-induced dissociation mass spectra (ESI-MS-MS) of Se-containing positive molecular ions of a selenometabolite (selenoneine) in hawksbill turtle liver. The dissociation of the Se-containing molecular ion of the selenometabolite at *m*/*z* 278 was induced with 15 (A), 25 (B), and 35 (C) eV collision energies, and fragment ions were detected with the second mass spectrometer. The assignment of fragments obtained is shown.

Selenoneine is specifically detected in marine biota. In fact, although substantial amounts of selenoneine are detected in the livers of marine turtles, selenoneine in the livers of terrestrial turtles is below the detection limit (Anan *et al.*, 2011b). Those findings suggest that selenoneine is transferred among marine organisms *via* the food web to result in its ubiquitous distribution in the marine ecosystem. However, the origin of selenoneine is still unclear.

The antioxidative effects of selenoneine have been investigated as well. Indeed, its structure indicates that selenoneine is a potential antioxidant (Yamashita *et al.*, 2010). In addition, it is speculated that selenoneine plays a role in the detoxification of inorganic and methyl mercury (Anan *et al.*, 2011a; Yamashita *et al.*, 2013b). Further studies are needed to confirm this speculation.

Selenoneine is also detected in blood samples from a fish-eating population (Klein *et al.*, 2011; Yamashita *et al.*, 2013a). *Se*-Methyl-selenoneine is identified in a fish-eating population as well, suggesting that the redox activity of selenoneine is likely reduced in human, *i.e.*, the toxicity of selenoneine is masked by replacing its selenyl/selenone group with a methyl group (Klein *et al.*, 2011).

9.3 Conclusion

As mentioned above, Se-containing compounds having carbon–Se covalent bond(s), *i.e.*, selenometabolites, are formed in animals and plants. Hence, the identification of selenometabolites is a means to determine the metabolic pathway of Se in animals and plants. Further improvement of the sensitivity of organic mass spectrometry is expected to lead to the complete identification of selenometabolites, thereby paving the way to revealing the entire metabolic pathway of Se.

Summary Points

- This chapter focuses on the identification of selenium-containing metabolites (selenometabolites) in animals and plants by mass spectrometry.
- Inductively coupled plasma mass spectrometry (ICP-MS) is the most sensitive and robust technique for the detection of selenium in biological samples.
- The hyphenated technique consisting of HPLC and ICP-MS is the most frequently used to detect and screen for selenometabolites in animals and plants.
- Mass spectrometers having a softer ionization source than ICP are utilized for the identification of unknown selenometabolites.
- Major urinary selenometabolites are identified as selenosugars by mass spectrometry. The most abundant and important selenosugar is *Se*-methylseleno-*N*-acetylgalactosamine. As minor selenosugars, *Se*-methylseleno-*N*-acetylglucosamine and *Se*-methylseleno-galactosamine are also detected.
- Unique selenium-containing amino acids (selenoamino acids) are also identified in specific animals and plants.

Key Facts of Selenometabolites

1. Selenium is excreted into urine as selenosugars when the ingested amount of selenium is small. Selenosugars are supposed to be *de novo* synthesized in the selenium-specific metabolic pathway because no sulfur analogs of selenosugars have been identified so far.
2. Trimethylselenonium is also excreted into urine when exceedingly large amounts of selenium are ingested.
3. In plants that have a high capacity for selenium metabolism, unique selenoamino acids, such as *Se*-methylselenocysteine, γ-glutamyl-*Se*-methylselenocysteine, and selenohomolanthionine, are *de novo* synthesized.
4. Selenoneine (2-selenyl-*N*,*N*,*N*-trimethyl-L-histidine) is specifically accumulated in marine animals by biomagnification.

5. Selenium analogs of thioamino acid derivatives, such as *Se*-adenosylselenomethionine, *Se*-adenosylselenohomocysteine, and *Se*-methylselenomethionine, are detected in animal and plant samples. This indicates that selenomethionine shares the same metabolic pathway, in part, with methionine.

Definitions and Explanations of Key Terms

Speciation. Speciation analysis is the analytical activity of identifying and/or measuring the quantities of one or more individual chemical species in a sample.

Hyphenated technique. This technique is developed from the coupling of a separation technique and an online spectroscopic detection technology.

Photodiode array detector (PDA). PDA is useful for recording the full UV-Vis absorption spectra of samples that are rapidly passing through a sample flow cell of HPLC.

Selenosugar. This is a major urinary selenium metabolite, and consists of selenium and a sugar moiety which is *N*-acetylated or nonacetylated hexosamine.

Wheat germ extract. This is prepared by grinding wheat germ in an extraction buffer followed by centrifugation to remove cell debris. The supernatant is applied to chromatography that separates endogenous amino acids and plant pigments from the extract. The extract is also treated with micrococcal nuclease to destroy endogenous mRNA.

List of Abbreviations

DMSe	Dimethylselenide
ESI-MS-MS	Electrospray ionization tandem mass spectrometry
γGluMeSeCys	γ-Glutamyl-*Se*-methylselenocysteine
ICP-MS	Inductively coupled plasma mass spectrometry
MeSeCys	*Se*-Methylselenocysteine
Met	Methionine
MMSe	Monomethylselenol
MeSeMet	*Se*-Methylselenomethionine or selenomethionine methylselenonium
SeAH	*Se*-Adenosylselenohomocysteine
SeAM	*Se*-Adenosylselenomethionine
SeCys	Selenocysteine
SeHCys	Selenohomocysteine
SeHLan	Selenohomolanthionine
MeSeGalNAc	*Se*-Methylseleno-*N*-acetyl-galactosamine
MeSeGalNH$_2$	*Se*-Methylseleno-galactosamine
MeSeGlcNAc	*Se*-Methylseleno-*N*-acetyl-glucosamine
SeMet	Selenomethionine
TMSe	Trimethylselenonium

References

Anan, Y., Tanabe, S. and Ogra, Y., 2011a. Comparison of selenoneine found in marine organisms with selenite in the interaction with mercury compounds *in vitro*. *Journal of Toxicological Sciences*. 36: 725–731.

Anan, Y., Ishiwata, K., Suzuki, N., Tanabe, S. and Ogra, Y., 2011b. Speciation and identification of low molecular weight selenium compounds in the liver of sea turtles. *Journal of Analytical Atomic Spectrometry*. 26: 80–85.

B'Hymer, C. and Caruso, J. A., 2006. Selenium speciation analysis using inductively coupled plasma-mass spectrometry. *Journal of Chromatography A*. 1114: 1–20.

Beilstein, M. A., Vendeland, S. C., Barofsky, E., Jensen, O. N. and Whanger, P. D., 1996. Selenoprotein W of rat muscle binds glutathione and an unknown small molecular weight moiety. *Journal of Inorganic Biochemistry*. 61: 117–124.

Bierla, K., Szpunar, J., Yiannikouris, A. and Lobinski, R., 2012. Comprehensive speciation of selenium in selenium-enriched yeast. *Trends in Analytical Chemistry*. 41: 122–132.

Block, E., Glass, R. S., Jacobsen, N. E., Johnson, S., Kahakachchi, C., Kamiński, R., Skowrońska, A., Totoe Boakye, H., Tyson, J. F. and Uden, P. C., 2004. Identification and synthesis of a novel selenium–sulfur amino acid found in selenized yeast: rapid indirect detection NMR methods for characterizing low-level organoselenium compounds in complex matrices. *Journal of Agricultural and Food Chemistry*. 52: 3761–3771.

Chassaigne, H., Vacchina, V. and Lobinski, R., 2000. Elemental speciation analysis in biochemistry by electrospray mass spectrometry. *Trends in Analytical Chemistry*. 19: 300–313.

Far, J., Preud'homme, H. and Łobiński, R., 2010. Detection and identification of hydrophilic selenium compounds in selenium-rich yeast by size exclusion-microbore normal-phase HPLC with the on-line ICP-MS and electrospray Q-TOF-MS detection. *Analytica Chimica Acta*. 657: 175–190.

Gammelgaard, B., Grimstrup Madsen, K., Bjerrum, J., Bendahl, L., Jøns, O., Olsen, J. and Sidenius, U., 2003. Separation, purification and identification of the major selenium metabolite from human urine by multi-dimensional HPLC-ICP-MS and APCI-MS. *Journal of Analytical Atomic Spectrometry*. 18: 65–70.

Ghasemi, E., Sillanpaa, M. and Najafi, N. M., 2011. Headspace hollow fiber protected liquid-phase microextraction combined with gas chromatography-mass spectroscopy for speciation and determination of volatile organic compounds of selenium in environmental and biological samples. *Journal of Chromatography A*. 1218: 380–386.

Juresa, D., Blanusa, M., Francesconi, K. A., Kienzl, N. and Kuehnelt, D., 2007. Biological availability of selenosugars in rats. *Chemico-Biological Interactions*. 168: 203–210.

Klein, M., Ouerdane, L., Bueno, M. and Pannier, F., 2011. Identification in human urine and blood of a novel selenium metabolite, Se-methylselenoneine, a potential biomarker of metabolization in mammals of the naturally occurring selenoneine, by HPLC coupled to electrospray hybrid linear ion trap-orbital ion trap MS. *Metallomics*. 3: 513–520.

Kobayashi, Y., Ogra, Y., Ishiwata, K., Takayama, H., Aimi, N. and Suzuki, K. T., 2002. Selenosugars are key and urinary metabolites for selenium excretion within the required to low-toxic range. *Proceedings of the National Academy of Sciences of the United States of America*. 99: 15932–15936.

Kotrebai, M., Tyson, J. F., Block, E. and Uden, P. C., 2000. High-performance liquid chromatography of selenium compounds utilizing perfluorinated carboxylic acid ion-pairing agents and inductively coupled plasma and electrospray ionization mass spectrometric detection. *Journal of Chromatograpy A*. 866: 51–63.

Lee, A., Lin, Z. Q., Pickering, I. J. and Terry, N., 2001. X-ray absorption spectroscopy study shows that the rapid selenium volatilizer, pickleweed (*Salicornia bigelovii* Torr.), reduces selenate to organic forms without the aid of microbes. *Planta*. 213: 977–980.

Lin, S. Y., Wang, G. R., Huang, Q. P. and Liu, C. Y., 2006. Speciation of selenium compounds by open tubular capillary electrochromatography-inductively coupled plasma mass spectrometry. *Electrophoresis*. 27: 4257–4265.

Lu, Y. and Pergantis, S. A., 2009. Selenosugar determination in porcine liver using multidimensional HPLC with atomic and molecular mass spectrometry. *Metallomics*. 1: 346–352.

McSheehy, S., Yang, W., Pannier, F., Szpunar, J., Lobinski, R., Auger, J. and Portin-Gautier, M., 2000. Speciation analysis of selenium in garlic by two-dimensional high-performance liquid chromatography with parallel inductively coupled plasma mass spectrometric and electrospray tandem mass spectrometric detection. *Analytica Chimica Acta*. 421: 147–153.

Montes-Bayón, M., Yanes, E. G., Ponce de León, C., Jayasimhulu, K., Stalcup, A., Shann, J. and Caruso, J. A., 2002. Initial studies of selenium speciation in *Brassica juncea* by LC with ICPMS and ES-MS detection: an approach for phytoremediation studies. *Analytical Chemistry*. 74: 107–113.

Ogra, Y., Ishiwata, K., Takayama, H., Aimi, N. and Suzuki, K. T., 2002a. Identification of a novel selenium metabolite, Se-methyl-*N*-acetyl-selenohexosamine, in rat urine by the HPLC-ICP MS and ESI MS/MS methods. *Journal of Chromatography B: Analytical Technologies in the Biomedical and Life Sciences*. 767: 301–312.

Ogra, Y., Ishiwata, K., Takayama, H., Aimi, N. and Suzuki, K. T., 2002b. Identification of a novel selenium metabolite, Se-methyl-*N*-acetylselenohexosamine, in rat urine by high-performance liquid chromatography-inductively coupled plasma mass spectrometry and electrospray ionization tandem mass spectrometry. *Journal of Chromatography B: Analytical Technologies in the Biomedical and Life Sciences*. 767: 301–312.

Ogra, Y., Kitaguchi, T., Ishiwata, K., Suzuki, N., Iwashita, Y. and Suzuki, K. T., 2007. Identification of selenohomolanthionine in selenium-enriched Japanese pungent radish. *Journal of Analytical Atomic Spectrometry*. 22: 1390–1396.

Ogra, Y., Kitaguchi, T., Ishiwata, K., Suzuki, N., Toida, T. and Suzuki, K. T., 2009. Speciation of selenomethionine metabolites in wheat germ extract. *Metallomics*. 1: 78–86.

Ogra, Y., Kitaguchi, T., Suzuki, N. and Suzuki, K. T., 2008. *In vitro* translation with [^{34}S]-labeled methionine, selenomethionine and telluromethionine. *Analytical and Bioanalytical Chemistry*. 390: 45–51.

Pedrero, Z. and Madrid, Y., 2009. Novel approaches for selenium speciation in foodstuffs and biological specimens: a review. *Analytica Chimica Acta*. 634: 135–152.

Pickering, I. J., Wright, C., Bubner, B., Ellis, D., Persans, M. W., Yu, E. Y., George, G. N., Prince, R. C. and Salt, D. E., 2003. Chemical form and distribution of selenium and sulfur in the selenium hyperaccumulator *Astragalus bisulcatus*. *Plant Physiol*. 131: 1460–1467.

Ruiz Encinar, J., Ouerdane, L., Buchmann, W., Tortajada, J., Łobinski, R. and Szpunar, J., 2003. Identification of water-soluble selenium-containing proteins in selenized yeast by size-exclusion-reversed-phase HPLC/ICPMS followed by MALDI-TOF and electrospray Q-TOF mass spectrometry. *Analytical Chemistry*. 75: 3765–3774.

Ruiz Encinar, J., Połatajko, A., Szpunar, J. and Łobiński, R., 2004. Identification of non-peptide species in selenized yeast by MALDI mass spectrometry using post-source decay and orthogonal Q-TOF detection. *Analyst*. 129: 846–849.

Templeton, D. M., Ariese, F., Cornelis, R., Danielsson, L., Muntau, H., van Leeuwen, H. P. and Łobiński, R., 2000. Guidelines for terms related to chemical speciation and fractionation of elements. Definitions, structural aspects, and methodological approaches (IUPAC Recommendations 2000). *Pure and Applied Chemistry*. 72: 1453–1470.

Yamashita, M., Yamashita, Y., Ando, T., Wakamiya, J. and Akiba, S., 2013a. Identification and determination of selenoneine, 2-selenyl-*N*α,*N*α,*N*α-trimethyl-L-histidine, as the major organic selenium in blood cells in a fish-eating population on remote Japanese islands. *Biological Trace Element Research*. 156: 36–44.

Yamashita, M., Yamashita, Y., Suzuki, T., Kani, Y., Mizusawa, N., Imamura, S., Takemoto, K., Hara, T., Hossain, M. A., Yabu, T. and Touhata, K., 2013b. Selenoneine, a novel selenium-containing compound, mediates detoxification mechanisms against methylmercury accumulation and toxicity in zebrafish embryo. *Marine Biotechnology*. 15: 559–570.

Yamashita, Y., Yabu, T. and Yamashita, M., 2010. Discovery of the strong antioxidant selenoneine in tuna and selenium redox metabolism. *World Journal of Biological Chemistry*. 1: 144–150.

Yamashita, Y. and Yamashita, M., 2010. Identification of a novel selenium-containing compound, selenoneine, as the predominant chemical form of organic selenium in the blood of bluefin tuna. *Journal of Biological Chemistry*. 285: 18134–18138.

CHAPTER 10

Bioavailability of Selenium and Selenium Compounds in Foodstuff by In Vivo Testing

JORGE MOREDA-PIÑEIRO*[a] AND ANTONIO MOREDA-PIÑEIRO[b]

[a]Department of Analytical Chemistry, Faculty of Sciences, University of A Coruña, Campus da Zapateira, s/n, 15071 – A Coruña, Spain; [b]Department of Analytical Chemistry, Nutrition and Bromatology, Faculty of Chemistry, University of Santiago de Compostela, Avenida das Ciencias, s/n, 15782 – Santiago de Compostela, Spain
*E-mail: jorge.moreda@udc.es

10.1 Introduction

The ingestion of several essential/toxic substances by humans occurs mainly *via* food. The knowledge of the actual forms and the total content of essential/toxic substances in foodstuffs do not provide enough information for assessing the beneficial effects or the human dietary risks after food consumption. Understanding the amount of the species that reaches the systemic circulation after ingestion and that is available to perform its biological activity or toxic effect (bioavailability) is therefore required. Regarding trace elements, the bioavailability of the different species of an essential/toxic element is also important because the benefits or the damages are

Food and Nutritional Components in Focus No. 9
Selenium: Chemistry, Analysis, Function and Effects
Edited by Victor R. Preedy

Published by the Royal Society of Chemistry, www.rsc.org

dependent on the chemical form of the element that is absorbed. In addition, the new eating habits and the increase in the consumption of functionalized foodstuffs to improve the content of some essential species in the diet of several countries (*i.e.* Northern Europe is selenium deficient), requires the toxicology and bioavailability assessment of those nutrients/essential substances.

Due to selenium being an essential element for humans, the determination of selenium and selenium species in foodstuffs and the establishment of bioavailability studies of these species are of special concern. Useful information about selenium bioavailability in foodstuffs can be found in recent reviews (Finley, 2006; Intawongse and Dean, 2006; Fairweather-Tait *et al.*, 2010; Moreda-Piñeiro *et al.*, 2011; Thiry *et al.*, 2012).

10.1.1 Bioaccessibility/Bioavailability/Bioactivity Definitions

Bioaccessibility, bioavailability and bioactivity are often used as synonymous terms although they refer to different concepts. When dealing with human nutrition, bioaccessibility indicates the maximum fraction of a nutrient or substance in food that is theoretically soluble from its matrix in the gastrointestinal (GI) tract (bioaccessible fraction), and that is therefore available for subsequent processes of absorption through the intestinal mucosa (Ruby *et al.*, 1999). The nonaccessible fraction cannot be absorbed through the intestinal membrane and is directly excreted without having affected the body mechanisms (Figure 10.1).

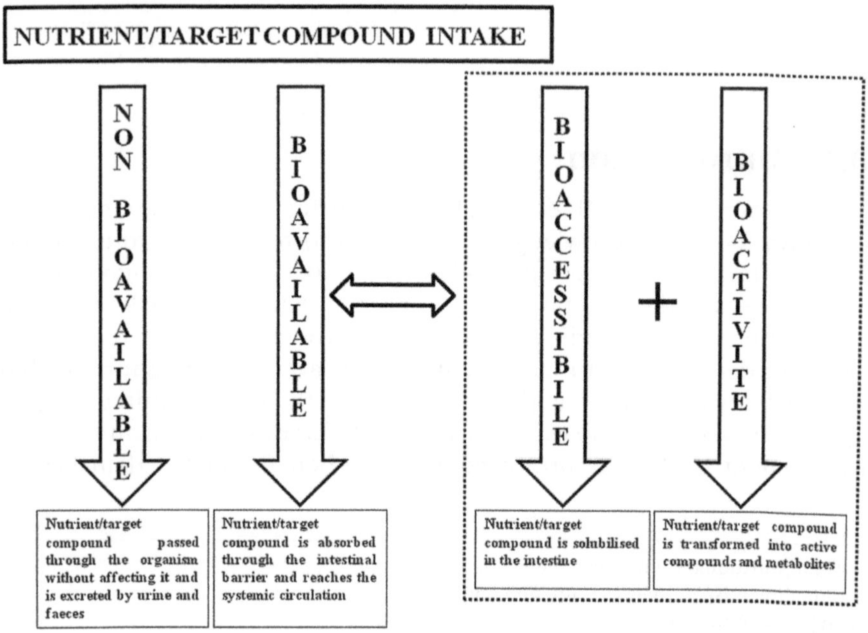

Figure 10.1 Definitions of nonbioavailable, bioavailable, bioaccessible and bioactive fractions. Note that bioaccessibility plus bioactivity is bioavailability.

Bioavailability refers to the fraction of substance released from food that is absorbed through the intestinal barrier (bioavailable fraction), passes into the bloodstream or an organ, and is available to promote its action in the exposed organism (Ruby *et al.*, 1999). Other authors consider the bioavailable fraction as the quantity of substance that is absorbed and reaches the systemic circulation, that is distributed to organs and tissues and that is transformed into a biochemically active form and that effectively is used by the organism (bioactive fraction) (Schümann *et al.*, 1997). The bioactive fraction of a target compound is the one that is converted into biologically active metabolites (selenoproteins, *i.e.* glutathione peroxidase, GPx; thioredoxin reductase, TrxR; and selenoprotein P, SeP; in the case of selenium) from that compound. Bioavailability includes the entire sequence of events that take place during the digestive transformation of food into material that can be assimilated by the body, the absorption/assimilation into the cells of the intestinal epithelium, and lastly, the presystemic metabolism (both intestinal and hepatic) (Fernández-García *et al.*, 2009). Thus, bioavailability is defined as a sum of bioaccessibility and bioactivity (Figure 10.1).

Finally, the concept of bioactivity includes events linked to how the compound is transported and reaches the target tissue, how it interacts with biomolecules, the metabolism or biotransformation that it may undergo, and the generation of a biomarker and the physiological response it causes (Fernández-García *et al.*, 2009). As can be seen, bioaccessibility, bioavailability and bioactivity correspond to the different level of interaction that substances can have with the organism.

The bioaccessibility/bioavailability/bioactivity of a substance is directly affected by the type of food and/or the composition of the food matrix, and by the synergies and antagonisms that may be established among the several components. In addition, the GI conditions, food processing, compound species distribution and special physiological conditions of the consumer also influence the bioavailability (Figure 10.2).

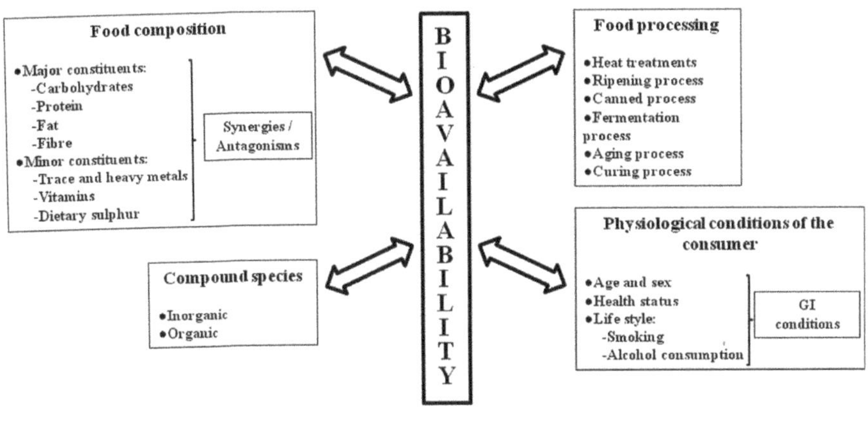

Figure 10.2 Factors that affect bioavailability of a target compound.

10.1.2 Bioavailability Assessment: *In Vivo* Tests

The assessment of substances bioavailability in foodstuff can be performed by using several analytical approaches: *in vivo* and *in vitro* studies. Both approaches present strengths and drawbacks that are summarized in Figure 10.3.

The theoretical and practical aspects of *in vivo* tests will be described in this Chapter. In addition, *in vivo* tests applied for selenium bioavailability assessment in foodstuff will also be summarized in this chapter. *In vitro* tests performances and application for selenium bioavailability assessment in foodstuff will be described and summarized in Chapter 11.

An *in vivo* test is performed by dosing experimental humans (*in vivo* human) or animals (*in vivo* animal) with several concentrations of the target compound (isotopically labeled or nonisotopically labeled). Foodstuffs with high target compound concentration or fortified food enriched with the target compound is generally supplied to subjects (volunteer people or animals) in *in vivo* tests (Figure 10.4). The determination of the target concentration in several tissues and further balance studies are then performed, which allows the determination of the absorbed amount of nutrients or their metabolites.

Tissue analysis is generally developed by monitoring the increase in plasma/serum concentration of the nutrient/metabolite. In studies dealing with selenium bioavailability the measure of GPx and TrxR activity and SeP concentration, induced by certain food or selenium species doses, the metabolite is assessed/monitored.

Balance studies are performed by determining the absorbed amounts (apparent absorption and retention) after measuring the difference

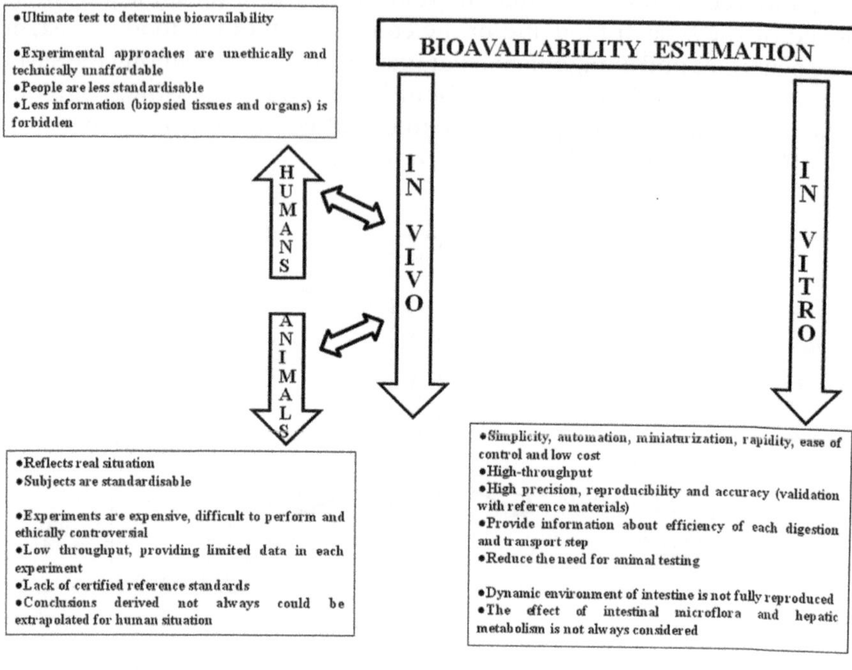

Figure 10.3 Strengths and drawbacks of *in vivo* human, *in vivo* animal and *in vitro* tests are summarized.

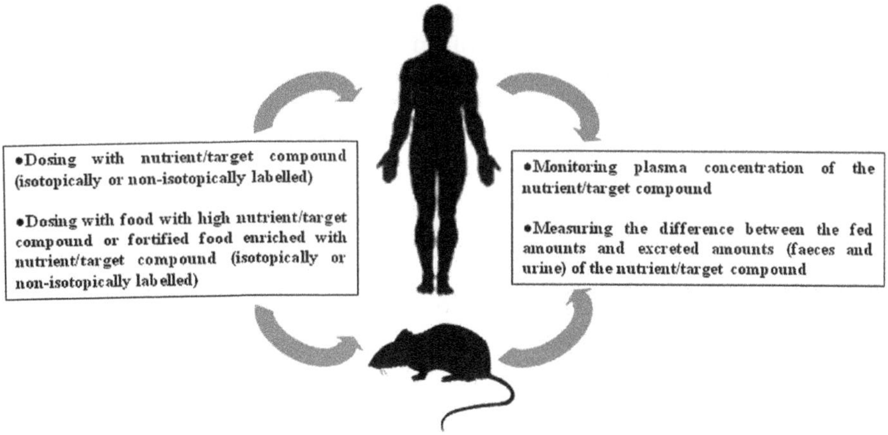

Figure 10.4 Schematic diagram of *in vivo* approaches in human and animal tests procedures are summarized.

between the fed amounts from food and the excreted amounts of the nutrient/metabolite in feces and urine. Apparent absorption and retention of nutrients/target compounds can be calculated by using the following equations (eqn (10.1) and (10.2)):

$$\text{Apparent}_{\text{absorption}} = \frac{\text{Conc}_{\text{intake}} - \text{Conc}_{\text{infaeces}}}{\text{Conc}_{\text{intake}}} \times 100 \tag{10.1}$$

$$\text{Retention} = \frac{\text{Conc}_{\text{intake}} - \text{Conc}_{\text{infeces}} - \text{Conc}_{\text{inurine}}}{\text{Conc}_{\text{intake}} - \text{Conc}_{\text{infeces}}} \times 100 \tag{10.2}$$

where $\text{Conc}_{\text{intake}}$ is the concentration of target compound intake and $\text{Conc}_{\text{infeces}}$ and $\text{Conc}_{\text{inurine}}$ are the concentrations of target compound in feces and urine, respectively. The knowledge of nutrient/metabolite concentration in feces allows for bioacccessibility estimation. In addition, bioavailability estimation can be achieved knowing nutrient/metabolite concentrations in urine and tissues.

Other methods for assessing *in vivo* animal tests consist of performing animal feeding in parallel with the selected foodstuff and with a reference selenocompound (selenite (Se(IV)) and selenomethionine (SeMet) mainly) during the repletion period. The results obtained when feeding with the tested food are then compared to those obtained when feeding with the single selenocompound. The comparison was done by using eqn (10.3) (slope-ratio assay) after achieving the slopes of the regression lines generated by:

(1) dose–response relationship of a tested food item and;
(2) dose–response relationship of the reference selenium chemical pure.

$$\text{Ratio} = \frac{S_{\text{food}}}{S_{\text{selenocompound}}} \times 100 \tag{10.3}$$

where S_{food} and $S_{selenocompound}$ are the slopes of regression lines dose–response relationship for foodstuff and selenium chemical pure form, respectively. The resulting ratio can then be used to quantitatively compare the bioavailability of selenium in different foods.

10.1.3.1 *In Vivo Human*

Although *in vivo* human studies are the ultimate test to determine bioavailability of nutrients or bioactive compounds, compared to *in vivo* animal or *in vitro* tests, some experimental approaches are unethically and technically unaffordable (Figure 10.3). People are not as standardizable as animals or cells and they cannot release as much information as animals that are usually sacrificed at the end of the experiment for a further analysis of biopsied tissues and organs (Thiry *et al.*, 2012).

10.1.3.2 *In Vivo Animal*

In vivo studies using animal models were the first proposed assays and nowadays they are still the most common models for predicting bioavailability. A variety of animal models have been used in bioavailability studies, including primates, swine, ewes, chickens, rodents, rabbits and fish. However, monkeys, swines and rats are the most commonly animal models used (Figure 10.5).

Although *in vivo animal* tests reflect the real situation, several drawbacks are associated with *in vivo* animal approaches. Testing with animals is expensive, difficult to perform, and ethically controversial, it provides limited data in each experiment and doubts of overall contribution of factors involved on bioaccessibility. The lack of certified reference standards to compare data among studies/laboratories also limits the quality of *in vivo* tests. In addition, the relevance for human situation derived from *in vivo* animal studies may not always be extrapolated (Fernández-García *et al.*, 2009) (Figure 10.3).

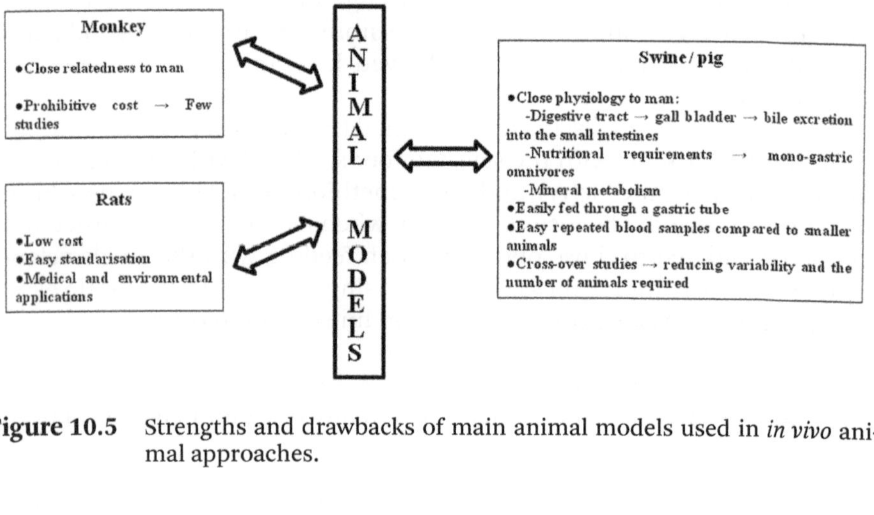

Figure 10.5 Strengths and drawbacks of main animal models used in *in vivo* animal approaches.

As an alternative, the bioavailability studies have also been estimated through various *in vitro* methods (see Chapter 11).

10.2 *In Vivo* Testing for Assessing Selenium Species Bioaccessibility/Bioavailability

Several *in vivo* human and animal tests have been used to evaluate the impact of selenium in foodstuffs on the organism. First, selenium-enriched food is directly dosed (after controlled oral intake) to subjects directly or after subjects were selenium depleted by using a diet with low selenium content. Secondly, selenium bioavailability was assessed by:

(1) Using foodstuff intrinsically labeled with stable selenium isotopes. Foodstuffs can be labeled with stable selenium isotopes in order to differentiate the selenium contained in the tested food items from the selenium coming from other constituents of the diet or from endogenous excretions.
(2) Measuring changes in plasma/blood or tissue selenium concentrations (restoration of plasma or tissue selenium levels in selenium-depleted subjects).
(3) Measuring selenium levels in feces and urine (absorption/retention studies, see eqn (10.1) and (10.2)).
(4) Measuring changes of activity or concentration in several functional biomarkers of selenium status (such as SeP, GPx and TrxR) in selenium-depleted humans or animals (restoration of selenoprotein P and GPx activities in selenium-depleted subjects).
(5) Functional assessing of the ability of selenium to ameliorate selenium-responsive disease conditions in humans or animals.

Among many other selenoenzymes, GPx is often monitored to estimate selenoenzyme activity because it is the most abundant selenoenzyme and also, its activity is easy to quantify. SeP is also a good functional biomarker due to there are measured in readily accessible tissues (blood) and appear to reach a plateau after 2–4 weeks of the supplementation period, and they are well correlated with plasma selenium across a wide range of selenium status (Fairweather-Tait *et al.*, 2010).

10.2.1 *In Vivo* Tests in Humans

Several *in vivo* tests have been developed with human for assessing selenium bioavailability estimation in foodstuffs: vegetables (mushroom, broccoli) (Mutanen, 1986; Finley, 1999), cereals (selenium-rich wheat, wheat biscuits, rice) (Levander *et al.*, 1981; Levander *et al.*, 1983; Thompson *et al.*, 1985; Mutanen, 1986; Van der Torre *et al.*, 1991; Meltzer *et al.*, 1993; Djujić *et al.*, 2000; Hawkes *et al.*, 2003; Kirby *et al.*, 2008), fish and seafood (Levander *et al.*, 1981; Meltzer *et al.*, 1993; Hagmar *et al.*, 1998; Bügel *et al.*, 2001; Fox *et al.*, 2004), yeast (Fox *et al.*, 2004), selenized yeast (Levander *et al.*, 1983; Bogye *et al.*, 1988; Nève *et al.*, 1988; Alfthan *et al.*, 1991; Xia *et al.*, 1992; Thompson *et al.*, 1993; Alfthan *et al.*, 2000; Larsen *et al.*, 2004; Bügel *et al.*, 2008; Bost and Blouin,2009) and selenized tablets (Finley *et al.*, 1999; Mahmoud, 2012), milk (Chen *et al.*, 2004), milk-based infant formulas (Van Dael *et al.*, 2002), Brazil nuts (Thompson *et al.*, 2008), eggs (Sirichakwal *et al.*, 1985), meat (beef (Van der Torre *et al.*, 1991; Hawkes *et al.*, 2003), pork (Bügel *et al.*, 2004) and chicken (Christensen *et al.*, 1983) meats), and self-selected diets (Levander *et al.*, 1984; Luo *et al.*, 1985).

Some conclusions can be derived from these studies:

(a) Selenium levels in plasma and/or whole blood increased significantly after supplementation with rich selenium foods or selenium enriched foodstuffs. High increases were reached after supplementation with wheat (Thompson *et al.*, 1985; Mutanen, 1986; Van der Torre *et al.*, 1991; Djujić *et al.*, 2000; Kirby *et al.*, 2008), yeast and selenized tablets (Levander *et al.*, 1983; Alfthan *et al.*, 1991; Xia *et al.*, 1992; Thompson *et al.*, 1993; Alfthan *et al.*, 2000; Larsen *et al.*, 2004; Bügel *et al.*, 2008; Bost and Blouin, 2009; Mahmoud, 2012), rice (Hawkes *et al.*, 2003) and Brazil nuts (Thompson *et al.*, 2008). Fecal and urinary selenium excretion was also increased during post-supplementation.

(b) Levels of plasma selenium, GPx and SeP reached after supplementation were significantly dependent on the selenium intake level and the amount of ingested foodstuff. The amount of foodstuffs is highly correlated with plasma selenium, SeP and GPx levels (Hagmar *et al.*, 1998).

(c) The presence of different selenium species in foodstuffs determines the selenium level increase in plasma/whole blood. Plasma selenium concentrations after supplementation with selenium-enriched wheat were found to increase from 122 $\mu g\ L^{-1}$ to 194 $\mu g\ L^{-1}$ after 6 months; in contrast, supplementation with process-fortified selenium biscuits showed only a small increase in mean total selenium plasma concentrations mean (122 $\mu g\ L^{-1}$ at 0 months to 140 $\mu g\ L^{-1}$ at 4 months). The difference in total selenium plasma concentrations may be due to the presence and bioavailability of different selenium species in fortified wheat (the main selenium species is SeMet) and process-fortified biscuits (the main selenium species is selenomethionine-Se-oxide, SeOMet) (Kirby *et al.*, 2008). The bioavailability in mushroom (Mutanen, 1986) is different from that found in wheat (Thompson *et al.*, 1985; Mutanen, 1986; Van der Torre *et al.*, 1991; Djujić *et al.*, 2000; Kirby *et al.*, 2008) and yeast (Levander *et al.*, 1983; Alfthan *et al.*, 1991, 2000; Xia *et al.*, 1992; Bügel *et al.*, 2008; Bost and Blouin, 2009). Most of the selenium in wheat and yeast is thought to exist as organoselenium compounds (SeMet), while 80% of selenium in mushrooms has been shown to exist as nonprotein compounds. SeMet is a good source of selenium that is both deposited in tissues as well as being available for GPx. The different forms of selenium also explain the low selenium bioavailability in mushrooms.

(d) Plasma/whole blood selenium levels remained elevated during the postsupplementation period.

(e) A slight plasma and platelet GPx activity enhancement was found after supplementation, except for yeast (Alfthan *et al.*, 1991) and wheat and fish (Meltzer *et al.*, 1993). The activity remained slightly elevated for several weeks during the postsupplementation period.

(f) Supplementation with selenium-rich food (*i.e.* shrimps) in addition to their habitual diet and with fortified foodstuffs increased the apparent selenium absorption and retention (Bügel *et al.*, 2001).

(g) Food processing and cooking could also affect selenium bioavailability. The apparent selenium absorption of selenium in cooked trout fish (87.8%) and the salted trout fish (90.4%) (Fox *et al.*, 2004) was significantly different in freeze-dried ground tuna (62%) (Levander *et al.*, 1981). However, cooking did not affect the apparent selenium absorption or retention in fish (Fox *et al.*, 2004).

(h) The composition of foodstuffs also affected significantly selenium bioavailability. High apparent selenium absorption in pork meat (94%) (Bügel *et al.*, 2004), milk-based infant formulas (73–97%) (Van Dael *et al.*, 2002), yeast (54–89%) (Fox *et al.*, 2004; Bügel *et al.*, 2008), fish (62–90.4%) (Levander *et al.*, 1981; Fox *et al.*, 2004) and eggs (54.1–85%) (Sirichakwal *et al.*, 1985) were assessed, with values significantly different from selenium absorbed in broccoli (72%) (Mutanen, 1986) or chicken meat (71–72%) (Christensen *et al.*, 1983). Despite of the high selenium bioavailability in pork, the plasma selenium concentration and the red blood cell GPx activity were not affected by the diet intervention (Bügel *et al.*, 2004). In the same way, selenium bioavailability in milk was also high (estimated fractional absorption between 61 and 77%) (Chen *et al.*, 2004).

Regardless of the high selenium bioavailability assessed in fish and seafood (Levander *et al.*, 1981; Fox *et al.*, 2004), Meltzer *et al.* (1993) reported a low selenium bioavailability in selenium-rich fish (trout, mackerel, wolffish, rosefish, and haddock) in humans. Several reasons that explain the low selenium plasma/whole blood and low GPx increase were reported by the authors:

(1) Selenium was not well absorbed, due to interactions between the mercury and arsenic content in fish and selenium. The mercury and arsenic intake in selenium-rich fish used in this study (18 and 806 µg per day for mercury and arsenic, respectively), which made it probable that some interaction takes place. However, some authors reported that mercury content in erythrocytes is correlated with both selenium in plasma, SeP and GPx, which minimizes the antagonistic effect of mercury (Hagmar *et al.*, 1998).

(2) Heavy metals and selenium interactions occurred in forms like selenium compounds or seleno–sulfur bonds, which are poorly available.
(3) The high methionione (Met) and polyunsaturated fatty acids content in selenium-rich fish species reduces deposition of Met-bound Se in tissues and increased the oxidant stress, thereby compromising selenium status by increasing the turnover of selenium through GPx. However, other studies showed that the resulting differences in fatty acid composition of the human diets, with a higher content of polyunsaturated fatty acids appeared not to affect selenium absorption or retention (Bügel *et al.*, 2004).

On the other hand, the low bioavailability of selenium in mushroom could be due to the high intake of cadmium (45 µg per day) and mercury (20 µg per day) content in food. Cadmium and mercury would have complexed selenium into a nonavailable form already in the gastrointestinal tract, decreasing the GPx activity (Mutanen, 1986).

Finally, it has to be remarked that the results obtained are only valid under the established conditions valid for the examined group (healthy subjects with a well-defined basal selenium status, and selenium-deficient subjects). Individual selenium requirement actually depends on many factors that have not yet been considered in these studies, such as pro- and antioxidant status.

10.2.2 *In Vivo* Tests in Animals

Similarly to *in vivo* human tests, animals are first made selenium deficient with a low selenium diet in *in vivo* animal tests before feeding with the food of interest. During the period of repletion, the extent and the rapidity of the selenium status recovery are measured. The percentage of selenium concentration/GPx and TrxR activities increase in several tissues is then achieved. In addition, the apparent absorption (eqn (10.1)) and slope-ratio assay (eqn (10.3)) are also estimated.

Several *in vivo* animal studies have been carried out to assess the bioavailability of selenium from naturally selenium-rich or selenized foodstuffs: wheat, oat, vegetables, seafood, algae, yeast, milk, Brazil nuts, and meat. Rat, fish, chicken and ewes are the main animal models used for selenium bioavailability assessment.

10.2.2.1 *Selenium Bioavailability in Vegetables*

Several vegetables and selenized vegetables such as peas (Yan and Johnson, 2011), soybean and soybean products (Gabrielsen and Opstvedt, 1980; Marks and Mason, 1993; Yan *et al.*, 2009, 2010; Jang *et al.*, 2010), broccoli (Finley, 1998; Finley *et al.*, 2000), mushroom (Chansler *et al.*, 1986; da Silva *et al.*, 2010), garlic, cabbage and ramps (Whanger *et al.*, 2000; Seo *et al.*, 2008) have been used as selenium sources in several *in vivo* animal approaches.

Selenium in naturally produced high-selenium yellow peas and oats (harvested in selenized soils) is highly bioavailable (Yan and Johnson, 2011). The overall selenium bioavailability was approximately 88% and 92% higher in yellow peas and oats, respectively, than selenium bioavailability when testing SeMet solutions.

The bioavailability of selenium in soybean and soybean products was found to be very high (Marks and Mason, 1993; Yan *et al.*, 2009, 2010; Jang *et al.*, 2010). On the contrary, Gabrielsen and Opstvedt (1980) reported very low selenium bioavailability ratios in soybean and corn gluten (18% and 26% for soybean and corn gluten, respectively). Selenium in soy was as bioavailable as sodium selenite in the reference diet. In addition, selenium in soy grown with sodium selenite was as bioavailable for rats as selenium in soy grown with sodium selenate. These data indicate that soy protein provided a bioavailable source of dietary selenium to rats and that the form of selenium available in plants did not influence the selenium bioavailability of the resulting soy-based product (Marks and Mason, 1993). Finally, high bioavailability was also obtained for protein isolate from soybean (101%) and tofu (94%) (Yan *et al.*, 2010).

Selenium in high-selenium broccoli was much less effective than selenite, selenate and SeMet in restoring kidney and plasma selenium concentrations, and GPx activity (Finley, 1998; Finley *et al.*, 2000, 2004). Selenium in high-selenium broccoli is absorbed, distributed, and retained in a different manner than selenium from selenite, selenate and SeMet solutions (Finley *et al.*, 2000).

Mushrooms have a very low capacity in stimulating selenium levels and GPx activity, which suggests that mushrooms may contain slightly available selenium species such as methylated compounds (Chansler *et al.*, 1986). However, the selenium concentration in the plasma of rats fed with selenized mushrooms was higher than in rats fed with a normal diet containing selenate (da Silva *et al.*, 2010). These results showed that selenium-enriched mushrooms could increase the selenium bioavailability of this food (da Silva *et al.*, 2010).

Selenium in selenized garlic, cabbage and ramps (Whanger *et al.*, 2000; Seo *et al.*, 2008) has been found to be equally bioavailable to selenium in selenite and selenate solutions when bioavailability was based on repletion of tissue selenium concentrations and GPx activity (Seo *et al.*, 2008). Bioavailability studies with rats indicated that selenium in ramps was 15–28% more available for regeneration of GPx activity than inorganic selenium as selenite (Whanger *et al.*, 2000). Thus, selenium in Se-enriched garlic, cabbage and ramp is highly bioavailable and can potentially be beneficial in enhancing selenium status and GPx activity.

10.2.2.2 Selenium Bioavailability in Cereals

Selenium bioavailability in selenized and nonselenized wheat grain, wheat flour, wheat bread and bran and corn gluten (Gabrielsen and Opstvedt, 1980; Douglass *et al.*, 1981; Alexander *et al.*, 1983; House and Welch, 1989; Reeves *et al.*, 2005, 2007) has also been assessed. Selenium bioavailability in wheat and selenized wheat is similar to that obtained when using SeMet. This could be due to SeMet represents almost the totality of the selenium species in wheat grain. Several studies have shown that selenium in wheat is well retained by various body tissues (Douglass *et al.*, 1981; House and Welch, 1989), especially when it is milled in meal because this process disrupts the fibrous cell walls that can impede selenium bioaccessibility (Reeves *et al.*, 2007). GPx activity induced by selenized wheat also seems to be satisfactory (Reeves *et al.*, 2007). In addition, similar selenium bioavailability was found for whole wheat flour, whole wheat bread or bran (Alexander *et al.*, 1983). Similarly, the ability of selenium in buckwheat bran to restore selenium-dependent enzymes activity and tissue selenium concentrations in selenium-deficient rats was 70–80% as high as selenite or SeMet (Reeves *et al.*, 2005). Finally, the restoration of liver TrxR activity for selenium in buckwheat bran was only 40% as high as selenite and 70% as high as SeMet (Reeves *et al.*, 2005).

10.2.2.3 Selenium Bioavailability in Fish

The biological availability in high-selenium seafood such as crab, oyster, shrimp and Baltic herring was close to that observed when administrating selenite solutions (the criterion used was either plasma Se level or plasma GPx activity) (Mutanen *et al.*, 1986a). When using the criterion of increased liver GPx activity, herring had a biological availability comparable to that observed for selenite solutions; whereas, crab and oyster were distinctly inferior in this regard (Mutanen *et al.*, 1986a). Some studies have reported that selenium bioavailability in fish is low, especially when compared to that obtained with wheat (Douglass *et al.*, 1981; Alexander *et al.*, 1983). Tuna increased selenium concentration in the body tissues, and RBC was similar or slightly lower than that obtained for selenite solutions and wheat (Douglass *et al.*, 1981; Alexander *et al.*, 1983). Concerning GPx induction, tuna was less efficient than both wheat and selenite solution (Douglass *et al.*, 1981; Alexander *et al.*, 1983; Wen *et al.*, 1997), which suggests that selenium in wheat is more available than in tuna. Selenium in salmon and shrimps was found to be better, or equally, absorbed and retained than when administrating inorganic selenium (Mutanen *et al.*, 1986a; Ørnsrud and Lorentzen, 2002). Finally, a high GPx activity increase (~100–540) was reported in processed anchovy (Haratake *et al.*, 2007).

10.2.2.4 Selenium Bioavailability in Algae

The bioavailability of selenium in selenium-rich Spirulina was less effective than sodium selenite and SeMet solutions in restoring selenium concentration in the liver but not in kidney. Similarly, selenium in Spirulina was less effective than the other forms of selenium in restoring GPx activity, except in plasma and red blood cells for which no differences were noted among the three selenium sources (Cases *et al.*, 2001). However, selenium in the retentate fraction of Spirulina (resulting from ultrafiltration of the supernatant through a 30 kDa exclusion membrane) was highly bioavailable. The retentate fraction allowed full replenishment of selenium concentration in liver and kidney (as did sodium selenite) and GPx activity in liver, kidney, plasma, and erythrocytes (Cases *et al.*, 2002).

10.2.2.5 Selenium Bioavailability in Yeast

Selenium yeast increased selenium status to a similar extent to SeMet. In addition, selenium in yeast was almost twice as bioavailable as selenium in selenite and selenate for restoration of depleted GPx activity (Smith and Picciano, 1987; Wang and Lovell, 1997; Yoshida *et al.*, 1999; Mateo *et al.*, 2007; Wang and Xu, 2008; Rider *et al.*, 2009, 2010; Jang *et al.*, 2010; Hall *et al.*, 2012; Briens *et al.*, 2013).

10.2.2.6 Selenium Bioavailability in Milk

Cow and goat milk and selenized milk have also been tested (Mutanen *et al.*, 1986b; Alférez *et al.*, 2003; Barrionuevo *et al.*, 2003). Selenited-barley milk was more available than selenite milk when the plasma selenium level was used as the response criterion. Otherwise, the bioavailability of selenium in several milks was not different when plasma or liver GPx activities were used as the response criteria. Overall bioavailability for selenite milk, selenite barley milk, and American milk was only slightly lower than that that observed for selenite. These findings show that milk can be a readily available source of dietary selenium (Mutanen *et al.*, 1986b).

Freeze-dried goat and cow milk was also used to evaluate selenium bioavailability. The apparent absorption and retention of selenium were higher for a goat-milk-based diet than for diets with cow milk (Alférez *et al.*, 2003; Barrionuevo *et al.*, 2003).

10.2.2.7 Selenium Bioavailability in Brazil Nuts

Based on selenium contents in plasma and liver, and also on experiments regarding plasma and liver GPx activity restoring in rodents, selenium can be considered as fully available in Brazil nuts (Chansler *et al.*, 1986).

10.2.2.8 Selenium Bioavailability in Meat

Concerning beef, pork, chicken and lamb, the GPx activity and the increase of selenium concentration in several tissues is similar to those observed when using a selenite reference diet (Shi and Spallholz, 1994*a*, 1994*b*; Wen *et al.*, 1997). The bioavailability of selenium achieved was 77, 77, 80, 86 and 58% for veal, chicken, beef, pork and lamb, respectively. However, selenium in beef, veal, chicken, pork and lamb was not sufficient to restore liver and muscle selenium after nine weeks of recovery following a six-week period of selenium depletion (Wen *et al.*, 1997). In addition, the bioavailability of selenium in various commercial portions of beef, liver, striploin, round, shoulder and brisket was also assessed (Shi and Spallholz, 1994*a*, 1994*b*). The GPx activity was 108, 105, 106, 106 and 103% for liver, striploin, round, shoulder and brisket, respectively. The experimental results suggest that all cuts of beef appear to be highly bioavailable sources of dietary selenium when compared with selenite or SeMet (Shi and Spallholz, 1994*a*, 1994*b*). Finally, regarding sheep meat, selenium levels and GPx activity were found to be higher when feeding rats with the sheep muscle than when feeding with the liver and hemoglobin (Butler *et al.*, 1991).

10.2.2.9 Effect of Food Processing on Selenium Bioavailability

Food processing and cooking could also affect selenium bioavailability. Thus, the bioavailability of raw, precooked and canned tuna, and whole-wheat flour, whole-wheat bread and bran was assessed (Alexander *et al.*, 1983). No differences were found among the three tuna products and among the three wheat products tested. There was a significant increase in the liver selenium content of rats fed with canned tuna and in kidney, blood and muscle of rats fed with canned tuna in comparison to rats fed with the same levels of selenium in raw or precooked tuna. Although selenium levels were different when feeding with cooked, raw, precooked and canned tuna, the similar GPx activity values measured did not allow conclusive results regarding selenium bioavailability for cooked and uncooked tuna to be obtained. Thus, food processing does not appear to affect selenium availability, but selenium appears to be more available in wheat than in tuna (Alexander *et al.*, 1983). Similarly, the bioavailability of selenium in raw and cured selenized-salmon fillets was assessed in selenium-deficient rats (Ørnsrud and Lorentzen, 2002). The data for fractional apparent absorption and fractional retention showed higher bioavailability in raw salmon than for cured salmon. These results suggested that curing salmon altered the utilization of selenium, but further work is necessary to clarify the effect of processing on fish fillets with regard to the bioavailability of selenium (Ørnsrud and Lorentzen, 2002).

10.2.2.10 Effect of Sample Composition on Selenium Bioavailability

The composition of foodstuffs also significantly affected selenium bioavailability. Selenium retention increased in iron-deficient diets (Díaz-Castro *et al.*, 2011). This contributes to keep the enzymatic antioxidant activity of GPx at normal levels and to store the increase of selenium in the organs, especially in kidney.

Similarly, zinc had an antagonistic effect on selenium absorption by selenium-depleted rats (House and Welch, 1989). The antagonistic effect occurred between natural forms of zinc and selenium at concentrations potentially encountered in wheat grain.

Heavy metals such as mercury (Ralston *et al.*, 2008) and dietary sulfur (Met) (Waschulewski and Sunde, 1988) tend to reduce selenium bioavailability in *in vivo* studies with rats. Methylmercury–selenocysteine complexes appear likely to impair selenium bioavailability, interrupting synthesis of selenium-dependent enzymes. Therefore, selenoenzymes may be the molecular target of methylmercury toxicity. Sulfur competes with selenium for absorption and utilization.

The effect of dietary fat on the availability of selenium was investigated in chicks fed either 4% or 20% butter, olive oil, rape oil, corn oil or sunflower oil in the diet (Mutanen and Mykkănen, 1984). Plasma GPx levels increased with increasing proportions of the polyunsaturated fatty acids in the diet. Increasing the amount of fat from 4 to 20% significantly enhanced the GPx activity in the groups receiving butter and olive oil, but had no effect in animals fed with unsaturated fats. The absorption of selenium tended to be lower in chicks fed with corn oil and sunflower oil than that observed in animals receiving butter in their diet. On the other hand, the type of dietary fat did not appear to affect the absorption of the orally administered selenite (Mutanen and Mykkănen, 1984).

The potential interaction between selenium and vitamin E in juvenile hybrid striped bass was also evaluated. Whole-body selenium and whole-body selenium retention were linearly influenced by sodium selenite and selenomethionine. However, there was no significant effect of vitamin E on selenium availability (Jaramillo *et al.*, 2009).

Finally, the effect of ethanol consumption affects selenium bioavailability in rat and in their progeny. Ethanol decreased selenium retention, affecting the tissue selenium deposits. Their progeny's weight and oxidation balance was committed. The effects of ethanol were caused by a direct alcohol-generated oxidation action (Ojeda *et al.*, 2009).

Summary Points

- This chapter focuses on selenium and selenium species bioavailability in foodstuffs.
- *In vivo* (human and animal) bioavailability assays for releasing selenium and selenium species in food samples is summarized.
- *In vivo* human bioavailability is assessed by apparent absorption/retention assays.
- *In vivo* animal bioavailability is assessed by apparent absorption/retention, slope ratio and apparent digestability assays.
- Selenium species bioavailability depends on foodstuff processing and cooking.
- Selenium species bioavailability depends on foodstuff composition.

Key Facts of *In Vivo* Testing for Bioavailability Assessment

1. *In vivo* testing involves subjects (volunteer people or animals) to bioavailability assessment.
2. First, subjects are depleted by using a diet with low concentration of target compounds.
3. Target compound levels in tissues are monitored.
4. Secondly, subjects are repleted by using a diet enriched with target compounds.
5. The increase of target compound levels in tissue and further balance studies allows for bioavailability estimation.
6. Target compounds, isotopically labeled or nonisotopically labeled, are generally used in these approaches.

Definitions of Words and Terms

Bioaccessibility. This refers to the fraction of an administered dose of a substance that is soluble in the gastrointestinal tract and that is therefore available to reach the systemic circulation.

Bioactivity. This refers to the fraction of an administered dose of a substance that is absorbed in the gastrointestinal tract, reaches the systemic circulation and is distributed to organs and tissues, that is transformed into a biochemically active form and that effectively is used by the organism.

Bioavailability. This refers to the fraction of an administered dose of a substance that reaches the systemic circulation.

Depletion period. This is the time period (usually several weeks) during which the living organism (animals or people) is fed with a diet depleted with the target substance. During this time period, target substance concentration is decreasing in the body tissues.

GPx. Glutathione peroxidase (GPx) is a selenium-containing enzyme which plays a crucial role in protecting cells from damage by free radicals. The measurement of GPx activity in blood and tissues are an index of selenium status in the organism.

***In vivo* tests.** These involve experimentation using living organism; human (*in vivo* human) and animals (*in vivo* animals).

Plateau period. During the repletion period, the target substance concentration in the body tissues is increased until a plateau value and then remains constant (plateau period).

Repletion period. This is the time period (usually several weeks) during which the living organism (animals or people) is fed with a diet enriched with the target substance. During this time period, the target substance concentration is increasing in the body tissues.

SeP. Selenoprotein P is a selenocysteine-containing protein with antioxidant properties. Levels of SeP in blood and tissue are an index of selenium status in the organism.

TrxR. Thioredoxin reductase (TrxR) is a selenocysteine-containing flavoprotein that plays a key role in regulating cellular redox homeostasis. The measurement of TrxR activity in blood and tissue are an index of selenium status in the organism.

Abbreviations and Acronyms

GI	Gastrointestinal
GPx	Glutathione Peroxidase
Met	Methionine
RBC	Red Blood Cell
Se(IV)	Selenite
Se(VI)	Selenate
SeCys	Selenocysteine
SeMet	Selenomethionine
SeOMet	Selenomethionine-Se-oxide
SeP	Selenoprotein P
TrxR	Thioredoxin reductase

References

Alexander, A. R., Whanger, P. D. and Miller, L. T., 1983. Bioavailability to rats of selenium in various tuna and wheat products. *Journal of Nutrition*. 113: 196–204.

Alférez, M. J. M., López-Aliaga, I., Barrionuevo, M. and Campos, M. S., 2003. Effect of dietary inclusion of goat milk on the bioavailability of zinc and selenium in rats. *Journal of Dairy Research*. 70: 181–187.

Alfthan, G., Aro, A., Arvilommi, H. and Huttunen, J. K., 1991. Selenium metabolism and platelet glutathione peroxidase activity in healthy Finnish men: effects of selenium yeast, selenite, and selenate. *American Journal of Clinical Nutrition*. 53: 120–125.

Alfthan, G., Xu, G.-L., Tan, W.-H., Aro, A., Wu, J., Yang, Y.-X., Liang, W.-S., Xue, W.-L. and Kong, L.-H., 2000. Selenium supplementation of children in a selenium-deficient area in China. Blood selenium levels and glutathione peroxidase activities. *Biological Trace Element Research*. 73: 113–125.

Barrionuevo, M., López-Aliaga, I., Alférez, M. J. M., Mesa, E., Nestares, T. and Campos, M. S., 2003. Beneficial effect of goat milk on bioavailability of copper, zinc and selenium in rats. *Journal of Physiology and Biochemistry*. 59: 111–118.

Bogye, G., Alfthan, G. and Machay, T., 1988. Bioavailability of enteral yeast-selenium in preterm infants. *Biological Trace Element Research*. 65: 143–151.

Bost, M. and Blouin, E., 2009. Effect of supplementation with Se-enriched yeast and factors influencing Se concentration in plasma of transplant recipients. *Journal of Trace Elements in Medicine and Biology*. 23: 36–42.

Briens, M., Mercier, Y., Rouffineau, F., Vacchina, V. and Geraert, P. A., 2013. Comparative study of a new organic selenium source v. seleno-yeast and mineral selenium sources on muscle selenium enrichment and selenium digestibility in broiler chickens. *British Journal of Nutrition*. 110: 617–624.

Bügel, S. H., Sandström, B. and Larsen, E. H., 2001. Absorption and retention from shrimps in man. *Journal of Trace Elements in Medicine and Biology*. 14: 198–204.

Bügel, S., Sandtröm, B. and Skibsted, L. H., 2004. Pork meat: a good source of selenium? *Journal of Trace Elements in Medicine and Biology*. 17: 307–311.

Bügel, S., Larsen, E. H., Sloth, J. J., Flytlie, K., Overvad, K., Steenberg, L. C. and Moesgaard, S., 2008. Absorption, excretion, and retention of selenium from a high selenium yeast in men with a high intake of selenium. *Food and Nutrition Research*. 52: 1–8.

Butler, J. A., Deagen, J. T., Van Ryssen, J. B. J., Rowe, K. E. and Whanger, P. D., 1991. Bioavailability to rats of selenium in ovine muscle, liver and hemoglobin. *Nutrition Research*. 11: 1293–1305.

Cases, J., Vacchina, V., Napolitano, A., Caporiccio, B., Besançon, P., Lobinski, R. and Rouanet, J. M., 2001. Selenium from selenium-rich spirulina is less bioavailable than selenium from sodium selenite and selenomethionine in selenium-deficient rats. *Journal of Nutrition*. 131: 2343–2350.

Cases, J., Wysocka, I. A., Caporiccio, B., Jouy, N., Besançon, Szpunar, J. and Rouanet, J. M., 2002. Assessment of selenium bioavailability from high-selenium spirulina subfractions in selenium-deficient rats. *Journal of Agricultural and Food Chemistry*. 50: 3867–3873.

Chansler, M. W., Mutanen, M., Morris, V. C. and Levander, O. A., 1986. Nutritional bioavailability to rats of selenium in Brazil nuts and mushrooms. *Nutrition Research*. 6: 1419–1428.

Chen, J., Lindmark-Månsson, H., Drevelius, M., Tidehag, P., Hallmans, G., Hertervig, E., Nilsson, Å. and Åkesson, B., 2004. Bioavailability of selenium from bovine milk as assessed in subjects with ileostomy. *European Journal of Clinical Nutrition*. 58: 350–355.

Christensen, M. J., Janghorbani, M., Steinkes, F. H., Istfan, N. and Young, V. R., 1983. Simultaneous determination of absorption of selenium from poultry meat and selenite in young men: application of a triple stable-isotope method. *British Journal of Nutrition*. 50: 43–50.

Díaz-Castro, J., Ojeda, M. L., Alférez, M. J. M., López-Aliaga, I., Nestares, T. and Campos, M. S., 2011. Se bioavailability and glutathione peroxidase activity in iron deficient rats. *Journal of Trace Elements in Medicine and Biology*. 25: 42–46.

Djujić, I. S., Josanov-Stankov, O. N., Milovac, M., Jankvić, V. and Djermanović, V., 2000. Bioavailability and possible benefits of wheat intake naturally enriched with selenium and its products. *Biological Trace Element Research*. 77: 273–285.

Douglass, J. S., Morris, V. C., Soares, J. H. and Levander, O. A., 1981. Nutritional availability to rats of selenium in tuna, beef kidney, and wheat. *Journal of Nutrition*. 111: 2180–2187.

Fairweather-Tait, S. J., Collings, R. and Hurst, R., 2010. Selenium bioavailability: current knowledge and future research requirements. *American Journal of Clinical Nutrition*. 91(suppl. 5): 1484S–1491S.

Fernández-García, E., Carvajal-Lérida, I. and Pérez-Gálvez, A., 2009. *In vitro* bioaccessibility assessment as a prediction tool of nutritional efficiency. *Nutrition Research*. 29: 751–760.

Finley, J. W., 1998. The absorption and tissue distribution of selenium from high-selenium broccoli are different from selenium from sodium selenite, sodium selenate, and selenomethionine as determined in selenium-deficient rats. *Journal of Agricultural and Food Chemistry*. 46: 3702–3707.

Finley, J. W., 1999. The retention and distribution by healthy young men of stable isotopes of selenium consumed as selenite, selenate or hydroponically grown broccoli are dependent on the isotopic form. *Journal of Nutrition*. 129: 865–871.

Finley, J. W., Duffield, A., Ha, P., Vanderpool, R. A. and Thompson, C. D., 1999. Selenium supplementation affects the retention of stable isotopes of selenium in human subjects consuming diets low in selenium. *British Journal of Nutrition*. 82: 357–360.

Finley, J. W., Davis, C. D. and Feng, Y., 2000. Selenium from high selenium broccoli protects rats from colon cancer. *Journal of Nutrition*. 130: 2384–2389.

Finley, J. W., Grusak, M. A., Keck, A. S. and Gregoire, B. R., 2004. Bioavailability of selenium from meat and broccoli as determined by retention and distribution of ^{75}Se. *Biological Trace Elements Research*. 99: 191–209.

Finley, J. W., 2006. Bioavailability of selenium from foods. *Nutrition Reviews*. 64: 146–151.

Fox, T. E., Van den Heuvel, E. G. H. M., Atherton, C. A., Dainty, J. R., Lewis, D. J., Langford, N. J., Crews, H. M., Luten, J. B., Lorentzen, M., Sieling, F. W., van Aken-Schneyder, P., Hoek, M., Kotterman, M. J. J., van Dael, P. and Fairweather-Tait, S. J., 2004. Bioavailability of selenium from fish, yeast and selenate: a comparative study in humans using stable isotopes. *European Journal of Clinical Nutrition*. 58: 343–349.

Gabrielsen, B. O. and Opstvedt, J., 1980. Availability of selenium in fish meal in comparison with soybean meal, corn gluten meal and selenomethionine relative to selenium in sodiumselenite for restoring glutathione peroxidase activity in selenium-depleted chicks. *Journal of Nutrition*. 110: 1096–1100.

Hagmar, L., Persson-Moschos, M., Åkesson, B. and Schütz, A., 1998. Plasma levels of selenium, selenoprotein P and glutathione peroxidase and their correlations to fish intake and serum levels of thyrotropin and thyroid hormones: a study on Latvian fish consumers. *European Journal of Clinical Nutrition*. 52: 796–800.

Hall, J. A., Van Saun, R. J., Bobe, G., Stewart, W. C., Vorachek, W. R., Mosher, W. D., Nichols, T., Forsberg, N. E. and Pirelli, J. G., 2012. Organic and inorganic selenium: I. Oral bioavailability in ewes. *Journal of Animal Sciences*. 90: 568–576.

Haratake, M., Takahashi, J., Ono, M. and Nakayama, M., 2007. An assessment of Niboshi (a processed Japanese anchovy) as an effective food source of selenium. *Journal of Health Science*. 53: 457–463.

Hawkes, W. C., Alkan, F. Z. and Oehler, L., 2003. Absorption, distribution and excretion of selenium from beef and rice in healthy North American men. *Journal of Nutrition*. 133: 3434–3442.

House, W. A. and Welch, R. M., 1989. Bioavailability of and interactions between zinc and selenium in rats fed wheat grain intrinsically labeled with ^{65}Zn and ^{75}Se. *Journal of Nutrition*. 119: 916–921.

Intawongse, M. and Dean, J. R., 2006. *In vitro* testing for assessing oral bioaccessibility of trace metals in soil and food samples. *Trends in Analytical Chemistry*. 25: 876–886.

Jaramillo, F., Peng, L. and Gatlin, D. M., 2009. Selenium nutrition of hybrid striped bass (*Morone chrysops* × *M. saxatilis*) bioavailability, toxicity and interaction with vitamin E. *Aquaculture Nutrition*. 15: 160–165.

Jang, Y. D., Choi, H. B., Durosoy, S., Schlegel, P., Choi, B. R. and Kim, Y. Y., 2010. Comparison of bioavailability of organic selenium sources in finishing pigs. *Asian-Australasian Journal of Animal Sciences*. 23: 931–936.

Kirby, J. K., Lyons, G. H. and Karkkainen, M. P., 2008. Selenium speciation and bioavailability in biofortified products using species-unspecific isotope dilution and reverse phase ion pairing-inductively coupled plasma-mass spectrometry. *Journal of Agricultural and Food Chemistry*. 56: 1772–1779.

Larsen, E. H., Hason, M., Paulin, H., Moesgaard, S., Reid, M. and Rayman, M., 2004. Speciation and bioavailability of selenium in yeast-based intervention agents used in cancer chemoprevention studies. *Journal of AOAC International*. 87: 225–232.

Levander, O. A., Sutherland, B., Morris, V. C. and King, J. C. J., 1981. Selenium balance in young men during selenium depletion and repletion. *American Journal of Clinical Nutrition*. 34: 2662–2669.

Levander, O. A., Alfthan, G., Arvilommi, H., Gref, C. G., Huttunen, J. K., Kataja, M., Koivistoinen, P. and Pikkarainen, J., 1983. Bioavailability of selenium to Finnish men as assessed by platelet glutathione peroxidase activity and other blood parameters. *American Journal of Clinical Nutrition*. 37: 887–897.

Levander, O. A., Sutherland, B. and Morris, V. C., 1984. Dietary selenium levels needed to maintain balance in North American adults consuming self-selected. *American Journal of Clinical Nutrition*. 39: 809–815.

Luo, X., Wei, H., Yang, C., Xing, J., Qiao, C., Feng, Y., Liu, J., Liu, Z., Wu, Q., Liu, Y., Stoecker, B. J., Spallholz, J. E. and Yang, S. P., 1985. Selenium intake and metabolic balance of 10 men from a low selenium area of China. *American Journal of Clinical Nutrition*. 42: 31–37.

Mahmoud, W. A., 2012. Bioavailability of selenocysteine complex in ten healthy volunteers using flameless atomic absorption spectrophotometer. *Asian Journal of Chemistry*. 24: 5705–5707.

Marks, H. S. and Mason, A. C., 1993. Selenium bioavailability of soy-based diets in rats. *Journal of Nutritional Biochemistry*. 4: 523–527.

Mateo, R. D., Spallholz, J. E., Elder, R., Yoon, I. and Kim, S. W., 2007. Efficacy of dietary selenium sources on growth and carcass characteristics of growing-finishing pigs fed diets containing high endogenous selenium. *Journal of Animal Science*. 85: 1177–1183.

Meltzer, H. M., Bibow, K., Paulsen, I. T., Mundal, H. H., Norheim, G. and Holm, H., 1993. Different bioavailability in humans of wheat and fish selenium as measured by blood platelet response to increased dietary Se. *Biological Trace Element Research*. 36: 229–241.

Moreda-Piñeiro, J., Moreda-Piñeiro, A., Romarís-Hortas, V., Moscoso-Pérez, C., López-Mahía, P., Muniategui-Lorenzo, S., Bermejo-Barrera, P. and Prada-Rodríguez, D., 2011. *In vivo* and *in vitro* testing to assess the bioaccessibility and the bioavailability of arsenic, selenium and mercury species in food samples. *Trends in Analytical Chemistry*. 30: 324–345.

Mutanen, M. L. and Mykkänen, H. M., 1984. Effect of dietary fat on plasma glutathione peroxidase levels and intestinal absorption of ^{75}Se-labeled sodium selenite in chicks. *Journal of Nutrition*. 114: 829–834.

Mutanen, M., 1986. Bioavailability of selenium in mushrooms, *Boletus edulis*, to young women. *International Journal of Vitaminology and Nutrition Research*. 56: 297–301.

Mutanen, M. L., Koivistoinen, P., Morris, V. C. and Levander, O. A., 1986a. Nutritional availability to rats of selenium in four seafoods: crab (*Callinectes sapidus*), oyster (*Crassostrea virginica*), shrimp (*Penaeus duorarum*) and Baltic herring (*Clupea harengus*). *British Journal of Nutrition*. 55: 219–225.

Mutanen, M. L., Aspila, P. and Mykkänen, H. M., 1986b. Bioavailability to rats of selenium in milk of cows fed sodium selenite or selenite barley. *Annals of Nutrition and Metabolism*. 30: 183–188.

Nève, J., Van Erum, S., Drama, I. X. M., Chamart, S. and Vertongen, F., 1988. Selenium status in humans as investigated by the effects of supplementation with Se-enriched yeast tablets. In: Nève, J. and Favier, A. (ed.) *Selenium in Medicine and Biology*. Walter de Gruyter, New York, pp. 315.

Ojeda, M. L., Vázquez, B., Nogales, F., Murillo, M. L. and Carreras, O., 2009. Ethanol consumption by Wistar rat dams affects selenium bioavailability and antioxidant balance in their progeny. *International Journal of Environmental Research and Public Health*. 6: 2139–2149.

Ørnsrud, R. and Lorentzen, M., 2002. Bioavailability of selenium from raw or cured selenomethionine-enriched fillets of Atlantic salmon (*Salmo salar*) assessed in selenium-deficient rats. *British Journal of Nutrition*. 87: 13–20.

Ralston, N. V. C., Ralston, C. R., Blackwell, J. L. and Raymond, L. J., 2008. Dietary and tissue selenium in relation to methylmercury toxicity. *Neurotoxicology*. 29: 802–811.

Reeves, P. G., Leary, P. D., Gregoire, B. R., Finley, J. W., Lindlauf, J. E. and Johnson, L. K., 2005. *Journal of Nutrition*. 135: 2627–2633.

Reeves, P. G., Gregoire, B. R., Garvin, D. F., Hareland, G. A., Lindlauf, J. E., Johnson, L. K. and Finley, J. W., 2007. Determination of selenium bioavailability from wheat mill fractions in rats by using the slope-ratio assay and a modified torula yeast-based diet. *Journal of Agricultural and Food Chemistry*. 55: 516–522.

Rider, S. A., Davies, S. J., Jha, A. N., Fisher, A. A., Knight, J. and Sweetman, J. W., 2009. Supra-nutritional dietary intake of selenite and selenium yeast in normal and stressed rainbow trout (*Oncorhynchus mykiss*): implications on selenium status and health responses. *Aquaculture*. 295: 282–291.

Rider, S. A., Davies, S. J., Jha, A. N., Clough, R. and Sweetman, J. W., 2010. Bioavailability of co-supplemented organic and inorganic zinc and selenium sources in a white fishmeal-based rainbow trout (*Oncorhynchus mykiss*) diet. *Journal of Animal Physiology and Animal Nutrition*. 94: 99–110.

Ruby, M. V., Schoof, R., Brattin, W., Goldade, M., Post, G., Harnois, M., Mosby, D. E., Casteel, S. W., Berti, W., Carpenter, M., Edwards, D., Cragin, D. and Chappell, W., 1999. Advances in evaluating the oral bioavailability of inorganics in soil for use in human health risk assessment. *Environmental Science and Technology*. 33: 3697–3705.

Schümann, K., Classen, H. G., Hages, M., Prinz-Langenohl, R., Pietrzik, K. and Biesalki, H. K., 1997. Bioavailability of oral vitamins, minerals, and trace elements in perspective. *Arzneimittel-Forschung Drug Research*. 47: 369–380.

Seo, T. C., Spallholz, J. E., Yun, H. K. and Kim, S. W., 2008. Selenium-enriched garlic and cabbage as a dietary selenium source for broilers. *Journal of Medical Food*. 11: 687–692.

Shi, B. and Spallholz, J. E., 1994a. Selenium from beef is highly bioavailable as assessed by liver glutathione peroxidase (EC 1.11.1.9) activity and tissue selenium. *British Journal of Nutrition*. 72: 873–881.

Shi, B. and Spallholz, J. E., 1994b. Bioavailability of selenium from raw and cooked ground-beef assessed in selenium-deficient Fischer rats. *Journal of the American College of Nutrition*. 13: 95–101.

da Silva, M.-C. S., Naozuka, J., Oliveira, P. V., Vanetti, M. C. D., Bazzolli, D. M. S., Costad, N. M. B. and Kasuya, M. C. M., 2010. In vivo bioavailability of selenium in enriched *Pleurotus ostreatus* mushrooms. *Metallomics*. 2: 162–166.

Sirichakwal, P. P., Young, V. R. and Janghorbani, M., 1985. Absorption and retention of selenium from intrinsically labeled egg and selenite as determined by stable isotope studies in humans. *American Journal of Clinical Nutrition*. 41: 264–269.

Smith, A. M. and Picciano, M. F., 1987. Relative bioavailability of selenocompounds in the lactating rat. *Journal of Nutrition*. 117: 725–731.

Thiry, C., Ruttens, A., De Temmerman, L., Schneider, Y.-J. and Pussemier, L., 2012. Current knowledge in species-related bioavailability of selenium in food. *Food Chemistry*. 130: 767–784.

Thompson, C. D., Ong, L. K. and Robinson, M. F., 1985. Effects of supplementation with high-selenium wheat bread on selenium, glutathione peroxidase and related enzymes in blood components of New Zealand. *American Journal of Clinical Nutrition*. 41: 1015–1022.

Thompson, C. D., Robinson, M. F., Butler, J. A. and Whanger, P. D., 1993. Long-term supplementation with selenate and selenomethionine: selenium and glutathione peroxidase (EC 1.11.1.9) in blood components of New Zealand women. *British Journal of Nutrition*. 69: 577–588.

Thompson, C. D., Chisholm, A., McLachlan, S. K. and Campbell, J. M., 2008. Brazil nuts: an effective way to improve selenium status. *American Journal of Clinical Nutrition*. 87: 379–384.

Van Dael, P., Davidsson, L., Ziegler, E. E., Fay, L. B. and Barclay, D., 2002. Comparison of selenite and selenate apparent absorption and retention in infants using stable isotope methodology. *Pediatric Research*. 51: 71–75.

Van der Torre, H. W., Van Dokkum, W. T. M., Schaafsma, G., Wedel, M. and Ockhuizen, T., 1991. Effect of various levels of selenium in wheat and meat on blood Se status indices and on Se balance in Dutch men. *British Journal of Nutrition*. 65: 69–80.

Wang, C. and Lovell, R. T., 1997. Organic selenium sources, selenomethionine and selenoyeast, have higher bioavailability than an inorganic selenium source, sodium selenite, in diets for channel catfish (*Ictalurus punctatus*). *Aquaculture*. 152: 223–234.

Wang, Y. and Xu, B., 2008. Effect of different selenium source (sodium selenite and selenium yeast) on broiler chickens. *Animal Feed Science and Technology*. 144: 306–314.

Whanger, P. D., Ip, C., Polan, C. E., Uden, P. C. and Welbaum, G., 2000. Tumorigenesis, metabolism, speciation, bioavailability, and tissue deposition of selenium in selenium-enriched eamps (*Allium tricoccum*). *Journal of Agricultural and Food Chemistry*. 48: 5723–5730.

Waschulewski, I. H. and Sunde, R. A., 1988. Effect of dietary methionine on utilization of tissue selenium from dietary selenomethionine for glutathione peroxidase in the rat. *Journal of Nutrition*. 118: 367–374.

Wen, H. Y., Davis, R. L., Shi, B., Chen, J. J., Chen, L., Boylan, M. and Spallhol, J. E., 1997. Bioavailability of selenium from veal, chicken, beef, pork, lamb, flounder, tuna, selenomethionine, and sodium selenite assessed in selenium-deficient rats. *Biological Trace Element Research*. 58: 43–53.

Xia, Y., Zhao, Y., Zhao, X., Zhu, L. and Whanger, P. D., 1992. Metabolism of selenate and selenomethionine by a selenium-deficient population of men in China. *Journal of Nutritional Biochemistry*. 3: 202–210.

Yan, L., Graef, G. L., Reeves, P. G. and Johnson, L. K., 2009. Selenium bioavailability from soy protein isolate and tofu in rats fed a torula yeast-based diet. *Journal of Agricultural and Food Chemistry*. 57: 11575–11580.

Yan, L., Reeves, P. G. and Johnson, L. K., 2010. Assessment of selenium bioavailability from naturally produced high-selenium soy foods in selenium-deficient rats. *Journal of Trace Elements in Medicine and Biology*. 24: 223–229.

Yan, L. and Johnson, L. K., 2011. Selenium bioavailability from naturally produced high-selenium peas and oats in selenium-deficient rats. *Journal of Agricultural and Food Chemistry*. 59: 6305–6311.

Yoshida, M., Fukunaga, K., Tsuchita, H. and Yasumoto, K., 1999. An evaluation of the bioavailability of selenium in high-selenium yeast. *Journal of Nutritional Science and Vitaminology*. 45: 119–128.

CHAPTER 11

Bioavailability of Selenium and Selenium Compounds in Foodstuff by In Vitro Testing

JORGE MOREDA-PIÑEIRO*[a] AND
ANTONIO MOREDA-PIÑEIRO[b]

[a]Department of Analytical Chemistry, Faculty of Sciences, University of A Coruña, Campus da Zapateira, s/n, 15071 – A Coruña, Spain; [b]Department of Analytical Chemistry, Nutrition and Bromatology, Faculty of Chemistry, University of Santiago de Compostela, Avenida das Ciencias, s/n, 15782 – Santiago de Compostela, Spain
*E-mail: jorge.moreda@udc.es

11.1 Introduction

As commented in a previous chapter (Chapter 10), the assessment of substances bioavailability in foodstuff can be performed by using *in vivo* and *in vitro* approaches. Strengths and drawbacks of *in vivo* and *in vitro* methods were summarized in Figure 10.2 (Chapter 10 regarding *in vivo* tests for selenium bioavailability assessment). In this chapter, theoretical and practical aspects of *in vitro* tests will be described. The application of *in vitro* approaches for selenium bioavailability assessment in foodstuff will be summarized and critically commented.

Food and Nutritional Components in Focus No. 9
Selenium: Chemistry, Analysis, Function and Effects
Edited by Victor R. Preedy

Published by the Royal Society of Chemistry, www.rsc.org

11.2 Bioavailability Assessment: *In Vitro* Tests

In vitro methods simulate the physiologic conditions and the sequence of events that occur in the human gastrointestinal (GI) tract during human body digestion (Miller *et al.*, 1981; Ruby *et al.*, 1999). Physiological conditions such as temperature, shaking or agitation, pH, and enzyme/chemical composition of saliva, gastric and intestine juices are the main conditions simulated. The sequence of events simulated in this approach enclosed processing in the mouth, in the stomach and in the duodenum.

In the mouth, samples are mechanical ground in the presence of the α-amylase (an enzyme that initiates hydrolysis of carbohydrates) present in saliva. Due to the entire process in the mouth lasting from a few seconds to minutes, and due to the saliva's pH being close to neutral, a significant compound dissolution from food samples is not expected during this stage. It is therefore usually assumed that saliva has only a negligible effect on the level of mobilization of compounds (Intawongse and Dean, 2006) and in most of the cases, only simulated gastric and intestinal extractions are considered.

Simulating gastrointestinal digestion occurs in two steps:

(1) Gastric digestion; *i.e.* with pepsin at pH 2 followed by intestinal digestion (*i.e.* with amylase, pancreatin and bile salts at neutral pH). During this step food nutrients are affected by the digestion conditions (pH, enzymes), and by interactions with other food components (fiber, fat and protein), which could affect efficiency of the simulated digestion process. After this step a digesta or intestinal preparation is then obtained.
(2) Digesta assimilation by the intestinal mucosa. Different experimental options are available for this purpose, *i.e.* dialysis membrane and cell-culture-based models (Caco-2 cell model).

Most studies on the assessment of the bioaccessibility/bioavailability of nutrients/target compounds in foodstuffs are based on *in vitro* digestion with commercially available enzymes of nonhuman origin, such as different proteolytic enzymes, and bile salts. Due to possible target compound changes during extractions, the use of nonhuman enzymes would lead to us false conclusions. The use of human gastric (HGJ) and duodenal (HDJ) juices (obtained by aspirating juices and collecting on ice before freezing at −80 °C) should give precise, more reproducible and realistic information on foodstuffs degradation than those obtained after using single nonhuman enzymes. The use of a single purified enzyme rather than the use of complex biological mixtures offers, however, the advantage of facilitating standardization of the *in vitro* digestion models (Hur *et al.*, 2011).

Simulated GI digestion can be performed following a static or a dynamic model, see Figure 11.1.

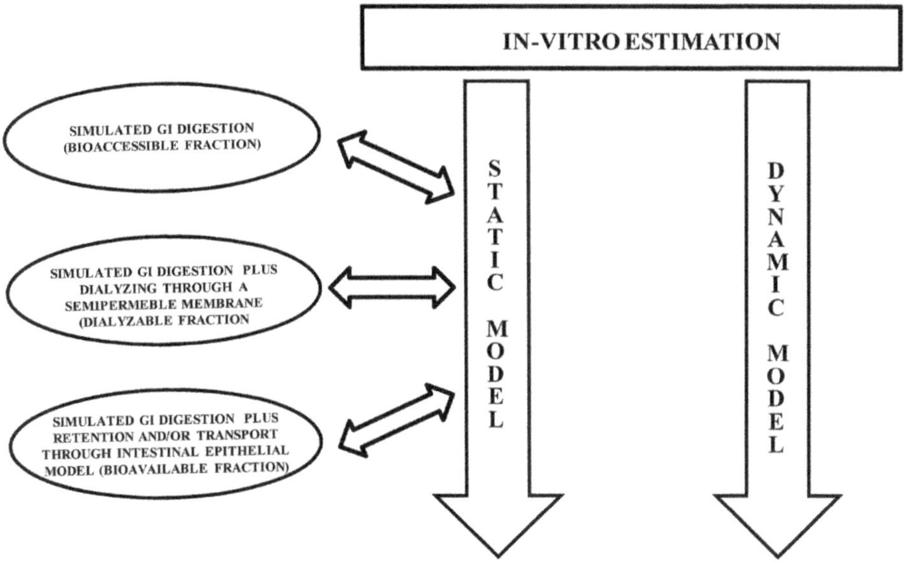

Figure 11.1 *In vitro* approaches classification.

11.2.1 *In Vitro* Static Models

Static models, also known as biochemical models, are defined as models where digestion products are not removed during the digestion process and that do not mimic the physical processes that occur *in vivo* (*i.e.* shear, mixing, hydration, changes in conditions over time, peristalsis, changes on viscosity, particle-size reduction, diffusion, and partitioning of nutrients) (Wickham *et al.*, 2009). These *in vitro* approaches are therefore oversimplified.

Static models involve homogenization of the food; acidification (with hydrochloric acid); addition of gastric and pancreatic enzymes and bile salts; simulating gastric and intestinal residence times whilst stirring at 37 °C, neutralization (with sodium carbonate or hydroxide). Commonly, the ratios of surfactants, enzymes and substrates are not physiological because the model is intended to cause exhaustive digestion (Wickham *et al.*, 2009).

There are several *in vitro* static approaches (Figure 11.1) such as:

(1) Measurements of the maximum soluble concentrations of the nutrient/target compound in the simulated GI solution after filtration or centrifugation (bioaccessible fraction). Figure 11.2 shows a schematic diagram of *in vitro* gastric and intestinal digestion to achieve the bioaccessible fraction. The bioaccessible fraction, expressed as a percentage, is calculated using the eqn (11.1):

$$\text{Bacc} = \frac{\text{Conc}_{\text{liquid extract}}}{\text{Conc}_{\text{food}}} \times 100 \quad (11.1)$$

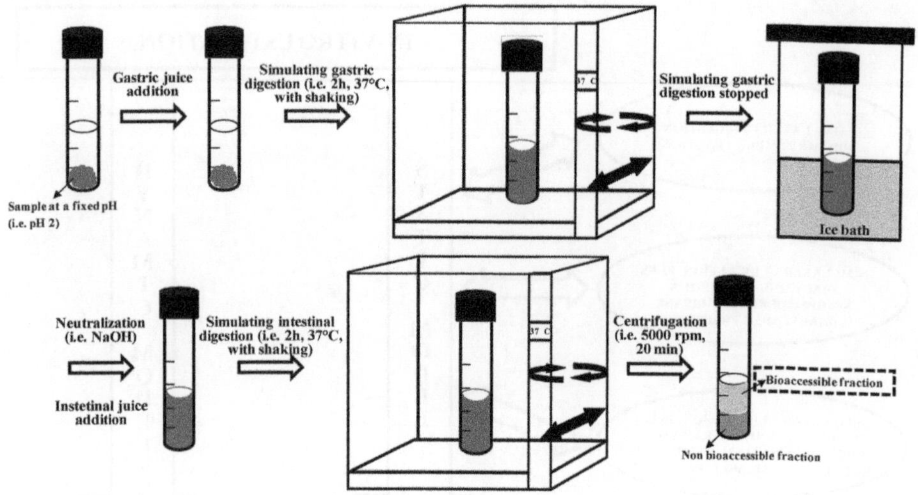

Figure 11.2 Schematic diagram of *in vitro* GI digestion.

where Bacc is the percentage of bioaccessibility, $Conc_{liquid\ extract}$ and $Conc_{food}$ are the nutrient/target compound concentrations after *in vitro* digestion and in a food sample before *in vitro* GI digestion, respectively.

(2) Measurement of the dialyzable fraction of the nutrient/compound that is capable of dialyzing (during intestinal digestion) through a semipermeable membrane with a specified pore size (dialysate or bioavailable fraction) at equilibrium or nonequilibrium conditions. Dialyzable fraction achievement is schematically shown in Figure 11.3. Several molecular weight cut-off (MWCO) dialysis membranes (1–10 kDa) are used for this purpose. Dialysis membranes are filled with an acceptor solution (*i.e.* sodium hydrogen carbonate or piperazine-*N*-*N*′-bis(2-ethanesulfonic acid buffer solution, PIPES) at the physiological pH of 7.0. The use of sodium hydrogen carbonate as an acceptor solution presents two drawbacks: first, the pH can vary during the intestinal digestion process, and secondly, a further acid–base titration is necessary to determine the required amount of sodium hydrogen carbonate to be added at the end of the simulated process to ensure a pH of 7.0 (Domínguez-González *et al.*, 2010). By contrast, the use of PIPES buffer solution keeps a physiological pH during intestinal digestion, improving the reproducibility of the *in vitro* digestion conditions (Haro-Vicente *et al.*, 2006). The bioavailable and dialyzed fractions, expressed as a percentage, are calculated using eqn (11.2) and (11.3):

$$\mathrm{Bav} = \frac{\mathrm{Conc}_{\mathrm{dialyzate}}}{\mathrm{Conc}_{\mathrm{food}}} \times 100 \tag{11.2}$$

$$\mathrm{Dial} = \frac{D_{\mathrm{dialyzate}}}{W_{\mathrm{food}} \times \mathrm{Conc}_{\mathrm{food}}} \times 100 \tag{11.3}$$

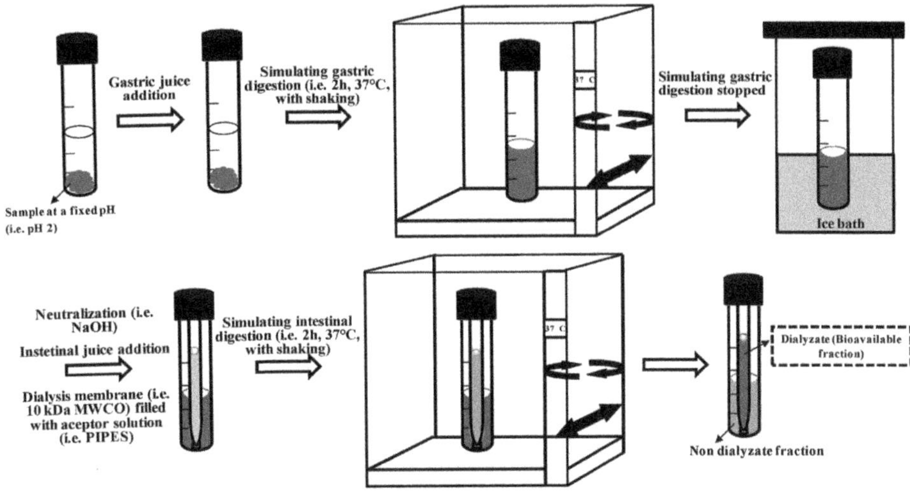

Figure 11.3 Schematic diagram of *in vitro* GI digestion plus dialyzing step.

where Bav and Dial are the percentages of bioavailability and dialyzability, respectively; $Conc_{dialyzate}$ is the nutrient/target compound concentration after *in vitro* digestion plus dialysis process, $D_{dialyzate}$ is total amount of dialyzed compound (expressed in mass) and W_{food} is the sample weight of sample used for intestinal digestion plus dialysis process.

(3) Measurement of the fraction of the nutrient/compound capable of be retained and/or transported through a solid or microporous supports (bioavailable fraction) in which human cells grown are incorporated (intestinal epithelial model) (Karlsson and Artursson, 1991). In this model a previous *in vitro* GI digestion is required. Several cell lines are available to mimic the intestinal mucosa. Caco-2 cell line (derived from human colon carcinoma cell) has similar structural and functional characteristics to those of colon enterocytes. The Caco-2 cell model is suitable to study cellular uptake, retention, transport and metabolism processes of target compounds (Zeng *et al.*, 2008) and offers a more reliable approximation to the *in vivo* situation in estimating bioavailability at the intestinal level.

Figure 11.4 shows a schematic diagram of a bicameral system used in transport and retention studies by using Caco-2 cell cultures; in which a monolayer of Caco-2 cells is grown on a filter separating two stacked well plates. The upper well (apical side) is filled with the donor solution (containing the digesta), and the bottom (basal side) is filled with the acceptor solution (minimum essential medium, MEM). Uptake (retention and transport percentages) by Caco-2 cells is calculated by eqn (11.4) and (11.5):

$$\text{Retention} = \frac{\text{Conc}_{\text{cells}}}{\text{Conc}_{\text{bioaccessible}}} \times 100 \qquad (11.4)$$

$$\text{Transport} = \frac{\text{Conc}_{\text{basal}}}{\text{Conc}_{\text{bioaccessible}}} \times 100 \tag{11.5}$$

where $\text{Conc}_{\text{cells}}$, $\text{Conc}_{\text{bioaccessible}}$ and $\text{Conc}_{\text{basal}}$ are target compound concentrations in monolayer Caco-2 cells, in the bioaccessible fraction after simulated GI digestion and in the basal side, respectively. The transport efficiency can also be measured by means of the apical to basolateral permeability (apparent permeability coefficient, P_{app}) according to eqn (11.6). P_{app} is expressed as cm s^{-1}.

$$P_{\text{app}} = \left(\frac{dC}{dt}\right) \times \left(\frac{V}{(A \times \text{Conc}_0)}\right) \tag{11.6}$$

where dC/dt is the change in concentration in the basolateral side over time, V is the volume of solution in the basolateral side, A is the membrane surface area, and Conc_0 is the initial concentration in the apical side.

The bioavailable fraction is considered to be the sum of, first, the fraction of soluble nutrient/target compound that was transported through a monolayer of Caco-2 cells in a bicameral culture system and; secondly, the fraction that was retained in cells during the process of absorption (eqn (11.7)).

$$\text{Bav} = \frac{\text{Conc}_{\text{basal}} + \text{Conc}_{\text{cells}}}{\text{Conc}_{\text{food}}} \times 100 \tag{11.7}$$

Cell-culture-based models have the main drawback that the analytical techniques used for target compound quantification must be highly sensitive and selective enough to avoid interferences from culture/transport media composition (Fernández-García *et al.*, 2009).

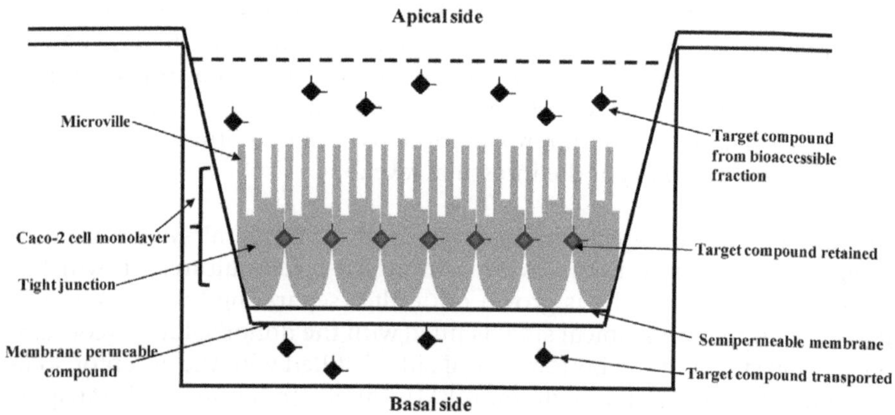

Figure 11.4 Schematic diagram of a bicameral system used in transport and retention studies by Caco-2 cell cultures during an *in vitro* approach.

11.2.2 *In Vitro* Dynamic Models

Dynamic models are validated systems of gastrointestinal tract to study physicochemical, enzymatic and microbial parameters of the digestion in a controlled *in vitro* setting. Parameters involved on human digestion such as body temperature, flow of saliva, gastric and pancreatic juices (including digestive enzymes), bile peristalsis for mixing and gastrointestinal transit times, regulation of gastric and intestinal pH, continuous removal of digested lipophilic and hydrophilic compounds are monitored and controlled. Physical conditions of the digesta changes over time (viscosity, particle size reduction, diffusion, creation of colloidal phases, partitioning of nutrient between phases) are also taking into account in dynamic models (Wickham *et al.*, 2009). Dynamic models are therefore designed to encompass the physical processing and temporal changes in the luminal conditions that mimic *in vivo* conditions. Dynamic models are flexible (can be adapted to mimic specific conditions related to age, type of meal (liquid, semisolid, and solid meals), and health or disease status), and they are accurate and precise assays. These models are usually computer controlled by software that includes monitoring of all parts in real time.

Several dynamic models such as the simulator of the human intestinal microbial ecosystem (SHIME) (Molly *et al.*, 1993) and TIM gastrointestinal (Minekus *et al.*, 1995) models that mimic *in vivo* physical processes, have been described and marketed by several manufactures.

The original SHIME model (Figure 11.5) consisted of five reactors that sequentially simulate the stomach, small intestine and the three regions of the large intestine, *i.e.* the ascending, transverse and descending colon. The SHIME model is usually inoculated with human fecal material to establish a microbial population that resembles that found in the human gastrointestinal tract. A careful control of the environmental parameters in these reactors allows the establishment of complex and stable microbial communities that are highly similar in both structure and function to the microbial community in the different regions of the human colon. A recent evolution of the

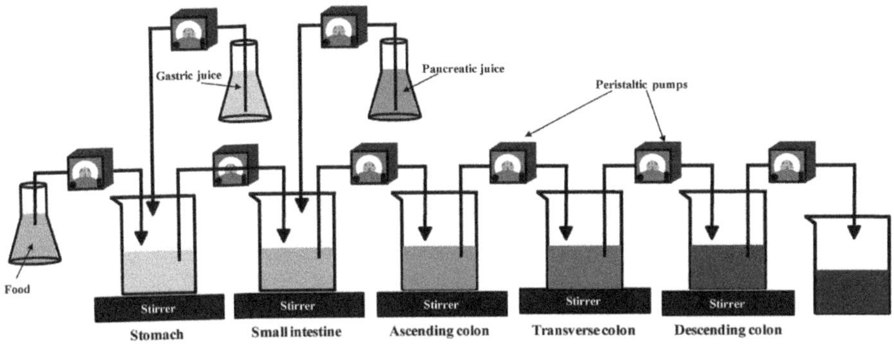

Figure 11.5 Schematic diagram of *in vitro* dynamic SHIME simulator.

SHIME is the TWINSHIME, in which two SHIME systems are run in parallel, and where all the environmental parameters are completely identical. This allows the study of two nutrients under identical conditions without interference of external parameters.

In vitro methods provide an effective approximation to *in vivo* scenarios and offer the advantages of high-throughput screening tools, simplicity (automation, and miniaturization), rapidity, ease of control, low cost, high precision and accuracy (validation and standardization with reference materials) and good reproducibility (Figure 10.2, (Chapter 10)). They provide information about efficiency of each digestion/absorption step and transport mechanisms; and, in addition, the use of laboratory animals is avoided, which connects with increasing interest in reducing the use of animals for testing. On the contrary, the dynamic environment of intestine is not fully reproduced with biochemical and cell-culture models, and, in addition, the effect of intestinal microflora and hepatic metabolism is not always considered in *in vitro* assays (Fernández-García *et al.*, 2009).

11.3 *In Vitro* Gastrointestinal Extraction Methods to Selenium Species Bioaccessibility/Bioavailability in Foodstuff

In this section, *in vitro* bioaccessibility/bioavailability studies of selenium in foodstuff will be summarized. Besides parameters related to the person itself (status, age, sex, lifestyle), selenium bioavailability is tightly related to selenium chemical forms, the presence of certain other components of the food matrix (or the ingestion of other nutrients with it) and the food processing (*i.e.* cooking treatments).

11.3.1 Oral Bioaccessibility/Bioavailability of Selenium in Food Samples

Several GI conditions (gastric and intestinal juices composition and incubation temperature and time), intestinal epithelial models (Caco-2 cell) and dynamic models (mainly SHIME) are extensively used.

In vitro studies are focused on assessing bioaccessible/bioavailable selenium species in foodstuff with high selenium content and selenium-fortified foodstuff:

(1) Selenium-fortified cereals such as wheat (Govasmark *et al.*, 2010), rice (Fang *et al.*, 2010; Jaiswal *et al.*, 2012; Wang *et al.*, 2013) and maize (Jaiswal *et al.*, 2012).
(2) Raw and cooked selenium-fortified vegetables such as radish (Pedrero *et al.*, 2006), garlic (Dumont *et al.*, 2006a), onion (Kápolna and Fodor,

2007), chives (Kápolna and Fodor, 2007), leek (Lavu *et al.*, 2012), broccoli (Zeng *et al.*, 2008) and mushroom (Turlo *et al.*, 2010; Bhatia *et al.*, 2013).

(3) Raw and cooked fish (Crews *et al.*, 1996; Shen *et al.*, 1997; Cabañero *et al.*, 2004, 2007; Calatayud *et al.*, 2012; Moreda-Piñeiro *et al.*, 2013a) and shellfish (Calatayud *et al.*, 2012; Moreda-Piñeiro *et al.*, 2013a).

(4) Raw and cooked seaweed (García-Sartal *et al.*, 2013; Moreda-Piñeiro *et al.*, 2013b).

(5) Yeast, selenized yeast, and selenized-yeast-based nutritional supplements (Dumont *et al.*, 2004; Hinojosa-Reyes *et al.*, 2006a, 2006b; Thiry *et al.*, 2013).

(6) Milk (human, cow, goat and sheep) (Shen *et al.*, 1996) and selenium-fortified milk (whole (Muñiz-Naveiro *et al.*, 2006; Zohoori *et al.*, 2009), skimmed (Zohoori *et al.*, 2009) and fermented (Alzate *et al.*, 2010) milk).

(7) Others foodstuffs such as Brazil nuts (Dumont *et al.*, 2006b; Gomes da Silva *et al.*, 2013), selenized-chicken eggs (Lipiec *et al.*, 2010), and chicken (Brandt-Kjelsen *et al.*, 2012; Govasmark *et al.*, 2010) and beef (Ramos *et al.*, 2012).

11.3.1.1 Selenium Bioavailability in Cereals

Selenium bioaccessibility was assessed in selenium-fortified wheat grain and bran after soil enrichment with selenium (seeding by using selenium-enriched fertilizers) (Govasmark *et al.*, 2010). Selenium yields were within the 50–75% range in wheat and bran when using a simulated gastric digestion. The bioaccessible ratio, however, did not increase after the simulated duodenal digestion. In addition, results from bioaccessibility studies using commercial enzymes and human gastric juices for wheat flour and bran digestion suggested that selenium bioavailability was overestimated when using commercial enzymes (Govasmark *et al.*, 2010).

The selenium bioaccessibility assessment in selenium-enriched rice (use of selenium-enriched fertilizers prepared from selenite (Fang *et al.*, 2010; Wang *et al.*, 2013) or by harvesting in seleniferous soils (Jaiswal *et al.*, 2012)) was also performed by using an *in vitro* GI digestion. Studies were also carried out for selenium-enriched maize (Jaiswal *et al.*, 2012). The results show that selenium content in rice was increased by up to 51 times, after foliar application of selenite-enriched fertilizers. No symptoms of selenium toxicity were observed in the plants (Wang *et al.*, 2013). The analysis of the bioaccessible extracts showed that low molecular weight (LMW) selenium-containing proteins, SeMet and selenocysteine (SeCys) were the major selenium-containing moieties found in the gastrointestinal extracts (Fang *et al.*, 2010; Wang *et al.*, 2013). In addition, selenium bioaccessible levels were higher in rice (65%) than in maize (Jaiswal *et al.*, 2012).

11.3.1.2 Selenium Bioavailability in Vegetables

Selenium-fortified vegetables have been extensively used for selenium *in vitro* bioaccessibility/bioavailability studies. The potential bioaccessibility of selenium and selenium species in radish grown in hydroponics media in the presence of inorganic selenium was assessed. Low selenium transformation into organic forms was observed in radish plants grown in Se(VI)-enriched culture media (Pedrero *et al.*, 2006). On the contrary, studies developed with plants exposed to selenite showed that percentages higher than 95% in the bioaccessibility fraction comprised selenocystine ($SeCys_2$), selenomethionine (SeMet), and Se-methylselenocysteine (SeMetSeCys). The concentrations of $SeCys_2$, SeMet and SeMetSeCys in raw samples remained almost unaltered after a simulated GI digestion, and the analysis of the GI digests showed that almost 100% of selenium in the raw plants was found in the bioaccessible fraction. Due to radish usually undergoing a short development cycle, these results suggest that radish enriched in selenite could be a good choice as an organoselenium supplement for the human diet and animal feed (Pedrero *et al.*, 2006).

Selenite and selenate were also added once to the soil on which the garlic samples were harvested (Dumont *et al.*, 2006a). Supplemented garlic samples were treated with saliva and simulated gastric and intestinal fluids, and the same selenium species observed, SeMet, SeMetSeCys and γ-glutamylselenomethylselenocysteine (γ-glutamyl-Se-Me-SeCys), as well as additional unknown selenium compounds, were assessed in saliva and in both simulated fluids. No data regarding bioaccessibility percentages for each selenium species were given (Dumont *et al.*, 2006a).

Selenium bioaccessibility in selenized green onion and chive using an *in vitro* GI digestion was also assessed. After green onions and chives had attained full growth, selenium supplementation was performed by using independent solutions containing selenite (Se(IV)), selenate (Se(VI)) and SeMet that were added daily for 2 weeks (Kápolna and Fodor, 2007). The amount of selenium accumulated in chive was one-fold higher than that observed in green onion. Results regarding selenium fortification with Se(VI) showed that 80–90% of the total selenium content (mainly as selenate) became bioaccessible in the simulated GI digests. Unknown selenium compounds were detected (2% of the bioaccessible total selenium). Otherwise, a lower selenium bioaccessible percentage was obtained in Se(IV)-fortified green onion (30%) and chives (12%), and also in SeMet-fortified green onion (20%) and chives (22%). Se(VI), Se(IV) and SeMet species were found and quantified in the bioaccessible fractions in Se(IV)-fortified green onion and chives. Certain unknown Se compounds (3.0%) were also detected. Concerning SeMet-fortified green onion and chives, Se(IV), SeMet, MeSeCys, and selenomethionine-Se-oxide (SeOMet) species were quantified in the GI digests. In addition, unknown selenium compounds were also observed (2–10% of the bioaccessible total selenium).

Studies regarding selenium bioavailability in seleno-broccoli extracts and in Se(IV), SeMSeCys, and SeMet aqueous standards were also performed by using Caco-2 cell models after simulated GI digestion (Zeng *et al.*, 2008). Results showed similar selenium bioavailability ratios in seleno-broccoli extracts and in SeMSeCys and SeMet aqueous standards.

The assessment of total selenium bioavailability was also established in selenium-enriched leek was studied (Lavu *et al.*, 2012). In this development, the colon digestion was performed by using a SHIME simulator. Selenium-enriched leek was obtained after using selenium-enriched fertilizers and harvesting in a green house. The lowest selenium uptake was observed when using sodium selenite-based fertilizers. When soil was amended with selenate-based fertilizers, a percentage of 55 ± 5% of total selenium was found as inorganic forms in leek; whereas, only 21 ± 8% was inorganic selenium when fertilizing with selenite-based fertilizers. Most of the selenium in the selenium-enriched leek was found to be bioaccessible in the stomach (~60%), and in the small intestine (~80%). Regarding simulated colon digestions, selenium bioaccessibility in selenium-enriched leek samples started to decrease to below 50% after a 24 h incubation period. The decrease in selenium bioaccessibility in the colon can probably be attributed to sorption phenomena between selenium and organic matter, or to uptake processes by the colon microbial biomass.

Finally, selenized mushroom (Shiitake mushroom (Turlo *et al.*, 2010) and oyster mushrooms (Bhatia *et al.*, 2013)) were also used for assessing selenium bioaccessibility. Shiitake and oyster mushrooms were cultivated in sodium selenite and selenic acid media (Turlo *et al.*, 2010); whereas, oyster mushrooms were grown on wheat straw from the seleniferous belt (Bhatia *et al.*, 2013). The selenium concentration in biofortified oyster mushrooms was 800 times higher when compared with control samples (use of wheat straw from nonselenium-rich areas) (Bhatia *et al.*, 2013). Bioaccessibility percentages were high in Shiitake mushrooms (60 and 80% from dried mycelium and mycelial lyophilisate, respectively) (Turlo *et al.*, 2010) and in oyster mushroom (Bhatia *et al.*, 2013) (75% of the selenium was found in the GI digest). Approximately 25% of the solubilized selenium was dissolved in the pancreatic juice (Turlo *et al.*, 2010). Se speciation in the gastrointestinal extract was dominated by SeMet, which accounted for 73% of the extracted selenium. A small SeMet percentage (2%) was present as SeOMet, and as other oxidized Se-species (Bhatia *et al.*, 2013). On the other hand, only 2% of the detected species was present as inorganic selenium (Se(IV)). A number of other low molecular weight selenocompounds generated during the gastrointestinal digestion (25% of the detected selenium-based chromatographic signals) remained unidentified (Bhatia *et al.*, 2013).

11.3.1.3 *Selenium Bioavailability in Fish and Shellfish*

Several studies regarding selenium bioavailability in fish and shellfish have shown that total selenium and selenium species contents in *in vitro* GI digests (bioaccessible fraction) in raw/cooked fish (Crews *et al.*, 1996;

Cabañero *et al.*, 2004, 2007; Calatayud *et al.*, 2012), and raw shellfish (Calatayud *et al.*, 2012) were dependent on the type of seafood (see Sections 11.3.2 and 11.3.3). The fish and shellfish under study were: cod, swordfish, sardine, tuna, anchovy, anglerfish, blue whiting, hake, salmon, small hake, sole, clam, cuttlefish, lobster, mussel, prawn, shrimp, and squid. Higher selenium solubility percentages in the GI supernatants in swordfish (76–96%) (Cabañero *et al.*, 2004, 2007; Calatayud *et al.*, 2012), anglerfish (100%) (Calatayud *et al.*, 2012), blue whiting (91%) (Calatayud *et al.*, 2012), hake and small hake (86–99%) (Calatayud *et al.*, 2012), salmon (100%) (Calatayud *et al.*, 2012) and sole (84%) (Calatayud *et al.*, 2012) than those obtained for sardine (53–83%) (Cabañero *et al.*, 2004, 2007; Calatayud *et al.*, 2012), tuna (50%) (Cabañero *et al.*, 2004, 2007) and cod (61.3%) (Crews *et al.*, 1996) were reported. In addition, high median bioaccessibility values were found for shellfish products (clam: 73%, cuttlefish: 98%, lobster: 72%, mussel: 100%, prawn: 72%, shrimp: 90%, and squid: 92%) (Calatayud *et al.*, 2012). Simulated human gastric and intestinal digestion led to the identification of SeMet in the GI digests in swordfish, sardine and tuna, and similar bioaccessible SeMet percentages (16, 14 and 19%, for swordfish, sardine and tuna, respectively) (Cabañero *et al.*, 2004, 2007). Data regarding bioaccessibility percentages for selenium species in other seafood types were not available (Calatayud *et al.*, 2012; Crews *et al.*, 1996).

In addition, *in vitro* bioavailability (bioavailable fraction) of total selenium and selenium species in several raw seafood (Moreda-Piñeiro *et al.*, 2013a, 2013b; Shen *et al.*, 1997) and processed fish (maatjesherring) (Shen *et al.*, 1997) has also been assessed by using a simulated gastric and intestinal digestion/dialysis method (Moreda-Piñeiro *et al.*, 2013a, 2013b). Seafood tested in these studies were: edible cockle, razor shell, variegated scallop, carpet-shell clam, shrimp, scallop, plaice, herring, hake, cod, anglerfish, Atlantic pomfret, poor cod, tuna, sardine, Atlantic mackerel, Atlantic horse mackerel, plaice, herring. SeMet was the major selenium species found in dialyzates in seafood (Moreda-Piñeiro *et al.*, 2013a) $SeCys_2$ was detected at low concentrations, while SeMeSeCys and inorganic selenium species (selenite and selenate) were not detected in the dialyzate (Moreda-Piñeiro *et al.*, 2013a). Low selenium bioavailability (percentages ranging from 4.0% to 13%) was found in white fish (hake, cod, poor cod), cold water fish (anglerfish, Atlantic pomfret, tuna, sardine, Atlantic mackerel, Atlantic horse mackerel) and mollusc (edible cockle, razor shell, variegated scallop, carpet-shell clam) (Moreda-Piñeiro *et al.*, 2013a, 2013b). Low bioavailability percentages were also assessed in plaice (25.8%), herring (21.3–31.3%), cod (16.7%) and shrimp (17.2%) (Shen *et al.*, 1997). The bioavailability percentages obtained in seafood are lower than data regarding bioaccessible total selenium in seafood, *i.e.* percentages from 50% to 100% in fish and shellfish (Crews *et al.*, 1996; Cabañero *et al.*, 2004, 2007; Calatayud *et al.*, 2012). These differences are attributed to the additional dialysis stage in the *in vitro* bioavailability methods, which improves the simulation of the intestinal digestion (a dialyzability stage is included in the simulation process), and leads therefore to lower bioavailable total selenium percentages. It can also be concluded that total selenium is highly bioaccessible (Crews *et al.*, 1996; Shen *et al.*, 1997; Cabañero *et al.*, 2004, 2007 Calatayud *et al.*, 2012;) but it is not easily dialyzed (bioavailable). These results are in agreement with those reported when including Caco-2 cells transport and retention simulation (Calatayud *et al.*, 2012). High bioaccessibility percentages (96%) for swordfish, but low bioavailability percentages have been reported (selenium transport from the food was less than 14%, and cell retention was between 8 and 12%) (Calatayud *et al.*, 2012).

11.3.1.4 Selenium Bioavailability in Edible Seaweed

Bioavailability of total selenium and selenium species in different raw and cooked edible seaweed (Dulse, Nori, Kombu, Wakame, Sea spaghetti, Sea lettuce, and canned/cooked seaweed) has also been assessed by using a simulated gastric and intestinal digestion/dialysis method (Moreda-Piñeiro *et al.*, 2013b; García-Sartal *et al.*, 2013). High bioavailability percentages, ranged from 48 to 100% for raw edible seaweed (Moreda-Piñeiro *et al.*, 2013b), and moderate dialyzable percentages for cooked edible seaweed (from 14.6 to 61.3%) were assessed (García-Sartal *et al.*, 2013). Dialyzability percentages were dependent on the type of seaweed studied. This may be due to the different seaweed composition (fat, protein, carbohydrates and fiber contents, among other dietary factors). In general, Wakame followed by Kombu was the seaweed exhibiting the highest selenium bioavailability percentages (see Section 11.3.3).

11.3.1.5 Selenium Bioavailability in Yeast and Yeast-Based Nutritional Supplements

Results from yeast (Dumont *et al.*, 2004), and selenized yeast (Hinojosa-Reyes *et al.*, 2006a, 2006b; Thiry *et al.*, 2013), and yeast-based nutritional supplements (Arkovital, Verdalia and Selenium-ACE) (Hinojosa-Reyes *et al.*, 2006a, 2006b), and selenate-based food supplements, and selenite-based food supplement showed that SeMet was the major compound identified in the GI extracts (Dumont *et al.*, 2004; Hinojosa-Reyes *et al.*, 2006a, 2006b; Thiry *et al.*, 2013); whereas, SeOMet was its main degradation product after medium and long-term sample storage (Dumont *et al.*, 2004; Hinojosa-Reyes *et al.*, 2006a, 2006b). Other minor selenium compounds were identified as $SeCys_2$, SeMeSeCys, and selenium low molecular weight compounds. Results showed that total selenium percentages of 55–89% (Dumont *et al.*, 2004; Hinojosa-Reyes *et al.*, 2006a, 2006b) and 57–100% (Thiry *et al.*, 2013) were extracted after the GI approach, but only 41% (Hinojosa-Reyes *et al.*, 2006a, 2006b) and 70% (Thiry *et al.*, 2013) was surprisingly quantified as free SeMet. Therefore, although this free selenoamino acid appears to be the most abundant selenium species in the GI digests, an important total SeMet fraction in the extracts still seems to be associated to selenium-containing peptides after the *in vitro* GI digestion. The remaining extracted selenium species (20%) were found to be in the form of low molecular weight compounds (Hinojosa-Reyes *et al.*, 2006a, 2006b). Concerning yeast-based nutritional supplements, low SeMet percentages (26–37%) were observed, and the remaining solubilized selenium consisted of unknown selenium compounds (probably selenium-containing peptides) (Hinojosa-Reyes *et al.*, 2006a, 2006b). The bioavailable fraction (Caco-2 cell model) was also achieved for selenium-enriched yeast, selenate-based food supplements, and selenite-based food supplements (Thiry *et al.*, 2013). Results proved the existence of selenium bioavailability differences among the three types of food supplements. These differences could partially be attributed to the selenium species profile because selenate-food supplements and selenite-food supplements contain exclusively Se(VI) and Se(IV), respectively, but speciation studies in the bioavailable fractions in selenized yeast showed the presence of SeMet, Se(IV), and SeMeSeCys. Overall selenium bioavailability in the three food supplements was low (7.0, 14 and 1.0% for selenium-enriched yeast, selenate-based food supplements, and selenite-based food supplements) (Thiry *et al.*, 2013). Among the three samples, the highest bioavailable selenium percentage was obtained in selenate-based food supplements, followed by selenized-yeast, and by selenite-based food supplements (selenium can be considered as nonbioavailable in the latter food supplement). These results confirm the low availability of inorganic selenium but were somewhat unexpected regarding the yeast-based food supplements since selenium-rich yeasts are usually reported to contain an important fraction of available selenium. SeMet was the species detected in the basolateral compartment and in cells after *in vitro* absorption (Thiry *et al.*, 2013). The bioavailable SeMet fraction accounted for 24% of the original SeMet content, and for approximately 35% of bioaccessible SeMet.

11.3.1.6 Selenium Bioavailability in Milk

Selenium bioavailablity in human milk and some common ruminant milk (cow, goat and sheep milk) was estimated by a simulated *in vitro* GI digestion procedure (Shen *et al.*, 1996). Dialysate percentages in human milk (11.1 ± 0.4%) were significantly higher than those observed for cow (6.8 ± 0.2%), goat (6.2 ± 0.2%) and sheep (~2%) milk. Reported studies showed that selenium availability (15.0 ± 0.2, 10.0 ± 0.9, 12.0 ± 0.4 and 2.4 ± 0.5% for human, cow, goat and sheep milk, respectively) increases on removing the milk fat fraction. In addition, the selenium availability may be related to the gastric digestibility of protein (Shen *et al.*, 1996) (see Sections 11.3.2 and 11.3.3).

Selenium bioavailability and bioaccessibility in fortified cow milk after cows were fed with forages supplemented with selenium species (SeMet and sodium selenite) (Muñiz-Naveiro *et al.*, 2006), and after adding sodium selenite to milk samples (Alzate *et al.*, 2010) was also assessed. Bioavailability percentages of approximately 11% were found for cows' milk obtained after supplementation of forage with 0.4–0.5 μg selenium per gram organic selenium without altering the original milk composition. Low bioavailable percentages (<10%) were obtained after supplementation with inorganic selenium (0.2–0.3 μg selenium per gram) (Muñiz-Naveiro *et al.*, 2006). By contrast, high bioaccessible percentages were found (results showed that 76 ± 3% selenium was extracted after GI digestion and 24 ± 6% remained in the insoluble residue) (Alzate *et al.*, 2010). As discussed above for seafood, it can also be concluded that selenium is highly bioaccessible but it is not easily bioavailable in milk. On the other hand, $SeCys_2$ and SeMetSeCys were the most abundant compounds in the bioaccessible fraction in selenium-fortified fermented milk. Two unknown selenium compounds were also identified in the bioaccessible fraction (Alzate *et al.*, 2010). Zohoori *et al.* (2009) studied the impact of adding fluoride on the *in vitro* bioaccessibility of selenium and other trace elements (iron, zinc, copper, chromium, and molybdenum) in pasteurized skimmed (0.3% fat) and whole (4% fat) milk samples. Although no data on bioaccessibility percentages of total selenium were given, the authors confirmed that selenium bioaccessibility was higher in skimmed milk than in whole milk. Finally, a clear relationship between selenium bioaccessibility and fluoride concentrations was not found.

11.3.1.7 Selenium Bioavailability in Brazil Nuts

When assessing bioaccessible selenium in Brazil nuts (Dumont *et al.*, 2006b; Gomes da Silva *et al.*, 2013), one of the products exhibiting the highest selenium content, $SeCys_2$ and SeMet were the main compounds found after simulated GI digestion. SeMet bioaccessible percentage was high, and it accounted for 74% of the total selenium (Gomes da Silva *et al.*, 2013).

11.3.1.8 Selenium Bioavailability in Eggs

Selenium bioavailability in eggs claimed by the producers to be fortified with health-promoting additives, including selenium, and eggs obtained in hens fed with organic selenium feed supplement (Selplex™) was also established (Lipiec *et al.*, 2010). The analysis of control and supplemented eggs showed that SeCys was the dominating selenoaminoacid in the yolk; whereas SeMet accounted for more than 50% of selenium in the white. The selenium supplementation in chicken affected principally the egg white, where both SeMet and SeCys concentrations were doubled. The forms of selenium initially present were degraded under the gastrointestinal conditions to low molecular weight peptides, aminoacids (SeMet and SeCys) and inorganic selenium. However, bioaccessibility percentage data for each selenium species were not given in the study.

11.3.1.9 Selenium Bioavailability in Chicken and Beef

The bioaccessibility of total selenium was studied in meat in chicken (fed with selenium-enriched wheat (Govasmark *et al.*, 2010; Brandt-Kjelsen *et al.*, 2012)) and in unaged and aged beef (Ramos *et al.*, 2012). The *in vitro* gastric digestion increased the selenium yield to 70% and to 65 and 62%

after digestion using commercial enzymes and human juices, respectively (Brandt-Kjelsen *et al.*, 2012). Results revealed that selenium proteins are subjected to enzymatic breakdown during GI digestion. After GI digestion, most selenium species were degraded to LMW compounds (<44 kDa) (Govasmark *et al.*, 2010; Brandt-Kjelsen *et al.*, 2012).

On the other hand, aging affected negatively the selenium content and the selenium bioaccessibility. Bioaccessibility percentages of 75–91% and 58–80% were achieved for unaged and aged meat, respectively (Ramos *et al.*, 2012). The main selenium form in beef was SeMet, which is much more bioavailable than other forms such as selenite (Ramos *et al.*, 2012).

11.3.2 Effect of Food Processing on Selenium Species Bioaccessibility/Bioavailability

Foodstuffs are usually processed in order to improve the palatability of the product. Food processing might affect the digestive enzymes efficiency for releasing nutrients from food due to sample composition could drastically be modified (see Section 11.3.3).

Heat treatments could produce changes in the total selenium content and in the level of the several selenium species. During heat and ripening processing, loss of water and protein degradation occurs. Protein degradation can improve protein digestibility, which will facilitate oligoelements (selenium) release and enhance the bioaccessibility ratio (Cabañero *et al.*, 2004). Therefore, selenium bioaccessible percentages in cooked tuna was slightly higher than those found in uncooked tuna (Cabañero *et al.*, 2004). The effect of cooking of Dulse, Nori, Kombu, Wakame, Sea spaghetti and Sea lettuce on the selenium bioavailablity ratios was also studied (García-Sartal *et al.*, 2013). Results showed that selenium was released into the cooking water during the heat treatment. In addition, low dialyzability percentages were achieved for cooked seaweed (García-Sartal *et al.*, 2013). Opposite results have been reported for canned seaweeds (mixture of Sea spaghetti and *Saccorhiza polyschides*). The dialyzability percentage obtained for canned seaweed (containing Sea Spaghetti) was higher (~100%) than the bioavailability percentages obtained for raw Sea spaghetti (~80%) (Moreda-Piñeiro *et al.*, 2013b). These results suggest that the cooking procedure affects the major nutrient composition and consequently, selenium bioavailability.

The bioavailability percentage in ripening herring (maatjesherring: herring ripened for a couple of days in oak barrels in a salty solution) was higher (31.3%) than those found in unripened herring (21.3%) (Shen *et al.*, 1997). Choosing a proper way in fishery products processing could improve the nutritive value. Milk fermentation procedure also improved the digestibility of milk proteins, which increased selenium availability in milk (Shen *et al.*, 1996). Finally, aging negatively affected selenium bioaccessibility in beef (Ramos *et al.*, 2012). Natural enzymes present in beef break down the fibrous tissues during the aging time; and, in addition, moisture is partially lost. Selenium bioaccessibility ratios ranged between 75 and 91% in unaged beef. After aging, selenium bioaccessibility was significantly lowered (65 to 78%).

11.3.3 Effect of Sample Composition on Selenium Species Bioaccessibility/Bioavailability

Food components can influence nutrients/target compound bioavailability in foods. Differences in food composition (carbohydrate, proteins, dietary fiber and fat contents) affect the capability of the enzymes during the *in vitro* methods for solubilizing selenium and other substances. In addition, the presence of other minor components, such as trace metals, can also modify selenium bioavailability.

11.3.3.1 Effect of Carbohydrate Content

The carbohydrate content in foodstuffs seems to affect strongly trace elements bioavailability. Findings showed that selenium bioavailability increases with the carbohydrate content in fish, shellfish and seaweed (Moreda-Piñeiro *et al.*, 2013b) due to carbohydrate is capable of forming micelles, which can increase the partition of hydrophobic molecules in the aqueous solution (Yu *et al.*, 2010; Moreda-Piñeiro *et al.*, 2013b). On the contrary, selenium-containing polysaccharides in mushrooms could not be solubilized after GI digestion due to selenium-enriched mycelia being associated with chitin-containing structures in cell walls (Bhatia *et al.*, 2013). The formation of selenium-containing polysaccharides might explain the low selenium bioavailability found in mushroom. Similar conclusions have been addressed from mushroom after *in vivo* human test (Mutanen, 1986) and in Brazil nuts and mushroom after *in vivo* animal tests (Chansler *et al.*, 1986).

11.3.3.2 Effect of Protein Content

Although there are few data about the effect of protein content on selenium bioavailability, some studies have indicated that selenium bioavailability decreases on increasing the protein content in foodstuff such as fish, shellfish and seaweed samples (Moreda-Piñeiro *et al.*, 2013b). This fact has been attributed to the effect caused by the ionic strength. Proteins are hydrolyzed to amino acids during the GI digestion. Most of these amino acids are soluble and carried positive or negative charges at the physiological pH used during the digestion. Therefore, these amino acids exist as a sort of salt that increases the ionic strength of the aqueous phase (GI digest). The increased ionic strength results in a lower solubility of the Se species (the "salting-out" effect) in the aqueous phase, and hence lower bioaccessibility (Yu *et al.*, 2010; Moreda-Piñeiro *et al.*, 2013b).

11.3.3.3 Effect of Fat Content

The fat content in foodstuffs does not affect either selenium bioavailability or increase selenium bioavailability. From fish and shellfish, selenium bioavailability is not affected by the fat content. This is not surprising because selenium species are poorly lipophilic molecules. Selenium species and fat are released from the food matrix during the GI digestion, and selenium species are not expected to be bound to the fat, even after fat emulsification by bile salts and the transfer of the emulsified fats to the liquid phase to become accessible (Moreda-Piñeiro *et al.*, 2013b). These results agree with *in vivo* tests with rats (Mutanen and Mykkănen, 1984). In addition, as previously mentioned, fat removal from milk (human, cow, goat and sheep milk) significantly increased selenium bioavailability (Shen *et al.*, 1996; Zohoori *et al.*, 2009). The negative influence of animal fat on selenium availability is unclear, but it was stated that selenium in plants is usually more available than selenium from animals (Hazell, 1985). Selenium mainly exists in protein-bound forms in milk, and almost 80% of selenium in cow's milk is associated with casein. It has been suggested that protein digestibility is the main determinant factor for selenium bioavailability in milk. Thus, the high selenium bioavailability in skimmed milk might be due to the better protein digestibility of skimmed milk (Zohoori *et al.*, 2009).

11.3.3.4 Effect of Fiber Content

The fiber (soluble fiber or nonsoluble fiber) content seems to affect selenium bioavailability differently. Findings when analyzing seaweed have demonstrated that selenium bioavailability is increased with increased soluble fiber concentrations (Moreda-Piñeiro *et al.*, 2013b). This result agrees with those reported by Vitali *et al.* (2008), which have attributed high mineral bioavailability ratios in vegetables with high soluble dietary fiber. On the other hand, *in vivo* test with rats showed that nondigestible fiber content in wheat and bran decreased selenium bioavailability (Reeves *et al.*, 2007). The low solubility of this fiber likely effectively encapsulates cellular proteins within intact cells in the bran fraction and renders them unavailable for digestion in the upper gut. Because most of the selenium is incorporated primarily into these proteins, the selenium itself becomes unavailable.

Summary Points

- This chapter focuses on selenium and selenium species bioavailability in foodstuff.
- *In vitro* bioavailability assays for releasing selenium and selenium species in food samples is summarized.
- *In vitro* bioavailability is assessed by bioaccessible and dialyzable fraction measurements.
- Caco-2 cell model allows the *in vitro* bioavailability assessment.

- Selenium species bioavailability depends on the foodstuff processing and cooking.
- Selenium species bioavailability depends on the foodstuff composition.

Key Facts of *In Vitro* Testing for Bioavailability Assessment

1. *In vitro* testing simulates the physiologic conditions and the events that occur in the human gastrointestinal tract during human digestion.
2. The sequence of events enclosed processing in the mouth, in the stomach and in the duodenum.
3. Bioaccessibility is assessed after the measurement soluble concentrations of the target compound in the simulated gastrointestinal solution after filtration or centrifugation.
4. Bioavailability is assessed after measurement of the dialyzable fraction.
5. The dialyzable fraction is the amount of the target compound that is capable of dialyzing (during intestinal digestion) through a semipermeable membrane with a specified pore size.
6. Bioavailability is also assessed after the measurement of the fraction of the target compound capable of be retained and/or transported through a solid or microporous supports in which grown human cells (Caco-2 cell) are incorporated (intestinal epithelial model).

Definitions of Words and Terms

Bioaccessibility. Bioaccessibility refers to the fraction of an administered dose of a substance that is soluble in the gastrointestinal tract and that is therefore available to reach the systemic circulation.

Bioavailability. Bioavailability refers to the fraction of an administered dose of a substance that reaches the systemic circulation.

Caco-2 cell model. This model is a human intestinal model of the intestinal barrier used to study (*in vitro*) the transport and retention of substances throughout the intestine.

Dialysis. This is the physical process for removing substances from a liquid sample by passage through a microporous semipermeable membrane (with a fixed molecular weight cutoff). Thus, substances larger than the pore size of the membrane may not pass through. Substances smaller than the molecular weight cutoff pass through the membrane and are dissolved in a capturing solution.

Fortified food. This is food enriched by the addition of micronutrients and essential elements. This fortification can be achieved by micronutrient supplementation of livestock or through micronutrient fertilization of crops.

***In vitro* tests.** These involve experimentation (under exhaustively controlled conditions) in a "test tube". Controlled conditions mimic real organism conditions.

PIPES. Piperazine-*N*-*N*′-bis(2-ethane-sulfonic acid) (PIPES) is a buffer solution with a pK_a value near the physiological pH. PIPES is usually used for culturing cells *in vitro* applications.

SeOMet. Selenomethionine-Se-oxide (SeOMet) is a degradation product of selenomethionine (SeMet) that is formed under *in vitro* tests.

SHIME. This is a simulator of the human intestinal microbial ecosystem that mimics *in vivo* physical and chemical processes of the human gastrointestinal tract.

TIM. TIM gastrointestinal models simulate the conditions of the gastrointestinal tract (body temperature, peristaltic movements, swallowing of saliva, secretion of gastric acids and enzymes, pancreatic juice and bile salts).

Abbreviations and Acronyms

DIW	Deionized Water
GI	Gastrointestinal
GPx	Glutathione Peroxidase
HDJ	Human Duodenal Juice
HGJ	Human Gastric Juice
LMW	Low Molecular Weight
MEM	Minimum Essential Medium
SeMeSeCys	Selenomethylselenocysteine
MWCO	Molecular Weight Cutoff
PIPES	Piperazine-*N*-*N*′-bis(2-ethane-sulfonic acid)
Se(IV)	Selenite
Se(VI)	Selenate
SeCys	Selenocysteine
$SeCys_2$	Selenocystine
SeMet	Semethionine
SeOMet	Selenomethionine-Se-oxide
SHIME	Simulator of the Human Intestinal Microbial Ecosystem
γ-Glutamyl-Se-Me-SeCys	γ-Glutamyl-selenomethylselenocysteine

References

Alzate, A., Pérez-Conde, M. C., Gutiérrez, A. M. and Cámara, C., 2010. Selenium-enriched fermented milk: a suitable dairy product to improve selenium intake in humans. *International Dairy Journal*. 20: 761–769.

Bhatia, P., Aureli, F., D'Amato, M., Prakash, R., Cameotra, S. S., Nagaraja, T. P. and Cubadda, F., 2013. Selenium bioaccessibility and speciation in biofortified *Pleurotus* mushrooms grown on selenium-rich agricultural residues. *Food Chemistry*. 140: 225–230.

Brandt-Kjelsen, A., Govasmark, E., Vegarud, G., Haug, A., Szpunar, J. and Salbu, B., 2012. In vitro digestion of selenium from seleniumenriched chicken. *Pure and Applied Chemistry*. 84: 249–258.

Cabañero, A. I., Madrid, Y. and Cámara, C., 2004. Selenium and mercury bioaccessibility in fish samples: an *in vitro* digestion method. *Analytical Chimica Acta*. 526: 51–61.

Cabañero, A. I., Madrid, Y. and Cámara, C., 2007. Mercury-selenium species ratio in representative fish samples and their bioaccessibility by an *in vitro* digestion method. *Biological Trace Elements Research*. 119: 195–211.

Calatayud, M., Devesa, V., Virseda, J. R., Barberá, R., Montoro, R. and Vélez, D., 2012. Mercury and selenium in fish and shellfish: occurrence, bioaccessibility and uptake by Caco-2 cells. *Food Chemical Toxicology*. 50: 2696–2702.

Chansler, M. W., Mutanen, M., Morris, V. C. and Levander, O. A., 1986. Nutritional bioavailability to rats of selenium in Brazil nuts and mushrooms. *Nutrition Research*. 6: 1419–1428.

Crews, H. M., Clarke, P. A., Lewis, D. J., Owen, L. M., Strutt, P. R. and Izquierdo, A., 1996. Investigation of selenium speciation in *in vitro* gastrointestinal extracts of cooked cod by high-performance liquid chromatography inductively coupled plasma mass spectrometry and electrospray mass spectrometry. *Journal of Analytical and Atomic Spectrometry*. 11: 1177–1182.

Domínguez-González, R., Romarís-Hortas, V., García-Sartal, C., Moreda-Piñeiro, A., Barciela-Alonso, M. C. and Bermejo-Barrera, P., 2010. Evaluation of an *in vitro* method to estimate trace elements bioavailability in edible seaweeds. *Talanta*. 82: 1668–1673.

Dumont, E., Vanhaecke, F. and Cornelis, R., 2004. Hyphenated techniques for speciation of Se in *in vitro* gastrointestinal digests of *Saccharomyces cerevisiae*. *Analytical and Bioanalytical Chemistry*. 379: 504–511.

Dumont, E., Ogra, Y., Vanhaecke, F., Suzuki, K. T. and Cornelis, R., 2006a. Liquid chromatography-mass spectrometry (LC-MS): a powerful combination for selenium speciation in garlic (*Allium sativum*). *Analytical and Bioanalytical Chemistry*. 384: 1196–1206.

Dumont, E., De Pauw, L., Vanhaecke, F. and Cornelis, R., 2006b. Speciation of Se in *Bertholletia excelsa* (Brazil nut): a hard nut to crack?. *Food Chemistry*. 95: 684–692.

Fang, Y., Catron, B., Zhang, Y., Zhao, L., Caruso, J. A. and Hu, Q., 2010. Distribution and *in vitro* availability of selenium in selenium-containing storage protein from selenium-enriched rice utilizing optimized extraction. *Journal of Agricultural and Food Chemistry*. 58: 9731–9738.

Fernández-García, E., Carvajal-Lérida, I. and Pérez-Gálvez, A., 2009. In vitro bioaccessibility assessment as a prediction tool of nutritional efficiency. *Nutrition Research*. 29: 751–760.

García-Sartal, C., Barciela-Alonso, M. C., Moreda-Piñeiro, A. and Bermejo-Barrera, P., 2013. Study of cooking on the bioavailability of As, Co, Cr, Cu, Fe, Ni, Se and Zn from edible seaweed. *Microchemical Journal*. 108: 92–99.

Gomes da Silva, E., Verola-Mataveli, R. L. and Zezzi-Arruda, M. A., 2013. Speciation analysis of selenium in plankton, Brazil nut and human urine samples by HPLC-ICP-MS. *Talanta*. 110: 53–57.

Govasmark, E., Brandt-Kjelsen, A., Szpunar, J., Bierla, K., Vegarud, G. and Salbu, B., 2010. Bioaccessibility of Se from Se-enriched wheat and chicken meat. *Pure and Applied Chemistry*. 82: 461–471.

Haro-Vicente, J. F., Martínez-García, C. and Ros, G., 2006. Optimisation of *in vitro* measurement of available iron from different fortificants in citric fruit Juices. *Food Chemistry*. 98: 639–648.

Hazell, T., 1985. Minerals in foods: dietary sources, chemical forms, interactions, bioavailability. *World Review of Nutrition and Dietetics*. 46: 1–123.

Hinojosa-Reyes, L., Ruiz-Encinar, J., Marchante-Gayón, J. M., García-Alonso, J. I. and Sanz-Medel, A., 2006a. Selenium bioaccessibility assessment in selenized yeast after "*in vitro*" gastrointestinal digestion using two-dimensional chromatography and mass spectrometry. *Journal of Chromatography A*. 1110: 108–116.

Hinojosa-Reyes, L., Marchante-Gayón, J. M., García-Alonso, J. I. and Sanz-Medel, A., 2006b. Application of isotope dilution analysis for the evaluation of extraction conditions in the determination of total selenium and selenomethionine in yeast-based nutritional supplements. *Journal of Agricultural and Food Chemistry*. 54: 1557–1563.

Hur, J. S., Lim, O. B., Decker, A. E. and McClemets, D. J., 2011. In vitro human digestion models for food applications. *Food Chemistry*. 125: 1–12.

Intawongse, M. and Dean, J. R., 2006. In vitro testing for assessing oral bioaccessibility of trace metals in soil and food samples. *Trends in Analytical Chemistry*. 25: 876–886.

Jaiswal, S. K., Prakash, R., Acharya, R., Nathaniel, T. N., Reddy, A. V. R. and Prakash, N. T., 2012. Bioaccessibility of selenium from Se-rich food grains of the seleniferous region of Punjab, India as analyzed by instrumental neutron activation analysis. *CyTA – Journal of Food*. 10: 160–164.

Kápolna, E. and Fodor, P., 2007. Bioavailability of selenium from selenium-enriched green onions (*Allium fistulosum*) and chives (*Allium schoenoprasum*) after '*in vitro*' gastrointestinal digestion. *International Journal of Food Sciences and Nutrition*. 58: 282–296.

Karlsson, J. and Artursson, P., 1991. A method for the determination of cellular permeability coefficients and aqueous boundary-layer thickness in monolayers of intestinal epithelial (Caco-2) cells grown in permeable filter chambers. *International Journal of Pharmaceutics*. 71: 55–64.

Lavu, R. V. S., Du Laing, G., Van De Wiele, T., Pratti, V. L., Willekens, K., Vandecasteele, B. and Tack, F., 2012. Fertilizing soil with selenium fertilizers: impact on concentration, speciation, and bioaccessibility of selenium in leek (*Allium ampeloprasum*). *Journal of Agricultural and Food Chemistry*. 60: 10930–10935.

Lipiec, E., Siara, G., Bierla, K., Ouerdane, L. and Szpunar, J., 2010. Determination of selenomethionine, selenocysteine, and inorganic selenium in eggs by HPLC-inductively coupled plasma mass spectrometry. *Analytical and Bioanalytical Chemistry*. 397: 731–741.

Molly, K., Van de Woestyne, M. and Verstraete, W., 1993. Development of a 5-step multichamber reactor as a simulation of the human intestinal microbial ecosystem. *Applied Microbiology and Biotechnology*. 39: 254–258.

Moreda-Piñeiro, J., Moreda-Piñeiro, A., Romarís-Hortas, V., Domínguez-González, R., Alonso-Rodríguez, E., López-Mahía, P., Muniategui-Lorenzo, S., Prada-Rodríguez, D. and Bermejo-Barrera, P., 2013a. In vitro bioavailability of total selenium and selenium species from seafood. *Food Chemistry*. 139: 872–877.

Moreda-Piñeiro, J., Moreda-Piñeiro, A., Romarís-Hortas, V., Domínguez-González, R., Alonso-Rodríguez, E., López-Mahía, P., Muniategui-Lorenzo, S., Prada-Rodríguez, D. and Bermejo-Barrera, P., 2013b. ICP-MS for the determination of selenium bioavailability from seafood and effect of major food constituents. *Microchemical Journal*. 108: 174–179.

Miller, D. D., Schricker, B. R., Rasmussen, R. R. and Van Campen, D., 1981. An *in vitro* method for estimation of iron availability from meals. *American Journal of Clinical Nutrition*. 34: 2248–2256.

Minekus, M., Marteau, P., Havenaar, R. and Huisintveld, J. H. J., 1995. A multicompartamental dynamic computer-controlled model simulating the stomach and small-intestine. *Alternatives to Laboratory Animals*. 23: 197–209.

Muñiz-Naveiro, O., Domínguez-González, R., Bermejo-Barrera, A., Bermejo-Barrera, P., Cocho, J. A. and Fraga, J. M., 2006. Study of the bioavailability of selenium in cows' milk after a supplementation of cow feed with different forms of selenium. *Analytical and Bioanalytical Chemistry*. 385: 189–196.

Mutanen, M. L. and Mykkănen, H. M., 1984. Effect of dietary fat on plasma glutathione peroxidase levels and intestinal absorption of ^{75}Se-labeled sodium selenite in chicks. *Journal of Nutrition*. 114: 829–834.

Mutanen, M., 1986. Bioavailability of selenium in mushrooms, *Boletus edulis*, to young women. *International Journal of Vitaminology and Nutrition Research*. 56: 297–301.

Pedrero, Z., Madrid, Y. and Cámara, C., 2006. Selenium species bioaccessibility in enriched radish (*Raphanus sativus*): a potential dietary source of selenium. *Journal of Agricultural and Food Chemistry*. 54: 2412–2417.

Ramos, A., Cabrera, M. C. and Saadoun, A., 2012. Bioaccessibility of Se, Cu, Zn, Mn and Fe, and heme iron content in unaged and aged meat of Hereford and Braford steers fed pasture. *Meat Science*. 91: 116–124.

Reeves, P. G., Gregoire, B. R., Garvin, D. F., Hareland, G. A., Lindlauf, J. E., Johnson, L. K. and Finley, J. W., 2007. Determination of selenium bioavailability from wheat mill fractions in rats by using the slope-ratio assay and a modified Torula yeast-based diet. *Journal of Agricultural and Food Chemistry*. 55: 516–522.

Ruby, M. V., Schoof, R., Brattin, W., Goldade, M., Post, G., Harnois, M., Mosby, D. E., Casteel, S. W., Berti, W., Carpenter, M., Edwards, D., Cragin, D. and Chappell, W., 1999. Advances in evaluating the oral bioavailability of inorganics in soil for use in human health risk assessment. *Environmental Science and Technology*. 33: 3697–3705.

Shen, L., Van Dael, P., Luten, J. and Deelstra, H., 1996. Estimation of selenium bioavailability from human, cow's, goat and sheep milk by an *in vitro* method. *International Journal of Food Sciences and Nutrition*. 41: 75–81.

Shen, L. H., Hoek-Van Nieuwenhuizen, M. and Luten, J. B., 1997. Speciation and *in vitro* bioavailability of selenium in fishery products. In: Luten, J. B., Barresen, T. and Oehlenschlinger, J. (ed.) *Seafood From Producer to Consumer, Integrated Approach to Quality*. Elsevier Science, pp. 653.

Thiry, C., Schneider, Y.-J., Pussemier, L., De Temmerman, L. and Ruttens, A., 2013. Selenium bioaccessibility and bioavailability in Se-enriched food Supplements. *Biological and Trace Elements Research*. 152: 152–160.

Turlo, J., Gutkowska, B., Herold, F., Klimaszewska, M. and Suchocki, P., 2010. Optimization of selenium-enriched mycelium of *Lentinula edodes* (Berk.) pegler as a food supplement. *Food Biotechnology*. 24: 180–196.

Vitali, D., Dragojević, I. V. and Šebečić, B., 2008. Bioaccessibility of Ca, Mg, Mn and Cu from whole grain tea-biscuits: Impact of proteins, phytic acid and polyphenols. *Food Chemistry*. 110: 62–68.

Wang, Y.-D., Wang, X. and Wong, Y.-S., 2013. Generation of selenium-enriched rice with enhanced grain yield, selenium content and bioavailability through fertilisation with selenite. *Food Chemistry*. 141: 2385–2393.

Wickham, M., Faulks, R. and Mills, C., 2009. In vitro digestion methods for assessing the effect of food structure on allergen breakdown. *Molecular Nutrition and Food Research*. 53: 952–958.

Yu, Y.-X., Li, J.-L., Zhang, X.-Y., Yu, Z.-Q., Van de Wiele, T., Han, S.-Y., Wu, M.-H., Sheng, G.-Y. and Fu, J.-M., 2010. Assessment of the bioaccessibility of polybrominated diphenyl ethers in foods and the correlations of the bioaccessibility with nutrient contents. *Journal of Agricultural and Food Chemistry*. 58: 301–308.

Zeng, H., Botnen, J. H. and Johnson, L. K., 2008. A selenium-deficient Caco-2 cell model for assessing differential incorporation of chemical or food selenium into glutathione peroxidase. *Biological Trace Elements Research*. 123: 98–108.

Zohoori, V., Seal, C. J., Moynihan, P. J., Steen, I. N. and Maguire, A., 2009. The impact of fluoridatedmilks on the availability of trace elements in milk. *Journal of the Science of Food and Agriculture*. 89: 595–602.

CHAPTER 12

Assay of Selenium in Dietary Supplements

KRZYSZTOF JANKOWSKI*[a], ANNA TYBURSKA-STANIEWSKA[b], AND MARTA JANKOWSKA[c]

[a]Department of Analytical Chemistry, Faculty of Chemistry, Warsaw University of Technology, ul. Noakowskiego 3, 00-664 Warsaw, Poland; [b]Institute of Applied Optics, ul. Kamionkowska 18, 03-805 Warsaw, Poland; [c]Department of Immunology, Biochemistry and Nutrition, Medical University of Warsaw, ul. Żwirki i Wigury 61, 02-091 Warsaw, Poland
*E-mail: kj@ch.pw.edu.pl

12.1 Introduction

Selenium is present in the environment in trace or ultratrace levels and is referred to as an essential microelement. It is a component of about 30 proteins, including several selenoenzymes, in particular, glutathione peroxidase that is important in the cellular detoxification of peroxides. Moreover, selenium protects from metabolic disorders and acts in synergy with vitamins A, E, C, β-carotene, and lycopene. Although the mechanism of such effects is not fully understood yet, recent studies shows that it can stop the progress of malicious tumors by promoting the formation of white blood cells, which destroy cancer cells.

Obviously, selenium is necessary for proper functioning of living organisms; however, when consumed in excess, it may cause disorders in some functions. A daily demand for selenium ranges from 6 to 21 μg for infants up

Food and Nutritional Components in Focus No. 9
Selenium: Chemistry, Analysis, Function and Effects
Edited by Victor R. Preedy

Published by the Royal Society of Chemistry, www.rsc.org

to 75 μg for pregnant and lactating women. The maximum dose for safe daily consumption is 400 μg and this includes selenium supplied both in food and in dietary supplements. In the case of additional selenium supplementation the maximum amount cannot exceed 200 μg (Institute of medicine, 2000). Selenosis symptoms are observed when the daily selenium consumption is greater than 900 μg per day. The toxic dose is usually between 3.2 and 6.7 mg selenium per 24 h.

The toxicity, bioavailability, and environmental mobility of selenium are very much dependent on its chemical form (Fox *et al.*, 2004). Plants absorb soluble inorganic selenium species from soil and thus become the main source of selenium in human and animal diet. Next, those inorganic forms are methylated into organic forms, mostly selenomethionine and selenocysteine. Organoselenium compounds supplied in food are better accumulated than inorganic ones. The amount of selenium absorbed by human organism can be assessed by its content in serum, urine, hair or nails (Morris and Crane, 2013). The measured content is directly proportional to the amount of selenium consumed. In serum the normal content should be ranged between 75 and 140 μg L^{-1}.

In most widely available foodstuffs the selenium content oscillates between several ng mL^{-1} in water and several hundred ng g^{-1} in fruits, vegetables and meat. The *World Health Organization (WHO)* recommendation suggests the maximum selenium concentration in drinking water to be 40 ng mL^{-1} assuming that one supplies about 10% of daily selenium dose by drinking water. In the European Union the maximum selenium concentration in drinking water is assumed to be 10 ng mL^{-1}. The selenium content in food products is different depending on the region of origin, and it is proportional to the local selenium content in soil. However, it was observed that in the countries in which the selenium content in soil is high, people may also suffer from selenium deficiency. The main reason for this is the consumption of preprocessed food. In particular, frying or baking reduces the amount of selenium in the final product by 40%. In addition, heavy metals present in food may significantly reduce the absorption rate, because they form sparingly soluble selenides. To reduce the selenium deficiency, it is necessary to provide additional supplementation by adding Se directly to animal feed, fertilizing the soils or orally administrating selenium compounds by humans. Dietary supplements are considered as food products that are a source of vitamins, minerals or other substances of nutritive properties. Dietary supplements available on the market are usually in a concentrated form of pills, capsules or liquids. The internal regulations of different countries and international organizations treat dietary supplements in various ways, considering them either as food or as pharmaceuticals. The *Codex Alimentarius* (international repertory of nutrition norms) defines dietary supplements as food. All activities regarding the above-mentioned nutrition regulations are conducted under the supervision of *Codex Alimentarius Commission*, which was established by the *Food and Agriculture Organisation* of the *United Nations* (FAO) and WHO. The members of *Codex Alimentarius Commission* include now the *European Community* as well as 180 countries from outside Europe. European Union member countries are obliged to adapt their national law to the EU law, which on the strength of *Codex Alimentarius Commission* describes which vitamins and minerals can be used in the production of dietary supplements, and allows for the use of selenium compounds in the production of dietary supplements; however, it does not define their maximum concentration in the final product. In practice, the amount of selenium ranges from several to several hundred micrograms per pill.

12.2 Basic Characteristics of Selenium-Containing Dietary Supplements

Oral preparations intended for selenium supplementation are available in the form of strengthening mixtures, vitamin–mineral compositions or products containing only one selected component. In such preparations, selenium is present in inorganic form as sodium selenite, or in the form of Se aminoacids, typically selenomethionine. Both inorganic and organic selenium compounds can be absorbed directly from the alimentary canal, and the most effective absorption is observed for selenomethionine. Living organisms can acquire up to 95% of selenium from selenomethionine, 60–85% from other organic selenium compounds, and only 45–70% of selenium from inorganic selenium. By bioavailability of selenium one understands the amount of selenium not excreted with urine. Comparison of different dietary supplements available on the market containing different active Se forms and different Se content per pill, is presented in Table 12.1.

The subtle difference between the selenium dose beneficial for health and the toxic amount, as well as the need for control of the Se content in food products, leads to the development of sensitive methods of selenium determination. Additionally, sensitive analytical methods are necessary to monitor selenium content in living organisms, which helps to correct a diet setup and the possible prescription of supplementation. The control of selenium content in supplemented food and dietary supplements on every stage of the production process is also of very high concern.

In the case of dietary supplements containing organic selenium, selenized yeast is typically added. In particular, yeast from *Saccharomyces* sp. is an effective selenium biosorbent with 75% of the accumulated selenium in the form of selenomethionine. In the case of yeast from *Saccharomyces cerevisiae*, 80% of selenium is bound by high molecular weight biopolymers including cell walls, mitochondria and proteins. Using modern technologies, one can produce selenium yeast preparations containing from 500 to 2500 mg of selenium per kg. Some of these technologies have been patented including Cerosel-50.

Apart from baker's yeast, other micro-organisms, such as bacteria, algae, or some other yeast species (including brewing yeast) gain a high interest as selenium biosorbents. However, the use of other micro-organisms for the production of dietary supplements is limited, because they accumulate much smaller amounts of selenium than yeast. Lactic acid fermentation bacteria can be effective in the production of Se-enriched

Table 12.1 Comparison and characterization of different dietary supplements containing selenium compounds.

Product name	Se content in tablet (μg)	Se form	Producer
SelenoPrecise	200	—	Pharma Nord
SelenoQ10	100	Selenium yeast	Pharma Nord
Spirulina Selenium	30	Algae *Spirulina*	BlueBio(Yantai) Bio-pharmaceutical
Selenium oral drops	150/3 drops	Na_2SeO_3	Clinicians
Selenium with A C & E	100	L-Selenomethionine	Holland & Barrett
Selenium	200	Selenium yeast	Healthspan
Centrum	25	Na_2SeO_3	Whitehall
Centrum Junior	15		
Centrum Silver	25		
Ekovit	50	Na_2SeO_3	Merck Darmstadt
Femibion	50	Selenium yeast	Merck Darmstadt
Full Spectrum	20	Selenium yeast	California Fitness
Integral	35	Selenium yeast	Merck Darmstadt
Selenium	100	L-Selenomethionine	Swanson
Rutinoscorbin Plus	8	Na_2SeO_3	GlaxoSmithKline
Selenium	100	Selenium yeast	Holistic Health
Xenosulf	45	Algae	Laboratoires Le Stum Xenosulf
Spirulina organic Green Food	12,5	Algae *Arthrospira platensis*	Swanson
Cerosel-50	50	—	Biochefa
Vigor Complete Vigor Senior	25	Na_2SeO_3	US Pharmacia
Multitabs Active	61	—	Ferrosan
Power health Se, Zn	100	Brewery yeast Selenium yeast	—

food and dietary supplements. *Lactobacillius* can accumulate 400 mg of selenium per 1 kg of dry biomass. The process of selenium accumulation by the micro-organisms runs according to one of the following two mechanisms: (1) extracellular bonding by active groups present in bacterial cell membranes or cell walls of yeast or fungi; or (2) intracellular selenium accumulation, which requires selenium to pass the barrier of a cell wall and cytoplasmic membrane. Various active groups of proteins, phospholipids and polysaccharides present in cell walls participate in extracellular selenium bonding.

12.3 Sample Pretreatment and Separation of Selenium

The analysis of dietary supplements may cause a lot of problems due to the chemical complexity of the sample matrices. Apart from the active substance, they contain many excipients, such as talc, silica, zinc oxide, magnesium stearate, as well as flavoring or coloring substances, emulsifiers, stabilizers and preservatives. Due to the large complexity and variety of the sample components, and the lack of highly selective instrumental methods for determining selenium, it is often necessary to preseparate the analytes. Therefore, the methods used for analysis of medicines and dietary supplements can be divided into direct methods and complex analytical procedures involving separation stages. Several routine procedures are discussed in (Goenaga-Infante *et al.*, 2008).

Wet digestion with concentrated acids, acid leaching, enzymatic digestion and alkaline sample decomposition procedures are recommended for sample pretreatment before the assay of selenium (Table 12.2). These procedures can be substantially improved by application of high-efficiency microwave and ultrasonic devices (Capelo *et al.*, 2006; Vale *et al.*, 2007). Closed-vessel microwave technology permits prevention against the loss of volatile selenium compounds.

Another critical point in the pretreatment stage is the separation of selenium from interfering components of dietary supplements including organic substances and transition metals. This is often accompanied by preconcentration of selenium. Inorganic selenium was preconcentrated on functionalized cellulose sorbent Cellex T with quaternary amine exchange groups. The analyte was retained on-column and next eluted with diluted hydrochloric acid. Additionally, selenium can be separated from even 1 million-fold excess of Fe, Cr, Cu and Zn that is beneficial for analysis of multimineral supplements (Pyrzyńska, 1995).

Another technique of analyte separation and preconcentration is solid-phase microextraction (SPME). In this solvent-free sampling technique, a fiber coated with polymer film is used. Two modes of SPME procedure can be distinguished: direct immersion microextraction (DI-SPME) based on fiber immersion in sample solution, and headspace microextraction (HS-SPME) based on volatile component extraction from the gaseous phase over the sample solution. When determining total selenium content, the use of alkaline solution of tetramethylammonium hydroxide (TMAH) for dietary

Table 12.2 Sample pretreatment and separation of selenium from the matrix of dietary supplements.

Operation	Option	Goal
Pretreatment	Wet digestion Microwave digestion Enzymatic digestion Alkaline dissolution Acid leaching Electrothermal vaporization	Decompose the matrix and/or release of Se species from the matrix
Separation and preconcentration	Solid-phase extraction Solid-phase microextraction Liquid extraction	Selective separation and collection of Se compound(s) from the bulk sample

supplement sample digestion reduces the loss of volatile Se compounds, decreases the transition-metal interference, and eliminates the use of acids potentially harmful for the apparatus. Selenium is converted into volatile selenium hydride, and after separation by HS-SPME, subsequent thermal desorption of the analyte is performed in the injector directly connected to the ICP-OES detection system (Tyburska and Jankowski, 2013).

12.4 Instrumental Techniques for Selenium Assay in Dietary Supplements

12.4.1 Electrochemical Methods

Electrochemical methods are widely used for determining both active substances and impurities, which may occur in pharmaceuticals at trace level. These methods offer several advantages including the ease of sample preparation, fast measuring and selectivity. Therefore, they are successfully used for the determination of selenium in dietary supplements. Either adsorptive stripping voltammetry (AdSV) or differential pulse cathodic stripping voltammetry (DPCSV) methods are most commonly used. AdSV has been applied for the determination of inorganic Se(IV) in pharmaceuticals using a controlled-growth mercury drop electrode. The LOD achieved was in the order of tens of pg mL^{-1}, which makes the method well suitable for routine determinations of selenium in pharmaceutical products. However, all of the electroanalytical methods, including AdSV, are prone to matrix interference and require a very thorough procedure for real samples. The results obtained with this method were comparable with those obtained by ICP-MS, and by electrothermal-atomic adsorption spectrometry (ETAAS) (Kowalczyk *et al.*, 2005). In differential pulse cathodic stripping voltammetry (DPCSV), the analyte is initially preconcentrated by electrolysis, while in AdSV this is done due to adsorption. The second stage of the measurement procedure is similar in both techniques, and takes advantage of the change of the electrode potential caused by the reduction of the accumulated depolarizer. Both the CSV and AdSV offer similar LODs, and can be used for the determination of trace selenium in pharmaceuticals. It is noteworthy that all of the electroanalytical techniques, including AdSV and DPCSV, are prone to matrix interferences. In some cases, the use of standard additions minimizes the influence of several microelements (such as Zn, Mo, Cr) and vitamins (Stoica *et al.*, 2002) on the determination of selenium. The disadvantage of both AdSV and CSV is that they require very precise optimization of measurement conditions to ensure repeatability, high sensitivity and selectivity of the determination.

The use of differential pulse polarography (DPP) for the determination of selenium in vitamin supplements seems to be more convenient because of its susceptibility to the presence of excess transition metals, such as Cu, Pb, Zn. However, compared to AdSV, the LOD for Se by DPP is higher, reaching 5 ng mL^{-1} (Kherwa and Dubey, 2013).

12.4.2 Spectroscopic Methods

Spectroscopic techniques are the most widely used for determining total selenium content in dietary supplements (Renard and Tompkins, 2002). Spectrophotometry in the UV-Vis range was very popular in the past decades. Nowadays, the significance of this method is marginal and it is hard to find any up-to-date reports regarding the determination of selenium in dietary supplements based on this technique. Obviously, the sensitivity of these techniques is too low for this purpose. However, hyphenated techniques including a preconcentration step with the use of cloud point extraction of selenium hydrophobic complexes into a surfactant-rich phase have been proposed for the determination of total selenium content in supplements containing inorganic or organic selenium. In all cases Se species were converted into Se(IV) before complexation. The method based on Se-dithizone complex with LOD of 4.4 ng mL^{-1} shows susceptibility to the presence of Cu and Fe.

Selenium was assayed in vitamin supplements using a yellow-colored complex of Se(IV) with 1-naphthyloamine-7-sulfonic acid (Cleve's acid) of maximum absorption at 350 nm (Pyrzyńska, 1997). Selenium can be determined over the range of 20–300 ng mL^{-1} of final solution after on-column preconcentration on functionalized cellulose sorbent Cellex T.

Apart from the above-mentioned spectrophotometric methods, several indirect methods for determination of selenium in dietary supplements have been developed based on oxidation of iodide with Se(IV). In the reaction between iodide and Se(IV) in acidic medium, iodine is formed that in turn interacts with starch, thereby forming a blue-colored adduct with maximum absorption at 570 nm. In the more sensitive method, thionine is used to form colored complex with the liberated iodine, which becomes violet with maximum absorption at 600 nm. Beer's law was obeyed between 0.1 and 0.5 μg mL^{-1} (Revanasiddappa and Kiran Kumar, 2002). Both methods were applied for assaying selenium in dietary supplements after typical sample pretreatment including acid digestion, conversion of all selenium species into Se(IV) and masking of interfering metals. However, the application of the methods is restricted by a relatively low selectivity and sensitivity. In particular, the determination of selenium in multivitamin and multimineral formulations should be validated due to possible interferences from excess Zn, Fe and Cu. Moreover, organic selenium may be lost in the sample preparation step because some organic forms are resistant to acid digestion and some others are volatile. In such cases, microwave digestion is recommended.

In acid media, Se(IV) can stimulate oxidative coupling reaction between 4-aminoantipyrine (4-AAP) and *N*-(naphthalene-1-yl)ethane-1,2-diamine-dihydrochlorine (NEDA), which provides spectrophotometric detection in an automated flow-injection procedure for indirect determination of selenium in pharmaceutical samples. When operating under the flow-injection regime, the interference from transition metals present in the sample solution can be minimized. A relatively high LOD might be considered as a main drawback of this method, thus large samples are required and acid digestion might be troublesome (Kumar *et al.*, 2008).

Atomic spectrometry, including absorption spectrometry, emission spectrometry and atomic fluorescence spectrometry, is widely used for determination of selenium in dietary supplements and pharmaceuticals. Depending on the type of atomizer one can distinguish between flame atomization (FAAS) and electrothermal atomization (ETAAS). The latter is more suitable for determination of trace selenium because of its high sensitivity. The FAAS technique may be applied for the analysis of dietary supplements of high selenium content (above 100 μg per g). However, the LOD for selenium can be lowered about ten times when hydride generation is used for sample introduction (Kozak *et al.*, 2008). Even lower detection limit can be achieved with the use of ETAAS (Vale *et al.*, 2010). The use of chemical modifiers in the ETAAS technique is crucial for determining selenium because certain selenium compounds are volatile and could be lost. The addition of a modifier to the sample increases the volatility of the matrix components and/or lowers the volatility of the analytes. It allows for the separation of selenium from other sample components. Another disadvantage is the necessity of conversion of all selenium compounds present in the sample into one chemical form before assaying of selenium. In the last decade a high-resolution continuum light source (HR-CS) was introduced to ETAAS that offered new analytical possibilities for this technique. The use of this system allowed for the determination of selenium at the level of several ng mL^{-1}. This can be done even in the presence of excessive iron, due to the high resolution of the optical system. The advantage of this method is the possibility of multielement determination of almost all mineral components including Cr, Cu, and Fe apart from Se (Krawczyk, 2014).

Another sensitive technique suitable for trace analysis is inductively coupled plasma optical emission spectrometry. The instrumental LOD for selenium by the ICP-OES with pneumatic nebulization is relatively high (10 ng mL^{-1}). However, introduction techniques based on the chemical vapor generation have been developed by taking advantage of the volatility of some selunium compounds. The application of the hydride generation (HG) technique lowers the LOD for Se by up to 100 times (Tyburska *et al.*, 2010). When applied for the analysis of dietary supplements, the main disadvantage of HG is the susceptibility to the presence of transition metals, which easily decompose selenium hydride. To minimize this effect one can apply a direct HG inside the spray chamber directly connected to the plasma torch (Tao and Sturgeon, 1999). The other possibility is to conduct the HG process from alkaline medium (Tyburska and Jankowski, 2013).

Also, the combination of HG with AFS was used for the determination of trace selenium in dietary supplements. The advantages of the AFS method include fair selectivity, a low noise level, and wide linear dynamic range extending to several orders of magnitude. The LOD for selenium is lower than 1 ng mL^{-1}. The analytical potential of HG-AFS has been substantially improved by using liquid chromatography for preseparation of selenium species (Viñas *et al.*, 2005; Stibilj *et al.*, 2010).

Solid-sampling techniques in conjunction with atomic spectrometry have been proposed for assaying selenium in dietary supplements with the aim to avoid the potential analyte losses during the sample preparation stage. The continuous powder introduction technique proved to be compatible with microwave-induced plasma optical emission spectrometry (MIP-OES). The analytical potential of this technique has been shown during the reliable analysis of various samples including dietary supplements. Selenium has been determined in dietary supplements at the level of 15–100 μg per g by using both matrix matching and particle-size matching for preparing calibration standards (Tyburska, 2011). Also, slurry sampling has been successfully applied and coupled with ETAAS (Krawczyk, 2014). Another technique employing solid-sampling methodology is total reflectance X-ray fluorescence spectrometry, which is a well-established method in medical diagnostics. Also, it has been applied as a fast screening method for determination of selenium in dietary supplements for samples containing as low as 8–30 μg mg^{-1} of Se (Stosnach, 2010).

Nowadays, the ICP-MS offers the lowest LODs at the level of pg mL^{-1}, and can be successfully used for assaying trace selenium in pharmaceutical samples (Ayouni *et al.*, 2007). The most important limitation of ICP-MS is that the most abundant selenium isotopes (^{80}Se, ^{78}Se, ^{76}Se) are burdened with polyatomic interferences generated by argon plasma gas. Elimination of the spectral interference can be achieved by choosing a different isotope, usually less abundant, with the expense of lower sensitivity. However, the use of either a high-resolution MS analyzer or collision chamber, placed before the entrance to the mass analyzer proved to be more effective. In such a chamber, polyatomic ions (for example $^{40}Ar^{40}Ar^{+}$, $^{38}Ar^{40}Ar^{+}$) are converted into neutral molecules or ions of different *m*/*z* through collisions and chemical reactions. Thus, the use of two main isotopes of selenium is possible for determining total selenium content in dietary supplements. Before the measurement, samples are usually digested with the aid of acids or selenium species are extracted with hot water. The effectiveness of hot water extraction is 80–95% for selenate and selenomethionine and only 10–25% for selenium present in dietary supplements as Se-enriched yeast. Different analytical methodologies have been examined for the determination of total selenium by ICP-MS including external calibration, standard additions or isotope dilution MS and show comparable results (Goenga-Infante *et al.*, 2008; Sannac *et al.*, 2009).

The comparison of the above-described analytical procedures for the assay of total selenium used in dietary supplement analysis is presented in Table 12.3.

12.5 Selenium Speciation in Dietary Supplements

Nowadays, apart from the necessity of the determination of total selenium content in dietary supplements, the information about its chemical form is also important due to differences in the bioavailability and toxicity of organic and inorganic selenium compounds. In this case speciation analysis

Table 12.3 Summary of analytical procedures used for the assay of total selenium in dietary supplements.

Sample (Se concentration)	Species	Separation	Analytical technique	DL (ng mL^{-1})	RSD (%)	References
Ekovit Rutinoscorbin Plus	Na_2SeO_3	• Extraction of Se(IV) with water • Controlled-growth mercury drop electrode	AdSV	0.05	3.5	Kowalczyk *et al.* (2005)
Optima forte (17 µg g^{-1}) Antioxidant (59 µg g^{-1}) Selenium Walmark (151 µg g^{-1}) Eurovita (13 µg g^{-1}) Orgasel (117 µg g^{-1})	Na_2SeO_3	• Microwave digestion: 6 mL HNO_3 + 1 mL H_2O_2 • Mercury drop electrode	DPCSV	0.1	4.8–8.8	Stoica *et al.* (2002)
Revup (150 µg per tablet) Fourts B (100 µg per tablet)	Na_2SeO_3	• Wet digestion: 5 mL HNO_3 • Leaching with H_2SO_4 • Neutralization NaOH	UV-Vis	10	0.3–0.6	Revanasiddappa and Kiran Kumar (2002)
Acetoxy (70 µg per tablet) Oxyban (150 µg per tablet) Recharje plus (60 µg per capsule) Revox (150 µg per capsule) Reconia (30 µg per tablet)	—	• Heating with 10 mL HNO_3 • Oxidative coupling reaction of 4-AAP and NEDA in the presence of Se(IV)	UV-Vis	500	2	Kumar *et al.* (2008)
Tablets A (30 mg kg^{-1}) Tablets B (8 mg kg^{-1})	—	• Preconcentration: (Triton X-100)	XRF	100–200	—	Stosnach (2010)

Se-enriched food supplements	—	• Microwave-assisted acid digestion • Anion exchange column: Hamilton PRP-X100	HPLC-ETAAS	Se(IV) 200 Se(VI) 265 SeMet 299 SeCys 297	—	Vale *et al.* (2010)
Multivitamin dietary supplement (MVDS-3)	—	• Microwave digestion (1 mL 30% H_2O_2, 4 mL conc. HNO_3 and 1.5 mL conc. HF)	HR-CS-GFAAS	40	7	Krawczyk (2014)
Centrum-Junior ABC Plus Bio-selenium + Zinc Unicap T; Unicap M Sport Fuel Triovit	Na_2SeO_3 Na_2SeO_3 Organoselenium Na_2SeO_3 — Selenium yeast	• Pressurized acid digestion (0.5 mL conc. H_2SO_4 + 1.5 mL conc. HNO_3 + 2 mL H_2O_2) • Oxidation with 0.15 mL V_2O_5 in H_2SO_4 • Prereduction to Se(IV) with 2.5 mL conc. HCl	HG-AFS	0.2	2	Stibilj *et al.* (2005)
Rutionoscorbin Plus (37 µg g^{-1})	Na_2SeO_3	• Grinding and blending with SiO_2	CPI-MIP-OES	5000	4	Tyburska (2011)
Bioselenium + Zn (100 µg per tablet) Selenoprecise (200 µg per tablet) Bioantioxidant (50 µg per tablet) Visio Ace (75 µg per tablet) Selenium Ace (100 µg per tablet)	SeMet Selenium yeast SeMet Selenium yeast Selenium yeast	• Extraction with 20 mL of water • Reversed-phase and ion-pair HPLC: column Nucleosil 120 A, C_{18}, 5 µm • Mobile phase:30 mM ammonium formate, 5% v/v methanol	ICP-MS	Se(VI) 0.082 Se(IV) 0.044 SeCys 0.042 SeMet 0.036 SeEt 0.058	—	Marchante-Gayon *et al.* (2000)

is recommended. Alternatively, the information about selenium content may contain its total concentration in the sample as well as its origin (for example, Se-enriched yeast), without revealing the chemical form. Here, only a brief summary of publications regarding selenium speciation in pharmaceuticals and dietary supplements will be given. From this point of view, analytical procedures allowing for determination of inorganic selenium and selenomethionine in rich sample matrices are required, preferably with the use of high-performance liquid chromatography inductively coupled plasma mass spectrometry (HPLC-ICP-MS).

Reversed-phase chromatography allows for the separation of both inorganic selenium and several organic forms including selenocysteine, selenomethionine, selenoethionine, selenourea, selenocystanine and trimethylselenium ion. Another separation technique is size-exclusion chromatography, which, contrary to the RP-LC, shows low matrix effects.

Due to the rich matrix of dietary supplements, several scenarios of sample pretreatment are possible, based on liquid extraction and enzymatic digestion. Sequential extraction with hot water, protease and nitric acid was used for selenium compounds determination in dietary supplements. The effectiveness of such extraction is 80% and 95% for selenate and selenomethionine, respectively (Viñas *et al.*, 2005). The use of Protease 14 or Proteinase K enzymes to sample digestion can help identifying selenomethinonine, its oxidized form and *S*-(methyl-seleno)cysteine (Capelo *et al.*, 2006; Vale *et al.*, 2007; Goenaga-Infante *et al.*, 2008). Alkaline dissolution with tetramethylammonium hydroxide (TMAH) may be also used for sample pretreatment leading to about 100% recovery of the total selenium present in selenium yeast.

γ-Glutamyl-Se-methyl-selenocysteine has been separated from selenium yeast extract by RP-HPLC and identified by ESI-MS/MS (Goenaga-Infante *et al.*, 2005). In the other study, a number of Se-enriched plants and yeast have been examined by HPLC-ICP-MS with a variety of Se species identified including selenomethionine in yeast and *Se*-methylselenocysteine in onion and garlic (Kotrebai *et al.*, 2000). Total selenium measurements and selenium speciation profiles obtained by ion-pair RP-HPLC-ICP-MS show major differences in speciation for selenomethionine, selenomethionine Se-oxide and *S*-(methyl-seleno)cysteine with respect to time and conditions of storage of nutritional supplements (Amoako *et al.*, 2007).

Apart from ICP-MS, ETAAS has also been successfully used for speciation studies of selenium (Vale *et al.*, 2007; Vale *et al.*, 2010). It is noteworthy that several routine procedures dedicated to selenium speciation analysis in dietary supplements have been subjected to interlaboratory comparison and metrological assessment (Goenaga-Infante *et al.*, 2008; Sannac *et al.*, 2009).

12.6 Conclusions

The increasing consumption of dietary supplements stimulates the development of analytical procedures for determining total selenium content in a variety of nutritional products. Analytical laboratories offer several reliable instrumental techniques or even routine analytical procedures enabling the determination of selenium in such materials with good accuracy and precision. ICP-MS is the most efficient tool, which when combined with microwave digestion or selected extraction procedures and with advanced calibration methodology including isotope dilution can serve for both the assay of total selenium content and, in combination with liquid chromatography, speciation studies covering a wide variety of selenium species. However, some problems remain unresolved including sample pretreatment strategy, selenium species-dependent recovery and elimination of matrix effects caused by organic substances and transition metals accompanying selenium. Because of the discrepancies between samples, the results obtained by commercial laboratories may be unreliable, especially for samples containing organoselenium forms. Selection of the appropriate method of sample pretreatment and determination of selenium is critical in the analysis of dietary supplements containing organoselenium compounds. However, the analysis of the formulations containing inorganic selenium seems to be reliable. The attractive features of certain hydride generation plasma spectrometric techniques, in terms of a relatively high tolerance toward the presence of elevated levels of transition metals, may help to improve the detection power of total Se in dietary samples.

Because bioavailability, and ultimately, toxicity, is strongly dependent upon the chemical form of selenium, total selenium content in dietary supplements may be uninformative regarding any positive and negative effects. Therefore, stricter control on the Se content in dietary supplements in terms of Se species will need to become mandatory, together with the information about total Se content.

Summary Points

- This chapter focuses on determination selenium in dietary supplements.
- Both organic and inorganic selenium forms are added in preparations, and the organic forms are better absorbed.
- A number of spectroscopic or electrochemical methods are recommended for the determination of trace selenium.
- Essential problems occurring during sample preparation and separation of selenium from the sample matrix are pointed out including selenium losses due to its volatility, species-dependent recovery of selenium from samples, and severe matrix effects from both organic and mineral matrix components.
- Today, ICP-MS is the most accurate and sensitive method for determination of total selenium content in dietary supplements, and when coupled with liquid chromatography it provides additional species-selective information.
- It is postulated that the information about selenium species present in the preparation must accompany the total selenium content for the dietary supplement characteristic to be sufficiently informative.

Key Facts

Key Facts of Inductively Coupled Plasma/Optical Emission Spectrometry (ICP-OES)

1. The ICP was developed for OES by Fassel in the USA and Greenfield in the UK in the mid-1960s.
2. A sample of complex matrix should be digested and pretreated before being measured by ICP-OES.
3. ICP-OES provides low ppb or sub-ppb-level LODs and is commonly used for multielemental analysis.

Key Facts of Inductively Coupled Plasma/Mass Spectrometry (ICP-MS)

1. The ICP was introduced as an ion source for MS by Gray and Date in the UK and Houk in the USA in the late 1970s.
2. A sample matrix should be removed or analytes should be separated from the matrix before the ICP-MS measurement to provide accurate results.
3. ICP-MS allows for the determination of metals and several nonmetals with atomic masses ranging from 7 to 250 amu.
4. ICP-MS provides low ppt or even sub-ppt-level LODs and is commonly used for multielemental analysis.

Key Facts of Electrothermal Atomic Absorption Spectrometry (ETAAS)

1. ETAAS or graphite-furnace AAS was pioneered by Lvov in Russia and Massmann in Germany in the late 1950s.
2. In ETAAS a transient signal is generated, the area of which is directly proportional to the mass of analyte.
3. In ETAAS, matrix effects are minimized with the use of specific chemical modifiers and after a controlled multistep heating procedure.
4. ETAAS provides low ppb or sub-ppb-level LODs for most elements.

Definitions of Words and Terms

Dietary supplement. A dietary supplement is a foodstuff, administrated to supplement the normal diet, which is a concentrated source of vitamins or minerals, or other substances with a nutritional or physiological effect, marketed in a form that allows the dosage including capsules, tablets, and other similar forms, sachets of powder, ampoules of liquid, drop-dispensing bottles, and other similar forms of liquids and powders designed to be eaten in small, measured amounts.

Biosorbent. Micro-organism possessing natural ability to absorb, metabolize and accumulate different selenium compounds that in turn increases the bioavailability of selenium.

Bioavailability. A fraction of the consumed amount of the nutritional component in a given chemical form being absorbed by living organisms.

Solid-phase microextraction (SPME). Separation and preconcentration technique of trace analytes based on the sorption of the analytes on a thin polymer layer coating a quartz or metallic microfiber.

Voltammetry. A group of electrochemical analytical techniques based on the measurement of electric current generated under selected conditions during electrochemical reaction in the system of working and reference electrode immersed in the sample solution containing the determined substance.

Inductively coupled plasma/optical emission spectroscopy (ICP-OES). An analytical technique of emission spectroscopy, in which the plasma generated in the electromagnetic field at radio frequencies, is the source of excited atoms of analytes. After excitation the atoms return to the ground state and emit the energy, specific to the given element that is detected.

Inductively coupled plasma/mass spectrometry (ICP-MS). An analytical technique of elemental mass spectrometry that enables determination of elements at ng L^{-1} levels. This is achieved by ionizing the sample in the radio frequency-operated plasma and then using a mass spectrometer to separate and quantify created ions.

Electrothermal–atomic absorption spectrometry (ETAAS). An analytical technique of atomic spectrometry in which sample constituents including an analyte are vaporized and atomized under controlled multistep heating conditions. Next, a monochromatic light emitted by the lamp is absorbed by the analyte atoms separated from the sample matrix, and the fraction of the intensity of the incident light that is being absorbed by the atoms corresponds to the analyte concentration in the sample.

Atomic spectrometry. A group of spectroscopic analytical techniques based on the partial conversion of the element determined into atomic or ionic state. The resulting atoms participate in different spectroscopic processes including absorption or emission of light, fluorescence or, if present in the form of ions, are separated according to their mass-to-charge ratio.

Hydride generation. A kind of sample introduction technique used in atomic spectrometry based on conversion of analytes into volatile hydrides.

Reversed-phase high-performance liquid chromatography (RP HPLC). A type of analytical separation technique of liquid sample components based on the differences in their partitioning between the stationary phase of lower polarity and mobile phase of higher polarity.

Chemical speciation. The presence of an element in the sample in different chemical forms.

Limit of detection. The value specific for each analyte determined by a given analytical method and defined as the lowest quantity of the analyte that can be identified using this method.

List of Abbreviations

4-AAP	4-Aminoantipyrine
AdSV	Adsorptive stripping voltammetry
AFS	Atomic fluorescence spectrometry
DI-SPME	Direct immersion solid-phase microextraction
DPCSV	Differential pulse cathodic stripping voltammetry
DPP	Differential pulse polarography
ESI-MS/MS	Electrospray ionization tandem mass spectrometry
ETAAS	Electrothermal atomic adsorption spectrometry
FAAS	Flame atomic absorption spectrometry
FAO	Food and Agriculture Organization
HG	Hydride generation
HPLC	High-performance liquid chromatography
HR-CS	High-resolution continuum-light source
HS-SPME	Headspace solid-phase microextraction
ICP-MS	Inductively coupled plasma mass spectrometry
ICP-OES	Inductively coupled plasma optical emission spectrometry
LOD	Limit of detection
MIP-OES	Microwave-induced plasma optical emission spectrometry
NEDA	*N*-(Naphthalene-1-yl)ethane-1,2-diamine dihydrochlorine
RP-HPLC	Reversed-phase high-performance liquid chromatography
TMAH	Tetramethylammonium hydroxide
TRXRF	Total reflectance X-ray fluorescence spectrometry
UV-Vis	Ultraviolet-visible
WHO	World Health Organization

References

Amoako, P. O., Kahakachchi, C. L., Dodova, E. N., Uden, P. C. and Tyson, J. F., 2007. Speciation, quantification and stability of selenomethionine, *S*-(methylseleno)cysteine and selenomethionine Se-oxide in yeast-based nutritional supplements. *Journal of Analytical Atomic Spectrometry*. 22: 938–946.

Ayouni, L., Barbier, F., Imbert, J. L., Lantéri, P. and Grenier-Loustalot, M. F., 2007. Speciation of selenium in a commercial dietary supplement by liquid chromatography coupled with inductively coupled plasma-mass spectrometry (ICP-MS). *Journal of Toxicology and Environmental Health, Part A*. 70: 735–741.

Capelo, J. L., Fernandez, C., Pedras, B., Santos, P., Gonzalez, P. and Vaz, C., 2006. Trends in selenium determination/speciation by hyphenated techniques based on AAS or AFS. *Talanta*. 68: 14421447.

Fox, T. E., Van den Heuvel, E. G. H. M., Atherton, C. A., Dainty, J. R., Lewis, D. J., Langford, N. J., Crews, H. M., Luten, J. B., Lorentzen, M., Sieling, F. W., Van Aken-Schneyder, P., Hoek, M., Kotterman, M. J. J., Van Dael, P. and

Fairweather-Tait, S. J., 2004. Bioavailability of selenium from fish, yeast and selenate: a comparative study in humans using stable isotopes. *European Journal of Clinical Nutrition*. 58: 343–349.

Goenaga-Infante, H., O'Connor, G., Rayman, M., Wahlen, R., Spallholz, J. E., Hearn, R. and Catterick, T., 2005. Identification of water-soluble gamma-glutamyl-Semethylselenocysteine in yeast-based selenium supplements by reversed-phase HPLC with ICP-MS and electrospray tandem MS detection. *Journal of Analytical Atomic Spectrometry*. 20: 864–870.

Goenaga-Infante, H., Sturgeon, R., Turner, J., Hearn, R., Sargent, M., Maxwell, P., Yang, L., Barzev, A., *et al.*, 2008. Total selenium and selenomethionine in pharmaceutical yeast tablets: assessment of the state of the art of measurement capabilities through international intercomparison CCQM-P86. *Analytical and Bioanalytical Chemistry*. 390: 629–642.

Institute of Medicine, 2000. Food and Nutrition Board. *Dietary Reference Intakes: Vitamin C. Vitamin E, Selenium, and Carotenoid*. National Academy Press, Washington, DC.

Kherwa, V. and Dubey, S., 2013. A simple method for voltammetric determination of trace amounts of selenium in diverse matrices. *Research Journal of Recent Sciences*. 2: 51–54.

Kotrebai, M., Birringer, M., Tyson, J. F., Block, E. and Uden, P. C., 2000. Selenium speciation in enriched and natural samples by HPLC-ICP-MS and HPLC-ESI-MS with perfluorinated carboxylic acid ion-pairing agents. *Analyst*. 125: 71–78.

Kowalczyk, P., Łozak, A. and Fijałek, Z., 2005. Determination of selenium in multicomponent pharmaceutical preperations. *Analytical Chemistry*. 50: 437–447.

Kozak, L., Rudnicka, M. and Niedzielski, P., 2012. Determination of inorganic selenium species in dietary supplements by hyphenated analytical system HPLC-HG-AAS. *Food Analytical Methods*. 5: 1237–1243.

Krawczyk, M., 2014. Determination of macro and trace elements in multivitamin dietary supplements by high-resolution continuum source graphite furnace atomic absorption spectrometry with slurry sampling. *Journal of Pharmaceutical and Biomedical Analysis*. 88: 377–384.

Kumar, K. S., Suvardhan, K. and Kang, S. H., 2008. Facile and sensitive determination of selenium(IV) in pharmaceutical formulations by flow injection spectrophotometry. *Journal of Pharmaceutical Sciences*. 97: 1927–1933.

Marchante-Gayon, J. M., Thomas, C., Feldmann, I. and Jakubowski, N., 2000. Comparison of different nebulisers and chromatographic techniques for the speciation of selenium in nutritional commercial supplements by hexapole collision and reaction cell ICP-MS. *Journal of Analytical Atomic Spectrometry*. 15: 1093–1102.

Morris, J. S. and Crane, S. B., 2013. Selenium toxicity from a misformulated dietary supplement, adverse health effects, and the temporal response in the nail biologic monitor. *Nutrients*. 5: 1024–1057.

Pyrzyńska, K., 1995. Separation of inorganic selenium species on anion-exchange resins. *Analyst*. 120: 1933–1936.

Pyrzyńska, K., 1997. Spectrophotometric determination of selenium with 1-naphthyloamine-7-sulfonic acid. *Analytical Sciences*. 13: 629–632.

Renard, N. E. and Tompkins, T. A., 2002. Evaluation of methods for total selenium determination in yeast. *Biological Trace Element Research*. 88: 185–191.

Revanasiddappa, H. D. and Kiran Kumar, T. N., 2002. Spectrophotometric determination of selenium by use of thionin. *Analytical and Bioanalytical Chemistry*. 374: 1121–1124.

Sannac, S., Pannier, F., Oster, C., Labarraque, G., Fisicaro, P. and Potin-Gautier, M., 2009. Validation of a reference measurement procedure for the assessment of selenomethionine in nutritional supplements. *Journal of Analytical Atomic Spectrometry*. 24: 237–241.

Stibilj, V., Smrkolj, P. and Krbavcic, A., 2010. Investigation of the declared value of selenium in food supplements by HG-AFS. *Microchimica Acta*. 150: 323–327.

Stoica, A. I., Babaua, G. R., Iorgulescu, E. E., Marinescu, D. and Baiulescu, G. E., 2002. Differential pulse cathodic stripping voltammetric determination of selenium in pharmaceutical products. *Journal of Pharmaceutical and Biomedical Analysis*. 30: 1425–1429.

Stosnach, H., 2010. Analytical determination of selenium in medical samples, staple food and dietary supplements by means of total reflection X-ray fluorescence spectroscopy. *Spectrochimica Acta, Part B: Atomic Spectroscopy*. 65: 859–863.

Tao, G. H. and Sturgeon, R. E., 1999. Sample nebulization for minimization of transition metal interferences with selenium hydride generation ICP-AES. *Spectrochimica Acta, Part B: Atomic Spectroscopy*. 54: 481–489.

Tyburska, A., 2011. Application of solid phase extraction and hydride generation for the determination trace selenium by selected optical emission spectrometry techniques. Ph.D. thesis, Warsaw University of Technology. (in Polish).

Tyburska, A. and Jankowski, K., 2013. Determination of selenium in dietary supplements by optical emission spectrometry after alkaline dissolution and subsequent headspace solid phase microextraction. *Journal of Pharmaceutical and Biomedical Analysis*. 74: 268–272.

Tyburska, A., Jankowski, K., Ramsza, A., Reszke, E., Strzelec, M. and Andrzejczuk, A., 2010. Feasibility study of the determination of selenium, antimony and arsenic in drinking and mineral water by ICP-OES using a dual-flow ultrasonic nebulizer and direct hydride generation. *Journal of Analytical Atomic Spectrometry*. 25: 210–214.

Vale, G., Pereira, S., Mota, A., Fonseca, L. and Capelo, J. L., 2007. Enzymatic probe sonication as a tool for solid–liquid extraction for total selenium determination by electrothermal-atomic absorption spectrometry. *Talanta*. 74: 198–205.

Vale, G., Rodrigues, A., Rocha, A., Rial, R., Mota, A. M., Goncalves, M. L., Fonseca, L. P. and Capelo, J. L., 2010. Ultrasonic assisted enzymatic digestion (USAED) coupled with high-performance liquid chromatography and electrothermal-atomic absorption spectrometry as a powerful tool for total selenium and selenium species control in Se-enriched food supplements. *Food Chemistry*. 121: 268–274.

Viñas, P., Lopez-Garcıa, I., Merino-Merono, B., Campillo, N. and Hernandez Cordoba, M., 2005. Determination of selenium species in infant formulas and dietetic supplements using liquid chromatography-hydride generation atomic fluorescence spectrometry. *Analytica Chimica Acta*. 535: 49–56.

CHAPTER 13

Selenium as a Modulator of Fourier Transform Infrared Spectra in Cancer

ANSHOO MALHOTRA[a] AND D. K. DHAWAN*[b]

[a]Department of Biophysics, Post Graduate Institute of Medical Education and Research (PGIMER), Chandigarh, India-160012;
[b]Department of Biophysics, Panjab University, Chandigarh, India-160014
*E-mail: dhawan@pu.ac.in

13.1 Introduction

A trace element is a micronutrient that is required in "trace amounts (<100 ppm)" for various cellular functions and in maintaining normal homeostasis in the mammalian system. Trace elements such as cobalt, copper, zinc, iron, molybdenum, iodine, manganese and selenium are considered essential as they are structural components of biomolecules and are needed for various catalytical functions. Another criterion for the essentiality of a trace element is their involvement in oxidation–reduction processes for the release of metabolic energy. Research is still continuing to discover new possible roles of essential trace elements in mammalian systems. The physiological importance of each of the essential trace elements is summarized in Table 13.1. Selenium, a trace element is required by humans in minute quantities and is a physiological constituent of selenoproteins that are part of various processes that include thyroid hormone metabolism,

Food and Nutritional Components in Focus No. 9
Selenium: Chemistry, Analysis, Function and Effects
Edited by Victor R. Preedy

Published by the Royal Society of Chemistry, www.rsc.org

Table 13.1 This table summarizes all essential trace elements and their respective physiological roles.

Sr. no.	Essential trace element	Role
1.	Iron	An integral component of hemoglobin that transports oxygen from the lungs to different parts of the body. In myoglobin, iron enables storage of oxygen in muscle cells. Iron is also part of many enzymes and is essential for growth, healing, immune function and synthesis of DNA
2.	Iodine	Iodine is essential for the formation of thyroid hormones T3, or triiodothyronine, and T4, or thyroxine
3.	Fluoride	It has a role in forming bones and teeth and may prevent bone loss
4.	Copper	Copper helps in production of energy from carbohydrates, protein and fat and is a good antioxidant. Copper is also essential for formation of bone, connective tissues and red blood cells
5.	Zinc	It is integral component of many enzymes, improves immunity and enables normal growth and development
6.	Chromium	Chromium is an important trace mineral that is necessary for normal functioning of insulin, a hormone that maintains blood sugar levels
7.	Manganese	Manganese not only helps in the formation of enzymes, but is also necessary for their activation. It works as an antioxidant, helps develop bones and heals wounds by increasing collagen production
8.	Molybdenum	Helps in some enzyme activation and enables normal cell function

DNA synthesis, reproductive health and protection from oxidative stress as well as infection (Sunde, 2012). This chapter presents a broad view of physiological importance of selenium in human and experimental biology with special emphasis on its chemopreventive as well as biophysical role in cancer.

13.1.1 Sources of Selenium

Selenium intake in humans is through diet as well as wheat and is the most important dietary source of selenium (Reilly, 2006). The presence of selenium in food products depends upon its levels in the soil that in turn is different from one region to another. So, food products have variable selenium contents according to geographical location, for example, the selenium content in the soil of the high plains of northern Nebraska and the Dakotas is very high, and the human population from this area has the highest selenium intakes in the United States (Longnecker *et al.*, 1991). Besides, other sources of selenium include meat, fish and poultry. Several food sources of selenium are listed in Table 13.2.

Table 13.2 Several food sources of selenium.

Sr. no.	Food source	Micrograms/serving
1.	Brazil nuts (1 ounce)	544
2.	Chicken (3 ounces), cottage cheese (1 cup)	22
		20
3.	Beef liver (3 ounces)	28
4.	Macaroni (1 cup)	37
5.	Egg boiled (1)	15
	Rice brown (1 cup)	19
6.	Bread whole wheat (1 slice)	13
	Baked beans (1 cup)	13
7.	Milk with 1% fat (1 cup)	8
	Yogurt (1 cup)	8
8.	Cashew nuts (1 ounce)	3
	Bread white (1 slice)	6
9.	Corn flakes (1 cup), green peas (1 cup), bananas	2
		2
10.	1 potato, peaches	1

13.1.2 Selenoproteins and their Biosynthesis

Synthesis of selenoproteins is the result of selenium metabolism that involves glutathione (GSH) as the core component that is responsible for various reduction reactions leading to production of hydrogen selenide (H_2Se) – the precursor of selenoproteins. This precursor then further provides active selenium for the production of selenoproteins (Mistry *et al.*, 2012). A brief flow diagram of selenium metabolism is provided in Figure 13.1. Other end products of selenium metabolism are the result of further methylation of hydrogen selenide (H_2Se) to methylselenide (CH_3SeH) which in turn gets methylated once to form dimethylselenide ($[CH_3]_2$ SeH) and twice to form trimethylselenide ($[CH_3]_3$ SeH), respectively. These methylated selenides are the end products of selenium metabolism that are excreted out *via* urine or feces from the human system. There are 25 known selenoprotein genes in total that code for different selenoproteins (Table 13.3). Glutathione peroxidase (Gpx1 to GPx6) and selenoproteins act as antioxidants and efficiently prevent oxidative stress by reducing hydrogen peroxide as well as other lipid peroxides to harmless products such as water (H_2O) and alcohol (OH) (Brigelius *et al.*, 2003). Out of these, antioxidant selenoprotein GPx4 is essential for the production of the male reproductive hormone testosterone as well as spermatozoa (Flohe, 2007). Selenium uptake through diet improves antioxidant efficacy of GPx, thereby prevents oxidative damage to the sperm membrane that otherwise leads to a decrease in sperm viability and would be one of the causes of male infertility (Safarinejad and Safarinejad, 2009). In females also, selenium deficiency has been observed to be one of the causes of infertility (Ozkaya *et al.*, 2010). Selenium deficiency is reportedly linked to pre-eclampsia that is a leading cause of maternal and prenatal deaths (Meplan *et al.*, 2008). Selenium deficiency has also been reported to exacerbate iodine deficiency leading to cretinism

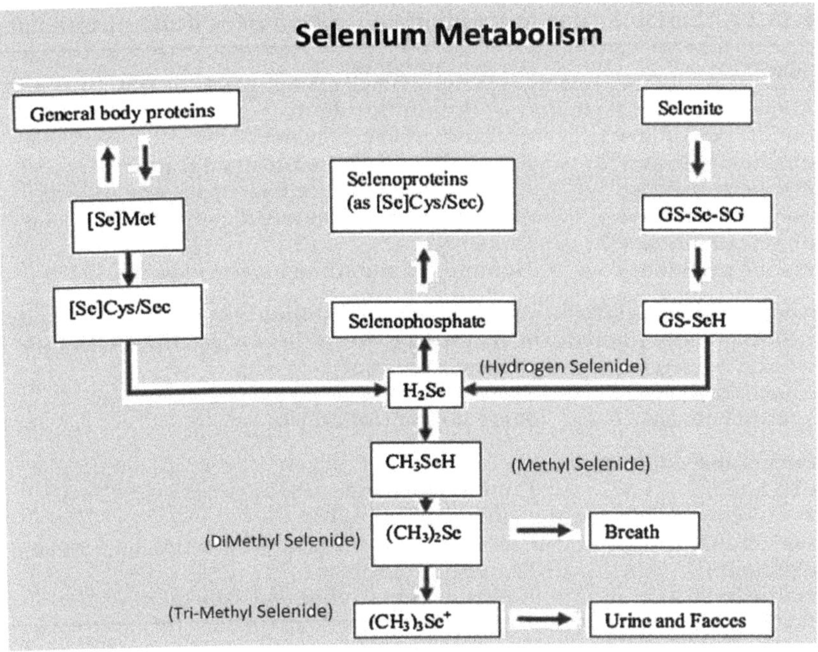

Figure 13.1 The diagram illustrates how selenoprotiens are produced from various sources *in vivo*. Briefly, GSH (glutathione) *via* reduction reactions converts selenite to hydrogen selenide (H_2Se) that is the ultimate precursor of selenium in active form for selenoproteins production. Further metabolism involves sequential methylation reactions leading to production of dimethyl selenide and trimethyl selenide. Abbreviations: Se (selenium), Cys (cysteine), Met (methionine), Sec (selenocysteine).

in infants (Sunde, 2012). Selenium is associated in reducing the deaths due to cardiovascular diseases by its ability to prevent lipid peroxidation, inflammation and aggregation of platelets. Cognitive impairment that is the result of decline in brain activity in elder people is linked with a decline in selenium concentrations (Shahar *et al.*, 2010). Selenium is also required for the optimum production and utilization of thyroid hormones as its deficiency can cause thyroid dysfunctions. The benefit of selenium supplementation during thyroid diseases is being investigated and success can be achieved in cases related to pregnant women with postpartum thyroiditis (Reid *et al.*, 2010). So, all these observations and research investigations showed physiological importance and wider application of this trace element in human health care.

13.1.3 Selenium – Its Role in Cancer

Cancer is considered as one of the most frequently faced public health problems in the world (Ao *et al.*, 2008). The key molecular events associated with the onset of carcinogenesis are somehow related to improper dietary habits. The search for finding new dietary chemopreventive agents is

Table 13.3 This table summarizes important selenoproteins and their functions.

Selenoproteins	Known function
Glutathione peroxidases: mainly work as antioxidants	
Glutathione peroxidase 1	Antioxidant in cell cytosol
Glutathione peroxidase 2	Antioxidant in gastrointestinal tract
Glutathione peroxidase 3	Antioxidant in extracellular space and plasma
Glutathione peroxidase 4	Antioxidant in membranes
Glutathione peroxidase 5	Unknown
Glutathione peroxidase 6	Homolog of glutathione peroxidase 1
Thioredoxin reductases: mainly work as dithiol disulfide oxoreductase, detoxify peroxides, reduce thioredoxin, maintain redox state of transcription factors	
Thioredoxin reductase 1	Present in cytosol
Thioredoxin reductase 2	Present in testis
Thioredoxin reductase 3	Present in mitochondria
Iodothryonine deiodinases	
Types D1 and D2	Converts thyroxine (T4) to bioactive 3,4,3-tri-iodothyronine (T3)
Types D1 and D3	Converts thyroxine (T4) to bioactive 3,3,5 reverse T3
Selenoprotein P	Se transport protein
Selenoprotein W	Antioxidant in cardiac and skeletal muscle

being given top priority by scientists to initiate the fight against this deadly disease. Selenium is one such chemopreventive agent that has been observed to show chemopreventive efficacy against cancer both in humans (Margaret *et al.*, 2009) as well as in animal models (Ganther, 1999). Selenium has found favor for its multimechanism approach against carcinogenesis which includes its ability to repair DNA (Sunde, 2006), to induce apoptosis in cancer cells (Rayman, 2012) and to regulate both endocrine and immune systems (Comb and Gray, 1998; Allen *et al.*, 2008). By virtue of these chemopreventive approaches, selenium has been reported to quench the risk of colorectal, prostate, lung, bladder, skin and gastric cancers to a significant extent (Dennert *et al.*, 2006).

The studies from our laboratory have shown the chemopreventive potential of selenium during colon carcinogenesis, induced by using dimethylhydrazine (DMH) as a carcinogen (Ghadi *et al.*, 2013). We have reported the chemopreventive efficacy of selenium against colon cancer by covering various biochemical, biophysical and histological indices in order to elucidate the comprehensive mechanism behind chemoprevention. We have seen reduced levels of selenium in colon tissues of DMH-treated rats. However, upon selenium supplementation, the levels of selenium were improved, which is a consequence of induction of antioxidant defense enzymes upon selenium supplementation (Ghadi *et al.*, 2013). We also evaluated modulatory effects of selenium on microviscosity and fluidity parameters on the brush border membrane of DMH-treated rats. Further, selenium proved its potential by mitigating ultrastructural changes in colonic histoarchitecture during colon carcinogenesis. All these studies from our research laboratory clearly show selenium to be an efficient anticancer trace element. In the next phase of selenium anticancer research, we conducted a study to explore the potential of Fourier transform infrared spectroscopy (FTIR), a biophysical technique to be used as an efficient biomarker for the detection of colon carcinogenesis. In the next section of this chapter, we elaborate this technique and explain its prospects as a diagnostic tool of cancer.

13.2 Fourier Transform Infrared (FTIR) Spectroscopy

Fourier transform infrared (FTIR) spectroscopy is basically a technological advancement of infrared spectroscopy (IR). IR in simple term means the interaction of a molecule with infrared radiation whereby the molecule is excited to a higher vibrational state by absorbing IR radiation. For the last couple of decades, FTIR – a biophysical technique – has remained favorable among scientists for material analyses. The reason for this is the fact that the infrared spectrum is a unique fingerprint of the composition of the material under analysis. Infrared spectrum basically represents absorption peaks that correspond to vibrations frequencies between the atomic bonds that actually make up the material under observation. As each material differs in its composition due to unique atomic bonds combinations, no two compounds can have exactly the same infrared (IR) spectrum. Practically, in IR spectroscopy, infrared radiation is passed through the sample and some energy of the IR radiation is absorbed by the sample and the rest is transmitted. The resultant spectrum shows the molecular absorption and transmission of IR, thereby providing a molecular fingerprint of the sample under investigation. So, this is an efficient analytical tool that helps scientists to study the composition of different compounds qualitatively. Moreover, in addition to the composition, the size of the peaks in the infrared spectrum can give an estimate of the amount of a particular material present in the compound of interest.

In order to study biological samples, an advancement in IR spectroscopy was needed, as the biological samples are much more complex compared to nonbiological materials. Moreover, a method that measures all IR frequencies simultaneously is required. Therefore, Fourier transform infrared spectroscopy was introduced that is more sensitive, faster, having greater optical throughput and is mechanically simple as compared to older IR spectroscopy. In this advanced technique, a very simple optical device called an "interferometer" is used, which allows the interaction of light from IR source with a beam splitter. The unique signal called an "interferogram" is generated from this interferometer, after interaction with the sample. The interferogram has all the infrared frequencies encoded in it and its measurement is done in a few seconds. In this way, the introduction of an interferometer in IR spectroscopy effectively reduces the time domain per sample

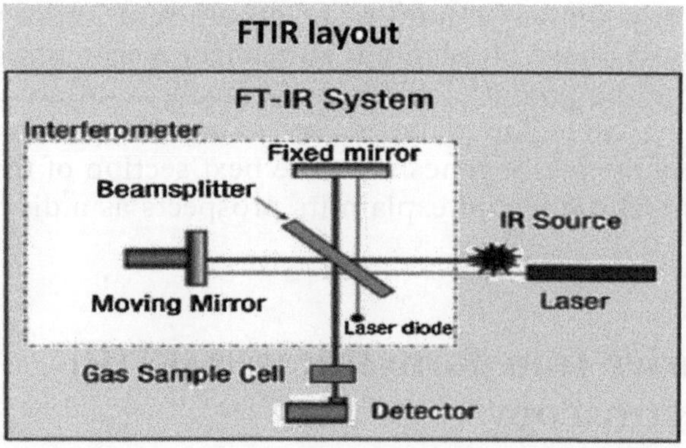

Figure 13.2 Optical diagram showing basic layout Fourier transform infrared spectrometer (FTIR) system.

from several minutes to a few seconds. So the use of interferometer allows the measurement of an interferogram that encodes all infrared frequencies coming from the sample simultaneously. The next step is the identification of peaks at each frequency or in other words, the frequency spectrum (plot of intensity with frequency) is required. To achieve this, the interferogram is to be decoded and the mathematical tool used for decoding is called a Fourier transformation, which is performed by a computer so as to provide the frequency spectrum for the sample under observation. The layout of the FTIR is given in Figure 13.2. Amazing speed as well as sensitivity benefits of the Fourier transform infrared (FT-IR) technique has brought significant practical advantages to infrared spectroscopy and made the use of infrared analysis virtually limitless.

13.2.1 FTIR – Applications in Medical Sciences

FTIR, due to its wide and quick applicability, has been used by various medical scientists, as an effective research analyzing tool in mammalian systems. A recent study by Wei (2013), utilized the FTIR technique to study cadmium-induced changes in the gonads of sea urchin. The effects of cadmium treatment on urchin gonads were investigated by monitoring the conformational changes in functional groups of proteins and lipids by using FTIR spectroscopy. Another study by Alvarez-Ordónez *et al.* (2010), used FTIR to understand the cellular changes at the molecular level in various biological samples. Infrared spectroscopy has also been used for tissue characterization as well as differentiation in diseased states (Muller *et al.*, 1999). Insull (2009) used this technique to analyze atheromatous plaques during arthrosclerosis. In this work, Fourier transform infrared spectroscopy was used to analyze the biochemical properties of human atherosclerotic plaques in order to understand how these plaques originated and matured.

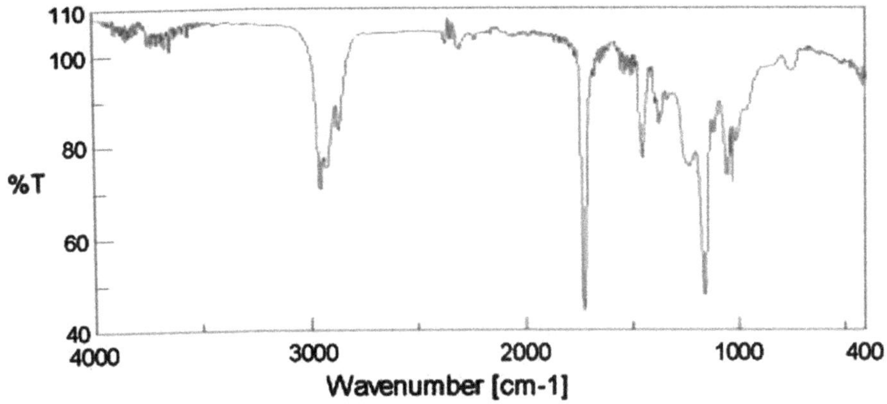

Figure 13.3 This shows an example to demonstrate how a FTIR spectrum looks. It shows wave numbers on the *x*-axis and percentage transmittance (*T*) on the *y*-axis.

13.2.2 FTIR Spectra

This section will help the reader to understand the basics of FTIR spectra in order to use them in biology. In its simplest form, we can say that an FTIR spectrum is presented by peaks between two axis that are the "*x*"-axis and the "*y*"-axis (see Figure 13.3). On the "*x*" axis, the wavelength (λ) or the wave number (v) is presented, whereas on the "*y*" axis, we have either absorption intensity (A) or % transmittance (T). The range in wave numbers is from 13 000 to 10 cm^{-1}, however, the wavelength ranges from 0.78 to 1000 μm. Wave numbers are basically the number of waves per unit length. In other words, wave number is directly proportional to the frequency as well as the energy of the IR radiation. Use of wavelength on the "*x*" axis in FTIR spectra is a thing of the past and presently wave numbers are used everywhere. Furthermore, on the "*y*" axis, % transmittance gets preference by modern FTIR instruments, as it provides better contrast in comparison to absorption intensities. Transmittance and absorption intensity are related by the equation given below. Transmittance is basically the ratio of intensity transmitted (I) by the sample under investigation to the intensity incident (I_0) on the sample. Absorbance (A) is the $\log_{10}$ of the reciprocal of the transmittance (T).

$$A = \log_{10}(1/T) = -\log_{10} T = \log_{10} I/I_0$$

13.2.2.1 *FTIR Spectra in Cancer*

Research emphasis can be focused on the use of biophysical techniques in the diagnosis of cancer as well as understanding the treatment response of chemotherapeutic drugs. In cancer diagnosis research, the prime reason for the shift of focus towards FTIR is understandable in the light of its noninvasive and reagent-free character that can observe changes at the molecular level in the biochemical composition of cells as well as tissues during carcinogenesis. In an FTIR spectrum, such changes can be visualized using a spectrum of wave numbers usually taken from the midinfrared range (4000 to 400 cm^{-1}) and the region of wave numbers from 1800 to 950 cm^{-1}, is promptly used for analyses in biomedical research and is popularly known as the fingerprint region.

FTIR has shown promise as a sensitive diagnostic tool to distinguish neoplastic cells from normal cells in different cancers that include colon, prostate, breast, cervical, gastric and esophageal cancer (Lewis *et al.*, 2010). The key areas where this technique can prove to be effective are differentiation of normal from cancerous tissue, differentiation between carcinogenesis and other pathological conditions with cancer-like manifestations. Furthermore, for the monitoring of efficacy of anticancer drugs, FTIR is proven to be effective and has also been used in tumor-grading studies. Lewis *et al.* (2010) has tested the possibility of using FTIR as a noninvasive diagnostic tool in sputum of lung cancer patients. They used the fingerprint region of wave numbers from 1800 to 950 cm^{-1} and found a total of 92 wave numbers that were different in IR absorbance in the IR spectra of sputum of lung cancer patients when compared with IR spectra of normal ones. Out of these 92 wave numbers, five peaks at wave numbers 964 cm^{-1}, 1024 cm^{-1}, 1411 cm^{-1}, 1577 cm^{-1} and 1656 cm^{-1} distinguished cancer spectra from normal spectra significantly. Jian-Sheng *et al.* (2003) utilized the FTIR technique in esophagus cancer for diagnostic purposes. In a recent study, Fourier transform infrared (FTIR) spectroscopy has been employed to study two cancer cell lines: the Henrietta Lacks (HeLa) human cervix carcinoma and 5637 human bladder carcinoma cell lines in order to investigate biochemical changes and differences in these cell lines by studying their FTIR spectra (Otero *et al.*, 2012).

13.3 Selenium and FTIR Spectra

In this section, we will discuss the efficacy of selenium as a modulator of FTIR spectra in a study from our own laboratory (Ghadi *et al.*, 2013) on an experimental model of colorectal carcinogenesis. Among all cancers in the world, colorectal carcinogenesis ranks third. To induce carcinogenesis, 1,2-dimethyl hydrazine (DMH) was used as a carcinogen in rats. The prime focus of our study was to investigate the modulatory potential of selenium, if any, on Fourier transform infrared (FTIR) spectra of brush border membranes of colons.

Fourier transform infrared spectroscopy provides a rapid and efficient option for detecting changes in biochemical composition of the cells at the molecular level that are of enormous significance especially during carcinogenesis. It involves computerized methodology that transforms the spectrum from optical path difference into the frequency domain and allows noise-free rapid measurements of complete regions of wave numbers. FTIR has the ability to measure complex molecular vibrational modes that contain valuable information on changes occurring in various biomolecules during cancer.

The prime reason for choosing selenium as a modulator is that many dietary compounds require investigations for their prominent roles in the chemoprevention of cancer. Selenium was the trace element of interest in our study as it has been reported to prevent carcinogenesis both in humans (Margaret *et al.*, 2009) as well as in animal models (Ganther, 1999). Moreover, another study from our laboratory showed the chemopreventive effects of selenium on antioxidant enzymes and histological changes during colon carcinogenesis (Ghadi *et al.*, 2009). The study indicated that selenium supplementation appreciably modulated the DMH-induced changes in the FTIR spectra in the brush border membranes of DMH-treated rats.

The FTIR results (Figures 13.4–13.6) showed prominent alterations in the biochemical composition of the cells, thereby indicating transformation of cells from normal to cancerous. The prime spectral change that was indicative of alteration in biochemical composition of cells was in the amide bands. In FTIR, different bands are characterized on the basis of various stretching vibrations around a specific bond. For example, Amide band I arises from the stretching vibrations of C=O hydrogen bonds and amide II band arises from C—N stretching as well as from CNH bending vibrations (Jagannathan *et al.*, 2001). In addition to this, these bands show peaks in spectra at their specific wave number that actually helps us to identify these bands. For instance amide I band is visible by peak at 1645 cm^{-1} and amide II band is visible by a peak at 1539 cm^{-1}. So, by observing these bands in FTIR spectra and by noting their peaks at specific wave numbers, one can easily indentify the groups present in the sample that in turn help us to investigate the nature of the compound. So in our study, the presence of both the amide bands in the FTIR spectra (Figure 13.4) signified the existence of protein contents in cells under observation, as these amide bands are indicative of proteins.

Furthermore, when we observed the FTIR spectra of DMH treated rats (Figure 13.5), it showed loss of peaks at 1645 cm^{-1} (indicative of amide band I) and at 1539 cm^{-1} (indicative of amide band II), which confirmed the decline in the protein contents of the cells. So, in this way, by just observing the amide band peaks in the FTIR spectra of normal control and DMH-treated rats, we could get important information of protein loss in cancer. The results observed in our study were in agreement with various other studies (Robbins *et al.*, 1994; Andronie *et al.*, 2011). We then further analyzed both the spectra and observed loss of peak at wave number 2851 cm^{-1} in the DMH-treated rats that, however, was present in the normal FTIR spectra. This peak is the result of stretching vibrations around CH_2, CH_3 groups and is indicative of creatine contents, as it is known that essential cellular metabolites like creatinines fall in this region of 2800–3500 cm^{-1}. So, the loss of this peak gave another important clue of loss of creatine contents during cancer that may have been used in containing hyperproliferation of cells during carcinogenesis, as creatine has been reported to inhibit tumor growth (Kristensen *et al.*, 1999). We then observed another peak at 1399 cm^{-1} in the normal spectra and found it to be missing in the spectra of DMH-treated rats. This wave number is specific for the collagen contents in

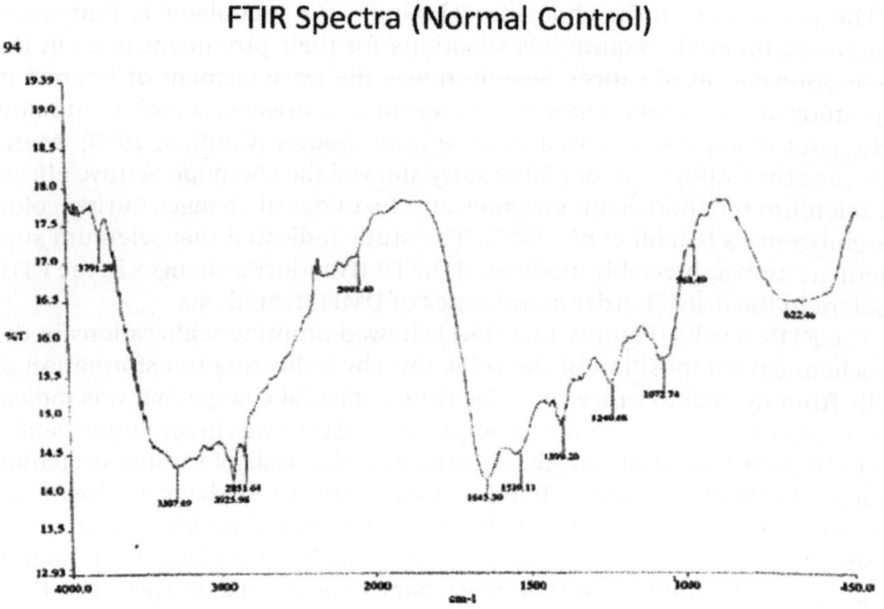

Figure 13.4 The above FTIR spectrum is of an experimental normal control rat confirming the presence of proteins, carbohydrates and other important biomolecules present in colon. *Original Source*: Ghadi *et al.*, 2013. *Nutrition and Cancer*. 65(1): 92–98 with permission from Taylor and Francis Ltd.

the cells. So, this observation suggested a decline in the collagen contents during carcinogenesis, as has also been shown in a previous study (Cheng *et al.*, 2009). Loss of collagen could be another factor contributing towards progression of carcinogenesis in the DMH-treated rats. Moreover, the presence of peaks at 1440 cm^{-1} and 1750 cm^{-1} in the spectra of DMH-treated rats signified an increase in the lipid contents that is in agreement with the observation by Andronie *et al.* (2011).

Administration of selenium to DMH-treated rats appreciably modulated the changes in the FTIR spectra (Figure 13.6), which included restoration of amide band at 1563 cm^{-1}, the collagen peak at 1322 cm^{-1} and the loss of the lipid peak at 1750 cm^{-1}. All these modulations in the biochemical composition of cancer cells by selenium had collectively contributed towards prevention of colorectal carcinogenesis. To further validate the above observation, our study also evaluated sialic acid levels, which is one of the prominent tumor markers of the colorectal carcinogenesis.

Earlier studies have reported sialic acid for its specificity and as a possible biomarker of colon carcinogenesis (Goodarzi *et al.*, 2005; Chadha and Dhawan, 2009). The reason for the above specificity is that sialic acid is one of the major constituents of cell membrane and minute surface changes in membrane can cause its release into the blood (Durgawale *et al.*, 2011).

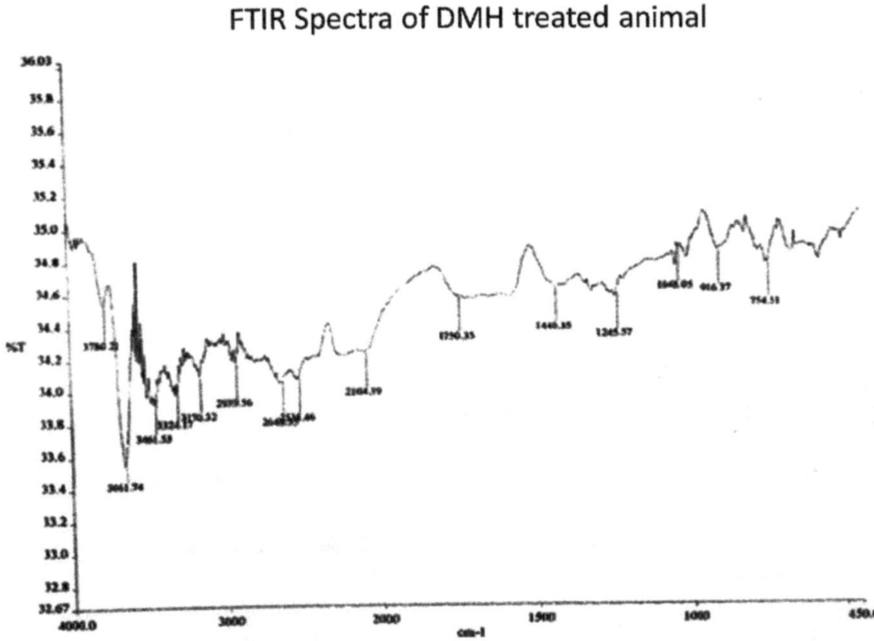

Figure 13.5 The above FTIR spectrum is of experimental colon cancer in rat showing absence/decrease in the peaks of important biomolecules in comparison to the spectra of control rat. *Original Source*: Ghadi *et al.*, 2013. *Nutrition and Cancer*. 65(1): 92–98 with permission from Taylor and Francis Ltd.

A recent study from our laboratory, has also observed changes in membrane parameters during colon carcinogenesis (Chadha and Dhawan, 2009). The observed increase in the total sialic acid in the sera of rats treated with DMH for 20 weeks is the result of altered microsurface changes on the membranes of the cells. Another plausible explanation would be altered carbohydrate metabolism in the cancer cells that is responsible for all the abnormal behavior, *i.e.* abnormal cell recognition, cell adherence, antigenicity and invasiveness (Varley, 1975). All these changes in the cell behavior contributed to hyperproliferation of cells that were also confirmed by changes in FTIR spectra of DMH-treated rats. Moreover, these changes in the cell behavior have also stimulated the release of glycoproteins in the blood that finally led to an increase in TSA levels in the serum of DMH-treated rats. There are reports that suggest a positive correlation of glycoproteins levels with malignancy (Hakomoris, 1984). Selenium supplementation to DMH-treated rats, efficiently moderated TSA levels in serum. The observed modulation in the sialic acid levels by selenium can be ascribed to its ability to maintain the membrane integrity either by a direct mechanism or indirectly by scavenging the free radicals. Furthermore, there have been reports that suggest that selenium is an integral component of selenoenzyme GPX and its metabolite selenodiglutathione

FTIR Spectra of DMH + Selenium treated animal

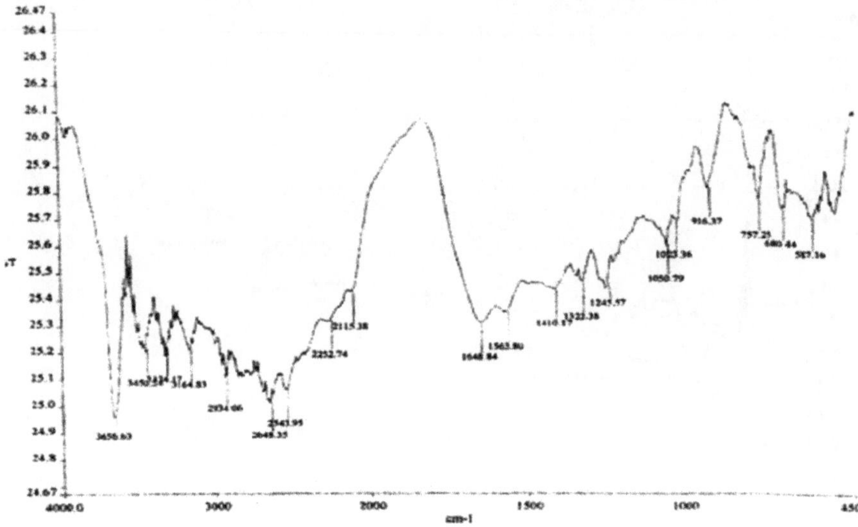

Figure 13.6 The above FTIR spectra is of experimental colon cancer in rat treated with selenium showing restoration of the peaks of important biomolecules confirming the efficacy of selenium as a protector during colon cancer. *Original Source*: Ghadi *et al.*, 2013. *Nutrition and Cancer*. 65(1): 92–98 with permission from Taylor and Francis Ltd.

inhibits hypercell proliferation as well as improves cell membrane integrity (Clement, 1998; Naithani *et al.*, 2008). So, it is clear from the above observations that FTIR has the potential to be used as a possible biophysical tool to diagnose cancer and modulation by selenium further affirms the above belief.

13.4 Future Prospects

The current need is an efficient, fast and convenient way to diagnose cancer. The FTIR technique is noninvasive and shall be effective in detecting cancer at early stages. The above facts substantiate its strength to be used as a possible tool in the diagnosis of cancer.

Furthermore, dietary supplementation of trace elements especially selenium should be considered with utmost importance in the path of chemoprevention against cancer. Selenium holds enormous potential to modulate the process of carcinogenesis at both the molecular and cellular levels.

Finally, we want to emphasize the need to explore further the potential of biophysical technique like FTIR in the early detection of cancer. Furthermore, for the prevention of cancer, proper diet consisting of all essential trace elements with special focus on selenium is essential.

Summary Points

- Trace elements are essential for the normal growth and metabolic functions of the body in order to attain normal homeostasis.
- Selenium is one of the essential trace elements and plays an important role in biochemical processes.
- Selenium is involved in physiological functions *via* selenoproteins – the key functional associates of this trace element.
- Selenium has been shown to exhibit anticancer effects during varied forms of cancers in studies on animals as well as human beings.
- Fourier transform infrared spectroscopy (FTIR) is the upcoming biophysical technique being explored the world over for its diagnostic efficacy in various pathological states including cancer.
- Selenium has the role of a modulator of FTIR spectra during colon carcinogenesis.

Key Facts

1. FTIR (Fourier transform infrared spectroscopy) is the improved form of an analytical technique called (IR) Infrared spectroscopy used widely to determine the structures of molecules with the molecules' characteristic absorption of infrared radiation.
2. Fourier transform is a mathematical tool, named after the French mathematician and physicist Jean Baptiste Joseph Fourier. It is used to transform a function.
3. It has applications in varied areas like organic synthesis, polymer science, petrochemical engineering, pharmaceutical industry and food industry.
4. FTIR spectrometers are third-generation infrared spectrometers having high accuracy with significantly higher signal-to-noise ratio.
5. FTIR is a new upcoming diagnostic biophysical tool for cancer.
6. The main advantages of FTIR over other techniques in cancer diagnosis lie in its faster speed, simplicity and noninvasive approach.

Definition of Words and Key Terms

Cancer. Cancer is a state when the natural control over division of cells is lost and it leads to formation of abnormal masses of cells that are called tumors.

Physiological role. This means the role being played by a particular biomolecule inside the mammalian system.

Chemoprevention. This is a way to reverse the process of carcinogenesis or cancer in a mammalian system by providing some natural dietary supplement.

Oxidative stress. This is basically the imbalance between the production of reactive oxygen species (ROS) and the antioxidant defense system comprising of enzymes that prevent the production of reactive oxygen species.

Lipid peroxidation. This is oxidative degradation of lipids that finally result in cell damage.

Inflammation. This is a natural defense of mammalian system in order to remove harmful stimuli like irritants, damaged cells, pathogens so as to start the process of healing.

Apoptosis. This is programmed cell death that enables removal of old worn-out defective cells from the mammalian system.

Immune system. This is defense of the mammalian system against any foreign substance/irritant/pathogen.

Dephophorylation. The removal of phosphate group from an organic compound by hydrolysis is called dephosphorylation. It is a physiological way to activate certain *in vivo* enzymes.

List of Abbreviations

FTIR	Fourier transform infrared spectroscopy
ppm	Parts per million
DNA	Deoxyribose nucleic acid
GSH	Reduced glutathione
GPx	Glutathione peroxidase
Se	Selenium
DMH	Dimethyl hydrazine
IR	Infrared

References

Allen, N. E., Appleby, P. N., Roddam, A. W., Tjonneland, A., Johnsen, N. F. and Overvad, K., 2008. Plasma selenium concentration and prostate cancer risk: results from the European prospective investigation into cancer and nutrition. *American Journal of Clinical Cancer*. 88: 1567–1575.

Alvarez-Ordónez, A., Halisch, J. and Prieto, M., 2010. Changes in Fourier transform infrared spectra of *Salmonella enterica serovars Typhimurium* and *Enteritidis* after adaptation to stressful growth conditions. *International Journal of Food Microbiology*. 142: 97–105.

Andronie, L., Pazaru, S. C., Cozar, O. and Domsa, I., 2011. FTIR spectroscopy for human colon tissue diagnostic. *Romanian Journal of Biophysics*. 21(2): 85–91.

Ao, P., Galas, D., Hood, L. and Zhu, X., 2008. Cancer as robust intrinsic state of endogenous molecular-cellular network shaped by evolution. *Medical Hypotheses*. 70(3): 678–684.

Brigelius, F. R., Banning, A. and Schnurr, K., 2003. Selenium-dependent enzymes in endothelialcell function. *Antioxidant Redox Signal*. 5: 205–215.

Chadha, V. D. and Dhawan, D. K., 2009. Membrane fluidity and surface changes during initiation of 1,2-dimethylhydrazine induced colon carcinogenesis: protection by zinc. *Oncology Research*. 18(1): 17–23.

Cheng, C., Xiong, W. and Tian, Y., 2009. Classification of rat FTIR colon cancer data using wavelets and BPNN. *Chinese Journal of Chemistry*. 27: 911–914.

Clement, I. P., 1998. Lessons from basic research in selenium and cancer prevention. *Journal of Nutrition*. 128(11): 1845–1854.

Comb Jr, G. F. and Gray, W. P., 1998. Chemopreventive agents: selenium. *Pharmacology & Therapeutics*. 79: 179–182.

Durgawale, P., Sontakke, A., Shukla, P. S., Durgawale, P. and Chogule, P., 2011. Altrations in total sialic acid (TSA), total proteins (TP) and TSA/TP ratio in cancer patients. *Biomedical Research*. 22(4): 492–494.

Flohe, L., 2007. Selenium in mammalian spermiogenesis. *Biological Chemistry*. 388: 987–995.

Ganther, H. E., 1999. Selenium metabolism, selenoproteins, and mechanisms of cancer prevention: complexities with thioredoxin reductase. *Carcinogenesis*. 20: 1657–1666.

Ghadi, F., Malhotra, A., Ramzani, A. and Dhawan, D. K., 2013. Chemopreventive effects of selenium on cancer marker indices and ultrastructural changes during 1,2 dimethylhydrazine induced colon carcinogenesis in rats. *Journal of Gastrointestinal Cancer*. 44(1): 54–59.

Ghadi, F. E., Ghara, A. R., Bhattacharya, S. and Dhawan, D. K., 2009. Selenium as a chemopreventive agent in experimentally induced colon carcinogenesis. *World Journal of Gastrointestinal Oncology*. 1(1): 74–81.

Ghadi, F. E., Malhotra, A., Ramzani, A. and Dhawan, D. K., 2013. Modulation of Fourier transform infrared spectra and total sialic acid levels by selenium during 1,2 dimethylhydrazine-induced colon carcinogenesis in rats. *Nutrition and Cancer*. 65(1): 92–98.

Goodarzi, M. T., Shafiei, M., Nomani, H. and Shahriarahmadi, A., 2005. Relationship between total and lipid bound sialic acid and some tumor markers. *Iranian Journal of Medical Sciences*. 30(3): 124–127.

Hakomoris, S., 1984. Tumor associated carbohydrate antigens. *Annual Review of Immunology*. 2: 103–126.

Insull, W., 2009. The pathology of atherosclerosis: plaque development and plaque responses to medical treatment. *American Journal of Medicine*. 122(1A): S3–S14.

Jagannathan, R., Ahmad, S. and Shaul, M., 2001. FTIR microscopic studies on normal, polyp and malignant human colonic tissues. *Subsurface Sensing Technologies and Applications*. 2(2): 99–117.

Jian-Sheng, W., Jing-Sen, S., Yi-Zhuang, X., Xiao-Yi, D., Li, Z., Jing, W., Li-Ming, Y., Shi-Fu, W. and Jin-Guang, W., 2003. FT-IR spectroscopic analysis of normal and cancerous tissues of esophagus. *World Journal of Gastroenterology*. 9(9): 1897–1899.

Kristensen, C. A., Askenasy, N., Jain, R. K. and Koretsky, A. P., 1999. Creatine and cyclocreatine treatment of human colon adenocarcinoma xenograsfts: 31P and 1H magnetic resonance spectroscopic studies. *British Journal of Cancer*. 79: 278–285.

Lewis, P. D., Lewis, K. E., Ghosa, R., Bayliss, S., Lloyd, A. J., Wills, J., Godfrey, R. and Kloer, P., 2010. Evaluation of FTIR spectroscopy as a diagnostic tool for lung cancer using sputum. *BMC Cancer*. 10: 640–650.

Longnecker, M. P., Taylor, P. R. and Levander, O. A., 1991. Selenium in diet, blood, and toenails in relation to human health in a seleniferous area. *American Journal of Clinical Nutrition*. 53: 1288–1294.

Margaret, P., Rayman, G. F., Combs, J. R. and David, J. W., 2009. Selenium and vitamin E supplementation for chemoprevention. *JAMA, the Journal of the American Medical Association*. 301(18): 1876.

Meplan, C., Crosley, L. K. and Nicol, F., 2008. Functional effects of a common single-nucleotide polymorphism (GPX4c718t) in the glutathione peroxidase 4 gene: interaction with sex. *American Journal of Clinical Nutrition*. 87: 1019–1027.

Mistry, H. D., Pipkin, F. B., Christopher, W. G. and Lucilla, P., 2012. Selenium in reproductive health. *American Journal of Obstetrics and Gynaecology*. 21–30.

Muller, G., Wasche, W., Bindig, U. and Liebold, K., 1999. IR-spectroscopy for tissue differentiation in the medical field. *Laser Physics*. 9(1): 348–356.

Naithani, R., Huma, L. C., Moriarty, R. M., McCormick, D. L. and Mehta, R. G., 2008. Comprehensive review of chemopreventive agents evaluated in experimental carcinogensis models and clinical trials. *Current Medicinal Chemistry*. 15(11): 1044–1071.

Otero, E. P., Eliel, G. S., Fonseca, J. M., Hickmann, J. M., Rodarte, R., Barreto, E. and Jalkanen, K. J., 2012. Study of cancer cell lines with Fourier transforms infrared (FTIR)/vibrational absorption (VA) spectroscopy. *Current Physical Chemistry*. 2.

Ozkaya, M. O., Naziroglu, M., Barak, C. and Berkkanoglu, M., 2010. Effects of multivitamin/mineral supplementation on trace element levels in serum and follicular fluid of women undergoing *in vitro* fertilization (IVF). *Biological Trace Element Research*. 139: 1–9.

Rayman, M. P., 2012. Selenium and human health. *Lancet*. 379: 1256–1268.

Reid, S. M., Middleton, P., Cossich, M. C. and Crowther, C. A., 2010. Interventions for clinical and subclinical hypothyroidism. *Cochrane Database Systematic Review*.

Reilly, C., 2006. *Selenium in Food and Health*. Springer, New York.

Robbins, S. L., Cotran, R. S. and Kumar, V., 1994. *Pathologic Basis of Diseases*. WB Saunders Co., Philadelphia.

Safarinejad, M. R. and Safarinejad, S., 2009. Efficacy of selenium and/or *N*-acetyl-cysteine for improving semen parameters in infertile men: a doubleblind, placebo controlled, randomized study. *Journal of Urology*. 181: 741–751.

Shahar, A., Patel, K. V., Semba, R. D. and Bandineli, S., 2010. Plasma selenium is positively related to performance in neurological task assessing coordination and motor speed. *Movement Disorders*. 25: 1909–1915.

Sunde, R. A., 2012. Selenium. In: *Modern Nutrition in Health and Disease*. 11th edn, Lippincott Williams and Wilkins, Philadelphia PA, 225 pp.

Sunde, R. A., 2006. Selenium. In: *Present Knowledge in Nutrition*. 9th edn, International Life Science Institute, Washington DC, 480 pp.

Varley, H., 1975. *Practical Clincial Biochemsitry*. 4th edn, Arnold Heinmann Pub. India, 172 pp.

Wei, Z., 2013. Use of FTIR to study cadmium induced changes strongylocentrotus nudus gonad. *Life Science Journal*. 10(1): 269–276.

CHAPTER 14

Selenium Status Assessment by Questionnaire in Clinical and Cardiovascular Studies

YAP-HANG CHAN*[a] AND HUNG-FAT TSE*[a]

[a]Cardiology Division, Department of Medicine, The University of Hong Kong, Rm 1928, Block K, Queen Mary Hospital, Hong Kong, China
*E-mail: hftse@hkucc.hku.hk, yhchanwill@gmail.com

14.1 Introduction: Need for Selenium Status Assessment Tools in Clinical Studies

The emerging role of selenium in cardiovascular and other chronic diseases has well been covered in other chapters. To study the effects of selenium in the wide-ranging number of clinical conditions, researchers and clinicians strived to explore the most definitive and feasible research designs and assessment methods. In observational clinical studies and population-based epidemiological studies, one of the primary key steps is to obtain accurate and reliable estimates of selenium exposure. Therefore, development of valid clinical tools for assessment of selenium status, which is mostly dominated by the dietary intake of selenium-containing food items, is essential. Even for interventional clinical studies, for example trials that investigate the vascular effects of selenium supplementation, it is important to assess baseline selenium status and dietary profile. The former is critical for assessing eligibility entry criteria for

Food and Nutritional Components in Focus No. 9
Selenium: Chemistry, Analysis, Function and Effects
Edited by Victor R. Preedy

Published by the Royal Society of Chemistry, www.rsc.org

subjects, in light of the preferential recruitment of subjects with baseline selenium deficiency for obvious methodological as well as ethical reasons. The latter is also important since selenium intake can be associated with other intercurrent dietary factors/patterns, and measurement and possibly statistical adjustments of these "noises", or confounders, are needed in subsequent data analysis to isolate the effects specifically attributable to selenium.

14.2 Clinical Assessment Tools for Long-Term Exposures

There are a number of clinical assessment tools of selenium exposures in the research studies of cardiovascular and other chronic diseases. Direct measurements of selenium levels can be performed in the blood, urine, toenail, and hair samples of subjects, and these will be discussed in other chapters. Along with these biomarker measurement modalities, questionnaire-based assessments are also increasingly used for the long-term intake estimation of selenium in clinical studies.

14.2.1 Food-Frequency Questionnaires (FFQ)

Dietary intake of selenium-containing food is the key determinant of selenium bioavailability in the body (Gonzalez *et al.*, 2006). Questionnaires that are specially designed to capture nutrients intake, commonly known as the food-frequency questionnaires (FFQ), (Karita *et al.*, 2003) contains a predefined number of food items, usually structured under several major food groups. The FFQ is usually administered to participants with the assistance of an interviewer. The number of food items varies, and how comprehensively a particular food group is interrogated in the FFQ may reflect the underlying focus and scope of interest of the research hypothesis tested. It has been shown that selenium intake as determined by FFQ correlates closely with blood selenium levels (Pestitschek *et al.*, 2013).

Such interviews typically lasts for at least 15 min to 1 h, where participants are asked to recall their individual dietary intake of each food item, both in terms of quantity of intake each time (*e.g.* portion size or estimated weight of food item), as well as the estimated frequency of intake over the prespecified period in the recent past. The definition of "recent past" is variable, but is often taken as about 3 months to 1 year, to allow a reasonably long period of estimate, but not to be excessively long such that recall of remote exposures would become difficult. In case-control studies, the assessed period is often defined as the recent past immediately before the diagnosed clinical condition. The estimation of food portion size is often assisted with the use of food demonstration models or standardized food photographs, to reduce interindividual variations in perceived quantity.

With the so-obtained intake estimates of both quantity and frequency of each specific food item, the average intake amount of each food item in the assessed period can then be calculated. Therefore, an FFQ is often semiquantitative (Stranges *et al.*, 2010). As selenium is a micronutrient derived from a diverse source of food items entailing seafood, crops, vegetables and animal proteins, an overall integrative estimate of selenium derived from all relevant food items is the most helpful in analyzing its effects in relation to cardiovascular biomarkers and clinical events.

This is achieved by coupling the semiquantitative intake estimates of each selenium-containing food item with the known per-weight selenium contents contained in such food items. Such latter information is available from the Food Composition Tables (FCT) (Garcia *et al.*, 2004), which are derived from prior standardized laboratory biochemical studies of commonly consumed food items at each jurisdiction, often performed by local health authorities or research institutions.

14.2.2 Other Questionnaire-Based Assessment Options

There are a few questionnaire-based assessment tools other than the FFQ. These include 24 h dietary recalls (Murphy *et al.*, 2002), and food diary/record (Hughes *et al.*, 2012). 24 h dietary recalls involve complete ascertainment of food intake in the immediate past 24 h through indepth interviews, as a rather recent snapshot but detailed account of nutrient intake of participants. To reduce the effect of intraindividual variability in dietary intake, multiple 24 h dietary recalls are often administered for each participant, usually separated by time intervals of days to months. A food diary/record is a complete, detailed written record of dietary intake by participants over a defined period, which is subsequently studied by researchers to formulate the dietary pattern and specific nutrients intake over that period. Both these methods are relatively labor intensive at the stage of study execution, requiring studies of often ample and open-ended type of data and derivation of analyzable estimates therefrom. In this chapter, we are interested to focus on the FFQ, which is a standardized tool that represents the relatively long-term exposure of selenium.

14.3 Comparison of FFQ with Clinical Biomarkers for Assessment of Selenium Status

As a questionnaire-based assessment, the FFQ has its pros and cons in comparison to various clinical biomarkers of selenium status. Such considerations are discussed below. We shall focus mainly on the clinical and cardiovascular research setting.

14.3.1 Duration of Exposure Assessed – Long Term *versus* Short Term

Clinical biomarkers of selenium, *e.g.* assessment of selenium level in blood (Laclaustra *et al.*, 2009), urine (Longnecker *et al.*, 1996), hair (Hong *et al.*, 2009), or nails (Kardinaal *et al.*, 1997), provide accurate quantitative measures of selenium status at a single snapshot. However, whether such levels reflect the long-term exposure of selenium, especially in subjects with more erratic changes in dietary habits, are less certain. In contrast, the FFQ aims to assess the relatively long-term dietary intake of selenium over a reasonably long period, which may be potentially more relevant in terms of causality studies of chronic diseases such as cardiovascular diseases where pathogenesis take months to years. The combined administrations of FFQ together with clinical biomarkers studies provide a more powerful assessment of both the short-term and long-term intake of selenium, and also allow mutual validation of study methods.

14.3.2 Categorized *versus* Continuous Estimates

Where clinical biomarkers provide precise continuous estimates of selenium level, FFQ is often semiquantitative. Even if the daily level of selenium intake can be estimated from FFQ by combining data with local FCT, the accuracy and precision of such fine estimates may be variable, as the methodology of FFQ is fundamentally limited by the recall nature. FFQ may be most useful for selenium-status assessment in terms of stratifying subjects into groups of relative selenium exposure, *e.g.* groups formulated based on tertiles, quartiles, quintiles, or even deciles. Based on our experience, FFQ is helpful to identify subjects with evident extremes of selenium deficiency, (Chan *et al.*, 2012) in the assessment of cardiovascular risk of high-risk patients in relation to systemic arterial function changes. These patients with selenium deficiency are also those who are most likely to have the potential for any benefit from interventions. As discussed in another chapter, addressing persons with selenium deficiency, rather than nontargeted screening or intervention, should be the focus of future cardiovascular studies on selenium.

14.3.3 Geographical Variations

The soil level of selenium varies substantially accordingly to geographical regions, *i.e.* higher in seleniferous areas (Longnecker *et al.*, 1991). This is a key challenge to the validity of using FFQ in the assessment of selenium status. As food composition depends upon local-specific selenium levels in crops, FCT in one region does not readily apply to another. There are two key implications that follow. First, FFQ used in selenium studies should be country/region specific to allow food contents of selenium to be compared based on identical food traditions and culture, and ideally identical/similar soil selenium concentrations. Secondly, studies in different geographical regions that use semiquantitative FFQ to assess selenium intake may not be directly comparable, unless differences in local food composition of selenium has been carefully taken into account. In comparison, clinical biomarkers of selenium do not pose geographical concerns once the laboratory measurements are standardized.

14.3.4 Seasonality

Studies showed that seasonal variations of selenium levels exist in diet (Watson and McDonald, 2007). These can be due to seasonal changes of dietary intake habits and food item availability, as well as seasonal variations in selenium concentrations in food items. In clinical studies such as those focusing on cardiovascular diseases, clinical outcomes may also exhibit seasonality. For example, vascular function parameters (Patel *et al.*, 2011) and the risk of cardiovascular events (Patel *et al.*, 2014) are known to be affected by seasonality. In such cases, such a seasonality factor should be adjusted for wherever possible, especially in large studies where enrollment of subjects spans throughout the year(s). This is the same for studies employing biomarkers of selenium for exposure assessment.

14.3.5 Assessment and Adjustment for Other Dietary Factors

The ability of the FFQ to assess not only selenium intake, but also other major domains of the dietary profile of participants is a major strength of this method. As selenium intake of participants could be associated with other dietary factors, these factors may potentially confound on the relations between selenium intake and clinical disease outcomes of interest. Thus, assessing and applying statistical adjustments for these factors is an important component of a well-performed clinical study. In the case of cardiovascular disease, dietary factors that must be considered at minimum should include total daily caloric intake, intake of carbohydrates, proteins, and fats, as well as dietary fiber and antioxidant vitamins. Although there are criticisms of FFQ in assessing total energy expenditure of subjects, in comparison to the gold standard by using doubly labeled water (DLW) (Kroke *et al.*, 1999), FFQ certainly stills provide a valuable assessment of the overall dietary pattern and major components that are not allowed by lone biomarkers assessment. The simultaneous inclusion of various techniques on top of the FFQ such as DLW will certainly strengthen the overall study design.

14.3.6 Recall Bias

A bias in selenium-exposure assessment is any systematic deviation from the true level of exposure. This impedes the accuracy of estimated selenium level in participants and hence adversely impacts on the assessment of any genuine associations, or the contrary, between selenium exposure and the clinical outcome.

Recall bias is an intrinsic deficit of any questionnaire-based assessments of exposures, in which participants may be preferentially prompted to recall and report prior selenium exposures disproportionate to their genuine exposure. This may happen as a consequence of self-awareness of current disease status; or due to disproportionate prompting by interviewers, and these may both happen in a subconscious and nondeliberate manner. This problem is particularly seen in case-control studies, where cases with known clinical outcomes are prone to overreporting of exposures. To minimize the effects of recall bias, assessment of past exposures is usually limited to the reasonably recent past. Although the definition of "recent past" varies, it is mostly ranges from 3 months to 1 year. Ideally, the exposure of interest, *i.e.* selenium, and the underlying study hypothesis, should be concealed from both the participants as well as the interviewers. The outcomes of the study should also be blinded to the assessment of exposures, and this is possible in prospective cohort studies where assessment of selenium status occurs at baseline; the occurrence of clinical outcomes only follows subsequently. The simultaneous use of biomarkers can provide opportunities for mutual validation of selenium exposures and partially mitigate this problem.

14.3.7 Noninvasiveness

Blood taking, or samples taken from the body such as hair or nail, may not be readily accepted in all participants. In this regard, the FFQ provides a noninvasive and valuable alternative for selenium assessment. This is particularly important in potentially vulnerable subjects including school children/adolescents and pregnant women.

14.3.8 Timeliness

At the patient recruitment level, the FFQ usually takes longer to complete compared to direct biospecimen collection. This is a potential limitation to its use in the busy clinical settings.

14.3.9 Selenium as an Acute Phase Reactant

Studies showed that selenium is an acute phase reactant, and its level fluctuates with recent acute insults such injury or inflammation (Nichol *et al.*, 1998). In this regard compared to the biomarkers, the FFQ is a relatively stable assessment tool free from effects of these short-term biological disturbances.

14.4 Summary and Future Research Prospects

Despite some of its drawbacks and criticisms, questionnaire-based assessment using FFQ presents a comprehensive, effective and noninvasive means of determining long-term selenium status in clinical studies of cardiovascular and other chronic diseases. Comparative features of FFQ and clinical biomarkers are summarized in Table 14.1. The use of prospective study designs and the blinding of exposure assessment to study outcomes or hypothesis can reduce recall biases. It is important to take into account seasonality and geographical variations in selenium levels in the analysis of FFQ-derived estimates. The assessment and adjustment for the major

Table 14.1 Comparative features of food-frequency questionnaires (FFQ) *versus* clinical biomarkers of selenium status in clinical and cardiovascular studies.

	FFQ	Clinical biomarkers	FFQ + clinical biomarkers
Exposure duration assessed	Long-term exposure	Short-term snapshot	Both long and short terms assessed
Strength of estimate method	Good at identifying extremes of selenium intake	Continuous estimates of selenium level	Crossvalidation possible
Assessment method affected by geographical variations	Affected	Not affected	Minimized by biomarker assessment
Seasonal variation	All		
Assessment and adjustment for other dietary factors	Yes	Possible but only for specific nutrients	Yes
Recall bias	Yes	Irrelevant	Effect minimized by crossvalidation
Subjected to reverse causality	Possible (in the form of recall bias for retrospective studies)	Possible (if measurement done after onset of disease)	Possible (if retrospective studies)
Noninvasiveness	Noninvasive	Invasive	Invasive
Timeliness	Requires 10 min to 1 h	Quick	Time and resource intensive
Affected by acute phase responses (*e.g.* infection, inflammation)	Irrelevant	Could be affected	Could be affected

dietary patterns and other key nutritional factors is a unique strength of FFQ not allowed by any other single means of biomarker measurements. The simultaneous use of questionnaire-based approaches with single or interval measurements of biomarkers can strengthen the exposure assessments and provide crossvalidation of different methods. Such a combined approach is the ongoing developmental trend in current research methodologies.

Furthermore, the recently identified genetic polymorphisms of mechanistic pathways related to selenium, *e.g.* variations of the selenoprotein S gene locus (Alanne *et al.*, 2007; Cox *et al.*, 2013), provides exciting opportunities of causal inference studies of the selenoprotein pathways and may allow the study of diet–gene interactions in the role of selenium in cardiovascular disease and other clinical conditions.

Summary Points

- This chapter focuses on the questionnaire-based assessment of selenium status in clinical studies of cardiovascular and chronic diseases.
- FFQ is a commonly used tool for assessing long-term selenium intake in clinical studies.
- FFQ and biomarkers of selenium status each has its own strengths and limitations. FFQ is a noninvasive tool compared to clinical biomarkers. It is designed to mainly determine long-term selenium intake and is good at identifying subjects with extremes of selenium intake. It also allows assessment of dietary patterns and other nutritional components, which make statistical adjustment for these factors possible.
- Selenium level in soil and food items varies by geographical regions. Selenium levels are also affected by seasonal variations. These should be taken into account in questionnaire-based assessment of selenium status.
- Recall bias is a key challenge to questionnaire-based assessment of selenium. The use of prospective study designs, focusing on exposures in the recent past, and the combined assessment approach together with biomarkers can help ameliorate the problem.

Definition of Key Words and Terms

Acute phase reactants. Circulating biomarkers that change in level during acute physiological stressors such as infection, inflammation, or tissue injuries. Selenium is one of these acute phase reactants.

Confounding. Simultaneous statistical dependence of any third variable with both selenium status and disease outcomes of interest, that causes a departure from the genuine relation between selenium and outcomes in clinical studies. This should be assessed and adjusted for in statistical analysis of clinical studies.

Doubly labeled water (DLW). A gold standard and noninvasive technique used to estimate the total energy expenditure of the body. The technique involves ingestion of a water sample that contains naturally occurring stable isotopes of hydrogen and oxygen, which helps to estimate the metabolic rate.
Food frequency questionnaire (FFQ). A structured, interviewer-administered or self-administered questionnaire that assesses dietary intake of participants in terms of quantity as well as frequency of intake, which usually results in a derived semiquantitative estimate. The FFQ is ideally designed to capture dietary pattern or the specific nutrients according to the research hypothesis tested.
Food composition table (FCT). A comprehensive database on the biochemical contents of commonly consumed food items at a defined jurisdiction. This is particularly important in the case of selenium assessment since soil selenium content varies greatly by region.
Gene–diet interactions. The interplay between genetic variations and dietary intake in determining the risk of cardiovascular and other chronic diseases. Increasingly this interplay has been recognized in numerous nutritional determinants of cardiovascular diseases. Such interactions may also have a potential role in selenium.
Recall bias. Systematic deviation from the true estimate of selenium exposure based on questionnaire assessment due to participant or interviewer factors. This is most commonly seen in case-control and retrospective studies where knowledge of disease status influences the behavior of participant reporting or interviewer inquiry.
Seleniferous region. Regions where the soil level of selenium is abundant.
Statistical adjustment. Statistical techniques to account for and mitigate the effect of confounders to generate an unbiased (ideally) estimate of the true relation of selenium exposure and clinical outcomes.
Vascular function. Surrogate estimate of vascular reactivity or severity of subclinical atherosclerosis, as a reflection of cardiovascular health. It also predicts the risk of future cardiovascular events.

Key Facts of Food-Frequency Questionnaire (FFQ)

The key facts of food-frequency questionnaire (FFQ), including the research tool characteristics, administration approach and research applications.

1. FFQ is a structured dietary exposures assessment tool, usually administered to participants with the assistance of an interviewer.
2. The number of food items in an FFQ and the comprehensiveness as to how a particular food group is interrogated depends on the underlying research hypothesis tested.
3. FFQ administration typically lasts for at least 15 min to 1 h, where participants are asked to recall their individual dietary intake of each food

item, both in terms of quantity of intake each time (*e.g.* portion size or estimated weight of food item), as well as the estimated frequency of intake over the prespecified period in the recent past.
4. The definition of "recent past" is variable, but is often taken as about 3 months to 1 year.
5. The estimation of food portion size is often assisted with the use of food demonstration models or standardized food photographs.
6. With the so-obtained intake estimates of both quantity and frequency of each specific food item, the average intake amount of each food item in the assessed period can then be calculated. Therefore, an FFQ is often semiquantitative.
7. Semiquantitative intake estimates of each selenium-containing food item can be coupled with data on the known per-weight selenium contents contained in such food items to generate an overall selenium intake estimate.
8. Selenium contents of each food item are available from the Food Composition Tables (FCT), which are derived from prior standardized laboratory biochemical studies of commonly consumed food items, often performed by local health authorities or research institutions.
9. It has been shown that selenium intake as determined by FFQ correlates closely with blood selenium levels.

List of Abbreviations

DLW Doubly Labeled Water
FCT Food Composition Table
FFQ Food-Frequency Questionnaire

References

Alanne, M., Kristiansson, K., Auro, K., Silander, K., Kuulasmaa, K., Peltonen, L., Salomaa, V. and Perola, M., 2007. Variation in the selenoprotein S gene locus is associated with coronary heart disease and ischemic stroke in two independent Finnish cohorts. *Human Genetics*. 122: 355–365.

Chan, Y. H., Siu, C. W., Yiu, K. H., Chan, H. T., Li, S. W., Tam, S., Cheung, B. M., Lau, C. P., Lam, T. H. and Tse, H. F., 2012. Adverse systemic arterial function in patients with selenium deficiency. *Journal of Nutrition, Health and Aging*. 16: 85–88.

Cox, A. J., Lehtinen, A. B., Xu, J., Langefeld, C. D., Freedman, B. I., Carr, J. J. and Bowden, D. W., 2013. Polymorphisms in the selenoprotein S gene and subclinical cardiovascular disease in the diabetes heart study. *Acta Diabetologica*. 50: 391–399.

Garcia, V., Rona, R. J. and Chinn, S., 2004. Effect of the choice of food composition table on nutrient estimates: a comparison between the British and American (Chilean) tables. *Public Health Nutrition*. 7: 577–583.

Gonzalez, S., Huerta, J. M., Fernandez, S., Patterson, E. M. and Lasheras, C., 2006. Food intake and serum selenium concentration in elderly people. *Annals of Nutrition & Metabolism*. 50: 126–131.

Hong, S. R., Lee, S. M., Lim, N. R., Chung, H. W. and Ahn, H. S., 2009. Association between hair mineral and age, BMI and nutrient intakes among Korean female adults. *Nutrition Research and Practice*. 3: 212–219.

Hughes, C. M., Woodside, J. V., McGartland, C., Roberts, M. J., Nicholls, D. P. and McKeown, P. P., 2012. Nutritional intake and oxidative stress in chronic heart failure. *Nutrition, Metabolism & Cardiovascular Diseases*. 22: 376–382.

Kardinaal, A. F., Kok, F. J., Kohlmeier, L., Martin-Moreno, J. M., Ringstad, J., Gomez-Aracena, J., Mazaev, V. P., Thamm, M., Martin, B. C., Aro, A., Kark, J. D., Delgado-Rodriguez, M., Riemersma, R. A., van't Veer, P. and Huttunen, J. K., 1997. Association between toenail selenium and risk of acute myocardial infarction in European men. The EURAMIC study. European Antioxidant Myocardial Infarction and Breast Cancer. *American Journal of Epidemiology*. 145: 373–379.

Karita, K., Sasaki, S., Ishihara, J., Tsugane, S. and JHPC, 2003. Validity of a self-administered food frequency questionnaire used in the 5-year follow-up survey of the JPHC study to assess selenium intake: comparison with dietary records and blood levels. *Journal of Epidemiology*. 13: S92–S97.

Kroke, A., Klipstein-Grobusch, K., Voss, S., Moseneder, J., Thielecke, F., Noack, R. and Boeing, H., 1999. Validation of a self-administered food-frequency questionnaire administered in the European Prospective Investigation into Cancer and Nutrition (EPIC) study: comparison of energy, protein, and macronutrient intakes estimated with the doubly labeled water, urinary nitrogen, and repeated 24 h dietary recall methods. *American Journal of Clinical Nutrition*. 70: 439–447.

Laclaustra, M., Navas-Acien, A., Stranges, S., Ordovas, J. M. and Guallar, E., 2009. Serum selenium concentrations and diabetes in U.S. adults: National Health and Nutrition Examination Survey (NHANES) 2003–2004. *Environmental Health Perspectives*. 117: 1409–1413.

Longnecker, M. P., Taylor, P. R., Levander, O. A., Howe, M., Veillon, C., McAdam, P. A., Patterson, K. Y., Holden, J. M., Stampfer, M. J., Morris, J. S., *et al.*, 1991. Selenium in diet, blood, and toenails in relation to human health in a seleniferous area. *American Journal of Clinical Nutrition*. 53: 1288–1294.

Longnecker, M. P., Stram, D. O., Taylor, P. R., Levander, O. A., Howe, M., Veillon, C., McAdam, P. A., Patterson, K. Y., Holden, J. M., Morris, J. S., Swanson, C. A. and Willett, W. C., 1996. Use of selenium concentration in whole blood, serum, toenails, or urine as a surrogate measure of selenium intake. *Epidemiology*. 7: 384–390.

Murphy, S. P., Wilkens, L. R., Hankin, J. H., Foote, J. A., Monroe, K. R., Henderson, B. E. and Kolonel, L. N., 2002. Comparison of two instruments for quantifying intake of vitamin and mineral supplements: a brief questionnaire versus three 24 hour recalls. *American Journal of Epidemiology*. 156: 669–675.

Nichol, C., Herdman, J., Sattar, N., O'Dwyer, P. J., St. J. O'Reilly, D., Littlejohn, D. and Fell, G., 1998. Changes in the concentrations of plasma selenium and selenoproteins after minor elective surgery: further evidence for a negative acute phase response? *Clinical Chemistry*. 44: 1764–1766.

Patel, Y. R., Han, K. A., Lteif, A. A., Wallace, J. P. and Mather, K. J., 2011. A cross-sectional evaluation of seasonality as a determinant of endothelial function. *Nitric Oxide: Biology and Chemistry*. 25: 282–287.

Patel, N. J., Pant, S., Deshmukh, A. J., Nalluri, N., Badheka, A. O., Shah, N., Chothani, A., Savani, G. T., Schwartz, C., Duvvuri, S., Bogin, M. and Vazzana, T. J., 2014. Seasonal variation of acute myocardial infarction related hospitalizations in the United States: perspective over the last decade. *International Journal of Cardiology*. 172: e441–442.

Pestitschek, M., Sonneck-Koenne, C., Zakavi, S. R., Li, S., Knoll, P. and Mirzaei, S., 2013. Selenium intake and selenium blood levels: a novel food frequency questionnaire. *Wiener Klinische Wochenschrift*. 125: 160–164.

Stranges, S., Sieri, S., Vinceti, M., Grioni, S., Guallar, E., Laclaustra, M., Muti, P., Berrino, F. and Krogh, V., 2010. A prospective study of dietary selenium intake and risk of type 2 diabetes. *BMC Public Health*. 10: 564.

Watson, P. E. and McDonald, B. W., 2007. Seasonal variation of nutrient intake in pregnancy: effects on infant measures and possible influence on diseases related to season of birth. *European Journal of Clinical Nutrition*. 61: 1271–1280.

Section III
Function and Effects

CHAPTER 15

Functional Foods Enriched in Selenium

TAMARA GARCÍA-BARRERA*[a], JOSÉ LUIS GÓMEZ-ARIZA[a], VERÓNICA GÓMEZ JACINTO[a], INÉS GARBAYO NORES[a], CARLOS VÍLCHEZ LOBATO[a], AND ZIVAN GOJKOVICA[b]

[a]Department of Chemistry and Materials Science, Faculty of Experimental Sciences, University of Huelva, Campus de El Carmen, 21007 Huelva, Spain; [b]Department of Food Technology and Biotechnology, Faculty of Chemistry, Brno University of Technology, 61200 Brno, Czech Republic
*E-mail: tamara@dqcm.uhu.es

15.1 Functional Foods, Nutraceuticals and Ready-to-Use Therapeutic Foods: Selenium-Enriched Foods

The economic, cultural and scientific development of our society has given rise to important changes in our food habits and life-style. In this sense, diets in developed countries are highly caloric, rich in saturated fats and sugars, while the consumption of complex carbohydrates and dietetic fiber is low. Diet is not the only factor that affects well-being and health, but is one of the most important, together with a decrease in physical activity, which has given rise to obesity problems, and as a consequence, heart diseases, diabetes and hypertension in the population.

Food and Nutritional Components in Focus No. 9
Selenium: Chemistry, Analysis, Function and Effects
Edited by Victor R. Preedy

Published by the Royal Society of Chemistry, www.rsc.org

The term "functional food" was first used in 1984 in Japan as a result of a study on the relationships between nutrition, sensory satisfaction, fortification and modulation of physiological systems in order to define those food products fortified with special constituents that possess advantageous physiological effects. In 1999 the European Commission Concerted Action on Functional Food Science in Europe (FUFOSE) stated some characteristics of functional foods, namely: (i) they should be in the form of normal foods and they must demonstrate their effects in amounts that can normally be expected to be consumed in the diet; (ii) a functional food can be a natural whole food, a food to which a component has been added, or a food from which a component has been removed by technological or biotechnological means; (iii) it can also be a food in which the nature of one or more components has been modified, or a food in which the bioavailability of one or more components has been modified, or any combination of these possibilities and (iv) a functional food may be targeted at the whole population or for particular groups, which may be defined, for example, by age or by genetic constitution. On the other hand, the International Life Sciences Institute of North America (ILSI) stated that a food can be regarded as "functional" if it is satisfactorily demonstrated to affect beneficially one or more target functions in the body, beyond adequate nutritional effects, in a way that is relevant to either an improved state of health and well-being and/or reduction of risk of disease, and they are not pills or capsules, but part of a normal food pattern. Functional foods present several objectives: they improve the general conditions of the body (*e.g.*, pre- and probiotics), decrease the risk of some diseases (*e.g.*, cholesterol-lowering products), and could be used for curing some illnesses (Siró *et al.*, 2008 and references therein). One of the main objectives of functional food science is to identify the beneficial interactions among a food, or specific ingredients, and one or more functions of the organism, obtaining, if possible, definitive proofs about the mechanisms involved in the interaction. This primary objective must be based on investigations *in vitro* or *ex vivo* in cellular lines or culture tissues; later in animal models and finally they must be corroborated in studies of observation or intervention in human (clinical trials). The rigorous design of a functional food needs to determine the biological activity at molecular level of their components and the bases of the disease (or diseases) considered as target. On the other hand, nutraceuticals are products elaborated from food, but presented as pill, powder, and other forms not related with food, which present demonstrated beneficial physiological properties or protection against chronic illness. Nutraceuticals from natural sources have been investigated for their putative chemopreventive and cancer therapeutic properties for the last few decades; they present pleiotropic effects and a relatively nontoxic nature (Go *et al.*, 2012). The ready-to-use-therapeutic food (RTUTF) can also contain nutraceutical compounds from vegetal origin (Espín *et al.*, 2007), specifically addressed to health condition support and therapy. The RTUTF are a mixture of nutrients designed and primarily addressed to the therapy of the severe acute malnutrition. The main ingredients of the formulation are powdered milk, peanuts butter, vegetal oil, sugar, and a mix of vitamins, salts, and minerals.

Selenium (Se)-containing foods can be considered as functional foods due to the health benefits of this element, and direct intake of Se supplements by humans has received considerable attention in recent years. Two types of multimicronutrients are commonly used such as: (i) multivitamins and multimineral preparations containing inorganic Se, other trace elements and vitamins, and (ii) supplements based on *Saccharomyces cerevisiae* yeast (Baker's yeast) (Roman *et al.*, 2014). Se supplementation in products for human nutrition has to be done with caution, having in mind the duality of Se action in human organism knowing that the upper limit of therapeutic Se dose (0.2 mg per day) is only a few times less than the potentially toxic dose (0.8–1.0 mg per day) (Rayman, 2004).

The selenized yeast (Se yeast) is particularly attractive due to its low cost, facility to grow under different conditions, and its ability to assimilate up to 3000 mg g^{-1} of Se starting from sodium selenite added to the growth medium. However, other Se-rich supplements can be proposed, like microalgae.

15.2 Selenium Biotransformation in Microalgae and Yeast

The peculiarity of Se is its high chemical and physical similarity to S and for this reason Se compounds follow the same metabolic routes as S compounds (Birringer *et al.*, 2002). Se can substitute S in amino acids and in other metabolites, however, a complete substitution is not possible, owing to the higher reactivity of Se, which is also the reason for its toxicity.

The analogies between Se and sulfur (S) metabolism have been demonstrated by the fact that all small organic Se compounds in plants, yeasts, bacteria or animals are isologs of S compounds, and the enzymes involved in sulfate metabolism and in the transulfuration pathway typically do not discriminate between S and Se compounds (Birringer *et al.*, 2002).

15.2.1 Selenized Yeast as Selenium Source

Se yeast produced by fermenting *Saccharomyces cerevisiae* in a Se-rich media, is a recognized source of organic food-form selenium. Se yeast is a common form of Se used to supplement the dietary intake of this important trace mineral, that present some beneficial aspects like the high content in proteins and the possibility to produce food from yeast inside of plants. Se yeast is currently the primary Se dietary supplement, where Se is present mostly as selenomethionine (SeMet), but several minor organic species have been reported, such as methylselenol, selenocysteine (SeCys), selenohomocysteine (Se-homocysteine), selenoadenosine and others at trace levels (Roman *et al.*, 2014).

In this process, virtually all of the Se structurally substitutes for S in the amino acid methionine (Met) thus forming SeMet *via* the same pathways and enzymes that are used to form S-containing Met (Figure 15.1). Owing to its similarity to S-containing methionine, SeMet is taken up nonspecifically and

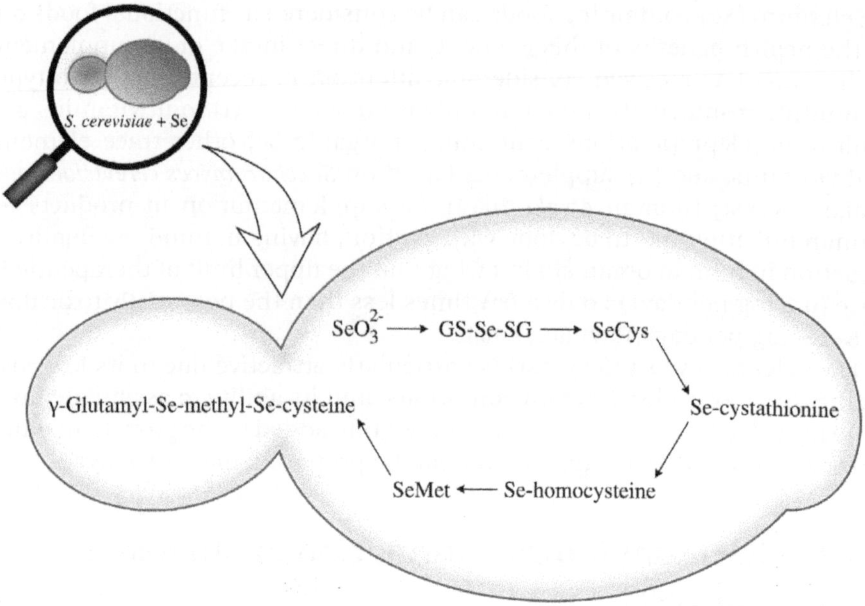

Figure 15.1 Schematic overview of metabolic transformation of compounds that have been identified in Se-enriched yeast extracts. GS-Se-SG – selenodiglutathione; SeCys – selenocysteine; SeMet – selenomethionine.

becomes part of yeast protein. It is this metabolic route (Figure 15.1) that makes Se yeast valuable in animal and human nutrition, since it offers the same organic form of Se produced by food-chain autotrophs (*i.e.*, most plants and certain blue-green algae) (Schrauzer, 2006).

By growing in Se-enriched media, yeast can accumulate large amounts of Se and incorporate them into organic Se-containing compounds, mainly SeMet (Rayman, 2004). Inorganic selenite as a low bioavailable toxic component can be converted to safer highly bioactive species with improved nutritional properties. *S. cerevisiae* is the only yeast strain that has been used by manufacturers for production of Se yeast. The ability of *S. cerevisiae* to transform inorganic selenium into organoselenium compounds (especially SeMet) is known and it is dependent on growth conditions. The main advantages of using Se yeast are illustrated in Figure 15.2.

15.2.2 Microalgae as Selenium-Containing Functional Food

Algae are photosynthetic organisms, which possess reproductive simple structures. These organisms constitute a total of 25–30 000 species, with a great diversity of forms and sizes, and that can exist from unicellular microscopic organisms (microalgae) to multicellular of great size (macroalgae). Microalgae are autotrophic micro-organisms having extremely high photosynthetic efficiency and are valued as rich source of lipids, hydrocarbons,

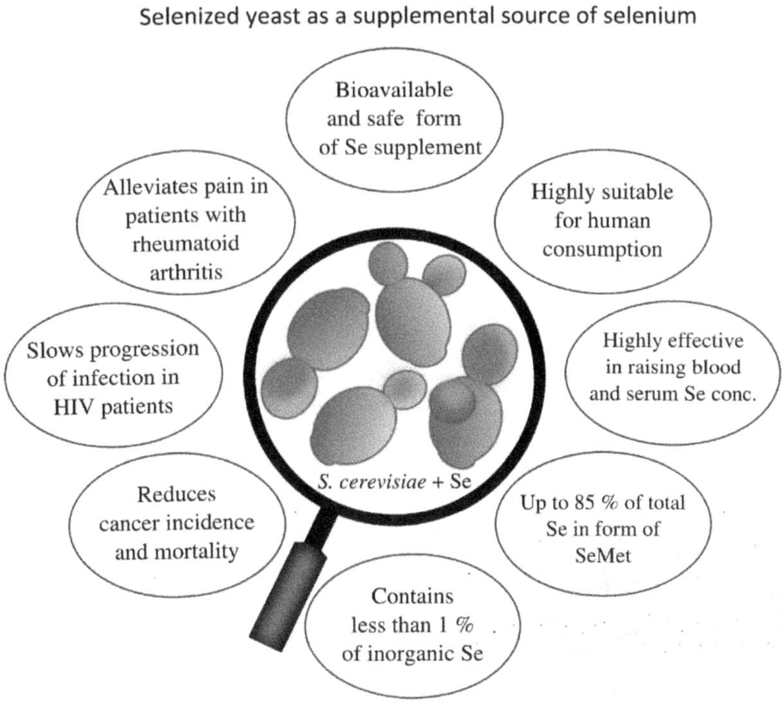

Figure 15.2 Chemical characteristics and main advantages of selenized yeast used as functional food.

and complex oils for biodiesel besides being an invaluable source of bioethanol, biomethane, and biohydrogen. Moreover, microalgae are a rich source of bioactive compounds with high interest in food and human health. In fact, some algae are organisms that live in complex habitats submitted to extreme conditions (for example, changes of salinity, temperature, nutrients, UV-vis irradiation), therefore, they must adapt rapidly to the new environmental conditions to survive, producing a great variety of secondary (biologically active) metabolites, which cannot be found in other organisms such as antioxidant complexes that are highly effective in dissipating antioxidant species.

In this way, algae and microalgae can be considered as genuine natural reactors being, in some cases, a good alternative to chemical synthesis for certain compounds. Several microalgae species have been used as functional foods like *Chlorella* spp., *Dunaliella* spp. and *Scenedesmus* spp., as well as the cyanobacteria *Spirulina* spp. (Kay, 1991) and some of them have also been proposed as food supplements due to the high iodine and Se accumulation capacity (Gómez-Jacinto *et al.*, 2010).

Microalgae can readily intake Se salts from aquatic environments or culture medium and transform it into various organic Se compounds. Algal metabolism of Se in the form of selenate is schematically presented in

Figure 15.3. Upon an ion's uptake *via* sulfate transporters and delivery to chloroplast, selenate is further reduced to selenite in a two-step reaction (Pilon-Smits and Quinn, 2010). First, selenate is activated by coupling to ATP giving adenosine phosphoselenate (APSe) that is further reduced to selenite by the action of glutathione containing adenosine phosphosulfate enzyme. Selenite is further reduced to selenide mediated by sulfite reductase. Selenide is consequently enzymatically coupled to *O*-acetylserine to form selenocysteine (SeCys). SeCys is converted to SeMet in multiple enzymatic reactions *via* Se-cystathionine and Se-homocysteine (Pilon-Smits and Quinn, 2010). The resulting SeMet can be accumulated intracellularly or incorporated nonspecifically into proteins or methylated

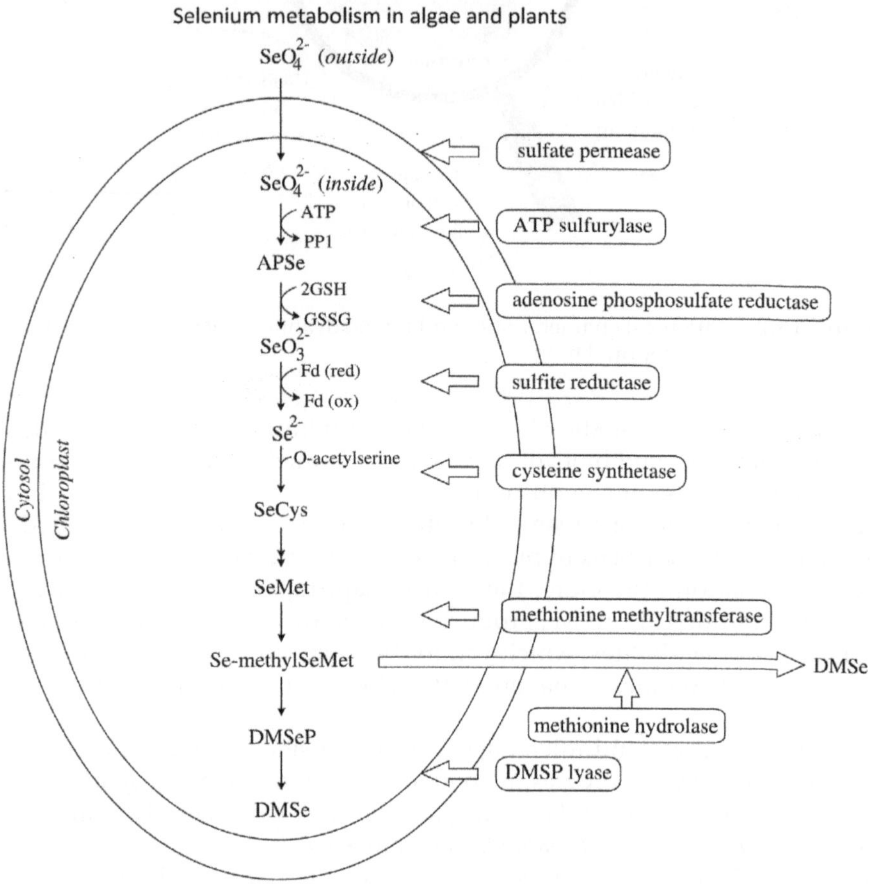

Figure 15.3 Schematic overview of selenium metabolism in algae and plants (Pilon-Smits and Quinn, 2010). APSe – adenosine phosphoselenate, SeCys selenocysteine, SeMet selenomethionine, DMSeP – dimethylselenoproprionate, DMSe dimethylselenide. GSH/GSSG – gutathione (reduced/oxidated form); Fd – ferredoxin.

to selenomethylselenocysteine(SeMeSeCys) that can be further metabolized to volatile dimethylselenide (DMSe) (Pilon-Smits and Quinn, 2010).

Once inside the cell, inorganic Se is metabolized to seleno-aminoacids with SeMet as the prevailing specie (Gojkovic *et al.*, 2013). *Chlorella sorokiniana*, accumulated up to 0.140 mg per g DW of SeMet cultivated for 120 h in a batch and up to 0.35 mg per g DW SeMet cultivated continuously in a photobioreactor (Gojkovic *et al.*, 2013). The fact that SeMet is the main Se form accumulated by microalgae intracellularly is of great importance, knowing that SeMet is one of the precursors of methylselenol, which is a potent anti-carcinogen that inhibits angiogenesis in tumor tissue and thus interrupts tumor nutrition. At the same time, microalgae are able to incorporate part of the Se into proteins specifically *via* SeCys insertion machinery and thus synthesizes various selenoenzymes. There are dozens of such selenoproteins identified in green algae *Chlamydomonas reinhardtii* (Novoselov *et al.*, 2002). This makes Se-enriched microalgae biomass an excellent carrier of highly bioavailable Se forms suitable for food supplementation.

Several exposure experiments to selenite have been described in the literature related to marine unicellular algae, like *Dunaliella primolecta*, *Porphyridium cruentum* and *Chlorella* sp. Se is present at the different biochemical components such as proteins, aminoacids (*Dunaliella primolecta* and *Chlorella* sp.) and probably, lipids and soluble carbohydrates (*Dunaliella primolecta* and *Chlorella* sp.). In *Chlorella sorokiniana* three major amino acids were found in the protein hydrolyzates: selenocysteic acid, SeMeSeCys and Se-methylselenocysteine selenoxide. On the other hand, the prevailing fraction of Se analogs in the free amino acids and in the proteins of *Chlorella sorokiniana* is a mixture containing SeCys, $SeCys_2$ and SeMet (Stadtman, 1980). In microalgae, Se can be incorporated into enzymes in the form of selenoaminoacids, especially, SeMet, as functional groups of glutathione peroxidase, which can eliminate hydrogen peroxide, preventing the damage of cellular membranes (Mattoo and Edelman, 1987). However, the excess of Se in the culture medium produces toxic effects, especially in relation to inorganic S assimilation mechanisms that cause damage in algae growth.

Culture conditions can increase the accumulation of SeMet, which improve the volatilization, or the formation of SeCys, increasing the incorporation of Se into proteins. Oxidative stress also positively influences Se intake by microalgae, since it provokes the overexpression of antioxidant systems, some of them requiring Se as cofactor to develop their function (*i.e.* glutathione peroxidase).

15.3 Nutritional Significance and Toxicological Aspects of Selenium Species

Se is an essential trace metalloid and the only one for which incorporation into proteins is genetically encoded, being a constituent of the 21st amino acid SeCys. Twenty-five selenoproteins have been identified in the human proteome, whereas only a few of them have been functionally characterized (Roman *et al.*, 2014). The molecules of interest include selenoproteins, like glutathione peroxidase and selenoprotein P, in which SeCys is specifically incorporated into their active sites, and selenoalbumin (SeAlb), not considered as a selenoprotein because the element is not specifically incorporated into the albumin moiety. Se has antioxidant properties and it is involved in cancer chemoprevention. It is needed to produce triiodothyronine required for healthy brain and bone development, normal growth and thermoregulation and at the proper levels, Se has been found to be necessary for proper immune-system functioning since it inhibits the progression of human immunodeficiency virus to acquired inmunodeficiency syndrome. Figure 15.4 summarizes the beneficial aspects of Se in human health.

Dietary Se intakes depend on regional Se availability and the types of foods consumed. The daily recommended Se amount is 0.04 and 0.03 mg per day, for adult men and women, respectively (WHO/FAO/IAEA, Geneva 1996). However, due to the narrow range between Se beneficial and toxic levels, both effects can be produced. Lower quantities can produce cardiovascular risk disease and other degenerative diseases, while high intake can produce hair and nail brittleness and loss, gastrointestinal disturbances, skin rash, garlic breath, fatigue, irritability and nervous system abnormalities. However,

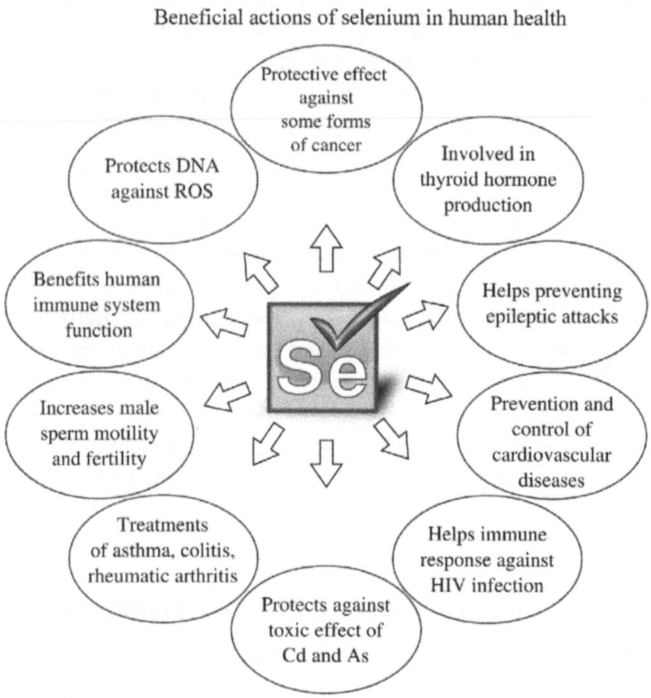

Figure 15.4 Schematic representation of beneficial actions of selenium in human health.

there is some controversy related to the upper limits for Se intake, and while some authors do not report any adverse level for an intake <0.8 mg per day for adults, others report selenosis with Se intakes ≥0.85 mg per day (Roman *et al.*, 2014). In this context, the US Environmental Protection Agency defined an intake limit of 1.262 mg per day as the reference at which clinical selenosis appears, while the tolerable upper intake level established by the European Community is 0.3 mg per day (Rayman, 2004).

On the other hand, it is well known that inorganic forms of Se are more acutely toxic than organic forms such as Se yeast, in which SeMet is the main specie, accounting for 54–74% of the total Se (Rayman, 2004). The LD_{50} in rats for Se yeast has been established by several authors at 0.0373 mg g^{-1} compared with 0.0127 mg g^{-1} for sodium selenite demonstrating that Se yeast is considerably less acutely toxic than sodium selenite. Histological examination of livers of animals fed Se yeast evidenced up to 50% greater deposition of Se, but there was no corresponding toxicity against that produced in rats fed selenite at levels of 0.016 mg g^{-1} diet over an 8 week period (severe hepatotoxicity, cardiotoxicity and splenomegaly) (Rayman, 2004). On the other hand, it has been suggested that organic forms of Se may be more toxic during long-term consumption due to the rapid incorporation into tissue proteins rather than be excreted. Then, the EC Scientific Committee on Food expressed concern in 2002 that organic Se, for example SeMet or Se yeast, could build up in body tissues to toxic levels. However, a large number of studies state that Se yeast can be administrated in doses as high as 0.8 mg per day for lengthy periods without any toxic effects, which led to the conclusion that there cannot be a continuing rise in tissue Se and that it is not more hazardous than supplementation with inorganic Se. In this sense, the conversion of organic Se into hydrogen selenide maybe an important regulator of Se bioavailability that may protect against excessive incorporation of Se into proteins and prevents toxicity mediated through reactive oxygen species from excessive intake (Rayman, 2004 and references therein).

15.4 Selenium Bioaccessibility and Bioavailability in Food

Due to the narrow range between Se beneficial and toxic levels, it is crucial to establish the Se bioavailability/bioaccessibility of a food in order to guarantee its required assimilation, since usually only a fraction is absorbed and transformed into a biologically available form.

The term bioavailability has been defined by several authors, but sometimes they interpret it differently. Some authors define this term as the fraction of an ingested element that is absorbed through the intestinal barrier and passes into the bloodstream or an organ, but others consider this fraction as the quantity of this element that is effectively used by the organism and transformed into a biochemically active form, while another possible definition of bioavailability is the fraction of an element that is absorbed and that reaches the systemic circulation in order to be distributed to organs and tissues, where it can eventually become bioactive. This fraction of Se has been estimated mainly using the activity of glutathione peroxidase (Pedrero and Madrid, 2009 and references therein), but also its concentration in different organs (Pedrero and Madrid, 2009 and references therein). On the other hand, the bioaccessible fraction is the fraction that is soluble in the intestine and is therefore available for subsequent processes of absorption through the intestinal mucosa (Thiry *et al.*, 2012 and references therein). In summary, the Se-bioavailable fraction can be considered as the quantity of Se absorbed through the intestinal barrier and reaching the systemic circulation; the Se-bioaccessible fraction as Se that is soluble in the intestine and the Se-bioactive fraction as the quantity of Se transformed into active selenometabolites (Thiry *et al.*, 2012).

In vivo bioaccessibility studies are both expensive and laborious, and the possibility of measuring certain parameters during the experiments is often limited. This soluble fraction of the element can be estimated by *in vitro* simulation of the gastrointestinal digestion, which usually includes two steps: a gastric digestion with pepsin at pH 2, followed by an intestinal digestion with amylase, pancreatin and bile salts at neutral pH.

Se bioavailability is affected by its chemical form and generally, organic compounds of Se are more bioavailable than the inorganic forms, since they have noticeable differences in the metabolism. Specifically, selenocompounds identified in plants include selenate, selenite, SeCys, Se-Met, Se-homocysteine, SeMeSeCys, γ-glutamil-selenocystathionine, SeMet selenoxide, γ-glutamyl-Se-methylselenocysteine (GGSeMCys), selenocysteineselenic acid, Se-proponylselenocysteine, dimethyl diselenide, selenosinigrin, selenopeptide and selenowax. On the other hand, selenocompounds in animal tissues are SeCys, selenotrisulfides of cystine, selenate and selenite. Other factors such as total protein, fat and the presence of heavy metals has been also described to affect Se bioaccessibility (Pedrero and Madrid, 2009 and references therein). Approximately 80% of dietary Se is absorbed, although it depends on the type of food consumed, the chemical specie and the Se status of the subject (Pedrero and Madrid, 2009 and references therein).

15.4.1 Selenium Bioavailability in Fish

Considering the bioavailability of Se in fish, the food source and chemical species are important. Tuna has been extensively studied since its ability to increase Se concentration in body tissues and red blood cells seems similar or slightly lower than wheat. Cabañero assessed Se bioaccessibility of swordfish, sardine and tuna using an *in vitro* approach (Cabañero *et al.*, 2004). Se bioaccessibility is different depending on the type of fish and higher Se-solubility percentages were observed in the intestinal digestion supernatants obtained from swordfish and sardine (76% and 83%, respectively) compared with those obtained from tuna (50%). Simulated human gastric and intestinal digestion led to the identification of SeMet in the intestinal digests from the three fishes, and similar bioaccessible percentages were observed for SeMet (16%, 14% and 19%, for swordfish, sardine and tuna, respectively). However, fish is a poor source of available Se, due in part to its high Hg content and other heavy metals, which bind to Se forming insoluble inorganic complexes.

15.4.2 Selenium Bioavailability in Meat

Meat is the most important Se dietary source since Se trends to be present in relatively high concentration and highly bioavailable. Several studies have shown that Se bioavailability in meat is high because Se forms in foods of animal origin are mostly SeCys and SeMet (Dumont *et al.*, 2006b). SeMet is known to be highly bioavailable and it is absorbed in the small intestine, being incorporated into the long-term body reserves. Se bioavailability was evaluated in Se-deficient rats fed veal, chicken, beef, pork and lamb, and the results shown that the activity of glutathione peroxidase (GPx) in liver is around 80% for all the species except for lamb (58%) (Wen *et al.*, 1997).

15.4.3 Selenium Bioavailability in Milk and Derived Products

The studies carried out with milk reveal that another important factor that influences Se assimilation is food composition. In this sense, Shen *et al.* demonstrated by an *in vitro* digestion that selenium bioavailability was higher in human milk than in cow, goat and sheep milk, which can be related with the protein composition (Shen *et al.*, 1996). Cabrera *et al.* reported a mean of Se absorbable fraction of 80 ± 10% in dairy products such as yogurt, custard, cream cheese, curd, crème caramel, ice-cream, and condensed milk (Cabrera *et al.*, 1996).

15.4.4 Selenium Bioavailability in Se-Fortified Foodstuffs

It is possible to find at the literature studies about the bioaccessibility and bioavailability of Se from several fortified foodstuffs like raw and cooked vegetables (*i.e.* radish, onion, chives and garlic) and nuts, raw and cooked fish, yeast and yeast-based nutritional supplements, wheat, and, chicken eggs. Fortification of these materials has usually been performed with several Se species, (*e.g.*, selenite, selenate and SeMet) and foodstuffs containing high Se concentrations (*e.g.*, Brazil nuts, yeast and fish) have also been studied.

Se bioaccessibility assessment in selenized yeast was carried out by Hinojosa (Hinojosa-Reyes *et al.*, 2006). The study shown that approximately 90% of total Se is in the bioaccessible fraction, but only 40% was quantified as SeMet. This SeMet recovery percentage obtained after gastrointestinal digestion, is noticeable lower than that in the proteolytic extract of Se yeast, of about 45% of total Se. This fact was attributed by the authors to the incomplete hydrolysis of the Se-containing peptides present in this matrix.

Pedrero *et al.* assessed the potential bioavailability of Se and Se species from radish (*Raphanus sativus*). The concentrations of $SeCys_2$, SeMet and SeMetSeCys in fresh samples remained almost unaltered after a simulated intestinal gastric digestion, and the analysis of the intestinal gastric digests showed that almost 100% of Se in the fresh plants is found in the bioaccessible fraction (Pedrero *et al.*, 2006). Dumont *et al.* treated garlic (*Allium sativum*) samples with simulated gastric and intestinal fluids, and measured 73% GGSeMCys, 13% SeMet, 4% γ glutamyl-selenomethionine, 3% SeMeSeMet, 2% selenate, 0.5% SeCys and 0.5% selenocystathionine (Dumont *et al.*, 2006a).

Kápolna *et al.* used *in vitro* intestinal gastric digestion to assess the bioavailability of Se from selenized green onion and chive samples (Kápolna and Fodor, 2007). After Se fortification with selenate, 80–90% of the total Se content (mainly as selenate) became bioaccessible in the simulated intestinal gastric digests and several Se unknown compounds were detected, which accounts approximately for 2% of the total bioaccessible Se content. Otherwise, lower Se bioaccessible percentages were obtained from Selenite-fortified green onion (30%) and chives (12%), and from SeMet-fortified green onion (20%) and chives (22%). In addition, selenate, selenite and SeMet species were quantified in the bioaccessible fraction of selenite-fortified green onion and chives. Concerning SeMet-fortified green onion and chives, selenite, SeMet, MeSeCys, and selenomethionine oxidase (SeOMet) species were quantified in the intestinal gastric digests.

Studies carried out with mushrooms (Bhatia *et al.*, 2013) showed that total Se bioaccessible is about 75% of the total Se. The main compound extracted by both gastric and intestinal fluids was SeMet (73%), but selenite was also present. Finally, Govasmark *et al.* assessed the bioaccessibility of Se species in Se-fortified wheat (Govasmark *et al.*, 2010). The Se yield is 70% in wheat when using simulated gastric digestion and the bioaccessible fraction did not increment when using simulated duodenal intestinal.

15.5 Antagonistic and Synergistic Interactions of Selenium with Other Elements

Se interacts with many elements, usually blocking the toxic effects of others. In this way, selenium and zinc counteract the toxic effect of cadmium. This interaction has been studied in cultured cells (Se/Cd) and rat kidney (Se/Cd, Zn/Cd). On the other hand, it has been shown that cadmium and other elements, ameliorated the frequency of selenium-induced cataracts and that tungsten, bismuth, germanium and antimony (as the trichloride administrated in the diet but not as sodium antimoniate) showed a partial effect against selenium toxicity (W/Se, Bi/Se, Ge/Se, Sb/Se) (García-Barrera *et al.*, 2012 and references therein).

Se also interacts with nickel and experiments carried out with Winstar rats conclude that the deleterious effects of $NiCl_2$ on the reproduction is antagonized by Na_2SeO_3 (Se/Ni).

15.5.1 Selenium Against Mercury Toxicity

The antagonistic interaction between Hg and Se was first reported in 1967 in an experiment with rats treated with mercury chloride and selenite (Parizek and Ostadalova, 1967). Then, other experiments confirmed this finding in tuna fish and other animals. Thus, in general, the simultaneous administration of selenite counteracts the negative impacts of the exposure to inorganic Hg, particularly in relation with neurotoxicity, fetotoxicity cardiovascular diseases. However, some authors suggest further studies to deepen the insight into this potential interplay in cardiovascular diseases, especially in relation to fish consumption. Also, quails fed a $MeHg^+$-containing diet survived longer when tuna with high levels of Se was coadministrated. In humans, selenite and SeMet dependent protection against Hg-induced apoptosis and growth inhibition in human cells has been observed. However, other studies reveals that inorganic Se is ineffective in preventing most of the $MeHg^+$-induced brain biochemical alterations and that it is also toxic alone. The Se to Hg molar ratio is very important, as described later in the present chapter and in this way it has been observed that Hg in molar excess over Se was a stronger inducer of metallothionein levels in trout (García-Barrera *et al.*, 2012 and references therein).

As previously stated, Se can also counteract the toxicity of $MeHg^+$. In an *in utero* study on mice with $MeHg^+$ and Se, the group that was given the lowest amount of Se and the highest dose of $MeHg^+$ was mostly adversely affected in neurobehavioral outcome. In rodents, antioxidant nutrients such as Se and vitamin E in the diet may alter reproductive and developmental toxicity induced by $MeHg^+$ Se has also been shown to reduce Hg bioavailability and trophic transfer in aquatic ecosystems. A great number of studies have been carried out related to the protective influence of the selenocompounds against $MeHg^+$ toxicity, especially SeMet. Also, exposition of $MeHg^+$ in rats resulted in a significant increase of urinary porphyrins and a decrease in motor activity that was counteracted by SeMet (García-Barrera *et al.*, 2012 and references therein).

15.5.2 The Interaction of Selenium and Iodine in Human Health: Food Supplemented with Iodine and Selenium

Iodine is well known to be crucial for animal metabolism through its development in hormones structure and functions. Since Se is a redox component of the enzyme thioredoxin reductase, the deficiency of this element causes metabolic disorders because this enzyme, together with GPx, is required for the immune system and anticarcinogenic function (Navarro and Lopez, 2000).

Se is specifically incorporated as SeCys into deiodinases, like in type I deiodinase, which is needed to convert thyroxine into the more active triiodothyronine (Beckett *et al.*, 1987), thus the interaction between these elements is established. In addition, the thyroid gland contains higher selenium concentrations than other organs and the deficiency of this element can contribute to endemic cretinism (Sánchez, 2009). The correlation between iodine and the Kashin–Beck disease, has also been established, concluding that rats fed with Se and iodine can decrease necrosis of the chondrocytes and dysfunctions of bone development similar to pathological changes in the growth plate and metaphysis (Yao *et al.*, 2012). Some investigators report a positive correlation between dietary intake of Se and iodine and the concentration of these trace elements in maternal milk, while others have been unable to demonstrate such a relationship.

15.5.2.1 Iodine Speciation in Microalgae

There is not much information about iodine species and their concentrations in microalgae, but it is believed that high iodine concentrations are toxic to algal growth. However, several species like *Bellerochea* sp., *Chlamydomonas reinhardtii*, *Dunaliella tertiolecta*, *Isochrysis galbana*, *Skeletonema costatum* and *Tetraselmis* sp. present high iodine tolerance, and other species like *Emiliania huxleyi*, *Gephyrocapsa oceanica*, show positive growth responses (Koji and Shiraiwa, 2012). Conversely, brown and red macroalgae are known to accumulate high concentrations of iodine, particularly the brown alga *Laminaria digitata* with a maximum accumulation percentage close to 5% (dry weight), which has the role of antioxidant (Küpper *et al.*, 2008). Other species like the brown alga *Petalonia fascia* requires iodine for growth, morphogenesis, and reproduction. Iodine is also required for the growth of the brown alga *Ectocarpus fasciculatus* and the red alga *Polysiphonia urceolata*. On the other hand, growth inhibition by high iodine concentrations has been reported in the brown alga *E. fasciculatus* and in the red alga *Goniotrichum elegans*. Brown algae are also known to transform iodate to iodide and release stored iodine as iodide, volatile iodocarbons, and molecular iodine.

The content of total iodine in water-soluble extracts and organic nonwater leachable has been measured in several macroalgae. In *Sargassum*, only 40% of iodine is water leachable, mainly as inorganic iodide and iodate. Iodine is mainly bonded to proteins, and to a lesser extent, to polyphenols, pigments and polysaccharides (algin and fucoidan). The microalga *Chlorella vulgaris* has been studied by Bourianová *et al.* for iodine accumulation and later use as food supplement since they can accumulate up to 500 mg of I per kg, mainly as water-soluble iodine (90%), in the form of iodide (Bourianová *et al.*, 2005). The water-soluble fraction isolated from *C. vulgaris* stuff contains 66.4% of iodine, mainly distributed between I^- (75%) and IO_3^- (25%), and several low-abundant iodine species, while the iodine macromolecular fraction represents about 13% of total iodine in the microalgae (Gómez-Jacinto *et al.*, 2010).

Summary Points

- Selenium-containing foods can also be considered as functional foods due to the health benefits of this element.
- Selenized yeast is particularly attractive due to its low cost, facility to grow under different conditions, and its ability to assimilate up to 3000 mg g^{-1}.
- Microalgae can also be used as functional food due to the high iodine- and selenium-accumulation capacity.
- Due to the narrow range between selenium beneficial and toxic levels, it is crucial to establish the selenium bioavailability/bioaccessibility of a food in order to guarantee its required assimilation.
- Selenium bioavailability is affected by its chemical form and generally, organic compounds of selenium are more bioavailable than the inorganic forms.
- There is much chemical and physical similarity to S and for this reason Se compounds follow the same metabolic routes as S compounds.
- All small organic Se compounds in plants, yeasts, bacteria or animals are isologs of sulfur compounds, and the enzymes involved in sulfate metabolism and in the transulfuration pathway typically do not discriminate between S and Se compounds.
- Selenium interacts with many elements (*e.g.* with mercury), usually blocking the toxic effects of others.
- Iodine is crucial for animal metabolism through its development in hormones structure and functions.
- Selenium plays a dual and a complex role in the thyroid hormone system.
- The thyroid gland contains higher selenium concentrations than other organs and the deficiency of this element produces endemic cretinism.
- Marine sea-foods contain the highest amount of iodine (0.0016–0.0032 mg g^{-1}), with seaweed and sea-salt being important sources.
- Selenium and iodine bioavailability can be affected by other components present in foods such as proteins, thiocyanates and their precursors such as glycosinolates.

Key Facts of *Saccharomyces cerevisiae*

1. *Saccharomyces cerevisiae* is an attractive model organism due to the fact that its genome has been sequenced, its genetics are easily manipulated, and it is very easy to maintain.
2. *Saccharomyces cerevisiae* is a yeast with high protein content that is related to the ability of incorporating Se.
3. The production of Se-enriched yeast is more manageable than the production of Se-enriched plants.
4. Selenized yeast is particularly attractive due to its low cost, facility to grow under different conditions, and its ability to assimilate up to 3000 mg g^{-1}.
5. *Saccharomyces cerevisiae* has a great capacity to transform inorganic selenium into organoselenium compounds, especially SeMet.

Definitions of Words and Terms

Bioaccessibility. The fraction of an element that is soluble in the intestine.
Bioactivity. The quantity of an element transformed into active selenometabolites.
Bioavailability. The quantity of an element absorbed through the intestinal barrier and reaching the systemic circulation.
Functional foods. Foods that affect beneficially one or more target functions in the body, beyond adequate nutritional effects, in a way that is relevant to either an improved state of health and well-being and/or reduction of risk of disease, and they are not pills or capsules, but part of a normal food pattern.
Nutraceuticals. Products elaborated from food, but presented as pill, powder, and other forms not related with food, which present demonstrated beneficial physiological properties or protection against chronic illness.
Ready-to-use-therapeutic food (RTUTF). Mixture of nutrients designed and primarily addressed to the therapy of the severe acute malnutrition.
Species (of an element). Specific form of an element defined as to isotopic composition, electronic or oxidation state, complex or molecular structure.

List of Abbreviations

APSe	Adenosine phosphoselenate
DMSe	Dimethylselenide
FAO	Food and Agriculture Organization
FUFOSE	European Commission Concerted Action on Functional Food Science in Europe
GGSeMCys	γ-Glutamyl-Se-methyl-selenocysteine
GPx	Glutathione peroxidase
IAEA	International Atomic Energy Agency
ILSI	International Life Sciences Institute of North America
Met	Methionine
$MeHg^+$	Methylmercury
RTUTF	Ready-to-Use-Therapeutic Food
SeAlb	Selenoalbumin
SeCys	Selenocysteine
$SeCys_2$	Selenocystine
Se-homocysteine	Selenohomocysteine
SeMeSeCys	Selenomethyl selenocysteine
SeMeSeMet	Selenomethyl selenomethionine
SeMet	Selenomethionine
SeOMet	Selenomethionine oxidase
Se yeast	Selenized yeast
WHO	World Health Organization

References

Beckett, G. J., Beddows, S. E., Morrice, P. C., Nicol, F. and Arthur, J. R., 1987. *Biochemical Journal*. 248: 443.
Bhatia, P., Aureli, F., D'Amato, M., Prakash, R., Cameotra, S. S., Nagaraja, T. P. and Cubadda, F., 2013. *Food Chemistry*. 140: 225.
Birringer, M., Pilawa, S. and Flohe, L., 2002. *Natural Product Reports*. 19: 693.
Bourianová, I., Machát, J., Niedová, E., Doucha, J. and Kanický, V., 2005. *Chemical Listy*. 99: 273.
Cabañero, A. I., Madrid, Y. and Camara, C., 2004. *Analytica Chimica Acta*. 526: 51.
Cabrera, C., Lorenzo, M. L., De Mena, C., López, M. C., 1996. *International Journal of Food Science and Nutrition*. 47: 331.
Dumont, E., Ogra, Y., Vanhaecke, F., Suzuki, K. T. and Cornelis, R., 2006a. *Analytical and Bioanalytical Chemistry*. 384: 1196.
Dumont, E., Vanhaecke, F. and Cornelis, R., 2006b. *Analytical and Bioanalytical Chemistry*. 385: 1304.
Espín, J. C., García-Conesa, M. T. and Tomás-Barberán, F. A., 2007. *Phytochemistry*. 68: 2986.
García-Barrera, T., Gómez-Ariza, J. L., González-Fernández, M., Moreno, F., García-Sevillano, M. A. and Gómez-Jacinto, V., 2012. *Analytical and Bioanalytical Chemistry*. 403: 2237.
Go, V. L. W., Harris, D. M. and Srihari, P., 2012. *Nutraceuticals and Cancer*. 1: 1.
Gojkovic, Z., Garbayo-Nores, I., Gomez-Jacinto, V., García-Barrera, T., Gómez-Ariza, J. L., Márová, I. and Vílchez-Lobato, C., 2013. *Process Biochemistry*. 48: 1235.
Gómez-Jacinto, V., Arias-Borrego, A., García-Barrera, T., Garbayo, I., Vílchez, C. and Gómez-Ariza, J. L., 2010. *Pure and Applied Chemistry*. 82: 473.
Govasmark, E., Brandt-Kjelsen, A., Szpunar, J., Bierla, K., Vegarud, G. and Salbu, B., 2010. *Pure and Applied Chemistry*. 82: 461.
Hinojosa-Reyes, L., Marchante-Gayón, J. M., García-Alonso, J. I. and Sanz-Medel, A., 2006. *Journal of Agricultural and Food Chemistry*. 54: 1557.
Kápolna, E. and Fodor, P., 2007. *International Journal of Food Sciences and Nutrition*. 58: 282.
Kay, R. A., 1991. *Critical Reviews in Food Science and Nutrition*. 30: 555.
Koji, I. and Shiraiwa, Y., 2012. *Procedia Environmental Sciences*. 15: 34.
Küpper, F. C., Carpenter, L. J., McFiggans, G. B., Palmer, C. J., Waite, T., Boneberg, E. M., Woitsch, S., Weiller, M., Abela, R., Grolimund, D., Potin, P., Butler, A., Luther, G. W., Kroneck, P. M., Meyer-Klaucke, W. and Feiters, M. C., 2008. *Proceedings of the National Academy of Sciences*. 13: 6954.
Mattoo, A. K. and Edelman, M., 1987. *Proceedings of the National Academy of Sciences of the United States of America*. 84: 1497.
Navarro, A. M. and Lopez, M. M. C., 2000. *Science of the Total Environment*. 249: 347.
Novoselov, S. V., Rao, M., Onoshko, N., Zhi, H. and Kryukov, G. V., 2002. *European Molecular Biology Organization Journal*. 21: 3681.

Parizek, J. and Ostadalova, I., 1967. *Experientia*. 23: 142.
Pedrero, Z., Madrid, Y. and Cámara, C., 2006. *Journal of Agricultural and Food Chemistry*. 54: 2412.
Pedrero, Z. and Madrid, Y., 2009. *Analytica Chimica Acta*. 634: 135.
Pilon-Smits, E. A. H. and Quinn, C. F., 2010. Selenium metabolism in plants. In: Hell, R. and Mendel, R. R. (ed.) *Cell Biology of Metals and Nutrients, Plant Cell Monographs*. Springer-Verlag, Berlin Heidelberg.
Rayman, M. P., 2004. *British Journal of Nutrition*. 92: 557.
Roman, M., Jitaru, P. and Barbante, C., 2014. *Metallomics*. 6: 25.
Sánchez, A., 2009. *Glánd Tir Paratir*. 18: 40.
Schrauzer, G., 2006. *Pure and Applied Chemistry*. 78: 105.
Shen, L. H., Van Dael, P., Luten, J. and Deelstra, H., 1996. *International Journal of Food Sciences and Nutrition*. 47: 75.
Siró, I., Kápolna, E., Kápolna, B. and Lugasi, A., 2008. *Appetite*. 51: 456.
Stadtman, T. C., 1980. *Annual Reviews in Biochemistry*. 49: 93.
Thiry, C., Ruttens, A., De Temmerman, L., Schneider, Y.-J. and Pussemier, 2012. *Food Chemistry*. 130: 767.
Wen, H. Y., Davis, R. L., Shi, B., Chen, J. J., Chen, L., Boylan, M. and Spallholz, J. E., 1997. *Biological Trace Element Research*. 58: 43.
WHO/FAO/IAEA, 1996. *Trace Elements in Human Nutrition and Health*. World Health Organization, Geneva.
Yao, Y. F., Pei, F. X., Li, X. B., Yang, J., Shen, B., Zhou, Z. K., Li, L. and Kang, P. D., 2012. *Biological Trace Element Research*. 146: 199.

CHAPTER 16

Selenium in Parenteral Nutrition

MALKI MILLER[a] AND SATHYAPRASAD BURJONRAPPA*[b]

[a]Brooklyn College, City University of New York, Department of Nutrition, Maimonides Medical Center, 4802 10th Avenue, Brooklyn, New York, USA; [b]Department of Surgery, Albert Einstein College of Medicine. Montefiore Medical Center, 111 E 210th St, Bronx, NY 10467.
*E-mail: sathyabc@yahoo.com

16.1 A Historical Overview

The early stages of total parenteral nutrition (TPN), introduced in the 1960s as an alternate mode of nutrient provision in the setting of limited gastrointestinal tract function, consisted of carbohydrate, protein, lipid, electrolytes and rudimentary forms of the vitamins. Little attention was given to trace elements and to selenium in particular, a mineral whose biological function was first being studied in animals at that time. Reports of trace element deficiencies prompted the first guidelines to add copper, zinc, chromium and manganese to parenteral nutrition formulations, a recommendation made in 1979 by the Nutrition Advisory Group of the American Medical Association. That same year, the first cases of symptomatic selenium deficiencies in humans were described. The initial account was the landmark publication reporting the association between selenium (Se) deficiency and Keshan disease, a form of cardiomyopathy endemic to Se-poor regions in China, while the second case manifested as muscular pain in conjunction with low plasma Se and

Food and Nutritional Components in Focus No. 9
Selenium: Chemistry, Analysis, Function and Effects
Edited by Victor R. Preedy

Published by the Royal Society of Chemistry, www.rsc.org

erythrocyte GSHP-x (glutathione peroxidase) activity in a woman maintained on selenium-free TPN for one month.

Since then, multiple reports have described Se deficiencies in patients receiving chronic TPN, characterized by skeletal myopathy, elevated creatine kinase, cardiomyopathy, congestive heart failure, poor immunity, encephalopathy, and whitened nail beds, all symptoms being reversed with Se supplementation (Table 16.1). In fact, these reports of clinical compromise in the scenario of low Se status first established the essentiality of this trace element in human health. Subsequently in 1984, the first American guidelines were published recommending routine Se addition to parenteral formulation for patients with prolonged TPN dependency. Our growing appreciation of the role Se plays in human health, the significance of oxidative stress in critically ill patients requiring intravenous nutrition support, and the emerging antioxidant role of Se in modulating this response has led to the evolution

Table 16.1 Reports of Se deficiencies in patients receiving parenteral nutrition. Summary of selected reports of clinical Se deficiencies reported on patients receiving long-term PN are shown. Information was obtained from reported cases in the literature from 1979 to the present. TPN: total parenteral nutrition; Se: selenium.

Year	Clinical Summary	Symptoms
1979	Adult with small intestinal perforation, multiple fistulae on TPN for 1 month.	Muscle pain; inability to walk
1983	Adult with multiple intestinal fistulae, on TPN for 18 months.	Low serum Se, erythrocyte glutathione peroxidase activity
1984	Adult with Crohn's disease, on home TPN for 8 years.	Fatal cardiomyopathy
1986	Adult with short bowel syndrome, on home TPN for 4 years.	Muscle weakness, low glutathione peroxidase
1987	Adult with Crohn's disease, on home TPN for 5 years.	Cardiomyopathy
1988	2 year old child with microvillous atrophy, on TPN since birth.	Regression of motor skills, skeletal myopathy
1989	Adult with Crohn's disease, on home TPN	Cardiomyopathy
1989	Adult with multiple gastrointestinal fistulae, on TPN for 1 month.	Muscle pain, elevated heart rate
1992	Adult with small bowel resection on TPN for 3 years.	Whitened nail beds
1994	Adult with Crohn's disease, on long-term TPN.	Impaired vision, encephalopathy
2003	Adult with Crohn's disease, on home TPN	Discoloration of nail beds, myopathy
2005	18 month old child on TPN for 5 months for juvenile polyposis.	Dry skin, alopecia
2006	Child with lymphangiomatosis.	Myopathy
2007	Adult with history of bariatic malabsorptive surgery.	Congestive heart failure
2007	5 infants On long-term TPN	Growth retardation, alopecia, pseudoalbinism

of these recommendations over the following three decades of nutrition research from the initial guidelines designed with the purpose of deficiency prevention to current research treating this trace element as a pharmaconutrient with potential to significantly affect clinical outcome.

16.2 Physiology and Biological Function

Selenium is a component of at least 25 selenoenzymes. One of the most important of the selenoproteins is glutathione peroxidase, which plays an important role in maintaining cell membrane integrity by serving as a scavenger of free radicals. The antioxidant activity is promoted by the breakdown of hydrogen peroxide and lipid hydroperoxides; reduced glutathione (GSH), the tripeptide located in cell membranes, is converted to its oxidized form (GS-SG) in the process by the combination of two glutathione molecules. Other important selenoenzymes include iodothyronine deiodinase that converts metabolically inactive tetraiodothyronine (T_4) to triiodothyronine (T_3), and thioredoxin reductase, an enzyme that regulates the redox state of cells. Immunoglobulin antibody function and cell-mediated immunity are also influenced by selenium and a higher incidence of infectious complications has been noted in neonates on selenium-free TPN protocols (Darlow and Austin, 2003).

Selenium is best absorbed in its organic form as selenomethionine; inorganic selenium (of the form found in TPN) is poorly retained compared to the organic forms. Prior to incorporation into selenoproteins, selenomethionine is converted to selenocysteine. Inorganic selenium is bound to low-density lipoproteins (LDL) and very low-density lipoproteins (VLDL), although the exact plasma selenium transport mechanism has yet to be identified, and then incorporated directly into selenocysteine.

In the body, extracellular Se is bound to selenoprotein-P that serves as the predominant Se carrier (60%). The rest of the plasma selenium is bound to glutathione peroxidase (30%), albumin (6–10%), and free selenium (<1%). In selenium-depleted patients, glutathione peroxidase activity was optimized earlier than selenoprotein-P concentrations, suggesting that selenoproteins are a better marker of selenium status. Selenium excretion occurs predominantly at the renal level; the kidneys excrete 75% of absorbed selenium (or 60% of dietary selenium), with the rest being excreted by the fecal route (Greene *et al.*, 1988).

16.3 Dosage Guidelines

The adult dietary reference intakes for Se, based on the oral intakes of Se that have been shown to maintain optimal GSHP-x activity, vary greatly by location given the different Se contents of soil; some areas may require as little as 20 mcg per day, while areas of New Zealand, traditionally known for their low selenium soil content and subsequent poor basal Se status, may require as high as 90 mcg per day. Consequently, establishing standardized guidelines

Table 16.2 Dosage guidelines for parenteral selenium. Current guidelines for intravenous dosing of Se in PN are shown, including clinical conditions that may increase or decrease parenteral Se requirements. Se: selenium; PN: parenteral nutrition. Source: Vanek, V. W., Borum, P., Buchman, A. *et al.*, 2012. A.S.P.E.N. position paper: recommendations for changes in commercially available parenteral multivitamin and multitrace element products. *Nutrition in Clinical Practice*. 27: 440.

	Parenteral dosage	Increased requirements	Decreased requirements
Adult	60–100 mcg per day	Sepsis, burns, postsurgery, SBS, diarrhea, CRRT	Oliguric renal failure (non-CRRT)
Pediatric	1–3 mcg per kg (up to 100 mcg per day)		
Preterm infant	2–7 mcg per kg		

for parenteral Se dosing is difficult (Table 16.2). Indeed, recommendations have varied substantially over the years, reflecting this lack of homogeneity in baseline Se status of different populations. The 2012 guidelines for parenteral Se dosing by the American Society for Parenteral and Enteral Nutrition are 60–100 mcg per day for adults under basal conditions. This number – increased from prior dietary reference intake guidelines – is extrapolated from the dietary reference intakes established in different parts of the world, reflecting a range of requirements, and has also been validated by studies on home TPN patients who were able to maintain normal serum Se levels at these dosages. Since selenium is excreted primarily *via* the kidneys, this dosage should be decreased in the case of renal impairment to prevent selenium toxicity, except in patients undergoing continuous renal replacement therapy (CRRT) where selenium losses in the ultrafiltrate may be significant (Vanek *et al.*, 2012).

For the pediatric population, a dose of 1–3 mcg per kg per day up to a maximum of 100 mcg per day, the limit set for adults, is sufficient to maintain normal Se levels. In infancy, parenteral intake of 2–3 mcg per kg per day are recommended by the American Society for Parenteral and Enteral Nutrition and the European Society of Pediatric Gastroenterology, Hepatology and Nutrition. This is based on estimates of the Se content of mature breast milk at 6–28 mcg L^{-1} in the United States and Europe, translating into an average Se consumption of 2.5 mcg per kg per day in normal breastfed infants; assuming an 80% absorption rate, these numbers would place the absolute Se requirement during infancy at 2 mcg per kg per day (Greene *et al.*, 1988).

Preterm infants given this dosage of parenteral Se are able to maintain normal plasma Se levels. Some studies have supported the use of higher doses in this population, in the range of 5–7 mcg per kg per day. A meta-analysis associated higher Se supplementation with a lower rate of sepsis, although most of these data were taken from areas with poor Se soil content and thus may not be able to be extrapolated to other regions where maternal intakes of Se, and thus Se status of the infant at birth, are likely better (Darlow and Austin, 2003).

16.4 A Focus on Critical Care

Guidelines for parenteral Se dosing were established based on maintaining normal plasma Se and/or GSHP-x activity in stable patients on long-term parenteral nutrition support. The complex physiological dynamics of critical care, however, alter both Se utilization and endogenous Se losses, potentially impacting basal Se balance. The metabolic milieu of the critically ill patient is characterized by catabolism, oxidative stress, and inflammation, increasing the demand for Se in its capacity for endogenous antioxidant activity (Figure 16.1). Paradoxically, selenium bioavailability during this time is compromised due to trace element redistribution from the circulation to tissues involved in protein synthesis, increased capillary leakage, hemodilution from fluid resuscitation, and exaggerated urinary losses resulting from higher selenoprotein catabolism. Indeed, plasma selenium levels are often reduced in patients in the intensive care unit, with low selenium status correlated with poor clinical outcomes that can be reversed with restoration of proper selenium balance (Table 16.2). With these factors in mind, parenteral Se dosing may need to be adjusted in the following clinical scenarios.

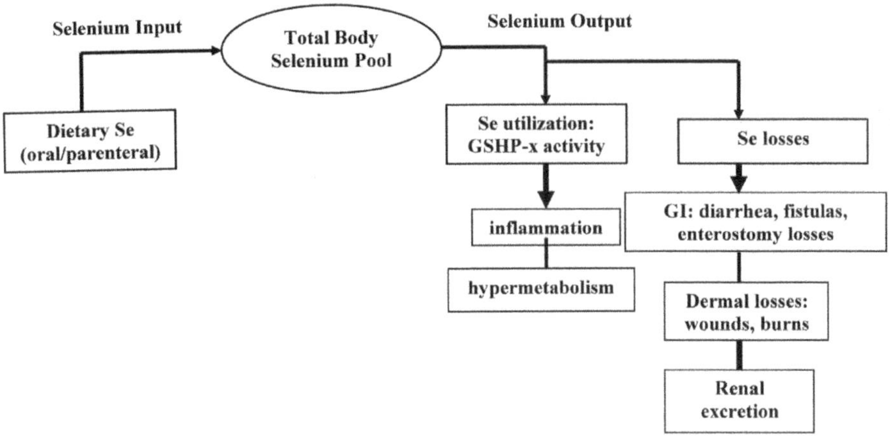

Figure 16.1 Alterations in selenium balance in critical care. Selenium balance in critical illness is shown. During hospitalization, there is often a decrease in Se intake due to poor dietary intake and/or inadequate parenteral Se dosing. Selenium output is usually higher due to accelerated endogenous Se utilization with increased GSHP-x activity associated with both inflammation and the hypermetabolic catabolic state. Additionally, endogenous Se losses are often elevated *via* the intestine, wounds, and higher urinary excretion related to increased selenoprotein catabolism. The simultaneous decrease in Se intake along with increased Se output leads to a condition of negative Se balance in the ICU setting that has been correlated with increased morbidity. Se: selenium; GSHP-x: glutathione peroxidase; ICU: intensive care.

16.4.1 Sepsis

The inflammatory cascade response during acute respiratory distress syndrome (ARDS) and systemic inflammatory response syndrome (SIRS) generates free radicals that upregulate the gene expression of TNF-alpha, interleukins and other proinflammatory cytokines. Selenium supplementation in supra-physiological doses inhibits this cascade reaction, resulting in suppressed inflammation and a lower C-reactive protein. Additionally, intravenous inorganic Se exerts a transient pro-oxidant effect before incorporation into selenoproteins; while this is clinically insignificant in normal doses usually supplied parenterally, megadoses of parenteral Se administered in bolus form may have a direct apoptotic effect on viral or bacterial cells as well as on activated inflammatory cells.

The effects of supraphysiological doses of parenteral selenium, in the form of sodium selenite, as a therapeutic strategy to improve the clinical course in septic patients has been studied in several randomized controlled trials and meta-analyses. Doses ranging from 350–2000 mcg per day, usually supplied as a loading dose within the first few days of admission to the intensive care unit and then a lower maintenance dose provided as a continuous infusion for several weeks thereafter, have been associated with improved morbidity and mortality and improvements in blood pressure, cardiac index, stroke volume index and overall hemodynamic status (Manzanares and Hardy, 2009). Studies, however, are heterogeneous in terms of dose and mode of infusion; additionally, some reports researched parenteral Se as monotherapy, while others included the mineral in an "antioxidant cocktail" containing glutamine, vitamin C, beta carotene, zinc, and/or vitamin E. Thus, while American and European guidelines for parenteral nutrition in sepsis recommend higher than baseline doses of Se, the optimal adjustment of parenteral Se to pharmaconutrient levels remains controversial.

16.4.2 Burns

Hypermetabolism and tissue catabolism following thermal injury, combined with exaggerated cutaneous selenium losses through wound exudates, predisposes the burn victim towards negative selenium balance. Decreased plasma selenium and GSHP-x activity are observed immediately after burn injuries, reductions that then persist for several weeks thereafter, along with increased urinary selenium excretion, likely indicative of catabolic losses. Topical silver nitrate commonly used on burns may have an antagonistic effect with selenium, further diminishing its antioxidant capacity.

Descriptive studies on selenium balance following thermal injury report persistently lower plasma selenium in burn patients, despite intakes resembling the recommended dietary allowance currently established for a healthy population. Plasma Se correlates negatively with total burn size, and better Se status is associated with lower rates of infection. Disruptions in Se homeostasis prompted recommendations to increase parenteral Se dosing in the setting of thermal injury, likely in the range of 315–380 mcg per day, although whether higher doses lead to improved outcomes has not yet been studied.

16.4.3 Gastrointestinal Surgery

A constellation of factors in the postoperative period alter normal selenium requirement. First, the metabolic insult of surgery alone, with its associated hypermetabolism and free-radical generation, increases Se utilization. High endogenous Se losses in fistulae, stoma effluent, and chylous fluid, as well as poor intestinal absorption of enteral selenium, similarly affect parenteral selenium requirement. In animals, intestine-specific glutathione peroxidase activity in the epithelial lining modulates local inflammation. Potential consequences of suboptimal activity in patients with underlying gastrointestinal inflammatory-related conditions such as irritable bowel disease, colon cancer, or necrotizing enterocolitis – conditions that are associated with low selenium status – are currently being investigated (Shenkin, 2009).

Deficiencies in this population were among the first selenium deficiencies described (see Table 16.1), likely due to this increased Se requirement, exaggerated Se losses, and chronic dependency on selenium-free parenteral nutrition. Even with adequate parenteral supplementation, case reports describe deficiencies attributed to high endogenous losses in gastrointestinal secretions, with stool mass and length of small bowel resection strong predictors of poor selenium status in patients with gastrointestinal disease or short bowel syndrome (SBS). Careful attention to selenium dosing, with periodic monitoring of selenium status, is recommended.

16.5 Pharmacology

For intravenous use, Se can be prepared as selenomethionine, and inorganically as selenious acid or sodium selenite. Selenomethionine is the form in which dietary Se enters the circulation and thus is most physiological although it is currently not available for commercial use. Selenomethionine undergoes conversion to selenocysteine and is then incorporated into body proteins. Although some studies showed the benefit of selenomethionine over selenious acid and sodium selenite in correcting Se status, parenteral Se in its inorganic forms have proven efficacious in reversing symptoms of Se deficiency as well, although they exert a transient pro-oxidant effect until incorporated into selenoproteins.

Selenious acid and sodium selenite are compatible with most TPN additives. Although selenite may be reduced to its elemental form by vitamin C in an acidic environment, this reaction is slowed by amino acids and other buffering agents and thus is generally not a concern in typical TPN formulations. The selenite form may also generate reactive oxygen species when interacting with sulfhydryl compounds such as cysteine and glutathione, and thus other forms of intravenous Se may be preferred when these additives are infused in the TPN solution (Terada *et al.*, 1999).

Parenteral Se is available separately or as part of a multitrace-element preparation. Trace-element packages commercially available (Table 16.3) may not provide adequate selenium when dosed with consideration to avoid

Table 16.3 Commercially available trace element packages. Parenteral trace element packages currently available in North America and Europe are shown, along with the Se content. Selenium is also available separately for intravenous injection. Source: Vanek, VW, Borum, P, Buchman, A. *et al.* 2012. A.S.P.E.N. position paper: recommendations for changes in commercially available parenteral multivitamin and multitrace element products. *Nutrition in Clinical Practice*. 27: 440.

Trace element package	Distributor	Standard dose	Se content, mcg (per standard dose)
Adults			
Multitrace-4	American Regent	1 ml	0
4-Trace elements	Hospira	5 ml	0
Multitrace-5	American Regent	1 ml	20
Multitrace-5 concentrate	American Regent	1 ml	60
Additrace	Fresenius Kabi	10 ml	32
Decan	Baxter & Laboratoires Aguettant	40 ml	70
Tracutil	B. Braun	10 ml	24
Pediatrics/Neonatal			
Multitrace-4 neonatal	American Regent	1 ml	0
Multitrace-4 pediatric	American Regent	1 ml	0
Trace elements injection 4-pediatric	American Regent	1 ml	0
Peditrace	Fresenius Kabi	1 ml	2
Inzolen-Infantibus sine NaK	Kohler	1 ml	0
Oligoelements pediatrique	Aguettant	1 ml	5

overload of other mineral components and may necessitate additional selenium provision as a separate infusion. Trace contamination of TPN additives may provide some source of Se as well, although these amounts are likely minimal and clinically insignificant. The inclusion of Se in parenteral formulations is especially important when parenteral lipids are administered, since Se deficiencies may induce lipid peroxidation, which has a damaging effect on cell membrane integrity. This is true in particular for patients with existing vitamin E deficiencies and also when using lipid formulations that do not contain additional vitamin E.

16.6 Monitoring Selenium Status

Selenium deficiency was recognized to be the causative factor in Keshan disease, a fatal cardiomyopathy affecting young women and children in China. Similar presentations were noted in TPN-dependent patients prior to the era of micronutrient replacement. Deficiency can develop within 6 weeks of providing selenium-free nutrition support. Kashin–Beck's disease is a chondrodystrophy also attributed to Se deficiency. Other manifestations of Se deficiency include muscle pain and tenderness, macrocytosis, and pigmentary changes to nails and hair. Increased erythrocyte fragility in response to oxidant stress has been reported in preterm infants with Se deficiency. Toxicity symptoms include nausea, vomiting, nail discoloration and brittleness, hair loss, fatigue, irritability and garlic odor in breath.

Blood and erythrocyte GSHP-x activity is thought to have a good correlation with whole blood Se. However, it is now known that enzyme activity may be influenced by exposure to oxygen and steroids and may not accurately reflect Se levels. Whole-blood Se or plasma Se reflects short-term changes in Se homeostasis; plasma selenium levels are a better indicator of Se stores in blood-transfusion recipients. Whole blood Se levels depend on dietary Se intake and range from 1.26–4.31 μmol L^{-1}. Plasma levels in North America average 1.27 μmol L^{-1}, whereas in new Zealand levels are only 0.51 μmol L^{-1}, reflecting the wide variation in different geographical areas (Shenkin, 2009).

There is generally no need to monitor Se levels in short-term TPN use. Long-term TPN requirement, use of higher pharmacological doses of Se, and specific disease states such as renal failure or short bowel syndrome warrant closer blood/plasma level monitoring.

16.7 Summary

The introduction of TPN created a scenario in which selenium-free nutrition was provided to malnourished patients with likely poor baseline trace element status. Combined with a paradoxically high Se utilization due to underlying hypermetabolic states and exaggerated gastrointestinal selenium excretion, this clinical situation created a model of selenium deficiency in humans for the first time that greatly enhanced our understanding of the trace element's role in human physiology.

Current guidelines suggest selenium supplementation in parenteral nutrition to be initiated in patients needing greater than 4 weeks of TPN support, with earlier supplementation considered for those at high risk for deficiency such as preterm infants and those in catabolic states or with documented high selenium losses. Although deficiency is rare in the era of routine micronutrient replacement, optimal dosing to not simply prevent deficiencies but to optimize human health and clinical status remains an exciting area to be explored.

Summary Points

- This chapter focuses on the use of selenium in parenteral nutrition.
- The first recommendations to add selenium to parenteral nutrition formulations were in 1984, after multiple reports described selenium deficiencies in patients receiving parenteral nutrition. The essentiality of selenium as a trace element was established with these reports.
- Parenteral dosage recommendations for selenium have evolved from these initial guidelines to reflect newer research on the functions of

selenium in the human body in health and disease conditions, for the adult, pediatric and neonatal population.

- Parenteral selenium is given inorganically as either sodium selenite or selenious acid. Organic selenomethionine can also be administered intravenously although this preparation is not widely available.
- Selenium losses may be increased with gastrointestinal fluid losses (*e.g.*, diarrhea, fistulas, and ostomy effluent) and dermal losses in burn injury. Endogenous Se utilization may be increased in clinical situations that promote hypermetabolism and oxidative stress since antioxidant activity is increased (*e.g.*, sepsis, burns, postsurgery). Parenteral dosing needs should be adjusted to reflect these changes in total body selenium balance.
- The use of pharmacological doses of parenteral Se in septic patients to help improve clinical outcomes is currently under investigation.
- Symptoms of selenium deficiency in parenterally nourished patients include skeletal myopathy, cardiomyopathy, encephalopathy, elevated creatine kinase, poor immunity, and whitened nail beds.
- Selenium should be added routinely to all parenteral nutrition formulations for patients requiring intravenous nutrition support for greater than 4 weeks (sooner if higher Se losses or utilization is suspected).
- Plasma Se and/or blood and erythrocyte GSHP-x activity should be monitored for patients on prolonged TPN or in patients with an underlying disease pathology that may alter Se requirement.

Key Facts

Key Facts of the Development of Parenteral Nutrition

The key facts of the development of parenteral nutrition, from its rudimentary stages of intravenous glucose infusion alone to its current formulation (Dudrick and Palesty, 2011). TPN: total parenteral nutrition; Se: selenium; GSHP-x: glutathione peroxidase.

1. 1896: First intravenous infusion of glucose in humans
2. 1918: Intravenous infusion of orange juice in a human to treat scurvy
3. 1935: Intravenous infusion of cottonseed oil emulsion in humans
4. 1939–1940: Use of intravenous protein (hydrolyzed casein and crystalline amino acids) in humans
5. 1944: First demonstration of the use of dextrose, insulin and protein *via* a peripheral vein in surgical patients; intravenous infusion of dextrose, amino acids, fat and saline in a surgical infant.
6. 1945: Intravenous infusion of dextrose, hydrolyzed protein, and fat *via* a peripheral vein in humans
7. 1961: Development of the first stable intravenous lipid emulsion
8. 1968: First report of long-term growth and development in an infant nourished by total parenteral nutrition; first report of an adult nourished *via* total parenteral nutrition for six months

9. 1979: Development of revised parenteral vitamin packages in response to reports of deficiencies.
10. 1979: First recommendations to add zinc, chromium, copper and manganese to parenteral nutrition formulations in response to reports of deficiencies.
11. 1979: First case report of Se deficiency in a woman maintained on selenium-free TPN for one month, manifested as muscular pain in conjunction with low plasma Se and erythrocyte GSHP-x activity. In this same year, the relationship between Se status and Keshan disease was first described.
12. 1984: First recommendations to add Se to total parenteral nutrition formulations in response to reports of deficiencies.
13. 1985–2014: Developments of alternative parenteral amino acid and lipid formulations, adjusted trace mineral and vitamin recommendations for TPN and exploration of the use of pharmacological doses of parenteral Se in critical illness for therapeutic use.

Key Facts of Parenteral Nutrition

The key facts of parenteral nutrition: its indications for use, benefits and potential complications. PN: parenteral nutrition.

1. PN is the intravenous infusion of nutrients, including dextrose, lipids, amino acids, fluid, electrolytes, vitamins and trace minerals.
2. When PN is infused *via* a central vein, it is referred to as "total parenteral nutrition" (TPN) since most of the nutrition needs can be met. When a peripheral vein is used, the term is "peripheral vein nutrition" or "partial parenteral nutrition" (PPN), since limits in osmolality, dextrose concentration and calcium provision usually do not allow for total nutrition needs to be met intravenously.
3. PN is used commonly in conditions that compromise the function of the gastrointestinal tract, such as short bowel syndrome, bowel obstruction, Crohn's disease, ulcerative colitis, and ischemic bowel disease.
4. While on PN, blood glucose, triglycerides, and electrolytes need to be monitored to evaluate adequacy and tolerance of the infusion.
5. Complications of prolonged PN use include line sepsis, thrombosis, and liver failure. Prolonged PN use can cause liver damage that may require transplantation if severe. Osteopenia is also a common complication of chronic TPN since calcium requirements are often limited to prevent precipitation with phosphate into an insoluble calcium phosphate salt.
6. Although life saving, PN is not considered the "ideal" mode to provide nutrition and the gastrointestinal tract should be used for feeds when medically feasible.

Definitions

Parenteral nutrition. The intravenous administration of nutrients, through a central or peripheral vein. This mode of nutrition is done to bypass the normal route of nutrition *via* the gastrointestinal tract when the intestinal tract is nonfunctional.

Keshan disease. A form of cardiomyopathy characterized by necrosis of heart tissue caused by the replacement of normal cardiac muscle tissue with fibrous tissue, leading to cardiogenic shock and/or congestive heart failure. This disease was first described in the Keshan region of China and was attributed to selenium deficiency.

Pharmaconutrient. A nutrient that exerts pharmacological effects in the body during disease processes and can be used therapeutically to improve clinical outcomes. Pharmaconutrients such as selenium, arginine, glutamine, fish oil, vitamin C, and vitamin E are among those studied for their impact on the underlying metabolic and immunologic processes in critical illness.

ARDS. ARDS is an abbreviation for acute respiratory distress syndrome, a form of pulmonary disease characterized by lung injury, inflammation, and respiratory failure. This potentially fatal condition may be caused by sepsis, burns, or other etiologies, and is usually treated with mechanical ventilation.

SIRS. SIRS is an abbreviation for systemic inflammatory response syndrome, a physiological state in which the body is in a state of inflammatory response frequently due to infection. SIRS can lead to multiorgan failure and is characterized by elevated heart and respiratory rates, altered body temperature and abnormal immune cell numbers.

Kashin-Beck's disease. A chronic degenerative joint condition characterized by osteoarthropathy and gradual deformation of the bones and joints. Kashin-Beck disease is endemic to central region of China, Siberia and Korea; although its pathogenesis is multifactorial, selenium deficiency is thought to contribute. The disease is named after two doctors, Nikolai Kashin and Evgeny Beck, who first described these bone abnormalities in patients of Russian origin.

Trace elements. A mineral that is needed by the body in minute amounts, in concentrations below 1 part per million.

Cardiomyopathy. A weakness of the cardiac muscle, leading to peripheral edema and shortness of breath, a condition that can ultimately result in cardiac failure.

Oxidative stress. A physiological imbalance where the systemic production of reactive oxygen species overwhelms the body's antioxidant ability, resulting in free-radical damage.

Selenoprotein-P. A glycoprotein synthesized mostly by the liver that serves as the main plasma selenoprotein. Its role is thought to involve extracellular antioxidant activity and Se transport.

List of Abbreviations

TPN	Total parenteral nutrition
GSHP-x	Glutathione peroxidase
GSH	Glutathione
Se	Selenium
CRRT	Continuous renal replacement therapy
ARDS	Acute respiratory distress syndrome
SIRS	Systemic inflammatory response syndrome
VLDL	Very low density lipoprotein

References

Darlow, B. A. and Austin, N., 2003. Selenium supplementation to prevent short-term morbidity in preterm neonates. *Cochrane Database of Systematic Reviews*. CD003312.

Dudrick, S. J. and Palesty, A., 2011. Historical highlights of the development of total parenteral nutrition. *Surgical Clinics of North America*. 91: 693–717.

Greene, H. L., Hambidge, K. M., Schanler, R. and Tsang, R. C., 1988. Guidelines for the use of vitamins, trace elements, calcium, magnesium, and phosphorus in infants and children receiving total parenteral nutrition: report of the Subcommittee on Pediatric Parenteral Nutrient Requirements from the Committee on Clinical Practice Issues of the American Society for Clinical Nutrition. *American Journal of Clinical Nutrition*. 48: 1324–1342.

Manzanares, W. and Hardy, G., 2009. Selenium supplementation in the critically ill: posology and pharmacokinetics. *Current Opinion in Clinical Nutrition and Metabolic Care*. 12: 273–280.

Shenkin, A., 2009. Selenium in intravenous nutrition. *Gastroenterology*. 137: S561–S569.

Terada, A., Yoshida, M., Seko, Y., Kobayashi, T., Yoshida, K., Nakada, M., Nakada, K., Echizen, H., Ogata, H. and Rikihisa, T., 1999. Active oxygen species generation and cellular damage by additives of parenteral preparations: selenium and sulfhydryl compounds. *Nutrition*. 15: 651–655.

Vanek, V. W., Borum, P., Buchman, A., Fessler, T. A., Howard, L., Jeejeebhoy, K., Kochevar, M., Shenkin, A. and Valentine, C., 2012. A.S.P.E.N. position paper: recommendations for changes in commercially available parenteral multivitamin and multitrace element products. *Nutrition in Clinical Practice*. 27: 440–491.

CHAPTER 17

Biofortification of Fruit and Vegetables with Selenium

MARIAN McKENZIE[a], ROSS LILL[a], STEPHEN TROLOVE[b], AND DAVID BRUMMELL*[a]

[a]Plant & Food Research, Private Bag 11600, Palmerston North 4442, New Zealand; [b]Plant & Food Research, Private Bag 1401, Havelock North 4157, New Zealand
*E-mail: david.brummell@plantandfood.co.nz

17.1 Introduction

Agronomic biofortification with Se refers to the addition of low concentrations of Se either to specialist fertilizers, to the water used to feed the crop, or to foliar sprays. Sodium selenate is most commonly used due to the poor transport of sodium selenite from the root to the shoot. In most countries, the primary intention is to supplement forage or pasture for animal feed, or to enhance the Se content in wheat grains and flour (Broadley *et al.*, 2006). However, in Finland where natural levels of Se in the soil are very low, more general soil Se supplementation has raised Se levels in a range of horticultural products (Broadley *et al.*, 2006). Producing a wider variety of crops containing adequate levels of Se may be an important way of raising dietary Se intakes, and thus increasing the Se status of populations of other countries with low-Se soils, such as New Zealand.

Fruit and vegetables generally contain low levels of dietary Se (on a fresh weight [FW] basis), relative to cereals, meat and fish. In both animals and

Food and Nutritional Components in Focus No. 9
Selenium: Chemistry, Analysis, Function and Effects
Edited by Victor R. Preedy

Published by the Royal Society of Chemistry, www.rsc.org

plants, a proportion of the Se assimilated as inorganic selenate is eventually incorporated into the selenoamino acids selenomethionine (SeMet) and selenocysteine (SeCys), which can be incorporated into proteins (Figure 17.1). However, while animals produce SeCys as the major selenocompound, plants mainly produce SeMet (Whanger, 2002). Free SeMet and proteins containing SeMet have a higher bioavailability than SeCys, and thus plants are better sources of dietary Se than meat or fish (Combs, 2001).

The recommended dietary allowance (RDA) for Se varies by country, but is usually 55–75 μg per day (Combs, 2001; Broadley *et al.*, 2006). For dietary biofortification, the serving size of a vegetable or fruit and the target Se content need to be carefully considered. If a Se content benefit is claimed, consumers need to obtain a supplementation of Se that is biologically significant (for example one-quarter to one-half of the RDA, or 20–35 μg). If it is assumed that consumers are already achieving the RDA of 55–75 μg per day and that some consumers may eat more than one serving in a day, this suggests an upper limit of approximately half the RDA per serving (35 μg). A daily intake of 200–300 μg Se has been reported to be effective in cancer prevention, but consumers should not exceed the tolerable upper intake level of 400 μg per day that can lead to selenosis, although this limit may be excessively conservative (Combs, 2001).

This review will cover the deliberate biofortification of the edible portions of fruit and vegetables with Se for human health benefits, rather than the effects of growing horticultural crops on soils possessing naturally high Se concentrations. For biofortification of wheat, various cereals and legumes, see other chapters in this volume.

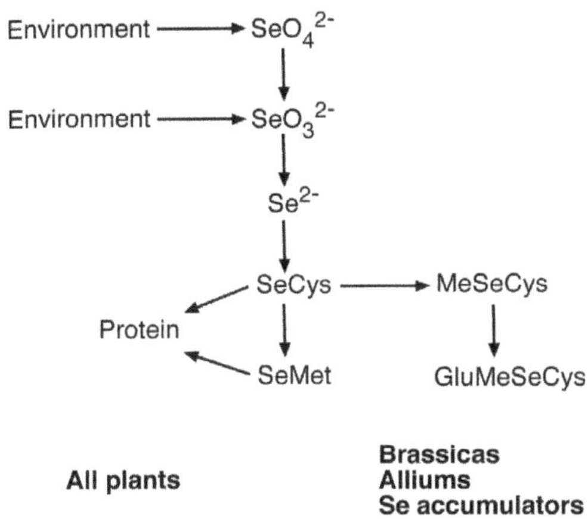

Figure 17.1 Simplified scheme showing the pathway for assimilation of inorganic selenate or selenite into organic selenocompounds in plants. Due to the presence of nonspecific methyltransferases most plants can form low amounts of methylselenocysteine (MeSeCys), but only some families can produce this methylated amino acid derivative at high levels. Some MeSeCys is converted to the dipeptide GluMeSeCys (γ-glutamyl methylselenocysteine). Figure modified from McKenzie *et al.* (2009).

17.2 Brassica Crops

Many members of the *Brassica* genus are natural accumulators of Se and are able to withstand relatively high tissue concentrations of Se (up to 1000 mg per kg dry weight [DW]) without apparent ill effects. The ability to take up and store excess Se results from the Brassicaceae's biochemistry, whereby SeCys is methylated to produce methylselenocysteine (MeSeCys) (Figure 17.1). Unlike SeCys and SeMet, MeSeCys is not incorporated into the plant's proteins and therefore does not contribute to Se toxicity symptoms. The biosynthesis of MeSeCys is catalyzed by a single enzyme, selenocysteine methyltransferase (SMT), and occurs in many members of the Brassicaceae (Table 17.1) and Alliaceae (Section 17.3). In broccoli, the *SMT* gene is inducible by selenate (Lyi *et al.*, 2005). The *SMT* gene from the Se-hyperaccumulator *Astragalus bisulcatus* has been used as a transgene to convert non-Se accumulating plant species into Se accumulators able to increase their Se content up to 6-fold that of controls, with up to 25% of the Se incorporated into MeSeCys (McKenzie *et al.*, 2009; Brummell *et al.*, 2011).

MeSeCys and its derivative, γ-glutamyl methylselenocysteine (GluMeSeCys), have greater bioefficacy in preventing the proliferation of cancer cells than other Se-containing compounds (Whanger, 2002). In the body, MeSeCys is efficiently converted to methylselenol, a nontoxic compound with demonstrated

Table 17.1 Reported contents of methylselenocysteine (MeSeCys) and Se in edible *Brassica* species following Se biofortification. Different species and treatment regimes result in considerable variation.

Crop	Se application method	MeSeCys content	Total Se content	References
Indian mustard (*B. juncea*)	Hydroponic culture of seedlings in 150 μM selenite for 7 days	5 μg per g FW	200 μg per g DW	LeDuc *et al.* (2006)
Broccoli (*B. oleracea* var. *Italica*)	Hydroponic culture of mature plants in 20 μM selenate	Up to 1.5 μmol per g FW in floret tissue	1200 μg per g DW in floret tissue	Lyi *et al.* (2005)
Broccoli (*B. oleracea* var. *Italica*)	Soil fertilization of mature plants with 100 mL 1.5 mM selenate twice weekly for six weeks	108 μg per g DW	489 μg per g DW	Avila *et al.* (2013)
Broccoli (*B. oleracea* var. *Italica*)	Hydroponic culture of sprouts in 25 μM selenate for 7 days	70 μg per g DW	77 μg per g DW	Avila *et al.* (2013)
Kale (*B. oleracea* Acephala Group)	Hydroponic culture of seedlings in up to 45 μg Se per mL for 15 days	34.7% of total Se content	Up to 400 μg per g DW	Maneetong *et al.* (2013)

anticancer properties including the induction of apoptosis and the reduction of angiogenesis in cancer tissues. Studies conducted on animal cancer models found that animals fed with Se-enriched broccoli had significantly less development of precancers compared with control animals fed nonenriched broccoli and inorganic Se supplements (Finley *et al.*, 2001). This strongly suggests the importance of this organic form of Se in cancer prevention. Therefore, the majority of the research on biofortification of vegetable crops has focused on plants of this genus.

Three methods of biofortification are commonly used to enrich the Se content of the brassicas: hydroponic culture, soil fertilization and foliar spraying. Hydroponic culture has been used most often, and generates greater Se content than the other two methods. This is presumably due to the constant exposure of the root system to the Se-enriched nutrient solution, and to the lack of Se–soil interactions that reduce the bioavailability of added Se. Se concentrations up to 2000 μg per g DW have been reported in the shoots of Indian mustard, rapid-cycling brassica and broccoli plants cultured in 26, 48 and 20 μM sodium selenate, respectively, for approximately two weeks (Kopsell and Randle, 1999; Ximénez-Embún *et al.*, 2004; Ramos *et al.*, 2011b). In general, the quantity of Se taken up decreases with shorter culture time, and is not necessarily compensated for by increasing Se concentration (Lyi *et al.*, 2005).

The Se source also affects Se accumulation. Brassicas exposed to sodium selenite have a much reduced Se content compared with those fed sodium selenate (Ximénez-Embún *et al.*, 2004; Lyi *et al.*, 2005). Root uptake of selenate occurs by high-affinity sulfate transporters, and is more efficient than uptake of selenite, which is believed to occur by phosphate transporters. Furthermore, selenite is converted to organoselenium compounds in the root, while selenate is transported *via* the xylem to the shoot where it is further metabolized and distributed rapidly throughout the plant (White and Broadley, 2009).

While hydroponic culture allows tight control of Se content, it is comparatively expensive commercially and can result in large volumes of selenized waste solution requiring careful and costly disposal. Consequently, research focused on the commercial production of Se-enriched brassicas has tended to utilize selenate application by enriched fertilizer solutions, either to solid plant growth media such as soil or sand, or in the form of foliar sprays.

For broccoli, the application of Se to the soil is generally achieved by regular addition of sodium selenate solution around the base of the stem, commencing at the appearance of immature head tissue at the apical meristem and continuing during head expansion. Application of 10–20 mL sodium selenate solutions of up to 5.2 mM every second day over 14 days resulted in the production of head material containing Se up to 880 μg per g DW (Lee *et al.*, 2005). Assuming that broccoli is 10% DW, a single serve (100 g FW) of this material would allow the ingestion of a Se dose well exceeding the tolerable upper intake level of 400 μg per day, and could be highly toxic. Nevertheless, these studies indicate such enrichment methods are more than adequate for the Se-biofortification of broccoli. We have shown that the application of much lower concentrations of sodium selenate results in the production of head material containing 50 μg of Se per 100 g serve, providing a sustainable increased dose of Se to the daily diet.

Foliar spraying of Se to young broccoli grown in the field has also been reported (Hsu *et al.*, 2011). A single foliar application of 2 mg sodium selenate per plant at the early stages of head development resulted in the production of floret material containing 10 μg Se per g DW, with about 25% of the Se found in the head tissue. This application would be appropriate for human ingestion as part of the diet, providing 100 μg Se in a 100 g FW serving of broccoli.

In addition to the ability to produce MeSeCys, members of the Brassicaceae are well known for their capacity to produce glucosinolates, sulfur (S)-based compounds believed to be involved in protection of the plant from herbivory and bacterial and fungal disease. The breakdown products of the glucosinolates have powerful anticarcinogenic effects and play an important role in human health. Therefore, there is interest in producing brassicas containing high levels of both glucosinolates and beneficial Se-based compounds. However, S and Se are chemically similar and compete for uptake and metabolism in plants, with many S-assimilatory enzymes being unable to distinguish between the two elements. Therefore, it is possible that Se-enrichment could affect S uptake and metabolism. Indeed, Se enrichment of rapid-cycling brassica and broccoli resulted in significant reductions in glucosinolate production in leaves and floret material (Charron *et al.*, 2001; Robbins *et al.*, 2005). However, the amount of Se applied was far greater than would be acceptable for a crop aimed at human consumption. When Se enrichment was provided at appropriate levels, no changes in glucosinolate concentrations have been observed even though the MeSeCys content increased (Hsu *et al.*, 2011).

Se enrichment of brassicas also resulted in the production of several selenoglucosinolates: glucoselenoiberverin, glucoselenoerucin and glucoselenoberteronin (Matich *et al.*, 2012). In these compounds, the S atom usually donated by methionine during glucosinolate biosynthesis was replaced by a Se atom, presumably due to incorporation of SeMet. These novel compounds and their breakdown products have been detected in broccoli, cauliflower and forage rape (Matich *et al.*, 2012), and in broccoli their breakdown products were found at concentrations up to 1300% of their S analogs. The bioefficacy of these novel compounds is currently unknown.

Despite abundant research showing that Se enrichment of edible brassicas can be achieved at concentrations suitable for human consumption and without decreasing concentrations of other health-related compounds, such as glucosinolates, there appear to be no examples of fresh Se-biofortified brassica products currently on the market. However, several nutraceutical products containing extracts from Se-enriched broccoli and broccoli sprouts are available. This could be due to the inherent difficulties of producing fresh products with a Se content guaranteed to be within a particular range. It is also possible that the generation of Se-containing volatiles (Matich *et al.*, 2009), which are unpleasantly odoriferous, could preclude the commercial production of packaged brassica goods, particularly salad greens that spend several days sealed in a bag and could produce relatively large amounts of volatile compounds.

17.3 Alliums

The alliums provide a range of flavorsome vegetables rich in S compounds. They are also notable as Se accumulators and have been a focus for Se-enrichment studies. There are considerable differences in the extent of Se accumulation amongst the allium species (Table 17.2). Garlic, leek and chives accumulate high concentrations of Se in response to enrichment of the growing medium, whereas onion bulbs and green onion accumulate only one-tenth as much.

Se enrichment can be achieved by foliar spray containing sodium selenate or sodium selenite (Kápolna *et al.*, 2012). The spray concentration must be selected to avoid damage to the foliage and associated plant stunting, which for a selenate spray occurred at 100 μg Se per L for onions. Notably, selenite could be sprayed at this concentration without plant damage. Multiple sprays are likely to be required to achieve useful enrichment, and a surfactant may be needed to penetrate the waxy cuticle of crop leaves. Bulb enrichment levels reported were modest, at 4 μg Se per g DW (Kápolna *et al.*, 2012).

Additions to the growing medium, both soil and hydroponic nutrient solution, provide for a more substantial accumulation of Se in the crop. In onions, sodium selenate at 2 mg per L of nutrient solution resulted in enriched bulbs with 71 μg Se per g DW (Kopsell and Randle, 1997b). Leaves were considerably

Table 17.2 Se accumulation in Se-biofortified *Allium* crops. The different treatments and species studied mean that direct comparisons are not possible, but considerable variation is observed.

Crop	Enrichment	Selenium content (μg Se per g DW)	References
Garlic bulb (*A. sativa*)	Soil, selenate, 8.4 μg Se per g soil	850	Larsen *et al.* (2006)
Leek (*A. ampeloprasum*)	Soil, selenate, 3.8 μg Se per g soil	982	Lavu *et al.* (2012)
Chives (*A. schoenoprasum*)	Pots watered 7 times over 14 days with a total of 2000 μg Se as selenate	613	Kápolna and Fodor (2007)
Onion bulb (*A. cepa*)	Hydroponic, selenate, 836 μg Se per L nutrient solution	85	Kopsell and Randle (1997a)
Green onion (*A. fistulosum*)	Pots watered 7 times over 14 days with a total of 2000 μg Se as selenate	66	Kápolna and Fodor (2007)

more enriched, with 208 µg Se per g DW. Adding Se salts to potting soil resulted in high concentrations of Se in leek and garlic (Table 17.2).

Sodium selenate is more readily taken up by alliums than is sodium selenite, but it also appears to be more toxic (Kápolna *et al.*, 2012). In a hydroponic system, 5 mg Se per L resulted in reduced leaf and root growth in onions (Wróbel *et al.*, 2004). Leaves contained 601 and 154 µg Se per g DW, and bulbs 51 and 16 µg Se per g DW for selenate and selenite, respectively. Green onions and chives grown in a hydroponic system showed stronger enrichment with selenate than selenite, with SeMet showing equivalent enrichment to selenite (Kápolna and Fodor, 2007).

Se is converted by alliums from the mineral form into a variety of organic compounds. Selenate is most readily taken up, but is not completely converted, with 30% (in chives) and 60% (in green onion) remaining as selenate (Kápolna and Fodor, 2007). Selenite, however, appears to be completely converted to organic forms. In onion bulbs, foliar spray with selenate resulted in greater amounts of inorganic Se (16% of total Se as selenate) than in those sprayed with selenite (1% as selenate), with no selenite detected (Kápolna *et al.*, 2012). The principal organic form was GluMeSeCys (80–90% of total Se) with lesser amounts of SeMet (5%) and MeSeCys (3%) for both selenate and selenite foliar sprays (Kápolna *et al.*, 2012). There is considerable variation in content of both GluMeSeCys and the selenoamino acids between and within species (Table 17.3; Pyrzynska, 2009).

17.4 Leafy Vegetables

Se-enriched leafy greens have been produced using a range of techniques: hydroponics (Ramos *et al.*, 2011a), aeroponics (Mazej *et al.*, 2008), application of foliar sprays (Zhang *et al.*, 2004), or soil applications (Mechora *et al.*, 2011). Zhang *et al.* (2004) sprayed a 200–400 µg per g solution of sodium selenite onto lettuce, spinach or Chinese cabbage seven days before harvest to biofortify with Se, and produced a Chinese cabbage vegetable juice drink containing 3 µg Se per g. The production of Se-biofortified fresh leafy vegetables became a commercial reality in February 2010, with Marks & Spencer (UK) selling spinach that was claimed to contain 10 times more Se than standard

Table 17.3 Main Se-containing compounds reported for Se-biofortified *Allium* crops (as percentage of total Se, drawn from Pyrzynska, 2009). GluMeSeCys, γ-glutamyl methylselenocysteine; MeSeCys, methylselenocysteine; SeCys, selenocysteine; SeMet, selenomethionine.

Crop	GluMeSeCys	MeSeCys	SeCys	SeMet
Garlic (*A. sativa*)	8–85	3–60	0.5–6	1–18
Chives (*A. schoenoprasum*)	—	3–48	2–42	0–5
Onion (*A. cepa*)	0–63	1–5	0.1–1	0.2–10
Shallot (*A. ascalonicum*)	66	5	—	—
Ramp (*A. tricoccum*)	1.5–3	34–44	—	0–21

spinach (http://www.express.co.uk/news/uk/156791/Super-spinach-is-set-to-go-on-sale). Se accumulation in leafy greens differed with genotype (*e.g.*, different accessions of lettuce showed a two-fold difference in Se accumulation; Ramos *et al.*, 2011a), which means that production systems would have to be adjusted to each cultivar since precise control of Se content is required.

Biofortification with Se has significant effects on plant metabolism. The addition of Se to leafy vegetables improved the storage life in most seasons, presumably due to the antioxidant properties of Se. The addition of 500 and 1000 µg Se per L (as selenate) to lettuce and chicory decreased ethylene production and improved the retention of chlorophyll during storage (Malorgio *et al.*, 2009). The supply of low concentrations of Se to plants also increased yield in some instances (Malorgio *et al.*, 2009; Rios *et al.*, 2009; Ramos *et al.*, 2011a). This growth increase was only exhibited by some cultivars (Ramos *et al.*, 2011a), and was thought to be due to Se enhancing the antioxidant system at low concentrations (5–20 µM), but acting as a pro-oxidant at higher concentrations (Rios *et al.*, 2009). Se addition has also been shown to increase nitrogen metabolism in lettuce (Rios *et al.*, 2010).

17.5 Shoots and Sprouts

Sprouts and microgreens are well suited to Se biofortification because they can be produced hydroponically in a closed system. This eliminates the need for the application of a large amount of Se fertilizer in the field, which may cause environmental pollution. If carefully managed, sprouts can be produced in a system where almost all of the Se is taken up by the sprouts, leaving very little in the waste water (Trolove *et al.*, unpublished). Sprouts also have the benefit that all of the plant is eaten, particularly if the husks can be removed before Se addition, which eliminates the need to dispose of seleniferous plant material.

The uptake of Se by sprouting seeds differs among plant species. In general, Fabaceae take up less Se than other plant families, while Brassicaceae and Apiaceae take up more (Sugihara *et al.*, 2004; Frias *et al.*, 2010). Different plant species also show a range in tolerance to Se toxicity: sprout species showing increasing tolerance to selenate were broccoli < mung bean < onion (Arscott and Goldman, 2012). Sprouts have the ability to accumulate high concentrations of Se (Arscott and Goldman, 2012), which could potentially be toxic to humans if large amounts were eaten; therefore care must be taken when producing sprouts for human consumption.

Sprouts are a rich source of antioxidants, and biofortification with Se further increases their antioxidant capacity (Frias *et al.*, 2010). Much of the interest in producing Se-enriched plants lies in their ability to incorporate Se into organic compounds that are powerful antioxidants, or that may have other health benefits. The supply of Se to sprouts may have quite different effects on the production of these organic Se compounds, compared with biofortification of mature plants. For example, selenate addition exerted a minimal effect on glucosinolate concentration in sprouts, but reduced glucosinolate concentration of broccoli florets by approximately 36% (Avila *et al.*, 2013). This is because glucosinolates are pre-existing in the seeds, and are not made *de novo* by the sprouting tissues. A higher percentage conversion of selenate to MeSeCys was shown by broccoli sprouts (~40%) than by florets (~10%), although since total Se concentration was greater in the florets there was more MeSeCys per g DW than in sprouts. In broccoli sprouts, more Se was taken up with selenate than with selenite, although both resulted in sprouts with the same MeSeCys concentration. This finding differs from mature broccoli, where selenate was more effective at stimulating MeSeCys production (Lyi *et al.*, 2005).

17.6 Fruit

Fruit are a more difficult vehicle than vegetables for dietary Se supplementation due to the length of time of fruit development, the size and architecture of the plants, the high concentrations of Se that can accumulate in the nonedible plant parts and variability between individual fruit due to their location on the plant. This can give rise to quality control concerns, since not every individual fruit can be tested for Se content, and fruit must meet claimed minimum dietary Se levels yet not approach concentrations that could exceed the tolerable upper intake level if consumers ate several fruit.

Tomato is the most studied fruit species. In a commercial setting, Se can be applied as sodium selenate either continuously at low concentrations in the watering medium from the seedling stage onwards (*e.g.*, 1.26 mM, Pezzarossa *et al.*, 1999), or as a single addition of higher concentration to the drip irrigation system (7.5 g per ha, Golubkina *et al.*, 2003). The application regime needs to be carefully optimized, since Se uptake varies with cultivar and is dependent on the S supply to the plant (Pezzarossa *et al.*, 1999). Foliar sprays applied prior to fruiting are also efficacious (Golubkina *et al.*, 2003; Schiavon *et al.*, 2013). Supplementation of Se content in tomato fruit to dietary levels (20–40 μg Se per average fruit of 100 g FW) is easily achievable (Figure 17.2). Indeed, Marks & Spencer (UK) marketed a Se-biofortified tomato in 2011 (http://health.marksandspencer.com/our-health-ranges/hot-topics-aug-11).

It has been reported that Se concentrations in tomato skin are several fold higher than in the flesh, although Se in the skin may be less soluble and less bioavailable (Golubkina *et al.*, 2003). Foliar spray with selenate altered the profiles of several compounds with potential nutraceutical value, with increases in chlorogenic acid in the flesh and kaempferol, quercetin and naringenin chalcone in the skin, but decreases in several phenolic acids (Schiavon *et al.*, 2013). However, the concentrations of Se accumulating in leaves of tomato are 10- to >100-fold greater than those in the fruit (Carvalho *et al.*, 2003; Brummell *et al.*, 2011; Schiavon *et al.*, 2013), creating a potential problem with regards to the disposal of the Se-contaminated nonedible parts. This inefficient use of the applied Se suggests that biofortification of vegetables, where most of the plant is edible (such as spinach, lettuce or sprouts), is a more economical practice than fruit.

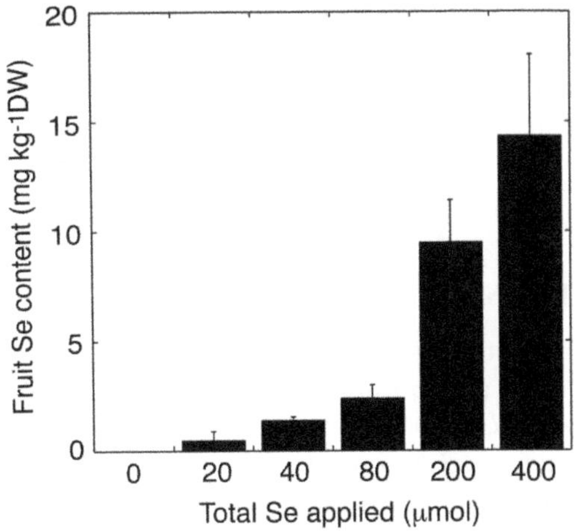

Figure 17.2 Se content of ripe tomato fruit at harvest after watering the plants with sodium selenate solution at various concentrations. Sodium selenate (20 mL per application) was applied to the soil just before watering twice per week for 10 weeks, beginning at fruit set, to give the total amount provided as shown. Three replicate fruit from each of trusses 1, 2 and 5 were collected, the fruit pericarp was freeze dried, powdered and assayed for Se content (mean ± SD, $n = 9$). A fruit Se content of 3 mg per kg DW equates to approximately 25 μg Se per 100 g FW of fruit pericarp.

There are few studies on other fruit species. Strawberry plants grown in soil supplemented with inorganic Se produced fruit with a Se content twice that of control fruit but, as with tomato, excessive concentrations of Se were present in the nonedible portions of the plant (Carvalho *et al.*, 2003). For fruit trees, both foliar and fruit application of sodium selenate have been trialed (Pezzarossa *et al.*, 2011). In pear, spraying directly onto the fruit was much more effective than spraying onto the leaves, whereas in peach the two methods were equivalent. Levels achieved were approximately 8 μg Se per pear fruit and 2 μg Se per peach fruit. In addition to a nutritional benefit, Se-biofortified fruit of both species showed delayed softening, thus prolonging shelf life.

One way to make Se consumption safer is to modify the form in which it is ingested. The organic Se compound MeSeCys not only lacks the potential toxicity properties of inorganic Se (since MeSeCys can be excreted in breath and urine), but it appears to possess powerful anticancer properties (Finley *et al.*, 2001). MeSeCys is produced at substantial levels in brassicas and alliums grown in seleniferous soils (see Sections 17.2 and 17.3), as well as in Se-hyperaccumulators that safely store excess Se as MeSeCys. All these plants possess the SMT enzyme, which is lacking in most horticultural species. Overexpression in tomato of an *Astragalus bisulcatus SMT* transgene resulted in a substantial conversion of inorganic Se to MeSeCys in the fruit (up to ~25% after accounting for losses during analysis), yet despite this, high concentrations of unconverted inorganic Se remained (Brummell *et al.*, 2011). This was due to the relatively high levels of Se that needed to be provided to the plants in order to divert Se into the organic pathway. Further work is needed to increase the conversion rate of inorganic Se to MeSeCys in such plants. MeSeCys is stable to heat, and survived processing of the fruit into juice (Brummell *et al.*, 2011). The Se content of tomato juice can be accurately controlled by mixing juice from high Se-containing fruit with juice produced from low-Se fruit.

17.7 Root Crops

There are preliminary studies on Se biofortification for several root crops, including radish, carrot and potato. Se supplementation of soil for growing radishes (Carvalho *et al.*, 2003) and foliar spray of carrots (Kápolna *et al.*, 2009) resulted in increased Se contents in the edible portions, but much greater Se accumulation in nonedible parts (leaves, stems). In carrots, foliar sprays of 100 µg Se per mL as selenate or selenite resulted in an accumulation of Se in the root at 2 µg Se per g DW (equivalent to ~24 µg Se per 100 g FW serving). Speciation showed that roots contained inorganic Se, SeMet and GluMeSeCys. However, at a spray concentration of 100 µg Se per mL, up to 80 µg Se per g DW accumulated in the leaves, where damage was visible as necrotic spots.

Application of sodium selenate to potato plants grown in pots of sand resulted in Se-biofortified tubers, although Se accumulation was substantially greater in the leaves, roots and stolons (Turakainen *et al.*, 2004). Of the Se present in the tubers, 49–65% was incorporated into protein (Turakainen *et al.*, 2006). Approximately 30% was in each of soluble protein, insoluble residue (proteins and polysaccharides) and as an inorganic form (probably selenate), and about 10% was as free selenoamino acids. These proportions remained approximately constant during storage of tubers up to 12 months, during which Se losses were minimal. Treatment of plants 8 weeks before harvest with a foliar spray of sodium selenate (10 mg Se per L) also resulted in Se-enriched tubers, with approximately 30% of the Se in tubers being incorporated into protein (Cuderman *et al.*, 2008).

Se-enriched potatoes have been released for sale in several countries, including Britain, Italy and Ireland. The "Selena" Se-enriched potatoes of Peter Keogh and Sons Ltd. sold in Ireland contained ~35 µg Se per 250 g serving (http://www.potatogrower.com/2009/04/selenium-enriched-potatoes-released-in-ireland/), which is approximately half of the RDA.

17.8 Mushrooms

Mushrooms are recognized as Se accumulators, although the efficiency of accumulation varies with species. Biofortification of cultivated species with Se has attracted attention not only to the major food crops but also to species grown for medicinal purposes.

Enrichment of the widely grown button mushroom (*Agaricus bisporus*) by addition of selenite through the irrigation water provided a Se concentration in the crop of 53 µg per g DW (Spolar *et al.*, 1999). Inclusion of this in the diet fed to rats significantly reduced chemically induced precancerous adducts in mammary tissue.

Oyster mushroom (*Pleurotus ostreatus*) is cropped on waste agricultural materials in Brazil, including coffee husks (da Silva *et al.*, 2012). Addition of selenite to coffee husks at 3.2 μg Se per g husks yielded mushrooms containing 58 μg Se per g DW. Mushrooms grown at the highest enrichment rate (102 μg Se per g husks) accumulated 858 μg Se per g DW. However, adverse growth effects were reported above 12.8 μg Se per g husks with smaller caps and larger stipes, and higher levels of enrichment than this resulted in mushrooms with an unpleasant smell. Enrichment of another species of oyster mushroom (*P. florida*) was achieved by growing the mushroom on wheat straw from seleniferous soils in Punjab, India (Bhatia *et al.*, 2013). When grown on straw containing 28 μg Se per g DW, the mushroom crop was enriched to 141 μg Se per g DW. The recommended daily intake (55 μg Se) would be gained from 5–7 g FW of these mushrooms.

Shiitake mushrooms (*Lentinula edodes*) have been enriched in Se by including selenite in the cold-shock water used to initiate the fruiting bodies after spawn run (Nunes *et al.*, 2012). Mushroom growth was affected at higher selenite concentrations, delaying veil formation. At selenite concentrations in the cold-shock water of 0.96 mM and above, no caps formed. Accumulation of Se at 0.64 mM selenite was 170 μg Se per g DW of mushroom.

SeMet was identified as the major organic form of Se in mushrooms, either free or associated with long chain molecules (Ogra *et al.*, 2004; Bhatia *et al.*, 2013).

17.9 Conclusions and Future Prospects

Research has shown that Se biofortification of various fruit and vegetable species is easily achievable to within recommended dietary levels. However, numerous technical and safety issues must be overcome. Se as selenate is a relatively expensive chemical that is toxic at high levels, and exposure of workers involved in the treatment of the plants must be avoided. Other serious issues include the potential accumulation of Se in soils used for repeated production of Se-enriched crops, the safe disposal of Se-contaminated water from hydroponics facilities, and the safe disposal of nonedible Se-containing plant parts remaining after harvest.

Only a few Se-enriched commercial products have so far been released onto the market, notably Se-enriched spinach, potato and tomato. Consumer awareness of the benefits of increased dietary Se consumption and the price premiums charged will determine if they become successful. For fruit, the inefficient use of the applied Se due to excessive accumulation of Se into nonedible parts suggests that biofortification of vegetables, where most of the plant is edible (such as spinach or sprouts), is a more economical practice. With these, if they are sold as bagged products the accumulation of potentially unpleasant Se volatiles might be a concern.

For the future, a great attraction of producing second-generation Se-enriched plants is due to the ability of particular plant species to incorporate

Se into organic compounds that are powerful antioxidants or anticarcinogens, or have other health benefits. However, plant metabolism is both complex and cultivar specific, and the effects of Se biofortification on the accumulation of other beneficial compounds, such as glucosinolates, must be investigated and specific treatment protocols designed. Studies aimed at improving the content of MeSeCys in particular crops have shown that the conversion of inorganic Se to MeSeCys is currently not efficient enough, with concentrations of unconverted selenate remaining too high.

The greatest concern of course is product safety. Se is beneficial at low levels, but is toxic and can be lethal at high doses. Both agricultural workers and consumers must be protected by the usc of rigorous application protocols and extensive product testing. Due to the inherently higher variability in Se content between individual fruit and vegetables compared with the more easily controlled Se enrichment of wheat flour, strict quality control is essential before a product enters the market.

Summary Points

- Se is essential for human health, and may play an additional role in preventing certain cancers.
- Plants are the source of Se in the diet, whether eaten directly by humans or through the animal food chain.
- A high proportion of many populations do not obtain sufficient Se in their diet, and it is important to increase the variety of foods available that are sources of Se.
- Plant crops are an attractive dietary source due to the high bioavailability of Se from plants, but fruit and vegetables generally contain low levels of Se unless biofortified.
- Plants produce organic forms of Se, which are a good source of Se for the human body.
- Research has shown that Se supplementation to dietary levels is possible in a range of vegetable and fruit species.
- Se biofortification of plants must be optimized carefully, and depends on the sulfur status of the plant, cultivar, maturity at application, time and frequency of application, and the tissue to be harvested.
- Se biofortification can cause positive or negative changes in the profiles of other compounds, some of which may also be important for human health (phenolic acids, flavonoids, and glucosinolates).
- In many plants, Se supplementation can result in much higher levels of Se in the nonedible parts (leaves and shoots) than in the edible fruit or tubers.
- Few Se-supplemented products have been produced so far, but examples of Se-supplemented commercial products include spinach, potato and tomato.
- Due to the potentially hazardous consequences of overexposure to Se, safe protocols for agricultural workers and extensive product testing to protect consumers are essential.

Key Facts of Human Se Nutrition

1. Se is a trace element essential for human health.
2. Se deficiency occurs in many countries due to low Se levels in the soil and thus in the crops grown in them (Finland, New Zealand, parts of China, Eastern Europe, Russia, Africa, *etc.*).
3. Symptoms of Se deficiency include juvenile cardiomyopathy (Keshan disease), chondrodystrophy (Kaschin–Beck disease), and reduced functionality of the immune system, thyroid and liver.
4. Some enzymes in the human body are selenoproteins and have an absolute requirement for Se; a dietary intake of ~40 µg Se per day is necessary to maintain their function.
5. Higher Se intake of 200–300 µg per day has been shown to provide protection against many cancers.
6. Some organic forms of Se, particularly methylselenocysteine (MeSeCys), appear to have powerful anticancer properties.
7. Cereals, meat, fish, shellfish and Brazil nuts are good dietary sources of Se.
8. Fruit and vegetables are poor sources of Se unless grown in supplemented soils, but Se from plants is more bioavailable than that from meat or fish.
9. Excessive intake of Se causes toxicity symptoms (selenosis).

Key Facts of Se Metabolism in Plants

1. Se in the soil is usually present as selenate or selenite.
2. Se uptake and assimilation occurs by enzymes of the sulfur (S) assimilation pathway, since most plant enzymes cannot discriminate between compounds containing S and Se.
3. Se is not an essential micronutrient for plants, and at high concentrations can interfere with S metabolism.
4. Selenate is translocated well from the roots to the shoots, whereas selenite is not.
5. Selenate is reduced to selenite before entering the organic pathway.
6. Se can replace S in the normally sulfur-containing amino acids cysteine and methionine, to produce the selenoamino acids selenocysteine and selenomethionine.
7. Selenoamino acids can be incorporated into proteins, where they may detrimentally affect enzyme stability or function.
8. Plants generally accumulate a proportion of Se as selenomethionine, which has high bioavailability to humans.
9. Plants of the Brassicaceae (broccoli, cabbage, *etc.*) and alliums (onions, garlic), as well as several nonedible species of plants adapted to live on highly seleniferous soils (Se hyperaccumulators), possess the enzyme selenocysteine methyltransferase (SMT).
10. The SMT enzyme specifically methylates SeCys to MeSeCys, which prevents it being incorporated into proteins.

11. Accumulating MeSeCys rather than SeCys is a way for these plants to safely store excess Se in a form that does not damage enzymes or metabolism.

Definitions of Words and Terms

Alliums. Members of the *Allium* genus, such as onion (*Allium cepa*), garlic (*A. sativa*) and leek (*A. ampeloprasum*).

Biofortification. The purposeful application of Se to plants to enhance the Se content of edible parts for human nutrition, as opposed to growing plants in soils naturally high in Se.

Brassicas. Members of the Brassicaceae (also known as Cruciferae) family. Species include broccoli (*Brassica oleracea* var. *Italica*), cabbage (*Brassica oleracea*) and rape (*Brassica napus*).

Glucosinolates. S- and N-based glucosides generated from amino acids in a multistep enzymatic process. Produced naturally by the brassicas. Breakdown products of these compounds have been shown to have anti-cancer properties in humans.

MeSeCys. Methylselenocysteine, sometimes called Se-methylselenocysteine. The methylated form of SeCys. Produced naturally by brassicas, alliums and Se hyperaccumulators when grown in the presence of Se (as selenate, selenite or certain organoselenium compounds), but at very low amounts in other crops.

Recommended dietary allowance (RDA, also called DRI, dietary reference intake). The amount of an essential nutrient required in the diet to maintain good health. For Se varies by country; *e.g.*, for adults 55 μg per day for both women and men (USA), or 60 μg per day for women and 75 μg per day for men (UK).

Se hyperaccumulators. A type of plant that can survive on highly seleniferous soils. Since the plants cannot prevent the uptake of Se, they convert and safely store it at high concentrations as methylselenocysteine, an amino acid derivative that is not incorporated into proteins. Examples include *Astragalus bisulcatus* and *Stanleya pinnata*.

Selenoamino acids. Se analogs of two normally sulfur-containing amino acids, selenomethionine (SeMet) and selenocysteine (SeCys). These can either be incorporated into proteins, or methylated as a way of reducing their incorporation into proteins.

Selenosis. Poisoning due to excessive intake of Se. Symptoms include hair and nail loss, skin lesions, damage to the nervous, immune and reproductive systems, convulsions, paralysis, gastrointestinal and circulatory disturbances.

List of Abbreviations

DW	Dry weight
GluMeSeCys	γ-Glutamyl methylselenocysteine
FW	Fresh weight

MeSeCys	Methylselenocysteine
RDA	Recommended dietary allowance
SeCys	Selenocysteine
SeMet	Selenomethionine
SMT	Selenocysteine methyltransferase

Acknowledgements

We thank Matthew Norris and Jovyn Ng for helpful comments on the manuscript.

References

Arscott, S. and Goldman, I., 2012. Biomass effects and selenium accumulation in sprouts of three vegetable species grown in selenium-enriched conditions. *HortScience*. 47: 497–502.

Avila, F. W., Faquin, V., Yang, Y., Ramos, S. J., Guilherme, L. R. G., Thannhauser, T. W. and Li, L., 2013. Assessment of the anticancer compounds Se-methylselenocysteine and glucosinolates in Se-biofortified broccoli (*Brassica oleracea* L. var. *italica*) sprouts and florets. *Journal of Agricultural and Food Chemistry*. 61: 6216–6223.

Bhatia, P., Aureli, F., D'Amato, M., Prakash, R., Cameotra, S. S., Nagaraja, T. P. and Cubadda, F., 2013. Selenium bioaccessibility and speciation in biofortified *Pleurotus* mushrooms grown on selenium-rich agricultural residues. *Food Chemistry*. 140: 225–230.

Broadley, M. R., White, P. J., Bryson, R. J., Meacham, M. C., Bowen, H. C., Johnson, S. E., Hawkesford, M. J., McGrath, S. P., Zhao, F. J., Breward, N., Harriman, M. and Tucker, M., 2006. Biofortification of UK food crops with selenium. *Proceedings of the Nutrition Society*. 65: 169–181.

Brummell, D. A., Watson, L. M., Pathirana, R., Joyce, N. I., West, P. J., Hunter, D. A. and McKenzie, M. J., 2011. Biofortification of tomato (*Solanum lycopersicum*) fruit with the anticancer compound methylselenocysteine using a selenocysteine methyltransferase from a selenium hyperaccumulator. *Journal of Agricultural and Food Chemistry*. 59: 10987–10994.

Carvalho, K. M., Gallardo-Williams, M. T., Benson, R. F. and Martin, D. F., 2003. Effects of selenium supplementation on four agricultural crops. *Journal of Agricultural and Food Chemistry*. 51: 704–709.

Charron, C. S., Kopsell, D. A., Randle, W. M. and Sams, C. E., 2001. Sodium selenate fertilisation increases selenium accumulation and decreases glucosinolate concentration in rapid-cycling *Brassica oleracea*. *Journal of the Science of Food and Agriculture*. 81: 962–966.

Combs, G. F., 2001. Selenium in global food systems. *British Journal of Nutrition*. 85: 517–547.

Cuderman, P., Kreft, I., Germ, M., Kovačevič, M. and Stibilj, V., 2008. Selenium species in selenium-enriched and drought-exposed potatoes. *Journal of Agricultural and Food Chemistry*. 56: 9114–9120.

Finley, J. W., Ip, C., Lisk, D. J., Davis, C. D., Hintze, K. J. and Whanger, P. D., 2001. Cancer-protective properties of high-selenium broccoli. *Journal of Agricultural and Food Chemistry*. 49: 2679–2683.

Frias, J., Gulewicz, P., Martinez-Villaluenga, C., Peñas, E., Piskula, M. K., Kozlowska, H., Ciska, E., Gulewicz, K. and Vidal-Valverde, C., 2010. Changes in nutritional value and cytotoxicity of garden cress germinated with different selenium solutions. *Journal of Agriculture and Food Chemistry*. 58: 2331–2336.

Golubkina, N. A., Zhumaev, A. A. and Dem'yanova-Roi, G. B., 2003. Pattern of selenium distribution in tomato *Lycopersicum esculentum* Mill. *Biology Bulletin*. 30: 468–471.

Hsu, F. C., Wirtz, M., Heppel, S. C., Bogs, J., Kramer, U., Khan, M. S., Bub, A., Hell, R. and Rausch, T., 2011. Generation of Se-fortified broccoli as functional food: impact of Se fertilization on S metabolism. *Plant Cell and Environment*. 34: 192–207.

Kápolna, E. and Fodor, P., 2007. Bioavailability of selenium from selenium-enriched green onions (*Allium fistulosum*) and chives (*Allium schoenoprasum*) after '*in vitro*' gastrointestinal digestion. *International Journal of Food Sciences and Nutrition*. 58: 282–296.

Kápolna, E., Hillestrøm, P. R., Laursen, K. H., Husted, S. and Larsen, E. H., 2009. Effect of foliar application of selenium on its uptake and speciation in carrot. *Food Chemistry*. 115: 1357–1363.

Kápolna, E., Laursen, K. H., Husted, S. and Larsen, E. H., 2012. Bio-fortification and isotopic labelling of Se metabolites in onions and carrots following foliar application of Se and ^{77}Se. *Food Chemistry*. 133: 650–657.

Kopsell, D. A. and Randle, W. M., 1997a. Short-day onion cultivars differ in bulb selenium and sulfur accumulation which can affect bulb pungency. *Euphytica*. 96: 385–390.

Kopsell, D. A. and Randle, W. M., 1997b. Selenate concentration affects selenium and sulfur uptake and accumulation by 'Granex 33' onions. *Journal of the American Society for Horticultural Science*. 122: 721–726.

Kopsell, D. A. and Randle, W. M., 1999. Selenium accumulation in a rapid-cycling *Brassica oleracea* population responds to increasing sodium selenate concentrations. *Journal of Plant Nutrition*. 22: 927–937.

Larsen, E. H., Lobinski, R., Burger-Meyer, K., Hansen, M., Ruzik, R., Mazurowska, L., Rasmussen, P. H., Sloth, J. J., Scholten, O. and Kik, C., 2006. Uptake and speciation of selenium in garlic cultivated in soil amended with symbiotic fungi (mycorrhiza) and selenate. *Analytical and Bioanalytical Chemistry*. 385: 1098–1108.

Lavu, R. V. S., Du Laing, G., Van de Wiele, T., Pratti, V. L., Willekens, K., Vandecasteele, B. and Tack, F., 2012. Fertilizing soil with selenium fertilizers: impact on concentration, speciation, and bioaccessibility of selenium in leek (*Allium ampeloprasum*). *Journal of Agricultural and Food Chemistry*. 60: 10930–10935.

LeDuc, D. L., AbdelSamie, M., Montes-Bayon, M., Wu, C. P., Reisinger, S. J. and Terry, N., 2006. Overexpressing both ATP sulfurylase and selenocysteine methyltransferase enhances selenium phytoremediation traits in Indian mustard. *Environmental Pollution*. 144: 70–76.

Lee, J., Finley, J. W. and Harnly, J. A., 2005. Effect of selenium fertilizer on free amino acid composition of broccoli (*Brassica oleracea* cv. Majestic) determined by gas chromatography with flame ionization and mass selective detection. *Journal of Agricultural and Food Chemistry*. 53: 9105–9111.

Lyi, S. M., Heller, L. I., Rutzke, M., Welch, R. M., Kochian, L. V. and Li, L., 2005. Molecular and biochemical characterization of the selenocysteine Se-methyltransferase gene and Se-methylselenocysteine synthesis in broccoli. *Plant Physiology*. 138: 409–420.

Malorgio, F., Diaz, K. E., Ferrante, A., Mensuali-Sodi, A. and Pezzarossa, B., 2009. Effects of selenium addition on minimally processed leafy vegetables grown in a floating system. *Journal of the Science of Food and Agriculture*. 89: 2243–2251.

Maneetong, S., Chookhampaeng, S., Chantiratikul, A., Chinrasri, O., Thosaikham, W., Sittipout, R. and Chantiratikul, P., 2013. Hydroponic cultivation of selenium-enriched kale (*Brassica oleracea* var. alboglabra L.) seedling and speciation of selenium with HPLC-ICP-MS. *Microchemical Journal*. 108: 87–91.

Matich, A. J., McKenzie, M. J., Brummell, D. A. and Rowan, D. D., 2009. Organoselenides from *Nicotiana tabacum* genetically modified to accumulate selenium. *Phytochemistry*. 70: 1098–1106.

Matich, A. J., McKenzie, M. J., Lill, R. E., Brummell, D. A., McGhie, T. K., Chen, R. K. Y. and Rowan, D. D., 2012. Selenoglucosinolates and their metabolites produced in *Brassica* spp. fertilised with sodium selenate. *Phytochemistry*. 75: 140–152.

Mazej, D., Osvald, J. and Stibilj, V., 2008. Selenium species in leaves of chicory, dandelion, lamb's lettuce and parsley. *Food Chemistry*. 107: 75–83.

McKenzie, M. J., Hunter, D. A., Pathirana, R., Watson, L. M., Joyce, N. I., Matich, A. J., Rowan, D. D. and Brummell, D. A., 2009. Accumulation of an organic anticancer selenium compound in a transgenic Solanaceous species shows wider applicability of the selenocysteine methyltransferase transgene from selenium hyperaccumulators. *Transgenic Research*. 18: 407–424.

Mechora, S., Stibilj, V., Radescek, T., Gaberscik, A. and Germ, M., 2011. Impact of Se(VI) fertilization on Se concentration in different parts of red cabbage plants. *Journal of Food, Agriculture and the Environment*. 9: 357–361.

Nunes, R. G. F. L., da Luz, J. M. R., Freitas, R., Higuchi, A., Kasuya, M. C. M. and Vanetti, M. C. D., 2012. Selenium bioaccumulation in shiitake mushrooms: a nutritional alternative source of this element. *Journal of Food Science*. 77: C983–C986.

Ogra, Y., Ishiwata, K., Encinar, J. R., Lobinski, R. and Suzuki, K. T., 2004. Speciation of selenium in selenium-enriched shiitake mushroom, *Lentinula edodes*. *Analytical and Bioanalytical Chemistry*. 379: 861–866.

Pezzarossa, B., Piccotino, D., Shennan, C. and Malorgio, F., 1999. Uptake and distribution of selenium in tomato plants as affected by genotype and sulphate supply. *Journal of Plant Nutrition*. 22: 1613–1635.

Pezzarossa, B., Remorini, D., Gentile, M. L. and Massai, R., 2011. Effects of foliar and fruit addition of sodium selenate on selenium accumulation and fruit quality. *Journal of the Science of Food and Agriculture*. 92: 781–786.

Pyrzynska, K., 2009. Selenium speciation in enriched vegetables. *Food Chemistry*. 114: 1183–1191.

Ramos, S. J., Rutzke, M. A., Hayes, R. J., Faquin, V., Guilherme, L. R. G. and Li, L., 2011a. Selenium accumulation in lettuce germplasm. *Planta*. 233: 649–660.

Ramos, S. J., Yuan, Y. X., Faquin, V., Guilherme, L. R. G. and Li, L., 2011b. Evaluation of genotypic variation of broccoli (*Brassica oleracea* var. *italica*) in response to selenium treatment. *Journal of Agricultural and Food Chemistry*. 59: 3657–3665.

Rios, J. J., Blasco, B., Cervilla, L. M., Rosales, M. A., Sanchez-Rodriguez, E., Romero, L. and Ruiz, J. M., 2009. Production and detoxification of H_2O_2 in lettuce plants exposed to selenium. *Annals of Applied Biology*. 154: 107–116.

Rios, J. J., Blasco, B., Rosales, M. A., Sanchez-Rodriguez, E., Leyva, R., Cervilla, L. M., Romero, L. and Ruiz, J. M., 2010. Response of nitrogen metabolism in lettuce plants subjected to different doses and forms of selenium. *Journal of the Science of Food and Agriculture*. 90: 1914–1919.

Robbins, R. J., Keck, A. S., Banuelos, G. and Finley, J. W., 2005. Cultivation conditions and selenium fertilization alter the phenolic profile, glucosinolate, and sulforaphane content of broccoli. *Journal of Medicinal Food*. 8: 204–214.

Schiavon, M., dall'Acqua, S., Mietto, A., Pilon-Smits, E. A. H., Sambo, P., Masi, A. and Malagoli, M., 2013. Selenium fertilization alters the chemical composition and antioxidant constituents of tomato (*Solanum lycopersicum* L.). *Journal of Agricultural and Food Chemistry*. 61: 10542–10554.

da Silva, M. C. S., Naozuka, J., da Luz, J. M. R., de Assuncao, L. S., Oliveira, P. V., Vanetti, M. C. D., Bazzolli, D. M. S. and Kasuya, M. C. M., 2012. Enrichment of *Pleurotus ostreatus* mushrooms with selenium in coffee husks. *Food Chemistry*. 13: 558–563.

Spolar, M. R., Schaffer, E. M., Beelman, R. B. and Milner, J. A., 1999. Selenium-enriched *Agaricus bisporus* mushrooms suppress 7,12-dimethylbenz[*a*]anthracene bioactivation in mammary tissue. *Cancer Letters*. 138: 145–150.

Sugihara, S., Kondo, M., Chihara, Y., Yuji, M., Hattori, H. and Yoshida, M., 2004. Preparation of selenium-enriched sprouts and identification of their selenium species by high-performance liquid chromatography-inductively coupled plasma mass spectrometry. *Bioscience Biotechnology and Biochemistry*. 68: 193–199.

Turakainen, M., Hartikainen, H. and Seppänen, M. M., 2004. Effects of selenium treatments on potato (*Solanum tuberosum* L.) growth and concentrations of soluble sugars and starch. *Journal of Agricultural and Food Chemistry*. 52: 5378–5382.

Turakainen, M., Hartikainen, H., Ekholm, P. and Seppänen, M. M., 2006. Distribution of selenium in different biochemical fractions and raw darkening degree of potato (*Solanum tuberosum* L.) tubers supplemented with selenate. *Journal of Agricultural and Food Chemistry*. 54: 8617–8622.

Whanger, P. D., 2002. Selenocompounds in plants and animals and their biological significance. *Journal of the American College of Nutrition*. 21: 223–232.

White, P. J. and Broadley, M. R., 2009. Biofortification of crops with seven mineral elements often lacking in human diets – iron, zinc, copper, calcium, magnesium, selenium and iodine. *New Phytologist*. 182: 49–84.

Wróbel, K., Wróbel, K., Kannamkumarath, S. S., Caruso, J. A., Wysocka, I. A., Bulska, E., Światek, J. and Wierzbicka, M., 2004. HPLC-ICP-MS speciation of selenium in enriched onion leaves – a potential dietary source of Se-methylselenocysteine. *Food Chemistry*. 86: 617–623.

Ximénez-Embún, P., Alonso, I., Madrid-Albarrán, Y. and Cámara, C., 2004. Establishment of selenium uptake and species distribution in lupine, Indian mustard, and sunflower plants. *Journal of Agricultural and Food Chemistry*. 52: 832–838.

Zhang, M., Li, C. L. and Cao, P., 2004. Effects of processing conditions of the green-leafy vegetable juice enriched with selenium on its quality stability. *Journal of Food Engineering*. 62: 393–398.

CHAPTER 18

Biofortification of Legumes with Selenium in Semiarid Conditions

M. J. POBLACIONES*[a], S. RODRIGO[a], AND O. SANTAMARÍA[a]

[a]Department of Agronomy and Forest Environment Engineering, University of Extremadura, Avda. Adolfo Suárez s/n, 06007 Badajoz, Spain
*E-mail: majops@unex.es

18.1 Selenium

Selenium (Se) is an essential micronutrient in animal and human nutrition which is involved in important regulatory and protective mechanisms. Selenium deficiency can cause many different health disorders including oxidative stress-related conditions, epilepsy, fertility reduction, immune deficiency and an increased risk of cancer (Rayman, 2012). This is why Se content in the human diet has become a topic of interest in public health systems around the world.

Although the European Recommended Dietary Allowance (RDA) of Se for humans is about 55 µg Se per day (Elmadfa, 2009), several authors go further and after carrying out clinical trials, have recommended a regular oral dose of 200 µg per day of Se to reduce the incidence of certain cancers, cardiomyopathy, free-radical-induced diseases and to protect against HIV (Reid *et al.*, 2008). In global terms and according to such recommendations, between 500 and 1000 million people around the world have a daily intake

Food and Nutritional Components in Focus No. 9
Selenium: Chemistry, Analysis, Function and Effects
Edited by Victor R. Preedy

Published by the Royal Society of Chemistry, www.rsc.org

of Se below the recommended dose (Li *et al.*, 2008). In European countries (Table 18.1), about 20% of the population has shown an inadequate Se intake (Roman-Viñas *et al.*, 2011).

Se deficiency in livestock is also common. It causes diseases such as white-muscle disease in cattle and sheep, hepatosis dietetica in pigs, pancreatic fibrosis and exudative diathesis (Hawkesford and Zhao, 2007). For the livestock, a minimum daily intake of 200–300 μg of Se per kg is required to prevent Se-deficiency-associated diseases in sheep and cattle, respectively (National Research Council, 1996).

Taking into account that diet is the principal route for the daily Se intake for animals and humans, the following strategies could be considered to remedy such deficiency (Zhao and McGrath, 2009):

- Fortification: addition of Se to food during processing.
- Supplementation: intake of Se supplements, *e.g.* in the form of pills. This is usually problematic because of adverse secondary effects, inadequate supply; need to hold a daily regime and expensive costs.
- Agronomic biofortification: increasing the Se concentrations in the edible portions of crop plants through the use of fertilizers.
- Genetic biofortification: increasing the Se concentrations in the edible portions of crop plants through crop selection or breeding.

Because Se is mainly accumulated in amino acids (Thavarajah *et al.*, 2008, 2010), foodstuffs with high proportion of proteins, like meat or fish, present the highest concentration of Se (Table 18.2). This is why legumes may also present a high potential of accumulation. However, total Se contained in plant-derived products is completely dependent of the initial Se concentration into the soil and of the capability of the plant species to uptake and accumulate the Se absorbed (Hawkesford and Zhao, 2007). Then, the application of Se fertilizer could increase the available Se into the soil, and consequently plants could increase the Se concentration in the edible parts.

However, not only the total Se content in the grain is important, the chemical Se species formed in seeds is also vital in terms of nutritional benefits. Selenium is an essential constituent of selenoproteins. In humans, the nutritional functions of Se are achieved by 25 selenoproteins, which are thought to discharge the nutritional essentiality of Se (Kryukov *et al.*, 2003).

Table 18.1 Daily Se intakes for several Mediterranean countries.

Country	Se intake	Source
Greece	39.3	Pappa *et al.* (2006)
Turkey	30	Foster and Sumar (1997)
Spain	32.3	Díaz-Alarcón *et al.* (1996a)
Portugal	32–43	Ventura *et al.* (2009)
Italy	39–48	Elmadfa (2009)
Croatia	33.9	Matec *et al.* (2000)
Egypt	49	Hussein and Bruggeman (1999)

According to Rayman *et al.* (2008), Se in foods can be found in the following chemical forms:

- Selenomethionine (SeMet): found in plant sources (especially cereals), yeast, and other selenium supplements. It is incorporated nonspecifically into body proteins in place of methionine. This form is considered very bioavailable and easily incorporated for humans and animals.
- Selenocysteine (SeCys): found mainly in animal foodstuffs (from their selenoproteins). It has been reported to be absorbed very well (Institute of Medicine, 2000).
- Selenoneine (2-selenyl-N_α,N_α,N_α-trimethyl-L-histidine): newly discovered as the major selenium compound in fish like tuna and mackerel; lower concentrations in squid, tilapia, pig, and chicken. It has strong radical-scavenging activity.
- Se-methylselenocysteine (MeSeCys) and γ-glutamyl-Se-methylselenocysteine (γ-glutamyl-MeSeCys): found in plant sources such as garlic, onions, broccoli, and in selenium-enriched yeast. It is metabolized to methyl selenol, which is thought to have anticancer effects.
- Sodium selenite and selenate: components of dietary supplements; selenite occasionally appears in water supplies. Some selenate is found in fish and plant sources (*e.g.*, cabbage). Almost 100% of selenate is absorbed; however, a significant fraction is lost in the urine. Regarding selenite, although a 50% is absorbed, it shows better retention than selenate in the organism.

Organic Se species are more bioavailable than inorganic Se species (Sors *et al.*, 2005). Organic Se compounds such as SeMet, MeSeCys and γ-glutamyl-MeSeCys are effective chemoprotective agents (Sors *et al.*, 2005; Rayman, 2012). SeMet has been confirmed to be the major Se species contained in the grain of many cereals and legumes (Li *et al.*, 2008; Hart *et al.*, 2011; Rayman, 2012; Poblaciones *et al.*, 2014a,b), and in chickpea sprouts (Zhang *et al.*, 2012).

Table 18.2 Selenium concentrations (μg per kg fresh weight) for major classes of foods from several Mediterranean countries.

Food	Portugal[a]	Spain[b]	Greece[c]	Croatia[d]	Egypt[e]
Fruit and vegetable	1.7–25	0.3–30	2–14	6–66	1–3
Cereals	18–40	2–78	6–132	38–114	7–37
Legumes	241–448	18–269	24–444	—	580
Milk	18–163	9.1–178	13–27	19–53	—
Meat and fish	88–737	28–1543	49–507	53–859	53
Eggs	173–331	—	73–172	53–177	168

[a]Ventura *et al.* (2009).
[b]Díaz-Alarcón *et al.* (1994a,b, 1996a,b).
[c]Bratakos and Ioannou (1991).
[d]Klapec *et al.* (1998).
[e]Hussein and Bruggeman (1999).

18.2 Importance of Legumes in Cropping Systems

Grain legumes should be important components of the cropping systems. The principal benefit attributed to legumes in crop rotations is the fixation of atmospheric nitrogen to the soil that will then be partly available to the subsequent crop. Several studies have showed a protein proportion increase in the subsequent crop due to this N fixation. Another important benefit is the break of the cereal crop disease cycles, the improvement of soil structure and the increase of water infiltration (Ruisi *et al.*, 2012). For these reasons, the promotion of legume crops should be an essential practice for sustainable agricultural systems, as it is also recognized by the European Commission in its Common Agricultural Policy by means of several specific payments for farmers.

The most cultivated legumes in the world are beans (*Phaseolus vulgaris* L.), peas (*Pisum sativum* L.), chickpeas (*Cicer arietinum* L.) and lentils (*Lens culinaris* L.), with a total cultivation area of 28.8, 12.1, 6.4 and 4.3 mill ha respectively (FAOSTAT, 2013). In the Mediterranean area, the countries with the greatest area harvested are Turkey, Egypt and Greece for beans; Spain and Morocco for peas; Turkey, Syria, Morocco, Algeria and Spain for chickpea and Turkey, Morocco and Spain for lentils (FAOSTAT, 2013).

Despite all the advantages of the rotation with legumes cropping system, in the European Mediterranean region, the cultivated area has decreased by 11% between 2000 and 2005. Since 2005, its area has remained relatively stable, with an estimated area of 667 000 ha in 2010. Spain is the European country with a highest cultivation area with approximately 30 700 ha (FAOSTAT, 2013).

18.3 Legumes in the Mediterranean Diet

The traditional Mediterranean diet is characterized by primarily plant-based foods, such as fruits and vegetables, whole grains, high intakes of plant proteins as legumes and nuts; butter has been replaced by healthy fats, such as olive oil; red meat intake is limited to a few times a month, eating fish and poultry more frequently, at least twice a week. The alcohol intake is moderate, and it uses herbs and spices instead of salt to flavor foods (Fung *et al.*, 2009). The diet also recognizes the importance of being physically active and enjoying meals with family and friends. The Mediterranean diet has been associated with decreased mortality (Mitrou *et al.*, 2007; Sofi *et al.*, 2008).

Although different regions in the Mediterranean basin have their own diets, it is appropriate to consider these as variants of a single entity, the Mediterranean diet. The dietary pattern is based on olive oil, with very well known health benefits, a consumption of large quantities of vegetables in salads and equally large quantities of legumes in the form of cooked foods. Other essential components of Mediterranean diet are wheat and grapes and their derivate products. However, the Mediterranean diet has evolved over a long time. While the proportion of animal–vegetables products was 15–85% in 1961, it was 26–74% in 2003 (Elmadfa, 2009). Specifically, the consumption of pulses in the Mediterranean countries decreased from 5–10 kg per capita per year to 4–6 kg per capita per year (Elmadfa, 2009). Even though pulses are still an essential component of the Mediterranean diet and they are consumed in large amounts.

18.4 Selenium Biofortification in the Main Legume Crops under Mediterranean Conditions

Agronomic biofortification with Se has been practiced in Finland since the mid-1980s in their major food crops (Eurola *et al.*, 1991), and later in New Zealand (Curtin *et al.*, 2006), China (Combs, 2001), UK (Broadley *et al.*, 2010), Australia (Lyons *et al.*, 2003), *etc.* Those Se biofortification programs have been carried out mainly under North and Central Europe conditions, especially in Finland and the UK. However, due to the different soil, climatic and cropping conditions, the application of such experience to Mediterranean conditions and with crops different from those tested, may be different.

Semiarid Mediterranean conditions are characterized by low rainfall, ranging between 450 and 600 mm, with a strong seasonal and interannual variation, with at least 65% falling in the winter half of the year and very dry summers. Sun intensity is high. Summers are warm to hot, and winters are cool but mild, with one month averaging below 59 °F (15 °C). Temperature, rainfall and solar radiation are the climatological factors most influencing plant growth. Rainfall fluctuations and dry spells have a negative effect on the production of rainfed crops.

These Mediterranean conditions result in an important reduction in the grain yield of most rainfed crops in comparison with that of Oceanic climates. For example, in 2012, grain yield for bread-making wheat in Spain was on average 2600 kg per ha, while in Germany it was 7300 kg per ha (FAOSTAT, 2013). However, the lack of water and the high temperatures during grain filling have a favorable effect on the grain protein content, and it usually is much higher than in plants growing under Atlantic or Oceanic conditions (Rao *et al.*, 1993). This is a very important point because the accumulation of Se in the grain has been shown to be very positively correlated with the proportion of proteins in the grain (Thavarajah *et al.*, 2008, 2010). Therefore, considering that legumes grain is the most important source of plant proteins, in addition to the fact that the conditions of the Mediterranean climate seems to increase the already high protein proportion, it is quite reasonable to suppose a high potential of legumes for agronomic Se biofortification in semiarid Mediterranean conditions.

In general terms, according to our knowledge there are very few studies dealing with Se biofortification programs on legume crops, and very few of them have been performed under Mediterranean conditions. Therefore, in our opinion there is an important gap in Se-biofortification knowledge regarding legumes, and especially under Mediterranean conditions. Additional efforts should be paid to clarify the many aspects that remain unknown. Most of the studies performed so far have been focused on beans, peas, chickpeas and lentils.

18.4.1 Beans

Common beans (*Phaseolus vulgaris*) are the most important grain legume crop in the Mediterranean region, and Turkey is the major Mediterranean beans producer and consumer country. World beans production is approximately 23 mill tons in 28.8 mill ha (FAOSTAT, 2013). There are some studies dealing with Se concentrations in beans in Mediterranean countries, with ranges between 152 and 268 μg Se per kg FW (fresh weight) (Díaz-Alarcón *et al.*, 1996b) or 15 to 204 μg Se per kg DW (dry weight) (Matos-Reyes *et al.*, 2010) in Spain; 183 to 448 μg per kg FW in Portugal (Ventura *et al.*, 2007); 24 to 143 μg Se per kg FW in Greece (Bratakos and Ioannou, 1991; Pappa *et al.*, 2006); 50 μg per kg DW in Turkey (Tuzen and Soylak, 2007) or only 7 μg per kg FW in Egypt (Hussein and Bruggeman, 1999). Although beans are the most important legume crop in terms of extension and production, its potential to be used in Se biofortification programs has been very scarcely studied and no study has been carried out under Mediterranean conditions.

In the greenhouse, Smrkolj *et al.* (2007) studied in four bean varieties the influence of two application methods on the Se accumulation in the grain and its speciation. The application methods were: (i) two foliar applications of 10 mg L^{-1} at the beginning of flowering with a 10 day interval between treatments, and (ii) soaking the seeds during 12 h in a solution containing 10 mg Se per L and after that they were hydroponically cultivated. Foliar sprayed beans accumulated 4 times more Se than soaked seeds. Regarding the Se-species, 85% of soluble selenium was found in foliar treatment plants and 65% in soaked treatment, mainly as Se-methionine and seleno-methyl-seleno-cysteine.

More recently, Hermosillo-Cereceres *et al.* (2013) studied the application of different doses of sodium selenate and sodium selenite (0–10–20–40–80–160 μM) that were added to a nutrient solution for 40 days in a greenhouse experiment. They found that the best doses were 40 and 20 μM of sodium selenite and sodium selenate, as they promoted growth in terms of biomass production and boosted the bioavailable concentration of Se in the bean fruit.

18.4.2 Peas

Peas crop (*Pisum sativum*) is one of the most cultivated legumes in Mediterranean areas. World peas production is approximately 9.8 mill tons. Spain, where the cultivated area has significantly increased by more than four times over the last decade, is the main producer with about 140 000 tons (FAOSTAT, 2013). Peas are mainly used for animal feeds (about 85% of the total production) as dried peas, but they are also harvested for humans (about 15% of the total production) as market peas. Despite its importance, there are very few studies dealing with Se biofortification in peas, and most of them were conducted in greenhouse or under temperate conditions. Those studies were mainly performed in Slovenia (Smrkolj *et al.*, 2006), Canada (Gawalko *et al.*, 2009; Thavarajah *et al.*, 2010) and USA (Yan and Johnson, 2011). Only the study carried out by Poblaciones *et al.* (2013) was conducted under Mediterranean climate.

Based on the information given by Poblaciones *et al.* (2013), which studied the effect of two foliar Se fertilizers (sodium selenate and sodium selenite) applied at five rates (0–10–20–40–80 g Se per ha), Se applied as sodium selenate was much more effective than sodium selenite in boosting Se in the peas grain. Regarding doses, those authors stated that there was a highly significant linear relationship between the Se contained in grain and the Se applied, regardless of the Se application form (Figure 18.1). However, they suggested the lowest dose of sodium selenate (10 g Se per ha) to be enough to increase to a great extent the content of Se in the grain, close to that recommended. An intake of 100 g peas from a Se-fertilized crop at that dose may provide about 90% of the 200 μg Se per day recommended to optimize the Se status for humans (Reid *et al.*, 2008), and between 60 and 90% for the livestock, with a minimum daily intake of 200–300 μg of Se per kg (National Academy of Sciences, 1983). At that dose, grain yield and crude protein proportion was not affected by the Se application. This is an important point to avoid impediments of farmers to successfully implement a potential biofortification program in the future. In terms of Se accumulation in the grain, peas showed a much higher capacity than other crops like cereals. As expected, the higher protein proportion prompted a higher Se accumulation. In comparison with other legumes, Se increments were similar to those found in other legumes like soybean (Gupta and MacLeod, 1994) after a Se fertilization of 10 g Se per ha as foliar application; or like beans (Hermosillo-Cereceres

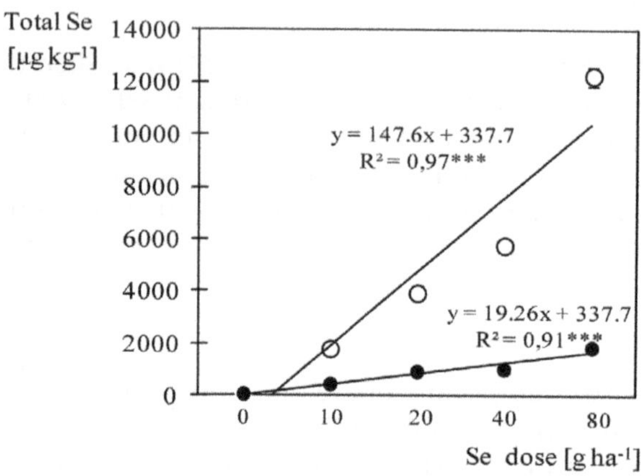

Figure 18.1 Total Se content in grain as μg Se per ha expressed as a linear relationship of the dose (0, 10, 20, 40 and 80 g per ha) for each Se form (sodium selenate empty marker and sodium selenite full marker). Vertical bars on the markers represent ± standard errors. Level of significance: $^{***}p < 0.001$ (Poblaciones *et al.*, 2013).

et al., 2013) after a soil Se application at several doses. However, as traditionally the grain yield (expressed as kg grain per ha) in peas is greater than in other legume crops in Spain (Ministerio de Agricultura, Alimentación y Medio Ambiente, 2011), the total Se accumulated in grain, expressed as μg Se per ha might be higher in peas than in other legume crops. Therefore, those authors showed the suitability of peas to be used in Se-biofortification programs.

18.4.3 Chickpeas

Chickpea (*Cicer arietinum*) is another important grain legume crop in the Mediterranean region. In Mediterranean countries, like Spain, the chickpea intake has been estimated at more than 5 kg per year per capita (Elmadfa, 2009). World chickpea production is approximately 9.4 mill tons, with India the main producer with about 74% of this total (FAOSTAT, 2013). The Indian production is mainly of Desi type, while Kabuli type is preferred by the European countries population. The bibliography indicates that in non- Se fertilized crop, the Se concentration in the chickpea grain in Mediterranean areas ranges between 94 μg per kg FW and 270 μg per kg DW (Díaz-Alarcón *et al.*, 1996b; Matos-Reyes *et al.*, 2010; Poblaciones *et al.*, 2014a). Considering those initially high values in crops growing in soils with total Se concentrations considered deficient (Hawkesford and Zhao, 2007), chickpeas, among legumes, seem to have a high potential of Se accumulation in the grain. Despite its potential, the information found in the scientific literature is scarce and mainly focused on conditions different from the Mediterranean climate (Srivastava *et al.*, 2012; Thavarajah and Thavarajah *et al.*, 2012; Zhang *et al.*, 2012). Under Mediterranean conditions, the current information is based on the studies carried out by Poblaciones *et al.* (2014a).

As in the case of peas, a significant linear regression was found between the total Se concentration in grain and the dose applied (Figure 18.2), which was in line with that found for other crops (Table 18.3). The recommended fertilization according to those authors (Poblaciones *et al.*, 2014a) was with sodium selenate at a dose of 10 mg Se per ha. In comparison to the study focused on peas carried out by these authors (Poblaciones *et al.*, 2013), in the chickpea study went further and analyzed the Se chemical species contained in the grain. As mentioned above, the knowledge of the chemical form in the Se fraction accumulated in the grain is of very high interest because bioavailability is more correlated to Se chemical form (Institute of Medicine, 2000). In this regard, the main component found was SeMet, with a proportion of 51–91%. This percentage was in agreement with the speciation of Se in several cereals like wheat, barley or rice (Li *et al.*, 2008) or in grain legumes like lentils (Thavarajah *et al.*, 2008). In other grain crops, like wheat, several other organic forms of Se were found (Hart *et al.*, 2011), however, it was not the case of chickpea. Perhaps these other forms of Se organic remained in the proportion of Se not recovered by HPLC. It is also important to highlight from the study by Poblaciones *et al.* (2014a) that a substantial, and not expected, proportion of the Se was recovered in inorganic forms,

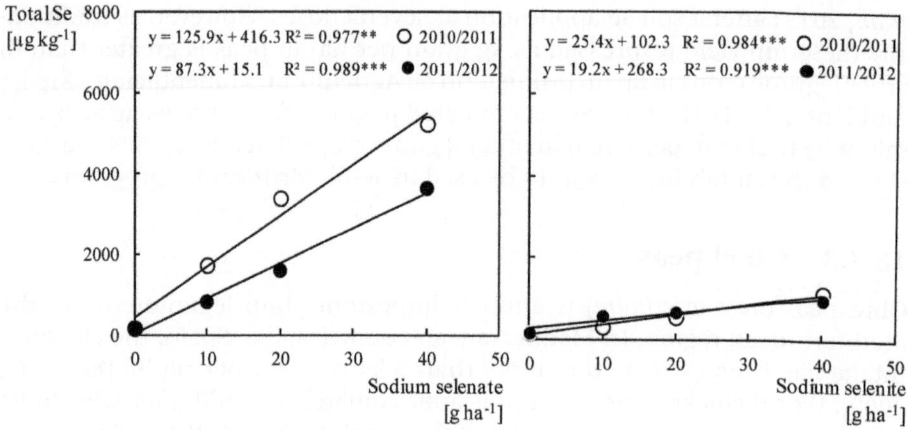

Figure 18.2 Total Se content in chickpea grain as μg Se per ha expressed as a linear relationship of the dose (0, 10, 20 and 40 g per ha) for each Se form, sodium selenate at left and sodium selenite at right (year 2010/2011 open circles and year 2011/2012 closed circles). Level of significance: $^{**}P \leq 0.01$; $^{***}P \leq 0.001$ (Poblaciones *et al.*, 2014a).

Table 18.3 Linear regressions of the relationships between Se concentrations in harvested parts of crops *versus* the amounts of Se(IV) or Se(VI) applied to soils (x; in g per ha) in Badajoz, and the relative bioavailability factor (slope with Se VI applied/slope with Se(IV) applied).[a]

	Se(IV)		Se(VI)		
Crop	Regression	R^2	Regression	R^2	Factor
Barley (2011)	$10.4x + 113$	0.995	$55.3x + 93$	0.997	5.3
Barley (2012)	$6.1x + 58.1$	0.951	$32.9x - 42.8$	0.996	5.4
Durum wheat (2011)	$38.1x - 3.6$	0.985	$137.6x + 8.3$	0.995	3.6
Durum wheat (2012)	$11.1x + 55.3$	0.995	$35.4x + 49.6$	0.995	3.2
Durum pasta (2012)	$7.34x + 13.6$	0.938	$30.2x - 22.5$	0.989	4.1
Bread wheat (2011)	$30.5x - 105$	0.866	$79.0x + 7.7$	0.993	2.6
Bread wheat (2012)	$8.45x + 70.4$	0.925	$37.5x - 12.2$	0.987	4.4
Chickpea (2011)	$25.4x + 102$	0.984	$125.9x + 416$	0.977	5.0
Chickpea (2012)	$19.2x + 268$	0.919	$87.3x + 15.1$	0.989	4.5
FieldPea (2011)	$19.3x + 337$	0.910	$147.6x + 337$	0.970	7.6

[a]McGrath *et al.* (2013).

with the highest proportion in the control. This finding was discussed by the author who concluded that the Se added was likely more easily transformed to organic forms into the plant than the Se taken directly from the soil by the roots.

Poblaciones *et al.* (2014a) also found an increase from 51% of SeMet to 63 to 91% when sodium selenite or sodium selenite was applied (Table 18.4). This range was in line with the found by Zhang *et al.* (2012) in chickpeas sprouts grown in different sodium selenite solutions. Therefore, considering that the

Table 18.4 Total Se obtained by ICP and HPLC in samples from each treatment in the 2011/2012 growing season, and Se recovered (expressed as %) in the HPLC in relation to the total Se obtained by ICP. Percentage of each Se species obtained from the total Se fraction (by HPLC) for each fertilization treatment.[a]

Treatment	Total Se ICP ($\mu g\ kg^{-1}$)	Total Se HPLC ($\mu g\ kg^{-1}$)	Se recovered (%)	Selenium species (% of total Se HPLC)		
				SeMet	Selenite	Selenate
0 g per ha	269.5	240.4	89.2	50.6	39.8	9.6
10 g per ha selenite	660.4	552.6	83.7	63.2	29.6	7.2
20 g per ha selenite	845.2	626.1	74.1	74.0	20.7	5.3
40 g per ha selenite	703.7	492.6	70.0	70.6	26.5	2.9
10 g per ha selenate	805.6	471.6	58.6	84.8	8.5	6.7
20 g per ha selenate	1051.2	699.0	66.5	83.6	11.0	5.4
40 g per ha selenate	2924.4	1676.2	57.3	90.9	5.3	3.8

[a]Poblaciones *et al.* (2014a).

SeMet proportion was higher in the grain from plots fertilized with sodium selenate, the already recommended dose of 10 g per ha might be again the most suitable; not only in terms of efficiency in the Se grain accumulation, but also in the chemical form of this Se, as SeMet has been regarded as one of the most bioavailable Se forms (Kirby *et al.*, 2008).

In the case of chickpea (Poblaciones *et al.*, 2014a), as well as in the case of peas (Poblaciones *et al.*, 2013), Se fertilization did not cause any negative effect in grain yield or other quality-related parameters like protein proportion. This fact, in addition to the high potential of Se accumulation in the grain, also indicates good suitability of chickpea to be used in Se biofortification programs under Mediterranean conditions.

18.4.4 Lentils

Lentils (*Lens culinaris*), which are also very important in the human diet in Mediterranean countries, are mainly cultivated in Turkey, Morocco and Spain. World lentil production is approximately 4.6 mill tons distributed as follows: Turkey about 440 000 tons, Spain 29 000 tons and Morocco 22 000 tons (FAOSTAT, 2013). In several studies conducted in Mediterranean countries, the Se concentration in the lentils grain, without any Se fertilization, has been estimated to range between 21 μg per kg FW in Spain (Díaz-Alarcón *et al.*, 1996b), 113 to 443 μg per kg FW in Greece (Bratakos and Ioannou, 1991; Pappa *et al.*, 2006) and 580 μg per kg FW in Egypt (Askar and Bielig, 1983).

In a non-Mediterranean country like Canada, lentils have shown a high potential for Se uptake and later accumulation in the grain (Thavarajah *et al.* 2007, 2008, 2011a,b). In these conditions Se concentrations in the grain have ranged between 425 and 673 μg Se per kg DW; finding most of the Se (over 70%) in the SeMet form. These values, obtained without any Se fertilization, are higher than those found in Mediterranean countries due to the higher concentrations of Se in the Canadian soils (37–301 μg Se per kg).

18.5 Other Considerations for Se Agronomic Biofortification under Semiarid Conditions

Evaluation of the Se accumulation potential under Mediterranean conditions, with a hot climate with rainfall scarce and irregular, is completely necessary. This is an important issue because, in other crops like cereals, low water availability during the growing season has been shown to prompt a low Se uptake, and therefore a low grain Se accumulation (Rodrigo *et al.*, 2013). The same significant and positive correlation between grain Se concentration and the amount of precipitation during the growing season has been found by Johnson (1991). In addition to that, precipitation occurred during the days before Se fertilization has been inversely correlated with Se uptake and later accumulation when applied as foliar spray (Rodrigo *et al.*, 2014). Therefore, the irregularity in precipitations typical of Mediterranean conditions seems to give rise to important differences in Se uptake and grain Se accumulation after fertilization. Consequently, under such conditions, special attention to rainfall, especially during the days prior to fertilizer application, should be paid in order to obtain Se-biofortification programs that are as effective as possible.

Another important aspect, also evaluated by Thavarajah *et al.* (2008) in the Canadian studies mentioned above, is the effect of the manufacturing and cooking processes on Se accumulation and speciation. In this regard, they found that cooking lentils in boiling water, also the most typical way of cooking chickpeas in the Mediterranean region, did not change the total Se content. There was a loss of Se from lentils to the liquid broth, but as lentils and broth are consumed together as a whole food source, the Se concentration and speciation may remain almost intact. However, the effect of cooking in the total Se and speciation should be determined in further studies, as great variations could be found between crops, the Se form used in the fertilization, and method of cooking.

To conclude, another important point that must also be studied is the introduction of Se in the food chain by increasing Se in the pasture species for forage production. Consequently, Se concentration in livestock milk and meat would be increased and the related Se-deficiency disorders such as white muscle disease would be reduced. Taking into account that extensive pastures are a mix of grasses, legumes and other species, the potential uptake and later accumulation of Se must to be different between them.

Some studies about this have been carried on, but in temperate climates. So, information regarding Se biofortification in the biodiversity of pastures to conserve them could be another important point of research.

Summary Points

- Legumes crops, whose products make up a major part of the Mediterranean diet, should be essential components of the rotations in rainfed cropping systems.
- The lack of water and the high temperatures during grain filling typical of the Mediterranean conditions produce reductions in the grain yield but have a favorable effect on the grain protein content.
- As a consequence, as the Se accumulation in the grain has been shown to correlate positively with protein proportion, legumes may present a high potential of Se accumulation in the grain under Mediterranean conditions.
- The Se biofortification experiments conducted on legumes under Mediterranean conditions have shown a significant linear regression between the total Se concentration in grain and the dose applied, regardless of the Se form used.
- However, a foliar application of 10 g per ha of sodium selenate at the start of the flowering state would be enough to increase to a great extent the content of Se in the grain, close to the recommendations.
- Under Mediterranean conditions, the moment of Se application might have major importance in the later accumulation in the grain, but due to the rainfall before application, rather than to the plant growth stage.
- In legumes grain, Se is mainly accumulated as selenomethionine (SeMet), which is one of the most bioavailable Se chemical species for humans.
- In the Se fertilization under Mediterranean climate, the Se added by foliar application is likely more easily transformed to organic forms into the plant than the Se taken directly from the soil by the roots.
- In legumes, Se fertilization under Mediterranean conditions did not cause any negative effect in grain yield or other quality-related parameters like protein proportion. This is an important point to avoid impediments of farmers to successfully implement a potential biofortification program in the future.
- However, in legume crops, the water availability seems to be positively correlated with the Se accumulation in grain. Therefore, due to the irregularity in precipitation typical of the Mediterranean climate, special attention to rainfall should be paid in order to obtain an accumulation as effective as possible.
- In conclusion, legumes show a great aptitude to be used in Se biofortification programs under rainfed conditions and could be a potentially good way of introducing Se into the human diet.

References

Askar, A. and Bielig, H. J., 1983. Selenium content of food consumed by Egyptians. *Food Chemistry*. 10: 231–234.

Bratakos, M. S. and Ioannou, P. V., 1991. Selenium in human milk and dietary selenium intake by Greeks. *Science of the Total Environment*. 1991: 101–107.

Broadley, M. R., Alcock, J., Alford, J., Cartwright, P., Foot, I., Fairweather-Tait, S. J., *et al.*, 2010. Selenium biofortification of high-yielding winter wheat (*Triticum aestivum* L.) by liquid or granular Se fertilization. *Plant and Soil*. 332: 5–18.

Combs, G. F., 2001. Selenium in global food systems. *British Journal of Nutrition*. 85: 517–547.

Curtin, D., Hanson, R., Lindley, T. N. and Butler, R. C., 2006. Selenium concentration in wheat (*Triticum aestivum* L.) grain as influenced by method, rate, and timing of sodium selenate application. *New Zealand Journal of Crop Horticultural Science*. 34: 329–339.

Díaz-Alarcón, J. P., Navarro-Alarcón, M., López-Martínez, M. C. and de la Serrana, H. L., 1994a. Determination of selenium in fresh fish from Southeastern Spain for calculation of daily dietary Intake. *Journal of Agricultural and Food Chemistry*. 42: 334–337.

Díaz-Alarcón, J. P., Navarro-Alarcón, M., de la Serrana, H. L. and López-Martínez, M. C., 1994b. Determination of selenium levels in vegetables and fruits by hydride generation atomic absorption spectrometry. *Journal of Agricultural and Food Chemistry*. 42: 2848–2851.

Díaz-Alarcón, J. P., Navarro-Alarcón, M., de la Serrana, H. L. and López-Martínez, M. C., 1996a. Determination of selenium in meat products by hydride generation atomic absorption spectrometry-selenium levels in meat, organ meats, and sausages in Spain. *Journal Agricultural and Food Chemistry*. 44: 1494–1497.

Díaz-Alarcón, J. P., Navarro-Alarcón, M., de la Serrana, H. L. and López-Martínez, M. C., 1996b. Determination of selenium in cereals, legumes and dry fruits from southeastern Spain for calculation of daily dietary intake. *The Science of the Total Environment*. 184: 183–189.

Elmadfa, I., 2009. *The European Nutrition and Health Report*. Forum of Nutrition, Vienna, vol. 62, 412 pp.

Eurola, M. H., Ekholm, P. I., Ylinen, M. E., Koivistoinen, P. E. and Varo, P. T., 1991. Selenium in Finish food after beginning the use of selenate-supplemented fertilizers. *Journal of the Science of Food and Agriculture*. 56: 57–70.

FAOSTAT, 2013. Food and Agriculture Organization of the United Nations. Available online at: http://faostat3.fao.org/faostat-gateway/go/to/download/Q/*/E.

Foster, L. H. and Sumar, S., 1997. Selenium in health and disease: a review. *Critical Reviews in Food Science and Nutrition*. 37: 211–228.

Fung, T. T., Rexrode, K. M., Mantzoros, C. S., Manson, J. E., Willett, W. C. and Hu, F. B., 2009. Mediterranean diet and incidence of and mortality from coronary heart disease and stroke in women. *Circulation*. 119: 1093–1100.

Gawalko, E., Garrettb, R. G., Warkentinc, T., Ning Wanga, T. and Richtera, A., 2009. Trace elements in Canadian field peas: a grain safety assurance perspective. *Food Additives & Contaminants, Part A: Chemistry, Analysis, Control, Exposure & Risk Assessment*. 26: 1002–1012.

Gupta, U. C. and MacLeod, J. A., 1994. Relationship between soybean seed selenium and harvested grain selenium. *Canadian Journal of Soil Science*. 79: 221–223.

Hart, D. J., Fairweather-Tait, S. J., Broadley, M. R., Dickinson, S. J., Foot, I., Knott, P., *et al.*, 2011. Selenium concentration and speciation in biofortified flour and bread: retention of selenium during grain biofortification, processing and production of Se-enriched food. *Food Chemistry*. 126: 1771–1778.

Hawkesford, M. J. and Zhao, F. J., 2007. Strategies for increasing the selenium content of wheat. *Journal of Cereal Science*. 46: 282–292.

Hermosillo-Cereceres, M. A., Sanchez-Chavez, E., Guevara-Aguilar, A., Munoz-Marquez, E. and Garcia-Banuelos, M. L., 2013. Biofortification and distribution patterns of selenium in bean: response to selenate and selenite. *Journal of Food and Agriculture Environment*. 11(2): 421–426.

Hussein, L. and Bruggeman, J., 1999. Selenium analysis of selected Egyptian foods and estimated daily intakes among population group. *Food Chemistry*. 67: 527–532.

Institute of Medicine, 2000. *Dietary Reference Intakes; Vitamin C, Vitamin E, Selenium and Carotenoids*. National Academies Press, Washington DC.

Johnson, L., 1991. Trends and annual fluctuations in selenium concentrations in wheat grain. *Plant and Soil*. 138: 67–73.

Kirby, J. K., Lyons, G. H. and Karkkainen, M. P., 2008. Selenium speciation and bioavailability in biofortified products using species-unspecific isotope dilution and reverse phase ion pairing-inductively coupled plasma-mass spectrometry. *Journal of Agricultural and Food Chemistry*. 56: 1772–1779.

Klapec, T., Mandic, M. L., Grgic, J., Primorac, L. J., Ikic, M., Lovric, T., Grgic, Z. and Herceg, Z., 1998. Daily dietary intake of selenium in eastern Croatia. *Science of the Total Environment*. 214: 127–136.

Kryukov, G. V., Castellano, S., Novoselov, S. V., *et al.*, 2003. Characterization of mammalian selenoproteomes. *Science*. 300: 1439–1443.

Li, H. F., McGrath, S. P. and Zhao, F. J., 2008. Selenium uptake, translocation and speciation in wheat supplied with selenate or selenite. *New Phytologist*. 178: 92–102.

Lyons, G., Stangoulis, J. and Graham, R., 2003. High-selenium wheat: Biofortification for better health. *Nutrition Research Review*. 16: 45–60.

Matec, M., Blanusa, M. M. and Grgic, J., 2000. Determination of the daily dietary selenium intake in Croatia. *European Food Research Technology*. 210: 155–160.

Matos-Reyes, M., Cervera, M., Campos, R. and de la Guardia, M., 2010. Total content of As, Sb, Se, Te and Bi in Spanish vegetables, cereals and pulses and estimation of the contribution of these foods to the Mediterranean daily intake of trace elements. *Food Chemistry*. 122: 188–194.

McGrath, S. P., Poblaciones, M. J. and Rodrigo, S. M., 2013. Biofortification of field crops with selenium in Mediterranean conditions. In: *Selenium in the Environment and Human Health*. CRC Press, pp. 115–116.

Ministerio de Agricultura, Alimentación y Medio Ambiente, 2011. Anuario de Estadística. http://www.magrama.gob.es/estadistica/pags/anuario/2011.

Mitrou, P. N., Kipnis, V., Thiebaut, A. C., Reedy, J., Subar, A. F., Wirfalt, E., Flood, A., Mouw, T., Hollenbeck, A. R., *et al.*, 2007. Mediterranean dietary pattern and prediction of all-cause mortality in a US population: results from the NIH-AARP Diet and Health Study. *Archives of Internal Medicine*. 167: 2461–2468.

National Academy of Sciences–National Research Council, 1983. *Selenium Nutrition*. NAS–NRC, Washington, DC.

National Research Council, 1996. *Selenium in Nutrition*. National Academy Press, Washington, DC.

Pappa, E. C., Pappas, A. C. and Surai, P. F., 2006. Selenium content in selected foods from the Greek market and estimation of the daily intake. *Science of the Total Environment*. 372: 100–108.

Poblaciones, M. J., Rodrigo, S. M. and Santamaria, O., 2013. Evaluation of the potential of peas (*Pisum sativum* L.) to be used in selenium biofortification programs under Mediterranean conditions. *Biological Trace Element Research*. 151: 132–137.

Poblaciones, M. J., Rodrigo, S. M., Santamaria, O., Chen, Y., and McGrath, S. P., 2014a. Selenium accumulation and speciation in biofortified chickpea (*Cicer arietinum* L.) under Mediterranean conditions. *Journal of the Science of Food and Agriculture*. 94: 1101–1106.

Poblaciones, M. J., Rodrigo, S. M., Santamaria, O., Chen, Y. and McGrath, S. P., 2014b. Agronomic selenium biofortification in *Triticum durum* under Mediterranean conditions: from grain to cooked pasta. *Food Chemistry*. 146: 378–384.

Rao, A. C. S., Smith, J. L., Jandhyala, V. K., Papendick, R. I. and Parr, J. F., 1993. Cultivar and climatic effects on protein content of soft white winter wheat. *Agronomy Journal*. 85: 1023–1028.

Rayman, M. P., Infante, H. G. and Sargent, M., 2008. Food-chain selenium and human health: spotlight on speciation. *British Journal of Nutrition*. 100: 238–253.

Rayman, M. P., 2012. Selenium and human health. *Lancet*. 379: 1256–1268.

Reid, M. E., Duffield-Lillico, A. J., Slate, E., Natarajan, N., Turnbull, B., Jacobs, E., *et al.*, 2008. The nutritional prevention of cancer: 400 Mcg per day selenium treatment. *Nutrition and Cancer*. 60(2): 155–163.

Rodrigo, S., Santamaria, O., Lopez-Bellido, F. J. and Poblaciones, M. J., 2013. Agronomic selenium biofortification of two-rowed barley under Mediterranean conditions. *Plant, Soil and Environment*. 59: 115–120.

Rodrigo, S., Santamaria, O. and Poblaciones, M. J., 2014. Selenium application timing: influence in wheat grain and flour selenium accumulation under Mediterranean conditions. *Journal of Agricultural Science*. 6(3). DOI: 10.5539/jas.v6n3p23.

Roman-Viñas, B., Ribas-Barba, L., Ngo, J., Gurinovic, M., Novakovic, R., Cavelaars, A., *et al.*, 2011. Projected prevalence of inadequate nutrient intakes in Europe. *Annuals of Nutrition and Metabolism*. 59: 84–95.

Ruisi, P., Giambalvo, D., Di Miceli, G., Frenda, A. S., Saia, S. and Amato, D., 2012. Tillage effects on yield and nitrogen fixation of legumes in Mediterranean conditions. *Agronomy Journal*. 104: 1459–1466.

Smrkolj, P., Germ, M., Kreft, I. and Stibilj, V., 2006. Respiratory potential and Se compounds in pea (*Pisum sativum* L.) plants grown from Se-enriched seeds. *Journal of Experimental Botany*. 57(4): 3595–3600.

Smrkolj, P., Osvald, M. and Osvald, J., 2007. Selenium uptake and species distribution in selenium-enriched bean (*Phaseolos vulgaris* L.) seeds obtained by two different cultivations. *European Food Research and Technology*. 225: 233–237.

Sofi, F., Cesari, F., Abbate, R., Gensini, G. F. and Casini, A., 2008. Adherence to Mediterranean diet and health status: meta-analysis. *British Medical Journal*. 337: 344.

Sors, T. G., Ellis, D. R. and Salt, D. E., 2005. Selenium uptake, translocation, assimilation and metabolic fate in plants. *Photosynthesis Research*. 86: 373–389.

Srivastava, A., Pathania, D., Swain, K. K., Ajith, N., Acharya, R., Reddy, A. V. R. and Nayyar, H., 2012. Application of INAA for phyto-accumulation study of selenium by chickpea plant. *Journal of Radioanalysis Nuclear Chemistry*. 294: 315–318.

Thavarajah, D., Vandenberg, A., George, G. N. and Pickering, I. J., 2007. Chemical form of selenium in naturally selenium-rich lentils (*Lens culinaris* from Saskatchewan. *Journal of Agricultural and Food Chemistry*. 55: 7337–7341.

Thavarajah, D., Ruszkowski, J. and Vandenberg, A., 2008. High potential for selenium biofortification of lentils (*Lens culinaris* L.). *Journal of Agricultural and Food Chemistry*. 56: 10747–10753.

Thavarajah, D., Warkentin, T. and Vandenberg, A., 2010. Natural enrichment of selenium in Saskatchewan field peas (*Pisum sativum* L.). *Canadian Journal of Plant Science*. 90(4): 383–389.

Thavarajah, D., Thavarajah, P., Sarker, A., Materne, M., Vandemark, G., Shrestha, R., Idrissi, O., Hacikamiloglu, O., Bucak, B. and Vandenberg, A., 2011a. A global survey of effects of genotype and environment on selenium concentration in lentils (*Lens culinaris* L.): implications for nutritional fortification strategies. *Food Chemistry*. 125: 72–76.

Thavarajah, D., Thavarajah, P., Wejesuriya, A., Rutzke, M., Glahn, R. P., Combs, G. F. and Vandenberg, A., 2011b. The potential of lentils (*Lens culinaris* L.) as a whole food for increased selenium, iron, and zinc intake: preliminary results from a 3 year study. *Euphytica*. 180: 123–128.

Thavarajah, D. and Thavarajah, P., 2012. Evaluation of chickpea (*Cicer arietinum* L.) micronutrient composition: Biofortification opportunities to combat global micronutrient malnutrition. *Food Research International*. 49: 99–104.

Tuzen, M. and Soylak, M., 2007. Evaluation of trace element contents in canned foods marketed from Turkey. *Food chemistry*. 102: 1089–1095.

Ventura, M. G., Freitas, M. C., Pacheco, A. M. G., van Meerten, T. and Wolterbbek, H. T., 2009. Selenium content in selected Portuguese foodstuffs. *European Food research Technology*. 224: 395–401.

Ventura, M. G., Freitas, M. C. and Pacheco, A. M. G., 2007. Determination of selenium daily intakes in two small groups of the Portuguese population by replicate sample neutron activation analysis. *Journal of Radioanalytical and Nuclear Chemistry*. 281: 193–196.

Yan, L. and Johnson, L. K., 2011. Selenium bioavailability from natural produced high-selenium peas and oats in selenium-deficient rats. *Journal of Agricultural and Food Chemistry*. 59: 6305–6311.

Zhang, L., Li, Q., Yang, X. and Xia, Z., 2012. Effects of sodium selenite and germination on the sprouting of chickpeas (*Cicer arietinum* L.) and its content of selenium, formononetin and biochanin A in the sprouts. *Biological Trace Element Research*. 146: 376–380.

Zhao, F. J. and McGrath, S. P., 2009. Biofortification and phytorremediation. *Current Opinion in Plant Biology*. 12: 373–380.

CHAPTER 19

Selenium in Saliva and Impact on Health

EBRU EMEKLI-ALTURFAN*[a], AYŞEN YARAT[a], AND SERAP AKYÜZ[b]

[a]Department of Biochemistry, Faculty of Dentistry, Marmara University, Guzelbahce Buyukciftlik Sok. No: 6 Nisantasi, Istanbul, Turkey;
[b]Department of Pediatric Dentistry, Faculty of Dentistry, Marmara University, Nisantasi, 34365 Istanbul, Turkey
*E-mail: eiemekli@marmara.edu.tr

19.1 Saliva and Salivary Components

Saliva is a clear and exocrine fluid that is slightly acidic. Due to its critical roles for the maintenance of oral health saliva may be referred as the most substantial body fluid. The terms whole or total saliva refer to a complex mixture of fluids from major and minor salivary glands, from gingival crevicular fluid, oral mucosa transudate, nasal cavity and pharynx mucous containing oral bacteria and food debris (Edgar, 1992; Humphrey and Williamson, 2001).

Saliva has various functions including digestion, taste, protection and lubrication, dilution and cleaning. Moreover, the buffering capacity and antibacterial properties of saliva are of vital importance for the maintenance of oral health. Salivary components are required to perform all these functions. Saliva is mainly composed of water (99.5%). Proteins, immunoglobulins, enzymes, mucins, and nitrogenous products, such as urea and ammonia

Food and Nutritional Components in Focus No. 9
Selenium: Chemistry, Analysis, Function and Effects
Edited by Victor R. Preedy

Published by the Royal Society of Chemistry, www.rsc.org

are also present in saliva. Saliva contains electrolytes such as potassium, calcium, sodium, magnesium, bicarbonate and phosphates (Humphrey and Williamson, 2001; Zahir and Sarkar, 2006).

The evaluation of the components of saliva and salivary flow provides important information to clinicians for the assessment and monitoring some diseases. The changes in the composition of saliva may lead to abnormalities in oral functions as several diseases have been related with the altered saliva composition (Shatzman and Herkin, 1981; Emekli-Alturfan *et al.*, 2008).

There is an increasing awareness of the need for trace elements for a healthy life. Accordingly, during the last two decades interest in trace element research in medicine, toxicology, dentistry, nutrition and environmental studies has increased progressively. With the developments in instrumentation the limits for determining trace elements have been lowered to the nanogram or even picogram range. Thanks to these advances research on trace elements in the body fluids including saliva has also increased. Organic (0.3%) and inorganic (0.2%) trace elements in saliva are necessary to maintain the functional integrity of the oral cavity. The relationship between dental caries and salivary trace element composition has been investigated by several researchers (Duggal *et al.*, 1991; Borella *et al.*, 1994). Selenium is one of the essential trace elements found in saliva. Functional and structural effects of selenium are necessary for maintaining oral health and salivary selenium content has been shown to be affected during various diseases (Hadjimarkos and Shearer, 1971; Kuntsal *et al.*, 2003; Abbas *et al.*, 2011).

19.2 Salivary Selenium

The relationship between dental caries and salivary elements were not among the most interesting topics in oral health for many years. On the other hand, as an exception fluoride in saliva attracted attention and many studies mainly focused on fluoride in teeth and saliva (Buttner, 1963). After the discovery that fluoride decreases the prevalence of caries, researchers began to explore the possible beneficial effects of other trace elements on caries.

Selenium concentrations have been evaluated in different organs of the humans such as; hair, nails, blood, urine and saliva and varying concentrations have been reported (Hojo, 1987). Accordingly, selenium concentrations have been reported to be in saliva at the very least and breast milk, plasma and erythrocytes followed respectively (Figure 19.1). Using a fluorimetric method the average saliva concentration has been reported as 2.3 ng mL^{-1} (Hojo, 1987). In order to understand the relation between selenium and dental caries selenium concentration in saliva was evaluated in saliva for the first time in 1971 (Hadjimarkos and Shearer, 1971). By a fluorometric method they evaluated selenium levels in samples obtained from 26 children (11–12 years old) attending a public school in Oregon. The authors reported that saliva samples contained selenium in amounts varying between 1.1 and 5.2 ppb with a mean concentration of 3.1 ppb. The concentration of salivary selenium for boys and girls were similar. On the other hand, salivary selenium concentration has been reported as 2 ng mL^{-1} in Mumbai adult population (Raghunath *et al.*, 2002). Benson (2013) evaluated the concentration of selenium in saliva and hair of children (age 6–9) going to school in Western Cape in South Africa. Selenium levels have been reported as 0.85 ± 0.8 μg kg^{-1} for hair and 0.42 ± 0.7 μg L^{-1} for saliva and no significant difference has been found between genders. Results of studies reporting salivary selenium concentrations are given in Table 19.1. Moreover, a strong correlation has been reported between selenium in hair and saliva.

Selenium levels in saliva have been investigated in some different physiological and pathological conditions (Figure 19.2). Selenium levels in the serum and saliva of patients with renal and ureteric stones have

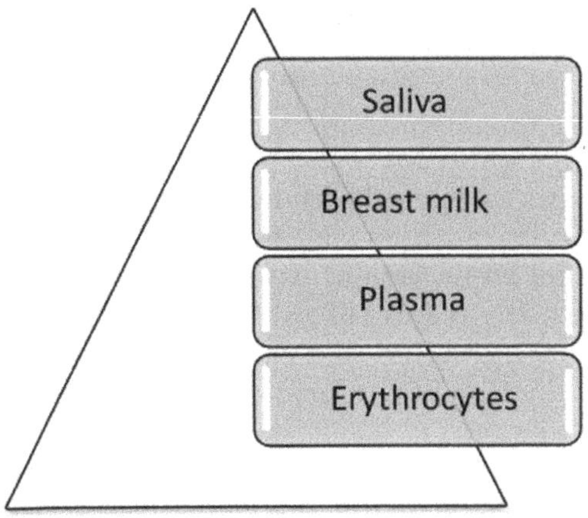

Figure 19.1 Selenium distribution in saliva, breast milk, plasma and erythrocytes. Selenium concentrations have been reported to be lowest in saliva and breast milk, plasma and erythrocytes followed, respectively.

Table 19.1 Selenium concentrations in saliva. Salivary selenium concentrations evaluated by Hadjimarkos and Shearer (1971), Hojo (1987), Raghunath *et al.* (2002) and Benson (2013) are 3.1 ppb, 2.3 ng mL^{-1}, 2 ng mL^{-1} and 0.42 μg L^{-1} respectively.

	Salivary selenium concentrations
Hadjimarkos and Shearer (1971) by fluorometric method	1.1–5.2 ppb (mean 3.1 ppb)
Hojo (1987) by fluorometric method	2.3 ng mL^{-1}
Raghunath *et al.* (2002) by differential pulse cathodic stripping voltammetry	2 ng mL^{-1}
Benson (2013) by atomic absorption spectrophotometer	0.42 μg L^{-1}

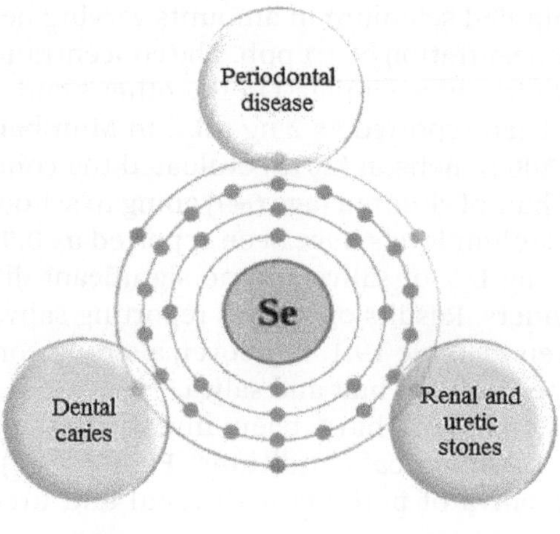

Figure 19.2 Salivary selenium levels have been related with periodontal disease and renal and uretic stone formation, whereas dietary selenium has been related with dental caries.

been evaluated and significant lower serum and saliva selenium levels in stone former patients compared with healthy controls have been reported (Abbas *et al.*, 2011). The authors suggested the use of salivary selenium as a biomarker in renal stone former patients. The findings of Abbas *et al.* are compatible with the results of Sakly of *et al.* suggesting selenium as an inhibitor of stone formation that can be stuck onto the crystal surface and may inhibit induction of new crystals, growth and aggregation (Sakly *et al.*, 2003).

In another study the authors examined whether physical exercise changed the composition of trace elements in whole saliva (Chicharro *et al.*, 1999). The most striking findings of the study were the increase in Na and Mg and a decrease in Mn levels after exercise. Resting and postexercise selenium levels were 0.91 and 0.64 mg L^{-1}, respectively, and no significant change was observed between them. Furthermore, it was suggested that adults with a greater periodontal disease incidence had higher levels of resting salivary Fe, Sc, Mn, Cr, Co,Cu, Se, Ag, and Hg concentrations and lower Zn concentrations (Zaichk and Bagirov, 1994).

19.3 Selenium and Teeth

The first studies on the selenium content of human teeth were conducted by Hadjimarkos and Bonhorst (1959). The authors suggested selenium as a normal constituent of enamel and dentine. They reported that the selenium content of primary teeth was higher than that of permanent teeth and observed no definite pattern showing an increase in the selenium content of enamel with increase in age.

Later, a study was designed to investigate the distribution of injected selenium in the hard and soft tissues of mother rats and their pups (Shearer and Hadjimarkos, 1973). The authors evaluated development and uptake of selenium by the molar teeth and the incorporation of selenium into the protein fraction of enamel and dentine in the developing rat molar teeth. They reported the accumulation of selenium both in the enamel and dentine fractions of the developing teeth and the erupted mature teeth. They also reported the incorporation of selenium as the selenium analogs selenocystine and selenomethionine into the protein fraction of enamel and dentine in the developing teeth.

It was then reported in 1979 that selenium concentrations in successive layers of bovine incisor enamel did not present a concentration gradient and no increase was observed with age. The authors also suggested that selenium was taken up into the continuously growing rat incisor proportional to the amount of selenium present in diet (Johnson and Shearer, 1979).

As the effects of selenium on promoting teeth caries have been linked to the degree it is incorporated into the enamel, the incorporation of selenium into oral tissues such as salivary glands and saliva has also been evaluated. Similar selenium activity has been shown in the salivary glands. On the other hand, although selenium is not accumulated in the salivary glands, salivary glands have been shown to be permeable to this element (Claycomb *et al.*, 1961).

Selenium has been shown to be deposited in all tissues with the largest amount in the lung and kidney and the lowest levels in teeth (molars and incisors) after the radioactive Se^{75} injection (Claycomb *et al.*, 1961). Moreover, Navia suggested that selenium was abundantly found in the pulp as well as the extravascular areas of teeth, but also in lower concentrations in the mineralized portions of the teeth (Navia, 1970).

19.4 Selenium and Dental Caries

Selenium was long considered as a potent toxic substance and the studies in the 1930s and 1960s were concerned with its toxicity. In particular, grazing animals eating selenium-accumulator plants of the genus *Astragalus* during periods of aridness or desert regions of western United States and China have been considered as the major target for its toxicity (Rosenfeld and Beath, 1964). Smith and Westfall were the first to report a high incidence of "poor teeth" in seleniferous areas (Smith and Westfall, 1937).

In the studies conducted in the United States many regional variations were reported in the prevalence of dental caries. The first that comes to mind are the geological conditions including soil and water that vary from area to area. However, water is not the major source for trace metals as it has been shown that compared with foodstuffs drinking water is not a significant source of selenium intake in humans (Rosenfeld and Beath, 1964). Except for fluoride it has been suggested that water supplies do not contribute significant amounts of trace elements to the daily diet (Hadjimarkos *et al.*, 1952; Rosenfeld and Beath, 1964).

The initial studies conducted in Oregon in United States, demonstrated a direct relationship between selenium intake and caries rates. In a series of regional epidemiological studies in 2069 Oregon children between the ages of 14 and 16, different caries prevalence was reported (Hadjimarkos *et al.*, 1952). In areas where selenium toxicity was prevalent one of the most frequent symptoms of the disease examined among children was a high caries prevalence (Hadjimarkos *et al.*, 1952; Hadjimarkos, 1969). A direct relationship has been reported by the authors between the prevalence of dental caries and the urinary selenium concentrations of the subjects. They concluded that selenium may render the teeth more susceptible to caries attack and suggested the need for further studies to understand the selenium and dental caries relationship. On the other hand, this relationship has not been substantiated by the investigation carried out in New Zealand (Cadell and Cousins, 1960). The authors suggested that there was no correlation between the proportion of available tooth surfaces decayed and filled and the total urinary selenium excreted in 24 h. They concluded that the significant differences in selenium excretion between groups was not associated with any significant difference in caries prevalence. Another study was conducted in the Western United States, in Wyoming. Wyoming has high-seleniferous areas where selenium toxicity in farm animals has been reported, children born in these areas were shown to be more susceptible to caries than those living in the control areas of the state (Hadjimarkos, 1969).

The reason for the increased susceptibility to dental caries due to selenium consumption has been explained by Shearer and Hadjimarkos as inhibition of mineralization by interfering with crystal nucleation due to the incorporation of selenium into the protein fraction of enamel (Shearer and Hadjimarkos, 1973).

19.5 Cariogenic Effect of Selenium in Laboratory Animals

After the demonstration of the relationship between the susceptibility to dental caries and selenium in humans researchers focused on the carogenicity of selenium in experimental animals. However, the results were contradictory. English suggested that thiouracil or organic selenium led to hypoplastic changes in the enamel of female dogs (English, 1949). In 1951, intraperitoneal sodium selenite administration to adult rats was reported to be uneffective on dentition; however, a significant toxic action has been observed when 0.5 and 1.0 mg of selenium per kg body weight were given (Wheatcroft *et al.*, 1951). At this level, the authors reported a trend toward increased caries incidence. Moreover, it has been suggested that selenium suppressed thyroid activity which may be compared with thiouracil, that increases caries (Muhler and Shafer, 1954). The activity of the thyroid gland has been associated with dental caries incidence in the rat, as injected thyroxine significantly decreased dental caries, whereas radiothyroidectomy significantly increased it (Muhler *et al.*, 1956). Later in 1957, the effect of selenium on dental caries was investigated in rats by feeding sodium selenite as a constituent of the diet. They reported that the selenium did not change the dental caries when compared to a control group treated with the same diet without added selenium. Moreover, the histology of the salivary glands was not changed by selenium administration (Muhler and Shafer, 1957). Buttner fed rats diets containing sodium selenite (5–10 ppm) in their drinking water and obtained 38 and 54% increase in carious lesions with decreased reproduction and decreased weight gain (Buttner, 1963). The effect of 4.56 ppm dietary sodium selenium on dental caries of rats was investigated by Claycomb *et al.* and they found no significant difference in caries incidence between test and control groups for a period of 100 days (Claycomb *et al.*, 1965). The effects of 4 ppm selenium as Na_2SeO_3 in the drinking water or in the diet were evaluated and it was reported that selenium had no effect except upon sulcus lesions in groups fed selenium in the drinking water (Navia *et al.*, 1968).

According to the results of both epidemiological and animal studies outlined here, it may be suggested that selenium has a slight caries-promoting property. On the other hand, further research is necessary to elucidate the mechanism of action of selenium on dental caries and to differentiate its toxic effects from physiological effects.

19.6 The Role of Selenium in the Antioxidant Activity of Saliva

Oxidative damage caused by free oxygen radicals plays a primary or secondary role in the pathology of many diseases. In recent years, studies on the antioxidant capacity of saliva have shown that some diseases may affect the antioxidants in saliva. In particular, the imbalance between the antioxidants and free radicals may lead to some inflammatory oral pathologies.

Emekli-Alturfan *et al.* have investigated the antioxidant activity of saliva in patients with periodontitis. In this study increased salivary glutathione levels in the periodontitis group compared with the control group were reported. Accordingly due to high glutathione levels lipid peroxidation decreased in the periodontitis group. On the other hand, as an index of increased inflammation salivary sialic acid levels were significantly higher in this group (Emekli-Alturfan *et al.*, 2013). The authors have also evaluated lipid peroxidation in patients with breast cancer and reported a significant reduction in total protein, pH and lipid peroxidation when compared with the healthy controls (Emekli-Alturfan *et al.*, 2008). In this study antioxidants in saliva were not evaluated, nevertheless decreased lipid peroxidation was suggested as a defense mechanism against oral complications of cancer and cancer therapy.

Selenium is in the structure of the antioxidant enzymes glutathione peroxidase (GSH-Px) and thioredoxin reductase that catalyze the reduction of hydrogen peroxide and organic hydroperoxide (Hojo, 1987) (Figure 19.3). Therefore selenium prevents the cell membrane from oxidative damage, reduces the risk of cancer and slows the ageing process (Schlag *et al.*, 1978; Mates *et al.*, 2012). Selenium has been suggested to be protective in cancer and the activity of selenium as a chemopreventive agent has been attributed

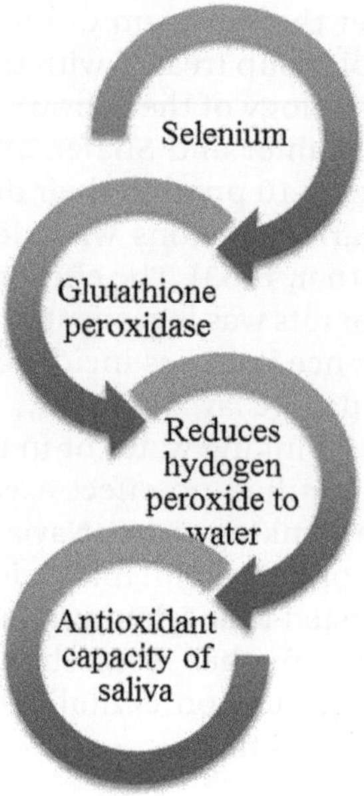

Figure 19.3 The contribution of selenium to the antioxidant capacity of saliva. Selenium is in the structure of glutathione peroxidase enzyme that reduces hydrogen peroxide to water and contributes to the antioxidant capacity of saliva.

to its antioxidant effects. It has been reported that serum selenium levels significantly decreased in oral cancer and precancer groups compared with the control (Khannaa *et al.*, 2013). Selenium-dependent GSH-Px activity was detected for the first time in human mixed saliva by Hojo (1987). In this study a significant positive correlation between GSH-Px and selenium contents has been reported.

The effects of selenium during liquid diet consumption have been investigated. Consumption of liquid diet for a long time has been shown to be responsible for causing degenerative changes on submandibular glands (Scott *et al.*, 1990). Kuntsal *et al.* investigated the effects of selenium supplementation against the degeneration of submandibular glands due to liquid diet consumption. By ultrastructural examination the authors reported that cellular integrity was protected and the organelles were intact and suggested that selenium protected the cell membrane integrity and prevented free-radical formation (Kuntsal *et al.*, 2003).

Although the antioxidant capacity of saliva has been shown in many studies, the studies investigating the role of salivary selenium in the antioxidant capacity are scarce in the literature. Therefore, further research is necessary emphasizing the contribution of selenium to the antioxidant activity.

19.7 Conclusion

Although selenium is known to be an important trace element with antioxidant activity its role in saliva and oral health have been overlooked for many years. Besides some very early studies investigating the carcinogenicity of this element, recent research performed with advanced methods is very limited. It can be suggested that there is an urgent need for new studies on this issue to illuminate the conflicting results existing in the literature.

Summary Points

- This chapter focuses on the effects of selenium in saliva.
- Salivary research has gained importance due to its easy availability with noninvasive methods.
- The antioxidant content of saliva protects oral mucosa against oxidative damage.
- Selenium is in the structure of antioxidant enzymes glutathione peroxidase and thioredoxin reductase, contributes to the antioxidant activity of the organism, however, contradictory results were reported on the effect of selenium on dental caries.
- Selenium concentrations have been evaluated in hair, nails, blood, urine and saliva and varying concentrations have been reported.
- Selenium has been suggested as a normal constituent of enamel and dentine.
- More studies are needed to illuminate the effects of salivary selenium on oral health.

Key Facts of Salivary Selenium

1. Selenium concentrations have been reported to be lowest in saliva and milk, plasma and erythrocytes followed, respectively.
2. The concentration of salivary selenium for boys and girls have been reported to be similar.
3. A strong correlation has been reported between selenium in hair and saliva.
4. Significant lower serum and saliva selenium levels have been reported in renal stone former patients compared with healthy controls and salivary selenium has been suggested to be used as a biomarker in renal stone former patients.
5. No significant change has been found in resting and postexercise selenium levels.
6. It was suggested that adults with a greater periodontal disease incidence had higher levels of resting salivary selenium levels.

Key Facts of Selenium and Teeth

1. Selenium content of primary teeth has been reported to be higher than that of permanent teeth.
2. The selenium content of enamel does not increase with age.
3. There are reports in the literature referring to a high incidence of caries activity in seleniferous areas.
4. Contradictory results have been reported in laboratory animals about the cariogenic effects of selenium.
5. The antioxidant activity of selenium in saliva needs to be illuminated with new research in the area.

Definitions of Words and Terms

Antioxidant. This is a molecule that inhibits the oxidation and terminates the chain reactions by removing free-radical intermediates.

Dental caries. This is an infection with bacterial origin causing demineralization and destruction of the hard tissues of the teeth. They are also known as tooth decay or a cavity.

Dentine. As a calcified tissue dentine is one of the major components of teeth along with enamel, cementum, and pulp.

Enamel. This is the hard mineralized surface of teeth that forms the visible part of the tooth covering the crown.

Glutathione peroxidase. This is the enzyme protecting the organism from oxidative damage by reducing lipid hydroperoxides.

Immunoglobulin. This is a Y-shaped protein produced by B cells used by the immune system to identify and neutralize foreign objects called antigens.

Lipid peroxidation. This is the term referring to the oxidative degradation of lipids in cell membranes by the free radicals.

Mucin. These are high molecular weight, glycosylated proteins.

Periodontitis. This is a set of inflammatory diseases affecting the periodontium.

Permanent teeth. The 32 teeth of the second dentition beginning to appear at about 6 years of age.

Primary teeth. These are the 20 teeth of the first dentition that are replaced by the permanent teeth, also known as deciduous teeth.

Saliva. This is the fluid present in the mouths of organisms, secreted by the salivary glands containing 99.5% water and electrolytes, mucus, glycoproteins, enzymes, and antibacterial compounds.

Seleniferous areas. These are areas containing selenium.

Sialic acid. *N*-Acyl derivative of neuraminic acid that is present in many polysaccharides, glycoproteins and glycolipids.

Thioredoxin reductase. This is the enzyme that catalyzes the reduction of thioredoxin.

List of Abbreviations

GSH-Px Glutathione peroxidase
ppb Parts per billion

References

Abbas, W. A., Fadil, A. and Al-Khazraji, S. K., 2011. Evaluation of selenium and chloride levels in both sera and saliva samples in renal stone former patients. *American Journal of Plant Sciences*. 10(2): 158–164.

Benson, O., 2013. Selenium in saliva and hair of school going children in Ceres district of the Western Cape South Africa. *Journal of Basic and Applied Scientific Research*. 3(3): 246–254.

Borella, P., Fantuzzi, G. and Aggazzotti, G., 1994. Trace elements in saliva and dental caries in young adults. *Science of the Total Environment*. 153: 219–224.

Buttner, W., 1963. Action of trace elements on the metabolism of fluoride. *Journal of Dental Research*. 42: 453–460.

Cadell, P. B. and Cousins, F. B., 1960. Urinary selenium and dental caries. *Nature*. 185: 863–864.

Chicharro, J. L., Serrano, V., Ureña, R., Gutierrez, A. M., Carvajal, A., Fernández-Hernando, P. and Lucía, A., 1999. Trace elements and electrolytes in human resting mixed saliva after exercise. *British Journal of Sports Medicine*. 33: 204–207.

Claycomb, C. K., Sorenson, F. M., Gatewood, D. C., Jump, E. B. and Weaver, M. E., 1961. Further studies on the presence of Se75 in rat saliva and teeth after intracardiac injection of radioactive sodium selenite. *Journal of Dental Research*. 40: 504–510.

Claycomb, C. K., Summers, G. W. and Jump, E. B., 1965. Effect of dietary selenium on dental caries in Sprague-Dawley rats. *Journal of Dental Research*. 44: 826.

Duggal, M. S., Chawla, H. S. and Curzon, M. E., 1991. A study of the relationship between trace elements in saliva and dental caries in children. *Archives of Oral Biology*. 36: 881–884.

Edgar, W. M., 1992. Saliva: its secretion, composition and functions. *British Dental Journal*. 172: 305–312.

Emekli-Alturfan, E., Demir, G., Kasikci, E., Tunali-Akbay, T., Pisiriciler, R., Caliskan, E. and Yarat, A., 2008. Altered biochemical parameters in the saliva of patients with breast cancer. *Tohoku Journal of Experimental Medicine*. 214(2): 89–96.

Emekli-Alturfan, E., Yarat, A., Çalışkan-Ak, E., Pisiriciler, R., Kuru, B. and Noyan, U., 2013. Determination of storage time of saliva samples obtained from patients with and without chronic periodontitis for the comparison of some biochemical and cytological parameters. *Journal of Clinical and Laboratory Analysis*. 27(4): 261–266.

English, J. A., 1949. Experimental effects of thiouracil and selenium on the teeth and jaws of dogs. *Journal of Dental Research*. 28: 172–194.

Hadjimarkos, D. M., 1969. Selenium: a caries-enhancing trace element. *Caries Research*. 3(1): 14–22.

Hadjimarkos, D. M. and Bonhorst, C. W., 1959. The selenium content of human teeth. *Oral Surgery, Oral Medicine, Oral Pathology*. 12(1): 113–116.

Hadjimarkos, D. M. and Shearer, T. R., 1971. Selenium concentration in human saliva. *American Journal of Clinical Nutrition*. 24: 1210–1211.

Hadjimarkos, D. M., Storvicka, C. A. and Remmerta, L. F., 1952. Selenium and dental caries: an investigation among school children of Oregon. *The Journal of Pediatrics*. 4: 451–455.

Hojo, Y., 1987. Selenium and glutathione peroxidase in human saliva and other human body fluids. *Science of the Total Environment*. 65: 85–94.

Humphrey, S. P. and Williamson, R. T., 2001. A review of saliva: normal composition, flow, and function. *Journal of Prosthetic Dentistry*. 85: 162–169.

Johnson, J. R. and Shearer, T. R., 1979. Selenium uptake into teeth determined by fluorimetry. *Journal of Dental Research*. 58(8): 1836–1839.

Khannaa, S., Udasb, A. C., Kumarb, G. K., Suvarna, S. and Karjodkar, F. R., 2013. Trace elements (copper, zinc, selenium and molybdenum) as markers in oral submucous fibrosis and oral squamous cell carcinoma. *Journal of Trace Elements in Medicine and Biology*. 27(4): 307–311.

Kuntsal, L., Firat, D. and Sirin, Y., 2003. Prevention of liquid-diet-induced damages on submandibular glands by selenium supplementation in rats. *Tohoku Journal of Experimental Medicine*. 201(3): 191–199.

Mates, J. M., Segura, J. A., Alonso, F. J. and Marquez, J., 2012. Oxidative stress in apoptosis and cancer: an update. *Archives Toxicology*. 86(11): 1649–1665.

Muhler, J. C., Bixer, D. and Shafer, W. C., 1956. Effect of replacement therapy on dental caries experience of radiothyroidectomized rats. *Proceedings of the Society for Experimental Biology and Medicine*. 93: 328–330.

Muhler, J. C. and Shafer, W. C., 1954. The effect of feeding desiccated thyroid and thiouracil on dental caries in rats. *Science*. 119: 687–689.

Muhler, J. C. and Shafer, W. G., 1957. The Effect of selenium on the incidence of dental caries in rats. *Journal of Dental Research*. 36(6): 895–896.

Navia, J. M., 1970. Effects of minerals on dental caries. In: Harris, R. S. (ed.) *Dietary Chemicals vs. Dental Caries*. American Chemical Society, vol. 94, pp. 139–140.

Navia, J. M., Menaker, L., Seltzer, J. and Harris, R. S., 1968. Effect of NA_2SeO_3 supplemented in the diet or the water on dental caries of rats. *Federation Proceedings*. 27: 676.

Raghunath, R., Tripathi, R. M., Mahapatra, S. and Sadasivan, S., 2002. Selenium levels in biological matrices in adult population of Mumbai, India. *Science of the Total Environment*. 285: 21–27.

Rosenfeld, I. and Beath, O. A., 1964. *Selenium-Geobotany, Biochemistry, Toxicity and Nutrition*. Academic Press, New York.

Sakly, R., Chaouch, A., el Hani, A. and Najjar, M. F., 2003. Effects of intraperitoneally administered vitamin E and selenium on calcium oxalate

renal stone formation: experimental study in rat. *Annales d'Urologie (Paris)*. 37(2): 47–50.
Schlag, P., Seeling, W., Merkle, P. and Betzler, M., 1978. Changes of serum Zn in breast cancer. *Langenbeck's Archives of Surgery*. 2: 129–133.
Scott, J., Berry, M. R., Gunn, D. L. and Woods, K., 1990. The effects of a liquid diet on initial and sustained, stimulated parotid salivary secretion and on parotid structure in the rat. *Archives of Oral Biology*. 35(7): 509–514.
Shatzman, A. R. and Henkin, R. I., 1981. Gustin concentration changes relative to salivary zinc and taste in humans. *Proceedings of the National Academy of Sciences of the United States of America*. 78: 3867–3871.
Shearer, T. R. and Hadjimarkos, D. M., 1973. Comparative distribution of 75 Se in the hard and soft tissues of mother rats and their pups. *Journal of Nutrition*. 103: 553–559.
Smith, M. I. and Westfall, B. B., 1937. Further field studies on the selenium problem in relation to public health. *United States Public Health Reports*. 52: 1375–1384.
Wheatcroft, M. G., English, J. A. and Schlack, C. A., 1951. Effects of selenium on the incidence of dental caries in white rats. *Journal of Dental Research*. 30: 523–524.
Zaichk, V. E. and Bagirov, S. T., 1994. The chemical element content of mixed unstimulated saliva in periodontal diseases. *Stomatologiia (Mosk)*. 73: 8–11.
Zahir, S. and Sarkar, S., 2006. Study of trace elements in mixed saliva of caries free and caries active children. *Journal of Indian Society of Pedodontics and Preventive Dentistry*. 24: 27–29.

CHAPTER 20

Selenohomolanthionine and Its Potential Use in Medicine and Nutrition

YASUMITSU OGRA*[a]

[a]Department of Toxicology and Environmental Health, Graduate School of Pharmaceutical Sciences, Chiba University, 1-8-1 Inohana, Chuo, Chiba 260-8675, Japan
*E-mail: ogra@chiba-u.jp

20.1 Identification of Selenohomolanthionine and Its Derivatives in Biota

Selenium (Se) is a nonessential element in plants. However, certain plants, such as garlic, onion, shallot, Indian mustard, and radish, are Se accumulators. Radish (*Raphanus sativus*) belongs to family Brassicaceae and is an important food consumed worldwide. A pungent variant of Japanese radish (*R. sativus* L. cv. "Yukibijin") is favored for use as a condiment. The compounds responsible for the pungent flavor are isothiocyanates, such as allyl isothiocyanate and benzyl isothiocyanate, which are generated from the hydrolysis of glycosinolates with myrosinase. Thus, the pungent variant is suggested to have higher glycosinolate content and/or higher myrosinase activity than regular radish (Hara *et al.*, 2001). These thio compounds show unique biological and pharmacological effects and have already been used as a supplement and a food additive (Winther and Nielsen, 2006). Because

Food and Nutritional Components in Focus No. 9
Selenium: Chemistry, Analysis, Function and Effects
Edited by Victor R. Preedy

Published by the Royal Society of Chemistry, www.rsc.org

Se and sulfur (S) are same-group elements, plants having high S content are expected to have high Se-assimilating ability. Indeed, Se-accumulating plants, such as garlic and Indian mustard, contain a large amount of S and produce unique thiometabolites (Block *et al.*, 2004). Japanese pungent radish fortified with Se (selenized) may contain unique Se metabolites. In this regard, selenized Japanese pungent radish was cultivated and examined for its ability to produce novel Se compounds (Ogra *et al.*, 2007).

The inductively coupled plasma mass spectrometer (ICP-MS) is a powerful instrument for elemental speciation owing to its high sensitivity and specificity when used in combination with such separation instruments as a high-performance liquid chromatograph (HPLC) and a gas chromatograph. HPLC-ICP-MS is the technique of choice for Se speciation, particularly in screening for Se compounds in samples (Wróbel *et al.*, 2004). The assignment of Se compounds by HPLC-ICP-MS is performed by matching the retention times of samples with those of authentic standards (Pedrero *et al.*, 2006). However, this technique is inappropriate and insufficient for the assignment of Se compounds whose standards are not available. Electrospray ionization mass spectrometry (ESI-MS) coupled with HPLC is another technique for elemental speciation. In particular, a tandem mass spectrometer (ESI-MS-MS) is superior to ICP-MS in terms of identification of unknown Se compounds/metabolites in biological samples (Ogra, 2008).

It was reported that 68.5% of total Se in Japanese pungent radish fortified with sodium selenate was water extractable. Speciation analysis by HPLC-ICP-MS revealed that the water extract contained selenate (around 45% of total Se), *Se*-methylselenocysteine (MeSeCys, around 11%), an unknown Se compound (around 5%), and others (around 1.2%). The unknown Se compound was identified as SeHLan by ESI-MS-MS.

Łobiński's group reported the existence of SeHLan and its derivative in selenized yeast extract (Dernovics *et al.*, 2009). Although Se is also a nonessential element in yeast, yeast can produce several unique Se metabolites due to its high capacity for Se metabolism. The group identified 2,3-dihydroxypropionyl-SeHLan (2,3-DHP-SeHLan) by ESI-quadrupole time-of-flight tandem mass spectrometry (ESI-Q-TOF MS-MS) and ESI-hybrid linear ion trap/Orbitrap MS^n, in addition to SeHLan (Far *et al.*, 2010). The structures of SeHLan and 2,3-DHP-SeHLan are depicted in Figure 20.1.

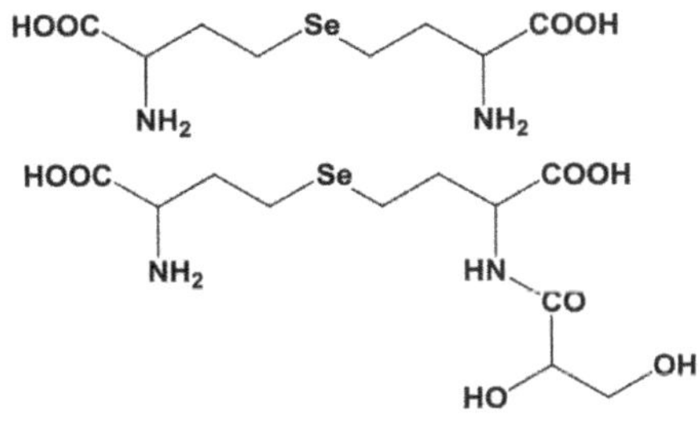

Figure 20.1 Structures of selenohomolanthionine and its derivative, 2,3-dihydroxypropionyl-selenohomolanthionine.

20.2 Biosynthetic Pathway of Selenohomolanthionine in Plants

The biosynthetic pathway of SeHLan is proposed as depicted in Figure 20.2. Although homoserine is suggested to be activated by the phosphorylation, acetylation or succinylation of its γ-OH group, it has been reported that phosphorylation is the dominant activation process in plants (Hacham *et al.*, 2003). *O*-phosphohomoserine is supposed to act as an important metabolic intermediate and to undergo three metabolic routes. The first route is the formation of selenocystathionine with selenocysteine (SeCys). The second route involves the reaction of *O*-phosphohomoserine with the reduced form of Se, selenide, which results in the synthesis of selenohomocysteine (SeHCys), and the third route is the cystathionine γ-synthase-catalyzed reaction of *O*-phosphohomoserine with SeHCys to form SeHLan (Krömer *et al.*, 2006). SeHCys is predominantly metabolized to SeHLan rather than SeMet. SeMet can be incorporated into a peptide sequence in the same manner as methionine (Met) in the plant translation machinery (Ogra *et al.*, 2008). Hence, the preferred metabolism to SeHLan is explainable by the fact that plants avoid translating proteins containing SeMet instead of Met.

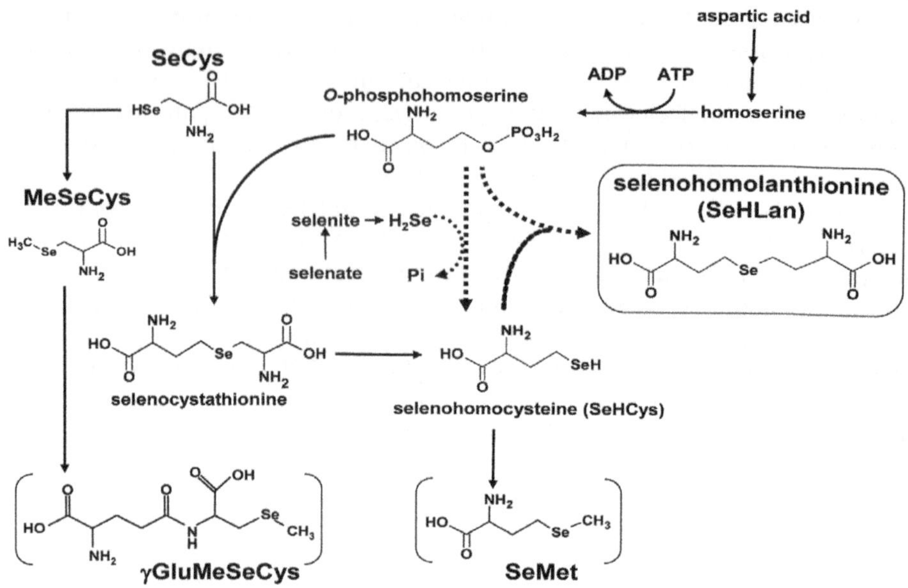

Figure 20.2 Proposed extended pathway for *trans*-selenation in selenized Japanese pungent radish. Proposed pathways for *de novo* synthesis of SeHLan are indicated by dotted lines. Se metabolites in brackets were not detected in this study but are important in other plants. γ-GluMeSeCys, γ-glutamyl-*Se*-methylselenocysteine; H_2Se, selenide; MeSeCys, *Se*-methylselenocysteine; SeCys, selenocysteine; SeHCys, selenohomocysteine; SeHLan, selenohomolanthionine; SeMet, selenomethionine.

20.3 Metabolic Pathway of Selenohomolanthionine in Animals

The distribution and metabolism of SeHLan were compared with those of SeMet in rat (Anan *et al.*, 2011). SeHLan or SeMet was injected intravenously at the bolus dose of 1.0 mg Se per kg body weight. SeMet was preferably accumulated in the pancreas, which increased the serum amylase level, an index of pancreatic damage (Tsuji *et al.*, 2010). In contrast, SeHLan was preferably accumulated in the kidneys, which increased serum creatinine level, an index of kidney damage. SeHLan clearly showed different pharmacokinetics and toxicity from SeMet by bolus administration: whereas SeMet was preferably distributed to the pancreas, SeHLan was accumulated in the kidneys. These specific distributions are reflected by their toxicities, *i.e.*, SeMet and SeHLan caused pancreatic and kidney damage, respectively. On the other hand, the levels of two major urinary selenometabolites, *Se*-methylseleno-*N*-acetyl-galactosamine (methyl-2-acetamido-2-deoxy-1-seleno-β-D-galactopyranoside, selenosugar) and trimethylselenonium ion (TMSe), were comparable between SeHLan- and SeMet-administered rats, suggesting that there may be no differences in the efficiency of metabolism of those two selenoamino acids into urinary selenometabolites despite the difference in distribution. SeHLan is expected to be a potential source of Se without inducing pancreatic damage. The kidney-specific toxicity of SeHLan may be avoided when SeHLan is administered at a lower dose than that used in the present study.

The proposed metabolic pathway of SeMet and SeHLan in mammals is summarized in Figure 20.3 (Ogra and Anan, 2009). It is suggested that

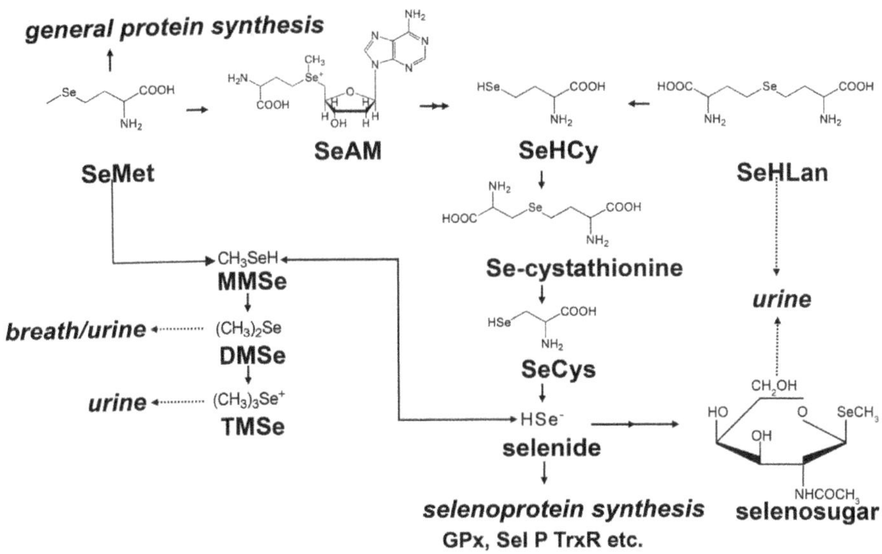

Figure 20.3 Proposed metabolic pathway of SeHLan and SeMet in animal cells.

SeMet participates in three metabolic pathways. SeMet mimics Met and shares metabolic pathways with Met. First, SeMet is used for peptide synthesis in place of Met because the translation machinery of animals cannot distinguish SeMet from Met. The pathways mentioned above may not contribute to Se utilization (selenoprotein synthesis). Secondly, SeMet is utilized for the biosynthesis of *Se*-adenosylselenomethionine (SeAM), an Se analog of *S*-adenosylmethionine (Ogra *et al.*, 2009). Then, SeAM is converted into SeCys *via* *Se*-adenosylselenohomocysteine (SeAH), SeHCys, and selenocystathionine in the *trans*-selenation pathway. The third route is a more specific pathway to SeMet than Met, *i.e.*, it involves the cleavage of methylselenol from SeMet by α,γ-lyase (Okuno *et al.*, 2005). The cleaved methylselenol is demethylated to form selenide or further methylated to form dimethylselenide and TMSe for the excretion of Se (Ohta and Suzuki, 2008). In contrast to SeMet, the metabolic pathway of SeHLan is simpler, *i.e.*, SeHLan is utilized only in the *trans*-selenation pathway for selenoprotein synthesis.

20.4 Antiseptic Effect of Selenohomolanthionine

Sodium selenite is used for the amelioration of sepsis and septic stress (Forceville, 2007). Although sodium selenite is more toxic than organic Se species, its antiseptic effect is promising when used at a bolus dose (Wang *et al.*, 2008). In clinical protocol, sodium selenite is intravenously administered to patients at doses of 4000 μg on the first day and 1000 μg per day on the subsequent nine days. These doses are extremely high compared to the recommended daily allowance of Se (30–50 μg per day). Thus, a safer and more effective Se compound than selenite is desired for clinical use. As mentioned above, SeHLan is the less toxic and more nutritive chemical form of Se. Then, the toxic and therapeutic effects of SeHLan in mouse were evaluated and compared with those of known Se species, such as selenite and SeMet. The therapeutic effect of Se compounds was determined from the extent of amelioration of septic stress in mice administered lipopolysaccharide (LPS).

The survival rate of the mice was 0.1 at 72 h after the injection of LPS alone at the dose of 15 mg kg^{-1} body weight (Figure 20.4). (Anan and Ogra, 2013) Sodium selenite, SeMet, and SeHLan improved the survival rate of mice injected with the above dose of LPS; in particular, SeHLan improved the survival rate from 0.1 to 0.5. Some mechanisms underlying the amelioration of septic stress by selenite were proposed. For instance, selenite acts as a redox modulator against oxidative damage induced by septic stress (Forceville, 2008). Selenite may boost the immune system (Rayman, 2012). Selenite recovers the amount of selenoprotein P (Sel P) that is reduced by septic stress (Hollenbach *et al.*, 2008). However, these mechanisms cannot explain unequivocally the effect of SeHLan and SeMet on septic stress. SeHLan and SeMet are redox-stable compounds in organisms. Although the

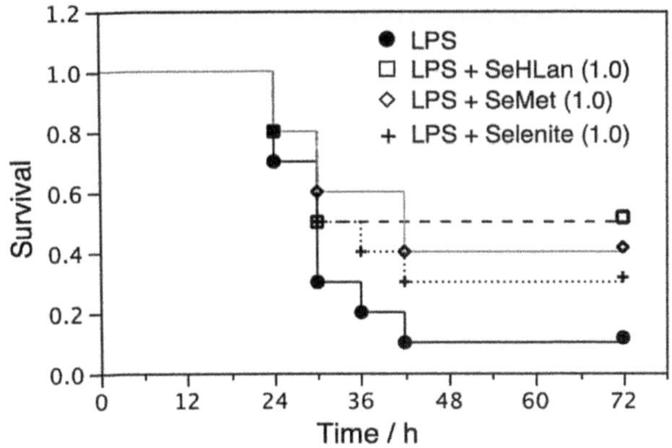

Figure 20.4 Protective role of Se compounds against LPS toxicity. Male ICR mice were intraperitoneally injected with LPS at the dose of 15 mg kg^{-1} body weight, and then intravenously injected with sodium selenite, SeMet, and SeHLan at the dose of 1.0 mg Se per kg body weight. The number of surviving mice was counted every 6 h over a 72 h period. Numbers in parenthesis show the dose in mg Se per kg body weight.

antioxidative effect of Sel P was proposed, the activity of Sel P as an antioxidant was lower than those of other Se-containing antioxidative enzymes, such as glutathione peroxidase and thioredoxin reductase (Takebe *et al.*, 2002). Thus, the actual biological function of Sel P is still controversial. Further studies are needed to clarify the precise mechanism of action of SeHLan and SeMet against septic stress.

Consequently, SeHLan more effectively improved the survival rate lowered by the LPS-induced septic stress than selenite and SeMet at the nontoxic dose of 1.0 mg Se per kg body weight. Overall, the results indicate that SeHLan can be used safely and effectively to treat sepsis in the clinical setting in place of selenite.

Summary Points

- This chapter focuses on selenohomolanthionine (SeHLan).
- Selenohomolanthionine (SeHLan) is the common name of 4,4′-selenobis[2-aminobutanoic acid] or *Se*-(3-amino-3-carboxypropyl)-homocysteine.
- SeHLan was first identified in an extract of Japanese pungent radish (*Raphanus sativus* L. cv. "Yukibijin") fortified with selenate, an inorganic form of selenium.
- 2,3-Dihydroxy-propionyl-selenohomolanthionine (2,3-DHP-SeHLan), a derivative of SeHLan, was identified in selenized yeast extract.
- SeHLan is less toxic than inorganic Se, and its tissue distribution differs from that of selenomethionine (SeMet) in experimental animals.
- SeHLan exhibited an antiseptic effect in mouse.

Key Facts of Selenohomolanthionine

1. SeHLan can be *de novo* synthesized in plants that have low Se-assimilating ability.
2. In these plants, inorganic selenium, such as selenate, is preferably converted into SeHLan rather than SeMet, thereby reducing the toxicity of inorganic selenium.
3. SeHLan can be metabolized in animals in the same manner as SeMet.
4. In contrast to SeMet, SeHLan does not accumulate in the pancreas.
5. SeHLan is able to reduce toxicity caused by septic stress.

Definitions and Explanations of Key Terms

Speciation. Speciation analysis is the analytical activity of identifying and/or measuring the quantities of one or more individual chemical species in a sample.

Selenosugar. This is a major urinary selenium metabolite, and consists of selenium and a sugar moiety which is *N*-acetylated or nonacetylated hexosamine.

List of Abbreviations

2,3-DHP-SeHLan	2,3-Dihydroxy-propionyl-selenohomolanthionine
DMSe	Dimethylselenide
γGluMeSeCys	γ-Glutamyl-*Se*-methylselenocysteine
MeSeCys	*Se*-Methylselenocysteine
Met	Methionine
MMSe	Monomethylselenol
SeAH	*Se*-Adenosylselenohomocysteine
SeAM	*Se*-Adenosylselenomethionine
SeCys	Selenocysteine
SeHCys	Selenohomocysteine
SeHLan	Selenohomolanthionine or 4,4′-selenobis[2-aminobutanoic acid] or Se-(3-amino-3-carboxypropyl)-homocysteine
Selenosugar	*Se*-Methylseleno-*N*-acetyl-galactosamine or methyl-2-acetamido-2-deoxy-1-seleno-β-D-galactopyranoside
SeMet	Selenomethionine
TMSe	Trimethylselenonium

References

Anan, Y., Mikami, T., Tsuji, Y. and Ogra, Y., 2011. Distribution and metabolism of selenohomolanthionine labeled with a stable isotope. *Analytical and Bioanalytical Chemistry*. 399: 1765–1772.

Anan, Y. and Ogra, Y., 2013. Toxicological and pharmacological analysis of selenohomolanthionine in mice. *Toxicology Research*. 2: 115–122.

Block, E., Glass, R. S., Jacobsen, N. E., Johnson, S., Kahakachchi, C., Kaminski, R., Skowronska, A., Boakye, H. T., Tyson, J. F. and Uden, P. C., 2004. Identification and synthesis of a novel selenium-sulfur amino acid found in selenized yeast: rapid indirect detection NMR methods for characterizing low-level organoselenium compounds in complex matrices. *Journal of Agricultural and Food Chemistry*. 52: 3761–3771.

Dernovics, M., Far, J. and Łobiński, R., 2009. Identification of anionic selenium species in Se-rich yeast by electrospray QTOF MS/MS and hybrid linear ion trap/orbitrap MS^n. *Metallomics*. 1: 317–329.

Far, J., Preud'homme, H. and Łobiński, R., 2010. Detection and identification of hydrophilic selenium compounds in selenium-rich yeast by size exclusion-microbore normal-phase HPLC with the on-line ICP-MS and electrospray Q-TOF-MS detection. *Analytica Chimica Acta*. 657: 175–190.

Forceville, X., 2007. Effects of high doses of selenium, as sodium selenite, in septic shock patients a placebo-controlled, randomized, double-blind, multi-center phase II study – selenium and sepsis. *Journal of Trace Elements in Medicine and Biology*. 21(suppl. 1): 62–65.

Forceville, X., 2008. The need for precise dose information of sodium selenite in the SIC study and rectification of GPx-3 plasma concentration. *Critical Care Medicine*. 36: 656–657.

Hacham, Y., Gophna, U. and Amir, R., 2003. In vivo analysis of various substrates utilized by cystathionine γ-synthase and *O*-acetylhomoserine sulfhydrylase in methionine biosynthesis. *Molecular Biology and Evolution*. 20: 1513–1520.

Hara, M., Eto, H. and Kuboi, T., 2001. Tissue printing for myrosinase activity in roots of turnip and Japanese radish and horseradish: a technique for localizing myrosinases. *Plant Science*. 160: 425–431.

Hollenbach, B., Morgenthaler, N. G., Struck, J., Alonso, C., Bergmann, A., Kohrle, J. and Schomburg, L., 2008. New assay for the measurement of selenoprotein P as a sepsis biomarker from serum. *Journal of Trace Elements in Medicine and Biology*. 22: 24–32.

Krömer, J. O., Heinzle, E., Schröder, H. and Wittmann, C., 2006. Accumulation of homolanthionine and activation of a novel pathway for isoleucine biosynthesis in *Corynebacterium glutamicum* McbR deletion strains. *Journal of Bacteriology*. 188: 609–618.

Ogra, Y., 2008. Integrated strategies for identification of selenometabolites in animal and plant samples. *Analytical and Bioanalytical Chemistry*. 390: 1685–1689.

Ogra, Y. and Anan, Y., 2009. Selenometabolomics: identification of selenometabolites and specification of their biological significance by complementary use of elemental and molecular mass spectrometry. *Journal of Analytical Atomic Spectrometry*. 24: 1477–1488.

Ogra, Y., Kitaguchi, T., Ishiwata, K., Suzuki, N., Iwashita, Y. and Suzuki, K. T., 2007. Identification of selenohomolanthionine in selenium-enriched

Japanese pungent radish. *Journal of Analytical Atomic Spectrometry*. 22: 1390–1396.

Ogra, Y., Kitaguchi, T., Ishiwata, K., Suzuki, N., Toida, T. and Suzuki, K. T., 2009. Speciation of selenomethionine metabolites in wheat germ extract. *Metallomics*. 1: 78–86.

Ogra, Y., Kitaguchi, T., Suzuki, N. and Suzuki, K. T., 2008. *In vitro* translation with [^{34}S]-labeled methionine, selenomethionine and telluromethionine. *Analytical and Bioanalytical Chemistry*. 390: 45–51.

Ohta, Y. and Suzuki, K. T., 2008. Methylation and demethylation of intermediates selenide and methylselenol in the metabolism of selenium. *Toxicology and Applied Pharmacology*. 226: 169–177.

Okuno, T., Motobayashi, S., Ueno, H. and Nakamuro, K., 2005. Identification of mouse selenomethionine alpha, gamma-elimination enzyme: cystathionine gamma-lyase catalyzes its reaction to generate methylselenol. *Biological Trace Element Research*. 108: 245–257.

Pedrero, Z., Madrid, Y. and Cámara, C., 2006. Selenium species bioaccessibility in enriched radish (*Raphanus sativus*): a potential dietary source of selenium. *Journal of Agricultural and Food Chemistry*. 54: 2412–2417.

Rayman, M. P., 2012. Selenium and human health. *Lancet*. 379: 1256–1268.

Takebe, G., Yarimizu, J., Saito, Y., Hayashi, T., Nakamura, H., Yodoi, J., Nagasawa, S. and Takahashi, K., 2002. A comparative study on the hydroperoxide and thiol specificity of the glutathione peroxidase family and selenoprotein P. *Journal of Biological Chemistry*. 277: 41254–41258.

Tsuji, Y., Mikami, T., Anan, Y. and Ogra, Y., 2010. Comparison of selenohomolanthionine and selenomethionine in terms of selenium distribution and toxicity in rats by bolus administration. *Metallomics*. 2: 412–418.

Wang, Z., Forceville, X., Van Antwerpen, P., Piagnerelli, M., Ahishakiye, D., Macours, P., De Backer, D., Neve, J. and Vincent, J. L., 2008. A large bolus, but not a continuous infusion, of sodium selenite improves outcome in peritonitis. *Shock*. 32: 140–146.

Winther, M. and Nielsen, P. V., 2006. Active packaging of cheese with allyl isothiocyanate, an alternative to modified atmosphere packaging. *Journal of Food Protection*. 69: 2430–2435.

Wróbel, K., Wróbel, K., Kannamkumarath, S. S., Caruso, J. A., Wysocka, I. A., Bulska, E., Swiatek, J. and Wierzbicka, M., 2004. HPLC-ICP-MS speciation of selenium in enriched onion leaves – a potential dietary source of Se-methylselenocysteine. *Food Chemistry*. 86: 617–623.

CHAPTER 21

Therapy with Selenium Cocktails and Co-use of Lycopene and Selenium

LETTERIA MINUTOLI[a], FRANCESCO SQUADRITO*[a], DOMENICA ALTAVILLA[b], AND HERBERT MARINI[a]

[a]Department of Clinical and Experimental Medicine, University of Messina, AOU Policlinico "G. Martino", Torre Biologica, 5th floor, Via C. Valeria Gazzi, 98125 Messina, Italy; [b]Department of Paediatric, Gynaecological, Microbiological and Biomedical Sciences, University of Messina, Italy
*E-mail: Francesco.Squadrito@unime.it

21.1 Selenium

21.1.1 General Aspects

Selenium (Se) is a mineral naturally present in many foods, and needed in the diet on a daily basis (Sunde 2006, 2012). It is also available in multivitamin/multimineral supplements and as a standalone supplement (Sunde, 2006). Plant foods are the major dietary sources of Se and the amount of Se in diet depends by several factors like soil Se concentrations, soil pH, geographic location, types and amounts of food consumed, *etc.*

Soil contains inorganic forms of Se, selenites and selenates that plants convert to organic forms, selenocysteine and selenomethionine and their methylated derivatives (Sunde, 2006; Terry and Diamond, 2012). Specifically,

Food and Nutritional Components in Focus No. 9
Selenium: Chemistry, Analysis, Function and Effects
Edited by Victor R. Preedy

Published by the Royal Society of Chemistry, www.rsc.org

Se occurs in foods as selenomethionine and the highest food sources of Se include Brazil nuts, fish, cereals, wheat germ, soybean and sunflower seeds (Sunde, 2012).

In the human body, skeletal muscle represents the major site of Se storage; moreover, elevated concentrations of Se are found in the liver, kidneys and thyroid gland (Terry and Diamond, 2012). Se is usually integrated into proteins to form selenoproteins (*i.e.* glutathione peroxidases, thioredoxin reductases, and iodothyronine deiodinases) which are involved in several biological functions as reproduction, thyroid hormone metabolism, DNA synthesis, and also show antioxidant and anti-inflammatory properties (Sunde, 2012).

The most commonly used measures of Se status are plasma and serum Se concentrations. Analyses of hair or nail Se content can be used to check longer-term Se intakes. Quantification of one or more selenoproteins is also employed to assess Se status in humans.

Intake recommendations for Se are provided in the Dietary Reference Intakes developed by the Food and Nutrition Board at the Institute of Medicine of the National Academies (National Research Council, 2000; Office of Dietary Supplements, 2013).

21.1.2 Selenium Deficiency

Human Se deficiency may occur in some countries (*e.g.* China, Tibet, and Siberia) where soil concentration of Se is low. There is evidence that Se deficiency in combination with additional stress factors may contribute to development of a form of heart disease (Keshan disease) (Chen, 2012), hypothyroidism (Zimmerman and Kohrle, 2002), osteoartritis (Kashin–Beck disease) (Jirong *et al.*, 2012), male infertility and a weakened immune system.

Overall, Se deficiency is rare but there are some groups of people at risk of Se inadequacy; patients living in Se-deficient regions or undergoing kidney dialysis or living with HIV or people who received total parenteral nutrition are typical examples.

21.1.3 Selenium and Cancer

Se might play a role in the prevention of cancer. Epidemiological evidence suggests that Se status is inversely associated with the risk to develop colorectal, prostate, lung, bladder, skin, esophageal, and gastric cancers (Dennert *et al.*, 2011). In humans, it has been shown that taking a daily supplement containing 200 μg of Se could reduce the risk of developing prostate, lung, and colorectal cancer (Clark *et al.*, 1996). Furthermore, it is reported that the men with the highest Se levels had a lower incidence of advanced prostate cancer (Yoshizawa *et al.*, 2005). The SU.VI.MAX study also reported that, after more than 7 years of follow-up, among men with normal prostate-specific antigen who had taken a combination of vitamin E, vitamin C, β-carotene, Se, and zinc there was a significant reduction in the rate of prostate cancer (Meyer *et al.*, 2005).

However, to date, taking into account that other clinical trials evaluating Se supplementation for cancer prevention have yielded conflicting results, the overall evidence is limited and not conclusive. Consequently, more research is needed to confirm the relationship between Se concentrations and cancer risk and to determine whether Se supplements can help prevent any form of cancer.

21.1.4 Selenium and Other Human Diseases

Because selenoproteins help prevent the oxidative modification of lipids, reducing inflammation and showing antiaggregant effects (Rayman, 2012), it has been suggested that Se supplementation could reduce the risk of cardiovascular disease or death associated with cardiovascular disease. However, the role of Se in cardiovascular disease is uncertain. The limited clinical-trial evidence, to date, does not support the use of Se supplements for preventing heart disease.

Se supplements are also investigated in order to prevent or treat cognitive decline in elderly people because it is well known that serum Se concentrations decline with age (Akbaraly *et al.*, 2007). Consequently, decreases in antioxidant activity of Se could negatively impact memory or other cognitive functions. However, regarding this specific topic, clinical evidence is insufficient to determine whether Se supplements can support an improvement of brain functions.

As before mentioned, Se has important functions in thyroid hormone synthesis and metabolism. Epidemiological evidence supports a relationship between Se levels and thyroid gland function. The SU.VI.MAX study indicated, particularly in women, an inverse relationship between serum Se concentrations and thyroid volume and echostructure, in people with mild iodine deficiency (Derumeaux *et al.*, 2003). Moreover, a Cochrane review indicated that selenomethionine supplementation is a promising strategy in pregnant women with hypothyroidism, especially for reducing postpartum thyroiditis (Reid *et al.*, 2010). However, considering that other studies investigating Se supplementation in patients with thyroid disease have obtained mixed results, additional research is needed to determine whether Se supplements can be used to prevent or treat thyroid disease.

21.1.5 Selenium Toxicity

High blood levels of Se can cause selenosis, a syndrome characterized by gastrointestinal symptoms, hair loss, white blotchy nails, garlic breath odor (caused by methylated Se), fatigue, irritability, and peripheral neuropathy (Koller and Exon, 1986). However, Se toxicity is generally rare. Brazil nuts could cause selenium toxicity if consumed regularly. The Institute of Medicine of the National Academy of Sciences has set a tolerable upper intake level for Se (400 μg per day) for adults to prevent the risk of developing selenosis (Table 21.1).

Table 21.1 This table lists the main features and therapeutic potential of use of selenium.

General aspects	Biological effects	Selenium-related pathologies
It is a mineral naturally present in many foods and needed in the diet on daily basis.	It is usually integrated into proteins to form selenoproteins that are involved in reproduction thyroid hormone metabolism, DNA synthesis, and also show antioxidant and anti-inflammatory properties.	Selenium deficiency: Keshan disease, hypothyroidism, osteoarthritis (Kashin–Beck disease), male infertility and a weakened immune system.
Highest food sources include Brazil nuts, fish, cereals, wheat, germ, soyabean and sunflower seeds.	It might play a role in the prevention of cancer.	Selenium deficiency: patients undergoing kidney dialysis or living with HIV or people who received total parenteral nutrition.
It is also available in multivitamin/multimineral supplements and as a standalone supplement.	Se supplementation: • Could reduce the risk of cardiovascular disease or death associated with cardiovascular disease. • Could be useful to prevent or treat cognitive decline in elderly people. • Could be a promising strategy in pregnant women with hypothyroidism.	High blood levels of selenium can cause selenosis.

21.2 Lycopene

21.2.1 General Aspects

Lycopene (Ly), a non-provitamin A carotenoid, is considered to be the major active component in tomatoes (Stahl *et al.*, 1992) although apricots, guava, watermelon, papaya, and pink grapefruit are also significant sources. It is twice as effective as β-carotene and 10-fold more active than α-tocopherol as an antioxidant and anti-inflammatory compound (Di Mascio *et al.*, 1989). Ly is not a necessary nutrient. However, like other substances found in fruits and vegetables, it may be very important for optimal human health (Linus Pauling Institute Micronutrient Research for Optimum Health Micronutrient Information Center).

Studies that looked at Ly levels in the blood found that levels were higher after people ate cooked tomatoes than after they ate raw tomatoes or drank tomato juice. Moreover, adding to Ly-rich food a small amount of oil or fat increases the amount of Ly absorbed by the intestines. Ly is also available in soft-gel capsule and liquid supplements. Much evidence indicates that Ly exhibits multiple biological functions, such as antioxidant activity, antiangiogenic, anti-inflammatory and anticancer properties. The amount of Ly found helpful in studies generally fell in the range of 4 to 8 mg, daily.

21.2.2 Lycopene and Cancer

In recent years, the role of the Ly in preventing cancer has been an interesting area of research and many investigators focused their attention particularly on prostate cancer.

Ly is known to accumulate in the prostate gland in high concentrations and is found in human semen. The abundance of Ly in prostatic tissue is indirectly implicated in the chemoprevention of pathologies likely to affect the prostate gland in the elderly male, such as benign prostatic hyperplasia (BPH) as well as prostatic cancer. In BPH these actions are thought to be mediated through various mechanisms including inhibition of cell growth in normal prostatic epithelial cells and/or induction of apoptosis in hyperplastic prostatic tissue (Bowen *et al.*, 2002; Obermüller-Jevic *et al.*, 2003).

Additionally, several mechanisms of action are implicated in the ability of Ly to prevent the development and progression of prostate cancer, including reduction of oxidative DNA damage in prostatic tissue, initiating upregulation of gap-junction proteins to enable improved intercellular communications and a reduction of local androgen signaling (Siler *et al.*, 2004). Moreover, Se and Ly supplementation reduced cisplatin-induced testicular toxicity, improved testicular function and prevented cisplatin-related injury to the rat testes by suppression of oxidative stress (Salem *et al.*, 2012).

Evidence from case-control and cohort studies suggests that high levels of dietary Ly intake are associated with a lower risk of prostate cancer and cell proliferation (Etminan *et al.*, 2004; Giovannucci, 2005). The underlying mechanism may be inhibition of 5-α reductase and interleukin-6 signaling, as demonstrated in benign prostate tissue of rats.

The Health Professionals Follow-Up Study showed a strong inverse relationship between the risk of developing prostate cancer and increased consumption of a tomato-enriched diet (Giovannucci *et al.*, 1995). Accordingly, in 2007, the World Cancer Research Fund reported that Ly, in the context of a high fruit and vegetable intake, may be useful in reducing the risk of cancer, including for prostate cancer. Other studies, however, found no link between tomato products or other Ly-rich foods and cancer (World Cancer Research Fund, 2007).

21.2.3 Lycopene and Other Human Diseases

Several observational studies suggest that foods containing Ly may help prevent macular degeneration, cataracts and cardiovascular disease. Moreover, another study evaluated the possible benefits of lycopene for pregnant women in reducing the risk of pre-eclampsia (Sharma *et al.*, 2003). In addition, use of lycopene appeared to help prevent inadequate growth of the fetus.

In a double-blind, placebo-controlled study, it has been shown that Ly was more effective than placebo for reducing signs and symptoms of leukoplakia (Singh *et al.*, 2004). Moreover, Ly has also shown interesting therapeutic activity for other pathological mouth conditions such as oral submucous fibrosis (Kumar *et al.*, 2007) and gingivitis (Chandra *et al.*, 2007). Finally, weaker evidence suggests that Ly or a standardized tomato extract containing Ly might be helpful for treating a number of medical conditions, including hypertension, breast cancer and male infertility, and for preventing heart disease, sunburn, and testicular damage caused by the cancer chemotherapy drug adriamycin.

Overall, results of clinical studies are not easily comparable because the researchers have used: (i) tomato or tomato-based foods or Ly supplements, other combinations with different dosage regimens as the active treatment; (ii) different design of the study, and (iii) the trials were of different duration. Since tomatoes also contain vitamins, potassium, and other carotenoids and antioxidants, it may be that other compounds in tomatoes may account for some of the protective effects ascribed to Ly and mentioned before. Moreover, it is also possible that different genetics, lifestyles and habits of populations examined could influence the positive results obtained. Here too, although Ly represents a promising functional natural pigment, additional research is needed to determine whether Ly supplements can be useful to prevent or treat different types of human diseases (Table 21.2).

21.3 Co-use of Selenium, Lycopene and *Serenoa repens*

Phytotherapeutic supplements, mainly based on Saw Palmetto-derived *Serenoa repens*, are numerous and used frequently. *Serenoa repens* is the sole species currently classified in the genus *Serenoa*. It has been known by a number of synonyms, including *Sabal serrulatum* (this term is particularly used in alternative medicine).

A recent paper suggested a prominent growth of prostate following testosterone administration and a consequent increase in its weight, with the typical histological features of BPH (Altavilla *et al.*, 2011). The combined treatment with Se-Ly-*Serenoa repens* was more effective than *Serenoa repens* alone in preventing BPH, then suggesting that Se and Ly further increase *Serenoa repens* efficacy in BPH. It has been supposed that prostate growth inhibition by Se-Ly-*Serenoa repens* association was stimulated through caspase-9 as well as an independent mechanism involving the proapoptotic Bax and the antiapoptotic Bcl-2 gene. Moreover, it has been shown that, during testosterone-induced prostate growth, there is an overexpression of the EGF that was prevented by treatment with Se-Ly-*Serenoa repens* combination (Altavilla *et al.*, 2011).

These data are also supported by previous findings in the bladder-obstruction model, in which a significant reduction of inflammatory status (including tumor necrosis factor-α, an important BPH inflammatory marker) has been observed (Bonvissuto *et al.*, 2011). Overall, in the same experimental approach it has been shown that a co-use of Se-Ly-*Serenoa repens* is more effective than *Serenoa repens* alone in reducing prostate inflammatory response, growth factor expression, oxidative stress and histological features (Bonvissuto *et al.*, 2011). The effects of *Serenoa repens*, Se and Ly, alone or in association, on the expression of four IAPs proteins cIAP1, cIAP2, NAIP and survivin, are currently under investigation in rats with experimental testosterone-dependent BPH. Our preliminary observations suggest that the enhancing efficacy of the association might be a direct consequence of a more pronounced ability of this combination of reawakening the apoptosis program and identifying in IAPs the molecular target. Furthermore, among the three compounds, Ly appears to give the major contribution in

Table 21.2 This table lists the main features and therapeutic potential of use of lycopene.

General aspects	Biological effects
Lycopene is not a necessary nutrient.	It exhibits antioxidant activity, anti-angiogenic, and anti-inflammatory properties.
Lycopene, a non-provitamin A carotenoid, is the red pigment of tomatoes although apricots, guava, watermelon, and pink grapefruit are also significant sources.	It might play a role in the prevention of cancer.
It is available in soft-gel capsule and liquid supplements.	It may help prevent macular degeneration, cataracts and cardiovascular disease.
	Possible benefits for pregnant women.
	Interesting therapeutic activity for pathological mouth conditions.

maximizing the effects of Se-Ly-*Serenoa repens*-induced activation of the programmed cell death (Minutoli *et al.*, 2013).

The efficacy and safety of *Serenoa repens* plus Se and Ly (Profluss) *versus* *Serenoa repens* alone were evaluated in patients suffering from category IIIa chronic prostatitis/chronic pelvic pain syndrome (CP/CPPS) (Morgia *et al.*, 2010). One hundred and two patients with IIIa CP/CPPS were enrolled and randomized into two groups each to receive *Serenoa repens* alone or in combination for 8 weeks. Evaluation was based on results of the National Institutes of Health–Chronic Prostatitis Symptom Index (NIH–CPSI), IPSS, maximum peak flow rate (MPFR), and PSA measurements at baseline and at weeks 4 and 8 after the end of treatment. The primary endpoint was a >50% reduction in NIH–CPSI score. Secondary endpoints evaluated were MPFR, IPSS, PSA and white blood cell count. This study confirms that Se-Ly-*Serenoa repens* association is safe and well tolerated and it improves symptoms associated with IIIa CP/CPPS (Morgia *et al.*, 2010).

Additionally, Morgia and coworkers (Morgia *et al.*, 2013) showed that the combined use of Se-Ly-*Serenoa repens* may have an anti-inflammatory activity that could be of interest in the treatment of prostatic chronic inflammation in BPH and/or prostatic intraepithelial neoplasia/atypical small acinar proliferation patients (Tables 21.3 and 21.4).

Table 21.3 This table lists the main features and therapeutic potential of use of *Serenoa repens.*

General aspects	Biological effects
Serenoa repens is commonly known as Saw Palmetto. It is particularly abundant in the Southeastern United States.	It exhibits anti-inflammatory and cell growth modulation effects.
The fruits of the Saw Palmetto are highly enriched with fatty acids and phytosterols. It is also available in supplements.	*Serenoa repens* extract has been promoted as useful for people with benign prostatic hyperplasia/prostate cancer.

Table 21.4 This table lists the biological properties and therapeutic potential of use of selenium, lycopene and co-use of selenium, lycopene and *Serenoa repens.*

Selenium	Lycopene	Selenium, lycopene and *Serenoa repens*
Antioxidant and anti-inflammatory properties	Antioxidant, antiangiogenic and anti-inflammatory properties	Anti-inflammatory and antioxidant properties
Antiaggregant effects	Prevention of macular degeneration, cataracts and cardiovascular disease	Treatment of BPH
Prevention of cancer	Prevention of cancer	Antiproliferative properties

21.4 Conclusions

To date, results of clinical studies investigating the effects of selenium as well as lycopene are not easily comparable and additional research is needed to determine whether these nutraceutics, alone or in association, can help prevent or treat different types of human diseases. *Serenoa repens* is frequently associated with selenium and lycopene to maximize their therapeutic effects. It has been reported that the selenium-lycopene-*Serenoa repens* association has greater and enhanced anti-inflammatory activity. In addition, the selenium-lycopene-*Serenoa repens* combination more efficiently inhibits prostate enlargement by inhibiting growth factors expression and inducing cell programmed death. Positive results by co-use of selenium, lycopene and *Serenoa repens* are observed in patients suffering from category IIIa chronic prostatitis/chronic pelvic pain syndrome. Moreover, the combination of selenium, lycopene and *Serenoa repens* may have interesting therapeutic activity in the treatment of prostatic chronic inflammation in BPH and/or prostatic intraepithelial neoplasia/atypical small acinar proliferation patients.

Summary Points

- Selenium might play a role in the prevention of cancer and other human diseases.
- Clinical evidence investigating the effects of selenium on human health is limited and not conclusive.
- Much evidence indicates that lycopene exhibits multiple biological functions, such as antioxidant, antiangiogenic, anti-inflammatory and anticancer properties.
- Results of clinical studies investigating the effects of lycopene are not easily comparable and additional research is needed to determine whether lycopene, alone or in association, can help prevent or treat different types of cancer in humans.
- Phytotherapeutic supplements, mainly based on Saw Palmetto-derived *Serenoa repens*, are numerous and used frequently.
- The combined treatment with selenium, lycopene and *Serenoa repens* was experimentally more effective than *Serenoa repens* alone in preventing BPH, then suggesting that selenium and lycopene further increase *Serenoa repens* efficacy in BPH.
- Experimental findings in the bladder-obstruction model, support the use of selenium-lycopene-*Serenoa repens* association in the treatment of BPH and in prevention of prostate cancer.
- Among the three compounds, lycopene appears to give the major contribution in maximizing the effects of therapeutic cocktail through the induced activation of the programmed cell death.
- The association of selenium, lycopene and *Serenoa repens* is safe and well tolerated and it improves symptoms associated with chronic prostatitis/chronic pelvic pain syndrome category IIIA in patients.

- The combination of selenium, lycopene and *Serenoa repens* may be useful in the treatment of prostatic chronic inflammation in BPH and/or prostatic intraepithelial neoplasia/atypical small acinar proliferation patients.

Key Facts of Selenium, Lycopene and *Serenoa repens*

1. The micronutrient selenium might play an important role in human health.
2. Selenium deficiency is rare but there are some groups of people at risk of selenium inadequacy.
3. Lycopene is not a necessary nutrient. However, it may be very important for optimal health in humans.
4. The role of lycopene in preventing cancer has been an attractive area of research, as well as observational studies suggest that foods containing lycopene may help prevent several human diseases.
5. *Serenoa repens* is one of several phytotherapeutic agents available for the treatment of BPH/prostate cancer.
6. The association of selenium, lycopene and *Serenoa repens* is safe and well tolerated.
7. Selenium and lycopene further increase *Serenoa repens* efficacy in BPH and/or in other medical conditions.
8. Lycopene is crucial in maximizing the effects of this therapeutic cocktail.

List of Abbreviations

BPH Benign prostatic hyperplasia/hypertrophy is a histologic diagnosis that refers to the nonmalignant proliferation of smooth muscle and epithelial cells of the prostate.

CP/CPPS Chronic prostatitis/chronic pelvic pain syndrome is a syndrome characterized by pelvic pain and painful ejaculation, associated with urinary symptoms and/or sexual dysfunction. CP/CPPS affects about 10–15% of the male population: it is associated with a high rate of mental disorders, such as anxiety and depression, and has a dramatic effect on quality of life (QoL). The etiology of the syndrome is still poorly understood.

Ly Lycopene a non-provitamin A carotenoid, is considered to be the major active component in tomatoes, although apricots, guava, watermelon, papaya, and pink grapefruit are also significant sources.

MPFR Maximum peak flow rate.

NIH–CPSI National Institutes of Health–Chronic Prostatitis Symptom Index is a formally developed and psychometrically validated instrument for the evaluation of CP/CPPS symptoms. It contains 13 items that are scored in three discrete domains: pain, urinary symptoms, and the impact on quality of life (QoL).

PSA Prostate-specific antigen. It is often elevated in the presence of prostate cancer or other prostate disorders. PSA is not a unique indicator of prostate cancer, but may also detect prostatitis or benign prostatic hyperplasia.

Se Selenium is a mineral naturally present in many foods, and needed in the diet on a daily basis. It is also available in multivitamin/multimineral supplements and as a standalone supplement.

References

Akbaraly, T. N., Hininger-Favier, I., Carrière, I., Arnaud, J., Gourlet, V., Roussel, A. M. and Berr, C., 2007. Plasma selenium over time and cognitive decline in the elderly. *Epidemiology*. 18: 52–58.

Altavilla, D., Bitto, A., Polito, F., Irrera, N., Marini, H., Arena, S., Favilla, V., Squadrito, F., Morgia, G. and Minutoli, L., 2011. The combination of *Serenoa repens*, selenium and lycopene is more effective than *Serenoa repens* alone to prevent hormone dependent prostatic growth. *Journal of Urology*. 186: 1524–1529.

Bonvissuto, G., Minutoli, L., Morgia, G., Bitto, A., Polito, F., Irrera, N., Marini, H., Squadrito, F. and Altavilla, D., 2011. Effects of *Serenoa repens*, lycopene and selenium, on proinflammatory phenotype activation: an "*in vitro*" and "*in vivo*" comparison study. *Urology*. 77: 248. e9–e16.

Bowen, P., Chen, L., Stacewicz-Sapuntzakis, M., Duncan, C., Sharifi, R., Ghosh, L., Kim, H. S., Christov-Tzelkov, K. and van Breemen, R., 2002. Tomato sauce supplementation and prostate cancer: lycopene accumulation and modulation of biomarkers of carcinogenesis. *Experimental Biology and Medicine (Maywood)*. 227: 886–893.

Chandra, R. V., Prabhuji, M. L., Roopa, D. A., Ravirajan, S. and Kishore, H. C., 2007. Efficacy of lycopene in the treatment of gingivitis: a randomised, placebo-controlled clinical trial. *Oral Health and Preventive Dentistry*. 5: 327–336.

Chen, J., 2012. An original discovery: selenium deficiency and Keshan disease (an endemic heart disease). *Asia Pacific Journal of Clinical Nutrition*. 21: 320–326.

Clark, L. C., Combs Jr, G. F., Turnbull, B. W., Slate, E. H., Chalker, D. K., Chow, J., Davis, L. S., Glover, R. A., Graham, G. F., Gross, E. G., Krongrad, A., Lesher Jr, J. L., Park, H. K., Sanders Jr, B. B., Smith, C. L. and Taylor, J. R., 1996. Effects of selenium supplementation for cancer prevention in patients with carcinoma of the skin. A randomized controlled trial. Nutritional prevention of cancer study group. *Journal of the American Medical Association*. 276: 1957–1963.

Dennert, G., Zwahlen, M., Brinkman, M., Vinceti, M., Zeegers, M. P. and Horneber, M., 2011. Selenium for preventing cancer. *Cochrane Database of Systematic Review*. CD005195.

Derumeaux, H., Valeix, P., Castetbon, K., Bensimon, M., Boutron-Ruault, M. C., Arnaud, J. and Hercberg, S., 2003. Association of selenium with thyroid volume and echostructure in 35- to 60-year-old French adults. *European Journal of Endocrinology*. 148: 309–315.

Di Mascio, P., Kaiser, S. and Sies, H., 1989. Lycopene as the most efficient biological carotenoid singlet oxygen quencher. *Archives of Biochemistry and Biophysics*. 274: 532–538.

Etminan, M., Takkouche, B. and Caamano-Isorna, F., 2004. The role of tomato products and lycopene in the prevention of prostate cancer: a meta-analysis of observational studies. *Cancer Epidemiology Biomarkers Prevention*. 13: 340–345.

Giovannucci, E., Ascherio, A., Rimm, E. B., Stampfer, M. J., Colditz, G. A. and Willett, W. C., 1995. Intake of carotenoids and retinol in relation to risk of prostate cancer. *Journal of the National Cancer Institute*. 87: 1767–1776.

Giovannucci, E., 2005. Tomato products, lycopene and prostate cancer: a review of the epidemiological literature. *Journal of Nutrition*. 135: 2030S–2031S.

Jirong, Y., Huiyun, P., Zhongzhe, Y., Birong, D., Weimin, L. and Ming, Y., 2012. Sodium selenite for treatment of Kashin–Beck disease in children: a systematic review of randomised controlled trials. *Osteoarthritis Cartilage*. 20: 605–613.

Koller, L. D. and Exon, J. H., 1986. The two faces of selenium-deficiency and toxicity are similar in animals and man. *Canadian Journal of Veterinary Research*. 50: 297–306.

Kumar, A., Bagewadi, A., Keluskar, V. and Singh, M., 2007. Efficacy of lycopene in the management of oral submucous fibrosis. *Oral surgery, oral medicine, oral pathology, oral radiology, and endodontics*. 103: 207–213.

Linus Pauling Institute Micronutrient Research for Optimum Health Micronutrient Information Center. Available at: http://lpi.oregonstate.edu/infocenter/phytochemicals/carotenoids/.

Meyer, F., Galan, P., Douville, P., Bairati, I., Kegle, P., Bertrais, S., Estaquio, C. and Hercberg, S., 2005. Antioxidant vitamin and mineral supplementation in the SU.VI.MAX trial. *International Journal of Cancer*. 116: 182–186.

Minutoli, L., Bitto, A., Squadrito, F., Marini, H., Irrera, N., Morgia, G., Passantino, A. and Altavilla, D., 2013. *Serenoa repens*, lycopene and selenium: a triple therapeutic approach to manage benign prostatic hyperplasia. *Current Medicinal Chemistry*. 20: 1306–1312.

Morgia, G., Mucciard,i, G., Galì, A., Madonia, M., Marchese, F., Di Benedetto, A., Romano, G., Bonvissuto, G., Castelli, T., Macchione, L. and Magno, C., 2010. Treatment of chronic prostatitis/chronic pelvic pain syndrome category IIIA with *Serenoa repens* plus selenium and lycopene (Profluss) versus *S. repens* alone: an Italian randomized multicenter-controlled study. *Urologia Internationalis*. 84: 400–406.

Morgia, G., Cimino, S., Favilla, V., Russo, G. I., Squadrito, F., Mucciardi, G., Masieri, L., Minutoli, L., Grosso, G. and Castelli, T., 2013. Effects of *Serenoa repens*, selenium and lycopene (Profluss®) on chronic inflammation associated with benign prostatic hyperplasia: results of "FLOG" (Flogosis and Profluss in Prostatic and Genital Disease), a multicentre Italian study. *International Brazilian journal of urology*. 39: 214–221.

National Research Council, 2000. *Dietary Reference Intakes for Vitamin C, Vitamin E, Selenium, and Carotenoids*, The National Academies Press, Washington, DC.

Obermüller-Jevic, U. C., Olano-Martin, E., Corbacho, A. M., Eiserich, J. P., van der Vliet, A., Valacchi, G., Cross, C. E. and Packer, L., 2003. Lycopene inhibits the growth of normal human prostate epithelial cell *in vitro*. *Journal of Nutrition*. 133: 3356–3360.

Office of Dietary Supplements National Institute of Health Dietary Supplement Fact Sheet: Selenium. Reviewed: July 02, 2013. Available at: http://ods.od.nih.gov/factsheets/selenium.

Rayman, M. P., 2012. Selenium and human health. *Lancet*. 379: 1256–1268.

Reid, S. M., Middleton, P., Cossich, M. C. and Crowther, C. A., 2010. Interventions for clinical and subclinical hypothyroidism in pregnancy. *Cochrane Database of Systematic Review*. CD007752.

Salem, E. A., Salem, N. A., Maarouf, A. M., Serefoglu, E. C. and Hellstrom, W. J., 2012. Selenium and lycopene attenuate cisplatin-induced testicular toxicity associated with oxidative stress in Wistar rats. *Urology*. 79: 1184. e1–e6.

Sharma, J. B., Kumar, A., Kumar, A., Malhotra, M., Arora, R., Prasad, S. and Batra, S., 2003. Effect of lycopene on pre-eclampsia and intra-uterine growth retardation in primigravidas. *International Journal of Gynecology and Obstetrics*. 81: 257–262.

Siler, U., Barella, L., Spitzer, V., Schnorr, J., Lein, M., Goralczyk, R. and Wertz, K., 2004. Lycopene and vitamin E interfere with autocrine/paracrine loops in the dunning prostate cancer model. *FASEB Journal*. 18: 1019–1021.

Singh, M., Krishanappa, R., Bagewadi, A. and Keluskar, V., 2004. Efficacy of oral lycopene in the treatment of oral leukoplakia. *Oral Oncology*. 40: 591–596.

Stahl, W., Schwarz, W., Sundquist, A. R. and Sies, H., 1992. *cis-trans* Isomers of lycopene and beta-carotene in human serum and tissues. *Archives of Biochemistry and Biophysics*. 294: 173–177.

Sunde, R. A., 2006. Selenium. In: Bowman, B. and Russell, R. (ed.) *Present Knowledge in Nutrition*. 9th edn, International Life Sciences Institute, Washington, DC, pp. 480–497.

Sunde, R. A., 2012. Selenium. In: Ross, A. C., Caballero, B., Cousins, R. J., Tucker, K. L. and Ziegler, T. R. (ed.) *Modern Nutrition in Health and Disease*. 11th edn, Lippincott Williams & Wilkins, Philadelphia, PA, pp. 225–237.

Terry, E. N. and Diamond, A. M., 2012. Selenium. In: Erdman, J. W., Macdonald, I. A. and Zeisel, S. H. (ed.) *Present Knowledge in Nutrition*. 10th edn, Wiley-Blackwell, Washington, DC, pp. 568–587.

World Cancer Research Fund/American Institute for Cancer Research, 2007. *Food, Nutrition, Physical Activity and the Prevention of cancer: A Global Perspective*, AICR, Washington, DC.

Yoshizawa, K., Willett, W. C., Morris, S. J., Stampfer, M. J., Spiegelman, D., Rimm, E. B. and Hercberg, S., 2005. Antioxidant vitamin and mineral supplementation in the SU.VI.MAX trial. *International Journal of Cancer*. 116: 182–186.

Zimmerman, M. B. and Kohrle, J., 2002. The impact of iron and selenium deficiencies on iodine and thyroid metabolism: biochemistry and relevance to public health. *Thyroid*. 12: 867–878.

CHAPTER 22

Selenium and Cancer

KATARZYNA JAWORSKA-BIENIEK[†*a], MARCIN LENER[†*a], MAGDALENA MUSZYŃSKA[a,b], PABLO SERRANO-FERNÁNDEZ[a], GRZEGORZ SUKIENNICKI[a], KATARZYNA DURDA[a], TOMASZ GROMOWSKI[a], SATISH GUPTA[a,c], JÓZEF KŁADNY[d], ANNA WIECHOWSKA-KOZŁOWSKA[e], TOMASZ GRODZKI[f], EWA JAWOROWSKA[g], JAKUB LUBIŃSKI[g], BARBARA GÓRECKA-SZYLD[h], GRAŻYNA WILK[h], TOMASZ HUZARSKI[a], TOMASZ BYRSKI[a], CEZARY CYBULSKI[a], JACEK GRONWALD[a], TADEUSZ DĘBNIAK[a], OLGIERD ASHURYK[a], ALEKSANDRA TOŁOCZKO-GRABAREK[a], ANTONI MORAWSKI[i,b], RODNEY J. SCOTT[j], ANNA JAKUBOWSKA[a,b], AND JAN LUBIŃSKI[a,b]

[a]Department of Genetics and Pathology, International Hereditary Cancer Center, Pomeranian Medical University, Szczecin, Poland; [b]Read-Gene SA, Grzepnica, Poland; [c]Postgraduate School of Molecular Medicine, Medical University of Warsaw, Poland; [d]Department of General and Oncological Surgery, Pomeranian Medical University, Szczecin, Poland; [e]Laboratory of Endoscopy, Division of Heath Care, Ministry of Internal Affairs and Administration in Szczecin, Poland; [f]Specialized Hospital for Lung Diseases, Szczecin-Zdunowo, Poland; [g]Department of Otolaryngology and Laryngological Oncology, Pomeranian Medical University, Szczecin, Poland; [h]Department of General and Dental Radiology, Pomeranian Medical University, Szczecin, Poland; [i]Institute of Inorganic Chemical Technology and Environmental Engineering, West Pomeranian University of Technology,

[†]These authors contributed equally to this work.

Food and Nutritional Components in Focus No. 9
Selenium: Chemistry, Analysis, Function and Effects
Edited by Victor R. Preedy

Published by the Royal Society of Chemistry, www.rsc.org

Szczecin, Poland; [j]Discipline of Medical Genetics, School of Biomedical Sciences, Faculty of Health, University of Newcastle and The Hunter Medical Research Institute, Newcastle, New South Wales, Australia
*E-mail: marcinlener@poczta.onet.pl, ka_jaworska@wp.pl

22.1 Introduction

Selenium is a micronutrient required for normal development and growth of the organism. In the body, selenium is incorporated into several proteins as selenocysteine. As part of selenoproteins selenium plays an enzymatic and structural function. It stimulates the immune system to increase the production of antibodies and results in the increased activity of immune cells (Combs *et al.*, 2001). It has been suggested that selenium may influence the risk of cancer not only by the enhancement of immune cell activity but also by protection against free radicals. The protective role of selenium against oxidative damage is related to its presence in the active center of antioxidant enzymes. One of these is glutathione peroxidase – GPX, which protects cellular components such as DNA (deoxyribonucleic acid) and lipids against the damaging action of H_2O_2 and of other organic peroxides produced as a result of the changes in the body (Zachara, 1992; Torun *et al.*, 1995).

22.2 Forms of Selenium and Its Metabolism in Organisms

Selenium is a microelement, which can exist in -II, -IV and -VI oxidation states (Borawska *et al.*, 2009). Experimental studies have revealed that the bioavailability of selenium is dependent on the chemical form of this element, affecting its intracellular distribution in the body. Selenium, delivered into the body from food or dietary supplements, occurs in organic forms such as selenomethionine (Se-Met) and selenocysteine (Se-Cys) and inorganic ones – selenites(IV) or selenates(VI) (Combs, 1997). In the body, metabolism of organic and inorganic forms of selenium is different but the final products of both metabolic transformations are selenides (H_2Se), which are directly used as a donor of selenium for the synthesis of selenoproteins. Selenium from exogenous Se-Cys is released directly as a selenide, whereas Se-Met (mainly occurring in vegetables), must be first converted to selenocysteine (a form not used in the protein synthesis) before it can become a source of selenides. In the case of inorganic forms of selenium (selenate, selenite) selenium is reduced to selenides through an intermediate form – selenodiglutathione (GS–Se–GS) using NADPH (Nicotinamide Adenine Dinucleotide Phosphate) and glutathione reductase. Selenides, as mentioned above, are the only active donors of selenium in the synthesis of amino acid selenocysteine. This amino acid is synthesized by incorporating selenium from the selenides to the amino acid serine by unique tRNASec (transfer ribonucleic acid specific for selenocysteine). Selenocysteine formed in this process is then used in protein biosynthesis. In studies on selenium metabolism it was observed that its methyl derivatives are removed from the body *via* the respiratory system feces or in urine (Combs and Gray, 1998; Figure 22.1).

Based on the above data on selenium metabolism it seems that methylated selenium compounds are not really useful in supplementation. Furthermore in our opinion, we find it also questionable to use Se-Met for supplementation. Se-Met can be incorporated into proteins nonspecifically, *i.e.* after

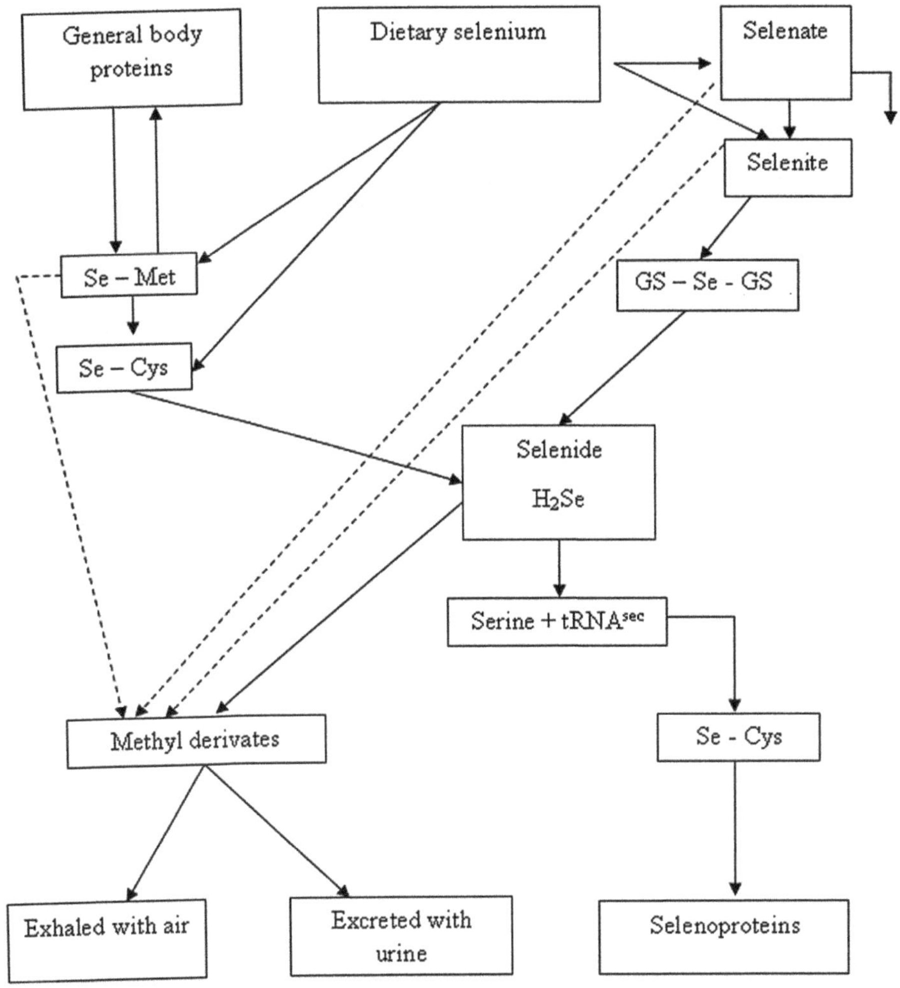

Figure 22.1 Scheme of selenium metabolism (Jacques, 2001 after modification; with Alltech Inc. permission). Scheme of selenium metabolism. Organic and inorganic forms of selenium compounds are transformed into the selenide. Se-Met can be converted to Se-Cys but also nonspecifically incorporated to nonselenoproteins, *e.g.* albumines.

long-term supplementation using Se-Met, nonphysiological selenoalbumin can be found in the peripheral blood (Schrauzer, 2000).

Thus, the use of supplements like selenium yeast, which are a mixture of different forms of selenium with a predominance of selenomethionine and methylated forms of selenium, may not be beneficial.

It is considered that inorganic selenium compounds showing greater bioavailability are much more efficiently absorbed by the body. Data reported in the literature indicate that inorganic selenium (mainly selenate(VI)) are not used immediately in selenoprotein synthesis but are metabolized in the liver (Jacques, 2001).

22.3 Medical Applications

22.3.1 Selenium and Cancer Mortality

Selenium is one of the trace elements, which has been the subject of interest of scientists, doctors and patients with cancer for many years. It was observed that there is a close correlation between the level of serum selenium and the risk of death regardless of the cause (Raymann, 2012a; Figure 22.2).

Association studies have shown that there is a strong correlation between the geographical distribution of selenium in soil and agricultural crops, amount of selenium consumed in the diet and cancer mortality. Schrauzer and colleagues in studies evaluating the content of selenium in the diet

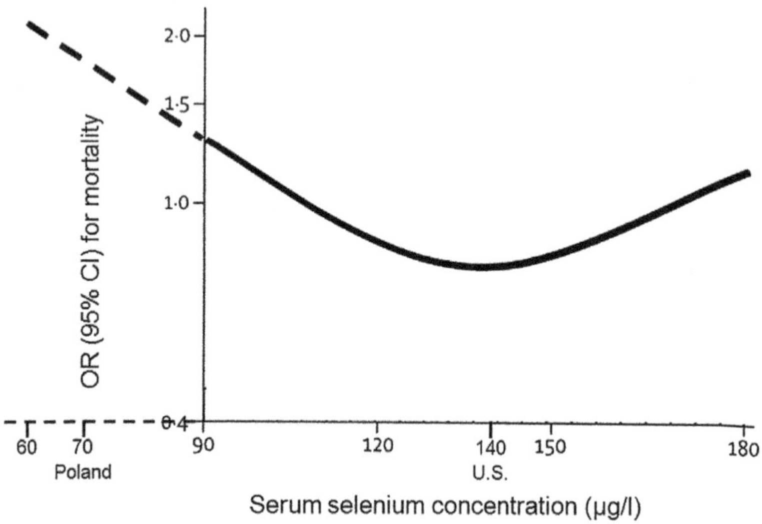

Figure 22.2 Adjusted hazard ratios for all-cause mortality depending on serum selenium concentration (Raymann, 2012a after modification; with Elsevier permission). The relationship between serum selenium concentration and all-cause mortality. The mean selenium levels for the US and Poland are ~140 and 70 μg L^{-1}, respectively.

comprising 27 countries have observed an inverse correlation between the amount of selenium intake and mortality caused by cancers of the lung, colon, rectum, prostate, breast and ovaries (Schrauzer *et al.*, 1977). Results of the Nutritional Prevention of Cancer (NPC) trial indicated that selenium supplementation can reduce not only the incidence but also mortality due to cancer (Duffield-Lillico *et al.*, 2002). It was found that in the group of participants, taking selenium supplements, cancer mortality was reduced by more than 40%. Data from meta-analysis of 49 prospective studies have also confirmed a reduction in cancer mortality by 45% in persons with an optimized selenium concentration (Dennert *et al.*, 2011).

22.3.2 Selenium as a Marker of Selection for Control Examinations in Surveillance

When the serum selenium level is examined in patients with newly diagnosed cancers the results can be influenced by the tumor itself. Therefore, retrospective case–control studies are not suitable to conclude that low selenium levels are the cause of disease, although such studies can be valuable in identifying whether selenium levels may be used as a useful marker in the selection of patients for disease surveillance.

There are literature data that clearly demonstrate that identification of low selenium levels can by useful in selecting for disease surveillance by such methods as computerized tomography for detection of the lung cancers and colonoscopy for detection of the early-stage colorectal cancers. More than 20 studies have shown that low serum selenium levels are related to an increased risk of lung cancer. Studies performed by our center, which evaluated the level of selenium in the serum of 86 patients with lung cancer and 86 healthy controls, has also shown a strong correlation between Se and cancer. Among lung-cancer cases, the mean selenium level was 63.2 $\mu g\ L^{-1}$, compared to a mean level of 74.6 $\mu g\ L^{-1}$ for their matched controls. A 10-fold reduced risk of lung cancer was also observed in individuals with serum selenium >80 $\mu g\ L^{-1}$ as compared to those patients whose serum Se levels were <60 $\mu g\ L^{-1}$ (Jaworska *et al.*, 2013). Between 2008 and 2011 in Szczecin – a single city of four hundred thousand inhabitants in North-West Poland – a programme for early detection of lung cancer was conducted. The enrolment criteria included both sexes, aged 55–65 years who had a history of at least 20 pack-years of tobacco smoking. The latest addition to the protocol of this programme in 2012 was the inclusion of a preselection of participants by measurement of their serum selenium levels. Only individuals with a low selenium level (<75 $\mu g\ L^{-1}$) were invited for computerized tomography scans. This new additional to the protocol resulted in more than doubling of the detection rate of lung cancers (Grodzki *et al.*, 2013).

There are also about 10 studies that suggest a protective effect of selenium on colorectal adenomas/carcinomas development. In a Polish–Estonian study that evaluated the effect of selenium on the risk of colorectal cancer in 169 cancer cases and 169 healthy controls, a more than 13-fold increased risk of colorectal cancer was found in patients with serum Se levels <40 μg L^{-1} compared to those patients whose serum Se levels were >72 μg L^{-1} (Lener *et al.*, 2013; Figure 22.3).

Generally, there appears to be little doubt that low selenium level is associated with increased risk of cancers, however, the optimal serum selenium levels can be different for particular populations/countries. For example, in the US the serum level of selenium associated with the lowest risk of colorectal adenomas is >150 μg L^{-1}. Such differences between Europe and America can be related to several genetic and environmental factors including exposure on chemical compounds (Ou *et al.*, 2012).

It is well recognized that for early colorectal cancer detection, colonoscopy is a well-established method for the detection of precancerous conditions and of early cancers. Screening studies using colonoscopy are not widely performed in the world as they are expensive and time consuming. It appears that preselection using the data based on the levels of serum selenium may be useful in increasing the cost effectiveness and specificity of detecting precancerous lesions occurring in the colon.

Further investigations are needed to determine the value of serum selenium levels as a selection marker for patient selection for prophylactic screening in other types of tumors such as cancers of the stomach, pancreas, and prostate.

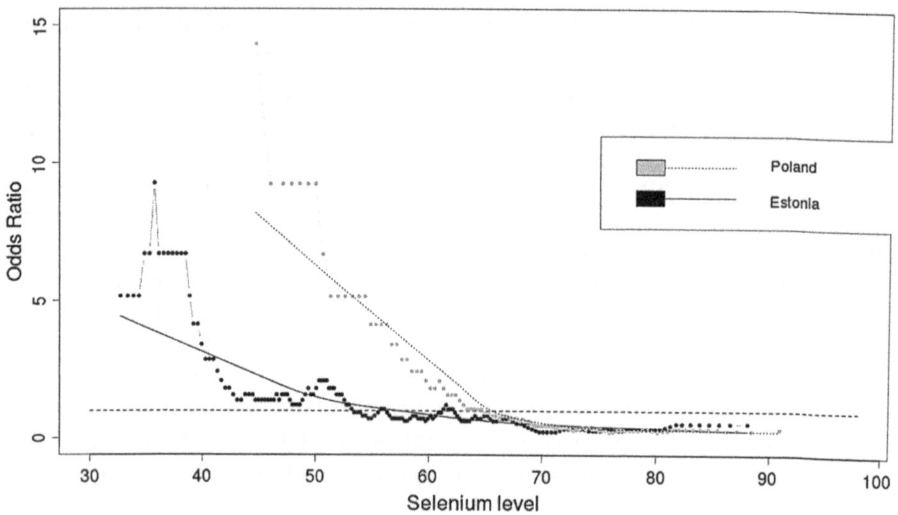

Figure 22.3 Selenium level and colorectal cancer risk (Lener *et al.*, 2013 after modification; with BMC series journals permission). The relationship between selenium concentration in blood serum and the risk of colorectal cancer is displayed in form of odds ratios at each selenium level, separately for patients from Estonia (black circles) and Poland (gray circles).

22.3.3 Selenium and Cancer Prevention

22.3.3.1 Prospective Observations

Data from a meta-analysis of 49 prospective studies revealed that overall the risk of cancer incidence was 31% lower and the risk of death from cancer was 45% lower in the highest category of selenium exposure than in the lowest. Reduction of the number of new cases of cancers with optimized selenium concentration was observed for cancers of the: bladder, lung, larynx, prostate, stomach and colon (Dennert *et al.*, 2011).

More specifically, several prospective studies have shown an inverse association of selenium concentration with cancer risk of: the prostate (van den Brandt *et al.*, 2003), lung (van den Brandt *et al.*, 1993; Knekt *et al.*, 1998), bladder (Zeegers *et al.*, 2002), oesophagus and gastric cardia (Mark *et al.*, 2000). One of the most extensively analyzed was the influence of selenium on prostate cancer risk. In a Dutch cohort study conducted on 540 prostate cancer patients and 1211 subcohort members among 58 279 men aged 55–69 years the prostate cancer risk was found to be reduced with higher levels of selenium (van den Brandt *et al.*, 2003). Selenium has been also shown to be associated with the risk of gastric and oesophageal cancers. The results of a study conducted on 590 oesophageal cancers, 402 gastric cardia cancers, 87 gastric noncardia cancers and 1062 control subjects have shown a significant association of higher serum selenium levels and a reduced risk of oesophageal and gastric cardia cancers (Mark *et al.*, 2000). The significant reduction in the risk of bladder and lung cancers with increasing selenium levels measured in toenails was found in a Dutch cohort study of 431 bladder cancer cases, 370 lung cancer cases against 2459 controls (van den Brandt *et al.*, 1993; Zeegers *et al.*, 2002). An analysis of 95 Finnish lung cancer cases and 190 matched controls have also shown an inverse association of higher serum selenium concentrations with lung cancer (Knekt *et al.*, 1998). In summary, the results of the above-mentioned prospective studies indicate that low selenium intake and/or levels in serum/plasma/toenail can be used as an important risk factor of several cancers.

22.3.3.2 Selenium Supplementation Trials

A meta-analysis of 9 randomized clinical trials, using selenium in the prevention of cancers, has shown a protective effect of dietary selenium on cancer incidence. Preventive effects were seen especially in populations with low baseline serum selenium concentrations (<125.6 mg L^{-1}) and reached a value of over 30% (Lee *et al.*, 2011).

The most fundamental results on the role of selenium supplementation in cancer prevention have been achieved in NPC (Nutritional Prevention of Cancer) and SELECT (Selenium and Vitamin E Cancer Prevention Trial) studies.

The NPC trial was a double-blind, placebo-controlled, randomized trial conducted with 1312 participants undertaken between the years 1983–1996.

Participants of this study received 200 μg of selenium (selenized yeast) per day or "placebo" and the length of follow-up observation was up to 8 years. The results of the study showed that the preventive effect of selenium supplementation depends on the initial (baseline) level of blood selenium concentration.

In the NPC study, a statistically significant decrease in the incidence of all cancers due to selenium supplementation has been demonstrated only for the subgroup of subjects with baseline selenium concentration <105 μg L^{-1}, while for individuals with high initial levels of serum selenium (>121.6 μg L^{-1}) supplementation was associated with an increased incidence of tumors (Duffield-Lillico *et al.*, 2002; Reid *et al.*, 2002; Duffield-Lillico *et al.*, 2003; Reid *et al.*, 2006). It was also observed that effective prevention by selenium supplementation depended not only on its initial level in the body but also on the cancer site. Selenium supplementation was associated with decreased risk of prostate, colon and lung cancers but there was a tendency for an increased risk of breast cancers.

Based on NPC results, in 2001, also in the US, Selenium and Vitamin E Cancer Prevention Trial (SELECT) was launched. It was a phase-3 randomized, placebo-controlled trial involving 35 535 men, investigating selenium supplementation (200 μg per day of L-selenomethionine), vitamin E (400 IU per day of all rac-α-tocopheryl acetate), or both for prostate cancer prevention (Lippman *et al.*, 2009; Klein *et al.*, 2011). Unfortunately, the results of the study showed that selenium supplementation did not reduce prostate cancer incidence. Despite the negative results of this trial it is absolutely incorrect to say that SELECT provides evidence for the lack of efficacy of selenium supplementation in prostate cancer prevention. Conversely, results of SELECT confirm the observations of NPC study that selenium supplementation strongly depends on the baseline selenium concentration and is beneficial only for individuals with low serum selenium levels (<106 μg L^{-1}) – the men who participated in the SELECT trial had selenium levels ~137 μg L^{-1}, which is too high to observe any protective effect of selenium supplementation. However, unpublished data from SELECT trial have also shown that men with serum Se <124 μg L^{-1} had a decreased risk of prostate cancer when supplemented with Se and only those with serum Se >124 μg L^{-1} had an increased risk (Raymann, 2012b; Rathner, 2014).

22.4 Summary

Recent studies evaluating the medical applications of selenium use has achieved significant progress mainly due to the observation that an optimal level of selenium is required and that this correlates with a reduction of overall mortality and risk of cancer as well. It has become clear that both too low and too high selenium levels are disadvantageous. Furthermore, it appears that the optimal selenium level is different among different populations. This could be related to selenium requirements as a result of environmental pollution, including chemical compounds, which are neutralized by selenium. It is also likely that the role of selenium in human health will be even greater if the genotypes, which are individualizing the optimal concentration of selenium for each person, are identified. There exists significant evidence that clearly demonstrates that optimization of selenium levels may improve the treatment of cancer and that selenium levels may also be used as a marker for high cancer risk patient groups, in the selection of persons for appropriate screening and early detection.

Summary Points

- It has become clear that both too low and too high selenium levels in the body are harmful.
- It appears that optimal selenium supplementation is population specific as the background selenium levels vary between populations.
- Studies evaluating medical applications of selenium use have achieved major progress mainly due to the observation that optimal levels of selenium are required and that this correlates with a reduction of overall mortality and risk of cancer as well.
- The use of selenium for human health will be even greater if the genotypes, which can be used to individualize the optimal levels of selenium for each person, are identified.
- The evidence accumulated to date provides a strong rationale for the optimization of selenium levels to improve the results of cancer treatment and that selenium levels can also be used as a marker to identify high cancer risk groups of patients that may be useful in the selection of persons for screening examinations.

Key Facts

1. It has been recognized that the content of micro-/macroelements in diet plays an important role in human health.
2. Deficiencies of micro-/macroelements, as well as their increased levels may affect the incidence of various diseases including cancer.
3. Optimal levels of some microelements, such as selenium, appear to protect against the development of various diseases.
4. Selenium levels can be used as a marker for identification of groups with high cancer risk.
5. Selenium supplementation can be beneficial in reducing cancer incidence and mortality provided it is offered for persons with selenium deficiency.

Definitions of Words and Terms

Supplementation. The use of micro- or macroelements in special types of food or pharmaceuticals in order to improve health, *e.g.* to make up its lack or deficiency.

Micronutrient. Any substance, such as a trace element or vitamin, essential for healthy growth and development, which is required in very small amounts.

Selenoproteins. A peculiar class of proteins containing selenium that is incorporated into the protein in the form of selenocysteine.

Trace elements. Chemical elements present in very small (trace) amounts in plants, animals and humans.

Free radicals. The molecule or ion containing at least one unpaired electron exhibiting paramagnetic properties and (usually) very high chemical activity; in living organisms, free radicals (oxygen compounds) play the role of reactive chemicals, moreover free radicals in the body are the cause of degradation of DNA and are considered to be a cause of aging.

Antioxidant. A group of chemical compounds that themselves perform activities in small concentrations that ameliorate the actions of free radicals; they are often present in food supplements and are widely regarded as preventive agents for numerous diseases.

Surveillance examination. Control examinations aimed to detect the earliest asymptomatic lesions – applied in oncology in order to detect early precancerous lesions and small cancers.

Antioxidant enzymes. Enzymes inactivating the reactive form of oxygen.

Bioavailability. An indicator of the efficiency of absorption into an organism of a particular substance.

Selenites, selenates, selenides. The inorganic forms of selenium having different oxidation states.

Adenomas. Benign epithelial tumors derived from glands.

Carcinomas. A group of malignant tumors that are derived from epithelial tissue.

Mortality. The ratio of deaths among all patients.

Prospective studies. Studies that are carried out before the occurrence of the final result, *e.g.*, studies evaluating the incidence of lung cancer among initially unaffected smokers; observation of specific groups of patients: exposed and not exposed to the pathogen or medical intervention beginning at a specified time and lasting for a period determined by the researcher.

Retrospective studies. Studies in which the final result occurred before the beginning of the study, *e.g.* the effect of smoking on the occurrence of lung cancer among people with newly diagnosed disease; the study is based on the collection of medical documentation, history of past exposure of patients on a putative factor in order to determine whether an effect relationship exists between the occurrence of a disease and the activity of the studied factor.

Case-control studies. An observational epidemiological study beginning with the identification of a group of patients with the disease or other clinical condition (case group) and an appropriate control group without the disease or clinical condition, but matched to the case group.

Meta-analyses. Analysis (mostly statistical) of available literature data, enriched by an analysis, conclusion and summary; a meta-analysis is a separate and full-fledged type of study.

Double-blind trial. Experimental technique used in many branches of science such as medical, biological or social studies. This method involves the planning of the experiment, in which neither participants nor the researchers had access to information that would be able to influence the outcome.

Placebo-controlled trial. The trial in which patients are randomly divided into two groups, one part of this group receives the test drug and the other the placebo.

Randomized trial. Experimental-clinical trial in which patients are divided randomly into case and control groups.

Follow-up observation. The time from the termination of the test intervention until the collection of the clinical data.

List of Abbreviations

DNA	Deoxyribonucleic acid
GPX	Glutathione peroxidase
H_2O_2	Hydrogen peroxide
Se-Met	Selenomethionine
Se-Cys	Selenocysteine
GS–Se–GS	Selenodiglutathione
NADPH	Nicotinamide adenine dinucleotide phosphate
tRNASec	Transfer ribonucleic acid specific for selenocysteine
NPC	Nutritional prevention of cancer
SELECT	Selenium and vitamin E cancer prevention trial

References

Borawska, M. H., Socha, K., Łazarczyk, B., Czyzewska, E., Markiewicz, R. and Darewicz, B., 2009. The effects of diet on selenium concentration in serum in patients with cancer. *Nutrition and Cancer*. 61: 629–633.

van den Brandt, P. A., Goldbohm, R. A., van't Veer, P., Bode, P., Dorant, E., Hermus, R. J. and Sturmans, F., 1993. A prospective cohort study on selenium status and the risk of lung cancer. *Cancer Research*. 53: 4860–4865.

van den Brandt, P. A., Zeegers, M. P., Bode, P. and Goldbohm, R. A., 2003. Toenail selenium levels and the subsequent risk of prostate cancer: a prospective cohort study. *Cancer Epidemiology, Biomarkers and Prevention*. 12: 866–871.

Combs Jr, G. F., 1997. Selenium and cancer prevention. In: Garewal, H. S. (ed.) *Antioxidants and Disease Prevention*. CRC Press, Boca Raton, New York, US, pp. 97–113.

Combs Jr, G. F. and Gray, W. P., 1998. Chemopreventive agents: selenium. *Pharmacology and Therapeutics*. 79: 179–192.

Combs Jr, G. F., Clark, L. C. and Turnbull, B. W., 2001. An analysis of cancer prevention by selenium. *Biofactors*. 14: 153–159.

Dennert, G., Zwahlen, M., Brinkman, M., Vinceti, M., Zeegers, M. P. and Horneber, M., 2011. Selenium for preventing cancer (review). *Cochrane Database of Systematic Reviews*. 11: CD005195.

Duffield-Lillico, A. J., Reid, M. E., Turnbull, B. W., Combs Jr, G. F., Slate, E. H., Fischbach, L. A., Marshall, J. R. and Clark, L. C., 2002. Baseline characteristics and the effect of selenium supplementation on cancer incidence in a randomized clinical trial: a summary report of the nutritional prevention of cancer trial. *Cancer Epidemiology, Biomarkers and Prevention*. 11: 630.

Duffield-Lillico, A. J., Dalkin, B. L., Reid, M. E., Turnbull, B. W., Slate, E. H., Jacobs, E. T., Marshall, J. R., Clark, L. C. and Nutritional Prevention of Cancer Study Group, 2003. Selenium supplementation, baseline plasma selenium status and incidence of prostate cancer: an analysis of the complete treatment period of the nutritional prevention of cancer trial. *BJU International*. 91: 608–612.

Grodzki, T., Wójcik, J., Jakubowska, A., Wójcik, R., Muszyńska, M., Kubisa, B., Jaworska-Bieniek, K., Pieróg, J., Gupta, S., Alchimowicz, J., Gromowski, T., Jankowski, H., Sukiennicki, G., Bielewicz, M. and Lubiński, J., 2013. International Association for the Study of Lung Cancer (IASLC). 15th World Conference on Lung Cancer, October 27th–30th, Sydney, Australia. Low selenium serum level is a good preselection factor for patients invited for low dose chest CT lung cancer screening programme. *Journal of Thoracic Oncology*. 8: 328.

Jacques, K. A., 2001. Selenium metabolism in animals: the relationship between dietary selenium form and physiological response. In: Lyons, T. P. and Jacques, K. A. (ed.) *Science and Technology in the Feed Industry. Proceedings of Alltech's 17th Annual Symposium*. Nottingham University Press, UK, pp. 319–348.

Jaworska, K., Gupta, S., Durda, K., Muszyńska, M., Sukiennicki, G., Jaworowska, E., Grodzki, T., Sulikowski, M., Woloszczyk, P., Wójcik, J., Lubiński, J., Cybulski, C., Dębniak, T., Lener, M., Morawski, A. W., Krzystolik, K., Narod, S. A., Sun, P., Lubiński, J. and Jakubowska, A., 2013. A low selenium level is associated with lung and laryngeal cancers. *PLoS One*. 8: e59051.

Klein, E. A., Thompson Jr, I. M., Tangen, C. M., Crowley, J. J., Lucia, M. S., Goodman, P. J., Minasian, L. M., Ford, L. G., Parnes, H. L., Gaziano, J. M., Karp, D. D., Lieber, M. M., Walther, P. J., Klotz, L., Parsons, J. K., Chin, J. L., Darke, A. K., Lippman, S. M., Goodman, G. E., Meyskens Jr, F. L. and Baker, L. H., 2011. Vitamin E and the risk of prostate cancer: the selenium and vitamin E cancer prevention trial (SELECT). *JAMA, the Journal of the American Medical Association*. 306: 1549–1556.

Knekt, P., Marniemi, J., Teppo, L., Heliövaara, M. and Aromaa, A., 1998. Is low selenium status a risk factor for lung cancer? *American Journal of Epidemiology*. 148: 975–982.

Lee, E. H., Myung, S. K., Jeon, Y. J., Kim, Y., Chang, Y. J., Ju, W., Seo, H. G. and Huh, B. Y., 2011. Effects of selenium supplements on cancer prevention: meta-analysis of randomized controlled trials. *Nutrition and Cancer*. 63: 1185–1195.

Lener, M. R., Gupta, S., Scott, R. J., Tootsi, M., Kulp, M., Tammesoo, M. L., Viitak, A., Metspalu, A., Serrano-Fernández, P., Kładny, J., Jaworska-Bieniek, K., Durda, K., Muszyńska, M., Sukiennicki, G., Jakubowska, A. and Lubiński, J., 2013. Can selenium levels act as a marker of colorectal cancer risk? *BMC Cancer*. 13: 214.

Lippman, S. M., Klein, E. A., Goodman, P. J., Lucia, M. S., Thompson, I. M., Ford, L. G., Parnes, H. L., Minasian, L. M., Gaziano, J. M., Hartline, J. A., Parsons, J. K., Bearden 3rd, J. D., Crawford, E. D., Goodman, G. E., Claudio, J., Winquist, E., Cook, E. D., Karp, D. D., Walther, P., Lieber, M. M., Kristal, A. R., Darke, A. K., Arnold, K. B., Ganz, P. A., Santella, R. M., Albanes, D., Taylor, P. R., Probstfield, J. L., Jagpal, T. J., Crowley, J. J., Meyskens Jr, F. L., Baker, L. H. and Coltmanr Jr, C. A., 2009. Effect of selenium and vitamin E on risk of prostate cancer and other cancers: the selenium and vitamin E cancer prevention trial (SELECT). *JAMA, the Journal of the American Medical Association*. 301(1): 39–51.

Mark, S. D., Qiao, Y. L., Dawsey, S. M., Wu, Y. P., Katki, H., Gunter, E. W., Fraumeni Jr, J. F., Blot, W. J., Dong, Z. W. and Taylor, P. R., 2000. Prospective study of serum selenium levels and incident esophageal and gastric cancers. *Journal of the National Cancer Institute*. 92: 1753–1763.

Ou, Y., Jiang, B., Wang, X., Ma, W. and Guo, J., 2012. Selenium and colorectal adenomas risk: a meta-analysis. *Nutrition and Cancer*. 64: 1153–1159.

Rathner, Z., 2014. Selenium and vitamin E supplementation above recommended dietary intake may increase prostate cancer risk. *JNCI, Journal of the National Cancer Institute*. 919: 677.

Raymann, M. P., 2012a. Selenium and human health. *Lancet*. 379: 1256–1268.

Raymann, M. P., 2012b. Selenium in cancer prevention. *International Conference. Clinical Genetics of Hereditary Cancers*. Pp. 30–31.08.2012, Szczecin, Poland.

Reid, M. E., Duffield-Lillico, A. J., Garland, L., Turnbull, B. W., Clark, L. C. and Marshall, J. R., 2002. Selenium supplementation and lung cancer incidence: an update of the nutritional prevention of cancer trial. *Cancer Epidemiology, Biomarkers and Prevention*. 11: 1285–1291.

Reid, M. E., Duffield-Lillico, A. J., Sunga, A., Fakih, M., Alberts, D. S. and Marshall, J. R., 2006. Selenium supplementation and colorectal adenomas: an analysis of the nutritional prevention of cancer trial. *International Journal of Cancer*. 118: 1777–1781.

Schrauzer, G. N., White, D. A. and Schneider, C. J., 1977. Cancer mortality correlation studies–III: statistical associations with dietary selenium intakes. *Bioinorganic Chemistry*. 7: 23–31.

Schrauzer, G. N., 2000. Selenomethionine: a review of its nutritional significance, metabolism, and toxicity. *Journal of Nutrition*. 130: 1653–1656.

Torun, M., Aldemir, H. and Yardim, S., 1995. Serum selenium levels in various cancer types. *Trace Elements and Electrolytes*. 12: 186–190.

Zachara, B. A., 1992. Mammalian selenoproteins. *Journal of Trace Elements and Electrolytes in Health and Disease*. 6: 137–151.

Zeegers, M. P., Goldbohm, R. A., Bode, P. and van den Brandt, P. A., 2002. Prediagnostic toenail selenium and risk of bladder cancer. *Cancer Epidemiology, Biomarkers and Prevention*. 11: 1292–1297.

CHAPTER 23

Selenium and Skin Cancer

PAMELA B. CASSIDY*[a], SANCY A. LEACHMAN[a], AND PHILIP J. MOOS[b]

[a]Department of Dermatology, Oregon Health & Science University, 3181 SW Sam Jackson Park Rd, Portland, OR, USA 97239; [b]Department of Pharmacology and Toxicology, College of Pharmacy, University of Utah, 30 S 2000 E, Room 201, Salt Lake City, UT, USA 84112
*E-mail: cassidyp@ohsu.edu

23.1 Etiology and Genetics of Skin Cancers

Melanoma is the most deadly of the skin cancers. There were an estimated 76 690 cases of melanoma and 9480 deaths caused by the disease in the US in 2013 (SEER, 2013). Although there have been significant recent advances in both targeted and immunotherapies for metastatic disease, the prognosis for Stage IV patients remains grim (Balch *et al.*, 2009). Early detection and avoidance of risk factors such as sun exposure are the most important strategies for decreasing mortality.

Melanoma arises from melanocytes, which are the pigment-producing cells located in the basal layer of the epidermis. Exposure to UV-light is the most significant environmental risk factor for melanoma. Other risk factors include fair skin and the inability to tan, red hair, freckles, a history of blistering sunburns, numerous and/or atypical nevi (moles), a family or personal history of melanoma, male gender and age greater than 55 years. Pooled analysis of risk phenotypes provide melanoma odds ratio (OR) estimates for

Food and Nutritional Components in Focus No. 9
Selenium: Chemistry, Analysis, Function and Effects
Edited by Victor R. Preedy

Published by the Royal Society of Chemistry, www.rsc.org

red hair (OR = 1.9), fair skin (OR = 1.7), high common nevus count (OR = 7.9) and ≥1 atypical nevi (OR = 4.0) (Chang *et al.*, 2009).

Germline mutations are associated with approximately 10% of melanomas diagnosed in Caucasian populations; of these 20–50% involve mutations at the *CDKN2A* locus on chromosome 9p, which harbors the genes encoding the cell-cycle regulators p16INK4a and p14ARF. These proteins are derived from differential splicing and the use of alternate reading frames within the same gene. Mutations are inherited in an autosomal dominant fashion, and penetrance varies from 58% in Europe to 76% in the United States to 91% in Australia (Pho *et al.*, 2006). This variation suggests that both geographical effects on exposure to sunlight and other modifying heritable factors may influence penetrance of susceptibility genes. One such modifying factor is the highly polymorphic melanocortin 1 receptor gene (*MC1R*, also known as the red hair gene) which is known to vary with ethnic background and has been shown to exacerbate risk associated with *CDKN2A* mutations (Box *et al.*, 2001). Numerous studies utilizing the techniques of genetic linkage, candidate gene analysis and genome wide association (GWAS) have identified other genes associated with melanoma susceptibility, many of which are associated with phenotypes of pigmentation and nevus count. These studies have been discussed recently in an excellent review (Marzuka-Alcala *et al.*, 2014).

Given the low prevalence (10%) of familial melanoma, the high degree of heritability for this skin cancer might not seem obvious especially in light of the significant impact of environmental factors such as sun exposure, on skin cancer risk. But in fact, twin studies estimate that 55% of the variation in melanoma risk is due to genetic effects (Shekar *et al.*, 2009). It has been argued that under a polygenic threshold model that because of melanoma's low incidence (21 per 100 000 men and women in the US, 57 per 100 000 in Australia) the majority of cases would be sporadic rather than familial if a large portion of the risk arises from common genetic variants with low risk in combination with common environmental factors (*i.e.* UV exposure) (Box *et al.*, 2001). The point of presenting this argument is that risk for melanoma varies greatly according to both environment and genetic makeup of the study population, and we believe that these factors likely contribute to the decidedly inconclusive cumulative results of the studies of the impact of selenium on risk for skin cancer (especially melanoma) discussed in the following sections.

The most common of the nonmelanoma skin cancers (NMSC), basal cell and squamous cell carcinoma (BCC and SCC, respectively) arise from keratinocytes in the epidermis. They represent one of the most common human neoplasms, although both are rarely metastatic and mortality is much lower (with the exception of cases of immune-suppressed patients) than is the case for melanoma (Dreno *et al.*, 2007; Rogers *et al.*, 2010). Nevertheless, NMSC can be extremely disfiguring and these diseases are a major burden both on the patients suffering from them and on healthcare systems world-wide. Like melanoma, UV exposure is the most common environmental risk factor for

NMSC. Many other risk factors are common to both classes of skin cancer and these include fair skin and the inability to tan, red hair, freckles, a history of blistering sunburns, and a family or personal history of melanoma. There are a number of inherited conditions that increase risk for NMSC, many of which are associated with impaired DNA repair machinery and pigmentation. For a review of genome-wide association studies of pigmentation and risk for NMSC see Meng *et al.* (2012).

There is correlative evidence to suggest that one link between UV radiation and both melanoma and NMSC is oxidative damage to skin cells. Multiple oxidizing species capable of damaging cellular components including proteins and DNA, are induced in the skin by UV; these include hydrogen peroxide, hydroxyl radical, superoxide, nitric oxide and oxidized lipids. UV-induced oxidizers can also damage skin cells by depleting vital intracellular reductants such as glutathione (GSH) (Schafer *et al.*, 2010). These reactive species are cleared from within cells by the action of multiple antioxidant enzymes including the glutathione peroxidases (GPxs), the glutathione S-transferases (GSTs), peroxiredoxins, superoxide dismutases and catalase. If high levels of oxidizers persist, mutations in DNA can result directly from oxidation of guanine residues and formation of 8-oxoguanine (8-OHdG), which may lead to $G \rightarrow T$ transversion mutations if not repaired by the DNA glycosylase OGG1 prior to DNA replication (Bruner *et al.*, 2000). Numerous selenium-containing enzymes are involved in repairing oxidative damage to proteins. These include the methionine sulfoxide reductases (MSRs) and thioredoxin reductases (TRs) (Arner, 2009). We will discuss the current literature on the involvement of each of the selenoproteins known to affect skin cancer and redox homeostasis in the skin in the following sections.

23.2 Dietary Selenium and Selenium Metabolism

Selenium (Se) is an essential micronutrient with a narrow range between deficient (<20 μg per day), essential (50–60 μg per day), and toxic doses (>400 μg per day) (Roman *et al.*, 2014). Se is incorporated *via* a specialized mechanism described below into selenoproteins in the form of selenocysteine (SeCys). Selenomethionine (SeMet) is sometimes incorporated nonspecifically into other proteins in place of methionine. Se occurs in the diet in both inorganic forms (selenite and selenate) and organic forms, principally SeMet, SeCys, selenocystine ($(SeCys)_2$), selenomethylselenocysteine (SeMCys) and selenodiglutathione ($Se(GS)_2$), see Roman *et al.* (2014) for a more detailed discussion. After absorption in the intestine, all of these forms that contribute to the biosynthesis of selenoproteins, including SeCys itself, must be converted first to hydrogen selenide. Selenide is the species commonly identified as the "central selenium pool" (Figure 23.1). Selenophosphate is synthesized from selenide, and selenocysteinyl-tRNA is the product of the reaction of selenophosphate with a phosphoseryl-tRNA. The selenocysteinyl-tRNA then recognizes a UGA codon (in most other cases interpreted as a stop codon) in mRNA in the context of a 3′-selenocysteine insertion sequence (SECIS), and

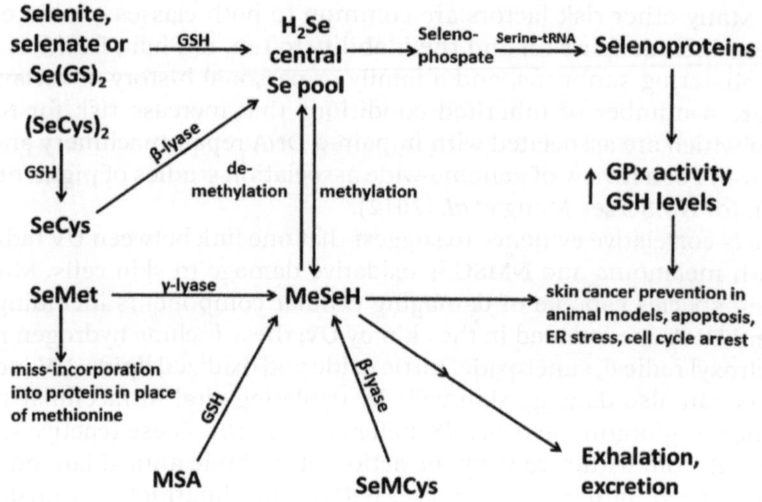

Figure 23.1 Metabolism and biological activities of dietary and synthetic selenocompounds. Dietary selenocompounds enter the central selenium pool after metabolism to give hydrogen selenide (H_2Se). This is the form used in the synthesis of selenoproteins, which can increase antioxidant capacity and prevent skin cancer. Some forms of selenium can be metabolized to methyl selenol (MeSeH) that is believed to have cancer-preventive activities. MeSeH is also a precursor the exhaled and excreted selenium species including di- and trimethylselenide as well as selenosugars.

SeCys is incorporated into a growing peptide in the same manner as other amino acids. Selenide is also the common precursor to secreted and exhaled forms of Se including di- and trimethylselenide and selenosugars (Turanov *et al.*, 2011).

Se-methylated species such as SeMet, SeMCys and the synthetic compound methylseleninic acid (MSA) can give rise to methyl selenol (MeSeH, Figure 23.1). Selenide can also be methylated, to give MeSeH. MeSeH is believed to mediate the cancer-preventive and antitumor effects observed in animal studies where supranutritional levels (>400 μg per day human equivalent dose) of Se have been administered (Wu *et al.*, 2005). Many effects have been attributed to MeSeH in tumors and cells treated *in vivo* with Se; these include induction of apoptosis, cell cycle arrest and endoplasmic reticulum (ER) stress.

23.3 Se and the Risk for Skin Cancer

Table 23.1 summarizes studies of the effects of dietary, environmental and supplemental Se on risk for skin cancers in humans. Interest in the effects of Se on skin-cancer risk was sparked by the report in 1984 of Clark *et al.* (1984) of an inverse association between plasma Se levels and risk for NMSC.

Table 23.1 Studies of selenium and risk for skin cancer in humans. Case control, interventional and cohort studies of the relationship between selenium intake and risk for skin cancer have been mixed.

Location/publication date	Study design	Study population	Outcome
US (Clark *et al.*, 1984)	Case-control	240 cases	Inverse association of plasma Se with risk for BCC and SCC
US (Breslow *et al.*, 1995)	Nested case-control	30 melanomas, 32 BCC, 37 SCC each matched with 2 controls	No association with serum Se and any skin cancer
Eastern US NPC study (Duffield-Lillico *et al.*, 2002)	Placebo-controlled intervention (4.5 years treatment with 5–10 years follow-up)	1312 patients with history of NMSC	Increased risk for SCC with supplementation of 200 µg Se per day
US (Karagas *et al.*, 1997)	Nested case-control	132 cases and 264 controls	No association with plasma selenium and SCC
Italy (Vinceti *et al.*, 1998)	Cohort	2065 people exposed to selenate (7-9X controls) in municipal water supply	Increased risk for melanoma in exposed cohort
Hawaii (Le Marchand *et al.*, 2006)	Case-control	278 cases and controls	No association with serum or toenail Se and melanoma
France (Dreno *et al.*, 2007)	Placebo-controlled intervention (3 years treatment with 2 years follow-up)	184 organ transplant patients	No reduction of skin cancers with supplementation of 200 µg Se per day
Australia (van der Pols *et al.*, 2009)	Prospective cohort	485 participants	Inverse association of serum selenium with risk for BCC and SCC
Western Washington State (Asgari *et al.*, 2009)	Prospective cohort	69 671 self-reporting consumers of multivitamins	No association of melanoma with ≤50 µg per day Se supplement
Northern Italy (Vinceti *et al.*, 2012)	Nested case-control	54 cases, 56 controls	Positive association of plasma Se with risk for melanoma, no association with toenail selenium

In their case-control study, the authors reported that the mean level of Se among patients with either BCC (n = 142), SCC (n = 48), or both (n = 50), was 0.141 μg g^{-1}. This was significantly lower than the mean plasma Se level of the 103 control subjects, which was 0.155 μg g^{-1}. The control groups were drawn from clinic patients at the same location. The logistic estimate of the odds ratio of skin cancer risk for the lowest *versus* the highest decile of Se for all cases combined *versus* the group of current patient controls was 4.39. This result was supported by an Australian study reported in 2009 (van der Pols *et al.*, 2009). In this community-based prospective cohort study the authors found that compared with participants with the lowest serum Se at baseline (32–79 ng mL^{-1}), those with the highest serum Se concentrations (103–221 ng mL^{-1}) had a lower incidence of BCC (RR 0.43, 95% CI 0.21–0.86) and SCC (RR 0.36, 95% CI 0.15–0.82). However, two additional (but smaller) case-control studies in the US in 1995 and 1997 failed to show any association between serum Se levels and NMSC incidence (Breslow *et al.*, 1995; Karagas *et al.*, 1997).

The observations of Clark *et al.* in 1984, were the impetus for the Nutritional Prevention of Cancer (NPC) study that tested the hypothesis that supplemental Se (200 μg per day in the form of selenized yeast) would prevent recurrence of NMSC cancer in a high-risk cohort of patients (Duffield-Lillico *et al.*, 2002). Results of the study showed that the hypothesis was not true; in fact the risk for SCC was increased (RR 1.25, 95% CI 1.03–1.51), the risk for BCC was increased but not statistically significant, (RR 1.09, 95% CI 0.94–1.26) and the excess risk for all NMSC was 17% (95% CI 2–34%) in the Se arm of the study. Another study in France reported in 2007 showed no benefit for similar supplementation with Se on risk for skin cancer in transplant patients who are at elevated risk by virtue of their compromised immune system (Dreno *et al.*, 2007). Secondary endpoint analysis in the NPC trial showed a significantly decreased risk for prostate cancer amongst patients treated with Se (RR 0.48, 95% CI 0.28–0.80, (Duffield-Lillico *et al.*, 2002) and this result was the basis of the SELECT trial. SELECT was a Phase III placebo-controlled trial of Se and vitamin E that showed no beneficial effect on incidence of prostate cancer in a trial involving more than 32 000 men (Platz and Lippman, 2009).

Studies of the effects of Se on risk for melanoma also have mixed results. A case-control study of Caucasians living in Hawaii found no association of serum or toenail Se with risk for melanoma, (Le Marchand *et al.*, 2006) but a smaller study from Northern Italy published in 2012 found a positive association between plasma, but not toenail, Se (Vinceti *et al.*, 2012). An unintended natural experiment on the effects of Se in drinking water took place in Italy from 1972–1988 when 2065 people were exposed to elevated levels of selenate from the municipal water supply (Vinceti *et al.*, 1998). The authors estimated that people consuming water from contaminated wells were exposed to approximately 10–20 μg Se per day in excess of that normally consumed by the Italian population in the diet. When the melanoma rates of the exposed cohort over a period of 11 years was compared to 130 000 people in the same location, the exposed cohort was found to have an elevated risk (RR 3.9, 95% CI 1.8–7.4). However, a study of 69 671 self-reported consumers of multivitamins including Se supplements performed in the western US found no association with risk for melanoma and consumption of ≤50 µg per day of Se (Asgari *et al.*, 2009).

These results illustrate the complexity of the problem of establishing optimal levels of Se intake for managing risk for skin cancer. Part of that difficulty likely lies in the genetics of skin cancer susceptibility that is dominated by numerous low penetrance genes that vary in their prevalence according to geographic location and ethnicity. Also, even though many of the studies attempt to control for these factors by taking into account phenotypes associated with susceptibility, some genetic factors influence risk without an overt phenotype. For example, homozygous or mixed heterozygous loss-of-function mutations in MC1R that confer a 4-fold increased risk for melanoma, generally cause a red-hair, fair-skinned phenotype in individuals for whom tanning is difficult or impossible. However, heterozygotes with one wild-type allele often have wild-type pigmentation, but they are still at 2-fold increased risk for melanoma and NMSC (Kanetsky *et al.*, 2010).

Another factor that likely contributes to the variable results found in the studies listed in Figure 23.1 is what has been referred to as a "U-shaped curve" (Figure 23.2) for cancer risk as a function of Se levels, *i.e.* both low and elevated Se levels may increase risk (Mayne *et al.*, 2012). This might explain the excess risk for NMSC seen in the Se arm of NPC trial and the variable results

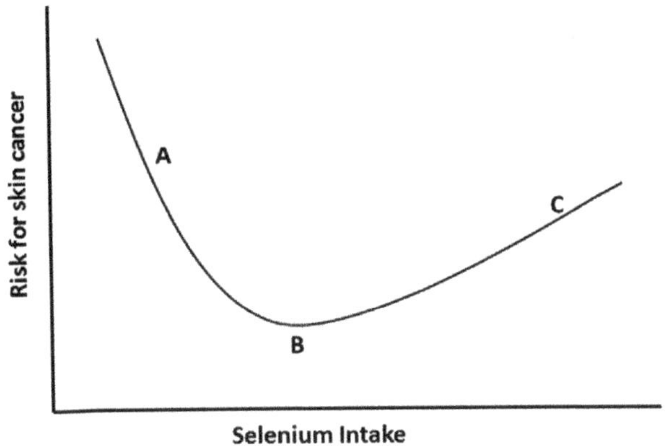

Figure 23.2 Hypothetical U-shaped curve for selenium intake and skin cancer risk. This curve represents the notion that there is an optimum selenium intake for lowering skin cancer risk; both too little and too much can increase risk. Results from epidemiological studies that found an elevated risk for skin cancer associated with low levels of plasma/serum selenium levels might be represented by points A (cases) and B (controls) (Clark *et al.*, 1984; van der Pols *et al.*, 2009) Points B and C represent hypothetical results for placebo and Se-treated groups, respectively, in an interventional study (Duffield-Lillico *et al.*, 2002) that found elevated risk in the Se arm.

reported in the case control studies. In a 2009 discussion of Se, genetic variation and the SELECT trial, Platz and Lippman (the PI for the SELECT trial) concluded that "At present, we do not know enough to determine how much selenium any man or woman should receive from the diet or a supplement"; (Platz and Lippman, 2009) this statement definitely applies to the question of the effects of Se on the risk for skin cancer at the present time. The discussion below will outline what is known about selenoproteins and the skin. We will highlight the ongoing research that promises to identify selenoprotein-related genetic loci that influence skin-cancer risk and biomarkers of redox homeostasis and carcinogenesis in the skin. These are the factors that will enable personalized assessments of optimal Se intake in the future.

23.4 Selenoproteins and the Skin

There are 25 selenoproteins in humans (Kryukov *et al.*, 2003). Each has at its active site(s) one or more selenocysteine residues. Many selenoproteins (all that we will discuss here) function as antioxidants. For an excellent general discussion of the biochemistry of the selenoproteins please see Roman *et al.* (2014). Selenoproteins are essential for proper keratinocyte function and skin development. Targeted disruption of the gene encoding the selenocysteine tRNA in keratinocytes eliminated the expression of all selenoproteins in the skin of transgenic mice. These mice had a runt phenotype, developed severe abnormalities in the hair and skin and died shortly after birth, possibly due to dysplastic mucosal and tongue epithelium that hampered food intake (Sengupta *et al.*, 2010).

23.4.1 Glutathione Peroxidases

The role(s) of 3 of the 5 Se-containing glutathione peroxidases (GPx1, GPx2 and GPx4) have been examined in the context of normal skin and/or skin cancers. These enzymes utilize the reducing power of GSH to transform hydrogen peroxide and lipid hydroperoxides to water or the corresponding alcohol (Brigelius-Flohe *et al.*, 2002). GPx1 is expressed in most tissues and it is very sensitive to dietary Se levels in that it is one of the first selenoproteins whose expression level is decreased under conditions of dietary insufficiency. GPx2 and GPx4 are expressed in cells of epithelial origin including the skin. UV-irradiation of keratinocytes, (Walshe *et al.*, 2007) melanocytes (Cassidy *et al.*, 2013) and whole-skin results in decreased total GPx activity. UV-irradiated $GPx2^{-/-}$ mice develop SCC faster and have a greater tumor burden than do wild-type mice, but $GPx1^{-/-}$ mice have a tumor profile identical to the wild-type mice in the same skin cancer model (Walshe *et al.*, 2007). A preliminary report describing the phenotype of transgenic mice with targeted disruption of GPx4 in keratinocytes has been published (Sengupta *et al.*, 2013). Perinatal mice had epidermal hyperplasia, dermal inflammatory infiltrate, dysmorphic hair follicles and alopecia. The moderation of the phenotype of GPx4 keratinocyte knockout mice compared to elimination of all selenoproteins could be due to compensatory elevation (and presumably overlapping functions) of GPx1 and TR1 in the GPx4 knockout mice. The authors did not report on the susceptibility of these animals to UV-induced skin cancer. They did find that feeding a COX-2 inhibitor to nursing mothers partially prevented the development of the abnormal phenotype in their pups; this is not unexpected given the known connection between GPx4, the metabolism of oxidized lipids, and the regulation of COX-2.

In humans, a polymorphism in the *GPX1* gene (Pro198Leu, rs 1050450) is associated with decreased incorporation of Se into the protein and lower GPx activity in tissue (He *et al.*, 2010). The association between this polymorphism and skin cancer risk was studied in a nested case-control study of Caucasians. Study participants included 218 cases of melanoma, 285 squamous cell carcinomas (SCC), 300 basal cell carcinomas (BCC) and 870 age-matched controls within the Nurses' Health Study (He *et al.*, 2010). After controlling for lifestyle factors including sun exposure and dietary carotenoid intake, the authors found that the *GPX1* Leu/Leu genotype was associated with a 2-fold increased risk for melanoma (RR 2.14; 95% CI, 1.22–3.72). No association was found between this polymorphism and risk for SCC or BCC.

23.4.2 Methionine Sulfoxide Reductase

A major effect of UV-induced oxidative stress is formation of sulfoxides in the sidechains of methionine residues in cellular proteins, which can cause disruption in protein structure and impairment of function. Methionine sulfoxide reductases A (MSRA) and B (MSRB) reduce the S- and R-sulfoxide diastereomers, respectively. One of the MSRB isoforms, MSRB1, is a selenoprotein (Roman *et al.*, 2014). Reduction of methionine sulfoxide residues by the MSRs can restore the function of oxidatively inactivated proteins. UV radiation significantly reduces the MSR activity in human epidermis, and MSRA immunostaining is significantly reduced in chronically sun-exposed skin compared to sun-protected skin (Picot *et al.*, 2007). Human melanocytes express both MSRA and MSRBs, and keratinocytes express MSRA (Schallreuter, 2006). All of the MSRs are tied to the thioredoxin reductase (TR)/thioredoxin (Trx) system because the disulfide or selenosulfide formed at the active sites as a result of the reduction of methionine sulfoxide is reduced (thereby reactivating the active site) by Trx (Roman *et al.*, 2014).

23.4.3 Thioredoxin Reductases

The human TR/Trx system enables the transfer of electrons from NADPH through TR to Trx. This system performs two electron reduction reactions and is integrated into the redox tone of cells through the reductive roles of the Trx-dependent peroxiredoxins and MSRs, as well as the direct reduction of other redox-sensitive signaling molecules important in health and disease states such as NF-κB, p53, ASK1, HIF-1α, and Ape/Ref-1 (Moos *et al.*, 2003; Lu & Holmgren, 2014). The TR/Trx system counters protein oxidation from reactive species like superoxide, hydrogen peroxide, and nitric oxide. TR also regulates other soluble redox regulators such as ascorbate.

In humans, there are three selenoprotein thioredoxin reductase genes; *TXNRD1* (which encodes the protein product TR1 also known as TrxR1) is the cytosolic isoform that has been associated with the inducible antioxidant response regulated by NRF2; *TXNRD2* (encoding TR3 or TrxR2) is the matrix-localized isoform that maintains redox function in the mitochondrion, and *TXNRD3* (encoding TR2 or TGR), which is a glutathione-thioredoxin reductase whose expression is limited to the testis. *TXNRD1* can generate multiple variants due to the complex structure of the 5′ end of the gene. The best characterized is the ~55 kDa protein that is known as the cytosolic TR1. Another isoform known as v3, is generated from an upstream start site and demonstrates membrane localization. In a broad evaluation of redox proteins in various human tissues, the protein products of *TXNRD1* and *TXNRD2* were analyzed by immunohistochemistry. TR1 and v3 are both expressed in the skin, but the staining of v3 is much stronger (Dammeyer & Arner, 2011). The expression of TR3 was found to be at relatively low levels.

The protective effects of Trx in the skin were shown in a recent study that found that human Trx expressed in a transgenic murine model or injected i.p. relieved UV-induced inflammation of the skin (Ono *et al.*, 2012). However, targeted removal of TR1 in keratinocytes did not cause any changes in skin phenotype or in the growth of the host (Sengupta *et al.*, 2013). The authors of the study suggest that TR1 may not be critical for skin function in the resting state or alternatively, its absence was compensated for by other proteins. They also state that additional studies with the TR1 knockout mice will be required to determine whether keratinocytes depend on this selenoprotein in order to function properly under conditions of stress (such as UV irradiation).

It is important to point out that the TR/Trx system has also been identified as an anticancer therapeutic target in the skin because Trx is associated with inhibition of apoptosis (Prickett *et al.*, 2014) and activation of PI3K/AKT signaling in melanoma (Meuillet *et al.*, 2004). In addition, several commonly used therapeutic agents, like cisplatin, cyclophosphamide, and doxorubicin appear to target TR as well as DNA, and it has been suggested that the off-target inhibition of TR may be responsible for some of the adverse effects of these agents (Powis and Kirkpatrick, 2007).

23.5 Studies of Se and Skin Cancer in Model Systems

Two studies have been performed using sodium selenite (Figure 23.3 and Table 23.2) in UV-induced models of NMSC. Both were done with hairless mice that were irradiated chronically over a period of 20 weeks. The first study used human-equivalent doses (HED) of 88–352 μg Se per day. The results showed a dose-dependent protection against skin cancer (Overvad *et al.*, 1985). The second study used much lower doses (9.7–49 μg per day HED) and found no effect on tumor formation, but did report a cessation

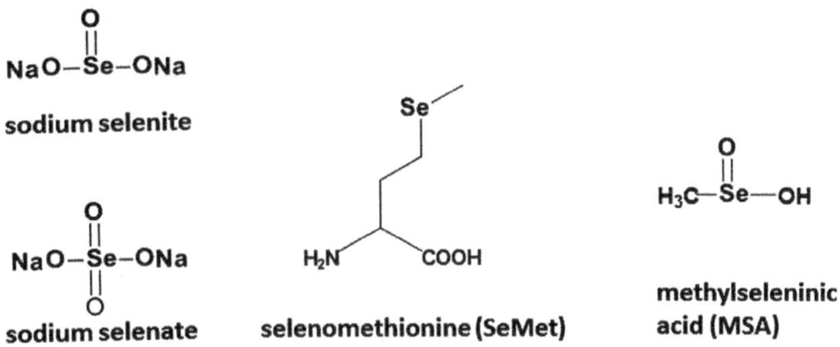

Figure 23.3 Structures of selenium species studied in animal models of skin cancer. Metabolites of these selenocompounds can either enter the central selenium pool, thereby supporting the synthesis of selenoproteins, and/or be used to generated MeSeH which is believed to have cancer-preventive activity.

of tumor growth in the Se-treated animals once UV irradiation was stopped (Pence *et al.*, 1994).

Selenite was used in two studies of melanoma metastasis in the B16F10 syngeneic model. In this model tumor cells are injected into the tail veins of mice, and the cells are then able to colonize the lungs. Both studies in this model found that Se treatment (both oral and i.p.) resulted in about 50% fewer and smaller lung tumors (Yan *et al.*, 1997; Song *et al.*, 2009). Mechanistic studies showed that Se treatment caused cell-cycle arrest and decreases in CDK2/4 mRNA in tumor cells.

There are 3 studies of the effects of selenomethionine (SeMet) on skin cancer in mice. Se from SeMet can enter the central Se pool by release of MeSH in a γ-lyase-catalyzed reaction, followed by demethylation (Figure 23.1). The effects of a topical cream containing SeMet were examined in a mouse model of UV-induced pigmentation and NMSC (Burke *et al.*, 2003). The authors found that SeMet treatment decreased both numbers of tumors and pigmentation, and they suggested that a reduction in UV-induced blistering of the skin was responsible for these effects. We tested a similar topical formulation in a mouse model of UV-induced melanoma and found that the time required for tumor formation was increased in the Se-treated animals (Cassidy *et al.*, 2013). We also found that while tumor numbers were similar in both Se and control arms, the rate of growth of the Se-treated tumors, once formed, was significantly greater than those treated with vehicle alone. A study with oral SeMet (the primary source of Se in high-Se soy protein) in the B16F10 system described above found that this source of Se also provided protection against lung metastases (Li *et al.*, 2004).

Methylseleninic acid (MSA, Figure 23.3) is a synthetic compound that reacts with GSH to give MeSeH (Figure 23.1). Se from MSA can also be incorporated into selenoproteins after demethylation of MeSeH to give selenide. We tested the effects of MSA on human melanomas both *in vitro* and in a xenograft model (Cassidy *et al.*, 2013). We found that daily oral administration of MSA

Table 23.2 Studies of selenium-containing compounds in skin cancer model systems. Selenium has beneficial activity in mouse models of skin cancer.

Compound, dose, delivery route	Model	Endpoints (results)	Mechanism
Selenite, oral (in water) 88–352 μg daily[a]	UV-irradiated hairless mice (Overvad *et al.*, 1985)	Dose dependent protection against tumor formation	None determined
Selenite, oral (in feed) 0 (deficient), 9.7, 49 μg daily[a]	UV-irradiated hairless mice, 20 weeks (Pence *et al.*, 1994)	No growth in tumors of Se-treated mice after cessation of UV	↑ GPx activity in skin of Se-treated mice
Selenite, oral and i.p., 194–388 μg daily[a]	B16F10 mouse melanoma syngeneic tail vein injection (Yan *et al.*, 1997; Song *et al.*, 2009)	~50% fewer and smaller lung metastases	Cell-cycle arrest, decrease in mRNA for CDK2/4
Hi-Se soyprotein (primarily SeMet), oral (in feed), 13 and 352 μg daily[a]	B16F10 mouse melanoma syngeneic tail vein injection (Li *et al.*, 2004)	↓ Lung metastasis in mice consuming high Se protein or SeMet supplements	None proposed
SeMet, topical	Mouse UV-induced pigmentation, NMSC (Burke *et al.*, 2003)	↓ Numbers of tumors, ↓ Pigmentation	↓ UV-induced blistering
SeMet, topical	HGF mouse UV-induced melanoma (Cassidy *et al.*, 2013)	↓ Time to first tumor, similar number of tumors	↓ UV-induced oxidative stress, ↑ growth of established tumors
MSA, oral	Mouse xenograft (Cassidy *et al.*, 2013)	↓ Tumor size	↓ SPARC secretion, ↑ cleavage of bid and apoptosis

[a]HED for 60 kg human.

reduced growth of tumors by 67% compared to tumors from animals treated with vehicle. *In vitro* studies showed that MSA decreased secretion of SPARC and increased levels of cleaved Bid and apoptosis.

There are literature reports of several additional classes of synthetic selenium-containing compounds that have been studied in animal and *ex vivo* models of skin cancer. To date none of these compounds has been shown to contribute to the central Se pool. Consequently, we believe that they should be considered a completely different class of agents from as those included in Figure 23.1.

Summary Points

- A substantial portion of the risk for skin cancer is inherited.
- UV light is the most important environmental risk factor for skin cancer.
- Some of the inherited risk factors, including polymorphisms in selenoproteins, affect the capacity of the skin to respond to UV-induced oxidative stress.
- The relationship between selenium consumption and risk for skin cancer is not well understood, and may be described by a "U-shaped curve" where risk is highest at both deficient and supranutritional levels.
- Studies with transgenic mice show that selenoproteins are essential for the function of healthy skin and for prevention of UV-induced skin cancer.

Key Facts of Skin Cancer

This lists statistics and risk factors for skin cancers.

1. Nonmelanoma skin cancers are the most common of the skin cancers, but melanoma is by far the most deadly, accounting for 80% of all skin cancer deaths.
2. Approximately 55% of melanoma risk is inherited.
3. Red hair, fair skin and the inability to tan increase an individual's risk for skin cancer.
4. Exposure to ultraviolet radiation from the sun is the most important environmental risk factor for both melanoma and nonmelanoma skin cancers.
5. In addition to direct damage to DNA, ultraviolet radiation may increase skin cancer risk *via* oxidative damage to both DNA and proteins.

Definitions and Explanations of Key Terms

Central selenium pool. The source of selenium used for selenoprotein biosynthesis, believed to be hydrogen selenide.

Familial melanoma. A melanoma that arises in a person carrying a high-penetrance mutation in a gene that predisposes to melanoma. Familial melanoma is diagnosed when a family has 3 or more melanomas (or a combination of melanomas and pancreas cancer) in first-degree relatives.

First-degree relative. A person who shares half of their genetic material with another individual, *i.e.* a parent, sibling or offspring.

GWAS. Genome-wide association study; a study of the association of genetic markers (in most cases single-nucleotide polymorphisms or SNPs) with incidence of a disease.

Keratinocyte. The cell type that makes up the majority of the epidermal layer of the skin. SCC and BCC arise from this cell type.

Melanocyte. A cell of neural crest origin found in the lowest layer of the epidermis in the skin. This is the cell type that produces melanins that give rise to pigmentation in the skin. They are also the cells from which melanoma arises.

Nevus (plural nevi). Also known as a mole, a benign growth usually in the skin, composed of melanocytes. Some melanomas may arise from what appears to be a nevus.

Relative risk. The factor by which an individual's risk for a disease is elevated (or decreased) when compared to that of the general population.

Selenoprotein. A protein in which selenocysteine has been incorporated into the polypeptide chain.

Sporadic. Having to do with a random or chance event. In skin cancer risk, sporadic tumors are thought to arise as a result of chance mutations instead of an inherited mutation.

List of Abbreviations

BCC	Basal cell carcinoma
CDKN2A	Cyclin-dependent kinase inhibitor 2A
GPx	Glutathione peroxidase
GSH	Glutathione
GWAS	Genome-wide association study
HED	Human-equivalent dose
i.p.	Intraperitoneal
MC1R	Melanocortin-1 receptor
MSA	Methylseleninic acid
MSR	Methionine sulfoxide reductase
NMSC	Nonmelanoma skin cancer
NPC	Nutritional prevention of cancer
p16ink4a	p16 protein inhibitor of cyclins 4 and 6
p14arf	p14 protein alternate reading frame
RR	Relative risk
SCC	Squamous cell carcinoma
SeCys	Selenocysteine
SeMet	Selenomethionine
SeMSec	Selenomethylselenocysteine
TR	Thioredoxin reductase
UV	Ultraviolet

References

Arner, E. S., 2009. *Biochimica et Biophysica Acta*. 1790: 495.

Asgari, M. M., Maruti, S. S., Kushi, L. H. and White, E., 2009. *Archives of Dermatology*. 145: 879.

Balch, C. M., Gershenwald, J. E., Soong, S. J., Thompson, J. F., Atkins, M. B., Byrd, D. R., Buzaid, A. C., Cochran, A. J., Coit, D. G., Ding, S., Eggermont, A. M., Flaherty, K. T., Gimotty, P. A., Kirkwood, J. M., McMasters, K. M., Mihm Jr, M. C., Morton, D. L., Ross, M. I., Sober, A. J. and Sondak, V. K., 2009. *Journal of Clinical Oncology*. 27: 6199.

Box, N. F., Duffy, D. L., Chen, W., Stark, M., Martin, N. G., Sturm, R. A. and Hayward, N. K., 2001. *American Journal of Human Genetics*. 69: 765.

Breslow, R. A., Alberg, A. J., Helzlsouer, K. J., Bush, T. L., Norkus, E. P., Morris, J. S., Spate, V. E. and Comstock, G. W., 1995. *Cancer Epidemiology, Biomarkers & Prevention*. 4: 837.

Brigelius-Flohe, R., Wingler, K. and Muller, C., 2002. *Methods in Enzymology*. 347: 101.

Bruner, S. D., Norman, D. P. and Verdine, G. L., 2000. *Nature*. 403: 859.

Burke, K. E., Clive, J., Combs Jr, G. F. and Nakamura, R. M., 2003. *Journal of the American Academy of Dermatology*. 49: 458.

Cassidy, P. B., Fain, H. D., Cassidy Jr, J. P., Tran, S. M., Moos, P. J., Boucher, K. M., Gerads, R., Florell, S. R., Grossman, D. and Leachman, S. A., 2013. *Nutrients*. 5: 725.

Chang, Y. M., Barrett, J. H., Bishop, D. T., Armstrong, B. K., Bataille, V., Bergman, W., Berwick, M., Bracci, P. M., Elwood, J. M., Ernstoff, M. S., Gallagher, R. P., Green, A. C., Gruis, N. A., Holly, E. A., Ingvar, C., Kanetsky, P. A., Karagas, M. R., Lee, T. K., Le Marchand, L., Mackie, R. M., Olsson, H., Osterlind, A., Rebbeck, T. R., Sasieni, P., Siskind, V., Swerdlow, A. J., Titus-Ernstoff, L., Zens, M. S. and Newton-Bishop, J. A., 2009. *International Journal of Epidemiology*. 38: 814.

Clark, L. C., Graham, G. F., Crounse, R. G., Grimson, R., Hulka, B. and Shy, C. M., 1984. *Nutrition and Cancer*. 6: 13.

Dammeyer, P. and Arner, E. S., 2011. *Biochimica et Biophysica Acta*. 1810: 111.

Dreno, B., Euvrard, S., Frances, C., Moyse, D. and Nandeuil, A., 2007. *European Journal of Dermatology*. 17: 140.

Duffield-Lillico, A. J., Reid, M. E., Turnbull, B. W., Combs Jr, G. F., Slate, E. H., Fischbach, L. A., Marshall, J. R. and Clark, L. C., 2002. *Cancer Epidemiology, Biomarkers & Prevention*. 11: 630.

He, C., Qureshi, A. A. and Han, J., 2010. *Journal of Dermatological Science*. 60: 54.

Kanetsky, P. A., Panossian, S., Elder, D. E., Guerry, D., Ming, M. E., Schuchter, L. and Rebbeck, T. R., 2010. *Cancer*. 116: 2416.

Karagas, M. R., Greenberg, E. R., Nierenberg, D., Stukel, T. A., Morris, J. S., Stevens, M. M. and Baron, J. A., 1997. *Cancer Epidemiology, Biomarkers & Prevention*. 6: 25.

Kryukov, G. V., Castellano, S., Novoselov, S. V., Lobanov, A. V., Zehtab, O., Guigo, R. and Gladyshev, V. N., 2003. *Science*. 300: 1439.

Le Marchand, L., Saltzman, B. S., Hankin, J. H., Wilkens, L. R., Franke, A. A., Morris, S. J. and Kolonel, L. N., 2006. *American Journal of Epidemiology*. 164: 232.
Li, D., Graef, G. L., Yee, J. A. and Yan, L., 2004. *Journal of Nutrition*. 134: 1536.
Lu, J. and Holmgren, A., 2014. *Free Radicals in Biology and Medicine*. 66: 75.
Marzuka-Alcala, A., Gabree, M. J. and Tsao, H., 2014. *Methods in Molecular Biology*. 1102: 381.
Mayne, S. T., Ferrucci, L. M. and Cartmel, B., 2012. *Annual Review of Nutrition*. 32: 369.
Meng, S., Zhang, M., Liang, L. and Han, J., 2012. *Pigment Cell & Melanoma Research*. 25: 612.
Meuillet, E. J., Mahadevan, D., Berggren, M., Coon, A. and Powis, G., 2004. *Archives of Biochemistry and Biophysics*. 429: 123.
Moos, P. J., Edes, K., Cassidy, P., Massuda, E. and Fitzpatrick, F. A., 2003. *Journal of Biological Chemistry*. 278: 745.
Ono, R., Masaki, T., Dien, S., Yu, X., Fukunaga, A., Yodoi, J. and Nishigori, C., 2012. *Journal of Dermatology*. 39: 843.
Overvad, K., Thorling, E. B., Bjerring, P. and Ebbesen, P., 1985. *Cancer Letters*. 27: 163.
Pence, B. C., Delver, E. and Dunn, D. M., 1994. *Journal of Investigative Dermatology*. 102: 759.
Pho, L., Grossman, D. and Leachman, S. A., 2006. *Current Opinion in Oncology*. 18: 173.
Picot, C. R., Moreau, M., Juan, M., Noblesse, E., Nizard, C., Petropoulos, I. and Friguet, B., 2007. *Experimental Gerontology*. 42: 859.
Platz, E. A. and Lippman, S. M., 2009. *Journal of Clinical Oncology*. 27: 3569.
Powis, G. and Kirkpatrick, D. L., 2007. *Current Opinion in Pharmacology*. 7: 392.
van der Pols, J. C., Heinen, M. M., Hughes, M. C., Ibiebele, T. I., Marks, G. C. and Green, A. C., 2009. *Cancer Epidemiology, Biomarkers & Prevention*. 18: 1167.
Prickett, T. D., Zerlanko, B., Gartner, J. J., Parker, S. C., Dutton-Regester, K., Lin, J. C., Teer, J. K., Wei, X., Jiang, J., Chen, G., Davies, M. A., Gershenwald, J. E., Robinson, W., Robinson, S., Hayward, N. K., Rosenberg, S. A., Margulies, E. H. and Samuels, Y., 2014. *Journal of Investigative Dermatology*. 134: 452.
Rogers, H. W., Weinstock, M. A., Harris, A. R., Hinckley, M. R., Feldman, S. R., Fleischer, A. B. and Coldiron, B. M., 2010. *Archives of Dermatology*. 146: 283.
Roman, M., Jitaru, P. and Barbante, C., 2014. *Metallomics*. 6: 25.
Schafer, M., Dutsch, S., auf dem Keller, U., Navid, F., Schwarz, A., Johnson, D. A., Johnson, J. A. and Werner, S., 2010. *Genes & Development*. 24: 1045.
Schallreuter, K. U., 2006. *Journal of Investigative Dermatology*. 126: 947.
SEER, 2013, http://seer.cancer.gov/statfacts/html/melan.html.
Sengupta, A., Lichti, U. F., Carlson, B. A., Ryscavage, A. O., Gladyshev, V. N., Yuspa, S. H. and Hatfield, D. L., 2010. *PLoS One*. 5: e12249.

Sengupta, A., Lichti, U. F., Carlson, B. A., Cataisson, C., Ryscavage, A. O., Mikulec, C., Conrad, M., Fischer, S. M., Hatfield, D. L. and Yuspa, S. H., 2013. *Journal of Investigative Dermatology*. 133: 1731.

Shekar, S. N., Duffy, D. L., Youl, P., Baxter, A. J., Kvaskoff, M., Whiteman, D. C., Green, A. C., Hughes, M. C., Hayward, N. K., Coates, M. and Martin, N. G., 2009. *Journal of Investigative Dermatology*. 129: 2211.

Song, H., Hur, I., Park, H. J., Nam, J., Park, G. B., Kong, K. H., Hwang, Y. M., Kim, Y. S., Cho, D. H., Lee, W. J. and Hur, D. Y., 2009. *Immune Network*. 9: 236.

Turanov, A. A., Xu, X. M., Carlson, B. A., Yoo, M. H., Gladyshev, V. N. and Hatfield, D. L., 2011. *Advances in Nutrition*. 2: 122.

Vinceti, M., Rothman, K. J., Bergomi, M., Borciani, N., Serra, L. and Vivoli, G., 1998. *Cancer Epidemiology, Biomarkers & Prevention*. 7: 853.

Vinceti, M., Crespi, C. M., Malagoli, C., Bottecchi, I., Ferrari, A., Sieri, S., Krogh, V., Alber, D., Bergomi, M., Seidenari, S. and Pellacani, G., 2012. *Tumori*. 98: 287.

Walshe, J., Serewko-Auret, M. M., Teakle, N., Cameron, S., Minto, K., Smith, L., Burcham, P. C., Russell, T., Strutton, G., Griffin, A., Chu, F. F., Esworthy, S., Reeve, V. and Saunders, N. A., 2007. *Cancer Research*. 67: 4751.

Wu, Y., Zhang, H., Dong, Y., Park, Y. M. and Ip, C., 2005. *Cancer Research*. 65: 9073.

Yan, L., Yee, J. A., McGuire, M. H. and Graef, G. L., 1997. *Nutrition and Cancer*. 28: 165.

CHAPTER 24

Selenium and Cancer Metastasis

YU-CHI CHEN[a] AND ANDREA M. MASTRO*[a]

[a]Department of Biochemistry and Molecular Biology, Eberly College of Science, The Pennsylvania State University, 107 Althouse Laboratory, University Park, Pennsylvania 16802, USA
*E-mail: a36@psu.edu

24.1 Selenium in Cancer

24.1.1 Overview

In recent decades, Se has garnered much attention because of its involvement in important biological functions, such as DNA synthesis, thyroid hormone metabolism, muscle regulation, antioxidant defense, fertility and reproduction (Mehdi *et al.*, 2013). The importance of Se has been further highlighted due to its participation in chemopreventive, anti-inflammatory and antiviral defense mechanisms (see Chen *et al.*, 2013b). Many of Se's functions are expressed through selenoproteins. Thus far, 30 have been indentified throughout 25 mammalian genes (Mehdi *et al.*, 2013). Some of these selenoproteins such as glutathione peroxidases (GPXs) and thioredoxin reductase (TXRs) have been extensively functionally characterized, but some require more investigation. Nevertheless, many selenoproteins are involved in antioxidant defense mechanisms. For instance, GPXs reduce hydrogen peroxide and organic hydroperoxides intracellularly and extracellulary, and protect

Food and Nutritional Components in Focus No. 9
Selenium: Chemistry, Analysis, Function and Effects
Edited by Victor R. Preedy

Published by the Royal Society of Chemistry, www.rsc.org

cells from oxidative damage (see Chen *et al.*, 2013b; Mehdi *et al.*, 2013). TXRs regulate the thioredoxin system and maintain the intracellular redox potential (Mehdi *et al.*, 2013). In addition, more recently, researchers reported that carefully designed phenolaminoethyl selenides can neutralize hydrogen peroxide directly and reduce the oxidative stress caused by lipopolysaccharide. Thus, it may be possible to use organoselenium compounds that are capable of forming a Se redox cycle as supplements to defend against oxidative stress (Cowan *et al.*, 2013). Taken together, selenoproteins and organoselenium can minimize the level of reactive oxygen species, a risk factor for DNA mutations and carcinogenesis (Loft *et al.*, 1996; Poulsen *et al.*, 1998). Because of its antioxidant nature, the relationship between selenium and cancer has been studied for decades, and selenium has been viewed as a potential therapeutic target. It is also noted that selenium levels in cancer patients have been reported to be lower than in their healthy counterparts (Surai, 2006). Furthermore, some alterations in selenoproteins are associated with cancers. For example, the loss of heterozygosity (LOH) of GPX1 has been documented in lung, breast, head and neck and colon cancers (see Hu *et al.*, 2003, 2004, 2005; Chen *et al.*, 2013b). The LOH of the 15 kDa selenoprotein (Sep15) has been linked to breast cancer (see Chen *et al.*, 2013b). Overexpression of TXR1 has been observed in many cancers, including nonsmall cell lung, thyroid, oral squamous, breast, colorectal, liver, gastric and cervical carcinoma (see Kawahara *et al.*, 1996; Grogan *et al.*, 2000; Kakolyris *et al.*, 2001; Matsutani *et al.*, 2001; Soini *et al.*, 2001; Raffel *et al.*, 2003; Hedley *et al.*, 2004; Turunen *et al.*, 2004; Lincoln *et al.*, 2010; Chen *et al.*, 2013b). It has been suggested that once tumorigenesis is initiated, the overexpression of the thioredoxin system may promote cancer progression by overcoming oxidative stress and increasing cell proliferation (Raffel *et al.*, 2003). Taken together, selenium controls the biosynthesis and function of selenoproteins, which guard against oxidative damage. In addition to regulating the redox-active proteins, monitoring their redox status and maintaining the intracellular redox balance, Se can protect against cancer by controlling inflammatory and immune responses, increasing DNA stability, dictating cell-cycle arrest or apoptosis, inhibiting cancer-cell migration and invasion, blocking angiogenesis, and upregulating phase II carcinogen-detoxifying enzymes (see Chen *et al.*, 2013b). Anticancer activity is not limited to selenoproteins. Some small Se compounds, including selenomethionine (SEM), methyl selenocysteine (MSC), methylseleninic acid (MSA) and methylselenol, have been reported to protect against cancer progression (Ganther, 1999; Chen *et al.*, 2013a,b).

24.1.2 Preclinical and Clinical Evidence for the Beneficial Effect of Se in Regard to Cancer

Clinical studies regarding the beneficial effect of dietary Se on cancer prevention are somewhat controversial. Initially, researchers reported a significant inverse association between Se content in forage crops or diets and overall cancer mortality (see Clark *et al.*, 1991; Chen *et al.*, 2013b). These data led to the Nutritional Prevention of Cancer (NPC) trial to evaluate the effect of increased Se uptake (in the form of Se-enriched yeast) on the recurrence of nonmelanoma skin cancer (see Chen *et al.*, 2013b). Although Se showed no protective effect on squamous or basal cell skin carcinoma, overall cancer mortality and the incidence of prostate, colorectal, lung and total cancers were significantly reduced (see Chen *et al.*, 2013b). A later analysis indicated that the incidences of prostate and total cancers in males were still significantly reduced (see Chen *et al.*, 2013b). These promising results led to the large Selenium and Vitamin E Cancer Prevention Trial (SELECT) to test the preventive effect of SEM and Vitamin E supplementation on prostate cancer (see Chen *et al.*, 2013b). Sadly, SELECT was terminated prematurely because of the absence of the positive outcome. Also long-term prevention was not observed either (see Chen *et al.*, 2013b). The inconsistency between these two trials suggests that researchers need to re-examine the methodology and strategies regarding the role of Se in cancer research. The Se baseline levels in the population, the different utilization rates among Se compounds, and the genetic polymorphisms in selenoproteins must be considered (see Chen *et al.*, 2013b). Results of other smaller studies focused on lung, bladder, colorectal, oesophageal, gastric, cardiac and thyroid cancers also suggested the beneficial effect of Se supplementation (see Chen *et al.*, 2013b). Taken together, it is proposed that increased Se uptake reduced the risk of cancers at least in populations with lower Se baselines.

The anticancer and chemopreventive effects of Se have been demonstrated more clearly in *in vitro* and preclinical animal models of liver, esophageal, pancreatic, prostate, colon and breast cancers (see Li *et al.*, 2008; Jung *et al.*, 2010; Chen *et al.*, 2013b). Mechanisms through which Se can suppress the incidence of primary tumor growth have already been reviewed (see Chen *et al.*, 2013b). Much of this work has been focused on the effect of Se on the initiating steps of cancer development. Because data from the SELECT trials indicated that the most protective effects of Se occurred on participants with low Se baselines prior to the intervention, it is difficult to determine the point to start Se supplementation in order to reduce cancer incidence. On the other hand, cancer metastasis is a late event and often the cause of death in cancer patients. If Se supplementation suppressed metastasis, it would greatly improve the quality of life and survival. Therefore, in this chapter, we will discuss current findings regarding the effects of Se on metastasis.

24.2 Se in Cancer Metastasis

24.2.1 Se and Cancer Metastasis

Compared to the large amount of published information concerning Se and cancer initiation, very little is known about the involvement of Se in metastasis. Nevertheless, there is credible evidence to suggest that Se can suppress/reduce metastasis. Recently, Chen *et al.* demonstrated that in a breast cancer cell, 4T1.2-mouse model, dietary Se compounds not only reduced the primary tumor growth but also affected spontaneous metastasis (Chen *et al.*, 2013a). Interestingly, this report drew attention to the differences among Se compounds, and the urgency to design or select appropriate Se molecules. Mice were fed diets supplemented with selenite, SEM or MSA for 3 months before cancer cell inoculation. Mice fed SEM had significantly smaller primary tumors and generally had the least number of cancer cells present in the secondary organs, including lung, heart, liver, kidney and bone. On the other hand, selenite supplementation provided only short-term suppression of primary tumor growth, and did not reduce cancer cells in secondary sites. Mice fed selenite, regardless of the amount of supplementation, displayed an increased tendency for kidney and bone metastasis.

Using an invasive prostate cancer cell PAIII rat model, Sinha *et al.* reported the antimetastatic effect of MSA (see Chen *et al.*, 2013b). PAIII cells were injected 2 weeks prior to daily MSA supplementation. There were significantly fewer foci present in the lungs of rats fed MSA. Although the inoculation of the cancer cells bypassed the establishment of primary tumors, and metastatic progression, the reduction of lung foci could be due to the downregulation of hypoxia-induced factor 1α (HIF-1α), which further inhibited angiogenesis by suppressing its downstream target, vascular endothelial growth factor (VEGF). In addition, GPX3 is often decreased in prostate cancer cells and can serve as a tumor suppressor (see Chen *et al.*, 2013b). Overexpression of GPX3 in PC3 prostate cancer cells led to reduced primary tumor growth with no metastasis when injected subcutaneously into SCID mice.

Selenoprotein, Sep 15, is highly involved in the progression of metastasis in colon cancer. It is often increased in colon cancer cell lines, and is essential to primary tumor growth and metastasis. For example, knockdown of Sep15 in murine colon cancer cells, CT26, led to a significant decrease (93% to 10%) in subcutaneous cancer formation as well as reduction in lung metastasis following tail-vein inoculation.

There is also evidence to suggest that the metastasis of melanoma can be controlled by Se. Dietary Se supplementation, *e.g.* SEM, selenite and even high-Se soy protein, for 2 weeks before and after the intravenous injection of murine melanoma cells, B16BL6 (see Chen *et al.*, 2013b) significantly reduced the metastatic incidence, the numbers of metastatic foci in the lung and the tumor size.

Several groups have reported the effectiveness of Se in inhibition of lung cancer metastasis in the Lewis lung carcinoma mouse model. A Se-yeast supplemented diet for 3 weeks before and after intramuscular injection of Lewis lung carcinoma cells (see Chen *et al.*, 2013b), reduced both the metastatic incidence and the number of metastatic lung foci. Yan *et al.* further demonstrated that although SEM is the main ingredient in Se-yeast, it may not necessarily be the most important in suppressing metastasis (see Chen *et al.*, 2013b). In a similar model, MSA supplementation resulted in significant reduction of lung metastasis (55% decrease intramuscularly and 20% decrease subcutaneously) while supplementation with SEM did not.

Taken together, there are many reports that Se can prevent the initiation of cancer but also suppress metastasis in breast, prostate, colon and lung cancers and melanoma. Both selenoproteins and Se compounds have been shown to have antimetastatic potential. However, these data also point out the diversity among Se compounds, and that different tissues may utilize different pathways to regulate cancer progression. Therefore, it is important to understand the underlying mechanisms that can be affected by Se in order to select the best Se molecules for therapeutic application.

24.2.2 Metastasis

Metastasis, a multistep event, starts from the infiltration of cancer cells into adjacent tissues and *trans*-endothelial migration to the circulation, followed by extravasation and reattachment of cancer cells in secondary sites. It ends with cancer cell proliferation and subsequent angiogenesis in metastatic organs (see Chen *et al.*, 2013b). In the following context, we discuss mechanisms reported to be involved in metastasis and the impact of Se on them. The effects of selenoproteins and Se molecules on different types of cancers are summarized in Tables 24.1–24.9.

24.2.3 The Effect of Se on Cell Integrity and Epithelial-Mesenchymal Transition (EMT)

In an *in vitro* study, with human metastatic breast cancer cells, MDA-MB-231, and nonmetastatic cells, MCF-7, selenate (50 ng mL^{-1}) (see Chen *et al.*, 2013b) was shown to increase transepithelial electronic resistance and decrease paracellular permeability to large molecules, and notably enhance cell–cell attachment and promote the insulation of cell colonies. The motility of MDA-MB-231 cells was highly reduced, and their ability to penetrate endothelial cell layers was impaired.

Table 24.1 Effects of Se compounds and selenoproteins on endothelial cells.

Se compounds/ selenoproteins	Studies	Cells	Function	References[a]
MSA	*In vitro*	HUVEC	↓ MMP-2, ↑ apoptosis	10, 36, 65
MeCN	*In vitro*	HUVEC	↓ MMP-2	10
Selenite	*In vitro*	HUVEC	↑ Apoptosis	65
	In vitro	Mammary endothelial cells	↓ VEGF	53
TXR1	*In vitro*	Mammary endothelial cells	TXR1 inhibition ↓ VEGF, cell migration, proliferation and tube formation	53

[a]Refers to the reference number in Chen *et al.* (2013b).

Table 24.2 Effects of Se compounds on breast cancer metastasis.

Se compounds	Studies	Cancer cells	Function	References[a]
MSA	*In vitro*	Human breast cancer cells, MDA-MB-468 and MCF-7	↓ VEGF	10
MSC	*In vitro*	Murine breast cancer cells, TM6	↓ Migration	64
	Rat	Carcinogen-induced breast cancer	↓ Angiogenesis	65
	Mice	Human breast cancer cells, MCF-7	↓ Angiogenesis	66
SEM	Mice	Murine breast cancer cells, 4T1.2	↓ Metastasis	Chen *et al.* (2013a)
Selenite	Rat	Carcinogen-induced breast cancer	↓ VEGF and angiogenesis	65
Selenate	*In vitro*	Human breast cancer cells, MDA-MB-231 and MCF-7	↑ Epithelial tight junction, ↓ motility and transendothelial invasion	38
Se-enriched garlic	Rat	Carcinogen-induced breast cancer	↓ VEGF and angiogenesis	65

[a]Refers to the reference number in Chen *et al.* (2013b).

Table 24.3 Effects of Se compounds and selenoproteins on prostate-cancer metastasis.

Se compounds/ selenoproteins	Studies	Cancer cells	Function	References[a]
MSA	*In vitro*	Human prostate cancer cells, DU145	↓ VEGF	10
	Mice	Human prostate cancer cells, DU145	↓ Tumor growth and angiogenesis	67
	Rat	Rat prostate cancer cells, PAIII	↓ HIF-1a, VEGF and metastatic lung foci	68
GPX3	*In vitro*	Human prostate cancer cells, PC3, DU145 and LNCaP	GPX3 overexpression ↓ cell invasion	60
	Mice	Human prostate cancer cells, PC3 with GPX3 overexpression	GPX3 overexpression ↓ primary tumor sizes and metastasis, ↑ survival	60

[a]Refers to the reference number in Chen *et al.* (2013b).

In contrast, Tsukamoto *et al.* reported that lower concentrations of selenate (0.04–0.16 mM, compared to IC_{50}: 0.88 mM) induced EMT in a colorectal cancer cell line, DLD-1, *in vitro* (Tsukamoto *et al.*, 2013). Twenty-four hours after treatment, DLD-1 cells detached from each other, became flat, and spindle-shaped and increased motility, migration and invasion. It was noted that selenate induced EMT through the AKT pathway.

Table 24.4 Effects of Se compounds and selenoproteins on melanoma metastasis.

Se compounds/ selenoproteins	Studies	Cancer cells	Function	References[a]
SEM	Mice	Melanoma	↓ Lung metastasis	50
Selenite	Mice	Murine melanoma cell, B16BL6	↓ Lung metastasis	51
	Mice	Murine melanoma cell, B16F10	↓ Lung metastasis	71 and 72
	In vitro	Murine melanoma cell, B16F10	↓ Cell migration, ↓ HIF-1a, VEGF, and IL-18	71 and 72
High-Se isolated soy proteins	Mice	Murine melanoma cell, B16BL6	↓ Lung metastasis	33

[a]Refers to the reference number in Chen *et al.* (2013b).

Table 24.5 Effects of Se compounds and selenoproteins on fibrosarcoma metastasis.

Se compounds/ selenoproteins	Studies	Cancer cells	Function	References[a]
MSA	*In vitro*	Human fibrosarcoma cell, HT1080	↓ Cell invasion, MMP-2 activation and MT1-MMP, ↑ TIMP-2	11
Methylselenol	*In vitro*	Human fibrosarcoma cell, HT1080	↓ Cell migration, invasion, and the expression and activity of MMP-2 and MMP-9, ↑ TIMP1 and TIMP2	54
Selenite	*In vitro*	Human fibrosarcoma cell, HT1080	↓ Cell migration, cell-ECM attachment, MMP-2, MMP-9 and uPA, ↑ TIMP-1	49

[a]Refers to the reference number in Chen *et al.* (2013b).

Table 24.6 Effects of Se compounds and selenoproteins on lung cancer metastasis.

Se compounds/ selenoproteins	Studies	Cancer cells	Function	References[a]
MSA	Mice	Lewis lung carcinoma cell	↓ Lung metastasis, plasma uPA and PAI-1	41
MSC	Mice	Human small cell lung cancer, H69	↓ Microvessel density, ↑ vascular maturation	55
	Mice	Human nonsmall epithelial lung carcinoma, A549	↑ Vascular maturation	55
TXR1	Mice	Lewis lung carcinoma cell	TXR1 inhibition ↓ lung metastasis	61
	In vitro	Lewis lung carcinoma cell	TXR1 inhibition ↓ HGF and OPN	61

[a]Refers to the reference number in Chen *et al.* (2013b).

Table 24.7 Effects of Se compounds and selenoproteins on colon-cancer metastasis.

Se compounds/ selenoproteins	Studies	Cancer cells	Function	References[a]
MSC	Mice	Human colon cancer cells, HCT-8, HT-29 and GEO	↓ Angiogenesis	45 and 46
Selenate	*In vitro*	Human colorectal cancer cell, DLD-1	↑ EMT	Tsukamoto *et al.* (2013)
GPX2	*In vitro*	Human colon adenocarcinoma cell, HT29	↓ Cell migration and invasion	39
SBP1	*In vitro*	Human colon cancer cell, HCT116	SBP1 overexpression ↓ cell migration	59
15-Sep	Mice	Murine colon cancer cells, CT26	Sep15 inhibition ↓ lung metastasis	48

[a]Refers to the reference number in Chen *et al.* (2013b).

Table 24.8 Effects of Se compounds and selenoproteins on liver-cancer metastasis.

Se compounds/ selenoproteins	Studies	Cancer cells	Function	References[a]
Selenite	Rat	Carcinogen-induced liver cancer	↓ Angiogenesis and angiogenic factors	52
Se-enriched malt	Rat	Carcinogen-induced liver cancer	↓ Angiogenesis and angiogenic factors	74
SBP1	*In vitro*	Human liver cancer cell, SMMC7721	SBP1 reduction ↑ cell migration and GPX1 activity	34

[a]Refers to the reference number in Chen *et al.* (2013b).

24.2.4 The Effect of Se on Cell Migration, Invasion and Extracellular Matrix (ECM) Degradation

In addition to selenate, MSC was also reported to inhibit breast cancer cell migration (see Chen *et al.*, 2013b). One hour of MSC pretreatment of murine transformed mammary epithelial cells, TM6, was sufficient to reduce by 44%, migration stimulated by insulin, epidermal growth factor (EGF) and serum.

In colorectal cancer, selenium-binding protein 1 (SBP1) and GPX2 also affect cell migration or invasion (see Chen *et al.*, 2013b). SBP1, which binds Se and is responsible for Se transport (see Chen *et al.*, 2013b), is a differentiation marker of colon cancer. Suppression of SBP1 occurs at the late stage and is an indicator of poor survival (see Chen *et al.*, 2013b). Pohl *et al.* provided evidence that SBP1 regulates tumor cell growth and progression (see Chen *et al.*, 2013b). By comparing several human colon cancer cell lines, they

Table 24.9 Effects of Se compounds and selenoproteins on head and neck squamous cell carcinoma metastasis.

Se compounds/ selenoproteins	Studies	Cancer cells	Function	References[a]
MSA	*In vitro*	Human head and neck squamous cell carcinoma, FaDu	↓ HIF-1α, and VEGF, ↑ prolyl hroxylases	43
MSC	Mice	Human head and neck squamous cell carcinoma, FaDu	↓ COX-2, iNOS, HIF-1α, VEGF and microvessel density, ↑ vascular maturation, drug delivery and distribution	42 and 75
	Mice	Human head and neck squamous cell carcinoma, A253	↓ Microvessel density, ↑ vascular maturation, drug delivery and distribution	55
TXR1	Clinically	Human oral squamous cell carcinoma	Correlated with lymph node metastasis and with the clinical stage	73

[a]Refers to the reference number in Chen *et al.* (2013b).

observed an inverse correlation between the methylation level of the SBP1 promoter and the expression of SBP1. Overexpression of SBP1 in human colon cancer cells, HCT116, with nondetectable endogenous SBP1 led to decreased cell proliferation, increased apoptosis and suppressed migration. On the other hand, GPX2 was often upregulated in colorectal cancer (see Chen *et al.*, 2013b). Stable knockdown of GPX2 in human colon adenocarcinoma cells, HT29, reduced tumor growth *in vivo* and decreased the capacity of HT29 to grow anchorage independently (see Chen *et al.*, 2013b). However, the lack of GPX2 enhanced cell migration and invasion through counteracting the function of cyclooxygenase-2 (COX-2), independently from its antioxidant property. Perhaps because of this unique mechanism, this regulation of migration and invasion was GPX2 specific.

The reduction of SBP-1 has been noted in several other cancers, including lung adenocarcinoma, ovarian, gastric, and liver. Moreover, it has been proposed that SBP1 is associated with liver-cancer metastasis (see Chen *et al.*, 2013b). SBP-1 was barely detectable in highly metastatic liver cell lines and minimally expressed in most slightly metastatic lines. Silencing of SBP-1 with RNAi inhibited cell migration but not the expression of GPX1. The combination of decreased SBP-1 and increased GPX1 positively correlated with the vascular invasion found in human hepatocarcinoma samples. Hydrogen peroxide induced the production of HIF-1α, which in turn increased the expression of SBP-1. These data suggest a feedback mechanism between SBP-1 and HIF-1α since hydrogen peroxide only induced HIF-1α when SBP-1 was present.

Sodium selenite ($NaSeO_3$) and sodium selenate ($NaSeO_4$) are the two most common inorganic Se compounds. They both affect fibrosarcoma but very differently (see Chen *et al.*, 2013b). Using human fibrosarcoma cells, HT1080, Yoon *et al.* reported that 10 μM selenite clearly decreased cell proliferation and viability, while no obvious cytotoxicity was observed under the 50 μM selenate treatment (see Chen *et al.*, 2013b). Treatment with selenite within the noncytotoxic range significantly reduced cell invasion but did not alter motility. Neither selenite nor selenate treatment affected cell–cell attachment in HT1080 cells. Only selenite reduced attachment to collagen. Cancer cells degrade ECM in order to invade surrounding areas. Selenite has been reported to decrease the degradation of ECM by the metalloproteinases (MMP) (see Chen *et al.*, 2013b). HT1080 cells constitutively express MMP-2, MMP-9 and urokinase-type plasminogen activator (uPA). Treatment of the cells with selenite but not selenate significantly reduced the activity of these proteins. Moreover, pretreatment with selenite counteracted the stimulatory effect of the MMP inducers, 12-*O*-tetradecanoylphorbol-13-acetate (TPA) and TNF-α. However, when selenite was added simultaneously with a MMP inhibitor, dexamethasone, the inhibition of MMP-9 was greater in combination than either treatment alone. There are two possible mechanisms why selenite inhibited the activity of MMP-9. First, selenite may reduce the protein level of MMP-9. Or, selenite may increase the level of the MMP-9 inhibitor, tissue inhibitor of metalloproteinase 1 (TIMP-1). Park *et al.* tested the effect of MSA on the inhibition of ECM degradation (see Chen *et al.*, 2013b), and found that MSA had no effect on the expression of pro-MMP-9 but significantly increased the level of pro-MMP-2 and decreased the active MMP-2. MMP-2 activation can be controlled by several regulators, including the initiator, membrane-type 1 matrix metalloproteinase (MT1-MMP) and the inhibitor, TIMP-2. TPA increased MT1-MMP and decreased TIMP-2. MSA neutralized TPA, suppressed the expression of MT1-MMP and enhanced the expression and activity of TIMP-2. Their data suggest that MSA reduced the activity of MMPs, leading to less ECM degradation and decreased cell invasion. MSA is the direct precursor of methylselenol that is considered to be the key molecule regulating the anticancer effects of Se. Others confirmed this protective effect of Se on ECM degradation (see Chen *et al.*, 2013b). Methylselenol generated by adding SEM and L-methionine-α-deamino-γ-mercaptomethane lyase (METase) in the HT1080 cell culture produced results similar to MSA treatment. Methylselenol increased the activity of pro-MMP2 and pro-MMP-9, decreased the active MMP-2 and increased the level of TIMP-1 and TIMP-2. This imbalance between prometastasis proteinases and antimetastasis molecules led to suppressed cell migration and invasion (Zeng *et al.*, 2006).

In addition to inducing apoptosis in human astrocytoma cells, IPSB-18, Rooprai *et al.* found that selenite altered the MMP profile extensively *in vitro* (see Chen *et al.*, 2013b). The expression of MMP-2, -9, -14, -15, -16, -24 and several inhibitors, TIMP-1, -2, -3, -4 were significantly reduced. In addition, epithelial growth factor receptor (EGFR), an often overexpressed contributor to cell proliferation, survival, motility and invasion of glioma (see Chen *et al.*, 2013b), was also reduced. Among all MMPs measured, only one, MMP-25, was elevated by selenite. Taken together, by reducing cell growth, inducing apoptosis and altering the ECM degradation efficiency, selenite likely suppressed the development of glioma.

Melanoma is the most fatal type of skin cancer due to its rapid metastasis. Se was shown to reduce the metastasis of the murine melanoma line, B16F10 (see Chen *et al.*, 2013b). Although the results indicate possible mechanisms for suppression, they also highlighted a potential difficulty; *i.e.* it may be difficult to reach the optimal dosage in one organ without causing damage or stress in others. They found that B16F10 cells had a very high tolerance to selenite *in vitro* (see Chen *et al.*, 2013b). More than 40 μM of selenite was required to bring about apoptosis as the result of G0/G1 cell-cycle arrest. Further investigation showed that selenite induced B16F10 cell-cycle arrest through a p53-independent p21Waf1 pathway. *In vivo*, when B16F10 cells were injected intraoptic venously along with 100 μM selenite treatment i.p. injection, lung metastasis was significantly reduced after 14 days (see Chen *et al.*, 2013b). It was interesting to note that although (20 μM) selenite did not affect cell viability, it significantly reduced cell migration (Song *et al.*, 2011). Taken together, selenite-reduced melanoma lung metastasis *in vivo* may be due to several mechanisms. At lower concentrations, selenite suppressed cell migration by blocking the expression of IL-18 and downregulating angiogenesis. At higher concentrations, selenite directly reduced cell number by inducing G0/G1 cell-cycle arrest in a p53-independent p21Waf1 pathway.

24.2.5 Angiogenesis

ECM degradation also plays a role in angiogenesis, an essential process in metastasis and development of solid tumors. Once cancer cells develop into a mass of a few mm^3, diffusion can no longer support their growth and survival without new blood vessels. It is noted that vascular tumor masses can exist dormant in asymptomatic cancer patients. In order for these small tumors to continuously develop, new blood vessels are required to ensure a constant supply of nutrients (Lu *et al.*, 2001). Generally, cancer cells secrete proangiogenic cytokines to initiate angiogenesis. Many cytokines or chemokines are involved, including VEGF, hepatocyte growth factor (HGF), axon guidance factors, interleukin-1 (IL-1), interleukin-6 (IL-6), interleukin-8 (IL-8), stromal cell-derived factor 1 (SDF-1), growth-regulated peptide-alpha/growth-regulated oncogene-1 (GRD-α), angiopioteins, osteopontin (OPN), and fibroblast growth factors (FGF) (see Chen *et al.*, 2013b). VEGF, which enhances endothelial cell proliferation and migration *in vitro* and blood-vessel permeability and angiogenesis *in vivo*, is considered the most important regulator of angiogenesis (Shweiki *et al.*, 1992; Neufeld *et al.*, 1999). Other molecules can affect angiogenesis by regulating VEGF expression, promoting EMT, inducing and activating MMPs, and enhancing endothelial cell proliferation. Endothelial cells respond to these stimulations, increase cell proliferation, enhance cell motility, and secret MMPs to degrade ECM. Taken together, both cancer cells and endothelial cells are important participants in the process of angiogenesis; therefore, treatments against both cells have the potential to interfere with angiogenesis and suppress metastasis.

24.2.6 The Effects of Se on ECM Degradation and Angiogenesis

Treatment of human umbilical vein endothelial cells (HUVEC) with a mixture of antioxidants containing Se did not affect cell proliferation but reduced migration and invasion (see Chen *et al.*, 2013b). Both ECM degradation and endothelial cell differentiation were significantly suppressed, but interestingly the differentiation was Se compound-specific (see Chen *et al.*, 2013b). The organic Se compound, methylselenocyanate (MeCN) and MSA significantly reduced MMP-2 through the MAPK1/2 pathway but the inorganic compound, selenite, did not (see Chen *et al.*, 2013b). The differences between Se compounds were further highlighted by Jiang *et al.* (see Chen *et al.*, 2013b). Both MSA and selenite induced apoptosis in HUVECs but through different pathways. MSA-induced apoptosis was caspase dependent and regulated by ERK1/2, AKT and p38 MAPK pathways, while selenite-induced apoptosis was caspase independent. In addition, Se can regulate the behaviors of endothelial cells and angiogenesis *via* selenoproteins (see Chen *et al.*, 2013b). For example, Streicher *et al.* reported that the expression of VEGF was increased when the activity of TXR was suppressed by 1-chloro-2,4-dinitro-benzene (DNCB) in primary bovine mammary endothelial cells (see Chen *et al.*, 2013b). The inhibition of TXR also resulted in the increase of the kinase insert domain-containing receptor (KDR) that can increase VEGF and promote endothelial cell proliferation and migration. However, specific inhibition of GPX did not produce the same effect. In conclusion, the cell proliferation, migration and differentiation of endothelial cells can be regulated both by Se molecules and selenoproteins. It is important to note the diverse effects among Se compounds and selenoproteins. The specific mechanisms involved in the regulation of endothelial cells by different Se materials need to be further investigated.

Because the most promising application of Se in cancer prevention was found in prostate cancer, the relationship between Se and prostate cancer has been extensively studied. It has been suggested that the basal Se level in humans was the critical factor affecting the preventive effect of Se (see Chen *et al.*, 2013b), and Se had a greater effect on more advanced prostate cancer (see Chen *et al.*, 2013b). Thus, it is likely that Se can suppress development of prostate-cancer metastasis. Jiang *et al.* reported that although in human prostate cancer cells, DU145, and human breast cancer cells, MDA-MB-468 (see below), the effect of MSA was both MAPK independent, the sensitivities to MSA of these cells were different (see Chen *et al.*, 2013b). A low concentration of MSA (2 μM) was sufficient to significantly reduce VEGF without causing apoptosis in DU145, but at least 5 μM MSA was required to have a notable inhibitory effect in MDA-MB-468 cells. Wang *et al.* (see Chen *et al.*, 2013b) demonstrated that MSA induced human microvascular endothelial cells, TIME G0/G1 cell-cycle arrest and apoptosis in a dose-dependent manner. MSA also inhibited prostate-cancer angiogenesis in mice.

To demonstrate the effect of Se on angiogenesis *in vivo*, Jiang *et al.* used Se garlic, selenite or MSC, in either conventional chemoprevention or acute exposure conditions (see Chen *et al.*, 2013b). In the conventional chemoprevention model, rats were injected intraperitoneally with 1-methyl-1-nitrosourea (MNU) to induce breast carcinoma. The Se-supplemented diets, Se garlic or selenite, started 7 days after the MNU and continued for 7 weeks. The microvessel density was significantly reduced by Se-garlic and selenite supplementation. Further examination indicated that small microvessels were most affected. The levels of VEGF expressed mainly by cancer epithelial cells were also decreased. Following acute exposure, the Se supplementation, MSC or selenite, also resulted in reduction of the microvessel density.

The effect of MSC on breast-carcinoma angiogenesis was confirmed with a MCF-7 xenograft model (see Chen *et al.*, 2013b). To summarize, several Se materials showed inhibitory effects on prostate and breast-carcinoma angiogenesis. However, these data point out the inconsistency between *in vitro* and *in vivo* studies. For example, selenite significantly reduced the microvessel density and suppressed VEGF *in vivo* (see Chen *et al.*, 2013b), but it failed to reduce the expression of VEGF by breast cancer cells *in vitro* (see Chen *et al.*, 2013b), which suggested that the regulation of Se in breast-cancer angiogenesis involves a complicated network of cells and the microenvironment.

In colon cancer, MSC has been shown to reduce both the tumor growth and the microvessel density *in vivo* (see Chen *et al.*, 2013b). Researchers implanted human colon cancer cells, HCT-8, HT-29 or GEO, into nude mice 3 or 6 days before the daily dose of 0.2 mg MSC was given for up to 16 days. At least 50% reduction in the microvessel density was found in all three xenograft models. MSC also enhanced the vascular maturation (see Chen *et al.*, 2013b) by nearly 50% in the HCT-8 model. The increased vascular maturation enhanced the tumor vascular permeability, reduced the tumor interstitial fluid pressure and improved the efficiency of drug delivery. These results indicated that MSC not only can work alone to reduce colon cancer growth and inhibit angiogenesis, but also it can be used together with other anticancer drugs.

In addition to MMPs, the urokinase plasminogen activator system (uPA, its receptor uPAR and plasminogen activator inhibitors (PAI)) regulate cancer development and metastasis. Plasmin triggers pro-uPA to bind uPAR and initiates the uPA activation. Active uPA turns the proenzyme, plasminogen, into the functional serine protease, plasmin that is involved in the ECM degradation process. Although the normal function of PAI-1 is to block plasmin-associated proteolysis, elevated plasma level of PAI-1 along with uPA and uPAR have been reported in several cancers. PAI-1 deficiency has been shown to decrease cell invasion and vascularization. Yan *et al.* showed that Se can target this system to downregulate the ECM degradation (see Chen *et al.*, 2013b). Mice bearing the Lewis lung carcinoma had significant elevated uPA levels in their plasma. MSA, but not SEM, caused a 50% reduction of the uPA plasma level, which suggested a lower activation of pro-uPA. Furthermore, MSA resulted in a 40% reduction of the plasma PAI-1 level. These data indicated that MSA manipulated the entire urokinase plasminogen activator system, and reduced ECM degradation and cell invasion. In addition, MSA directly affected angiogenesis in the Lewis lung carcinoma model by reducing the expression of several cytokines, including VEGF, FGF and platelet-derived growth factor-BB (PDGF-BB). Using a human small cell lung cancer, H69, and a human nonsmall epithelial lung carcinoma, A549, xenograft model, Rustum *et al.* demonstrated that MSC also reduced angiogenesis (see Chen *et al.*, 2013b).

As mentioned previously, TXR1 downregulated cell proliferation, migration and production of proangiogenic factors, VEGF and VEGFR, of mammary endothelial cells, which suggested that TXR1 is antimetastatic in breast cancer (see Chen *et al.*, 2013b). However, it was reported that TXR1 increased tumorigenesis and metastasis in the Lewis lung carcinoma (see Chen *et al.*, 2013b). Yoo *et al.* compared TXR1-knockdown Lewis lung carcinoma cells injected in the flank (primary tumors) with those injected in the tail vein (lung metastasis). In both scenarios, TXR1 deficiency significantly decreased the weight of primary tumor and completely eliminated lung metastasis. The changes in the tumor development may be explained by the fact that TXR1deficiency restored the morphology of the carcinoma cells to normal and reduced the expression of 2 proangiogenic factors, HGF and OPN. It is noted that in many cancer cell lines, the level of TXR1 is elevated (see Chen *et al.*, 2013b). Taken together, whether TXR1 is pro- or antimetastatic may be cancer or organ specific and the importance of TXR1 in tumor development warrants further investigation.

Liu *et al.* used a diethylnitrosamine-induced hepatocarcinoma rat model to investigate whether Se-enriched malt, SEM and selenite could reduce tumorigenesis and angiogenesis (see Chen *et al.*, 2013b). SEM showed the most promising results by decreasing mortality and the numbers of liver nodules. The reduction partially resulted from the decreased expression of several proangiogenic factors, including TNF-α, nitric oxide (NO), nitric oxide synthase (NOS), insulin growth factor II, VEGF and protein kinase Cα. Both Se-enriched malt and SEM increased the level of TNF-α at the initiation of carcinogenesis but decreased it in the later stage. On the other hand, selenite supplementation had much less effect on hepatocarcinoma angiogenesis.

Chintala *et al.* demonstrated that *in vitro*, MSA inhibited the proliferation of human clear cell renal cell carcinoma cells, RC2 as well as decreased the expression of HIF-1α and VEGF (see Chen *et al.*, 2013b). HIF-1α that is induced by hypoxia is a key regulator of oxygen availability and of VEGF expression. MSA may have increased the post-translational degradation of HIF-1α in a proteosome-independent, prolyl hydroxylase-dependent manner.

Organic Se compounds, MSC and MSA, have been reported to suppress angiogenesis in head and neck cell carcinoma. When exposed to hypoxia *in vitro*, MSA reduced the level of HIF-1α and VEGF in human head and neck squamous cell carcinoma cells, FaDu, (see Chen *et al.*, 2013b). Because the level of HIF-1α mRNA was not altered by MSA, some post-translational modification and degradation must be involved in decreasing the protein level of HIF-1α. In contrast to the mechanism regulating the level of HIF-1α in renal cells, RC2, the post-translational degradation of HIF-1α was proteosome dependent in FaDu cells (see Chen *et al.*, 2013b). In addition to HIF-1α, several important angiogenesis regulators, such as COX-2 and iNOS, have been found to be downregulated by MSC (see Chen *et al.*, 2013b). iNOS increased the production of NO that triggered the expression of COX-2, the main regulator in converting arachidonic acid into prostaglandins (PGs). One of these, PGE2, can increase VEGF and promote angiogenesis. Therefore, by reducing several key regulators of VEGF, MSC suppressed angiogenesis and inhibited tumor development. Bhattacharya *et al.* and Rustum *et al.*, respectively, used the Fadu and A253, xenograft models to demonstrate that MSC supplementation significantly reduced the microvessel density, and enhanced vascular maturation (see Chen *et al.*, 2013b). Similar to what transpired with lung cancer, improved vascular maturation significantly enhanced the vascular function, reduced tumor interstitial fluid pressure, and promoted the delivery and distribution of anticancer drugs (Rustum *et al.*, 2010).

24.3 The Future and Challenges of the Application of Se on Cancer and Metastasis Prevention

Results of many *in vitro* and animal studies support the beneficial chemopreventive and anticancer properties of Se. Although the epidemiological data and clinical trials do not fully correspond to these laboratory results, it is possible that the inconsistency is due to experimental design. For instance, the NPC trial was conducted with a relatively small population, and Se-enriched yeast was the source of Se. On the other hand, SELECT was a large phase III clinical trial using SEM. It is possible that although SEM is the major Se compound found in the Se-enriched yeast, other minor Se molecules may be more effective in cancer prevention. Also, the importance of the baseline Se level cannot be stressed enough. In the NPC trial, the greatest impact on prevention was found in the population with less than 106.4 ng mL^{-1} plasma, which implies the backgrounds and the physical profiles of participants were variable factors (see Chen *et al.*, 2013b). Se delivery is another variable. The most commonly used route is oral administration. Se molecules are normally digested and absorbed in the digestive tract, and transport to other organs. In a non-Se-deficient state, approximately 30% of total Se is stored in the liver, 30% in the muscle, 15% in the kidney, 10% in the plasma and other 15% in the rest of the body (see Chen *et al.*, 2013b). In a Se-deficient state, which often happens in cancer patients (Lopez-Saez *et al.*, 2003; Benderli Cihan *et al.*, 2011; Chen *et al.*, 2013b), some organs such as liver lose their Se content quickly, while other organs such as brain maintain Se. How can one ensure that the amount of Se reaches a protective level in one organ without reaching a cytotoxic level in other organs? Moreover, it is always a challenge to determine the optimal concentration and formula of the most effective Se compound since they affect the behaviors of cells through different mechanisms. Furthermore, the effects of certain Se compounds are cell specific. For example, selenate significantly enhanced cell–cell attachment, reduced cell motility, which could explain the reduction in migration and invasion in human breast cancer cells, MDA-MB-231 and MCF-7 (see Chen *et al.*, 2013b). On the other hand, only selenite but not selenate significantly reduced proliferation, attachment and invasion in human fibrosacoma cells, HT1080 (see Chen *et al.*, 2013b). These differences make it difficult to compare different Se compounds and types of cancers. Selenoproteins such as TXR1 also exhibited opposite effects in different situations; *i.e.* TXR1 negatively regulated cell proliferation, migration and the production of VEGF and VEGFR in breast endothelial cells and played an antimetastatic role (see Chen *et al.*, 2013b), while TXR1 significantly increased tumor growth and lung metastasis in the Lewis lung carcinoma (see Chen *et al.*, 2013b). Although these data are confusing, they also point out that our understanding of Se metabolism and the regulation of selenoproteins are far from complete. It is urgent to investigate the mechanisms behind these differences and design new strategies to apply Se in the prevention or treatment of cancer metastasis.

Summary Points

- Se has the potential to suppress cancer metastasis.
- Se can target multiple steps of metastasis, such as cancer-cell migration, invasion, ECM degradation and angiogenesis.
- Both selenoproteins and Se molecules have been shown to bring about a decrease in metastasis under specific conditions.
- Se utilizes many mechanisms to suppress metastasis. However, Se can inhibit metastasis efficiently in one cancer but not in another.
- Further investigation is needed to fully understand how Se suppresses metastasis, and to design optimal Se molecules for therapeutic applications.

Key Facts of Metastasis

1. Metastasis is the major cause of death in cancer patients.
2. Metastatic cancers are diagnosed as the same type as the original cancer; and they share some common characteristics. For example, when breast cancer metastasizes to the lung, the diagnosis is breast cancer lung metastasis, not lung cancer.
3. Different cancers have a predilection to spread to different organs. Overall, the skeleton, lung and liver are the most common sites of metastasis.
4. All cancers, including cancers of the blood and lymphatic system can metastasize to secondary organs.

Definitions

Loss of heterozygosity. The loss of an allele of a specific gene or a region on the chromosome.
Apoptosis. Programmed cell death that results in no damage to surrounding areas and no secretion of harmful molecules.
G0/G1 cell-cycle arrest. Several cell-cycle checkpoints ensure the accuracy and completion of cell division. Before the S phase, environmental conditions and other factors determine whether cells continue to divide or to exit the cell cycle and enter a quiescent stage, G0.
Metastasis. The spread of cancer cells from the original organ to other organs.
Epithelial-mesenchymal transition (EMT). Epithelial cells alter their cytoskeleton; lose cell polarity; decrease cell–cell adhesion; gain motility and change morphology to become mesenchymal cells.
Cell migration. A directional movement of cells from one place to another.
Cell invasion. Cancer cells penetrate the basement membrane to gain access to blood vessels or the lymphatic circulation.
Extracellular matrix. A network of proteins and glycosaminoglycans that surround cells to provide structural support and allow cell adhesion.
Angiogenesis. The formation and development of blood vessels.
IC_{50}. The half-maximum inhibitory concentration.

List of Abbreviations

Se	Selenium
GPX	Glutathione peroxidase
TXR	Thioredoxin reductase
Sep 15	15 kDa selenoprotein
SEM	Selenomethionine
MSC	Methyl selenocysteine
MSA	Methylseleninic acid
SELECT	Selenium and Vitamin E Cancer Prevention Trial
NPC	Nutritional prevention of cancer trial
HIF-1α	Hypoxia-induced factor 1α
VEGF	Vascular endothelial growth factor
EMT	Epithelial-mesenchymal transition
ECM	Extracellular matrix
EGF	Epidermal growth factor
SBP-1	Selenium-binding protein 1
COX-2	Cyclooxygenase-2
uPA	Urokinase-type plasminogen activator
TIMP	Tissue inhibitor of metalloproteinase
TPA	12-*O*-Tetradecanoyl-phorbol-13-acetate
MMP	Metalloproteinase
MT1-MMP	Membrane-type 1 matrix metalloproteinase

EGFR	Epithelial growth factor receptor
HGF	Hepatocyte growth factor
IL-1	Interleukin-1
IL-6	Interleukin-6
IL-8	Interleukin-8
SDF-1	Stromal cell-derived factor 1
FGF	Fibroblast growth factors
OPN	Osteopontin
GRO-α	Growth-regulated peptide-alpha/growth-regulated oncogene-1
MeCN	Methylselenocyanate
DNCB	1-Chloro-2,4-dinitro-benzene
KDR	Kinase insert domain-containing receptor
MNU	1-Methyl-1-nitrosourea
PAI	Plasminogen activator inhibitor
NO	Nitric oxide
NOS	Nitric oxide synthase
PG	Prostaglandin

References

Benderli Cihan, Y. *et al.*, 2011. Trace elements and heavy metals in hair of stage III breast cancer patients. *Biological Trace Element Research*. 144: 360–379.

Chen, Y. C. *et al.*, 2013a. Dietary selenium supplementation modifies breast tumor growth and metastasis. *International Journal of Cancer*. 133: 2054–2064.

Chen, Y. C. *et al.*, 2013b. Is selenium a potential treatment for cancer metastasis? *Nutrients*. 5: 1149–1168.

Clark, L. C. *et al.*, 1991. Selenium in forage crops and cancer mortality in U.S. counties. *Archives of Environmental and Occupational Health*. 46: 37–42.

Cowan, E. A. *et al.*, 2013. Cellular antioxidant activity of phenylaminoethyl selenides as monitored by chemiluminescence of peroxalate nanoparticles and by reduction of lipopolysaccharide-induced oxidative stress. *Enzyme and Microbial Technology*. 53: 373–377.

Ganther, H. E., 1999. Selenium metabolism, selenoproteins and mechanisms of cancer prevention: complexities with thioredoxin reductase. *Carcinogenesis*. 20: 1657–1666.

Grogan, T. M. *et al.*, 2000. Thioredoxin, a putative oncogene product, is overexpressed in gastric carcinoma and associated with increased proliferation and increased cell survival. *Human Pathology*. 31: 475–481.

Hedley, D. *et al.*, 2004. Up-regulation of the redox mediators thioredoxin and apurinic/apyrimidinic excision (APE)/Ref-1 in hypoxic microregions of invasive cervical carcinomas, mapped using multispectral, wide-field fluorescence image analysis. *American Journal of Pathology*. 164: 557–565.

Hu, Y. *et al.*, 2005. Allelic loss of the gene for the GPX1 selenium-containing protein is a common event in cancer. *Journal of Nutrition*. 135: 3021S–3024S.

Hu, Y. J. *et al.*, 2003. Role of glutathione peroxidase 1 in breast cancer: loss of heterozygosity and allelic differences in the response to selenium. *Cancer Research*. 63: 3347–3351.

Hu, Y. J. *et al.*, 2004. Allelic loss at the GPx-1 locus in cancer of the head and neck. *Biological Trace Element Research*. 101: 97–106.

Jung, H. J. *et al.*, 2010. Current issues of selenium in cancer chemoprevention. *Biofactors*. 36: 153–158.

Kakolyris, S. *et al.*, 2001. Thioredoxin expression is associated with lymph node status and prognosis in early operable non-small cell lung cancer. *Clinical Cancer Research*. 7: 3087–3091.

Kawahara, N. *et al.*, 1996. Enhanced coexpression of thioredoxin and high mobility group protein 1 genes in human hepatocellular carcinoma and the possible association with decreased sensitivity to cisplatin. *Cancer Research*. 56: 5330–5333.

Li, G. X. *et al.*, 2008. Superior *in vivo* inhibitory efficacy of methylseleninic acid against human prostate cancer over selenomethionine or selenite. *Carcinogenesis*. 29: 1005–1012.

Lincoln, D. T. *et al.*, 2010. Thioredoxin and thioredoxin reductase expression in thyroid cancer depends on tumour aggressiveness. *Anticancer Research*. 30: 767–775.

Loft, S. *et al.*, 1996. Cancer risk and oxidative DNA damage in man. *Journal of Molecular Medicine (Berlin)*. 74: 297–312.

Lopez-Saez, J. B. *et al.*, 2003. Selenium in breast cancer. *Oncology*. 64: 227–231.

Lu, J. *et al.*, 2001. Antiangiogenic activity of selenium in cancer chemoprevention: metabolite-specific effects. *Nutrition and Cancer*. 40: 64–73.

Matsutani, Y. *et al.*, 2001. Inverse correlation of thioredoxin expression with estrogen receptor- and p53-dependent tumor growth in breast cancer tissues. *Clinical Cancer Research*. 7: 3430–3436.

Mehdi, Y. *et al.*, 2013. Selenium in the environment, metabolism and involvement in body functions. *Molecules*. 18: 3292–3311.

Neufeld, G. *et al.*, 1999. Vascular endothelial growth factor (VEGF) and its receptors. *FASEB Journal*. 13: 9–22.

Poulsen, H. E. *et al.*, 1998. Role of oxidative DNA damage in cancer initiation and promotion. *European Journal of Cancer Prevention*. 7: 9–16.

Raffel, J. *et al.*, 2003. Increased expression of thioredoxin-1 in human colorectal cancer is associated with decreased patient survival. *Journal of Laboratory and Clinical Medicine*. 142: 46–51.

Rustum, Y. M. *et al.*, 2010. Architectural heterogeneity in tumors caused by differentiation alters intratumoral drug distribution and affects therapeutic synergy of antiangiogenic organoselenium compound. *Journal of Oncology*. 2010: 396286.

Shweiki, D. *et al.*, 1992. Vascular endothelial growth factor induced by hypoxia may mediate hypoxia-initiated angiogenesis. *Nature*. 359: 843–845.

Soini, Y. *et al.*, 2001. Widespread expression of thioredoxin and thioredoxin reductase in non-small cell lung carcinoma. *Clinical Cancer Research*. 7: 1750–1757.

Song, H. *et al.*, 2011. Selenium inhibits migration of murine melanoma cells *via* down-modulation of IL-18 expression. *International Immunopharmacology*. 11: 2208–2213.
Surai, P., 2006. *Selenium in Nutrition and Health*. Nottingham University Press, Nottingham.
Tsukamoto, T. *et al.*, 2013. Selenate induces epithelial-mesenchymal transition in a colorectal carcinoma cell line by AKT activation. *Experimental Cell Research*. 319: 1913–1921.
Turunen, N. *et al.*, 2004. Thioredoxin is associated with proliferation, p53 expression and negative estrogen and progesterone receptor status in breast carcinoma. *Acta Pathologica, Microbiologica et Immunologica Scandinavica*. 112: 123–132.
Zeng, H. *et al.*, 2006. The selenium metabolite methylselenol inhibits the migration and invasion potential of HT1080 tumor cells. *Journal of Nutrition*. 136: 1528–1532.

CHAPTER 25

The Selenium and Vitamin E Cancer Prevention Trial (SELECT): Prevention of Prostate Cancer Using Selenium and/or Vitamin E in the SELECT Cancer Prevention Trial

HOLLY L. NICASTRO[a] AND BARBARA K. DUNN*[b]

[a]Cancer Prevention Fellowship Program, Nutritional Science Research Group, Division of Cancer Prevention, National Cancer Institute, National Institutes of Health, 9609 Medical Center Dr #5E-572, Bethesda, MD 20852, USA; [b]Chemoprevention Agent Development Research Group, Division of Cancer Prevention, National Cancer Institute, National Institutes of Health, 9609 Medical Center Dr #5E-534, Bethesda, MD 20852, USA
*E-mail: dunnb@mail.nih.gov

25.1 Prostate Cancer

Prostate cancer continues to be a major public health burden, as it is the second leading cause of cancer death among men in the United States (US) and is the most commonly diagnosed noncutaneous cancer. One in 6 men in

Food and Nutritional Components in Focus No. 9
Selenium: Chemistry, Analysis, Function and Effects
Edited by Victor R. Preedy

Published by the Royal Society of Chemistry, www.rsc.org

the US is expected to be diagnosed with this disease in his lifetime. In 2013, an estimated 238 590 new cases were diagnosed in the US and approximately 29 720 men died of prostate cancer (Siegel *et al.*, 2013). When diagnosed in the localized or regional stages, as 81% and 12% of prostate cancers, respectively, are, the disease can be treated effectively with surgery or radiation, resulting in a five-year survival rate of 100%. Thus, prostate-cancer incidence far exceeds prostate cancer mortality. However, prostate cancer treatments often result in decreased quality of life due to side effects like declining urinary, bowel, or sexual functioning (reviewed in Nicastro and Dunn (2013)). When diagnosed in the distant stage (4% of all prostate cancers), the five-year survival rate drops to 28%. Current treatments for advanced prostate cancer are largely palliative. Most known risk factors for prostate cancer, including age, race, and genetic factors, are nonmodifiable, whereas modifiable risk factors include obesity, physical activity, and possibly dietary factors.

Prostate cancer screening, while once lauded for increasing early detection, is now controversial in large part due to overdiagnosis leading to overtreatment of nonfatal disease. In 2012, the United States Preventive Services Task Force changed its stance to strongly recommend against prostate-specific antigen (PSA) screening for prostate cancer, a screening modality that it once endorsed (Moyer, 2012). Because prostate cancer has a long natural history, mainly nonmodifiable risk factors, and an incidence rate that far exceeds the mortality rate, a focus on prevention overscreening or early detection appears to be a more appealing way to decrease the burden of this disease. Goals for prevention strategies should focus on reducing prostate-cancer incidence and delaying prostate cancer diagnosis until the individual succumbs to other causes.

Early efforts in prostate cancer chemoprevention focused on the use of the antiandrogens finasteride and dutasteride to decrease prostate-cancer incidence. However, two large multicenter randomized controlled clinical trials found that both agents slightly increased the rates of high-grade prostate cancers. Due to the concerns about increasing risks of high-grade prostate cancer, a US Food and Drug Administration (FDA) advisory panel voted overwhelmingly not to approve finasteride or dutasteride for prostate-cancer prevention (Theoret *et al.*, 2011). Concomitant with the interest and subsequent disinterest in antiandrogens, an independent approach to prostate-cancer chemoprevention involved nutritional agents, specifically vitamin E and selenium. Secondary analyses of other large-scale chemoprevention trials had suggested that these compounds may decrease the risk of prostate cancer (Clark *et al.*, 1996; Heinonen *et al.*, 1998). Further controlled intervention trials, human observational studies, and preclinical studies all provided evidence for potential chemopreventive efficacy of these compounds. It is important to keep in mind, however, that some preclinical studies suggested no benefit, but few of these were published (McCormick *et al.*, 2010; Nicastro and Dunn, 2013). Adding to the appeal of selenium and vitamin E, both agents are naturally occurring micronutrients essential to human health that have antioxidant activities. Thus, the Selenium and Vitamin E Cancer Prevention Trial (SELECT) was undertaken to test the chemopreventive efficacy of these micronutrients. In this chapter, we will describe the rationale, results, and implications of SELECT.

25.2 Selenium

25.2.1 Definition and Sources

Selenium, a nutritionally essential trace mineral, is found in soil in the inorganic forms of selenate (SeO_4^{2}) and selenite (SeO_3^{-2}). Plants convert selenium from these inorganic forms to organic forms, including the major form L-selenomethionine and, in lesser amounts, L-selenocysteine. Therefore, selenium concentrations in foods are dependent on the selenium content of the soil in which the foods or ingredients were grown. The western US has high soil selenium content, while portions of the eastern coastal US have very low soil selenium content. In the US, however, increasing fortification of the food system with selenium and increasing nationwide shipping of food products have lessened concerns that low selenium soil content might affect the selenium status of individuals compared to several years ago (Institute of Medicine (U.S.), 2000).

Selenium is found in the most abundant amounts in Brazil nuts, meats, fish, eggs, and cereals. Other sources of selenium include cruciferous vegetables, garlic, and mushrooms. Selenium is also found in dietary supplements, including multivitamin/multimineral supplements, as selenomethionine or as selenized yeast. Selenized yeast is baker's yeast grown in a selenium-rich medium. Commercially available selenized yeast can provide up to 2000 $\mu g\ g^{-1}$ selenium, over 90% of which is selenomethionine. Certain selenium supplements, particularly weight-loss products or infant formulas, contain sodium selenite or sodium selenate, though these inorganic forms are not highly bioavailable.

25.2.2 Metabolism and Biological Activities

A key role of dietary selenium is to provide the starting material for synthesis of selenoproteins. The major dietary forms of selenium including selenomethionine, selenocysteine, selenate, or selenite can all be metabolized into forms for selenoprotein synthesis. Selenomethionine can be nonspecifically incorporated into proteins in place of methionine or converted to selenocysteine *via* a trans-sulfuration pathway. Selenocysteine can be converted to hydrogen selenide, an intermediate fundamental to both selenocysteine incorporation into proteins and to selenium excretion. Selenate and selenite are reduced to hydrogen selenide by glutathione (Figure 25.1) (Hatfield and Gladyshev, 2009). *Via* conversion to hydrogen selenide, all dietary forms of selenium can therefore be used for selenoprotein synthesis.

Hydrogen selenide is converted to selenophosphate *via* selenophosphate synthetase. L-Selenocysteinyl-tRNA is then formed by the reaction of selenophosphate with L-seryl tRNA. L-Selenocysteine is considered the 21st amino acid. L-Selenocysteinyl-tRNA is inserted in selenoproteins during protein synthesis where specified by the UGA codon in mRNA. To date, twenty-five human selenoproteins have been identified, including: members of the glutathione peroxidase (GPx), thioredoxin reductase (TR) and thyroid hormone deiodinase families; selenophosphate synthetase; and several uncharacterized proteins (Davis *et al.*, 2012). A major role of these selenoproteins is in maintaining redox balance and proper cellular functioning.

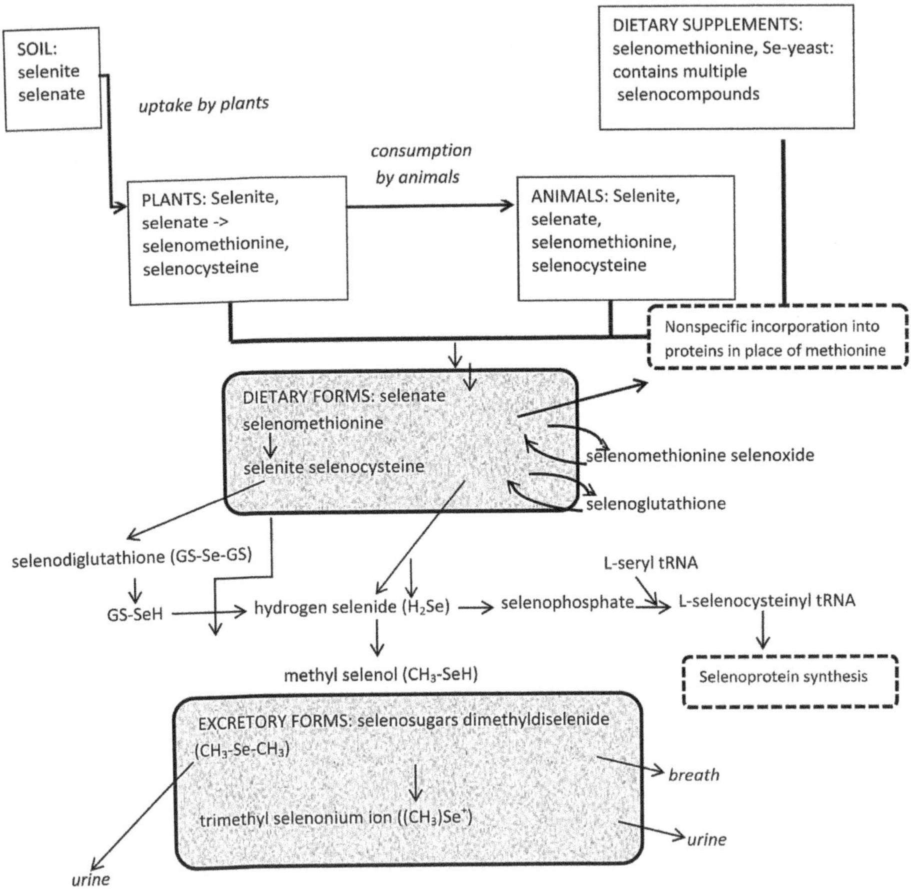

Figure 25.1 Selenium metabolism and distribution. Selenium enters the body as selenate, selenite, selenomethionine, or selenocysteine from plant, animal, or supplemental sources. Selenomethionine may be non-specifically incorporated into proteins in place of methionine, or all dietary forms of selenium may be converted to the key intermediate hydrogen selenide. Hydrogen selenide serves as a precursor to selenophosphate, which is loaded onto L-seryl tRNA and used for selenoprotein synthesis, or hydrogen selenide can be converted to the excretory forms. Adapted from Nicastro and Dunn (2013).

Hydrogen selenide can also be methylated to methyl selenol, or di-, or trimethylated for excretion. Dimethyl selenium is exhaled from the lungs following larger doses of selenium, but the trimethylselenonium ion and selenosugars are the main excretory forms of selenium and are excreted in the urine. The clinical efficacy of selenium, including its chemopreventive effects, derives from the activities of intracellular selenium metabolites, including methyl selenol (reviewed in Nicastro and Dunn (2013)).

25.2.3 Nutritional Requirements

Issues related to nutritional requirements for selenium have been discussed previously (Nicastro and Dunn, 2013). Adult men and women have a recommended dietary allowance (RDA) for selenium of 55 μg per day (Institute of Medicine (U.S.), 2000), based on the daily selenium intake necessary to achieve maximal activity of GPx-3, a selenoprotein that maintains redox balance by detoxification of hydrogen peroxide. Nevertheless, not all selenoproteins' activity levels would be maximal with the RDA intake level, leading others to suggest that an RDA of 80 μg per day for men may be more appropriate for achieving selenium balance. However, it is worth noting that the physiological implications for maximal activity levels of all selenoproteins that may be achieved by higher intakes of selenium remain unknown. The estimated average requirement (EAR) of selenium in adults is 45 μg per day. The EAR represents the average daily intake level estimated to meet the requirements of half of the healthy individuals in a group. In studies on the health effects of varying selenium intake in China, deficiency symptoms, including immunoincompetency, progression of viral infections, and reproductive symptoms, became apparent with intakes <11 μg per day. Conversely, symptoms of selenium toxicity, or selenosis, including hair and nail brittleness/loss, gastrointestinal disturbances, skin rash, garlic breath odor, fatigue, or irritability, appeared with 800 μg per day of selenium intake. Based on this observation and dividing by an uncertainty factor of 2, the tolerable upper limit (UL) for selenium intake is 400 μg per day (Institute of Medicine (U.S.), 2000).

Most adults in the United States have adequate selenium intake. Data from the National Health and Nutrition Examination Survey (NHANES) 2003–2006 showed that the usual intake of selenium for individuals over the age of 19 was 109 ± 1 μg per day from naturally occurring dietary sources and 126 ± 1 μg per day from naturally occurring sources plus supplements. Less than 1% of adults had intakes below the estimated average requirements (EAR) and even fewer had intakes above the tolerable UL (Fulgoni *et al.*, 2011).

NHANES data also show that 19 ± 1% of males reported taking a dietary supplement, including multivitamin/multimineral supplements, that contained selenium. Older men were more likely than younger men to take any dietary supplements, including those containing selenium. Interestingly, men who were selenium-supplement users had higher food selenium intake than men who did not use selenium supplements. This important observation suggests that those who took supplemental selenium did not do so in order to meet selenium nutritional requirements. In fact, 0.1 ± 0% of men in the analysis had selenium intake below the EAR. These NHANES analyses by and large demonstrate that men in the US have adequate selenium intake.

25.3 Vitamin E

25.3.1 Definition and Sources

Vitamin E is the collective name for a group of tocopherols that are the major lipid-soluble antioxidants in cell membranes and act as peroxyl and alkoxyl free-radical scavengers (reviewed in Dunn *et al.* (2010)). Naturally occurring vitamin E has eight different forms: the α-, β-, γ-, and δ-isomers of tocopherol and tocotrienol (Figure 25.2). The isomers have varying levels of biological activity, with α-tocopherol being the most active, and α-tocopherol is the only vitamin E isomer that is maintained in the plasma and is therefore the only

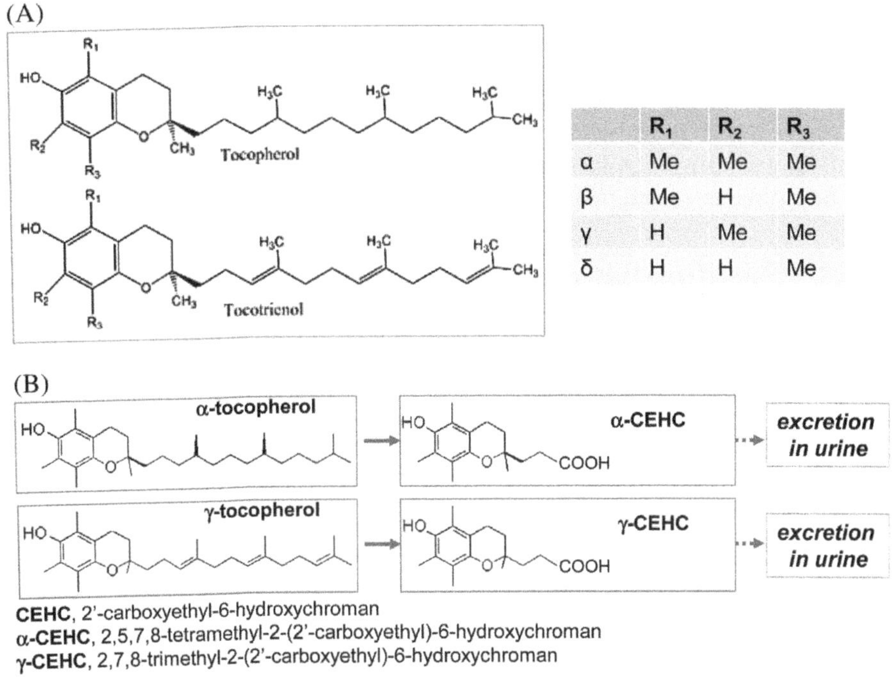

	R_1	R_2	R_3
α	Me	Me	Me
β	Me	H	Me
γ	H	Me	Me
δ	H	H	Me

Figure 25.2 (A) Naturally occurring forms of vitamin E (B) Vitamin E metabolites. (A) shows the 8 naturally occurring forms of vitamin E, the base molecules are tocopherol and tocotrienol. Each of these can be modified by 4 combinations (α, β, γ. δ) of the 3 R groups shown in the accompanying table. This gives rise to the 8 total number of forms. (B) shows the major vitamin E metabolites excreted in the urine. These are derived from 2 of the 8 forms shown in (A).

isomer that is recognized to meet human requirements. Furthermore, of the 8 possible stereoisomers of α-tocopherol, only four, the d-stereoisomers, are maintained in human tissues and can meet human dietary requirements (Institute of Medicine (U.S.), 2000).

The most abundant dietary sources of α-tocopherol include nuts, seeds, and oils. Other sources include green leafy vegetables or fortified cereal products. Vitamin E supplements may contain either α-tocopherol alone, mixed tocopherols, or mixed tocopherols and tocotrienols. Naturally occurring α-tocopherol exists as the D-stereoisomer, while the synthetic α-tocopherol present in dietary supplements or fortified foods is either all D-isomers or a 50:50 mixture of D- and L-stereoisomers (labeled "all-racemic" or "D, L-") (Institute of Medicine (U.S.), 2000). Therefore, those using all-racemic synthetic vitamin E to meet dietary requirements would need to double the amount of supplementation to achieve the desired α-tocopherol levels.

25.3.2 Metabolism and Biological Activities

Oral absorption of vitamin E is approximately 20–50%. Hepatic α-tocopherol transfer protein (α-TTP) transfers hepatic α-tocopherol into plasma lipoproteins for extrahepatic delivery and is therefore responsible for maintaining plasma α-tocopherol concentrations (Manor and Morley, 2007).

Vitamin E's major function is as a free-radical scavenger/antioxidant in the lipid phase of cell membranes. However, α-tocopherol exhibits other cellular activities, including regulation of cell growth, proliferation, apoptosis, adhesion, angiogenesis, and inflammation (reviewed in Azzi (2007)).

25.3.3 Nutritional Requirements

The RDA for vitamin E for adult men and women is 15 mg and the EAR is 12 mg. Vitamin E deficiency, which manifests as peripheral neuropathy, is exceedingly rare and occurs almost exclusively as a result of genetic abnormalities in α-TTP or as a result of protein-energy malnutrition. Vitamin E deficiency as a result of low dietary intake of vitamin E has never been clinically described (Institute of Medicine (U.S.), 2000). Rather, the RDA is based on amounts of vitamin E sufficient to prevent hydrogen peroxide-induced hemolysis. Little is known about the possible harmful effects of extremely high levels of vitamin E. However, even prior to the release of SELECT results, adverse events in recent clinical trials had begun to emerge, highlighting the need for more research into this area, particularly concerning long-term use of vitamin E supplements (The Alpha-Tocopherol Beta Carotene Cancer Prevention Study Group, 1994a). The tolerable upper limit for vitamin E intake is 1000 mg per day based on the lowest observed intake level that resulted in hemorrhagic events divided by uncertainty factors (Institute of Medicine (U.S.), 2000).

Because of the complexity inherent in the fact that vitamin E and α-tocopherol exist in many possible forms, a useful approach has been to express vitamin E intake and requirements in terms of biological activity. One international unit (IU) of vitamin E activity is defined as having the activity of 1 mg of all-racemic α-tocopherol acetate, 0.67 mg D-α-tocopherol, or 0.74 mg D-α-tocopherol acetate (Institute of Medicine (U.S.), 2000). The RDA of 15 mg corresponds to an RDA of 22.4 IU.

NHANES analyses (Fulgoni *et al.*, 2011) show that, in stark contrast to results with selenium, 93.3 ± 0.4% of people >2 years of age did not meet the vitamin E requirement from food in 2003–2006, with a usual intake of 6.5 ± 0.1 mg per day. Among users of dietary supplements containing vitamin E, 60.3 ± 0.8% did not meet vitamin E requirements, though the mean usual vitamin E intake in this group was 37.7 ± 1.4 mg per day. Only a very small percentage (0.1 ± 0.1%) exceeded the UL. Analyses from the same years showed that 29 ± 1% of individuals in the US, 33 ± 2% of males aged 51–70, and 40 ± 2% of males over the age of 71 used supplements containing vitamin E. Similar to selenium supplements, vitamin E supplements were more likely to be used by older individuals.

25.4 Rationale for SELECT

25.4.1 Prostate-Cancer Prevention by Selenium

25.4.1.1 Preclinical Studies

In vitro and animal studies have explored potential antitumorigenic mechanisms of selenium. Cellular functions affected by supplemental selenium include enhanced activity of selenoenzymes, increased carcinogen metabolism, augmented natural killer cell-dependent cytotoxicity, and perturbation of tumor cell growth and cell cycle progression. Selenite has been used in many of the preclinical studies assessing the chemopreventive efficacy and mechanisms of action of selenium. In a review of over 100 animal studies that investigated selenium's chemopreventive efficacy, investigators noted that selenium significantly reduced tumor incidence in two-thirds of the studies, and half of these studies showed tumor reductions of 50% or more at a variety of cancer sites (Combs and Gray, 1998).

25.4.1.2 Epidemiological Studies

Ecological analyses offered the first observational evidence suggesting that selenium had cancer preventive activity, pointing to an association between a higher incidence of certain cancers and mortality in selenium-deficient regions of the US compared to selenium-replete areas (reviewed in Nicastro and Dunn (2013)). In addition, multiple case-control studies have revealed that a general trend exists between higher selenium levels (as assessed by prediagnostic blood levels, serum levels, toenail selenium levels, or dietary selenium intake) and decreased total cancer incidence and mortality (reviewed in Combs and Gray (1998)). Some studies focusing on selenium in relation to prostate cancer support this same general association, while others do not. For example, a matched case-control study nested within the prospective US cohort Health Professionals Follow-Up Study showed the highest quintile of toenail selenium level to be associated with a decreased risk of advanced prostate cancer compared to the lowest quintile (OR: 0.49, 95% CI: 0.25–0.96, $P_{trend} = 0.11$) (Nicastro and Dunn, 2013). In contrast, no such association was observed in two nested case-control studies among men in the European Prospective Investigation into Cancer and Nutrition (EPIC) cohort, where investigators found that plasma selenium was not associated with prostate-cancer risk (Dunn *et al.*, 2010). Although the basis for these discrepancies remains unclear, the authors noted that the European cohort had substantially lower plasma selenium concentrations (mean: 70 μg L^{-1}) than those found in men in the US (>100 μg L^{-1}). The suggestion here is that the selenium levels of men in even the highest quintile in the EPIC cohort may have been below levels necessary for cancer prevention.

In a review of 17 cohort studies, 3 ecological studies, 14 case-control studies, and 1 clinical trial (NPC, see below) by the World Cancer Research Fund/American Institute for Cancer Research (WCRF/AICR), the conclusion was reached that selenium and foods containing selenium probably protect against prostate cancer. This conclusion was based on observed dose–response relationships as well as evidence for plausible mechanisms (reviewed in Dunn *et al.* (2010)). This report contained a meta-analysis that indicated a 5% decrease in the risk of prostate cancer and a 13% decrease in the risk of advanced or aggressive prostate cancer for every 10 ng mL^{-1} increase in plasma selenium or a 20% decrease in the risk of advanced or aggressive prostate cancer for every 100 ng g^{-1} increase in toenail selenium. Although this report was not released before the start of SELECT, many of the studies used in the systematic review and meta-analysis were published prior to the planning phase of SELECT.

25.4.1.3 *Clinical Trials*

Qidong and Linxian, China were the sites of some of the first large randomized trials to demonstrate cancer preventive activities of selenium in humans. The focus of these trials, however, were liver, gastric, and esophageal, *i.e.* not prostate, cancers. In Qidong, a selenium-deficient region, a community intervention trial showed that regions receiving salt supplemented with sodium selenite (providing 50–80 μg of selenium per day) experienced a decrease in incidence of primary liver cancer of 35.1% over 8 years compared to control regions (reviewed in Nicastro and Dunn (2013)). Following discontinuation of the intervention, when the selenized salt was no longer administered, primary liver cancer increased in incidence in the intervention regions. In the same region a clinical trial implemented in the same time period among 226 hepatitis B surface antigen positive participants showed that 200 μg selenium as selenized yeast daily for 4 years was associated with no cases of primary liver cancer among 113 individuals compared to 7/113 new cases in the placebo group.

The goal of the Linxian Nutritional Intervention Trials was to determine whether supplementation with multiple vitamins and minerals would decrease the risk of esophageal and gastric cardia cancers in individuals with esophageal dysplasia. The first trial randomized 29 854 participants 40–69 years old with esophageal dysplasia to receive one of eight combinations of vitamins and minerals. The group assigned to a daily dose of Factor D (50 μg selenium, 30 mg vitamin E, and 15 mg β-carotene daily) exhibited a decrease in cancer mortality of 13% and a decrease in stomach cancer mortality of 21% over 5 years compared to those not receiving Factor D (Blot *et al.*, 1993). The second trial included 3000 individuals with esophageal dysplasia who were randomized to receive either a multivitamin/multimineral supplement containing 12 micronutrients (including 50 μg selenium) and a separate 15 mg β-carotene supplement daily, or two placebos. At 6 years of follow-up, total cancer mortality decreased by 4%, gastric/esophageal cancer mortality decreased by 8%, and esophageal cancer mortality decreased by 16% in the supplement group compared to the placebo group. None of these decreases was statistically significant, however, and cancer incidence rates in the two groups were similar (Li *et al.*, 1993).

The Nutritional Prevention of Cancer (NPC) Trial (Clark *et al.*, 1996) was the main study that generated the hypothesis that selenium prevents prostate cancer, prompting the use of selenium in SELECT. The NPC trial was randomized, double blinded, and placebo controlled. The enrolled 1312 participants, from the eastern US and with a history of skin cancer, were randomized to receive either 200 μg of selenium in the form of selenized yeast or a placebo daily. The primary aim was to test the efficacy of selenium supplementation in preventing nonmelanoma skin cancer, while secondary endpoints included total cancer incidence, total cancer mortality, and mortality due to lung, prostate, and colorectal cancers. After 8271 person-years of follow-up, the selenium intervention did not favorably affect the primary endpoint of skin cancer, although a decrease in other cancer incidence and mortality became evident. Specifically, prostate cancer was reduced in the selenium group by 64% after 4.5 years and by 49% after 10 years (incidence RR: 0.51, 95% CI: 0.29–0.87), a decrease that was most pronounced in former smokers. In a stratified analysis selenium supplementation was shown to lower the incidence of prostate cancer in men in the lowest two tertiles of baseline selenium status (RR_{tert1}: 0.14, 95% CI: 0.02–0.59; RR_{tert2}: 0.39, 95% CI: 0.14–0.99), but not in those in the highest tertile of baseline selenium status (RR: 1.20, 95% CI: 0.50–2.97). A significant interaction between treatment group and baseline selenium status was therefore observed. Also, the reduction in prostate-cancer risk was also limited to men whose baseline PSA levels were ≤4.0 ng mL^{-1} (RR: 0.35, 95% CI: 0.13–0.87), although no significant interaction was observed between treatment and baseline PSA (Nicastro and Dunn, 2013). While this trial examined prostate-cancer incidence and mortality as a secondary endpoint, it greatly increased interest in prostate-cancer prevention by selenium, and also highlighted the point that only specific subgroups of men may benefit.

After the NPC trial, a small pilot study was initiated to test the effect of selenium supplementation on selenium levels in the prostate and seminal vesicles (Nicastro and Dunn, 2013). Men whose cancer was confined to the prostate ($n = 66$) and who were planning on surgical resection of the prostate were randomized to receive either 200 μg L-selenomethionine daily for 14–31 days or standard of care observation (control group) during the presurgical period. The concentrations of selenium in the prostate were 22% higher in the selenium than in the control group whereas selenium concentrations in the seminal vesicles were similar in both groups. The selective accumulation of supplemental selenium in the prostate provides additional biological plausibility for chemopreventive effects of selenium against prostate cancer.

25.4.2 Prostate-Cancer Prevention by Vitamin E

25.4.2.1 Preclinical Studies

Vitamin E's cellular functions extend beyond its antioxidant activities (reviewed in Dunn *et al.* (2010)). In prostate-cancer cells *in vitro*, α-tocopherol inhibited NF-κB, androgen receptor function, and expression of PSA, induced apoptosis *via* the mitochondrial caspase cascade, decreased vascular endothelial growth factor expression, and downregulated the proinflammatory cytokines IL-6 and IL-8. The compound also inhibited prostate-cancer cell invasiveness with concomitant reduction in secreted matrix metalloproteinase-9 but without alterations in cell cycle, cell adhesion, or cell motility. α-Tocopherol also inhibits expression of intercellular adhesion molecule-1 with concomitant decrease in the metastatic potential of PC-3 prostate-cancer cells. Additional mechanisms proposed to underlie the anticarcinogenic effect of vitamin E include blocking nitrosamine synthesis; inducing the detoxification enzyme nicotinamide adenine dinucleotide phosphate: quinine reductase; and inhibiting fatty acid metabolism, protein kinase C activity, and arachidonic acid and prostaglandin metabolism.

In animal studies, vitamin E has been shown to prevent various chemically induced tumors, some of which are hormonally mediated. In rats receiving various doses of chemotherapeutic agents, vitamin E has been shown to slow the growth of prostate cancer.

25.4.2.2 Epidemiological Studies

Many foods contain vitamin E but generally at low concentrations. This leads to difficulty in assessing dietary intake so that serum or plasma α-tocopherol concentrations are often used to assess vitamin E status. In observational studies a possible beneficial association between vitamin E status (α-tocopherol circulating levels) or intake and prostate-cancer risk has only inconsistently been demonstrated (reviewed in Dunn *et al.* (2010)). Prediagnosis serum or plasma vitamin E concentrations were shown to be lower years prior to prostate cancer diagnosis in cases compared to noncases in a number of positive studies. In contrast, as discussed in (Dunn *et al.*, 2010), other case-control and cohort studies have reported no association. One cohort analysis, nested within the Alpha-Tocopherol Beta-Carotene (ATBC) Cancer Prevention Trial (see below), showed no association between any baseline measures of vitamin E (serum α-tocopherol, dietary vitamin E) and prostate cancer, except when the analysis was limited to the intervention group. A subsequent case-control study, also nested within the ATBC, contained 100 randomly selected incident prostate cancer cases and 200 matched controls. This study showed that high baseline serum levels for both α- and γ-tocopherol were associated with decreased prostate-cancer risk (highest *vs.* lowest tertile for α-tocopherol, OR = 0.49, 95% CI = 0.24–1.01, $P_{trend} = 0.05$; for γ-tocopherol, OR = 0.57, 95% CI = 0.31–1.06, $P_{trend} = 0.08$); this association was stronger in the α-tocopherol-supplemented arm. Nevertheless, such associations have not been consistently supported by other observational analyses.

25.4.2.3 *Clinical Trials*

High-dose vitamin E supplements (above the 15 mg, or 33 IU, *all rac* α-tocopherol dietary level) have been tested in several randomized clinical trials. The results of these trials suggested benefits for noncancer endpoints, such as a decrease in coronary heart disease risk. With respect to cancer, the ATBC trial, which was conducted in Finland by a joint effort of the National Public Health Institute of Finland and the US National Cancer Institute, generated the most convincing evidence that vitamin E is associated with a decrease in prostate-cancer risk. This randomized, double-blind, placebo-controlled trial enrolled 29 133 male smokers 50 to 69 years old and compared α-tocopherol (50 mg synthetic DL-α-tocopherol acetate) daily and β-carotene (20 mg) daily alone or in combination (Heinonen *et al.*, 1998). The purpose of the ATBC trial was to determine whether these nutritional agents would reduce the risk of lung cancer in study participants. The primary endpoint, lung cancer incidence, was not affected by α-tocopherol acetate after 5–8 years of follow-up, while, paradoxically, the incidence of lung cancer increased among men receiving β-carotene. Prostate-cancer incidence, a prespecified secondary endpoint, showed a statistically significant 32% decrease (95% CI = 12–47, $p = 0.002$) with α-tocopherol. Of the 14 564 patients assigned to the α-tocopherol supplementation arms, 99 incident prostate cancers occurred compared with 147 in the 14 569 in the non-α-tocopherol arms. This preventive effect appeared to be stronger for clinically evident cases (stages B–D disease), with participants who received α-tocopherol showing a 40% (95% CI = −20 to −55) decrease in disease. Although based on fewer events, prostate-cancer mortality also showed a statistically significant decrease of 41% (95% CI = −1 to −64) among the 14 564 men taking vitamin E compared to the men not receiving vitamin E (Heinonen *et al.*, 1998). These findings for prespecified secondary endpoints offered strong support for testing vitamin E in a prospective clinical trial of prostate-cancer prevention.

25.4.3 Prostate-Cancer Prevention by Selenium and Vitamin E Combined

25.4.3.1 In Vitro Studies

Experimental evidence suggests that although selenium and vitamin E both have antiprostate-cancer effects, they function by modulating different cellular processes (reviewed in Dunn *et al.* (2010)). For example, α-tocopherol inhibits growth of several prostate-cancer cell lines, including the androgen-dependent LNCaP and the androgen-insensitive PC-3 and DU145 cell lines. The growth inhibitory effect that α-tocopherol has on LNCaP cells is associated with suppression of expression of the AR (androgen receptor) and PSA, and contrasts with the inhibitory effect of selenium on LNCaP cell growth, which is unrelated to the AR/PSA pathway. These differences in anticancer mechanisms are reflected in laboratory observations showing a substantially more pronounced effect of selenium and vitamin E together on molecular markers of anticancer cellular activity than that observed with either nutrient alone. In one study, vitamin E inhibited LNCaP growth by 47%, selenium by 37%, and the combination by 78%. The combination also induced significantly more apoptosis (37–43%) in prostate-cancer cells, which was accompanied by greater increases in the protein Bax/Bcl-2 ratio. In addition, reduction in PCNA protein levels, a marker of cell proliferation, was greater with the combination than with either nutrient alone. These additive anticancer effects supported the inclusion of a combined treatment arm in SELECT.

25.4.3.2 Preclinical Studies

Few selenium or vitamin E prostate-cancer prevention studies that used tumor endpoints in animal models have been conducted (Table 25.1). Only two studies were conducted with designs relevant to SELECT. These studies tested selenium and vitamin E alone or in combination on prostate-cancer incidence and multiplicity and reported null findings (McCormick *et al.*, 2010; Ozten *et al.*, 2010). Other *in vivo* selenium and prostate-cancer prevention studies either only used selenium in combination with other agents and not alone or used other forms of selenium such as methylseleninic acid or selenite (Wang *et al.*, 2009).

25.4.3.3 Clinical Studies

Interestingly, epidemiologic and clinical studies that examined the effects of selenium and vitamin E in combination have not always shown obvious benefit of the two agents together. The Vitamins and Lifestyle (VITAL) study, a prospective cohort study, showed no association of vitamin E and selenium with prostate-cancer risk, although greater long-term vitamin E supplementation was associated with reduced clinically relevant advanced disease (Peters *et al.*, 2008). The Supplementation en Vitamins et Mineraux

Table 25.1 *In vivo* animal studies that assessed prostate-cancer prevention by selenium. This table describes preclinical selenium chemoprevention trials with relevance to the SELECT study design. Only two preclinical studies tested the effects of L-selenomethionine and DL-α-tocopherol in rodent prostate-cancer models. These two reports were not published until after the design and implementation of SELECT and both reports showed null findings. Other studies compared the effects of L-selenomethionine and/or vitamin E in combination with lycopene, but not alone, or studied the effects of other forms of selenium.

Animal model (Ref.)	Agent and dose	Results
Studies with direct relevance to SELECT trial design		
N-Nitroso-*N*-methylurea (MNU) + testosterone-treated Wistar-Unilever rats (McCormick *et al.*, 2010)	• L-Selenomethionine (1.5 or 3 mg kg^{-1} diet) • DL-α-Tocopherol (4000 or 2000 mg kg^{-1} diet) • L-Selenomethionine (3 mg kg^{-1} diet) + DL-α-tocopherol (2000 or 5000 mg kg^{-1} diet) • Selenized yeast (target Se levels of 9 or 3 mg kg^{-1} diet) • Control	No group showed an effect on prostate-cancer incidence
Testosterone + estradiol-treated NBL rat (Ozten *et al.*, 2010)	• L-Selenomethionine (1.5 or 3.0 mg kg^{-1} diet) • DL-α-Tocopherol (4000 or 2000 mg kg^{-1} diet) • Control	No group showed an effect on prostate tumor incidence, multiplicity, or death
Studies on selenium in combination with other agents, including vitamin E		
Lady transgenic mice (Venkateswaran *et al.*, 2009)	• α-Tocopherol succinate (800 IU) + L-selenomethionine (200 μg) + lycopene (50 mg) • α-Tocopherol succinate (800 IU) + L-selenomethionine (200 μg) • Control	Both treatment groups showed increased survival ($p < 0.0001$) compared to control; neither group exhibited an effect on prostate tumor incidence
Lady transgenic mice (Venkateswaran *et al.*, 2004)	• α-Tocopherol succinate (800 IU) + L-selenomethionine (200 μg) + lycopene (50 mg) • Control	Treatment group exhibited four-fold decrease in prostate-cancer incidence compared to control animals ($p < 0.0001$)
Studies on other forms of selenium		
Transgenic adenocarcinoma mouse prostate (TRAMP) model (Wang *et al.*, 2009)	• Methylseleninic acid (3 mg selenium per kg body weight) for 10 weeks • Methylseleninic acid (3 mg selenium per kg body weight) for 16 weeks • Control	Methylseleninic acid groups showed decreased cancer-specific mortality compared to control group ($p_{10\ \text{weeks}} = 0.0078$, $p_{16\ \text{weeks}} = 0.0385$)
MNU + testosterone-treated Wistar rats (Bespalov *et al.*, 2011)	• Sodium selenite (4 mg L^{-1} in drinking water per day) • Control	Sodium selenite group showed no effect on prostate intraepithelial neoplasia, but exhibited decreased prostate cancer multiplicity by 44.6% compared to control group

Antioxidants (SU.VI.MAX) trial randomized 12 741 participants to placebo or to a supplement that included 30 mg of vitamin E and 100 µg of selenium along with 120 mg of ascorbic acid, 6 mg of β-carotene, and 20 mg of zinc. Results from an adjunct SU.VI.MAX study showed an overall moderate, non-significant reduction in prostate cancer rate with supplementation *vs.* placebo. However, this prostate-cancer incidence rate reduction was statistically significant in men with normal PSA (Meyer *et al.*, 2005).

25.5 SELECT

25.5.1 Rationale and Objectives

The Selenium and Vitamin E Cancer Prevention Trial (SELECT) was funded by the US National Cancer Institute (NCI) and implemented by the Southwest Oncology Group (SWOG). The primary objective of SELECT was to assess the efficacy of selenium and vitamin E alone and in combination on the incidence of prostate cancer. The study was also planned to address prespecified secondary endpoints: prostate cancer-free survival, all-cause mortality, incidence and mortality of other cancers including lung and colorectal cancer, overall cancer incidence and survival, and disease potentially impacted by chronic administration of selenium and/or vitamin E. Other areas of study were to include monitoring of serious cardiovascular events, assessment of quality of life, investigation of the relationship between serum micronutrient levels and prostate-cancer risk, and the evaluation of biological and genetic markers associated with the risk of prostate cancer (Lippman *et al.*, 2005).

25.5.2 Agent Formulations and Dose

L-Selenomethionine was chosen as the selenium formulation for SELECT despite the use of selenized yeast in the hypothesis generating NPC trial (Clark *et al.*, 1998). This recommendation from an NCI-sponsored panel of experts was based on large batch-to-batch variability and lack of commercial availability of selenized yeast. According to laboratory analyses, L-selenomethionine was the predominant selenium species in selenized yeast available at the time, so there was reason to believe that the daily dose of 200 µg was similar to the dose of 200 µg selenized yeast used in the NPC trial. However, the precise amount of selenium delivered by the selenized yeast remained unpredictable and variable (Lippman *et al.*, 2005). While one could conclude that 200 µg L-selenomethionine delivered more selenium than 200 µg selenized yeast, direct comparisons between the two doses were difficult to make.

Based on the association of long-term supplementation of the racemic mix of α-tocopherol, which includes the D- and L-isomers, with reduced prostate-cancer incidence in the ATBC trial, this formulation of vitamin E was chosen. The dose of 400 mg per day was selected based on its use in vitamin supplements, which suggested safety, and on its potential benefits for non-cancer diseases, including cardiovascular disease and Alzheimer's disease.

However, this chosen dose was eight times higher than that the 50 mg per day used in the ATBC study (The Alpha-Tocopherol Beta Carotene Cancer Prevention Study Group, 1994b).

25.5.3 Trial Design and Outcome Ascertainment

SELECT was a prospective, randomized, double-blind, placebo-controlled 2 × 2 factorial design clinical trial of selenium and vitamin E alone and in combination in eligible healthy men who were at elevated risk for prostate cancer. The men were randomized to receive daily oral doses of either 200 µg selenium plus placebo, 400 mg α-tocopherol plus placebo, 200 µg selenium plus 400 mg α-tocopherol, or two placebos. The study was planned to last for 12 years with 5 years for accrual and 7–12 years of intervention and follow-up. To address five predetermined comparisons, 32 400 men were required. The comparisons were: vitamin E *vs.* placebo, selenium *vs.* placebo, vitamin E plus selenium *vs.* placebo, vitamin E plus selenium *vs.* vitamin E alone, and vitamin E plus selenium *vs.* selenium alone. This sample size would provide adequate power to detect ≥25% decreases in the incidence of prostate cancer for selenium or vitamin E alone and an additional 25% decrease for selenium and vitamin E combined compared to either agent alone.

Routine clinical diagnostic evaluations, including yearly digital rectal exams (DRE) and serum PSA measurements, were used to assess prostate cancer. Biopsies were performed only at the discretion of study physicians or for participants with either suspicious DREs or elevated PSA levels. Unlike the previous NCI-sponsored prostate-cancer chemoprevention trial comparing finasteride to placebo, no end-of-study biopsies were implemented.

25.5.4 Cohort

Men were eligible for SELECT if they were at elevated risk of prostate cancer due to age and/or African ancestry. The age criteria for eligibility were ≥55 years for Caucasian men and ≥50 years for African American men. The rationale for this age difference is that African American men aged 50 to 55 have an incidence of prostate cancer that is comparable to that of Caucasian men aged 55 to 60. Other inclusion criteria were that men be healthy, and have a total PSA ≤ 4.0 ng mL^{-1}, a DRE not suspicious for cancer, no previous prostate cancer or high-grade prostate intraepithelial neoplasia, normal blood pressure, no current use of anticoagulation therapy, and be willing to stop taking off-study supplements. In a 3 year period, 35 533 eligible men from the US, Canada, and Puerto Rico were enrolled; this enrollment rate exceeded the goals for both number of participants and for anticipated length of the accrual period. The cohort included 21% minorities (12% African American, 7% Hispanic, and 2% other). Following randomization, known prostate-cancer risk factors, including age, race, baseline PSA, and smoking status were equally balanced among the four treatment groups (Lippman *et al.*, 2009).

25.5.5 Primary Endpoint Results

SELECT results were released in two separate reports, one in 2009 and one in 2011 (Lippman *et al.*, 2009; Klein *et al.*, 2011). Before publication of the results, in September 2008 an independent data and safety monitoring committee unanimously voted to discontinue the use of study supplements due to lack of evidence of benefit. The first SELECT results publication included data current as of October 23, 2008, the date on which study sites were advised to discontinue supplement administration. At that point, median follow-up time was 5.46 years and the range was 4.17–7.33 years. There were no significant differences in rates of prostate cancer among the four intervention arms (Table 25.2). The hazard ratio (HR) of prostate cancer for the vitamin E group was 1.13 (99% CI: 0.95–1.35), for selenium 1.04 (99% CI: 0.87–1.24), and for the combination group 1.05 (99% CI: 0.88–1.25), relative to the placebo group. The 13% increase in prostate-cancer risk in the vitamin E group was statistically nonsignificant ($p = 0.06$), but troubling nonetheless.

Data included in the second SELECT report had been collected through July 2011 and contained an additional 54 464 person-years of follow-up. During this additional follow-up time, study supplements were not used. An additional 521 prostate cancers were diagnosed: 113 in the placebo group, 147 in the vitamin E group, 143 in the selenium group, and 118 in the selenium + vitamin E group. There were no statistically significant differences in the rates of prostate-cancer detection in the selenium group *vs.* placebo or in the selenium + vitamin E group *vs.* placebo (Table 25.2); however, risk of prostate cancer in the vitamin E group was increased 17% compared to the placebo group, and in contrast to the 13% increase in the first report, now appeared to be significant (HR = 1.17, 99% CI: 1.004–1.36, $p = 0.008$). Cumulative incidence curves of prostate cancer by supplement group suggest that the increased risk in the vitamin E group began approximately 3–4 years after randomization (Nicastro and Dunn, 2013). Despite the increased risk in the vitamin E only group, there was no increased risk of prostate cancer for the vitamin E + selenium group.

Overall, in both reports, most prostate cancers were diagnosed by prostate biopsies administered due to abnormal PSA or DRE results, most were early stage and low Gleason grade, and stage, grade, and PSA levels did not differ by treatment group.

25.5.6 Secondary Endpoints, Adverse Outcomes, and Adherence

In addition to prostate-cancer incidence, SELECT investigators specified several *a priori* secondary endpoints. These included other cancers, including colorectal and lung cancer, total cancer incidence, cardiovascular events, diabetes, and deaths. No significant differences among supplement groups were observed for any of these endpoints (Table 25.3).

Table 25.2 First and second reports (Lippman *et al.*, 2009; Klein *et al.*, 2011) of the primary endpoint results from SELECT. This table shows the primary endpoint results from the first and second reports on SELECT. No significant differences in prostate-cancer incidence were reported in the first report, but in the 2011 report, the investigators showed that there was a 17% increase in prostate-cancer incidence in the group receiving vitamin E alone.[a]

	First report, October 2008				Second report, July 2011			
	Placebo ($n = 8696$)	Vitamin E ($n = 8737$)	Selenium ($n = 8752$)	Selenium + Vitamin E ($n = 8703$)	Placebo ($n = 8696$)	Vitamin E ($n = 8737$)	Selenium ($n = 8752$)	Selenium + Vitamin E ($n = 8702$)
Prostate cancer								
No. events	416	473	432	437	529	620	575	555
HR (99% CI)	1 (ref)	1.13 (0.95–1.35)	1.04 (0.87–1.24)	1.05 (0.88–1.25)	1 (ref)	1.17 (1.004–1.36)[a]	1.09 (0.93–1.27)	1.05 (0.89–1.22)
Method of diagnosis, *n* (%)								
Prostate biopsy	404 (97)	458 (97)	419 (97)	420 (96)	n.r.	n.r.	n.r.	n.r.
Other/unknown	12 (3)	15 (3)	13 (3)	17 (4)	n.r.	n.r.	n.r.	n.r.
Gleason score, *n* (%)								
2–6	240 (66)	249 (63)	217 (60)	220 (60)				
4–6					286 (69)	310 (67)	281 (64)	281 (63)
7	101 (28)	124 (31)	124 (34)	115 (32)	102 (24)	118 (25)	135 (31)	124 (28)
8–10	24 (7)	23 (6)	20 (6)	30 (8)	31 (7)	37 (8)	26 (6)	40 (9)
Not graded	51	77	71	72	110	155	133	110

[a] $p = 0.008$. n.r.: not reported.

Table 25.3 First and second reports (Lippman *et al.*, 2009; Klein *et al.*, 2011) of the secondary endpoint results and adverse outcomes from SELECT. This table shows the secondary endpoint results and adverse events from the first and second reports on SELECT. There were no significant differences in any endpoint/outcome among treatment groups. Numbers for specific types of cancers (lung, colorectal, other primary cancers, prostate cancers from Table 25.2) may not sum to number of individuals with cancer (described as "any cancer") due to potential multiple cancers per person.[a]

	First report, October 2008				Second report, July 2011			
Trial arm	Placebo ($n = 8696$)	Vitamin E ($n = 8737$)	Selenium ($n = 8752$)	Selenium + Vitamin E ($n = 8703$)	Placebo ($n = 8696$)	Vitamin E ($n = 8737$)	Selenium ($n = 8752$)	Selenium + Vitamin E ($n = 8702$)
Any cancer (including prostate cancer)								
No. events	824	856	837	846	1108	1190	1132	1149
HR (99% CI)	1 (ref)	1.03 (0.91–1.17)	1.01 (0.89–1.15)	1.02 (0.90–1.16)	1 (ref)	1.07 (0.96–1.19)	1.02 (0.92–1.14)	1.02 (0.92–1.12)
Lung cancer								
No. events	67	67	75	78	92	104	94	104
HR (99% CI)	1 (ref)	1.00 (0.64–1.55)	1.12 (0.73–1.72)	1.16 (0.76–1.78)	1 (ref)	1.11 (0.76–1.61)	1.02 (0.70–1.50)	1.11 (0.76–1.62)
Colorectal cancer								
No. events	60	66	63	77	75	85	74	93
HR (99% CI)	1 (ref)	1.09 (0.69–1.73)	1.05 (0.66–1.67)	1.28 (0.82–2.00)	1 (ref)	1.09 (0.72–1.64)	0.96 (0.63–1.46)	1.21 (0.81–1.81)
Other primary cancer								
No. events	306	274	292	290	579	570	557	594
HR (99% CI)	1 (ref)	0.89 (0.72–1.10)	0.95 (0.77–1.17)	0.94 (0.76–1.16)	1 (ref)	0.97 (0.83–1.14)	0.96 (0.83–1.13)	1.02 (0.92–1.14)
Diabetes								
No. events	669	700	724	660	869	918	913	875
HR (99% CI)	1 (ref)	1.04 (0.91–1.18)	1.07 (0.94–1.22)	0.97 (0.85–1.11)	1 (ref)	1.05 (0.93–1.17)	1.04 (0.93–1.17)	0.99 (0.89–1.12)
Cardiovascular (CV) events								
No. events	1050	1034	1080	1041	n.r.	n.r.	n.r.	n.r.
HR (99% CI)	1 (ref)	0.98 (0.88–1.09)	1.02 (0.92–1.13)	0.99 (0.89–1.10)				
CV events, grade ≥ 4								
No. events	n.r.	n.r.	n.r.	n.r.	969	909	939	943
HR (99% CI)					1 (ref)	0.93 (0.83–1.05)	0.97 (0.86–1.09)	0.97 (0.86–1.09)
Deaths, all cause								
No. events	382	358	378	359	564	571	551	542
HR (99% CI)	1 (ref)	0.93 (0.77–1.13)	0.99 (0.82–1.19)	0.94 (0.77–1.13)	1 (ref)	1.01 (0.86–1.17)	0.98 (0.84–1.14)	0.96 (0.82–1.12)

[a] n.r.: not reported; ref: reference value.

Investigators assessed adherence to study supplements by using pill counting and participant diaries for all participants and by measuring serum levels of selenium and cholesterol-adjusted α-tocopherol and γ-tocopherol levels in a bioadherence subcohort. Mean adherence by pill count was 83% at year 1 and 65% at year 5. Serum selenium and α-tocopherol levels rose only in participants randomized to receive those agents and not in the other groups, demonstrating good adherence and minimal drop-ins. Drop-in rates were also assessed *via* asking participants whether they took either supplement; rates were 3.1% or less for vitamin E and 1.8% or less for selenium (Lippman *et al.*, 2009).

25.6 Why Didn't SELECT Produce Expected Results?

25.6.1 Acceptance of Null Hypothesis

In SELECT, neither selenium nor vitamin E alone or in combination decreased the incidence of prostate cancer. Further, vitamin E significantly increased the incidence of prostate cancer among healthy men. Given the promising previous epidemiological and clinical studies, the findings from SELECT were surprising.

The simplest explanation of the results of SELECT is that neither agent has a potent preventive effect on prostate cancer and that the prior supportive data from randomized trials in humans merely reflected secondary endpoints that did not carry statistical validity. Laboratory data mimicking the design of SELECT showed no statistically significant reductions in prostate-cancer incidence with either selenium as L-selenomethionine or vitamin E alone or together in rodents (McCormick *et al.*, 2010). Nevertheless, interest remains in pursuing selenium as a potential cancer preventive agent for prostate as well as other cancers and alternative hypotheses to explain the null findings center on the agent formulation and dose that were chosen, the cohort, or the study design.

25.6.2 Agent

SELECT utilized a daily dose of L-selenomethionine of 200 μg per day, whereas the NPC trial used a dose of 200 μg per day of selenized yeast. An optimal dose of selenium for cancer prevention had not been established prior to SELECT, nor has one been established to date. It is likely that a narrow range of optimal doses exist, and these doses may depend on the baseline selenium status of the individual. Regardless, the SELECT dose was determined by an expert panel, and the 200 μg per day dose is much less controversial than the choice to use pure L-selenomethionine.

Organic selenium was chosen because while inorganic selenium had been shown to have better *in vitro* anticancer activities, these forms were linked to DNA single strand breaks. L-Selenomethionine was chosen over selenized yeast because of the batch-to-batch variability and lack of availability of the yeast form of selenium. Furthermore, L-selenomethionine is the predominant selenium species in the yeast, although selenized yeast contains numerous other selenocompounds with varying and possibly uncharacterized chemopreventive efficacy (discussed in Nicastro and Dunn (2013)). Selenocompounds, (including selenomethionine) that directly produce methyl selenol are more active than inorganic forms of selenium. Selenomethionine, however, is also nonspecifically incorporated into proteins in place of methionine, which diverts the selenium away from its potential active chemopreventive form(s) (Figure 25.1). Other selenocompounds like selenocysteine, selenite, or selenate, are not used nonspecifically in proteins and therefore are more likely to be converted to the potentially antitumorigenic metabolites *via* methyl selenol (Hatfield and Gladyshev, 2009).

In spite of the suspected differences in L-selenomethionine content between pure L-selenomethionine and selenized yeast, recent findings have indicated that selenomethionine and selenized yeast have similar biological activities in prostatic tissue of dogs (reviewed in Nicastro and Dunn (2013)). After seven months of supplementation with either selenomethionine or selenized yeast, there were no differences in levels of intraprostatic dihydrotestosterone and testosterone, dihydrotestosterone: testosterone ratio, DNA damage, proliferation, or apoptosis in the prostates of dogs according to formulation received. Further, there were no differences in toenail or intraprostatic selenium levels by formulation. Thus, while much controversy has surrounded the use of L-selenomethionine over selenized yeast in SELECT, it seems unlikely that the formulation of selenium used in SELECT is a major reason for the null results observed. Nevertheless, the effects of these different selenium formulations still need to be compared in humans.

25.6.3 Cohort

25.6.3.1 Baseline Selenium Status

The NPC trial recruited men from low environmental selenium areas in the eastern coastal US, whereas SELECT recruited across the US and Canada. This resulted in large discrepancies in baseline selenium status between men in the two trials. The mean baseline serum selenium concentration in SELECT was 135 ng mL^{-1}, compared to 114 ng mL^{-1} in NPC. This translated to 78% of the volunteers in SELECT having serum selenium levels that were above the lower two tertiles in NPC, and therefore above levels of the subgroup who saw protection from selenium in NPC (Dunn *et al.*, 2010).

As with many micronutrients, selenium status exhibits a U-shaped relationship with certain biomarkers or health outcomes. Such a relationship between toenail selenium concentration and prostatic DNA damage in elderly dogs paralleled data from studies in humans (reviewed in Dunn *et al.* (2010)). The optimal range of toenail selenium levels for prostate-cancer risk reduction in humans has been estimated to be 119–137 ng mL^{-1} and any additional selenium above this would either offer no benefit or cause harm (Chiang *et al.*, 2009). This proposed range is consistent with NPC results where protection was only seen in men with baseline serum selenium concentrations <123.2 ng mL^{-1}. Analysis of SELECT data stratified by baseline selenium status may identify a subpopulation of men who experienced a benefit from selenium. However, it is possible that the baseline serum selenium concentrations of men in even the lowest quantile in this cohort may be above concentrations at which supplemental selenium could offer protection. A stratified analysis has not been published to date.

The importance of baseline selenium status was likely not adequately recognized during planning of SELECT. A specified baseline selenium concentration was not established as an enrollment criterion, nor was a soil selenium-deficient region considered for the study location. Because SELECT was a very large trial, it was necessary to be inclusive with regards to geography in order to accrue an adequate number of participants. The study was therefore conducted through more than 400 clinical sites across the United States, Puerto Rico, and Canada. Further, due to increasing fortification of the food system with selenium in the years since NPC, it would be prohibitively difficult to ensure low baseline selenium status by limiting the study's geographical area to one with low soil selenium.

25.6.3.2 Genetics

Genetic factors were not a focus of NPC or SELECT, but they may nevertheless affect selenium status or an individual's response to selenium supplementation.

Polymorphisms in genes that encode either selenoproteins or proteins that function in selenium metabolism may influence outcomes (genetics studies are reviewed in Nicastro and Dunn (2013)). An example of a gene with variants that may influence the effect of selenium on cancer prevention is *SOD2*, which encodes the mitochondrial antioxidant enzyme manganese superoxide dismutase. The AA genotype of codon 16 (rs4880) of *SOD2* has been shown to confer a lower risk of total prostate cancer (RR: 0.3, 95% CI: 0.2–0.7) and of clinically aggressive prostate cancer (RR: 0.2, 95% CI: 0.1–0.5) among men with higher selenium levels compared to those with lower selenium levels. This protection was much weaker in men with VV or VA genotypes. Another example is seen in an analysis of prostate-cancer mortality in the Physicians' Health Study. In this study three different polymorphisms in the selenoprotein gene *SEP15* (rs479341, rs1407131, and rs561104) significantly affected survival time in men with prostate cancer. Depending on the polymorphism and the genotype, the survival time was either increased or decreased. An additional interaction was shown between selenium and the rs561104 polymorphism. High levels of selenium were associated with decreased prostate-cancer mortality only in those with the increased risk homozygous variant genotype and not in those with the wild-type genotype ($P_{\text{interaction}} = 0.02$). In a final example, the genotype of *GPX1*, which encodes the selenoprotein GPx1, was recently shown to be a determinant of selenium requirements. Among 161 men and women, those with the *GPX1* 679 (rs1050450) T/T genotype had significantly lower plasma selenium levels than those with the C/C genotype. The mean plasma selenium level in this study was 142.0 ng mL^{-1}, slightly above that of men in SELECT.

Because the polymorphisms in these and other genes can potentially help to determine levels of selenium necessary for selenium balance and outcomes in response to supplemental selenium, consideration of genotype may influence study results. Analysis of SELECT data according to participants' genotypes may elucidate relationships between supplemental selenium and prostate-cancer risk that were not evident in the SELECT population as a whole.

25.6.3.3 Age

Intervention in men older than 50 years of age may be too late, given the long natural history of prostate cancer (reviewed in Nicastro and Dunn (2013)). The critical window, if any, for prostate cancer chemoprevention, has not been clearly defined. In an analogous situation in breast cancer, also a hormonally driven cancer, an inverse association was observed between adolescent soy food intake and adult breast cancer, but not with adult soy intake and adult breast cancer. In rodents, selenite and other selenium metabolites inhibit carcinogen-induced mammary cancer during initiation *via* decreasing DNA damage, but selenium also has postinitiation activities and in fact is most efficacious when administered beginning during initiation and continuously thereafter (Combs and Gray, 1998). A clearer understanding of the timing by which selenium influences prostate carcinogenesis might help researchers to better predict an optimal age range at which selenium supplementation should take place. Regardless, results from NPC suggest that intervention in men over the age of 50 still has the potential to yield some anticancer benefit.

25.6.4 Design

In NPC, prostate-cancer incidence was a secondary outcome measure, and in SELECT, prostate-cancer incidence was a primary outcome measure. This important distinction may help to explain the different results. Both trials were adequately powered to detect differences of predetermined magnitudes in their respective primary outcomes only. In clinical trials with multiple outcomes, primary endpoints must be designated *a priori* in order to protect that endpoint from concerns about the observed results being due to chance due to multiple testing. Secondary endpoints do not have this protection. In NPC, results on skin cancer were protected, while results for prostate cancer were not. NPC results were particularly open to chance findings due to the small sample size of 64 prostate cancer cases in 1312 participants. Therefore, corroboration of the prostate-cancer incidence findings from NPC with a large randomized trial where prostate-cancer incidence was the primary outcome of interest was essential. Further, participants with a history of skin cancer were recruited in NPC. Because the primary endpoint was skin-cancer incidence, certain risk factors for prostate cancer, including PSA, were not considered as inclusion or exclusion criteria.

In addition, SELECT was designed to test the effects on prostate cancer after 7–12 years of vitamin E or selenium supplementation. One possible explanation for the apparent lack effect of either intervention is that a lag time exists before either a beneficial or harmful effect can be observed. Possibly a much longer period of time is required to see such an impact.

SELECT was not powered to address subgroup analyses. Regardless, findings according to quantile of baseline serum selenium concentration, genotype, or other risk factors could be enlightening and may help generate hypotheses regarding subpopulations that may benefit from selenium.

25.7 Implications and Future Directions

Following the discontinuation of supplement administration and publication of the primary data, SELECT transitioned into an observational cohort study, the SELECT Centralized Follow-up Study (SELECT CFU). As of December 2011, 17 761 participants, 58% of the 32 569 SELECT participants who were still alive and not refusing further contact, had enrolled in SELECT CFU (Goodman *et al.*, 2013). The SELECT biorepository, maintained by SWOG, contains over 100 000 banked tissue samples, including blood, tissue, and toenail samples. To date, at least 15 projects utilizing these samples have been approved, with continuing plans for additional projects.

Regarding vitamin E, the United States Preventive Services Task Force (USPSTF) issued a grade D recommendation against the use of vitamin E supplements for the prevention of cardiovascular disease or cancer. This recommendation is based on moderate certainty of lack of a net benefit of vitamin E supplementation.

Selenium, however, remains an area of interest for prostate-cancer chemoprevention, with caveats. One of the most important findings of SELECT was that there is a need to better understand selenium biology (Hatfield and Gladyshev, 2009). A clearer view of selenium's biological activities will aid researchers in choosing the appropriate doses and formulations of selenium, as well as the appropriate cohorts and study design. One promising area of selenium research is the ongoing characterization of selenium's anti-DNA damage activities (Chiang *et al.*, 2009). Furthermore, the tissue specificity, function, regulation, and enzyme kinetics of many of the 25 selenoproteins identified in humans remain uncharacterized (Davis *et al.*, 2012). Ongoing characterization of these proteins will contribute to a better understanding of selenium's mechanisms of action. Before any new large phase III clinical trials are undertaken, there is a need for smaller pilot human studies that will determine the proper doses and formulations of selenium and for further preclinical studies on selenium's mechanisms of action. It is also imperative to determine any subpopulations that may benefit from selenium supplementation. Conflicting or inconclusive results from clinical and epidemiological studies underscore the difficulty in making nutritional recommendations for cancer prevention to the general population and bring to light the need for studies using subgroups at greater risk or who may benefit more than others (Nicastro *et al.*, 2012). To date, the best evidence indicates that men with low serum selenium concentrations or those that live in selenium-deficient regions may represent the optimal cohort for studying prostate-cancer prevention by selenium, and more research is necessary to determine whether further refinement of eligibility criteria by age or selenoprotein genotype may also be useful.

Summary Points

- Prostate cancer is the most commonly diagnosed noncutaneous cancer in men in the United States.
- Investigators were interested in studying selenium and vitamin E as prostate-cancer prevention strategies, as they are naturally occurring essential micronutrients with promising preclinical, observational, and clinical data suggesting risk-reducing efficacy.
- The Selenium and Vitamin E Cancer Prevention Trial (SELECT) was conducted to test the efficacy of selenium and vitamin E alone and in combination on the incidence of prostate cancer.
- Supplemental selenium is available as L-selenomethionine or selenized yeast. The major component of selenized yeast is L-selenomethionine, but selenized yeast contains other components.
- The majority of men in the United States have adequate selenium status but inadequate vitamin E intake.
- SELECT was a prospective, randomized, double-blinded, placebo-controlled, 2 × 2 factorial design clinical trial of selenium and vitamin E alone or in combination in eligible healthy mean who were at elevated risk for prostate cancer.
- No significant differences in prostate-cancer incidence were reported in the first report of SELECT, but in the 2011 report, the investigators showed that there was a 17% increase in prostate-cancer incidence in the group receiving vitamin E alone.
- The null findings regarding selenium in SELECT may be partially attributable to the formulation of selenium chosen (selenized yeast), the relatively high baseline selenium levels in the trial cohort, age or genetic factors in the cohort, or the study design of SELECT.
- There is a need for a better understanding of selenium biology before additional clinical trials that test selenium efficacy are undertaken.

Key Facts

Key facts related to SELECT study design. SELECT: Selenium and Vitamin E Cancer Prevention Trial

1. Randomized: participants were randomly assigned to 1 of 4 possible treatment groups.
2. Prospective: participants free from prostate cancer were followed over time.

3. Double blind: participants did not know to which treatment group they were assigned.
4. Placebo-controlled: effects of the different treatments were compared to the effects of inert treatments.
5. 2 × 2 factorial design: two agents were tested alone and in combination.
6. Doses: 200 μg per day L-selenomethoinine and 400 mg per day racemic mix of α-tocopherol.
7. Sample size: 35 533 men.
8. Primary outcome: incident prostate cancer rates.
9. Intervention length: 7–12 years.
10. Predetermined comparisons: vitamin E *vs.* placebo, selenium *vs.* placebo, vitamin E plus selenium *vs.* placebo, vitamin E plus selenium *vs.* vitamin E alone, vitamin E plus selenium *vs.* selenium alone.

Definitions of Words and Terms

Cancer prevention refers to the prevention of initiation of the carcinogenic process as well as the halting, reversal, or slowing down of ongoing carcinogenesis so that the premalignant state never transforms into invasive cancer.

Chemoprevention is the prevention of disease before it becomes clinically evident by intervening with pharmaceutical agents/drugs. Some authors include intervention with nutritional agents in chemoprevention. In this chapter the term chemoprevention applies specifically to the prevention, or reduction in risk, of cancer.

EAR, or estimated average requirement is the intake of a nutrient that is expected to satisfy the needs of half of the individuals in a given age group.

Genetic polymorphisms are changes in the nucleotide sequence in genes that create novel alleles, or variant forms of the genes. Polymorphisms are common and occur with a certain minimum frequency, often defined as at least 1% of the alleles in a population. Unlike disease-causing mutations, polymorphisms often appear not to affect health. However, many may modify disease risk to a minor degree and may contribute to the overall risk of developing some of the common complex diseases like cancer. Similarly, some polymorphisms modify the ability of a drug to achieve the desired response (improvement in a disease endpoint).

The primary endpoint of a randomized clinical trial is the most important outcome that we are interested in We want to see if the intervention, whether a drug or a nutritional agent, leads to an improvement in the primary disease endpoint. In chemoprevention of cancer trials, this endpoint is usually the occurrence, or incidence, of cancer. The statistical design of a trial is based on its primary endpoint.

The randomized clinical trial is often considered the "gold standard" of study design for trials of drug interventions in humans. The randomization, whereby participants are randomly assigned to each of the 2 or more drug groups or placebo (the control) group, ensures that there is no bias in the assignments. This is the only way that any benefit derived from the drug intervention can be trusted as being real and not the result of a bias in the assignment of participants to the different groups.

The recommended daily allowance (RDA) of a dietary component, or nutrient, is the amount an individual is recommended to take on a daily basis to maintain health in each stage of life and for each gender. These RDA values are calculated based on Dietary Reference Indices. The US Department of Health and Human Services and US Department of Agriculture are involved in determining the RDA values.

SELECT is an acronym for the Selenium and Vitamin E Cancer Prevention Trial. This clinical trial, which was run by the Division of Cancer Prevention in the National Cancer Institute together with the Southwest Oncology Group (SWOG), tested the two nutrients in the title for their ability to reduce the risk of prostate cancer in a population of men at moderately increased risk by virtue of age.

Selenium is a trace element that is essential in the diet.

Vitamin E is a vitamin that is essential in the diet.

List of Abbreviations

AR	Androgen receptor
ATBC	The Alpha-Tocopherol Beta-Carotene Cancer Prevention Trial
BMI	Body mass index
DRE	Digital rectal exam
EAR	Estimated average requirement
EPIC	European Prospective Investigation into Cancer and Nutrition
FDA	Food and Drug Administration (US)
GPx	Glutathione peroxidases
IOM	Institute of Medicine
IU	International units
NHANES	National Health and Nutrition Examination Survey (US)
NPC	Nutritional prevention of cancer trial
OR	Odds ratio
PSA	Prostate-specific antigen
RDA	Recommended daily allowance
RR	Relative risk
SELECT	Selenium and Vitamin E Cancer Prevention Trial
SELECT CFU	SELECT Centralized Follow-Up study
SOD2	Mitochondrial superoxide dismutase gene
SU.VI.MAX	Supplementation en Vitamins et Mineraux Antioxidants trial
SWOG	SouthWest Oncology Group
TR	Thioredoxin reductase
UL	Upper limit
US	United States
USPSTF	United States Preventive Services Task Force

VITAL	Vitamins and Lifestyle study
WCRF/AICR	World Cancer Research Fund/American Institute for Cancer Research

References

Azzi, A., 2007. Molecular mechanism of alpha-tocopherol action. *Free Radical Biology Medicine*. 43: 16–21.

Bespalov, V. G., Panchenko, A. V., Murazovla, G. and Chepik, O. F., 2011. Influence of sodium selenite on carcinogenesis of the prostate and other organs induced by methylnitrosourea and testosterone in rats. *Voprosy Onkologii*. 57: 486–492.

Blot, W. J., Li, J. Y., Taylor, P. R., Guo, W., Dawsey, S., Wang, G. Q., Yang, C. S., Zheng, S. F., Gail, M. and Li, G. Y., 1993. Nutrition intervention trials in Linxian, China: supplementation with specific vitamin/mineral combinations, cancer incidence, and disease-specific mortality in the general population. *Journal of National Cancer Institute*. 85: 1483–1492.

Chiang, E. C., Shen, S., Kengeri, S. S., Xu, H., Combs, G. F., Morris, J. S., Bostwick, D. G. and Waters, D. J., 2009. Defining the optimal selenium dose for prostate cancer risk reduction: insights from the U-shaped relationship between selenium status, DNA damage, and apoptosis. *Dose Response*. 8: 285–300.

Clark, L. C., Combs, G. F., Turnbull, B. W., Slate, E. H., Chalker, D. K., Chow, J., Davis, L. S., Glover, R. A., Graham, G. F., Gross, E. G., Krongrad, A., Lesher, J. L., Park, H. K., Sanders, B. B., Smith, C. L. and Taylor, J. R., 1996. Effects of selenium supplementation for cancer prevention in patients with carcinoma of the skin. A randomized controlled trial. Nutritional prevention of cancer study group. *Journal of American Medical Association*. 276: 1957–1963.

Clark, L. C., Dalkin, B., Krongrad, A., Combs, G. F., Turnbull, B. W., Slate, E. H., Witherington, R., Herlong, J. H., Janosko, E., Carpenter, D., Borosso, C., Falk, S. and Rounder, J., 1998. Decreased incidence of prostate cancer with selenium supplementation: results of a double-blind cancer prevention trial. *British Journal of Urology*. 81: 730–734.

Combs Jr, G. F. and Gray, W. P., 1998. Chemopreventive agents: selenium. *Pharmacology and Therapeutics*. 79: 179–192.

Davis, C. D., Tsuji, P. A. and Milner, J. A., 2012. Selenoproteins and cancer prevention. *Annual Review of Nutrition*. 32: 73–95.

Dunn, B. K., Richmond, E. S., Minasian, L. M., Ryan, A. M. and Ford, L. G., 2010. A nutrient approach to prostate cancer prevention: the selenium and vitamin E cancer prevention trial (SELECT). *Nutrition and Cancer*. 62: 896–918.

Fulgoni 3rd, V. L., Keast, D. R., Bailey, R. L. and Dwyer, J., 2011. Foods, fortificants, and supplements: where do Americans get their nutrients? *Journal of Nutrition*. 141: 1847–1854.

Goodman, P. J., Hartline, J. A., Tangen, C. M., Crowley, J. J., Minasian, L. M., Klein, E. A., Cook, E. D., Darke, A. K., Arnold, K. B., Anderson, K., Yee, M., Meyskens, F. L. and Baker, L. H., 2013. Moving a randomized clinical trial into an observational cohort. *Clinical Trials*. 10(1): 131–142.

Hatfield, D. L. and Gladyshev, V. N., 2009. The outcome of selenium and vitamin E cancer prevention trial (SELECT) reveals the need for better understanding of selenium biology. *Molecular Interventions*. 9: 18–21.

Heinonen, O. P., Albanes, D., Virtamo, J., Taylor, P. R., Huttunen, J. K., Hartman, A. M., Haapakoski, J., Malila, N., Rautalahti, M., Ripatti, S., Maenpaa, H., Teerenhovi, L., Koss, L., Virolainen, M. and Edwards, B. K., 1998. Prostate cancer and supplementation with alpha-tocopherol and beta-carotene: incidence and mortality in a controlled trial. *Journal of National Cancer Institute*. 90: 440–446.

Institute of Medicine (U.S.), 2000. Panel on Dietary Antioxidants and Related Compounds. *Dietary Reference Intakes for Vitamin C, Vitamin E, Selenium, and Carotenoids: a report of the Panel on Dietary Antioxidants and Related Compounds, Subcommittees on Upper Reference Levels of Nutrients and of Interpretation and Use of Dietary Reference Intakes, and the Standing Committee on the Scientific Evaluation of Dietary Reference Intakes*, Food and Nutrition Board, Institute of Medicine. National Academy Press, Washington, D.C., USA, vol. 506, p. 506.

Klein, E. A., Thompson Jr, I. M., Tangen, C. M., Crowley, J. J., Lucia, M. S., Goodman, P. J., Minasian, L. M., Ford, L. G., Parnes, H. L., Gaziano, J. M., Karp, D. D., Lieber, M. M., Walther, P. J., Klotz, L., Parsons, J. K., Chin, J. L., Darke, A. K., Lippman, S. M., Goodman, G. E., Meyskens Jr, F. L. and Baker, L. H., 2011. Vitamin E and the risk of prostate cancer: the selenium and vitamin E cancer prevention trial (SELECT). *Journal of American Medical Association*. 306: 1549–1556.

Li, J. Y., Taylor, P. R., Li, B., Dawsey, S., Wang, G. Q., Ershow, A. G., Guo, W., Liu, S. F., Yang, C. S. and Shen, Q., 1993. Nutrition intervention trials in Linxian, China: multiple vitamin/mineral supplementation, cancer incidence, and disease-specific mortality among adults with esophageal dysplasia. *Journal of National Cancer Institute*. 85: 1492–1498.

Lippman, S. M., Goodman, P. J., Klein, E. A., Parnes, H. L., Thompson, I. M., Kristal, A. R., Santella, R. M., Probstfield, J. L., Moinpour, C. M., Albanes, D., Taylor, P. R., Minasian, L. M., Hoque, A., Thomas, S. M., Crowley, J. J., Gaziano, J. M., Stanford, J. L., Cook, E. D., Fleshner, N. E., Lieber, M. M., Walther, P. J., Khuri, F. R., Karp, D. D., Schwartz, G. G., Ford, L. G. and Coltman, C. A., 2005. Designing the selenium and vitamin E cancer prevention trial (SELECT). *Journal of National Cancer Institute*. 97: 94–9102.

Lippman, S. M., Klein, E. A., Goodman, P. J., Lucia, M. S., Thompson, I. M., Ford, L. G., Parnes, H. L., Minasian, L. M., Gaziano, J. M., Hartline, J. A., Parsons, J. K., Bearden, J. D., Crawford, E. D., Goodman, G. E., Claudio, J., Winquist, E., Cook, E. D., Karp, D. D., Walther, P., Lieber, M. M., Kristal, A. R., Darke, A. K., Arnold, K. B., Ganz, P. A., Santella, R. M., Albanes, D., Taylor, P. R., Probstfield, J. L., Jagpal, T. J., Crowley, J. J., Meyskens, F. L., Baker, L. H. and Coltman, C. A., 2009. Effect of selenium and vitamin E on risk of prostate cancer and other cancers: the selenium and vitamin E cancer prevention trial (SELECT). *Journal of American Medical Association*. 301: 39–51.

Manor, D. and Morley, S., 2007. The alpha-tocopherol transfer protein. *Vitamins and Hormones*. 76: 45–65.

McCormick, D. L., Rao, K. V., Johnson, W. D., Bosland, M. C., Lubet, R. A. and Steele, V. E., 2010. Null activity of selenium and vitamin E as cancer chemopreventive agents in the rat prostate. *Cancer Prevention Research (Phila)*. 3: 381–392.

Meyer, F., Galan, P., Douville, P., Bairati, I., Kegle, P., Bertrais, S., Estaquio, C. and Hercberg, S., 2005. Antioxidant vitamin and mineral supplementation and prostate cancer prevention in the SU.VI.MAX trial. *International Journal of Cancer*. 116: 182–186.

Moyer, V. A., 2012. Screening for prostate cancer: U.S. preventive services task force recommendation statement. *Annals of Internal Medicine*. 157: 120–134.

Nicastro, H. L., Trujillo, E. B. and Milner, J. A., 2012. Nutrigenomics and cancer prevention. *Current Nutrition Reports*. 1: 37–43.

Nicastro, H. L. and Dunn, B. K., 2013. Selenium and prostate cancer prevention: insights from the selenium and vitamin E cancer prevention trial (SELECT). *Nutrients*. 5: 1122–1148.

Ozten, N., Horton, L., Lasano, S. and Bosland, M. C., 2010. Selenomethionine and alpha-tocopherol do not inhibit prostate carcinogenesis in the testosterone plus estradiol-treated NBL rat model. *Cancer Prevention Research (Phila)*. 3: 371–380.

Peters, U., Littman, A. J., Kristal, A. R., Patterson, R. E., Potter, J. D. and White, E., 2008. Vitamin E and selenium supplementation and risk of prostate cancer in the vitamins and lifestyle (VITAL) study cohort. *Cancer Causes Control*. 19: 75–87.

Siegel, R., Naishadham, D. and Jemal, A., 2013. Cancer statistics, 2013. *Ca-A Cancer Journal for Clinicians*. 63: 11–30.

The Alpha-Tocopherol Beta Carotene Cancer Prevention Study Group, 1994a. The effect of vitamin E and beta carotene on the incidence of lung cancer and other cancers in male smokers. *New England Journal of Medicine*. 330: 1029–1035.

The Alpha-Tocopherol Beta Carotene Cancer Prevention Study Group, 1994b. The alpha-tocopherol, beta-carotene lung cancer prevention study: design, methods, participant characteristics, and compliance. *Annals of Epidemiology*. 4: 1–10.

Theoret, M. R., Ning, Y.-M., Zhang, J. J., Justice, R., Keegan, P. and Pazdur, R., 2011. The risks and benefits of 5α-reductase inhibitors for prostate-cancer prevention. *New England Journal of Medicine*. 365: 97–99.

Venkateswaran, V., Fleshner, N., Sugar, L. and Klotz, L., 2004. Antioxidants block prostate cancer in lady transgenic mice. *Cancer Research*. 64: 5891–5896.

Venkateswaran, V., Klotz, L., Ramani, M., Sugar, L. M., Jacob, L. E., Nam, R. K. and Fleshner, N. E., 2009. A combination of micronutrients is beneficial in reducing the incidence of prostate cancer and increasing survival in the Lady transgenic model. *Cancer Prevention Research (Philadelphia, PA)*. 2: 473–483.

Wang, L., Bonorden, M. J., Li, G. X., Lee, H. J., Hu, H., Zhang, Y., Liao, J. D., Cleary, M. P. and Lu, J., 2009. Methyl-selenium compounds inhibit prostate carcinogenesis in the transgenic adenocarcinoma of mouse prostate model with survival benefit. *Cancer Prevention Research (Phila)*. 2: 484–495.

CHAPTER 26

Selenium in Autoimmune Thyroiditis

D. NACAMULLI[a], D. PETRICCA[b], C. MIAN[a], AND C. BETTERLE*[a]

[a]Endocrinology Unit, Department of Medicine (DIMED), University of Padua, Via Ospedale n. 105, 35128 Padova, Italy; [b]Department of Medicine (DIMED), University of Padua, Via Ospedale n. 105, 35128 Padova, Italy
*E-mail: corrado.betterle@unipd.it

26.1 Introduction

In the last decade, several clinical studies have examined the effects of selenium (Se) supplementation in thyroid autoimmune diseases (TADs): Hashimoto thyroiditis (HT), postpartum thyroiditis (PTT), Graves' disease (GD) and Graves' orbitopathy (GO).

HT is a common autoimmune disease that mainly affects middle-aged women and its incidence tends to increase with age. Many individuals with thyroid autoantibodies progress to hypothyroidism (McLeod and Cooper, 2012). The onset of the disease depends on genetic factors, and individuals who express human leucocyte antigens (HLA) DR3 and DR5 are more prone to developing HT (Tomer, 2010). Several environmental factors, such as viral infections, iodine intake, iron and Se status, and immunotherapeutic agents are also thought to have a role in the onset of the disease (Gärtner and Gasnier, 2003; Köhrle, 2013). HT is characterized by infiltration of the thyroid by lymphocytes (both T and B cells), plasma cells and macrophages,

Food and Nutritional Components in Focus No. 9
Selenium: Chemistry, Analysis, Function and Effects
Edited by Victor R. Preedy

Published by the Royal Society of Chemistry, www.rsc.org

leading to the destruction of the gland and the presence of several circulating organ-specific autoantibodies, *i.e.* thyroid peroxidase (TPOAb), thyroglobulin (TgAb) and TSH-receptor antibody (TRAb). Given the pattern of cytokine release, HT is classified as a T helper-1 disorder (Krysiak and Okopien, 2011). $CD4^+$ T lymphocyte activation induces an increase in IFN-γ with a subsequent expression of class II MHC molecules on thyrocytes, and this triggers an inflammatory process that destroys the follicles (see Figure 26.1).

On ultrasound, the gland's destruction appears as varying degrees of hypoechogenicity, depending on the stage of the disease. The diagnosis of HT is consequently based on the presence of TPOAb and/or TgAb, and/or a hypoechogenic thyroid pattern on ultrasound. Serum TSH levels may remain within the reference range, or be more or less elevated, with serum FT4 concentrations within the reference range (in cases of subclinical hypothyroidism), or below its lower limit (in cases of overt hypothyroidism).

Women with TPOAb are known to be at risk of developing thyroid dysfunction during pregnancy and after delivery. PTT is due to an autoimmune destruction of the gland in the first postpartum year in women with TPOAb. It is caused by the immunological rebound that follows the partial immunosuppression induced by pregnancy (Negro *et al.*, 2007).

Like HT, GD is an autoimmune disease and more common among women between the ages of 20 and 50 years. Again like HT, the onset of GD is related

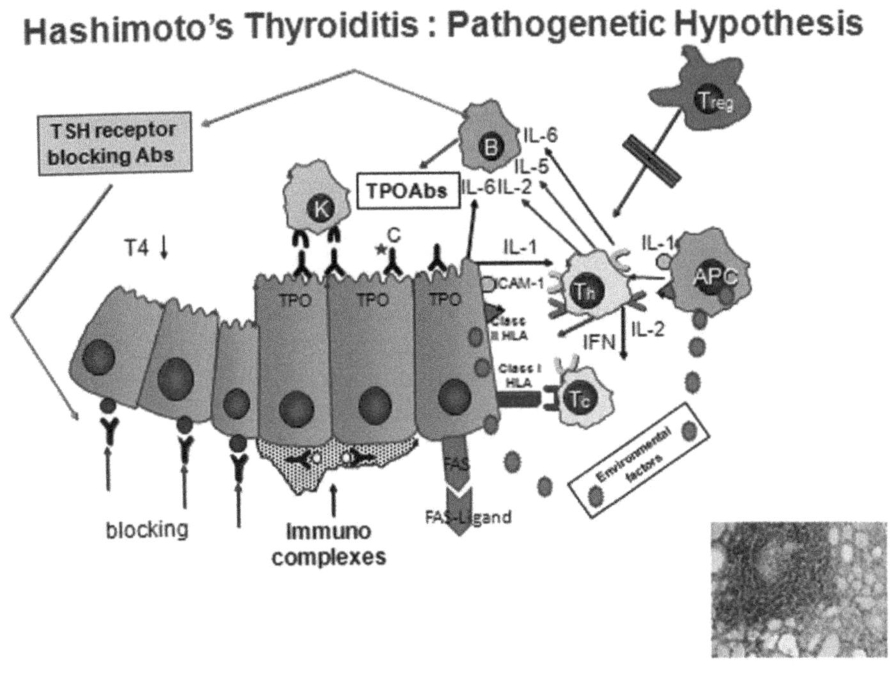

Figure 26.1 Pathogenetic hypothesis on Hashimoto's thyroiditis. Tc = T-cytotoxic, Th = T-helper, Treg = T-regulator, APC = antigen-presenting cells, B = B-lymphocytes, K = Killer-cells.

to a combination of genetic factors such as certain Class II HLA alleles and polymorphisms of the cytotoxic T-lymphocyte antigen-4 (CTLA-4) gene, and environmental factors such as iodine repletion in iodine-deficient areas and stress, whereas smoking is linked to GO but not to GD. TSH receptor (TSHR) is internalized and presented to T helper 2 cells by dendritic cells and macrophages. GD develops as a result of $CD4^+$ T cells losing their TSHR tolerance, which leads to the secretion of cytokines such as INF-γ and IL-2, and to the subsequent activation of B cells and TRAb secretion. TRAb cause the excessive thyroid hormone production responsible for the clinical manifestations of the disease and, by stimulating cyclic adenosine monophosphate, this leads to thyroid follicular cell proliferation and hyperplasia. TRAb bind to the TSHR in the orbital fibroblasts, giving rise to an inflammatory process that causes edema, extraocular muscle enlargement, and adipose tissue expansion, resulting in GO (Duntas, 2012).

In all these autoimmune thyroid diseases, an inflammatory process disrupts the antioxidant and oxidant balance, prompting an increased production of radical oxygen species (ROS) and cytokines, and thus causing oxidative stress and tissue injury. Glutathione peroxidase enzyme (GPX) was the first selenoprotein to be discovered in 1973 (Rotruck *et al.*, 1973). It plays an important part in the thyroid's antioxidant defense, together with other enzymatic defense systems comprising thioredoxin reductases (TRs), superoxide dismutase (SOD), and catalase (CAT) (Beckett and Arthur, 2005). Since the 1970s, at least 30 other selenoproteins have been identified. By lowering the propagation of free radicals and reactive oxygen species, and reducing the production of proinflammatory prostaglandins and leukotrienes (Rayman, 2008), Se-dependent enzymes are involved in several biological processes such as free-radical catabolism and carcinogenesis (Köhrle *et al.*, 2005). Iodothyronine selenodeiodinases are another important group of selenoproteins, responsible for activating and inactivating thyroid hormones in several extrathyroid tissues (Köhrle *et al.*, 2005; Köhrle and Gärtner, 2009). It has also recently been demonstrated that Se takes effect directly on the immune cells, targeting the T lymphocytes. A recent *in vitro* study provided evidence of Se having a dose-dependent inhibitory effect on the expression of HLA-DR molecules (Balázs and Kaczur, 2012).

Human Se intake, in the form of selenomethionine, selenocysteine and other sulfur amino acid analogs, depends on the diet. Organ meats and seafood are both rich in Se (Monograph, 2003). Adequacy of Se intake can be checked by testing for serum selenoprotein P (SePP), a hepatic selenoprotein responsible for transporting most of the Se in the blood to peripheral organs, or by measuring serum levels of glutathione peroxidase enzyme type 3 (GPX3). Plateau levels of GPX3 are reached with an intake of 125 μg of Se a day, while higher doses are needed to saturate SePP levels (Köhrle, 2013). An inverse relationship has been identified between TADs and Se plasma levels.

The thyroid gland contains the highest concentrations of Se in the body and, in contrast with other tissues, Se levels in the thyroid remain relatively stable even in cases of long-term nutritional deficiency, indicating that the thyroid is resistant to a scarce Se intake (Holben and Smith, 1999). Se in the thyroid gland is involved, as a selenoprotein component, in several processes for maintaining thyroid homeostasis, protecting thyrocytes against oxidative injury caused by H_2O_2, for example, and ensuring hormone balance (Bianco *et al.*, 2002).

Based on these premises, several authors have suggested in the last ten years that a poor Se status has a crucial role in generating and perpetuating the autoimmune process in TADs. Over the last decade, several clinical trials have assessed the effects of Se supplementation in individuals with TADs, obtaining inconclusive and partially contradictory results. These studies were conducted in European and Mediterranean regions where there is a mild shortage of Se in the soil, such as Germany, France, Italy, Turkey, and Croatia. The low Se status in these regions is due mainly to intensive farming methods used in the last 50–60 years, and partly to changes in eating habits (Rayman, 2000).

26.2 Selenium Supplementation in Hashimoto Thyroiditis

26.2.1 Effects of Selenium Supplementation on Thyroid Autoantibodies (Data Summarized in Tables 26.1 and 26.2)

In 2002, Gärtner *et al.* conducted a blinded, placebo-controlled prospective study on patients with HT in Germany. Seventy female patients with known HT and high TPOAb and/or TgAb were enrolled, and 36 were given 200 µg of sodium selenite for 3 months, while 34 received a placebo. All patients were taking L-thyroxine (L-T4) replacement therapy at a dosage planned to keep their basal TSH within the normal range. At the study baseline, the mean TPOAb concentrations were identical in the two groups. After 3 months, the mean TPOAb levels had dropped significantly to 63.6% ($p = 0.013$) in the Se group and these patients' quality of life had improved, whereas there was no change in autoantibody titers in the placebo group (Gärtner *et al.*, 2002). In a follow-up of this study (Gärtner and Gasnier, 2003), 13 patients continued to take Se supplementation for a further 6 months and achieved a further, significant reduction in their TPOAb concentrations ($p = 0.004$), while 9 stopped taking Se and experienced a significant increase in their TPOAb levels ($p = 0.017$); a significant decrease in TgAb was only seen in the latter control group.

In 2003, Duntas *et al.* performed a randomized, placebo-controlled prospective trial on patients with HT. This study was apparently conducted in a Se-sufficient area in the south of Greece. Sixty-five patients (56 of them women) were randomized to form two groups: the Se group ($n = 34$) took L-T4 and a daily tablet containing 200 µg of selenomethionine (Seme), while the placebo group ($n = 31$) was given L-T4 and no other treatment. The TPOAb

Table 26.1 Characteristics and results of trials demonstrating the effectiveness of Se in reducing TPOAb titer in subjects with TADs.

Studies	Patients	Type of study	L-T4 therapy	Type and dose of Se	Effects on autoantibodies	Follow-up
Gärtner *et al.* (2002)	HT from Germany	Prospective placebo controlled ($N = 70$ ♀)	Yes	Sodium selenite 200 μg	Decreased TPOAb (60%)	3 months
Gärtner and Gasnier (2003)	HT from Germany	Crossover study ($N = 47$ ♀)	Yes	Sodium selenite 200 μg	TPOAb further decreased ($p = 0.004$) in 13 pts who continued Se. Increased TPOAb levels in patients stopping Se	6 months
Duntas *et al.* (2003)	HT from Greece	Prospective placebo controlled ($N = 65$ ♀ and ♂)	Yes	Seme 200 μg	Decreased TPOAb levels (46 and 55%). No effects on TgAb levels.	3 months
Turker *et al.* (2006)	HT from Turkey	Prospective placebo controlled ($N = 88$ ♀)	Yes	Seme 200 μg and 100 μg	Decreased TPOAb levels (20–30%) and modest reduction in TgAb levels (5.2%) only in patients given 200 μg Se.	3 months
Mazokopakis *et al.* (2007)	HT from Greece	Prospective case-control ($N = 80$ ♀)	Yes and Not	Seme 200 μg	Decreased TPOAb levels (20%), increased TPOAb levels in patients stopping Se. No effect on TgAb levels.	6 months
Nacamulli *et al.* (2010)	HT from Italy	Prospective case-control ($N = 76$ ♀ and ♂)	Not	Sodium selenite 80 μg	Decreased TPOAb levels (30%). Decreased TgAb levels (14%).	12 months

Table 26.2 Characteristics and results of trials finding no effect of Se supplementation in reducing autoantibody titer in subjects with TADs.

Studies	Patients	Type of study	L-T4 therapy	Type and dose of Se	Effects on autoantibodies	Follow-up
Karanikas *et al.* (2008)	HT patients from Austria	Prospective placebo-controlled ($N = 36$ ♀ and ♂)	Yes	Sodium selenite 200 μg	No effect on TPOAb levels	3 months
Kvicala *et al.* (2009)	Elderly people from Southern Bohemia	Prospective placebo-controlled ($N = 253$ ♀ and ♂)	Not known	Yeast-bound Se 100 μg	TPOAb and TgAb decreased only in placebo group. No effects in subjects treated with Se.	12 months
Bonfing *et al.* (2010)	Children and adolescents from Germany	Non-blinded prospective case-control study ($N = 49$ ♀ and ♂)	Yes	Sodium selenite 100 and 200 μg	No effect on TPOAb levels. TgAb levels decreased significantly in subjects treated with L-T4 alone and in those treated with L-T4 + NaSe 200 μg	12 months
Onal *et al.* (2012)	Children and adolescents from Turkey	Prospective study ($N = 23$ ♀ and ♂)	No	Seme 50 μg	No effects on TPOAb and TgAb levels	3 months
Eskes *et al.* (2013)	HT adult women from the Netherlands	Placebo-controlled, double-blind study ($N = 61$ ♀)	No	Sodium selenite 200 μg	No effects on TPOAb in any group	At 3 and 6 months, and also 3 months after stopping sodium selenite treatment (*i.e.* at 9 months)

levels in the Se group had dropped by 46% after 3 months ($p < 0.0001$), and by 55.5% after 6 months ($p < 0.05$), and the patients taking Se reported an improvement in their well-being. In the group treated with L-T4 alone, there was only an overall 21% decrease in TPOAb levels after 3 months, which reached 27% at 6 months ($p < 0.001$). During this study, the TgAb titers did not change significantly in either the treated or the control group (Duntas *et al.*, 2003).

Turker performed a trial to assess the long-term effects (up to 9 months) of different doses of Seme (200/100 µg per day) in 88 female Turkish patients with HT. Turkey is a country with a mild Se deficiency. All patients were taking L-T4 to keep their serum TSH levels within the normal range. This study showed that Seme lowered the serum concentrations of TPOAb in patients with HT, but only when it was administered in doses higher than 100 µg of Se a day, which were needed to maximize GPX activity. In fact, 200 µg of Se a day had lowered the HT patients' serum concentration of TPOAb levels by 26% already after 3 months, and this trend continued over a further 6 months (with a further 23.7% drop), while lower doses had no such effect (the TPOAb titer increased significantly in the group treated with only 100 µg of Se a day). There was a mild reduction in TgAb in both groups, but it was only significant in the treated cases (5.2%) (Turker *et al.*, 2006).

Mazokopakis performed a trial on 80 Greek women with HT living in a region with a paucity of Se. These patients had TSH levels below 2 mIU L^{-1}, either spontaneously or maintained by administering L-T4 therapy. All patients took 200 µg of Seme orally for 6 months, then the patients were divided into two groups: 40 continued the Se treatment and 40 stopped. The investigation was performed over a period of 12 months and the results showed a significant, gradual reduction in TPOAb levels (−21%) starting after 3 months of Se supplementation. Extending the Se supplementation for another 6 months led to an additional 8% decrease, while the TPOAb levels rose again significantly (+4.8%) in patients who stopped taking Se. The TgAb levels showed no significant variation in either group. In this study, the authors do not say how many patients were on L-T4 therapy at the baseline and how many were not. They also fail to mention whether there were any differences in the outcome of patients who were or were not taking L-T4 as well as Se during the trial (Mazokopakis *et al.*, 2007).

Our own study (Nacamulli *et al.*, 2010) focused mainly on the effect of a year-long course of Se supplementation in Italian patients ($n = 76$; 65F) with early stage HT and a normal thyroid function or mild hypofunction (normal or slightly elevated TSH, FT4 within normal range). Patients were not taking L-T4 hormone replacement therapy. They were divided into two groups, one (30 cases) given no treatment, the other (46 cases) treated with 80 µg per day of sodium selenite in a single oral dose for 12 months. The patients enrolled for this study were iodine sufficient and there was no significant difference in urinary iodine concentration between the two groups. No significant changes in TPOAb or TgAb levels were recorded in either group after 6 months, but both values dropped significantly (by around 30% and 19%, respectively) after 12 months in the Se group, but not in the control group.

In contrast with previous trials, Karanikas conducted a controlled study in 2008 on a cohort of 36 Austrian HT patients with moderate HT receiving treatment with L-T4 and found no significant reduction in TPOAb levels after 3 months of Se supplementation with 200 μg of sodium selenite (Karanikas *et al.*, 2008).

The latest clinical works also produced negative results Kvicala gave 100 μg of yeast-bound Se to 126 of a sample of 253 elderly people living in rest homes in Southern Bohemia, and placebo to the remainder of the sample. A year later, only the group given no supplementation showed a significant reduction in their TPOAb and TgAb levels (Kvicala *et al.*, 2009).

Bonfig *et al.* and Onal *et al.* conducted two studies on children and adolescents (12.2 ± 2.2 years of age). The first, nonblinded study involved 16 males and 33 females with newly diagnosed HT and hypothyroidism in Germany. Patients were divided into 3 groups and treated with L-T4 alone (18 cases), L-T4 plus 100 μg of sodium selenite (13 cases), or L-T4 plus 200 μg of sodium selenite (18 cases). After 12 months, there was no significant decrease in the concentration of TPOAb in patients on either 100 or 200 μg of Se a day (Bonfig *et al.*, 2010). In a second, uncontrolled study, 23 euthyroid patients with HT were treated with 50 μg of Seme for 3 months, which led to no change in their TPOAb or TgAb levels (Onal *et al.*, 2012).

Finally, Eskes recently randomized 61 euthyroid (TSH 0.5–5.0 mU L^{-1}) adult women with TPOAb ≥ 100 kU L^{-1} to receive sodium selenite 200 μg daily (n = 30) or placebo (n = 31) for 6 months. At 3 and 6 months of follow-up, serum Se and SePP had only increased in the patients treated with Se, who subsequently returned to baseline levels after suspending the Se supplementation. On the one hand, no changes in TPOAb titer, serum TSH levels or quality of life were recorded during the follow-up in either the Se-treated group or the placebo-treated controls, and no differences emerged between the two groups (Eskes *et al.*, 2013).

All patients were not taking L-T4 hormone replacement therapy during the studies conducted by Nacamulli, Onal and Eskes.

26.2.2 Effects of Selenium Supplementation on Thyroid Echogenicity

In 2002 Gärtner observed a normalization of the ultrasound pattern in parallel with a decline in TPOAb levels in 9 (25%) of 36 AIT patients given Se therapy for 3 months (Gärtner *et al.*, 2002). In our study we found a significantly lower thyroid echogenicity after 6 months in both their patients treated with physiological doses of Se and in their controls, but after 12 months the ultrasound pattern was unchanged in the group treated with Se, whereas the untreated controls' thyroid echogenicity was further reduced (Nacamulli *et al.*, 2010).

26.2.3 Effects of Selenium Supplementation on Circulating Cytokine Production

Karanikas assessed cell-mediated events. Patients underwent intracellular cytokine detection in the CD4 and CD8 T cells of peripheral blood mononuclear cells by flow cytometry before and then 3 months after taking Se or a placebo, which showed no significant changes in the pattern of cytokine production by T lymphocytes after Se supplementation (Karanikas *et al.*, 2008).

In contrast, a subsequent randomized clinical trial including euthyroid women from Poland with untreated TADs demonstrated that HT patients' monocytes and lymphocytes released larger amounts of all the cytokines considered than those of control subjects. Levothyroxine reduced the release by monocytes of TNF-α, IL-1β, IL-6, and monocyte chemoattractant protein-1, while Seme inhibited the release by lymphocytes of IL-2, INF-γ and TNF-α, accompanied by a reduction in plasma C-reactive protein levels. There was a more marked decrease in cytokine release and plasma C-reactive protein levels when both drugs were administered together (Krysiak and Okopien, 2011).

26.2.4 Effects of Selenium Supplementation on TSH and Thyroid Hormones

Only two studies have assessed TSH and FT4 in HT patients not taking L-T4, *i.e.* a subgroup of patients in the Mazokopakis study took 200 μg per day of Se alone, and all the patients in our study taking 80 μg per day of Se therapy were given no hormone replacement therapy. In both cases, there were no changes in TSH levels or FT4 concentrations after 6 and 12 months of Se supplementation (Mazokopakis *et al.*, 2007; Nacamulli *et al.*, 2010). Nordio recently published a double-blind randomized controlled trial on 48 women with subclinical HT from Italy (with TSH levels between 4.01 and 9.99 mIU L^{-1} and with normal F-T4 levels). For 6 months, 24 patients were given Seme 83 μg a day, while the other 24 received the same Seme 83 μg a day plus myo-inositol 600 mg a day. None of the patients were taking hormone replacement therapy. This study demonstrated that Se supplementation alone could only reduce TPOAb (−44%) and TgAb (−48%) levels, whereas combining Seme with myo-inositol (a TSH second messenger) also promoted a significant reduction in the patients' increased TSH levels (−31% $p = 0.01$) (Nordio and Pajalich, 2013).

26.3 Selenium Supplementation in Postpartum Thyroiditis

Women with TPOAb and a normal thyroid function are at great risk of PTT and permanent hypothyroidism (Negro *et al.*, 2007). There is also a strong association between thyroid autoimmunity and miscarriage. Women with a history of recurrent abortions have revealed lower levels of Se in their hair than controls (Prummel *et al.*, 2004). Plasma Se levels drop significantly in the third trimester of gestation, returning to baseline concentrations after delivery, and this explains why pregnant women are exposed to Se deficiency. In a prospective, randomized, placebo-controlled study conducted in Italy, Se supplementation (in the form of Seme 200 μg per day) during and after pregnancy in 77 TPOAb-positive euthyroid women resulted in a lower incidence of PPT and permanent hypothyroidism than in a group of 74 TPOAb-positive women given a placebo (the incidence of PTT was 28.6% in the Se group *vs.* 48.6% in the placebo group, $p < 0.01$; and the incidence of permanent hypothyroidism was 11.7% in the Se group *vs.* 20.3% in the placebo group, $p < 0.01$) Administering Se in a group of pregnant women also prompted a significant decrease in their TPOAb titer and an improvement in the ultrasound patterns of their thyroid gland by the end of the postpartum period by comparison with a group given a placebo. The increase in TPOAb titer recorded in both groups after delivery in this study was significantly smaller after 12 months in the treated group. About 20% of the pregnant women with antibody positivity and 2.5% of those in the control group were placed on L-T4 therapy due to the onset of thyroid hypofunction during pregnancy (Negro *et al.*, 2007).

26.4 Selenium Supplementation in Graves' Disease and Graves' Orbitopathy

Vrca *et al.* (2004) evaluated the effect of supplementation with a fixed combination of antioxidant (vitamin C 200 mg and E 36 mg, β-carotene 6 mg and Se 60 μg) for 2 months in 57 patients with GD from Croatia. All patients were taking methimazole. In this trial patients who received supplementation with antioxidants in addition to methimazole (group A 29 subjects) obtained euthyroidism faster than the patients treated with methimazole alone (group B 28 subjects). A statistically significant increase of GPX activity in whole blood in both groups was assessed ($p = 0.01$) but a more significant increase occurred in group A compared to group B, 30 days after the commencement of therapy ($p < 0.01$).

Marcocci *et al.* (2011) enrolled 159 patients with mild Graves' orbitopathy (GO) in a randomized, double-blind, placebo-controlled trial designed to ascertain the effects of Se. The patients were given Se 100 μg twice daily, pentoxifylline 600 mg twice daily, or placebo orally for 6 months, and they were followed up for another 6 months after this treatment was withdrawn. After the first 6 months, Se treatment was associated with a better quality of life ($p < 0.001$), less eye involvement ($p = 0.01$) and a slower progression of GO ($p = 0.01$) by comparison with the use of a placebo. The findings after 12 months confirmed the results seen after 6 months. The authors concluded that administering Se significantly improved quality of life, reduced eye involvement, and slowed the progression of disease in patients with mild GO.

26.5 Discussion

Most of the studies considered here were conducted on patients with HT. Judging from the earlier reports, Se doses higher than 100 μg a day seemed to be needed to achieve a short-term reduction in TPOAb titers in patients also taking L-T4 to keep their TSH within the reference range. A daily dose of 200 μg of Se succeeded in suppressing autoimmune activity within 3 months, whereas lower doses failed to do so, or took longer to take effect. After suspending Se therapy, however, there was a prompt increase in TPOAb levels (Turker *et al.*, 2006). Autoantibody levels also seemed to drop more in cases with more severe thyroiditis, in terms of their baseline serum autoantibody levels. In our study, autoantibody levels dropped with 80 μg Se (albeit more slowly than when higher doses are used) even in patients with a moderate disease activity taking no L-T4 replacement therapy. The latency period in this case became 6–12 months, and the effect of supplementation was only fully evident after 12 months (Nacamulli *et al.*, 2010).

Subsequent studies did not reach the same positive results. An excessive advanced stage of HT might explain the ineffectiveness of Se supplementation particularly in elderly people (Vanderpump and Tunbridge, 2002). However, Se failed in reducing autoantibody titer even in case of an early stage of HT as was demonstrated in the studies of Bonfig and Onal (Bonfig *et al.*, 2010; Onal *et al.*, 2012). We can speculate that the negative findings obtained in children and adolescents might be due to the more active autoimmune process in this age group as seen for instance, in type I diabetes in which children with antibodies against islet cells develop a diabetes mellitus that requires insulin as soon as it has been diagnosed, whereas only one in two adults reach this step (Betterle *et al.*, 2002; Guglielmi *et al.*, 2012). This hypothesis appears to be in conflict with the negative findings obtained by Karanikas in a group of patients with a less active autoimmune disease (Karanikas *et al.*, 2008).

In most of the studies L-T4 treatment was administered to maintain TSH levels in the normal range but some patients had particularly low levels (between 0.3 and 2 mIU L^{-1}) (Duntas *et al.*, 2003; Turker *et al.*, 2006; Karanikas *et al.*, 2008). It is well known that L-T4 therapy is able to reduce the autoantibody titer by reducing antigen exposure both when administered in suppressive doses or even in replacement doses in subjects with elevated TSH (Romaldini *et al.*, 1996). Consequently L-T4 treatment might have affected the results generating a bias among the groups studied.

However, when only Se was administered and no L-T4 therapy was given to any of the patients enrolled, a reduction in thyroid autoantibodies and gland infiltration were observed in any case, while Se alone was not able to reduce TSH levels (Nacamulli *et al.*, 2010; Nordio and Pajalich 2013). This would be consistent with previous findings in HT patients and healthy elderly populations treated with Se at physiological and higher concentrations.

Moreover, only some studies compared the selenium plasma levels between the control group with the supplemented group (Gärtner *et al.*, 2002; Duntas *et al.*, 2003; Karanikas *et al.*, 2008; Kvicala *et al.*, 2009; Nordio and

Pajalich, 2013). Another reason to keep in mind that might explain the discrepancies in the outcomes of the various studies is that Se is not alone in performing the work of selenoproteins, iodine and iron play a part too; in fact, Se supplementation may even result in a worsening function in hypothyroid subjects with severe iodine deficiency (Cotemprè *et al.*, 1992). In addition, even a modest iodine excess can disrupt the function of antioxidant systems against reactive oxygen species in susceptible individuals (Rostami *et al.*, 2013). All studies but one failed in measuring the 24 h urinary iodine excretion (Nacamulli *et al.*, 2010). Iron influences thyroid hormone metabolism and response to iodine in patients with iodine-deficiency disorders. TPO, which is a hemoprotein, is less active than normal in iron-deficiency anemia (Zimmermann and Köhrle, 2002). We should remember that autoimmune gastritis, a disease with sideropenic anemia, is frequently associated with TADs (Toh, 2014). It is also important to bear in mind that CAT (which is crucial to peroxide inactivation) is a hemoprotein. No studies have been conducted to assess the efficacy of CAT in cases of severe hyposideremia.

Se-dependent enzymes may have effects on the immune system too. Karanikas *et al.* found no significant immunological changes, in terms of cytokine production patterns in peripheral T cells, after a short period of Se therapy (Karanikas *et al.*, 2008); this is probably due to a weak correlation between the cytokine patterns detected in the peripheral $CD4^+$ and $CD8^+$ T lymphocytes, and the immunological processes occurring in the thyroid gland in HT. In contrast, a paper by Krysiak and Okopien indicated that Se had an anti-inflammatory effect. LT-4 takes effect mainly by suppressing monocytes, whereas Seme normalizes lymphocyte cytokine release. The anti-inflammatory effect was strongest in patients receiving both treatments, and it correlated with a reduction in TPOAb titers. These authors suggested that the discrepancy *vis à vis* the previous paper may relate to differences in baseline disease activity and/or baseline Se levels (Krysiak and Okopien, 2011).

In HT, the recruitment of autoreactive B cells and the production of TPOAb and TgAb occur after $CD4^+$ T cell activation. TPO antibodies are IgG, IgG1 and IgG4 in particular, and they may have a role in the pathogenesis of the thyroid damage. In fact it has been demonstrated that they have a cytotoxic effects *in vitro via* an antibody-dependent cell-mediated cytotoxicity (ADCC) and, concerning only the subclass IgG1, also *via* a complement-mediated cytotoxicity (in fact, with respect to IgG1, IgG4 cannot fix complement). TPOAb can likewise inhibit the enzyme's action, depending on the TPO epitope recognized (Prummel and Wiersinga, 2005; Swain *et al.*, 2005). We have also taken into count that IgG1 and IgG4 have a different ability to determine cytotoxicity and interfere with thyroid hormone production. Li, in his paper, distinguished two different types of HT on the basis of plasma cell infiltrate, as assessed by immunohistochemistry on surgical material. Plasma cells showing a high IgG4/IgG ratio are indicative of a more aggressive and rapidly evolving disease (Li *et al.*, 2010). TgAb would also be able to act *via* an ADCC.

Whether the antibody level reduction obtained after Se supplementation, may have an effect on the progression of HT or not, remain a still open issue. Moreover, although antibody titer is important in predicting the evolution of

HT together with TSH, autoantibody levels and their variations after Se treatment are presumably not very reliable as an indication of the stage of the disease (Vanderpump and Tunbridge, 2002). A dissociation between cytotoxicity and antibody titer had already been reported in a study by Chiovato *et al.* (1993). In this context thyroid ultrasound appears to be a more precise indicator of the extent of anatomical damage and lymphocytic infiltrate (Yoshida *et al.*, 1985; Li *et al.*, 2010). In a study by Haibo conducted in a model involving NOD.H-2^{H4} mice with iodine-induced thyroiditis, animals were randomly divided into control, HT untreated, and HT with Se treatment groups. Mice not treated with Se showed a marked lymphocytic infiltrate, with some follicles being destroyed and the remainder appearing enlarged on histological examination. Administering Se led to a regression of the anatomical abnormalities and presumably reduced the hypoechogenicity at ultrasonography examination (Haibo *et al.*, 2010). In the first study by Gärtner, all the subjects enrolled also received ultrasound thyroid examination, and the ultrasonography findings were subjectively assessed by the sonographer. An improvement in echogenicity was reported in 9 treated and 2 untreated cases, and the antibodies become negative in all but one of them (Gärtner *et al.*, 2002). In the study by Negro, conducted in patients with postpartum thyroiditis, echogenicity was classified as normal, mild, moderate or advanced on the basis of the sonographer judgment without using a quantitative method to assess the gland's echogenicity. At the end of the observation period, the percentage of subjects with moderate and severe hypoechogenicity was unchanged in the group treated with Se and significantly higher in the placebo group. Whether the author found a parallel decrease of the autoantibody levels in patients with the unchanged ultrasonography pattern was not mentioned, however (Negro *et al.*, 2007). In contrast with the previous study in the ours, ultrasound was assessed objectively. In short, it is the only study that considered HT stage using ultrasound and more precisely it included only patients with initial thyroiditis. Like Negro, we demonstrated that the untreated patients' hypoechogenicity progressed with time, while it stopped after 6 months of treatment with selenium. Our finding may suggest the most eligible therapeutic application of Se because we might logically expect greater improvements in patients with only modest anatomical damage than in cases where the destruction has already become severe. In our study no correlation between the degree of hypoechogenicity and the antibody titer was found, neither at the baseline nor at subsequent follow-ups after 6 and 12 months. As in other studies, however, there was an apparent (though not statistically significant) correspondence between individuals with improved thyroid ultrasound findings and those with a reduction in the antibody titer (Nacamulli *et al.*, 2010).

It is well known that 10–25% of people (mainly women) with positive autoantibodies have no ultrasound abnormalities or impaired thyroid function. On the other hand, 5–10% of individuals with clear sonographic evidence of thyroiditis and hypothyroidism have no circulating antibodies. In the study by Negro *et al.* (2007), 2 women in the control group (without antibodies) needed therapy with L-T4 during their pregnancy to deal with the onset of thyroid failure, and 3 showed signs of PPT; we are not told whether the same women were involved, or what their ultrasound pattern was like. This report is consistent with previous studies indicating that up to one in three of all cases of PPT occur in women who have no TPOAb (Kuijpens *et al.*, 1998).

Some physical features, such as body surface area and body mass index, or smoking behavior, are believed to be able to change the thyroid's volume and echogenicity (Derumeaux *et al.*, 2003), but they were not taken into account in these studies.

The efficacy of Se supplementation is also thought to depend on the bioavailability of the compounds used. Seme is readily absorbed in the gastrointestinal tract and retained in the body significantly better than the inorganic forms, with a slower whole-body turnover. Despite evidence of this difference in the literature, however, the results of available studies indicate that both Seme and sodium selenite are effective in slowing the progression of HT (Duntas, 2010).

Only one study has been conducted on pregnant women, but Se seems to have an important role in thyroid gland homeostasis during pregnancy too. By maximizing GPX activity, Se supplementation may reduce intraglandular inflammatory activity and the incidence of PTT and overt hypothyroidism, especially postpartum. This is important because it would point to a potential strategy for preventing PTT in pregnant women (Negro *et al.*, 2007).

A low Se status has been observed in patients with newly diagnosed GD too, but only one study has been conducted to date on the effectiveness of supplementation with Se and other antioxidants in GD patients taking methimazole. The higher concentration of FT4 and FT3 in GD raises the basal metabolic rate, resulting in a greater total oxygen consumption and production of free radicals responsible for tissue damage. Such a pro-oxidative state and disturbed antioxidative defense suggest that it may be useful to provide Se supplementation in GD. Treatment with methimazole and Se succeed in inducing euthyroidism faster than when methimazole is administered alone (Vrca *et al.*, 2004). Another study was conducted on patients with GO: administering Se was seen to significantly improve quality of life, reduce eye involvement, and slow disease progression in patients with mild GO (Marcocci *et al.*, 2011).

Judging from the results of clinical studies published to date, the evidence to support or refute the efficacy of Se supplementation in TADs is incomplete. Further clinical studies are needed to better elucidate this issue. Such studies should:

(1) clarify the precise mechanism of action of selenoproteins in TADs, their effects on thyrocytes, and their role in modulating immune cells;
(2) compare the serum Se levels in patients with TADs against findings in controls, and compare 24 h urinary iodine excretion and serum iron levels in patients with HT against findings in controls;

(3) assess the effectiveness of different (low, medium, high) doses of Se, distinguishing between the effects of Se supplementation in patients with: (a) normal TSH; (b) subclinical hypothyroidism or hyperthyroidism; and (c) severe hypothyroidism or hyperthyroidism.

Summary Points

- The onset of TADs depends on genetic factors and environmental factors such as poor Se status.
- In all TADs an inflammatory process disrupts the antioxidant and oxidant balance leading to oxidative stress and tissue injury.
- Se in the thyroid gland is involved in several processes for protecting thyrocytes against oxidative injury.
- Several authors have suggested that a poor Se status has a key role in generating and perpetuating the autoimmune process in TADs with inconclusive and partially contradictory results, the majority of the studies were conducted in patients with HT.
- Factors that might explain the discrepancies among the studies are:

  1. The different activity and the stage of the disease, taking in count that there are different types of HT on the basis of plasma cell infiltrate.
  2. Different baseline Se levels.
  3. Different iodine and iron status that play an important role with Se in maintaining thyroid homeostasis.
  4. L-T4 replacement therapy and TSH levels.

- Thyroid ultrasound is more useful for assessing the stage of HT with respect to autoantibody titer.
- The precise role of selenoproteins in modulating immune cells must be better elucidated.
- Further clinical studies are needed to better clarify the efficacy of Se supplementation in TADs.

Key Facts

1. The thyroid is a gland present in the neck responsible of the production of thyroid hormones, especially in the form of thyroxine (FT4) that is metabolized in active hormone (triiodothyronine or FT3) in the peripheral tissues.
2. Hormonal output from the thyroid gland is regulated by thyroid-stimulating hormone (TSH) produced by the anterior pituitary, which itself is regulated by thyrotropin-releasing hormone (TRH) produced by the hypothalamus.
3. The thyroid hormones are involved in the growth and in cellular metabolism.

4. Hypothyroidism involves a group of thyroid diseases that cause the production of too little amount of thyroid hormones. It is characterized by a slowed body metabolism.
5. Hyperthyroidism involves a group of thyroid diseases that cause the production of thyroid hormones in too great an amount. It is characterized by an accelerated body metabolism.
6. The diagnosis of hypothyroidism and hyperthyroidism is easily made by thyroid function tests in patients with signs and symptoms of thyroid dysfunction.
7. In hypothyroidism FT4 is low and TSH is high, while a low TSH indicates the pituitary may be the cause of hypothyroidism. In hyperthyroidism FT3 and FT4 are high and TSH is low.
8. The treatment of hypothyroidism consists in replacing FT4 with a specific synthetic thyroid hormone (levothyroxine). The main treatment of hyperthyroidism consists in the administration of antithyroid medications such as methimazole, which inhibit the synthesis of thyroid hormone.
9. Antibodies are proteins secreted by the plasma B cells after their recognition of foreign antigens or autoantigens in the case of autoimmune diseases. They mainly act by attacking an antigen directly or by activating the complement system.
10. Ultrasonography is an ultrasound-based diagnostic imaging technique used for visualizing internal body structures. Ultrasound images are made by sending a pulse of ultrasound into tissue using an ultrasound transducer. Acoustic echoes are reflected from the tissues and are recorded and displayed as an image to the operator.
11. Oxidative stress reflects an imbalance between the production of reactive oxygen species and the cell's ability to readily detoxify the reactive intermediates or to repair the resulting damage.
12. Selenoproteins are proteins that include a selenocysteine amino acid residue in their structure. They have an important role in protecting against oxidative damage especially the thyroid gland.

Definitions of Key Terms

Hashimoto thyroiditis is an autoimmune thyroid disease characterized by the destruction of thyroid cells by cell- and antibody-mediated immune processes. It is the most common cause of hypothyroidism.

Graves' disease is an autoimmune thyroid disorder that results in the overproduction of thyroid hormones. Autoantibodies recognize and bind to the TSH receptor and activate the secretion of FT4 and FT3 from the gland. Graves' disease is the most common cause of hyperthyroidism.

Postpartum thyroiditis is thyroiditis that occurs in women after the delivery of a baby.

Grave's orbitopathy is an autoimmune inflammatory disorder affecting the orbit around the eye, characterized by bulging of the eyes. It is caused by autoantibodies that bind to tissue around the eye and occurs with hyperthyroidism.

Hypoechoic in ultrasonography, giving off few echoes. Tissues or structures that reflect relatively few of the ultrasound waves directed at them.

Human leukocyte antigen (HLA) is a set of leucocytes surface molecules that for example determine the compatibility of donors for organ transplant as well as one's susceptibility to an autoimmune disease *via* crossreacting immunization.

Cytokines are cell-signaling molecules that aid cell to cell communication in immune responses and stimulate the movement of cells towards sites of inflammation.

An **antioxidant** is a molecule that inhibits the oxidation of other molecules and protects the cell from the damage of free radicals that start chain reactions in the cell. Antioxidants terminate these chain reactions by removing free-radical intermediates and inhibit other oxidation reactions.

Trace element is a dietary mineral that is needed in very minute quantities and is essential for many physiological and biochemical processes.

Antibody-dependent cell-mediated cytotoxicity is a mechanism of cell-mediated immune defense in which specific antibodies bind membrane-surface antigens and the target cell is lysed by an effector cell of the immune system.

List of Abbreviations

ADCC	Antibody-dependent cell-mediated cytotoxicity
CAT	Catalase
CTLA-4	Cytotoxic T-lymphocyte antigen-4
GD	Graves' disease
GO	Graves' orbitopathy
GPX	Glutathione peroxidase enzyme
GPX3	Glutathione peroxidase enzyme type 3
HLA	Human leucocyte antigens
HT	Hashimoto thyroiditis
L-T4	L-thyroxine
PTT	Postpartum thyroiditis
ROS	Radical oxygen species
Seme	Selenomethionine
SePP	Serum selenoprotein P
SOD	Superoxide dismutase
TADs	Thyroid autoimmune diseases
TgAb	Thyroglobulin antibody
TPOAb	Thyroid peroxidase antibody
TRAb	TSH-receptor antibody
TRs	Thioredoxin reductases
TSHR	TSH receptor

References

Balázs, C. and Kaczur, V., 2012. Effect of selenium on HLA-DR expression of thyrocytes. *Autoimmune Diseases*. 2012: 374635.

Beckett, G. J. and Arthur, J. R., 2005. Selenium and endocrine systems. *Journal of Endocrinology*. 184: 455–465.

Betterle, C., Spadaccino, A. C., Presotto, F. *et al.*, 2002. The number of markers of pancreatic autoimmunity is proportional to the risk for type 1 diabetes mellitus in Italian and English patients with organ-specific autoimmune diseases. *Annals of the New York Academy of Sciences*. 958: 276–280.

Bianco, A. C., Salvatore, D., Gereben, B., Berry, M. J. and Larsen, P. R., 2002. Biochemistry, cellular and molecular biology, and physiological roles of the iodothyronine selenodeiodinases. *Endocrine Reviews*. 23: 38–89.

Bonfig, W., Gärtner, R. and Schmidt, H., 2010. Selenium supplementation does not decrease thyroid peroxidase antibody concentration in children and adolescents with autoimmune thyroiditis. *Scientific World Journal*. 10: 990–996.

Contempré, B., Duale, N. L., Dumont, J. E., Ngo, B., Diplock, A. T. and Vanderpas, J., 1992. Effect of selenium supplementation on thyroid hormone metabolism in an iodine and selenium deficient population. *Clinical Endocrinology*. 36(6): 579–583.

Chiovato, L., Bassi, P., Santini, F. *et al.*, 1993. Antibodies producing complement-mediated thyroid cytotoxicity in patients with atrophic or goitrous autoimmune thyroiditis. *Journal of Clinical Endocrinology and Metabolism*. 77(6): 1700–1705.

Derumeaux, H., Valeix, P., Castetbon, K. *et al.*, 2003. Association of selenium with thyroid volume and echostructure in 35- to 60-years-old French adults. *European Journal of Endocrinology*. 148: 309–315.

Duntas, L. H., 2012. The evolving role of selenium in the treatment of Graves' disease and ophthalmopathy. *Journal of Thyroid Research*. 2012: 736161.

Duntas, L. H., Mantzou, E. and Koutras, D. A., 2003. Effects of a six-month treatment with selenomethionine in patients with autoimmune thyroiditis. *European Journal of Endocrinology*. 148: 389–393.

Duntas, L. H., 2010. Selenium and the thyroid: a close-knit connection. *Journal of Clinical Endocrinology and Metabolism*. 95: 5180–5188.

Eskes, S. A., Endert, E., Fliers, E. *et al.*, 2013. elenite supplementation in euthyroid subjects with thyroid peroxidase antibodies. *Clinical Endocrinology*. DOI: 10.1111/cen.12284, [Epub ahead of print].

Gärtner, R., Gasnier, B. C. H., Dietrich, J. W., Krebs, B. and Angstwurm, M. W. A., 2002. Selenium supplementation in patients with autoimmune thyroiditis decreases thyroid peroxidase antibodies concentrations. *Journal of Clinical Endocrinology and Metabolism*. 87: 1687–1691.

Gärtner, R. and Gasnier, B. C. H., 2003. Selenium in the treatment of autoimmune thyroiditis. *Biofactors*. 19: 165–170.

Guglielmi, C., Palermo, A. and Pozzilli, P., 2012. Latent autoimmune diabetes in the adults (LADA) in Asia: from pathogenesis and epidemiology to therapy. *Diabetes/Metabolism Research and Reviews*. 2(28 suppl.): 40–46.

Haibo, X., Weiwei, W., Yuanbin, L. *et al.,* 2010. Selenium upregulates $CD4^+$ $CD25^+$ regulatory T cells in iodine-induced autoimmune thyroiditis model of NOD.H-2^{H4} mice. *Endocrine Journal*. 57(7): 595–601.

Holben, D. H. and Smith, A. M., 1999. The diverse role of selenium within selenoproteins: a review. *Journal of the American Dietetic Association*. 99: 836–843.

Karanikas, G., Schuetz, M., Kontur, S. *et al.,* 2008. No immunological benefit of selenium in consecutive patients with autoimmune thyroiditis. *Thyroid*. 18: 7–12.

Köhrle, J., 2013. Selenium and the thyroid. *Current Opinion in Endocrinology, Diabetes and Obesity*. 20(5): 441–448.

Köhrle, J., Jakob, F., Contempré, B. and Dumont, J. E., 2005. Selenium, the thyroid, and the endocrine system. *Endocrine Reviews*. 26: 944–984.

Köhrle, J. and Gärtner, R., 2009. Selenium and thyroid. *Best Practice & Research, Clinical Endocrinology & Metabolism*. 23: 815–827.

Krysiak, R. and Okopien, B., 2011. The effect of levothyroxine and selenomethionine on lymphocyte and monocyte cytokine release in women with Hashimoto's thyroiditis. *Journal of Clinical Endocrinology and Metabolism*. 96: 2206–2215.

Kuijpens, J. L., De Hann-Meulman, M., Vader, H. L., Pop, V. J., Wiersinga, W. M. and Drexhage, H. A., 1998. Cell-mediated immunity and postpartum thyroid dysfunction: a possibility for the prediction of disease? *Journal of Clinical Endocrinology and Metabolism*. 83(6): 1959–1966.

Kvicala, J., Hrda, P., Zamrazil, V., Nemecek, J., Hill, M. and Jiranek, V., 2009. Effect of selenium supplementation on thyroid antibodies. *Journal of Radioanalytical and Nuclear Chemistry*. 280: 275–279.

Li, Y., Nishihara, E., Hirokawa, M., Taniguchi, E., Miyauchi, A. and Kakudo, K., 2010. Distinct clinical, serological, and sonographic characteristics of Hashimoto's thyroiditis based with and without IgG4-positive plasma cells. *Journal of Clinical Endocrinology and Metabolism*. 95(3): 1309–1317.

Marcocci, C., Kahaly, G. J., Krassas, G. E., Bartalena, *et al.,* 2011. Selenium and the course of mild Graves' orbitopathy. *New England Journal of Medicine*. 364: 1920–1931.

Mazokopakis, E. E., Papadakis, J. A., Papadomanolaki, M. G. *et al.,* 2007. Effect of 12 months treatment with L-selenomethionine on serum anti-TPO levels in patients with Hashimoto's thyroiditis. *Thyroid*. 17: 609–612.

McLeod, D. S. and Cooper, D. S., 2012. The incidence and prevalence of thyroid autoimmunity. *Endocrine*. 42(2): 252–265.

Monograph. Selenium. *Alternative Medicine Review*, 2003, 8: 63–71.

Nacamulli, D., Mian, C., Petricca, D. *et al.,* 2010. Influence of physiological dietary selenium supplementation on the natural course of autoimmune thyroiditis. *Clinical Endocrinology (Oxford)*. 18: 535–539.

Negro, R., Greco, G., Mangieri, T., Pezzarossa, A., Dazzi, D. and Hassan, H., 2007. The influence of selenium supplementation on postpartum thyroid status in pregnant women with thyroid peroxidase autoantibodies. *Journal of Clinical and Endocrinology Metabolism*. 92: 1263–1268.

Nordio, M. and Pajalich, R., 2013. Combined treatment with myo-inositol and selenium ensures euthyroidism in subclinical hypothyroidism patients with autoimmune thyroiditis. *Journal of Thyroid Research.* DOI: 10.1155/2013/424163.

Onal, H., Keskindemirci, G., Adal, E., Ersen, A. and Korkmaz, O., 2012. Effects of selenium supplementation in the early stage of autoimmune thyroiditis in childhood: an open-label pilot study. *Journal of Pediatric Endocrinology and Metabolism.* 25(7–8): 639–644.

Prummel, M. F., Wiersinga, W. M., Schuetz, M., Wahl, K., Paul, M. and Kontur, S., 2004. Thyroid autoimmunity and miscarriage. *European Journal of Endocrinology.* 150: 751–755.

Prummel, M. F. and Wiersinga, W. M., 2005. Thyroid peroxidase autoantibodies in euthyroid subjects. *Best Practice & Research: Clinical Endocrinology & Metabolism.* 19(1): 1–15.

Rayman, M. P., 2008. Food-chain selenium and human health: emphasis on intake. *British Journal of Nutrition.* 100: 254–268.

Rayman, M. P., 2000. The importance of selenium to human health. *Lancet.* 356: 233–241.

Romaldini, J. H., Biancalana, M. M., Figueiredo, D. I., Farah, C. S. and Mathias, P. C., 1996. Effect of L-thyroxine administration on antithyroid antibody levels, lipid profile, and thyroid volume in patients with Hashimoto's thyroiditis. *Thyroid.* 6(3): 183–188.

Rostami, R., Aghasi, M. R., Mohammadi, A. and Nourooz-Zadeh, J., 2013. Enhanced oxidative stress in Hashimoto's thyroiditis: inter-relationships to biomarkers of thyroid function. *Clinical Biochemistry.* 46(4–5): 308–312.

Rotruck, J. T., Pope, A. L., Ganther, H. E., Swanson, A. B., Hafeman, D. G. and Hoekstra, W. G., 1973. Selenium: biochemical role as a component of glutathione peroxidase. *Science.* 179: 588–590.

Swain, M., Swain, T. and Mohanty, B. K., 2005. Autoimmune thyroid disorders – an update. *Indian Journal of Clinical Biochemistry.* 20(1): 9–17.

Toh, B. H., 2014. Diagnosis and classification of autoimmune gastritis. *Autoimmunity Reviews.* DOI: 10.1016/j.autrev.2014.01.048. [Epub ahead of print], pii: S1568-9972(14)00060-3.

Tomer, Y., 2010. Genetic susceptibility to autoimmune thyroid disease: past, present and future. *Thyroid.* 20: 715–725.

Turker, O., Kumanlioglu, K., Karapolat, I. and Dogan, I., 2006. Selenium treatment in autoimmune thyroiditis: 9-month follow-up with variable doses. *Journal of Endocrinology.* 190: 151–156.

Vanderpump, M. P. and Tunbridge, W. M., 2002. Epidemiology and prevention of clinical and subclinical hypothyroidism. *Thyroid.* 12(10): 839–847.

Vrca, V. B., Skreb, F., Cepelak, I., Romic, Z. and Mayer, L., 2004. Supplementation with antioxidants in the treatment of Graves' disease; the effect on glutathione peroxidase activity and concentration of selenium. *Clinica Chimica Acta.* 341(1–2): 55–63.

Zimmermann, M. B. and Köhrle, J., 2002. The impact of iron and selenium deficiencies on iodine and thyroid metabolism: biochemistry and relevance to public health. *Thyroid*. 12: 867–878.

Yoshida, A., Adachi, T., Noguchi, T., *et al.* 1985. Echographic findings and histological feature of the thyroid: a reverse relationship between the level of echo-amplitude and lymphocytic infiltration. *Endocrinologia Japonica*. 32(5): 681–690.

CHAPTER 27

Selenoproteins in Brain

ULRICH SCHWEIZER*[a] AND SANDRA SEEHER[a]

[a]Institut für Biochemie und Molekularbiologie, Rheinische Friedrich-Wilhelms-Universität Bonn, Nussallee 11, 53115 Bonn, Germany
*E-mail: uschweiz@uni-bonn.de

27.1 Selenoproteins in the Brain

27.1.1 Mice Deficient in Selenoprotein P Paved the Way to the Discovery that Selenium Is Essential for Brain Function

An essential role for selenoproteins in brain function was not evident until ten years ago. Although there was a suspicion that low selenium levels and resulting low cytosolic glutathione peroxidase levels may entail some sort of vulnerability towards traumatic, ischemic or toxic insults, if they are paralleled by the generation of reactive oxygen species, there was little evidence that brain selenium levels respond significantly to low selenium diets (Schweizer *et al.*, 2004a). The concept of brain protection by selenoproteins received initial support when the glutathione peroxidase 1 (*Gpx1*) gene was inactivated in mice. Although these mice did not normally suffer from brain damage, they turned out particularly sensitive towards ischemic and toxic insults of the brain (Schweizer *et al.*, 2004c). Thus, it was clear that the selenoprotein glutathione peroxidase does indeed play a protective role in the brain, although only upon above-mentioned challenges. However, brain Gpx1 activity does not appreciably drop during times of low dietary selenium intake.

Food and Nutritional Components in Focus No. 9
Selenium: Chemistry, Analysis, Function and Effects
Edited by Victor R. Preedy

Published by the Royal Society of Chemistry, www.rsc.org

When we inactivated the gene-encoding selenoprotein P (*Sepp*) in the mouse, we found that tissue selenium levels dropped even in the presence of adequate selenium in the diet (Schomburg *et al.*, 2003). Most importantly, $Sepp^{-/-}$ showed a drastic drop in brain selenium levels. Accordingly, selenoenzyme activities of Gpx1 and cytosolic thioredoxin reductase (Txnrd) were significantly reduced (Schomburg *et al.*, 2003). Thus, $Sepp^{-/-}$ mice represented the first animal model in which brain selenium and selenoenzyme levels could be significantly modulated. In parallel efforts, another group also inactivated the *Sepp* gene and reported similar findings (Hill *et al.*, 2003). Subsequent studies by both groups established that $Sepp^{-/-}$ mice suffer a number of neurological abnormalities when fed adequate or reduced dietary selenium levels, while the development of these neurological defects could be dose-dependently prevented by feeding the mice a high selenium diet or adding selenite into their drinking water (Hill *et al.*, 2004; Schweizer *et al.*, 2004b). The neurological phenotypes included epileptic seizures, a movement disorder with dystonic movements and a wide, shuffling gait in addition to a coordination defect as measured by the wire test or on the rotarod. Degeneration of axon tracts in the brain stem was also reported (Valentine *et al.*, 2008; Figure 27.1).

27.1.2 Selenoprotein P and Its Endocytic Receptors Rationalize a Model of Selenium Transport to the Brain

Studies on the expression pattern of SePP demonstrated high expression in liver and kidney and to a lower extent, in brain. Since SePP was initially found as the selenoprotein carrying most of plasma selenium, and because

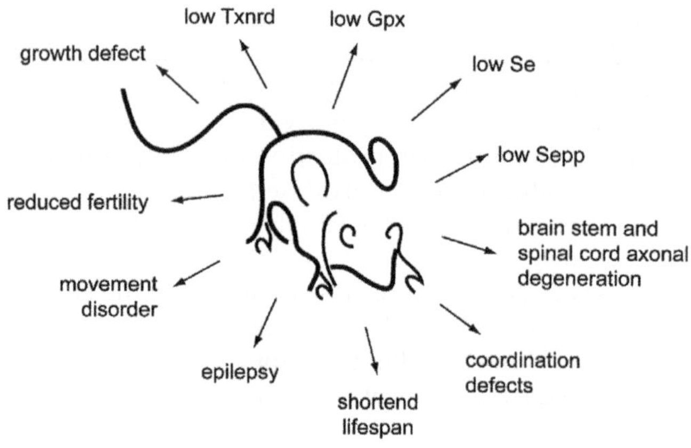

Figure 27.1 Phenotypes of selenoprotein P knockout mice. Deletion of the *Sepp* gene leads to disturbed selenium distribution as well as reduced selenoprotein expression and activity. *Sepp* knockout mice show growth defects, shortened lifespan and reduced fertility. Severe neuronal degeneration further leads to epileptic seizures, coordination defects, and movement disorder.

SePP contains 10–17 selenocysteine moieties per polypeptide chain depending on the species, it was regarded as a likely selenium-transport protein (Burk and Hill, 2005). We therefore inactivated SePP expression specifically in hepatocytes, thereby showing that plasma selenium levels are mainly determined by hepatic SePP expression (Schweizer *et al.*, 2005), a finding that was recently independently supported (Hill *et al.*, 2012). Moreover, we showed that liver-specific expression of SePP in *Sepp*$^{-/-}$ mice rescues plasma and brain selenium content and the neurological phenotype (Renko *et al.*, 2008). However, these studies also suggested that SePP expression in the brain likely contributes to maintaining brain selenium levels. In a landmark study, the apolipoprotein E receptor (ApoER2, Lrp8) was identified as a specific endocytic receptor of SePP (Olson *et al.*, 2007). Accordingly, *ApoER2*$^{-/-}$ mice suffer similar neurological damage as *Sepp*$^{-/-}$ mice (Burk *et al.*, 2007). Subsequently, megalin (Lrp2) was recognized as another SePP receptor expressed in kidney (Olson *et al.*, 2008) and we showed that inactivation of megalin in mice leads to urinary loss of selenium as a fragment of SePP that is apparently normally recovered along the renal proximal tubule epithelium *via* megalin (Chiu-Ugalde *et al.*, 2010). Moreover, movement coordination is moderately affected in megalin-deficient mice when fed a low-selenium diet (Chiu-Ugalde *et al.*, 2010). The current model of selenium transport in the mammalian body thus places SePP at center stage and expression of SePP receptors in different tissues, including the brain, determine a hierarchical distribution of selenium to organs (Schweizer, 2013). The brain contains relatively small amounts of selenium, but as detailed above, these are essential for its proper functioning. The brain is thus placed at the top of this hierarchy and any selenium entering the body under conditions of dietary selenium shortage are inserted into SePP in the liver and transported to the brain where ApoER2 and megalin mediate its uptake over the blood–brain barrier and into neurons. Selenium mobilized in other parts of the body may equally be distributed to the liver and sent to the brain. Biosynthesis of SePP in the kidney further supports this view. This mechanism is likely why feeding experimental animals with low selenium diets produced many peripheral defects, but apparently did not lead to brain defects that only became evident upon inactivation of SePP or ApoER2 (Figure 27.2).

27.1.3 Most Selenoproteins Are Expressed in Neurons

After establishing that selenium does play an essential role for brain function, it was important to define which selenoproteins are expressed in the brain and in which cell types. Since antibodies were only available for few selenoproteins and many of these were not suitable for immunohistochemical studies, several groups resorted to *in situ* hybridization to determine the expression patterns of selenoprotein mRNAs mostly in rodent brains. In a systematic bioinformatics approach, we used the recent first release of the Allen Brain Atlas and summarized the expression patterns of all selenoproteins in the mouse brain including several related genes (Zhang *et al.*, 2008).

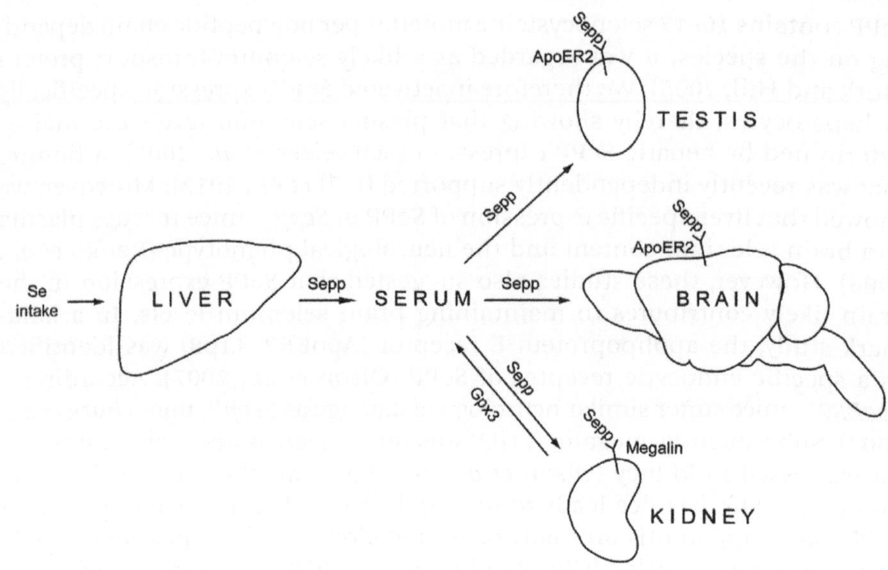

Figure 27.2 Model of selenium and selenoprotein P transport in mice. Se is taken up *via* the diet. In the liver, Se is incorporated as Sec into secreted Sepp and cellular selenoproteins. Sepp includes 10 Sec residues and therefore functions as Se storage and transport protein. It is distributed to the organs *via* the blood. The protein enters the cell by binding to specific endocytic receptors. The ApoER2 receptor is expressed in brain and testis, while the megalin receptor is expressed along the blood-brain barrier and in the kidneys. After proteolytic degradation, Sec/Se of Sepp is incorporated into new selenoproteins.

This study showed that most selenoproteins are expressed in the brain, selenoprotein expression is highest in neurons (with the exception of Sepp that was also localized in oligodendrocytes), and selenoprotein expression is particularly strong in several regions (*i.e.* cerebral cortex, cerebellar cortex, olfactory bulb, hippocampus). Studies following the developmental expression of selenoproteins demonstrated that most are expressed early on in the developing brain.

27.2 Functions of Selenoproteins in the Brain

The first, critical question regarding the role of selenium in the brain was whether selenium acts *via* selenoproteins or whether there is evidence for a selenoprotein-independent essential function of selenium in brain. We approached this question by targeted inactivation of selenoprotein biosynthesis in neurons by inactivation of the gene encoding $tRNA^{[Ser]Sec}$, *Trsp* (Wirth *et al.*, 2010). Since selenocysteine is synthesized attached to $tRNA^{[Ser]Sec}$ and because selenoprotein biosynthesis entirely depends on $tRNA^{[Ser]Sec}$, this approach selectively abrogated selenoprotein biosynthesis. As expanded below, the experiment suggested that essential selenium actions in the brain are mediated by selenoproteins.

27.2.1 Selenoproteins Are Essential for Neuronal Development and Survival

When cultured *in vitro*, neurons need protection from hyperoxia and profit from selenium supplementation (Roth *et al.*, 2010). Selenoprotein P has even been purified from plasma as a neurotrophic factor for cultured neurons (Kaufman and Barrett, 1983; Yan and Barrett, 1998). Selenium likely increases glutathione peroxidases in neurons and thus protects them from protein and lipid oxidation (Roth *et al.*, 2010). Inactivation of selenoprotein expression in cortical neurons greatly impairs their survival *in vivo* and *in vitro* (Wirth *et al.*, 2010). And lack of selenoprotein P or its receptor leads to axonal damage (Valentine *et al.*, 2008).

27.2.2 Inhibitory Interneurons of the Parvalbumin+ Class Are Most Sensitive to Impaired Selenoprotein Expression

An important question is which brain regions and brain-cell types are affected by selenoprotein deficiency. Hippocampus and cortical regions are likely target regions and epileptic seizures may originate there. Accordingly, we discovered in mice deficient in neuronal selenoprotein expression that a class of GABAergic inhibitory interneurons did not develop in the hippocampus and parts of the cerebral cortex (Wirth *et al.*, 2010). Interneurons are neurons that connect locally and do not project into other brain regions. Among the GABAergic inhibitory interneurons several classes exist and are histologically defined by expression of different neuropeptides or calcium-binding proteins. The three largest families of GABAergic cortical neurons are defined by expression of the small calcium-binding protein parvalbumin, by expression of the neurotransmitter neuropeptide Y, or somatostatin 14. These interneuron classes also differ in their developmental trajectory. In selenoprotein-deficient mice, specifically the parvalbumin-expressing interneurons fail to develop in the somatosensory cortex (Wirth *et al.*, 2010). Since these cells make up 60% of cortical GABAergic interneurons, their absence has significant consequences for brain function. Hyperexcitability or epilepsy as observed in mouse models with severely reduced numbers of parvalbumin+ interneurons is likely caused by their failure (Schomburg *et al.*, 2003; Schweizer *et al.*, 2004b; Seiler *et al.*, 2008; Carlson *et al.*, 2009; Wirth *et al.*, 2010). Interestingly, not all parvalbumin-expressing GABAergic neurons are affected by reduced selenoprotein expression, since respective cells in the globus pallidus or in several other brain regions are apparently normal. On the other hand, cerebellar Purkinje cells, which are parvalbumin-positive and GABAergic, degenerate in mice lacking neuronal selenoproteins (Schweizer and Schomburg, 2006). Interestingly, parvalbumin+

interneurons specifically express the SePP-receptor ApoER2 affording an explanation for their exquisite Se-dependence (Pitts *et al.*, 2012).

We have found that parvalbumin-positive interneurons are also affected in the cerebellar cortex, which is severely hypoplastic in mice lacking tRNA[Ser]Sec in neurons (Wirth and Schweizer, unpublished). Interestingly, Purkinje cells, a major cell type of the cerebellar cortex, are severely affected: these cells are GABAergic and express parvalbumin (Figure 27.3).

27.2.3 The Role of Glutathione Peroxidase 4 and Vitamin E

Glutathione peroxidase 4 is an essential selenoprotein in mice (Imai *et al.*, 2003; Yant *et al.*, 2003; Seiler *et al.*, 2008). Neuron-specific inactivation of Gpx4 leads to a phenotype similar to tRNA[Ser]Sec-inactivation in neurons, suggesting that lack of Gpx4 causes most of the damage in the tRNA knockout (Seiler *et al.*, 2008). Indeed, at least *in vitro*, vitamin E, a lipid-soluble peroxide

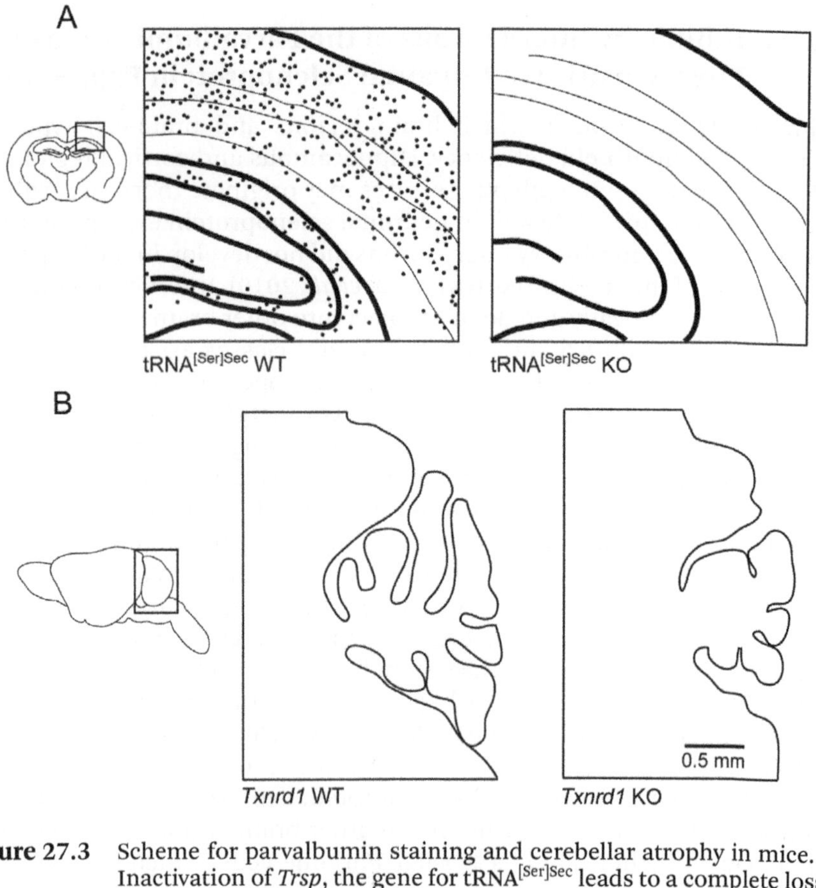

Figure 27.3 Scheme for parvalbumin staining and cerebellar atrophy in mice. (A) Inactivation of *Trsp*, the gene for tRNA[Ser]Sec leads to a complete loss of parvalbumin-positive interneurons (black dots in WT) in the knockout mice. (B) *Txnrd1* deletion causes severe cerebellar hypoplasia in mice.

scavenger, can rescue cultured *Gpx4*$^{-/-}$ neurons or cultured *Gpx4*$^{-/-}$ fibroblasts. In addition, vitamin E can also enhance survival of primary neurons in culture (Roth *et al.*, 2010).

27.2.4 Other Potentially Essential Selenoproteins

The only other selenoproteins that are known to be essential for embryonic development are the thioredoxin reductases (Txnrd). Inactivation of Txnrd1 in neurons leads to impaired movement coordination at advanced age (Figure 27.3). Inactivation of Txnrd1 in neural progenitors partially disrupts cerebellar development (Soerensen *et al.*, 2008). We have investigated parvalbumin-positive neuron numbers in several selenoprotein knockout mouse models (Sep15, Sepr, Sepk, Sepm, Sepn) without finding a phenotype (Figure 27.4). However, since the phenotype of neuron-specific tRNA$^{[Ser]Sec}$ knockouts is slightly more severe than that of *Gpx4*-knockouts, we assume that among the remaining selenoproteins which have not been knockout out yet, there may still be one that is essential for development – and possibly also important for brain function (Table 27.1).

Table 27.1 Phenotypes of Selenoprotein knockout mice.[b]

Selenoprotein	Brain phenotype
Selenoprotein P, Sepp	Complex
Gpx1	Only upon ischemic or toxic damage
Gpx2	None reported
Gpx3	None reported
Gpx4	Complex, neuronal survival, interneuron differentiation, retina
Txnrd1	Cerebellar development
Txnrd2	None reported
Txnrd3	Not done[a]
Dio1	None reported
Dio2	Endocrine, mild, retina, hearing
Dio3	Endocrine, interneurons, brain, hearing, retina
SelH, Seph	Not done
SelI, Sepi	Not done
SelK, Sepk	None reported
SelM, Sepm	None reported (Figure 27.4)
SelN, Sepn	None reported (Figure 27.4)
SelO, Sepo	Not done
SelR, MsrB1, Sepx	None reported
SelS, Seps, Vimp	Not done
SelT, Sept	None reported
SelV, Sepv	Not done
SelW, Sepw	Not done
Sel15, Sep15	None reported
Trsp hypomorphs	Complex, seizures, interneurons

[a]Some mouse models categorized as "not done" have been made, but no results are reported yet.
[b]Nomenclature is still inconsistent. While bacterial Sel-operon genes are denoted "SelA" *etc.*, human and mouse selenoproteins should be denoted "Sepp" (SelP could be confused with P-selectin).

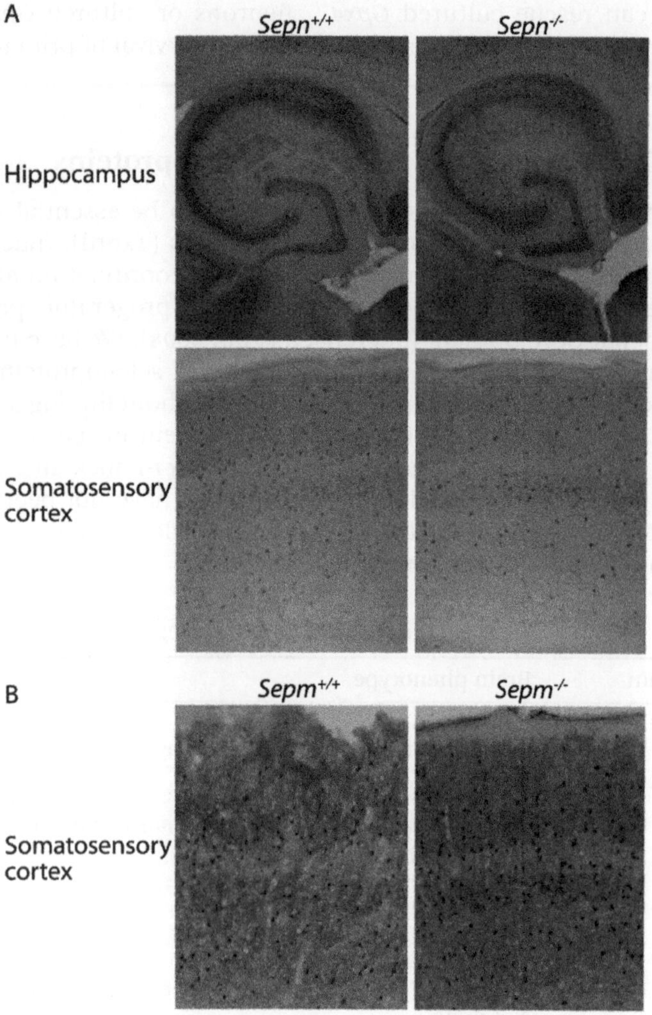

Figure 27.4 Immunohistochemical staining for parvalbumin in somatosensory cortex and hippocampus of mice. (A) The number of parvalbumin-positive interneurons is not reduced in *Sepn*$^{-/-}$ mice. (B) The number of parvalbumin-positive interneurons is not reduced in *Sepm*$^{-/-}$ mice.

27.3 Human Brain Disorders and Selenoproteins

Few reports clearly show that neurological disease in humans can be caused by lack of selenium. An interesting case report presented two children presenting with intractable seizures and low to undetectable plasma selenium levels (Ramaekers *et al.*, 1994). Both children responded positively to additional dietary selenium supplementation and it was concluded that the patients may suffer from an inborn error of selenium metabolism. This report

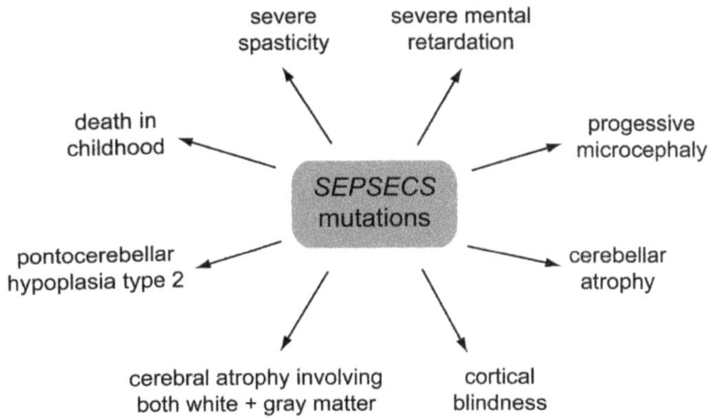

Figure 27.5 *SEPSECS* mutations cause PCCA in humans. The disease is characterized by cerebral and cerebellar atrophy and severe mental retardation. Affected children do not survive into childhood.

nicely complemented an earlier paper that also found an association of childhood seizures associated with low plasma Gpx activity (Weber *et al.*, 1991). At the time the earlier report was not received well, since a defect in Gpx activity did not appear to cause childhood seizures. In hindsight, the low plasma selenium is likely a result of a biosynthetic defect of selenoprotein expression – be it limited to SePP, its receptors or a general defect of selenoprotein deficiency. The phenotypic similarity with $Sepp^{-/-}$ mice is striking, to say the least.

27.3.1 Mutations in Selenocysteine Synthase in Humans Lead to Progressive Cerebello Cerebral Atrophy

Shortly after the identification of the gene-encoding selenocysteine synthase (*SEPSECS*), patients with autosomal recessive mutations in *SEPSECS* were identified (Agamy *et al.*, 2010). Affected children are diagnosed with Progressive Cerebello Cerebral Atrophy (PCCA) (Ben Zeev *et al.*, 2003). The patients suffer an ultimately fatal neurodevelopmental condition that also involves neurodegenerative aspects. Among other aspects, epilepsy and cerebellar hypoplasia are symptoms that also occur in mouse models with deficient selenoprotein expression. This report finally shows that the phenotypes identified earlier in transgenic mouse models are similarly found in the later discovered human syndrome (Figure 27.5).

27.3.2 Mutations in the *SECISBP2* Gene in Humans Lead to a Multisystem Disorder also Involving the Central Nervous System

The first mutations in *SECISBP2* that were identified in patients resulted in a rather mild syndrome of delayed growth and bone maturation (Dumitrescu *et al.*, 2005). The patients presented with high plasma thyroxine (T_4)

and inappropriately low plasma T_3, a constellation that likely underlies the growth phenotype. The unusual thyroid hormone constellation initially suggested defective activation of T_4 to T_3 through the action of deiodinases. Further analysis revealed that the deiodinase genes were not affected, but selenoprotein translation in general. The mutation in *SECISBP2* apparently renders translation of selenoprotein transcripts inefficient, affecting deiodinases, but also plasma selenoprotein SePP and GPx3 (Dumitrescu *et al.*, 2005). Neurological defects were not diagnosed in the first affected patients. Later reports on other patients affected by different mutations in *SECISBP2* demonstrated abnormal gait and possibly also mild neuropsychiatric findings (Azevedo *et al.*, 2010). More complete functional inactivity of SECISBP2 can reduce deiodinase 2 expression to a level that is insufficient to sustain normal development of the inner ear resulting in hearing impairment. Another patient with *SECISBP2* mutations exhibits a complex syndrome in addition to the hearing impairment (Schoenmakers *et al.*, 2010). This patient, however, does not show neurologic deficits. Taken together, *SECISBP2* mutations leading to a partial impairment of selenoprotein expression invariably involve abnormal thyroid hormone levels and may include additional symptoms, however, neuropsychiatric findings are not common in this syndrome.

27.3.3 Evidence that Selenoproteins Are Related to Major Neurodegenerative Disorders

So far, selenoprotein deficiency has not been causally linked to any one of the major neurodegenerative disorders. Clearly, "oxidative stress", *i.e.* increased levels of oxidative damage to cellular components like protein, DNA, and lipids, has been documented, for example, in Alzheimer's disease, Parkinson's disease, Huntington's disease, and amyotrophic lateral sclerosis (ALS). Since many selenoproteins reduce peroxides or oxidized proteins (*e.g.* glutathione peroxidases, methionine-sulfoxide reductase, thioredoxin reductases), it seems possible that their absence may enhance tissue damage in these diseases (Schweizer *et al.*, 2004a; Yoo *et al.*, 2010). Yet the known monogenetic forms of these diseases are not associated with selenoproteins. A unifying theme in age-related neurodegenerative disorders may be aberrant protein folding, as evidenced by formation of protein aggregates that are diagnostic hallmarks of neurodegenerative diseases like Alzheimer's (β-amyloid and neurofibrillary tangles) and Parkinson's (Lewy bodies). Dysfunction of protein folding pathways likely underlie formation of these aggregates (Kim *et al.*, 2013). Here, selenoproteins may as well modulate pathology, since several selenoproteins have been implicated in protein folding or quality control of unfolded proteins (*e.g.* SepS, SepM, Sep15). Again, no direct link has yet been shown that lack of selenoproteins impairs protein folding and thus causes protein aggregation.

Attempts have been made to correlate plasma or tissue selenium levels with neurodegenerative diseases or to reduce the incidence of age-related memory decline with selenium supplementation. Results have been inconclusive (Lopes da Silva *et al.*, 2013), possibly because brain selenium levels are independent of plasma selenium and only inborn errors of selenoprotein transport or biosynthesis may impact on brain selenoprotein levels. Their effects, however, are quite severe (*e.g.* PCCA) and may therefore affect the young. Most studies finding differences also were included small proband numbers and are likely underpowered. Selenoprotein P has been found associated with protein aggregates in brains of Alzheimer's disease patients (Bellinger *et al.*, 2008). Since SePP is expressed in many neurons, along axons, and is present in cerebrospinal fluid (Scharpf *et al.*, 2007), it is not clear whether this association is specific and related to pathology. Whether its reduction in demyelinating lesions in multiple sclerosis points to a role of selenoprotein P in the disease process is not known (Figure 27.6).

A prominent selenium intervention study in a genetic model of Alzheimer's disease demonstrated impressive improvements by selenate supplementation

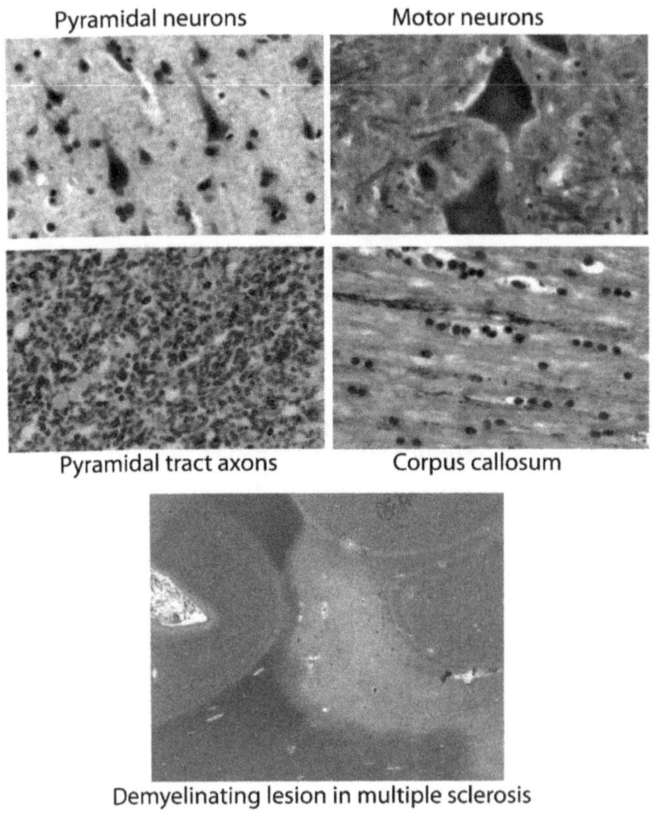

Figure 27.6 Immunohistochemical staining against selenoprotein P in the human brain. SePP expression on neurons, axons, and myelin sheaths. SePP expression is lost in demyelinating lesions in multiple sclerosis (M. Scharpf and U. Schweizer, unpublished).

(van Eersel *et al.*, 2010). However, selenate was suggested to act *via* stimulation of protein phosphatase 2 thereby reducing tau phosphorylation. The study did not show an effect on selenoprotein expression.

27.4 Conclusions

Many selenoproteins are expressed in the central nervous system. Several selenoproteins are essential for brain function, while other, highly expressed selenoproteins can be knocked out in mice without noticeable neurological defects. The presence of selenoproteins in neurons and white matter throughout the brain makes it tempting to assign specific roles to them in any disease process. The most specific effects of lack of selenoproteins (most likely of Gpx4) is impaired development and function of specific classes of GABAergic interneurons in somatosensory cortex, hippocampus, and cerebellar cortex. It thus seems prudent to focus on diseases in which these neurons are affected, *i.e.* schizophrenia and autism spectrum disorders.

Summary Points

- Mice deficient in selenoprotein P paved the way to the discovery that selenium is essential for brain function.
- Selenoprotein P and its endocytic receptors rationalize a model of selenium transport to the brain.
- Most selenoproteins are expressed in neurons.
- Selenoproteins are essential for neuronal development and survival.
- Inhibitory interneurons of the parvalbumin+ class are most sensitive to impaired selenoprotein expression.
- The role of glutathione peroxidase 4 and vitamin E.
- Other potentially essential selenoproteins.
- Mutations in selenocysteine synthase (*SEPSECS*) in humans lead to PCCA.
- Mutations in the *SECISBP2* gene in humans lead to a multisystem disorder also involving the central nervous system.
- There is evidence that selenoproteins are related to major neurodegenerative disorders.

Key Facts: Selenoprotein Biosynthesis

1. A UGA codon in the open reading frame is recoded as selenocysteine (Sec).
2. Sec is synthesized attached to its tRNA by selenocysteine synthase (SEPSECS).
3. The Sec insertion machinery involves a SECIS (selenocysteine insertion sequence) element, a stem-loop structure in selenoprotein mRNAs.
4. SECIS elements are located in the 3′-untranslated region and may be hundreds to thousands of bases away from the UGA codon.

5. The SECIS is recognized by the SECIS-binding protein 2 (SECISBP2).
6. SECISBP2 further interacts with EFSec, the Sec-specific translation elongation factor to promote Sec insertion instead of termination.

Key Facts: Essential Selenoproteins and Biosynthetic Cofactors for the Brain

1. Selenoprotein P is the selenium transport protein, which brings selenium from liver *via* plasma to brain. It is also produced within the brain.
2. ApoER2 and megalin are selenoprotein P endocytic receptors.
3. Glutathione peroxidase 4 is essential for neurons. Its absence reduces the number of parvalbumin-positive interneurons in hippocampus and some cortical regions.
4. Thioredoxin reductase 1 deficiency in neural precursors impacts on cerebellar development, while neuron-specific lack of this protein leads to age-related impairment of movement coordination.
5. Lack of $tRNA^{[Ser]Sec}$ or expression of a hypomorphic variant reduces selenoprotein expression, including glutathione peroxidase 4, and reduces the number of parvalbumin-positive interneurons.
6. Mutation of selenocysteine synthase causes progressive cerebello cerebral atrophy.

Key Facts: Mutations in Genes Affecting Selenoprotein Expression

1. Mutations in the selenoprotein N gene (*SEPN1*) were the first congenital defects of selenoprotein expression discovered.
2. Defects in selenoprotein N expression cause a type of muscular dystrophy.
3. Pathogenic mutations in *SEPN1* not only affect the open reading frame, but also the SRE and SECIS elements.
4. Mutations in the *SECISBP2* gene have been identified in patients from many different ethnic backgrounds and affect selenoprotein expression in a graded manner.
5. *SECISBP2* mutations lead to a multisystem disorder with symptoms ranging from retardation of growth and bone development, bilateral hearing loss, myopathy, azoospermia to photosensitivity.
6. Abnormalities in thyroid hormone levels (high T_4, low T_3, normal or slightly elevated TSH) are found in all affected patients and likely underlie many patient phenotypes.
7. Mutations in selenocysteine synthase cause the neurodegenerative disorder progressive cerebello cerebral atrophy (PCCA).
8. Patients afflicted with PCCA suffer a neurodevelopmental and neurodegenerative disease in infancy, which is fatal.
9. Symptoms of PCCA include epilepsy, blindness, cortical atrophy, cerebellar hypoplasia.

Definitions

3′-Untranslated region. Part of the mRNA downstream of the translational termination codon. In this region eukaryotic SECIS elements are located.

Elongation factor EFSec. A translation elongation factor specific for $tRNA^{[Ser]Sec}$ analogous to EF-1α.

Glutathione peroxidase (GPx). A selenoenzyme that reduces hydroperoxides using glutathione as electron donor. Glutathione peroxidase 4 is specific for lipid hydroperoxides.

Hypomorph. A term from genetics. It refers to a mutant (or its carrier) that is less active than the wildtype, but has not lost all activity.

Interneurons. Locally projecting neurons. Most inhibitory interneurons use γ-amino butyric acid (GABA) as a neurotransmitter.

Open reading frame is the protein coding part of a mRNA.

Multiminicore disease is one designation of myopathy caused by congenital defects in selenoprotein N expression.

Parvalbumin-positive interneurons express the small Ca^{2+}-binding protein parvalbumin. Among the GABAergic interneurons, this is the most abundant class of interneurons in the cortex. This class of interneurons is selectively affected by impaired selenoprotein expression.

Progressive cerebello cerebral atrophy is a neurodegenerative disorder based on mutations in the *SEPSECS* gene-encoding selenophosphate synthetase.

Selenoproteins are proteins containing the rare amino acid selenocysteine (Sec).

Selenoprotein P is a plasma selenoprotein and represents the major selenium transport protein in the body. It binds specifically to endocytic receptors establishing a hierarchy of selenium transport in the body.

SECIS elements are stem-loop structures in selenoprotein mRNAs that direct Sec insertion at UGA.

Thioredoxin reductase is a selenoenzyme in mammals. Reduced thioredoxin maintains the reduced state of enzymes and redox-regulated cytosolic or mitochondrial proteins.

$tRNA^{[Ser]Sec}$ is the tRNA that decodes UGA and brings Sec to the ribosome. It is initially charged with Ser and conversion to Sec occurs on the tRNA.

List of Abbreviations

ALS	Amyotrophic lateral sclerosis
ApoER2	Apolipoprotein E receptor
DNA	Desoxyribonucleic acid
EFSec	Translation elongation factor specific for $tRNA^{Sec}$
GABA	γ-Amino butyric acid
Gpx	Glutathione peroxidase
Lrp	Low-density lipoprotein receptor-related protein
PCCA	Progressive cerebello cerebral atrophy

Se	Selenium
Sec	Selenocysteine
SECIS	Selenocysteine insertion sequence
SECISBP2	SECIS-binding protein 2
SelX	Selenoprotein X (some instances) or *E. coli* gene in *Sel* operon
SEPSECS	Selenocysteine synthase; *O*-phosphoseryl-tRNA(Sec) selenium transferase
SepX	Selenoprotein X
SRE	Selenocysteine redefinition element
T3	Triiodothyronine; 3,3′,5-triiodo-L-thyronine
T4	Thyroxine; 3,3′,5,5′-tetraiodo-L-thyronine
tRNA$^{[Ser]Sec}$	Selenocysteine-specific transfer RNA
Trsp	Gene-encoding tRNA$^{[Ser]Sec}$
TSH	Thyroid-stimulating hormone; thyrotropin
Txnrd	Thioredoxin reductase

References

Agamy, O., Ben Zeev, B., Lev, D., Marcus, B., Fine, D., Su, D., Narkis, G., Ofir, R., Hoffmann, C., Leshinsky-Silver, E., *et al.*, 2010. Mutations disrupting selenocysteine formation cause progressive cerebello-cerebral atrophy. *American Journal of Human Genetics*. 87: 538–544.

Azevedo, M. F., Barra, G. B., Naves, L. A., Ribeiro Velasco, L. F., Godoy Garcia, C. P., de Castro, L. C., Amato, A. A., Miniard, A., Driscoll, D., Schomburg, L., *et al.*, 2010. Selenoprotein-related disease in a young girl caused by nonsense mutations in the SBP2 gene. *Journal of Clinical Endocrinology and Metabolism*. 95: 4066–4071.

Bellinger, F. P., He, Q. P., Bellinger, M. T., Lin, Y., Raman, A. V., White, L. R. and Berry, M. J., 2008. Association of selenoprotein P with Alzheimer's pathology in human cortex. *Journal of Alzheimer's Disease*. 15: 465–472.

Ben Zeev, B., Hoffman, C., Lev, D., Watemberg, N., Malinger, G., Brand, N. and Lerman-Sagie, T., 2003. Progressive cerebellocerebral atrophy: a new syndrome with microcephaly, mental retardation, and spastic quadriplegia. *Journal of Medical Genetics*. 40: e96.

Burk, R. F. and Hill, K. E., 2005. Selenoprotein P: an extracellular protein with unique physical characteristics and a role in selenium homeostasis. *Annual Review of Nutrition*. 25: 215–235.

Burk, R. F., Hill, K. E., Olson, G. E., Weeber, E. J., Motley, A. K., Winfrey, V. P. and Austin, L. M., 2007. Deletion of apolipoprotein E receptor-2 in mice lowers brain selenium and causes severe neurological dysfunction and death when a low-selenium diet is fed. *Journal of Neuroscience*. 27: 6207–6211.

Carlson, B. A., Schweizer, U., Perella, C., Shrimali, R. K., Feigenbaum, L., Shen, L., Speransky, S., Floss, T., Jeong, S. J., Watts, J., *et al.*, 2009. The selenocysteine tRNA STAF-binding region is essential for adequate selenocysteine tRNA status, selenoprotein expression and early age survival of mice. *Biochemical Journal*. 418: 61–71.

Chiu-Ugalde, J., Theilig, F., Behrends, T., Drebes, J., Sieland, C., Subbarayal, P., Köhrle, J., Hammes, A., Schomburg, L. and Schweizer, U., 2010. Mutation of megalin leads to urinary loss of selenoprotein P and selenium deficiency in serum, liver, kidneys and brain. *Biochemical Journal*. 431: 103–111.
Dumitrescu, A. M., Liao, X. H., Abdullah, M. S., Lado-Abeal, J., Majed, F. A., Moeller, L. C., Boran, G., Schomburg, L., Weiss, R. E. and Refetoff, S., 2005. Mutations in SECISBP2 result in abnormal thyroid hormone metabolism. *Nature Genetics*. 37: 1247–1252.
van Eersel, J., Ke, Y. D., Liu, X., Delerue, F., Kril, J. J., Gotz, J. and Ittner, L. M., 2010. Sodium selenate mitigates tau pathology, neurodegeneration, and functional deficits in Alzheimer's disease models. *Proceedings of the National Academy of Sciences of the United States of America*. 107: 13888–13893.
Hill, K. E., Wu, S., Motley, A. K., Stevenson, T. D., Winfrey, V. P., Capecchi, M. R., Atkins, J. F. and Burk, R. F., 2012. Production of selenoprotein P (Sepp1) by hepatocytes is central to selenium homeostasis. *Journal of Biological Chemistry*. 287(48): 40414–40424.
Hill, K. E., Zhou, J., McMahan, W. J., Motley, A. K., Atkins, J. F., Gesteland, R. F. and Burk, R. F., 2003. Deletion of selenoprotein P alters distribution of selenium in the mouse. *Journal of Biological Chemistry*. 278: 13640–13646.
Hill, K. E., Zhou, J., McMahan, W. J., Motley, A. K. and Burk, R. F., 2004. Neurological dysfunction occurs in mice with targeted deletion of the selenoprotein p gene. *Journal of Nutrition*. 134: 157–161.
Imai, H., Hirao, F., Sakamoto, T., Sekine, K., Mizukura, Y., Saito, M., Kitamoto, T., Hayasaka, M., Hanaoka, K. and Nakagawa, Y., 2003. Early embryonic lethality caused by targeted disruption of the mouse PHGPx gene. *Biochemical and Biophysical Research Communications*. 305: 278–286.
Kaufman, L. M. and Barrett, J. N., 1983. Serum factor supporting long-term survival of rat central neurons in culture. *Science*. 220: 1394–1396.
Kim, Y. E., Hipp, M. S., Bracher, A., Hayer-Hartl, M. and Hartl, F. U., 2013. Molecular chaperone functions in protein folding and proteostasis. *Annual Review of Biochemistry*. 82: 323–355.
Lopes da Silva, S., Vellas, B., Elemans, S., Luchsinger, J., Kamphuis, P., Yaffe, K., Sijben, J., Groenendijk, M. and Stijnen, T., 2013. Plasma nutrient status of patients with Alzheimer's disease: systematic review and meta-analysis. *Alzheimer's & Dementia*. 10(4): 485–502.
Olson, G. E., Winfrey, V. P., Hill, K. E. and Burk, R. F., 2008. Megalin mediates selenoprotein P uptake by kidney proximal tubule epithelial cells. *Journal of Biological Chemistry*. 283: 6854–6860.
Olson, G. E., Winfrey, V. P., Nagdas, S. K., Hill, K. E. and Burk, R. F., 2007. Apolipoprotein E receptor-2 (ApoER2) mediates selenium uptake from selenoprotein P by the mouse testis. *Journal of Biological Chemistry*. 282: 12290–12297.
Pitts, M. W., Raman, A. V., Hashimoto, A. C., Todorovic, C., Nichols, R. A. and Berry, M. J., 2012. Deletion of selenoprotein P results in impaired function of parvalbumin interneurons and alterations in fear learning and sensorimotor gating. *Neuroscience*. 208: 58–68.

Ramaekers, V. T., Calomme, M., Vanden Berghe, D. and Makropoulos, W., 1994. Selenium deficiency triggering intractable seizures. *Neuropediatrics*. 25: 217–223.

Renko, K., Werner, M., Renner-Muller, I., Cooper, T. G., Yeung, C. H., Hollenbach, B., Scharpf, M., Köhrle, J., Schomburg, L. and Schweizer, U., 2008. Hepatic selenoprotein P (SePP) expression restores selenium transport and prevents infertility and motor-incoordination in Sepp-knockout mice. *Biochemical Journal*. 409: 741–749.

Roth, S., Zhang, S., Chiu, J., Wirth, E. K. and Schweizer, U., 2010. Development of a serum-free supplement for primary neuron culture reveals the interplay of selenium and vitamin E in neuronal survival. *Journal of Trace Elements in Medicine and Biology*. 24: 130–137.

Scharpf, M., Schweizer, U., Arzberger, T., Roggendorf, W., Schomburg, L. and Köhrle, J., 2007. Neuronal and ependymal expression of selenoprotein P in the human brain. *Journal of Neural Transmission*. 114: 877–884.

Schoenmakers, E., Agostini, M., Mitchell, C., Schoenmakers, N., Papp, L., Rajanayagam, O., Padidela, R., Ceron-Gutierrez, L., Doffinger, R., Prevosto, C., *et al.*, 2010. Mutations in the selenocysteine insertion sequence-binding protein 2 gene lead to a multisystem selenoprotein deficiency disorder in humans. *Journal of Clinical Investigation*. 120: 4220–4235.

Schomburg, L., Schweizer, U., Holtmann, B., Flohé, L., Sendtner, M. and Köhrle, J., 2003. Gene disruption discloses role of selenoprotein P in selenium delivery to target tissues. *Biochemical Journal*. 370: 397–402.

Schweizer, U., 2013. Selenoprotein P. In: Kretsinger, R. H. U., Vladimir, N. and Permyakov, E. A., (ed.) *Encyclopedia of Metalloproteins*. Springer, New York, pp. 1941–1948.

Schweizer, U., Bräuer, A. U., Köhrle, J., Nitsch, R. and Savaskan, N. E., 2004a. Selenium and brain function: a poorly recognized liaison. *Brain Research Review*. 45: 164–178.

Schweizer, U., Michaelis, M., Köhrle, J. and Schomburg, L., 2004b. Efficient selenium transfer from mother to offspring in selenoprotein-P-deficient mice enables dose-dependent rescue of phenotypes associated with selenium deficiency. *Biochemical Journal*. 378: 21–26.

Schweizer, U., Schomburg, L. and Savaskan, N. E., 2004c. The neurobiology of selenium: lessons from transgenic mice. *Journal of Nutrition*. 134: 707–710.

Schweizer, U. and Schomburg, L., 2006. Selenium, selenoproteins, and brain function. In: Hatfield, D. L., Berry, M. J. and Gladyshev, V. N. (ed.) *Selenium: Its Molecular Biology and Role in Human Health*. Springer, New York, pp. 233–248.

Schweizer, U., Streckfuss, F., Pelt, P., Carlson, B. A., Hatfield, D. L., Köhrle, J. and Schomburg, L., 2005. Hepatically derived selenoprotein P is a key factor for kidney but not for brain selenium supply. *Biochemical Journal*. 386: 221–226.

Seiler, A., Schneider, M., Forster, H., Roth, S., Wirth, E. K., Culmsee, C., Plesnila, N., Kremmer, E., Radmark, O., Wurst, W., *et al.*, 2008. Glutathione peroxidase 4 senses and translates oxidative stress into 12/15-lipoxygenase dependent- and AIF-mediated cell death. *Cell Metabolism*. 8: 237–248.

Soerensen, J., Jakupoglu, C., Beck, H., Forster, H., Schmidt, J., Schmahl, W., Schweizer, U., Conrad, M. and Brielmeier, M., 2008. The role of thioredoxin reductases in brain development. *PLoS One*. 3: e1813.

Valentine, W. M., Abel, T. W., Hill, K. E., Austin, L. M. and Burk, R. F., 2008. Neurodegeneration in mice resulting from loss of functional selenoprotein P or its receptor apolipoprotein E receptor 2. *Journal of Neuropathology & Experimental Neurology*. 67: 68–77.

Weber, G. F., Maertens, P., Meng, X. Z. and Pippenger, C. E., 1991. Glutathione peroxidase deficiency and childhood seizures. *Lancet*. 337: 1443–1444.

Wirth, E. K., Conrad, M., Winterer, J., Wozny, C., Carlson, B. A., Roth, S., Schmitz, D., Bornkamm, G. W., Coppola, V., Tessarollo, L., *et al.*, 2010. Neuronal selenoprotein expression is required for interneuron development and prevents seizures and neurodegeneration. *FASEB Journal*. 24: 844–852.

Yan, J. and Barrett, J. N., 1998. Purification from bovine serum of a survival-promoting factor for cultured central neurons and its identification as selenoprotein-P. *Journal of Neuroscience*. 18: 8682–8691.

Yant, L. J., Ran, Q., Rao, L., Van Remmen, H., Shibatani, T., Belter, J. G., Motta, L., Richardson, A. and Prolla, T. A., 2003. The selenoprotein GPX4 is essential for mouse development and protects from radiation and oxidative damage insults. *Free RadicalBiology and Medicine*. 34: 496–502.

Yoo, M. H., Gu, X., Xu, X. M., Kim, J. Y., Carlson, B. A., Patterson, A. D., Cai, H., Gladyshev, V. N. and Hatfield, D. L., 2010. Delineating the role of glutathione peroxidase 4 in protecting cells against lipid hydroperoxide damage and in Alzheimer's disease. *Antioxidants and Redox Signaling*. 12: 819–827.

Zhang, Y., Zhou, Y., Schweizer, U., Savaskan, N. E., Hua, D., Kipnis, J., Hatfield, D. L. and Gladyshev, V. N., 2008. Comparative analysis of selenocysteine machinery and selenoproteome gene expression in mouse brain identifies neurons as key functional sites of selenium in mammals. *Journal of Biological Chemistry*. 283: 2427–2438.

CHAPTER 28

Selenoprotein H in Neuronal Cells

P. ANDY LI*[a], SURESH L. MEHTA[b], AND LI JING[c]

[a]Department of Pharmaceutical Sciences, Biomanufacturing Research Institute and Technological Enterprise (BRITE), North Carolina Central University, BRITE Building 2025, 1801 Fayetteville Street, Durham, NC 27707, USA; [b]Department of Neurosurgery, University of Wisconsin, School of Medicine and Public Health, 600 Highland Avenue, K4/870 CSC, Madison, WI 53792-8660, USA; [c]Department of Pathology, Ningxia Key Laboratory of Cerebrocranial Diseases, Incubation Base of National Key Laboratory, Ningxia Medical University, Yinchuan, China, 75004
*E-mail: pli@nccu.edu

28.1 Selenoprotein Biosynthesis

Like most trace elements, selenium functions through proteins of which it is a constituent. However, selenium is unique in that it is incorporated as the amino acid selenocysteine (Sec), the 21st amino acid. Sec is cotranslationally incorporated into selenoproteins at UGA codons that normally function as stop codons in most mRNAs (Bellinger *et al.*, 2009). Sec has a unique tRNA, tRNAsec, which is initially charged with serine. Selenium is phosphorylated by selenophosphate synthetase (SPS) and then added directly to the serine on tRNAsec by selenocysteine synthetase to produce Sec. The selenocysteine insertion sequence (SECIS) recodes the UGA stop codon for Sec insertion *via* recruitment of SECIS-binding protein 2 (SBP2). SBP2 in turn recruits the

Food and Nutritional Components in Focus No. 9
Selenium: Chemistry, Analysis, Function and Effects
Edited by Victor R. Preedy

Published by the Royal Society of Chemistry, www.rsc.org

Sec tRNA specific elongation factor (EFsec) to bind to tRNAsec, which then catalyzes selenocysteine incorporation at the ribosome. Among the 25 selenoproteins so far identified in the human genome, some of them, such as glutathione peroxidases (GPXs), thioredoxin reductases (TrxRs), phospholipid hydroperoxide glutathione peroxidase (PHGPX) and selenoprotein P (SelP), function as antioxidants, heavy-metal antidotes, and neural survival factors (for reviews, see Bellinger *et al.*, 2009; Reeves and Hoffmann, 2009). However, the functions of many recently identified selenoproteins including selenoprotein H (SelH) are still poorly understood.

28.2 The Functions of Selenoprotein H

SelH is a 14 kD protein located in most organs but predominantly in the brain and spleen (Novoselov *et al.*, 2007). SelH is one of the 25 so-far identified selenoproteins in humans. Like many selenoproteins, SelH contains a CXXU thioredoxin-like motif (where cysteine and Sec are separated by two other amino acids), indicative of a redox active protein. A significant subset of the mammalian selenoproteome has been associated with cellular detoxification functions (Yan and Barrett, 1998). Studies have shown that downregulation of *SelH* increases the sensitivity of mouse lung cancer LCC1 cells challenged by hydrogen peroxide (Novoselov *et al.*, 2007) and decreases cell viability and antioxidant concentrations in *Drosophila* (Morozova *et al.*, 2003). In contrast, upregulation of *SelH* results in increased levels of glutathione, total antioxidant capacity, and the mRNAs for several glutathione *S*-transferases (GSTs) in murine hippocampal neuronal HT22 cells treated with the glutathione depleting agent L-buthionine(*S*,*R*)-sulfoximine (Panee *et al.*, 2007). Overexpression of *SelH* protects cells against ultraviolet B (UVB) irradiation-induced damage by ameliorating superoxide production (Ben Jilani *et al.*, 2007).

Most interestingly, recent studies have suggested that SelH is a nuclear-localized DNA-binding protein that may exhibit transcription factor-like activity regulating genes involved in antioxidant protection and cellular detoxification (Novoselov *et al.*, 2007; Panee *et al.*, 2007). Transfection of SelH into HT22 cells increased the mRNA levels of glutamylcysteine synthetase (GCS), a rate-limiting enzyme for *de novo* GSH synthesis, and glutathione *S*-transferase-α2, -α4, and -ω1 subfamilies. Four putative nuclear localization signals, beginning at amino acids 3, 7, 71, and 101 in hSelH were identified using a protein subcellular localization prediction (PSORT) algorithm, and nuclear localization was demonstrated with GFP-tagged SelH. Using the Expert Protein Analysis System (ExPASy) analysis tool and alignment with ~380 known members of the high mobility group (HMG) family of DNA-binding proteins, Berry's group has identified a highly conserved AT-hook in SelH (Panee *et al.*, 2007). The AT hook is a DNA-binding protein motif consisting of a semiconserved 9 amino acid sequence. The nuclear DNA-binding property of SelH was further demonstrated by chromatin immunoprecipitation (ChIP) that reveals SelH binding to heat-shock elements (HSEs) and stress-response elements (STREs) (Panee *et al.*, 2007; Stoytcheva and Berry, 2009). *In silico* analysis of the genomic sequences of selenoproteins by the Berry lab revealed the presence of putative CREB, STAT, NF-κB and other nuclear transcription factor binding sites in the SelH promoter region (Stoytcheva and Berry, 2009).

28.3 Mitochondrial Protective Effects of Selenoprotein H

The mitochondrion is a key subcellular organelle responsible for sustaining life by producing energy through oxidative phosphorylation. During mitochondrial oxidation, small amount of free radicals are produced due to incomplete reduction of oxygen by electrons. Under physiological condition, these free radicals are degraded by superoxide dismutases, glutathione peroxidases and catalase into water and carbon dioxide. In stress condition, however, free radicals accumulate in the mitochondria either because of overproduction or reduced clearance. Mitochondrial accumulation of free radicals, together with mitochondrial calcium overload, can trigger the mitochondrial membrane to form a permeability transition pore, commonly referred to as MPTP, MTP, or MPT. The components of this mitochondrial megapore include cyclophilin D and adenine nucleotide translocase (ANT). The formation of MTPT allows mitochondrial proteins such as cytochrome c, small mitochondria derived activator of caspases (Smac/DIABLO), endonuclease G (EndoG), and apoptosis-inducing factor (AIF) to be released. Release of cytochrome c from the mitochondria to the cytosol forms an apoptosoma with procaspase-9 and apoptotic protease activating factor 1 (apaf-1), which subsequently activates caspase-9 and caspase-3. Caspase-3 can either directly translocate to the nucleus or causes DNA cleavage through activation of caspase-dependent DNase (CAD). Release of Smac/DIABLO to the cytosol directly binds and deactivates the inhibitor of apoptosis protein (IAP), a negative regulator of programmed cell death. Translocation of Endo G and AIF to nucleus causes chromosome and nuclear membrane damage. Since both cytochrome c and Smac/DIABLO induce apoptosis through the activation of caspases, these processes are also referred to as mitochondria-initiated, caspase-dependent cell-death pathways. The Endo G and AIF-mediated cell-death processes are referred as mitochondria-initiated, caspase-independent cell-death pathways.

Overexpression of SelH in murine hippocampal neuronal cell line HT22 cells blocks UVB-irradiation induced activation of caspase-3 and cell death (Mendelev *et al.*, 2009). Examination of caspase-3 upstream proteases reveals that SelH inhibits the cleavage of caspase-9 but not that of capase-8. Since caspase-8 mediates the mitochondria-independent cell-death pathway, these results suggest that SelH exerts its cellular protective function through blocking mitochondria-initiated cell-death pathway, but not the receptor-initiated pathway. Overexpression of SelH also stabilizes mitochondrial membrane potential and prevents UV stress-induced mitochondrial membrane depolarization (Figure 28.1), a marker of the MPTP formation (Mendelev *et al.*, 2009). It remains unknown whether SelH interacts with ANT, cyclophilin D, or Bcl-2 family proteins, which modulate the formation of the MPTP.

28.4 Effects of Selenoprotein H on Mitochondrial Biogenesis

Mitochondrial biogenesis refers to the process by which new mitochondria are formed. Activation of mitochondrial biogenesis is a novel promising target for protection against neurodegenerative diseases including cerebral stroke, Alzheimer's disease, Parkinson's disease, amyotrophic lateral sclerosis and other motor neuron disorders (Yin *et al.*, 2008; Vosler *et al.*, 2009).

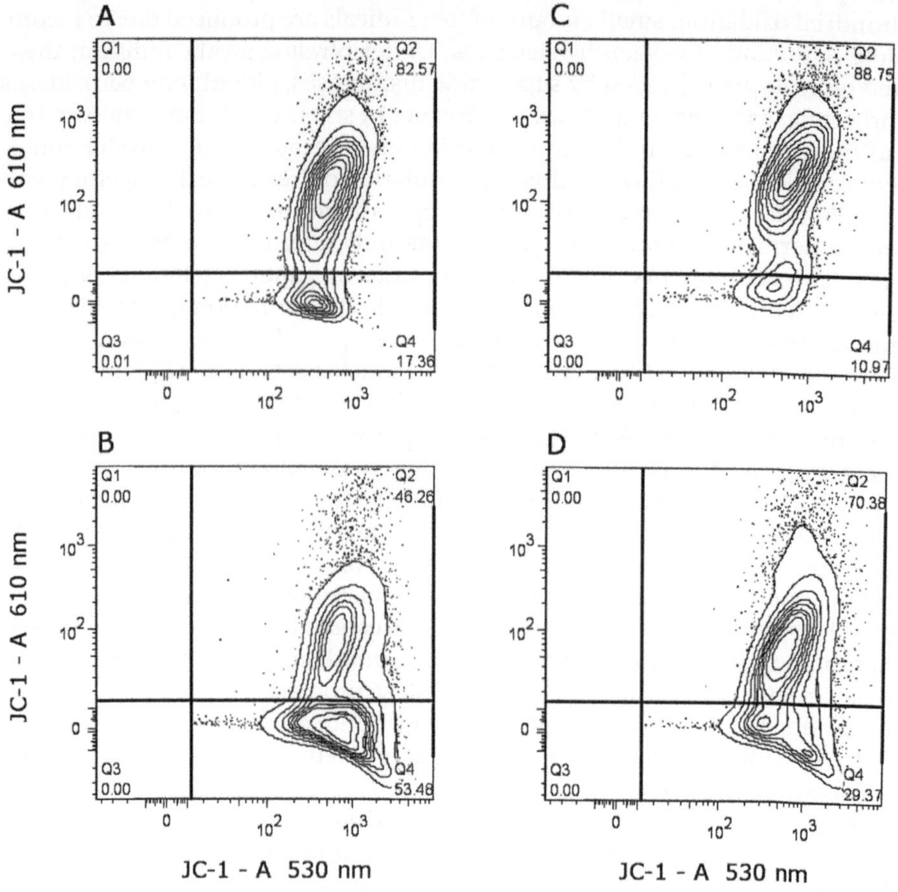

Figure 28.1 Overexpression of SelH prevented UVB-induced mitochondrial depolarization. Vector- and human SelH-transfected HT22 cells were exposed to at 7 J cm^{-2} UVB treatment and mitochondrial membrane potential was measured after 5 h using JC-1 fluorescent probe coupled with cytoflowmeter. UVB irradiation caused mitochondrial membrane depolarization in 53% of total cell population and overexpression of SelH reduced that population to 29%. (A) Vector-HT22 cells without UVB challenge; (B) vector-HT22 cells challenged with UVB; (C) SelH-HT22 cells without UVB; (D) SelH-HT22 cells with UVB challenge.

However, known compounds capable of activating mitochondrial biogenesis are few. Although both the mitochondrial and nuclear genomes control the transcription and translation of mitochondrial proteins, the majority of mitochondrial proteins are transcribed by the nuclear genome. Peroxisome proliferator-activated receptor (PPAR) gamma-1 coactivators alpha (PGC-1α) and nuclear respiratory factors (NRF1 and NRF2) are master regulators of mitochondrial biogenesis (Gromer *et al.*, 2005). PGC-1α is capable of binding to several nuclear receptors including NRF1 and NRF2, in addition to their interaction with PPAR. NRF1 and NRF2 directly regulate the expression of several nuclear-encoded electron-transport complex proteins, and indirectly regulate the three mitochondria-encoded cytochrome oxidase (COX) subunit genes by activating mitochondrial transcription factor A (Tfam), which is responsible for the transcription of nuclear-encoded mitochondrial proteins. These proteins include both structural mitochondrial proteins and those involved in mitochondrial DNA transcription, translation, and repair (Kelly and Scarpulla, 2004; Anne Stetler *et al.*, 2013; Wu *et al.*, 1999).

Overexpression of SelH significantly increases the protein levels of PGC-1α, NRF1, and Tfam, but not PGC-1β or NRF2, in the nuclear fraction. These stimulating effects of SelH on mitochondrial biogenesis regulators are not only observed in nonstressed cells, but also in UVB-stressed cells. Thus, SelH is capable of reinstalling the levels of these proteins that are suppressed by UVB-irradiation (Mendelev *et al.*, 2011). The observed effects are specifically induced by SelH. Knocking down of SelH by small interference RNA (miRNA) reverses the activating effect of SelH on mitochondrial biogenesis markers (Figure 28.2). Activation of the mitochondrial biogenesis signaling pathway may lead to increased mitochondrial protein synthesis, increased mitochondrial mass, and improved cellular respiratory function. These proved to be the case in SelH-transfected cells. Studies have reported that mitochondrial cytochrome c levels increased in the mitochondrial fraction, mitochondrial mass increased, and mitochondrial respiratory function increased almost 2-fold compared with vector-transfected control cells (Mendelev *et al.*, 2011). It is further revealed that SelH enhances the activities of mitochondrial electron-transport chain complexes I, II + III, and IV (Mehta *et al.*, 2013). These data suggest that SelH activates mitochondrial biogenesis and enhances mitochondrial respiratory function.

28.5 Selenoprotein H Regulates PGC-1α Upstream Signaling Pathways

As shown in Figure 28.3, the PGC-1α–NRF1 signaling pathway could be activated by the protein kinase A (PKA)–cAMP response element-binding (CREB) pathway, the calcium–calcium/calmodulin-dependent protein kinase (CaMK)–CREB pathway, the nitric oxide (NO)–cGMP pathway, and/or the AMP-activated protein kinase (AMPK) pathway (Scarpulla, 2008; Anne Stetler *et al.*, 2013). Our recent study shows that overexpression of SelH increases

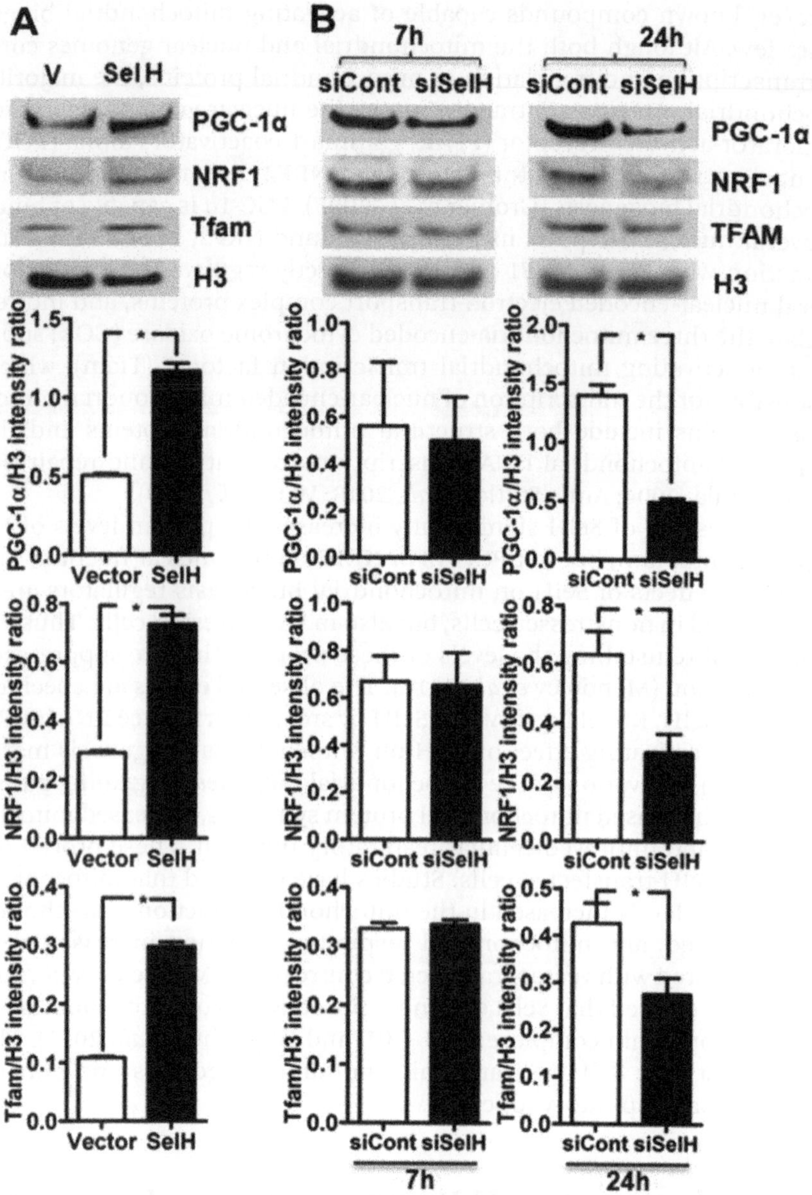

Figure 28.2 Effects of SelH transfection and SelH silencing on the protein levels of mitochondrial biogenesis regulators. (A) Protein levels of mitochondrial biogenesis regulators, PGC-1α, NRF1 and Tfam increased after SelH transfection whereas (B) silencing of SelH reduced the protein levels of PGC-1α, NRF1 and Tfam as compared to control in nuclear fraction. The data are collected from three independent experiments and presented as mean ± SD. *$P < 0.05$ *vs.* respective control. V, vector-transfected cells; SelH, SelH-transfected cells; H3, histone H3; siCont, control siRNA; siSelH, SelH siRNA; 7 h, 7 hours; 24 h, 24 hours. Reproduced with permission from Mehta *et al.* (2013).

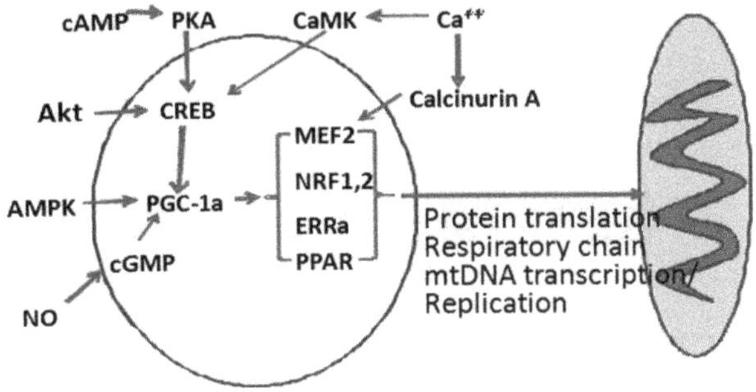

Figure 28.3 Signaling pathways that regulate mitochondrial biogenesis. PGC-1α and NRF1,2 are key transcription factors for mitochondrial biogenesis. The PGC-1α–NRF signaling pathway could be activated through the protein kinase A (PKA)–cAMP response element-binding (CREB) pathway, the Akt (also known as protein kinase B)–CREB pathway, the calcium–calcium/calmodulin-dependent protein kinase (CaMK)–CREB pathway, the nitric oxide (NO)–cGMP pathway, and/or the AMP-activated protein kinase (AMPK) pathway. Modified from Scarpulla (2008).

the levels of total and phosphorylated Akt, PKA and CREB (Mehta *et al.*, 2013). Knocking down of SelH by siRNA abolished the increases, suggesting the effects observed are specific to SelH (Figure 28.4). To ascertain the effect of SelH exerted on mitochondrial biogenesis by phosphorylation of CREB, CREB inhibitor naphthol AE-S phosphate was applied to SelH-transfected cell culture medium, and levels of PGC-1α were measured. Inhibition of CREB blocked the SelH-induced increase of PGC-1α (Mehta *et al.*, 2013), suggesting CREB is a key factor that mediates the mitochondrial effects of SelH. Other chemical compounds that inhibit AMPK [calcineurin, NF-κB, and nitric oxide synthase (NOS)] did not block the effects of SelH on the mitochondrial biogenesis signaling pathway.

It is likely that SelH exerts its mitochondrial effect through activation of the PKA–CREB–PGC-1α–NRF1 and Akt–CREB–PGC-1α–NRF1 pathways. In addition, SelH has been classified as a nuclear thioredoxin-like protein (Novoselov *et al.*, 2007). Thioredoxin 1 binds to nuclear factor (NF)-E2-related factor (Stoytcheva and Berry, 2009). Therefore, it is also possible that SelH binds to NF-E2 related factor that subsequently binds to NRF1 promoter region (Piantadosi *et al.*, 2008), and thereby, activates mitochondrial biogenesis. However, this remains to be examined.

28.6 Effects of Selenium on the Mitochondrial Function and Biogenesis

Selenium is a trace element necessary for normal cellular function in most animals including humans. It is integrated into the 21st amino acid selenocysteine, the latter being incorporated into selenoproteins. Among the 25 selenoproteins, GPX1, GPX3, TrxR1, TrxR2, TrxR3, SelH, SelP and SelW are

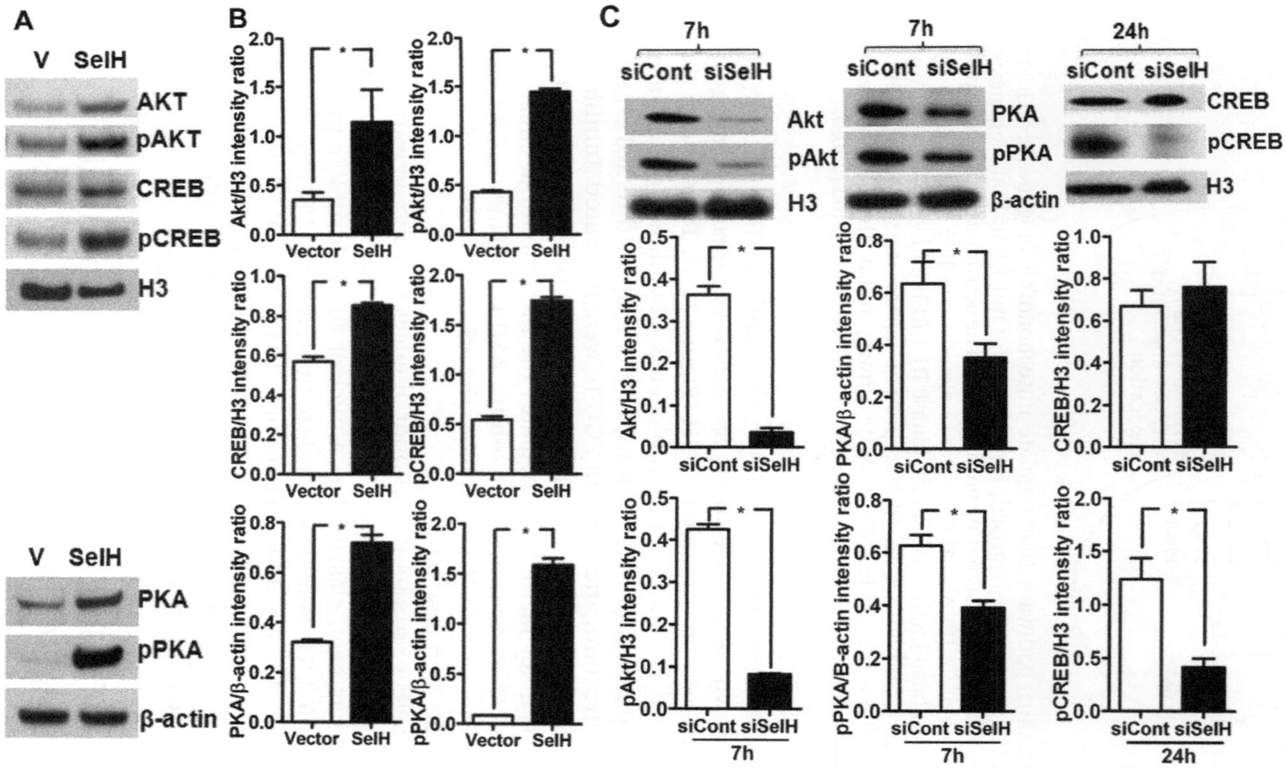

Figure 28.4 SelH overexpression increases and SelH siRNA decreases phosphorylation of Akt, PKA and CREB. (A) Protein and phosphorylation levels of Akt, PKA and CREB in vector- and SelH-transfected cells. Both, total protein and phosphoproduct of Akt (nuclear), PKA (cytosolic) and CREB (nuclear) increased in SelH cells as compared to vector-transfected cells. (B) Quantitative western analysis of Akt, PKA and CREB revealed that SelH transfection not only increased the protein levels of Akt, PKA and CREB but also significantly elevated the phosphorylation of these proteins. (C) SelH silencing decreased the total protein and phospho product of Akt, PKA and CREB in SelH-transfected cells. The data are collected from three independent experiments and presented as mean ± SD. $*P < 0.05$. SelH, SelH-transfected cells; H3, histone H3; siCont, control siRNA; siSelH, SelH siRNA; 7 h, 7 hours; 24 h, 24 hours. Reproduced with permission from Mehta *et al.* (2013).

sensitive to dietary selenium levels (Reeves and Hoffmann, 2009). It is commonly believed that selenium exerts most, if not all, of its function through the incorporation into selenoproteins. Selenium deficiency causes several disorders including Keshan disease, Kashin–Beck disease and hypothyroidism. Although selenium may cause cytotoxicity (Xiao *et al.*, 2006; Maraldi *et al.*, 2011), the majority of studies have shown that selenium supplementation prevents Keshan disease, increases the activity of deiodinase, protects cells from injuries induced by glutamate toxicity, oxidative stress, and inflammatory cytokines, and inhibits certain cancer cell growth (Papp *et al.*, 2007, 2010). We recently demonstrated that adding 100 nM sodium selenite to the culture medium increased the cell viability after glutamate exposure or hypoxic injury (Mehta *et al.*, 2012). The protective effect against glutamate toxicity was not only observed in selenite-pretreated HT22 cells but in 1 to 8 h of postglutamate exposed cells as well (Kumari *et al.*, 2012b). These protective effects are mediated by the abilities of selenium in reducing reactive oxygen free-radical accumulation, stabilizing mitochondrial membrane potential and preventing glutamate- or hypoxia-induced fall in activities of mitochondrial respiratory complexes (Mehta *et al.*, 2012; Kumari *et al.*, 2012b). In addition, incubation of HT22 cells with 100 nM selenite activates mitochondrial biogenesis. Thus, translational levels of PGC-1α and NRF1 significantly increased, as did two mitochondrial proteins cytochrome c and COX IV (Mendelev *et al.*, 2012). Similar to the effects of SelH overexpression, selenite supplementation enhances mitochondrial respiratory rate and improves the enzyme activity of mitochondrial respiratory complexes I–IV (Figure 28.5).

The mechanism that selenium promotes mitochondrial biogenesis is through activating the PGC-1α upstream factor Akt–CREB pathway. Selenium

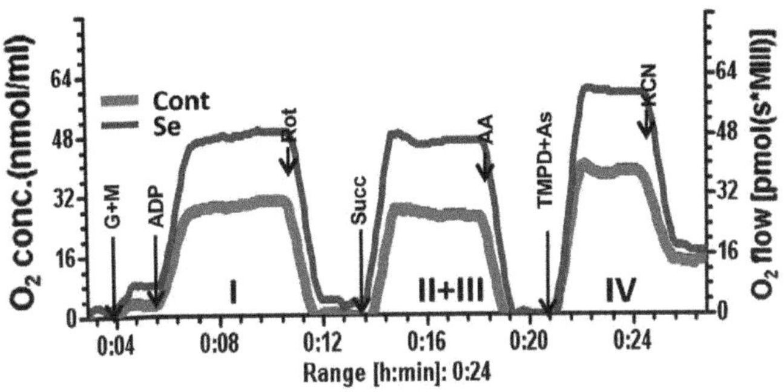

Figure 28.5 Selenite increases the activities of mitochondrial respiratory complexes. Polarograph recordings showing mitochondrial respiratory activities of complexes I, II + III and IV in control (red curve) and Se-treated (green curve) neurons. Oxygen consumption in digitonin permeabilized cells was measured using O_2K oxygraph with addition of multiple substrates and inhibitors. AA, antimycin A; As, ascorbate; G + M, glutamate and malate; KCN, potassium cyanide; Rot, rotenone; Succ, succinate; TMPD, *N*,*N*,*N*′,*N*′-tetramethyl-*p*-phenylenediamine dihydrochloride. Reproduced with permission from Mendelev *et al.* (2012).

modulates several cell signaling pathways. It has been reported that selenium causes phosphorylation of ERK1/2, p38 and JNK MAPK, deactivates apoptosis signaling-regulating kinase 1 (ASK1), phosphotidylinositol 3-kinase (PI3K)-Akt, and NF-κB pathways (Staplaton, 2000; Yoon *et al.*, 2002; Pillai *et al.*, 2012). Our study shows that incubation of HT22 cell with 100 nM selenite for 24 h increases the levels of phosphorylated Akt in both the cytosolic and nuclear fractions and CREB in the nuclear fraction, without influencing the total Akt and CREB levels (Mendelev *et al.*, 2012). An Akt inhibitor blocks the effects of selenite on CREB and PGC-1α, suggesting selenium activates mitochondrial biogenesis through phosphorylation of Akt–CREB pathway (Figure 28.6). In contrast, selenium did not increase the level of phospho-PKA. However, when an PKA inhibitor is applied to selenium-treated cells, the effects of selenium on CREB and PGC-1α were also blocked, suggesting the PKA–CREB pathway may also participate in mediating the mitochondrial biogenesis effect of selenium.

The effect of selenium on mitochondrial biogenesis was not only observed *in vitro* but *in vivo* in ischemic brain samples. Cerebral ischemia is a disease caused by occlusion of a cerebral blood vessel leading to cell death due to lack of glucose and oxygen supplies. Activation of mitochondrial biogenesis is one of the protective mechanisms that the ischemic tissue combats the devastating insult. Studies have shown that mRNA and protein levels of NRF1 increased significantly after ischemia and reperfusion up to 24 h (Yin *et al.*, 2008). Correspondingly, increased are mitochondrial protein COX IV, mitochondrial DNA content, mitochondrial enzyme synthase activity, and mitochondrial number.

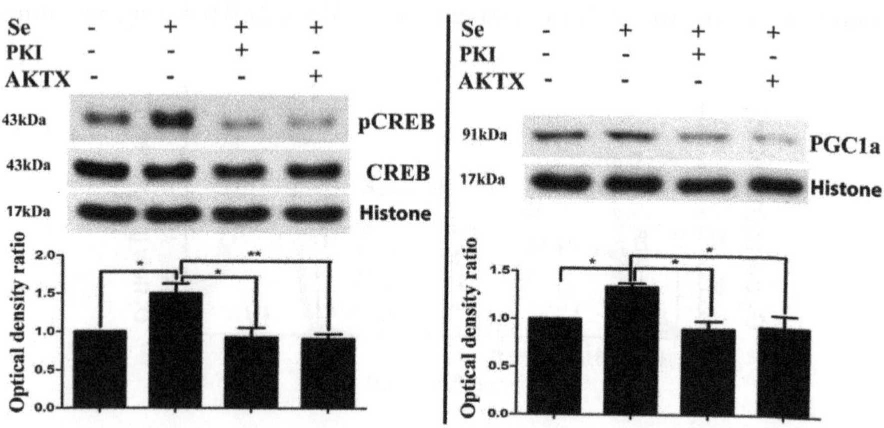

Figure 28.6 Effects of PKA and Akt inhibition selenium-induced phospho-CREB and PGC-1α. Cells were treated with 100 nM Selenite. PKA inhibitor 14–22 amide (PKI, 5 μM) or AKT inhibitor X (AKTX, 5 μM) were added to medium in the presence of selenium. Western blotting was performed after 5 h of incubation. *Upper panel*, representative Western blot bands. *Lower panel*, summarized bar graphs show band intensity presented as ratio of targeting protein over Histone H3. Values are means ± SD $*p < 0.05$ and $**p < 0.01$ by ANOVA followed by Tukey's test. Reproduced with permission from Mendelev *et al.* (2012).

Pretreatment with selenite (0.2 mg kg^{-1}, i.p.) for 1 week reduces the infarct volume from 36% in ischemic animals to 11% in selenite-pretreated ischemic animals (Mehta *et al.*, 2012). This *in vivo* neuroprotective effect of sodium selenite is associated with increased levels of PGC-1α and NRF1 prior to and after ischemia reperfusion injury (Mehta *et al.*, 2012).

28.7 Selenium Affects Mitochondrial Dynamics

Mitochondria are the crucial organelles of cells responsible for energy production and maintenance of calcium homeostasis as well as many other important biological functions. Any injury to mitochondria could lead to deleterious effects on overall physiology of the cell. Mitochondria in healthy cells resemble long filaments yet are remarkably dynamic, undergoing cycles of fission and fusion events. Fission processes are regulated by dynamin-related protein 1 (Drp1), mitochondrial fission 1 protein (Fis1 and endophilin B1 (Endo B1); while fusion processes are mediated by mitofusin-1 and -2 (Mfn1, -2) and optic atrophy 1 (OPA1) (Lee *et al.*, 2004). Disturbance in mitochondrial dynamics toward continuous fission may lead to MPTP formation, mitochondrial fragmentation, mitochondrial autophagy, activation of apoptosis, and neurodegeneration (Chan, 2006; Chen and Chan, 2009). Mitochondrial fusion, on the other hand, prevents depletion of metabolites or mitochondrial DNA, thereby protecting neurons (Lee *et al.*, 2004). Mitochondrial fission, triggered in the early stage of stroke and neurodegeneration, leads to ATP depletion, loss of mitochondrial DNA, respiratory deficits, increase in ROS and cell death by apoptosis, necrosis, or autophagy (Knott *et al.*, 2008). Drp1, which induces mitochondrial fission, has been reported to stimulate Bid-induced Bax oligomerization and cytochrome c release by promoting tethering and hemifusion of membranes (Montessuit *et al.*, 2010). Bax knockdown totally inhibited the mitochondrial accumulation of Drp 1 and Drp1 knockdown attenuated cell apoptosis (Wu *et al.*, 2011). Drp1/Bax activation therefore promotes mitochondrial fragmentation, reduces mitochondrial number, and activates apoptotic cell death.

Both *in vitro* exposure of neurons to glutamate, a key pathogenic factor mediating ischemic brain damage, and *in vivo* induction of transient cerebral ischemia in animals induces mitochondrial fragmentation. Mitochondrial fission proteins phospho-Drp1 and Fis 1 increased in cell lysates after 6 and 24 h of glutamate exposure. As a consequence, the number of neurons with fragmented mitochondria increased in glutamate-treated cells (Kumari *et al.*, 2012b). Similarly, Drp1 and mitochondrial fission/fusion index [= (Drp1 + Fis)/(Opa1 + Mfn1/2)] are increased after 6 or 72 h following 30 min of middle cerebral artery occlusion (Kumari *et al.*, 2012a). Selenite suppressed the glutamate-induced elevations of Drp1 and Fis1, reduced number of cells containing fragmented mitochondria in cultured HT22 neurons, and rebalanced mitochondrial fission and fusion markers in cerebral ischemic stroke animals (Kumari *et al.*, 2012a,b). A set of representative photomicrograms of glutamate-induced mitochondrial fission and the effects of selenium is given in Figure 28.7.

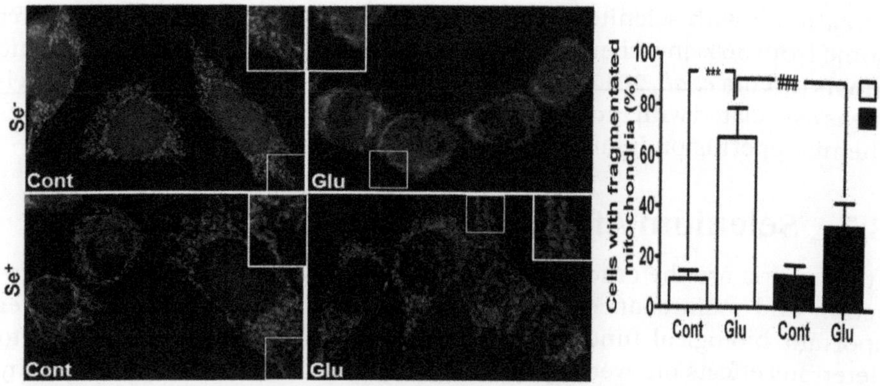

Figure 28.7 Selenium ameliorates glutamate-induced fragmentation. Exposure of HT22 cells to 4 mM glutamate for 24 h induced mitochondrial fragmentation as detected by MitoTracker®Red CM-XRos. Simultaneous application of 100 nM selenium significantly reduced glutamate-induced mitochondrial fragmentation. Data were collected from 5 experiments. Values are means ± SD. Two-way ANOVA test followed by post hoc Bonferroni's test was used for data analysis. $^{\#\#}p < 0.01$ and $^{***}p < 0.001$. Cont, control; Glu, glutamate; Se−, nonselenium treatment; and Se+, with selenium treatment. Red color indicated mitochondria whereas blue represents DAPI nuclear staining. Reproduced with permission from Kumari *et al.* (2012b).

Summary Points

- PGC-1α–NRF1 pathway is the major common pathway for mitochondrial biogenesis.
- Selenoprotein H activates mitochondrial biogenesis through Akt and PKA–CREB–PGC-1α–NRF1 pathway.
- Selenoprotein H enhances mitochondrial respiratory function and complex I–IV activities.
- Selenium protects neuronal cells against glutamate toxicity, hypoxia and cerebral ischemia.
- Selenium stimulates mitochondrial biogenesis through phosphorylation of Akt and CREB.
- Selenium maintains mitochondrial fission and fusion balance in glutamate-stressed cells and in ischemic animals.

Key Facts

Key Facts of PGC-1α in Relationship to Mitochondrial Biogenesis

1. Also known as PPARGC1A.
2. PGC-1α is a coactivating factor of PPAR.
3. PGC-1α is a key regulator for mitochondrial biogenesis.
4. PGC-1α activates NRF1.

Key Features of Mitochondrial Biogenesis

1. It is the processes of forming new mitochondria in the cell.
2. It can be activated by cellular stresses, environmental toxins and diseases.
3. It is regulated by both nuclear genome and mitochondrial genome, with the dominance of nuclear genome.
4. It is mainly regulated by PGC-1α–NRF1–Tfam pathway.
5. Increase of mitochondrial biogenesis is associated with enhanced mitochondrial oxidative phosphorylation.
6. Activation of mitochondrial biogenesis is a novel therapeutic target for mitochondria-associated cell death and mitochondria-related diseases.

Mitochondrial Dynamics

1. Mitochondria are dynamic organelles constantly in transition between fusion and fission.
2. Fusion forms mitochondrial tubular network; while fission divides one mitochondrion into two.
3. Fusion is the process by which 2 mitochondria fuse together to form a bigger mitochondrion.
4. Fusion facilitates content mixing as a form of complementation between 2 mitochondria and enhances energy production.
5. Fusion is controlled by Opa1, Mfn1 and Mfn2.
6. Fission releases degraded mitochondrial components and the latter is removed or recycled through mitochondrial autophagy.
7. Fission is associated with induction of apoptosis.
8. Fission is mediated by Drp1 and Fis 1

Definitions of Word and Terms

Autophagy. This is a process of cellular components degradation and recycling. In the beginning, a double-layered phagophore membrane forms, and then enwraps degraded cytosolic proteins, aged or damaged subcellualr organelles, and protein aggregates to form an autophagosome. Lysosome fuses with the autophagosome to become an autolysosome and cargo are lysed by lysosomal proteases.

Cerebral ischemia. This is a disease caused by significant narrowing or blockade of a blood vessel that carries blood to the brain tissue. In the brain areas where blood flow reduced to 0–20% of the normal level, neurons, glial cells and microvescular endothelial cells are all damaged forming an ischemic core. In the area where blood flow reduced to 20–40% of the control level, cellular functions are compromised yet able to produce enough energy to maintain their viability. This functionally compromised area is rcferred as penumbra.

Glutamate toxicity. Glutamate is an excitatory neurotransmitter in the central nervous system. It plays an important role in sending signals between neurons. When released in excessive amount, it can either cause overexcitation of the receiving neurons or calcium influx into the cells through binding to glutamate receptors. Intracellular calcium overload activates various lipases, proteinases and DNAses, causing cell damage. Calcium overload also triggers the formation of mitochondrial permeability transition pore (MPTP), which lead to activation of mitochondrial-initiated cell-death pathways.

Keshan disease. This is caused by lack of selenium in the body. It is prevalent in northeast of China where selenium content is low in foods. Clinically, Keshan disease presents as cardiac muscle necrosis, leading to cardiac malfunction, hypothyroidism and a compromised immune function.

Mitochondrial permeability transition. Also referred to as MPT, MPTP, and PTP. The MPTP opens occasionally to allow release of mitochondrial metabolic wastes under physiological condition. Increased free radicals and calcium in the mitochondria cause persistent opening of MPTP. It is believed that the MPTP is constituted by adenine nucleotide translocase (ANT) and cyclophilin D. Cyclosporine A is a potent MPTP inhibitor. Formation of the MPTP allows certain proapoptotic mitochondrial proteins to be release to either the cytoplasm or nucleus and activates cell-death pathways.

Mitochondria-initiated cell-death pathway. The MPTP formation causes releases of proapoptotic proteins from the mitochondria. These proteins mainly include cytochrome c, Smac/DIABLO, AIF and endonuclease G (EndoG). Release of cytochrome c to the cytosol causes downstream capase-9 and caspase-3 activation. Caspase-3 either directly translocates into the nucleus or through activation of caspase dependent DNase, which eventually causes DNA damage and cell death. Release of Smac/DIABLO into the cytosol inhibits the inhibitor of apoptosis proteins (IAPs), which leads to activation of caspase-9, -7, and -3. Releases of AIF and EndoG into the nucleus cause DNA damage. The cytochrome c and Smac/DIABLO-mediated cell-death pathways are sometime referred to as mitochondria-initiated, caspase-dependent cell-death pathways; and the AIF and Endo-mediated cell-death pathways are referred to as mitochondria-initiated, caspase-independent cell-death pathways.

Oxidative stress. This refers to excessive accumulation of free radicals in the cell. Free radicals are produced during cellular metabolic processes. Under normal circumstances, free radicals are detoxified by endogenous enzymes such as superoxide dismutase's, glutathione peroxidases and catalase. Some free radicals act as cellular messengers for signaling. However, accumulation of free radicals, either by increased production or decreased degradation, damage lipids, proteins and DNA.

Oxidative phosphorylation. This is a process during which ATP is produced. It involves 2 steps: oxidation and phosphorylation. In oxidation, electrons donated from NADH and FADH2 are transported from mitochondrial electron transport chain (ETC) complex I and II to III then IV. During this process, protons are pumped into the intramitochondrial membrane space to generate a proton chemoelectral potential across the inner mitochondrial membrane. The energy carried by these protons will be released to convert ADP to ATP when entering mitochondria through ETC complex V.

PSORT. This is an automated computer program that is used to predict subcellular localization of proteins. It uses the amino acids and origin of the protein as input to compare with the prestored sequences of known protein sorting signals. It reports possible localization of the input protein on candidate site.

Redox. This includes reactions that transfer electrons between atoms, molecules and ions, *etc.* The process involves reduction and oxidation reactions. Reduction decreases oxidation states that usually accomplished through receiving electrons and oxidation is an increase in oxidation state that is mostly achieved by losing electrons.

List of Abbreviations

AIF	Apoptosis-inducing factor
Akt	Also known as protein kinase B
AMPK	Adenosine monophosphate (AMP)-activated protein kinase
ANT	Adenine nucleotide translocase
apaf-1	Apoptotic protease activating factor 1
ASK1	Apoptosis signaling-regulating kinase 1
Bax	Bcl-2-associated X protein
Bcl-2	B-cell lymphoma 2
BID	BID BH3 interacting domain, a Bax-like BH3 protein
CAD	Caspase-dependent DNase
CaMK	Calcium–calcium/calmodulin-dependent protein kinase
ChIP	Chromatin immunoprecipitation
COX	Cytochrome c oxidase
COX IV	Cytochrome c oxidase or complex IV
CREB	3′-5′-Cyclic adenosine monophosphate (cAMP) response element-binding protein
Drp1	Dynamin-related protein 1
Endo B1	Endophilin B1
Endo G	Endonuclease G
ERK1/2	Extracellular signal-regulated kinases
ExPASy	The Expert Protein Analysis System
GCS	Glutamylcysteine synthetase
GFP	Green fluorescent protein
GPXs	Glutathione peroxidases
GSTs	Glutathione *S*-transferases

HSEs	Heat-shock elements
Fis 1	Fission 1, (mitochondrial outer membrane) homolog
IAP	Inhibitor of apoptosis protein
JNK	c-Jun N-terminal kinases
MAPK	Mitogen-activated protein kinase
Mfn1/2	Mitofusins 1 and 2
MTPT	Mitochondrial permeability transition pore
NF-κB	Nuclear factor kappa-light-chain-enhancer of activated B cells
NO	Nitric oxide
NOS	Nitric oxide synthase
NRF	Nuclear respiratory factors
OPA1	Optic atrophy 1
p38	p38 Mitogen-activated protein kinases
PGC-1	Peroxisome proliferator-activated receptor gamma-1 coactivator
PHGPX	Phospholipid hydroperoxide glutathione peroxidase
PI3K	Phosphotidylinositol 3-kinase
PKA	Protein kinase A
PSORT	Protein subcellular localization prediction
ROS	Reactive oxygen species
SBP2	SECIS-binding protein 2
SECIS	Selenocysteine insertion sequence
Sec	Selenocysteine
SelH	Selenoprotein H
SelP	Selenoprotein P
siRNA	Small interference RNA
Smac/DIABLO	Small mitochondria-derived activator of caspases/direct IAP-binding protein with low pI)
SPS	Selenophosphate synthetase
STAT	Signal transducers and activators of transcription pathway
STREs	Stress-response elements
Tfam	Transcription factor A, mitochondria; aka mtTFA
TrxRs	Thioredoxin reductases
UVB	Ultraviolet wave B

References

Anne Stetler, R., Leak, R. K., Gao, Y. and Chen, J., 2013. The dynamics of the mitochondrial organelle as a potential therapeutic target. *Journal of Cerebral Blood Flow Metabolism*. 33: 22–32.

Bellinger, F. P., Raman, A. V., Reeves, M. A. and Berry, M. J., 2009. Regulation and function of selenoproteins in human disease. *Biochemical Journal*. 422: 11–22.

Ben Jilani, K. E., Panee, J., He, Q., Berry, M. J. and Li, P. A., 2007. Overexpression of selenoprotein H reduces Ht22 neuronal cell death after UVB irradiation by preventing superoxide formation. *International Journal of Biological Sciences*. 3: 198–204.

Chan, D. C., 2006. Mitochondrial fusion and fission in mammals. *Annual Review of Cell and Developmental Biology*. 22: 79–99.

Chen, H. and Chan, D. C., 2009. Mitochondrial dynamics – fusion, fission, movement, and mitophagy – in neurodegenerative diseases. *Human Molecular Genetics*. 18: R169–R176.

Gromer, S., Eubel, J. K., Lee, B. L. and Jacob, J., 2005. Human selenoproteins at a glance. *Cellular and Molecular Life Sciences*. 62: 2414–2437.

Kelly, D. P. and Scarpulla, R. C., 2004. Transcriptional regulatory circuits controlling mitochondrial biogenesis and function. *Genes Development*. 18: 357–368.

Knott, A. B., Perkins, G., Schwarzenbacher, R. and Bossy-Wetzel, E., 2008. Mitochondrial fragmentation in neurodegeneration. *Nature Review Neuroscience*. 9: 505–518.

Kumari, S., Anderson, L., Farmer, S., Mehta, S. L. and Li, P. A., 2012a. Hyperglycemia alters mitochondrial fission and fusion proteins in mice subjected to cerebral ischemia and reperfusion. *Translational Stroke Research*. 3: 296–304.

Kumari, S., Mehta, S. L. and Li, P. A., 2012b. Glutamate induces mitochondrial dynamic imbalance and autophagy activation: preventive effects of selenium. *PLoS One*. 7: e39382.

Lee, Y. J., Jeong, S. Y., Karbowski, M., Smith, C. L. and Youle, R. J., 2004. Roles of the mammalian mitochondrial fission and fusion mediators Fis1, Drp1, and Opa1 in apoptosis. *Molecular Biology of the Cell*. 15: 5001–5011.

Maraldi, T., Riccio, M., Zambonin, L., Vinceti, M., De Pol, A. and Hakim, G., 2011. Low levels of selenium compounds are selectively toxic for a human neuron cell line through ROS/RNS increase and apoptotic process activation. *Neurotoxicology*. 32: 180–187.

Mehta, S. L., Kumari, S., Mendelev, N. and Li, P. A., 2012. Selenium preserves mitochondrial function, stimulates mitochondrial biogenesis, and reduces infarct volume after focal cerebral ischemia. *BMC Neuroscience*. 13: 79–90.

Mehta, S. L., Mendelev, N., Kumari, S. and Li, P. A., 2013. Overexpression of human selenoprotein H in neuronal cells enhances mitochondrial biogenesis and function through activation of protein kinase A, protein kinase B, and cyclic adenosine monophosphate response element-binding protein pathway. *International Journal of Biochemistry and Cell Biology*. 45: 604–611.

Mendelev, N., Mehta, S. L., Witherspoon, S., He, Q., Sexton, J. Z. and Li, P. A., 2011. Upregulation of human selenoprotein H in murine hippocampal neuronal cells promotes mitochondrial biogenesis and functional performance. *Mitochondrion*. 11: 76–82.

Mendelev, N., Witherspoon, S. and Li, P. A., 2009. Overexpression of human selenoprotein H in neuronal cells ameliorates ultraviolet irradiation-induced damage by modulating cell signaling pathways. *Experimental Neurology*. 220: 328–334.

Mendelev, N., Mehta, S. L., Idris, H., Kumari, S. and Li, P. A., 2012. Selenite stimulates mitochondrial biogenesis signaling and enhances mitochondrial functional performance in murine hippocampal neuronal cells. *PLoS One*. 7: e47910.

Montessuit, S., Somasekharan, S. P., Terrones, O., Lucken-Ardjomande, S., Herzig, S., Schwarzenbacher, R., Manstein, D. J., Bossy-Wetzel, E., Basañez, G., Meda, P. and Martinou, J. C., 2010. Membrane remodeling induced by the dynamin-related protein Drp1 stimulates Bax oligomerization. *Cell*. 142: 889–901.

Morozova, N., Forry, E. P., Shahid, E., Zavacki, A. M., Harney, J. W., Kraytsberg, Y. and Berry, M. J., 2003. Antioxidant function of a novel selenoprotein in *Drosophila melanogaster*. *Genes to Cells*. 8: 963–971.

Novoselov, S. V., Kryukov, G. V., Xu, X. M., Carlson, B. A., Hatfield, D. L. and Gladyshev, V. N., 2007. Selenoprotein H is a nucleolar thioredoxin-like protein with a unique expression pattern. *Journal of Biological Chemistry*. 282: 11960–11968.

Panee, J., Liu, W., Nakamura, K. and Berry, M. J., 2007. The responses of HT22 cells to the blockade of mitochondrial complexes and potential protective effect of selenium supplementation. *International Journal of Biological Sciences*. 3: 335–341.

Papp, L. V., Holmgren, A. and Khanna, K. K., 2010. Selenium and selenoproteins in health and disease. *Antioxidant and Redox Signaling*. 12: 793–795.

Papp, L. V., Lu, J., Holmgren, A. and Khanna, K. K., 2007. From selenium to selenoproteins: synthesis, identity, and their role in human health. *Antioxidant and Redox Signaling*. 9: 775–806.

Piantadosi, C. A., Carraway, M. S., Babiker, A. and Suliman, H. B., 2008. Heme oxygenase-1 regulates cardiac mitochondrial biogenesis *via* Nrf2-mediated transcriptional control of nuclear respiratory factor-1. *Circulation Research*. 103: 1232–1240.

Pillai, S. S., Sugathan, J. K. and Indira, M., 2012. Selenium downregulates RAGE and NFkappaB expression in diabetic rats. *Biological Trace Element Research*. 149: 71–77.

Reeves, M. A. and Hoffmann, P. R., 2009. The human selenoproteome: recent insights into functions and regulation. *Cellular and Molecular Life Science*. 66: 2457–2478.

Scarpulla, R. C., 2008. Transcriptional paradigms in mammalian mitochondrial biogenesis and function. *Physiological Review*. 88: 611–638.

Staplaton, S. R., 2000. Selenium: an insulin-mimetic. *Cellular and Molecular Life Sciences*. 57: 1874–1879.

Stoytcheva, Z. R. and Berry, M. J., 2009. Transcriptional regulation of mammalian selenoprotein expression. *Biochimica et Biophysica Acta*. 1790(11): 1429–1440.

Vosler, P. S., Graham, S. H., Wechsler, L. R. and Chen, J., 2009. Mitochondrial targets for stroke: focusing basic science research toward developing of clinically translatable therapeutics. *Stroke*. 40: 3149–3155.

Wu, S., Zhou, F., Zhang, Z. and Xing, D., 2011. Bax is essential for Drp1-mediated mitochondrial fission but not for mitochondrial outer membrane permeabilization caused by photodynamic therapy. *Journal of Cellular Physiology*. 226(2): 530–541.

Wu, Z., Puigserver, P., Andersson, U., Zhang, C., Adelmant, G., Mootha, V., Troy, A., Cinti, S., Lowell, B., Scarpulla, R. C. and Spiegelman, B. M., 1999. Mechanisms controlling mitochondrial biogenesis and respiration through the thermogenic coactivator PGC-1. *Cell*. 98: 115–124.

Xiao, R., Qiao, J. T., Zhao, H. F., Liang, J., Yu, H. L., Liu, J., Guo, A. M. and Wang, W., 2006. Sodium selenite induces apoptosis in cultured cortical neurons with special concomitant changes in expression of the apoptosis-related genes. *Neurotoxicology*. 27: 478–484.

Yan, J. and Barrett, J. N., 1998. Purification from bovine serum of a survival-promoting factor for cultured central neurons and its identification as selenoprotein-P. *Journal of Neuroscience*. 18: 8682–8691.

Yin, W., Signore, A. P., Iwai, M., Cao, G., Cao, Y. and Chen, J., 2008. Rapidly increased neuronal mitochondrial biogenesis after hypoxic-ischemic brain injury. *Stroke*. 39: 3057–3063.

Yoon, S. O., Kim, M. M., Park, S. J., Kim, D., Chung, J. and Chung, A. S., 2002. Selenite suppresses hydrogen peroxide-induced cell apoptosis through inhibition of ASK1/JNK and activation of PI3-K/Akt pathways. *FASEB Journal*. 16: 111–113.

CHAPTER 29

Current Clinical Perspectives of Selenium in Vascular Function and Cardiomyopathy

YAP-HANG CHAN*[a] AND HUNG-FAT TSE*[a]

[a]Cardiology Division, Department of Medicine, The University of Hong Kong, 19/F, Block K, Queen Mary Hospital, Hong Kong, China
*E-mail: hftse@hkucc.hku.hk, yhchanwill@gmail.com

29.1 Introduction

Derived mainly from the intake of seafood, crops, vegetables, and animal proteins, selenium as a trace element has received increasing research interests in recent years regarding its potential role in the risks of cardiovascular and various other chronic diseases (Fryer, 2000; Rayman, 2000). The main postulated mechanism of protection (or conversely, damage if intake is excessive) is *via* oxidative stress regulation. Experimental studies showed that selenium is involved in the synthesis of selenocysteine, a key constituent amino acid of selenoproteins, which helps regulate *in vivo* oxidative stress through the glutathione peroxidase and thioredoxin reductase and protects the vasculature against oxidative damage (Tappel, 1974). Low selenoprotein activity, as revealed by prior studies, is related to accelerated atherosclerosis (Yang *et al.*, 2011).

Food and Nutritional Components in Focus No. 9
Selenium: Chemistry, Analysis, Function and Effects
Edited by Victor R. Preedy

Published by the Royal Society of Chemistry, www.rsc.org

As selenium deficiency is highly reversible through dietary modifications, clarifying the potential role of selenium in cardiovascular health and disease will have important clinical as well as public health implications.

29.2 Selenium Deficiency and Relation to Cardiovascular Events

Supportive data that selenium has a protective role against cardiovascular diseases dated back to as early as the 1980s. An early prospective nested case-control study was performed, as part of a population-based cohort of 11 000 persons in Eastern Finland, a jurisdiction with extremely high risk of ischemic heart disease. It was found that a serum selenium level less than 45 μg L^{-1} was associated with excess risk of ischemic heart disease deaths by nearly twice after adjustment for other risk factors (Salonen *et al.*, 1982). In a subsequent Danish cohort study of middle-aged men (mean age 63 years), serum selenium level of less than 1 μg L^{-1} was independently associated with increased risk of incident ischemic heart disease by 55%, after adjustment for conventional risk factors (Suadicani *et al.*, 1992). There was another interesting multicenter case-control study of selenium status and risk of myocardial infarction across the regions of Europe and Israel, in which toenail selenium level was found inversely associated with risk of myocardial infarction, but only within the Germany center. Whether this was a genuine relationship or accounted for by residual confounding remained uncertain (Kardinaal *et al.*, 1997). In another small prospective nested case-control study, both reduced blood selenium and selenoprotein P were associated with increased risks of first incident stroke events (Koyama *et al.*, 2009), in support of the hypothesis that cardiovascular risk modulation by selenium status is not confined to the cardiac system but generalized across the atherosclerotic disease spectrum.

Other epidemiological studies yielded inconsistent results, some of which found no relationship between selenium levels and cardiovascular outcomes (Fryer, 2000; Rayman, 2000). Overall, a recent meta-analysis of 14 cohorts and 11 case-control studies concluded that 50% increase in selenium level was associated with 24% reduction of coronary heart disease risk (Flores-Mateo *et al.*, 2006).

It is interesting to note that selenium deficiency plays a role not only in the clinical occurrence of cardiovascular events, but also the extent of damage sustained during such episodes. In a recent clinical study of 100 patients presented with acute coronary syndrome, serum selenium level was inversely correlated with serum troponin I level on admission, a specific marker of myocardial necrosis, suggesting that selenium deficiency is a risk factor for potentiated myocardial damage during acute vascular insults (Bayir *et al.*, 2013). In the wider clinical setting, low blood selenium was also shown to be associated with increased operative risks and mortality in cardiac surgery (Koszta *et al.*, 2012).

29.3 Clinical Evidence of Selenium Impacting on Vascular Function

29.3.1 Clinical Observational Studies

To gain mechanistic insights into the observed links between selenium deficiency and adverse cardiac events, a number of studies have been carried out to elucidate the downstream mediating mechanisms beyond the biochemical level of oxidative stress regulation. In a recent cross-sectional vascular function study of 306 consecutive high-risk cardiovascular patients with established coronary/cerebrovascular diseases or type 2 diabetes, selenium deficiency based on dietary intake was found independently associated with worsened arterial stiffness, as indicated by increased brachial-ankle pulse wave velocity (PWV) (Chan *et al.*, 2012). The attributable increase in PWV was a clinically significant difference by more than 300 cm s^{-1}. Increased arterial stiffness is an independent predictor for both incident (Mattace-Raso *et al.*, 2006) and recurrent (Seo *et al.*, 2008) stroke, and predicts vascular complications in hypertensive (Laurent *et al.*, 2001; Boutouyrie *et al.*, 2002; Laurent *et al.*, 2003) and renal (Blacher *et al.*, 2001) patients. Its utility for cardiovascular risk prediction in the healthy population has also been proven beyond the paradigm of conventional risk factors (Sutton-Tyrrell *et al.*, 2005; Mitchell *et al.*, 2010). The implications of this study are that selenium deficiency may have an important role in the secondary progression of atherosclerosis in individuals with existing vascular disease or equivalent cardiovascular risk, and that such adverse arterial function may consequentially translate into clinically manifest cardiovascular events. Whether reversal of selenium deficiency in these patients may decelerate atherosclerotic progression or prevent cardiovascular events needs to be further studied in randomized controlled trials.

In another study of high-risk subjects, serum selenium was measured in 71 chronic hemodialysis patients who had coronary flow reserve determined as an index of endothelial function, using transthoracic Doppler echocardiography. It was shown that depressed serum levels of selenium was independently associated with reduced coronary flow reserve, suggesting a role of selenium deficiency in coronary endothelial dysfunction in hemodialysis patients (Atakan *et al.*, 2013). In another well-performed study of patients with chronic kidney disease (Yilmaz *et al.*, 2006), serum selenium decreases with stage of chronic kidney disease, and correlated positively with antioxidant enzymes including superoxide dismutase and glutathione peroxidase; but inversely with oxidative stress markers including malondialdehyde and asymmetric dimethylarginine (ADMA). Importantly, low serum selenium was related to worsened glomerular filtration rate and brachial flow-mediated vasodilatation (FMD), another key and independent predictor of cardiovascular events in persons with (Gokce *et al.*, 2003) or without (Shechter *et al.*, 2014) prior cardiovascular diseases.

In a further study of antioxidant status and carotid atherosclerosis among 1187 men and women aged 59–71 years who had no history of coronary or cerebrovascular disease (Bonithon-Kopp *et al.*, 1997), plasma selenium together with carotenoids and erythrocyte vitamin E levels were measured. Subclinical carotid atherosclerosis was assessed by vascular ultrasound examination, with the estimate of carotid intima-media thickness (IMT) as a structural indicator of vascular function. Carotid IMT was found to be negatively associated with erythrocyte vitamin E, but not with plasma selenium or carotenoids. Thiobarbituric acid-reactive substances (TBARS), a marker of lipid peroxidation, was significantly higher among men with carotid plaques compared to those without a plaque. In an exploratory analysis, this relation was strengthened in men with erythrocyte vitamin E, plasma selenium, and carotenoids that were below the lowest quartile, suggesting a potential interaction or effect modifying effect of selenium in such an association.

29.3.2 Clinical Interventional Studies

29.3.2.1 Setting the Scene – Literature Interpretation in Context

Despite the findings of adverse vascular function associated with deficient/inadequate selenium status from clinical observational studies, currently there is overall a lack of clinical trial evidence that supports any benefit of selenium supplementation for cardiovascular prevention. As we sail through the current literature, there are a number of key points we must note. First, few randomized controlled trials to date have been designed *a priori* and performed to specifically test the effects of selenium supplementation on vascular function or clinical cardiovascular events. Secondly, it is important to realize that the supplements used in current trials are often dietary interventions or composite supplements that contain other antioxidant vitamins, which may include vitamin E or C, beta-carotene, or zinc, rendering the specific effect of selenium supplementation unclear. Thirdly, observational and experimental studies indeed showed that the vascular effects of selenium are not linear, and in fact selenium status may exhibit a potentially U-shape in relation to cardiovascular effects. Indeed, the key beneficial vascular effects related to adverse selenium status from observational studies were amongst those with overt selenium deficiency, rather than derived from comparing subjects within the mediocre range of selenium intake. Thus, is it is pivotal for any clinical interventional studies to focus on reversing the selenium deficiency and test for the corresponding change in cardiovascular risks, rather than supplementing subjects with already adequate selenium status that could potentially dilute or reverse any overall beneficial effects – the potential detrimental risks of over-supplementation cannot be overstated.

29.3.2.2 Surrogate Vascular Endpoints

In a small, nonrandomized clinical trial among severely obese women, subjects received daily Brazil nuts for 8 weeks and cardiometabolic profiles were compared before and after intervention. It was shown that blood selenium increased, and this change was accompanied by improved lipid profile and reduced atherogenic risk as indicated by improved Castelli I and II indexes (Cominetti *et al.*, 2012). While this is an interesting study, the design was limited in terms of the small size, nonrandomized and uncontrolled nature, as well as the diverse ingredient contained in Brazil nuts other than selenium.

The SElenium Therapy in Coronary Artery disease Patients (SETCAP) Study was an excellently conducted clinical study on the cardiovascular effects of selenium among high-risk subjects (Schnabel *et al.*, 2008). The research group first showed in *in vitro* studies that selenium supplementation enhanced GPx-1 protein expression and activity dose dependently in human coronary artery endothelial cells. They then proceeded with a clinical trial that recruited 465 patients with coronary artery disease who had impaired endothelial function at baseline. Impaired endothelial function was defined as brachial FMD <8%. Patients were randomized to receive sodium selenite daily at doses of 200 μg or 500 μg, or matched placebo. Brachial FMD was assessed by ultrasonography, together with biochemical assessments at baseline and after 12 weeks. It was shown that selenium level was increased after selenium supplementation, with increased red-cell glutathione peroxidase 1 activity. A similar magnitude of treatment effect was seen at both supplementation dosage levels. Nevertheless, no significant changes were observed for endothelial dysfunction as measured by brachial FMD. In another randomized controlled trial comprising 70 patients with multiple cardiovascular risk factors (Shargorodsky *et al.*, 2010), patients who received antioxidant supplements twice daily (containing total selenium 200 μg per day, vitamins C and E and coenzyme Q10 for 6 months had significantly improved large and small arterial elasticity as measured by pulse wave contour analysis. There was also a decrease in blood pressure and favorable changes in glucose and lipid metabolism.

A few clinical studies further attempted to examine the vascular effects of selenium or antioxidant therapy among healthy subjects. A substudy (Zureik *et al.*, 2004) of the randomized, double-blind, placebo-controlled SU.VI.MAX Study recruited a group of apparently healthy individuals free of symptomatic chronic diseases, and randomly allocated them to receive either a daily composite supplement of antioxidants which included selenium (100 μg), vitamins C and E, beta-carotene and zinc, *versus* placebo over 7 years. Carotid IMT and presence of carotid plaque was determined by ultrasound examination. Carotid-femoral PWV was measured. It was found that PWV was significantly lower only among female participants receiving antioxidant supplement, but not in the overall sample. In fact, the percentage of participants identified with carotid plaques was alarmingly higher in the supplement *versus* the placebo group. Nevertheless, it should be emphasized that there was no baseline (and therefore no interval) assessment of vascular function, as the vascular assessments were only performed at the end of trial period as *ad hoc*, albeit blinded, investigations. The implication is that the vascular outcomes such conveyed cannot be fully interpreted as coming from a conventional interventional study. The specific effect of selenium as part of the composite supplement was also uncertain.

In another smaller, uncontrolled clinical trial among healthy subjects, 54 healthy North American men were administered selenium supplement 300 µg per day in the form of high-selenium yeast. Brachial FMD was measured at 24 and 48 weeks. While increased plasma and erythrocyte selenium levels were achieved, there was no significant change in brachial FMD. Nevertheless, these were healthy men who were likely to obtain adequate selenium intake from dietary sources (Hawkes and Laslett, 2009).

29.3.2.3 *Clinical Endpoints*

There were 2 major large randomized controlled trials available thus far to provide data on the effects of selenium supplementation on the risk of cardiovascular events. Both of these trials were not primarily designed to study cardiovascular outcomes.

The first one is the Nutritional Prevention of Cancer Trial (NPC Trial), which was originally designed to elucidate the role of nutritional prevention of cancer among 1312 participants with a history of basal or squamous cell carcinoma (study period 1983–1996). In the secondary exploratory analyses, the effects of selenium supplementation at 200 µg per day *versus* placebo (about 500 subjects per arm) were studied for clinical outcomes of incident cardiovascular disease and mortality, among participants free of cardiovascular disease at baseline (Stranges *et al.*, 2006). After 7.6 years of follow-up, there was no effect of selenium supplementation on all cardiovascular disease outcomes, myocardial infarction, stroke, or cardiovascular mortality. Furthermore, when the sample was stratified according to plasma selenium level by tertiles, the lack of effect of selenium supplementation on cardiovascular outcomes was similarly observed across all tertiles of selenium level. While this was an important attempt to dissect the effect of reversing selenium deficiency on cardiovascular outcomes, the small sample size after tertile stratification in a relatively low cardiovascular risk population may have limited the study power to detect potentially important differences. The generalizability of the study findings from the rather specific study sample to other populations of interest is also uncertain.

The largest study to date was the Selenium and Vitamin E Cancer Prevention Trial (SELECT), another study originally targeted at nutritional prevention of cancer, mainly prostate cancer. It was a multicenter, double-blinded, randomized controlled trial of 35 533 men from numerous centers scattered across the United States, Canada, and Puerto Rico. Participants were randomized to receive supplementations of selenium 200 µg per day (L-selenomethionine), vitamin E, selenium 200 µg per day plus vitamin E, *versus* placebo over a median follow-up of 5.5 years. The study showed that there was no effect of selenium supplementation in the risk of cardiovascular events in this relatively healthy male population.

29.3.2.4 Systematic Review

An earlier systematic review suggested that there is no clear benefit or harm of vitamins or mineral supplementation, including selenium, for the primary prevention of cardiovascular diseases or cancer (Fortmann *et al.*, 2013). A further Cochrane Database Systematic Review on selenium supplementation was recently published, which included 12 available randomized controlled trials on selenium supplementation for primary cardiovascular prevention, comprising a total of 19 715 participants in the randomized analyses (Brigo *et al.*, 2014). The strength of this review was the comprehensive inclusion but trials specifically testing selenium-only supplementation. Largest among the included trials were the SELECT and the NPC trials, which provided data on clinical events. The strengths and weaknesses of these two landmarks trials were already described in the aforementioned. In this systematic review, overall no significant effects of selenium supplementation on cardiovascular events (fatal or nonfatal) were observed (overall relative risk [RR] 1.03, 95% Confidence Interval [CI] 0.95–1.11). There was also no impact of selenium supplementation on cardiovascular or all-cause mortality. A small potentially increased risk of type 2 diabetes with selenium supplementation was seen (*a priori* specified), but this was not statistically significant (RR 1.06, 95% CI 0.97–1.15).

From the systematic review, it seems clear that selenium supplementation has no protective role for primary cardiovascular prevention in the healthy population. Yet importantly, as acknowledged in the review, the participants were relatively healthy subjects mostly residing in the United States and who were likely to have obtained adequate selenium from daily dietary intake. Thus, the key question that remains is whether reversing selenium deficiency in high-risk patients might result in cardiovascular benefits. Further studies should focus on high-risk subjects rather than persons of low cardiovascular risks. Also, one weakness of the studies included in the systematic review was the lack of inclusion of vascular function endpoints such as brachial FMD, PWV, or carotid IMT, which are important assessments from the mechanistic as well as risk prediction points of view.

29.4 Selenium in Other Endothelial Dysfunction-Related Clinical Conditions

Contemporary understanding of cardiovascular diseases and risk factors increasingly recognizes such entities as a continuous clinical spectrum (Dzau *et al.*, 2006), rather than discrete pathological entities – commonly characterized by disruption of the immunomodulatory, oxidative stress, neurohormonal, and prothrombotic mechanistic pathways, as well as vascular endothelial dysfunction. Adopting this concept, in dissecting the relations between selenium deficiency and endothelial dysfunction, one would also be interested to examine the wider spectrum of clinical conditions beyond conventional atherosclerotic diseases that are associated with endothelial dysfunction.

29.4.1 Rheumatoid Arthritis

One such condition is rheumatoid arthritis, which is linked through systemic inflammation to excessive cardiovascular risks. This is evidenced by impaired vascular endothelial function, accelerated atherogenesis and increased systemic arterial and coronary calcifications (Yiu *et al.*, 2011a), which eventually leads to higher risks of coronary artery disease, stroke and cardiovascular mortality (Yiu *et al.*, 2011b). In an interesting cross-sectional study, Pemberton *et al.* (2009) studied the relation of oxidative stress and inflammatory biomarkers to vascular function in 46 patients with rheumatoid arthritis *versus* age-matched controls. It was found that serum selenium was significantly lower in rheumatoid arthritis patients, and these patients had increased prevalence of ultrasound-ascertained carotid plaques, as well as adverse profiles of oxidative stress and inflammatory markers, together with increased vascular cell adhesion molecules (VCAM) and E-selectin. However, there was no report of direct relations between selenium level and these markers.

29.4.2 Pre-eclampsia

Pre-eclampsia is certainly another intriguing condition to explore in this context. Increasing numbers of studies showed that pre-eclampsia, characterized as gestational proteinuric hypertension, is a condition of generalized maternal vascular damage caused by systemic endothelial dysfunction that leads to various end-organ complications. Antiangiogenic factors released by the placenta have been postulated as a possible mechanism (Chambers *et al.*, 2001). Interestingly, studies observed that maternal oxidative stress was increased in pre-eclampsia, and such changes were related to reduced glutathione peroxidase activity and depleted selenium levels (Mistry *et al.*, 2008). A double-blind randomized controlled study investigated the effects of a low-dose selenium-enriched (60 μg per day) yeast supplement *versus* placebo in 230 primiparous pregnant women since 12 to14 weeks of gestation until delivery. As expected, plasma selenoprotein P was significantly higher among subjects in the intervention arm at the end of the study. Interestingly, serum soluble vascular endothelial growth factor receptor-1, which is an antiangiogenic factor associated with pre-eclampsia, was also significantly lower in the intervention *versus* placebo arm among subjects at the lowest quartile of baseline selenium status. Such exploratory findings and any benefit of selenium in preventing pre-eclampsia will warrant verification in future adequately powered trials (Rayman *et al.*, 2014).

29.5 Selenium Deficiency, Cardiomyopathy and Heart Failure

The links between selenium deficiency, cardiomyopathy and heart failure have been well recognized (de Lorgeril and Salen, 2000). Back in the 1930s there were reports of an endemic dilated cardiomyopathy in the Keshan region of Eastern China, where soil levels of selenium were low. Reported as the "Keshan disease", this endemic form of dilated cardiomyopathy presents with symptoms of acute/chronic congestive heart failure and cardiac arrhythmias, and is potentially fatal (Saliba *et al.*, 2010). It was found closely associated with selenium deficiency, and sodium selenite supplementation through dietary means was found to lower the occurrence of the disease (Lei *et al.*, 2009). In various clinical reports, selenium deficiency was also identified in idiopathic cases of heart failure, and reversal with supplementation led to resolution of heart-failure symptoms and also restoration of left ventricular function, as confirmed on echocardiography (Saliba *et al.*, 2010; Al-Matary *et al.*, 2013).

Indeed, there are a number of clinical conditions/factors that may contribute to selenium deficiency and in turn adversely affect cardiac function. For example, patients on ketogenic diet for epilepsy control are prone to develop selenium deficiency (Bergqvist *et al.*, 2003), and this may result in impaired myocardial function, clinical heart failure, electrocardiographic QT prolongation, and sudden cardiac deaths have also been reported (Best *et al.*, 2000; Bank *et al.*, 2008; Sirikonda *et al.*, 2012). Life-threatening cardiomyopathy has also been reported in patients who have undergone bariatric surgery and subsequently developed nutritional selenium deficiency (Boldery *et al.*, 2007). Similar cases have been encountered with intestinal malabsorption syndromes (Frustaci *et al.*, 2012). Furthermore, preclinical studies also suggested that selenium may play a role in the cardiac electrophysiology of diabetic cardiomyopathy, by restoring the prolongation of action potential at the cellular level, a key finding in diabetic cardiomyopathy (Ozturk *et al.*, 2013).

In areas where populations are more dependent on processed food consumption, such as the Saudi Arabian region, selenium deficiency may also be potentially more common (Al-Matary *et al.*, 2013). When encountering cases of idiopathic dilated cardiomyopathy/heart failure where symptoms remain significant and persistent, and that etiology remained unclear despite extensive investigations, a higher degree of clinical suspicion of selenium deficiency may be important. Confirming a serological deficient state of selenium and appropriate reversal by supplementation may result in clinical benefits (Saliba *et al.*, 2010; Marinescu and McCullough, 2011).

29.6 Genetic Polymorphisms of Selenoprotein: Cardiovascular Implications

A number of genetic studies have also explored the cardiovascular role of genetic polymorphisms linked to selenium and related oxidative stress regulatory pathways. For instance, variations of selenoprotein S gene locus has been associated with the risk of coronary heart disease and ischemic stroke in two independent Finish cohorts (Alanne *et al.*, 2007). In another study among patients with type 2 diabetes, genetic variants of selenoprotein S were further associated with subclinical atherosclerosis as measured by coronary calcification and carotid plaques, as well as self-reported prior history of cardiovascular diseases (Cox *et al.*, 2013). Furthermore, selenoprotein S genetic polymorphisms were also found to be associated with inflammatory cytokines including tumor necrosis factor-alpha and interleukin-1 beta, suggesting a direct mechanistic role of selenoprotein S in mediating inflammation (Curran *et al.*, 2005). Future genetic and genomic studies have the potential to further shed lights on the mechanistic pathways of selenium and selenoprotein in modulating the atherosclerotic process, and may as well provide a platform for genetic causal inference analysis at the population level.

29.7 The Selenium Paradox

While investigating for the potential beneficial effects of reversing selenium deficiency in susceptible patients, we should also cautiously adopt the precautionary principle in all clinical and research settings to *primum non nocere, i.e.* do no harm. Studies have identified potentially detrimental effects of high selenium levels on cardiovascular health, in terms of adverse changes in lipid profile (Stranges *et al.*, 2010), increased risk of diabetes (Laclaustra *et al.*, 2009), although these suggestions are unconfirmed. The potentially U-shaped association between selenium level and atherosclerosis (Bleys *et al.*, 2009) warrants special attention.

29.8 Current Recommendations and Future Directions

According to the current clinical evidence, we conclude that there is no evidence of any benefits of selenium supplementation for the primary prevention of cardiovascular diseases. Although there is no clear harm of selenium supplementation, there exists potential concerns of increased diabetes and adverse changes in lipid profile in nondiscriminate use of selenium supplements. Yet clinical studies also revealed that selenium deficiency may have a crucial role in vascular dysfunction and cardiovascular risks in susceptible patients. Any potential vascular protective effects of selenium are likely to be conferred to persons with selenium deficiency who are at high risk of cardiovascular events. Identifying and reversing selenium deficiency, rather than nontargeted universal supplementation should be the focus of future clinical trials. These trials should not only interrogate the hard clinical endpoints but also the mechanistic pathways that include vascular function surrogate assessments. The long-term side-effect profile of selenium supplementation should also be clarified. Before concrete clinical evidence of vascular benefits is available, routine selenium supplementation or food fortification should be not recommended.

Furthermore, the astute clinicians should be alerted to consider testing for selenium deficiency in patients presenting with unexplained cardiomyopathy or heart failure despite extensively negative etiological work-ups, especially in endemic regions of selenium deficiency or among those who are exposed to risk factors. These include restricted dietary intake pattern, dependency on processed food items, recent bariatric surgery, and gastrointestinal malabsorption syndromes. Prompt diagnosis and correction of selenium deficiency may reverse cardiac impairment and result in clinical improvements of heart failure. This clinical wisdom is most ideal to be tested in a well-designed randomized controlled trial.

Summary Points

- The focus of this chapter is the role of selenium in vascular function and cardiomyopathy based on current clinical studies.
- Selenium deficiency is independently associated with adverse systemic arterial function in patients with high cardiovascular risk. Whether reversal of selenium deficiency may give rise to clinical benefits needs to be investigated in randomized controlled trials.
- There is currently no evidence of benefit of selenium supplementation for the primary prevention of cardiovascular diseases.
- There are potential concerns of increased diabetes and adverse changes in lipid profile in subjects with high selenium levels, but these are not confirmed and need to be further investigated in future studies.
- Selenium deficiency may be an etiological factor in patients with unexplained cardiomyopathy or heart failure. If confirmed by a blood test, correction of selenium deficiency may reverse cardiac impairment and result in clinical improvements of heart failure.
- Risk factors of selenium deficiency include restricted dietary intake pattern, dependency on processed food items, recent bariatric surgery, and gastrointestinal malabsorption syndromes. Selenium levels are also affected by local soil selenium contents.

Key Facts

The Key Facts of Atherosclerosis, Including the Clinical Definition, Pathophysiological Features, Clinical Relevance, and Common Causes

1. Atherosclerosis is a chronic, degenerative pathophysiological process of blood vessels characterized by reduced vascular elasticity and formation of "fibrofatty" atherosclerotic plaques.
2. As atherosclerosis ensues, fibrofatty plaque progressively enlarges and cause luminal narrowing and impediment of blood flow.
3. When impediment of blood flow to target organs (such as the heart, brain, limbs) becomes so significant or when the plaques rupture causing acute obstruction to blood flow, symptomatic cardiovascular diseases result.
4. Causes of atherosclerosis are multifactorial and include high blood pressure, hyperlipidemia, diabetes, oxidative stress, inflammation, genetic predisposition, smoking, and a wide array of environmental/lifestyle factors.

The Key Facts of Heart Failure, Including Pathophysiological Features, Clinical Symptoms and Diagnosis

1. Heart failure is a clinical syndrome arising from insufficiency of circulatory output to maintain physiological needs of the body due to heart-function impairment.
2. Heart failure is characterized by symptoms of shortness of breath, tiredness, reduced exertional capacity, and features of gravity-dependent fluid retention in the body such as leg swelling.
3. Such heart-function impairment could be systolic (impaired cardiac contractility) or diastolic (impaired cardiac muscle relaxation); and could predominantly affect the left or right heart chambers, or both.
4. Diagnosis of heart failure is by clinical features and assisted by cardiac imaging such as echocardiography, which confirms the systolic or diastolic function impairment and may show abnormal dilatation (enlargement) of the heart ventricle(s).
5. Novel blood markers such as raised brain natriuretic peptide (BNP) and NT-proBNP (N-terminal prohormone of BNP) are also helpful in confirming the diagnosis.

Definitions of Words and Terms

Carotid intima-media thickness (IMT). Defined as the distance between the lumen-intima and media-adventitia border of the carotid arterial vascular wall with the use of electronic calipers on high-resolution ultrasonography, averaged by multiple measurements. It is a structural indicator of vascular function.

Cardiovascular continuum. The clinical concept where cardiovascular diseases and related risk factors are no discrete pathological entities, but are part of a continuous spectrum of risk. Persons at the early stages of the continuum have few/no risk factors and absence of atherosclerosis, where subclinical atherosclerosis ensues as the risk spectrum progresses, and persons at the advanced end of the spectrum are those with clinically manifest vascular events. This concept allows insights to be obtained into the mechanistic pathological changes that occur along the continuum, and provide conceptual basis for primary and secondary cardiovascular prevention.

Cohort studies. A prospective or retrospective (usually the former) analytic epidemiological study design, where baseline selenium exposure of a group of participants who share common experience is ascertained, and analyzed against cardiovascular surrogate or clinical endpoints at follow-up to elucidate etiological factors.

Endothelial dysfunction. Systemic impairment of vascular function that is dependent mainly *via* the nitric oxide synthase pathway. A precursor to atherosclerosis and is an independent predictor of cardiovascular events. It is believed to be a key mechanism of the cardiovascular continuum progression.

Flow-mediated vasodilatation (FMD). A noninvasive technique and surrogate estimate of vascular endothelial function determined by high-resolution ultrasound assessment. The patient is allowed to rest in the supine position, and the assessed arm is subjected to occlusion by blood-pressure cuff for a short duration, followed by sudden release that causes sudden blood flow along the brachial artery to impose shear stress and stimulate release of vasoactive nitric oxide. FMD is defined as the percentage dilatation of the diameter of the brachial artery.

Genetic polymorphisms. Genetic variations of a defined genetic locus, commonly described as single nucleotide polymorphisms.

Nested case-control studies. Studies that compare new-onset diagnosed cases of ongoing cohort studies with subjects who are free of the disease that are derived from the same cohort study, to elucidate etiological factors of the disease.

Pulse wave velocity (PWV). Velocity by which a pulse wave propagates along a defined functional path of vasculature. It is an estimate of arterial stiffness that indicates the severity of atherosclerosis.

Randomized controlled trial. This is the gold standard design of experimental clinical medicine where subjects are randomly allocated to the active intervention arm *versus* the control arm. Since all biases or confounding are effectively eliminated by random allocation provided that sample size is sufficiently large, any differences in clinical outcomes are due to the effect of the intervention.

Selenium paradox. This refers to the paradoxical phenomenon where selenium level exhibits a U-shaped relation with atherosclerosis, and suggests that adequate selenium may be protective against atherosclerosis in the lower ranges; where it may become detrimental to vascular heath if excessive.

List of Abbreviations

ADMA	Asymmetric Dimethylarginine
CI	Confidence Interval
FMD	Flow-Mediated Dilatation
IMT	Carotid Intima-Media Thickness
NPC Trial	Nutritional Prevention of Cancer Trial
PWV	Pulse Wave Velocity
RR	Relative Risk
SELECT	Selenium and Vitamin E Cancer Prevention Trial
SETCAP	SElenium Therapy in Coronary Artery disease Patients Study
TBARS	Thiobarbituric Acid-Reactive Substances
VCAM	Vascular Cell Adhesion Molecules

References

Alanne, M., Kristiansson, K., Auro, K., Silander, K., Kuulasmaa, K., Peltonen, L., Salomaa, V. and Perola, M., 2007. Variation in the selenoprotein S gene locus is associated with coronary heart disease and ischemic stroke in two independent Finnish cohorts. *Human Genetics*. 122: 355–365.

Al-Matary, A., Hussain, M. and Ali, J., 2013. Selenium: a brief review and a case report of selenium responsive cardiomyopathy. *BMC Pediatrics*. 13: 39.

Atakan, A., Macunluoglu, B., Kaya, Y., Ari, E., Demir, H., Asicioglu, E. and Kaspar, C., 2013. Decreased serum selenium levels are correlated with diminished coronary flow reserve among hemodialysis patients. *Biological Trace Element Research*. 155: 333–338.

Bank, I. M., Shemie, S. D., Rosenblatt, B., Bernard, C. and Mackie, A. S., 2008. Sudden cardiac death in association with the ketogenic diet. *Pediatric Neurology*. 39: 429–431.

Bayir, A., Kara, H., Kiyici, A., Ozturk, B. and Akyurek, F., 2013. Levels of selenium, zinc, copper, and cardiac troponin I in serum of patients with acute coronary syndrome. *Biological Trace Element Research*. 154: 352–356.

Bergqvist, A. G., Chee, C. M., Lutchka, L., Rychik, J. and Stallings, V. A., 2003. Selenium deficiency associated with cardiomyopathy: a complication of the ketogenic diet. *Epilepsia*. 44: 618–620.

Best, T. H., Franz, D. N., Gilbert, D. L., Nelson, D. P. and Epstein, M. R., 2000. Cardiac complications in pediatric patients on the ketogenic diet. *Neurology*. 54: 2328–2330.

Blacher, J., Guerin, A. P., Pannier, B., Marchais, S. J. and London, G. M., 2001. Arterial calcifications, arterial stiffness, and cardiovascular risk in end-stage renal disease. *Hypertension*. 38: 938–942.

Bleys, J., Navas-Acien, A., Laclaustra, M., Pastor-Barriuso, R., Menke, A., Ordovas, J., Stranges, S. and Guallar, E., 2009. Serum selenium and peripheral arterial disease: results from the national health and nutrition examination survey, 2003–2004. *American Journal of Epidemiology*. 169: 996–1003.

Boldery, R., Fielding, G., Rafter, T., Pascoe, A. L. and Scalia, G. M., 2007. Nutritional deficiency of selenium secondary to weight loss (bariatric) surgery associated with life-threatening cardiomyopathy. *Heart, Lung and Circulation*. 16: 123–126.

Bonithon-Kopp, C., Coudray, C., Berr, C., Touboul, P. J., Feve, J. M., Favier, A. and Ducimetiere, P., 1997. Combined effects of lipid peroxidation and antioxidant status on carotid atherosclerosis in a population aged 59–71 y: The EVA study. Etude sur le Vieillisement Arteriel. *American Journal of Clinical Nutrition*. 65: 121–127.

Boutouyrie, P., Tropeano, A. I., Asmar, R., Gautier, I., Benetos, A., Lacolley, P. and Laurent, S., 2002. Aortic stiffness is an independent predictor of primary coronary events in hypertensive patients: a longitudinal study. *Hypertension*. 39: 10–15.

Brigo, F., Storti, M., Lochner, P., Tezzon, F. and Nardone, R., 2014. Selenium supplementation for primary prevention of cardiovascular disease: proof of no effectiveness. *Nutrition, Metabolism & Cardiovascular Diseases*. 24: e2–e3.

Chambers, J. C., Fusi, L., Malik, I. S., Haskard, D. O., De Swiet, M. and Kooner, J. S., 2001. Association of maternal endothelial dysfunction with preeclampsia. *JAMA, the Journal of the American Medical Association*. 285: 1607–1612.

Chan, Y. H., Siu, C. W., Yiu, K. H., Chan, H. T., Li, S. W., Tam, S., Cheung, B. M., Lau, C. P., Lam, T. H. and Tse, H. F., 2012. Adverse systemic arterial function in patients with selenium deficiency. *Journal of Nutrition, Health & Aging*. 16: 85–88.

Cominetti, C., de Bortoli, M. C., Garrido Jr, A. B. and Cozzolino, S. M., 2012. Brazilian nut consumption improves selenium status and glutathione peroxidase activity and reduces atherogenic risk in obese women. *Nutrition Research*. 32: 403–407.

Cox, A. J., Lehtinen, A. B., Xu, J., Langefeld, C. D., Freedman, B. I., Carr, J. J. and Bowden, D. W., 2013. Polymorphisms in the selenoprotein S gene and subclinical cardiovascular disease in the diabetes heart study. *Acta Diabetologica*. 50: 391–399.

Curran, J. E., Jowett, J. B., Elliott, K. S., Gao, Y., Gluschenko, K., Wang, J., Abel Azim, D. M., Cai, G., Mahaney, M. C., Comuzzie, A. G., Dyer, T. D., Walder, K. R., Zimmet, P., MacCluer, J. W., Collier, G. R., Kissebah, A. H. and Blangero, J., 2005. Genetic variation in selenoprotein S influences inflammatory response. *Nature Genetics*. 37: 1234–1241.

Dzau, V. J., Antman, E. M., Black, H. R., Hayes, D. L., Manson, J. E., Plutzky, J., Popma, J. J. and Stevenson, W., 2006. The cardiovascular disease continuum validated: clinical evidence of improved patient outcomes: part I: pathophysiology and clinical trial evidence (risk factors through stable coronary artery disease). *Circulation*. 114: 2850–2870.

Flores-Mateo, G., Navas-Acien, A., Pastor-Barriuso, R. and Guallar, E., 2006. Selenium and coronary heart disease: a meta-analysis. *American Journal of Clinical Nutrition*. 84: 762–773.

Fortmann, S. P., Burda, B. U., Senger, C. A., Lin, J. S. and Whitlock, E. P., 2013. Vitamin and mineral supplements in the primary prevention of cardiovascular disease and cancer: an updated systematic evidence review for the U.S. Preventive Services Task Force. *Annals of Internal Medicine*. 159: 824–834.

Frustaci, A., Sabbioni, E., Fortaner, S., Farina, M., del Torchio, R., Tafani, M., Morgante, E., Ciriolo, M. R., Russo, M. A. and Chimenti, C., 2012. Selenium- and zinc-deficient cardiomyopathy in human intestinal malabsorption: preliminary results of selenium/zinc infusion. *European Journal of Heart Failure*. 14: 202–210.

Fryer, M. J., 2000. Selenium and human health. *Lancet*. 356: 943.

Gokce, N., Keaney Jr, J. F., Hunter, L. M., Watkins, M. T., Nedeljkovic, Z. S., Menzoian, J. O. and Vita, J. A., 2003. Predictive value of noninvasively determined endothelial dysfunction for long-term cardiovascular events in patients with peripheral vascular disease. *Journal of the American College of Cardiology*. 41: 1769–1775.

Hawkes, W. C. and Laslett, L. J., 2009. Selenium supplementation does not improve vascular responsiveness in healthy North American men. *American Journal of Physiology Heart and Circulatory Physiology*. 296: H256–H262.

Kardinaal, A. F., Kok, F. J., Kohlmeier, L., Martin-Moreno, J. M., Ringstad, J., Gomez-Aracena, J., Mazaev, V. P., Thamm, M., Martin, B. C., Aro, A., Kark, J. D., Delgado-Rodriguez, M., Riemersma, R. A., van 't Veer, P. and Huttunen, J. K., 1997. Association between toenail selenium and risk of acute myocardial infarction in European men. The EURAMIC study. European antioxidant myocardial infarction and breast cancer. *American Journal of Epidemiology*. 145: 373–379.

Koszta, G., Kacska, Z., Szatmari, K., Szerafin, T. and Fulesdi, B., 2012. Lower whole blood selenium level is associated with higher operative risk and mortality following cardiac surgery. *Journal of Anesthesia*. 26: 812–821.

Koyama, H., Abdulah, R., Ohkubo, T., Imai, Y., Satoh, H. and Nagai, K., 2009. Depressed serum selenoprotein P: possible new predicator of increased risk for cerebrovascular events. *Nutrition Research*. 29: 94–99.

Laclaustra, M., Navas-Acien, A., Stranges, S., Ordovas, J. M. and Guallar, E., 2009. Serum selenium concentrations and diabetes in U.S. adults: National Health and Nutrition Examination Survey (NHANES) 2003–2004. *Environmental Health Perspectives*. 117: 1409–1413.

Laurent, S., Boutouyrie, P., Asmar, R., Gautier, I., Laloux, B., Guize, L., Ducimetiere, P. and Benetos, A., 2001. Aortic stiffness is an independent predictor of all-cause and cardiovascular mortality in hypertensive patients. *Hypertension*. 37: 1236–1241.

Laurent, S., Katsahian, S., Fassot, C., Tropeano, A. I., Gautier, I., Laloux, B. and Boutouyrie, P., 2003. Aortic stiffness is an independent predictor of fatal stroke in essential hypertension. *Stroke*. 34: 1203–1206.

Lei, C., Niu, X., Wei, J., Zhu, J. and Zhu, Y., 2009. Interaction of glutathione peroxidase-1 and selenium in endemic dilated cardiomyopathy. *Clinica Chimica Acta*. 399: 102–108.

de Lorgeril, M. and Salen, P., 2000. Selenium and chronic heart failure. *Circulation*. 101: E74.

Marinescu, V. and McCullough, P. A., 2011. Nutritional and micronutrient determinants of idiopathic dilated cardiomyopathy: diagnostic and therapeutic implications. *Expert Review of Cardiovascular Therapy*. 9: 1161–1170.

Mattace-Raso, F. U., van der Cammen, T. J., Hofman, A., van Popele, N. M., Bos, M. L., Schalekamp, M. A., Asmar, R., Reneman, R. S., Hoeks, A. P., Breteler, M. M. and Witteman, J. C., 2006. Arterial stiffness and risk of coronary heart disease and stroke: the Rotterdam study. *Circulation*. 113: 657–663.

Mistry, H. D., Wilson, V., Ramsay, M. M., Symonds, M. E. and Broughton Pipkin, F., 2008. Reduced selenium concentrations and glutathione peroxidase activity in preeclamptic pregnancies. *Hypertension*. 52: 881–888.

Mitchell, G. F., Hwang, S. J., Vasan, R. S., Larson, M. G., Pencina, M. J., Hamburg, N. M., Vita, J. A., Levy, D. and Benjamin, E. J., 2010. Arterial stiffness and cardiovascular events. The Framingham heart study. *Circulation*. 121(4): 505–511.

Ozturk, N., Olgar, Y. and Ozdemir, S., 2013. Trace elements in diabetic cardiomyopathy: an electrophysiological overview. *World Journal of Diabetes*. 4: 92–100.

Pemberton, P. W., Ahmad, Y., Bodill, H., Lokko, D., Hider, S. L., Yates, A. P., Walker, M. G., Laing, I. and Bruce, I. N., 2009. Biomarkers of oxidant stress, insulin sensitivity and endothelial activation in rheumatoid arthritis: a cross-sectional study of their association with accelerated atherosclerosis. *BMC Research Notes*. 2: 83.

Rayman, M. P., Searle, E., Kelly, L., Johnsen, S., Bodman-Smith, K., Bath, S. C., Mao, J. and Redman, C. W., 2014. Effect of selenium on markers of risk of pre-eclampsia in UK pregnant women: a randomised, controlled pilot trial. *British Journal of Nutrition*. 1–13.

Rayman, M. P., 2000. The importance of selenium to human health. *Lancet*. 356: 233–241.

Saliba, W., El Fakih, R. and Shaheen, W., 2010. Heart failure secondary to selenium deficiency, reversible after supplementation. *International Journal of Cardiology*. 141: e26–e27.

Salonen, J. T., Alfthan, G., Huttunen, J. K., Pikkarainen, J. and Puska, P., 1982. Association between cardiovascular death and myocardial infarction and serum selenium in a matched-pair longitudinal study. *Lancet*. 2: 175–179.

Schnabel, R., Lubos, E., Messow, C. M., Sinning, C. R., Zeller, T., Wild, P. S., Peetz, D., Handy, D. E., Munzel, T., Loscalzo, J., Lackner, K. J. and Blankenberg, S., 2008. Selenium supplementation improves antioxidant capacity *in vitro* and *in vivo* in patients with coronary artery disease the SElenium Therapy in Coronary Artery disease Patients (SETCAP) study. *American Heart Journal*. 156(1201): e1201–e1211.

Seo, W. K., Lee, J. M., Park, M. H., Park, K. W. and Lee, D. H., 2008. Cerebral microbleeds are independently associated with arterial stiffness in stroke patients. *Cerebrovascular Diseases*. 26: 618–623.

Shargorodsky, M., Debby, O., Matas, Z. and Zimlichman, R., 2010. Effect of long-term treatment with antioxidants (vitamin C, vitamin E, coenzyme Q10 and selenium) on arterial compliance, humoral factors and inflammatory markers in patients with multiple cardiovascular risk factors. *Nutrition & Metabolism*. 7: 55.

Shechter, M., Shechter, A., Koren-Morag, N., Feinberg, M. S. and Hiersch, L., 2014. Usefulness of brachial artery flow-mediated dilation to predict long-term cardiovascular events in subjects without heart disease. *American Journal of Cardiology*. 113: 162–167.

Sirikonda, N. S., Patten, W. D., Phillips, J. R. and Mullett, C. J., 2012. Ketogenic diet: rapid onset of selenium deficiency-induced cardiac decompensation. *Pediatric Cardiology*. 33: 834–838.

Stranges, S., Marshall, J. R., Trevisan, M., Natarajan, R., Donahue, R. P., Combs, G. F., Farinaro, E., Clark, L. C. and Reid, M. E., 2006. Effects of selenium supplementation on cardiovascular disease incidence and mortality: secondary analyses in a randomized clinical trial. *American Journal of Epidemiology*. 163: 694–699.

Stranges, S., Laclaustra, M., Ji, C., Cappuccio, F. P., Navas-Acien, A., Ordovas, J. M., Rayman, M. and Guallar, E., 2010. Higher selenium status is associated with adverse blood lipid profile in British adults. *Journal of Nutrition*. 140: 81–87.

Suadicani, P., Hein, H. O. and Gyntelberg, F., 1992. Serum selenium concentration and risk of ischaemic heart disease in a prospective cohort study of 3000 males. *Atherosclerosis*. 96: 33–42.

Sutton-Tyrrell, K., Najjar, S. S., Boudreau, R. M., Venkitachalam, L., Kupelian, V., Simonsick, E. M., Havlik, R., Lakatta, E. G., Spurgeon, H., Kritchevsky, S., Pahor, M., Bauer, D. and Newman, A., 2005. Elevated aortic pulse wave velocity, a marker of arterial stiffness, predicts cardiovascular events in well-functioning older adults. *Circulation*. 111: 3384–3390.

Tappel, A. L., 1974. Selenium-glutathione peroxidase and vitamin E. *American Journal of Clinical Nutrition*. 27: 960–965.

Yang, S. J., Hwang, S. Y., Choi, H. Y., Yoo, H. J., Seo, J. A., Kim, S. G., Kim, N. H., Baik, S. H., Choi, D. S. and Choi, K. M., 2011. Serum selenoprotein P levels in patients with type 2 diabetes and prediabetes: implications for insulin resistance, inflammation, and atherosclerosis. *Journal of Clinical Endocrinology and Metabolism*. 96: E1325–E1329.

Yilmaz, M. I., Saglam, M., Caglar, K., Cakir, E., Sonmez, A., Ozgurtas, T., Aydin, A., Eyileten, T., Ozcan, O., Acikel, C., Tasar, M., Genctoy, G., Erbil, K., Vural, A. and Zoccali, C., 2006. The determinants of endothelial dysfunction in CKD: oxidative stress and asymmetric dimethylarginine. *American Journal of Kidney Diseases*. 47: 42–50.

Yiu, K. H., Wang, S., Mok, M. Y., Ooi, G. C., Khong, P. L., Lau, C. S. and Tse, H. F., 2011a. Relationship between cardiac valvular and arterial calcification in patients with rheumatoid arthritis and systemic lupus erythematosus. *Journal of Rheumatology*. 38: 621–627.

Yiu, K. H., Tse, H. F., Mok, M. Y. and Lau, C. S., 2011b. Ethnic differences in cardiovascular risk in rheumatic disease: focus on Asians. *Nature Reviews Rheumatology*. 7: 609–618.

Zureik, M., Galan, P., Bertrais, S., Mennen, L., Czernichow, S., Blacher, J., Ducimetiere, P. and Hercberg, S., 2004. Effects of long-term daily low-dose supplementation with antioxidant vitamins and minerals on structure and function of large arteries. *Arteriosclerosis, Thrombosis, and Vascular Biology*. 24: 1485–1491.

CHAPTER 30

Selenium and the Heart

JACOB JOSEPH*[a]

[a]Department of Medicine, VA Boston Healthcare, Harvard Medical School, 1400 VFW Parkway, West Roxbury, MA 02132, USA
*E-mail: jacob.joseph@va.gov

30.1 Introduction

Unlike other minerals that are important to cardiovascular health, selenium is unique in that it does not act as a cofactor or prosthetic group; rather it is directly incorporated into proteins termed selenoproteins as the unique amino acid selenocysteine (Joseph, 2013). Even though the functions of all these selenoproteins are not known, many of them, such as glutathione peroxidases (GPX) and thioredoxin reductases (TRR), are crucial for maintaining normal oxidant signaling and preventing oxidant stress in the myocardium. Since oxidant stress is implicated in the genesis and progression of myocardial and vascular diseases, selenium deficiency has been investigated as a risk factor for cardiovascular disease (CVD). Keshan disease (KD), a cardiomyopathy endemic to areas with low selenium content of soil, has been attributed to selenium deficiency. In addition, several studies have demonstrated an association of low blood selenium level with coronary artery disease; however, these associations are mostly observed in areas with lower soil selenium level, and studies have also shown a lack of association. Furthermore, two recent large-scale clinical trials of selenium supplementation in cancer prevention demonstrated that modest selenium supplementation increased the incidence of a major risk factor for CVD, *i.e.*, diabetes (Joseph, 2013; Joseph

Food and Nutritional Components in Focus No. 9
Selenium: Chemistry, Analysis, Function and Effects
Edited by Victor R. Preedy

Published by the Royal Society of Chemistry, www.rsc.org

and Loscalzo, 2013). Hence, it is clear that the exact mechanisms of selenium's actions and possible cardioprotection provided by modulating selenium levels are not clearly understood. This review will outline the current knowledge gained from preclinical studies of selenium and the heart, and propose novel avenues of preclinical research.

30.2 Selenium Deficiency and Cardiomyopathy

30.2.1 Keshan Disease

Keshan disease is a cardiomyopathy that is highly prevalent in areas of China that have a low selenium content in the soil. Clinical and epidemiologic studies suggest that selenium deficiency is a causative factor in this cardiomyopathy. Ge *et al.* described the histopathological features of KD in a postmortem analysis. In addition to multifocal myocardial necrosis and fibrous replacement of the myocardium, mitochondrial changes were also observed in this study (Ge *et al.*, 1983). A recent study examined the role of increased myocardial oxidant stress due to selenoprotein deficiency in KD in myocardial specimens obtained from subjects with KD and controls (Pei *et al.*, 2013). Immunohistochemical analysis showed that oxidant stress (assessed by expression of the DNA oxidation product 8-hydroxy-2-deoxy guanosine was increased in KD and associated with the degree of myocardial damage and that expression of the selenoproteins TRR-1 and GPX-1 was lower in KD. Conrad and colleagues examine the role of the selenoprotein thioredoxin reductase type 2 (TRR2) in cardiac development (Conrad *et al.*, 2004). In this study, a cardiac restricted deletion of TRR2 in mice resulted in a fatal cardiomyopathy similar to KD. Jia and colleagues examined the role of oxidant stress in KD utilizing combined selenium and vitamin E deficiency as a rodent model of KD (Jia *et al.*, 2011). Dehydroepiandrosterone, an adrenal androgenic steroid, reduced oxidant stress, matrix deposition, and activation of nuclear factor kappa-light-chain-enhancer of activated B cells (NFkB), and increase in the profibrotic cytokines TGF-β and connective tissue growth factor in this model. The benefits were attributed to the antioxidant effects of dehydroepiandrosterone. Since this model utilized combined deficiency of selenium and vitamin E, it is difficult to derive firm conclusions regarding a direct link between selenium deficiency and KD from this study.

Selenium deficiency has also been considered a predisposing factor to a second insult in the causation of KD. Based on an observed seasonal variation in the incidence of KD and isolation of viral isolates from myocardium of affected subjects, numerous preclinical studies have addressed whether KD results from an increased risk of viral myocarditis due to selenium deficiency (Beck *et al.*, 1994a). Beck and colleagues reported that a strain of coxsackievirus B3 that is not shown to cause myocarditis caused significant myocarditis-like changes in the myocardium of selenium-deficient mice compared to mice fed a normal diet, suggesting that the virus had mutated to a virulent genotype in selenium-deficient mice (Beck *et al.*, 1994a).

Further studies by Beck and colleagues showed that changes in the virus itself were responsible for worse myocarditis (as compared to immune changes) induced by selenium deficiency (Beck *et al.*, 1994b). Virus isolated from selenium-deficient mice induced severe myocarditis when inoculated into selenium-replete mice, while virus from selenium-replete mice did not cause worse myocarditis when injected into selenium-deficient mice. Sequence analysis of the different isolates revealed the presence of 6 nucleotide changes in the virulent strain obtained from selenium-deficient mice that were similar to known nucleotide differences between virulent and amyocarditic strains of coxsackievirus B3 (Beck *et al.*, 1995). Another study investigated the role of a specific selenoprotein in this process (Beck *et al.*, 1998). Wild-type and GPX deficient mice were inoculated with coxsackievirus B3. Myocarditis developed in GPX deficient mice, while wild-type mice were resistant to cardiac damage. Analysis of the viral genome showed point mutations in virus isolated from GPX deficient mice. One possible mechanism of this observed effect is the direct inhibition of coxsackievirus replication by selenium, a mechanism that has been demonstrated *in vitro* with exposure to high doses of selenite (5 μmol L^{-1}) (Cermelli *et al.*, 2002). This effect was not observed with selenate or selenomethionine, and selenite did not inhibit replication of two other virus strains. Since this effect was blocked by coincubation with the reducing agent dithiothreitol, selenite-induced oxidant stress was proposed by the investigators as the mechanism of action.

Epidemiologic and clinical studies of KD suggest that selenium deficiency is clearly linked to cardiomyopathy and heart failure. However, other known causes of cardiomyopathy may be acting synergistically with selenium deficiency in predisposing the myocardium to undergo adverse changes in structure and function. In support of this mechanism, an epidemiological study showed that annual incidence of KD decreased from 25.23 to 2.7 per 100 000 population with selenium supplementation (Cheng and Qian, 1990). The observation that incidence of KD decreased from 19.76 to 7.36 per 100 000 in a control population suggests that some of the decrease in incidence of KD may have been due to factors other than selenium. Further preclinical studies would be crucial to understand the precise role of selenium in modulating myocardial structure and function.

30.3 Selenium and Ischemic Myocardial Diseases

30.3.1 Myocardial Infarction

Tanguy and colleagues examined the effect of pretreatment with high selenium (1.5 mg kg^{-1}) or selenium-deficient diets on infarct size in rats subject to coronary ligation (Tanguy *et al.*, 2004). Supplemental selenium reduced infarct size, improved blood pressure after infarct, and maintained the ratio of reduced to oxidized glutathione compared to the low-selenium diet. Treatment for only 24 h with ebselen, a GPX mimetic, also reduced infarct size and myocardial dysfunction in rabbit model of experimental myocardial infarction (Baljinnyam *et al.*, 2006). These beneficial effects were associated with reduced levels of oxidant damage and markedly increased expression of heat-shock protein 72. Panicker *et al.* examined the effect of selenium in a nonischemic model of myocardial infarction, *i.e.*, isoproterenol infusion in rats (Panicker *et al.*, 2010), In this study the authors also observed less myocardial damage and oxidant stress with selenium supplementation. Nuclear translocation of the transcription factor NFkB, which is involved in expression of a number of proinflammatory molecules, was also reduced by selenium supplementation. The same group also demonstrated that, in this model of myocardial infarction, selenium supplementation led to a reduction in levels of the inflammatory mediator leukotriene B4, suggesting that the protective effects of selenium in myocardial infarction may be mediated *via* anti-inflammatory effects (Panicker *et al.*, 2012). Tanguy and colleagues reported that low selenium diet preceding myocardial infarction led to an increase in cardiac expression of the inflammatory cytokine tumor necrosis factor-alpha (TNF-α). In addition, the authors also found that selenium deficiency predisposed to dephosphorylation of a major gap junction protein, connexin-43 (Tanguy *et al.*, 2011).

30.3.2 Myocardial Ischemia–Reperfusion

Myocardial ischemia followed by reperfusion leads to release of reactive oxygen species and consequent cardiac dysfunction and arrhythmias. In a study utilizing an *ex vivo* model of 10 min of ischemia followed by reperfusion in isolated rat hearts, pretreatment of animals with selenium supplemented diets increased GPX activity and markedly reduced the occurrence of arrhythmias during reperfusion (Tanguy *et al.*, 1998). In another study in female rats, functional recovery following ischemia–reperfusion (I–R) was decreased and ultrastructural damage augmented by pretreatment of animals with a selenium-deficient diet for 10 weeks (Toufektsian *et al.*, 2000). Venardos and colleagues reported on the effects of various doses of selenium on myocardial oxidant stress following I–R. Selenium deficiency increased myocardial levels of protein carbonyls, lipid peroxidation, and I–R injury, while selenium supplementation had the opposite effect (Venardos *et al.*, 2004b). These findings were associated with an increase in GPX and TRR activity and gene expression with selenium supplementation, and the opposite effect in selenium deficiency (Venardos *et al.*, 2004a; Venardos *et al.*, 2005). As in the case with myocardial infarction, dephosphorylation of the gap junction protein connexin-43 was increased by selenium depletion during I–R injury, while selenium supplementation increased GPX activity and reduced connexin-43 dephosphorylation during myocardial I–R (Rakotovao *et al.*, 2005). Turan and colleagues have subsequently demonstrated that selenium's effects in reducing myocardial injury may involve deactivation of NFkB and prevention of nuclear translocation of this potent transcription factor during I–R (Turan *et al.*, 2005).

30.4 Doxorubicin (Adriamycin)-Induced Cardiomyopathy and Selenium

Van Vleet and Ferrans reported on the results of selenium supplementation in rabbits injected with adriamycin (also termed doxorubicin), a commonly used cancer chemotherapeutic agent associated with cardiomyopathy (Van Vleet and Ferrans, 1980). Selenium (sodium selenite 0.06 mg kg^{-1}) improved survival and reduced myocardial damage. A study by Jotti and colleagues examined the role of GPX enzymes in the selenium-induced cardioprotection in adriamycin toxicity (Jotti *et al.*, 1994). Interestingly, glutathione peroxidase activity in the heart was not different in adriamycin- or selenium-treated groups, suggesting restoration of GPX activity was not the mechanism of benefit of selenium. In contrast, Coudray *et al.* demonstrated that selenium supplementation reduced markers of oxidant stress and increased GPX activity in adriamycin treated rats (Coudray *et al.*, 1995). Another study demonstrated that the beneficial effects of selenium in adriamycin-induced myocardial dysfunction may be due to prevention of mitochondrial damage *via* restoring redox balance. Liu *et al.* examined the regulation of ATP-sensitive potassium channels, which may be involved in ischemia and cardiac excitability in selenium and adriamycin treated rats (Liu *et al.*, 2013). Cardiac expression of these channels were markedly reduced by selenium deficiency and adriamycin, and restored by selenium supplementation.

In contrast to the above reports, some studies have also reported a lack of effect of selenium status. Van Vleet *et al.* did not find a protective effect of selenium in a dog model of adriamycin induced cardiomyopathy (Van Vleet *et al.*, 1980). Similarly, Fischer and colleagues reported that severe selenium deficiency did not worsen doxorubicin-induced cardiomyopathy despite reducing GPX activity (Fischer *et al.*, 1992).

30.5 Selenium, Redox–Methylation Balance and the Heart

The lack of effect in clinical trials of selenium supplementation underscores the need to better elucidate the complex biology of selenium *in vivo*. Studies that examined the effect of selenium deficiency and supplementation have surprisingly found similar pathologic phenotypes. Novoselov and colleagues examined the effect of selenium-deficient and selenium-supplemented diets on liver carcinogenesis in a mouse model. Both deficiency and supplementation of selenium prevented carcinogenesis (Novoselov *et al.*, 2005). Another study showed that in normal mice, selenium supplementation led to an insulin-resistant state (Labunskyy *et al.*, 2011). In the same study, genetic modification to decrease selenoprotein synthesis also promoted abnormal glucose metabolism. These studies suggest the complex interaction of selenium status with biological processes *in vivo*, and the narrow range of selenium homeostasis outside of which selenium may induce pathology.

30.5.1 Selenium, *via* the Methionine–Homocysteine Cycle, Affects Redox and Methylation Potential

A potential explanation for these effects of selenium deficiency and supplementation lies in the close interaction of selenium with the methionine–homocysteine cycle *in vivo*. The methionine–homocysteine cycle is crucial to the proper functioning of redox mechanisms and methylation reactions in the body (Metes-Kosik *et al.*, 2012). Figure 30.1 illustrates the close interaction between selenium and the methionine–homocysteine cycle *in vivo*. The conversion of the sulfur amino acid methionine to homocysteine *via* intermediate metabolites provides methyl groups for all methylation reactions of the body except the remethylation of homocysteine itself back to methionine. Methylation reactions involve transfer of methyl groups to DNA, RNA and proteins, and are crucial for cellular homeostasis, gene

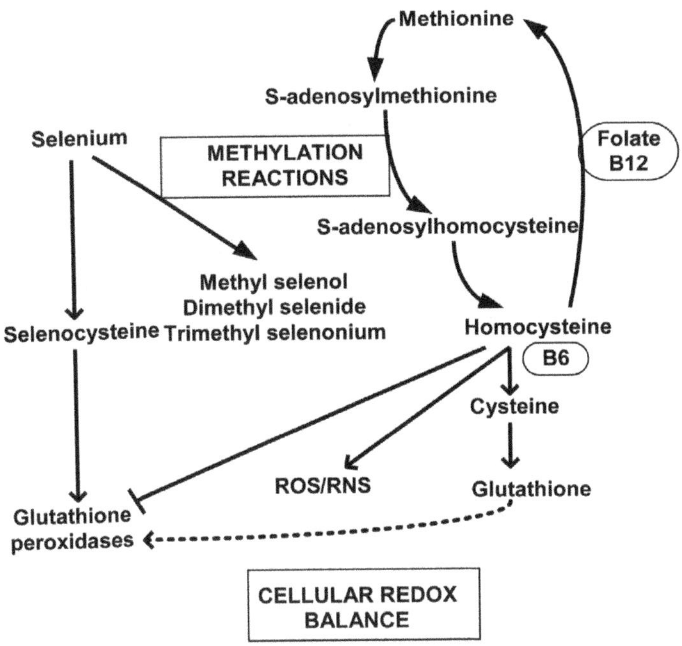

Figure 30.1 Selenium metabolism *in vivo* is intricately linked to the methionine–homocysteine cycle *via* effects on methylation and redox balance (reproduced with permission from Metes-Kosik *et al.*, 2012). Selenium itself undergoes methylation *in vivo*. The methionine–homocysteine cycle, *via* the intermediates *S*-adenosyl methionine and *S*-adenosyl homocysteine, regulates methylation reactions. Furthermore, homocysteine itself has a complex effect on redox balance: glutathione, the antioxidant substrate for glutathione peroxidases, is generated from homocysteine, while homocysteine has been shown to inhibit glutathione peroxidases, which are selenoproteins. ROS – reactive oxygen species; and RNS – reactive nitrogen species.

expression, and metabolism. Homocysteine is the precursor for the antioxidant molecules cysteine and glutathione (glutathione is the substrate for the GPX family of enzymes, which are selenoproteins); however, homocysteine can promote oxidant stress *via* direct generation of free radicals and by inhibition of GPX-1. Selenium, in addition to being a crucial component of antioxidant selenoproteins, also undergoes methylation *in vivo.* Preclinical studies have shown a substantive interaction of selenium and the methionine–homocysteine cycle *in vivo* (Uthus *et al.*, 2002; Uthus and Ross, 2007).

30.5.2 Selenium and Myocardial Dysfunction: Role of Redox–Methylation Balance

30.5.2.1 Selenium Deficiency and Supplementation Lead to Myocardial Fibrosis and Dysfunction

We examined the effects of selenium deficiency and modest selenium supplementation (0.5 mg per kg diet) on myocardial structure and function in normal mice (Metes-Kosik *et al.*, 2012). Figure 30.2 demonstrates the changes in myocardial structure induced by varying selenium in the diet. Both deficiency and supplemental selenium led to increased deposition of collagen in the myocardium, both around the smaller blood vessels and in between myocytes. This led to an increased stiffness of the myocardium and diastolic dysfunction in both selenium deficient and supplemented mice as shown in Figure 30.3. Contractile function of the heart was reduced in the mice fed the selenium-deficient diet compared to the other groups, suggesting that selenium deficiency may have affected the myocyte compartment in addition to matrix, and led to systolic dysfunction. Analysis of gene expression showed changes in expression of matrix genes in both selenium-deficient and -supplemented states; the pattern of changes were different in the presence of low and excess selenium.

30.5.2.2 Both Selenium Deficiency and Supplementation Affect Redox Balance and Epigenetic Mechanisms

In our study (Metes-Kosik *et al.*, 2012), plasma selenium was significantly lower in the selenium-deficient group; levels were not significantly higher in the selenium supplemented group compared to control diet (Table 30.1). Myocardial GPX activity increased with increasing levels of selenium in the diet, and myocardial expression of 4-hydroxynonenal, a marker of oxidant stress, decreased with increasing selenium in the diet. Plasma levels of isoprostanes, a marker of systemic oxidant stress, were significantly increased in the selenium deficient group compared to the other two groups. Cumulatively, these data suggested that increasing dietary selenium intake induced myocardial antioxidant mechanisms and reduced myocardial oxidant stress.

We further examined the changes in methylation balance and effects on gene methylation. There were no observed differences in the levels of the methyl donor *S*-adenosyl methionine. However, the level of *S*-adenosyl

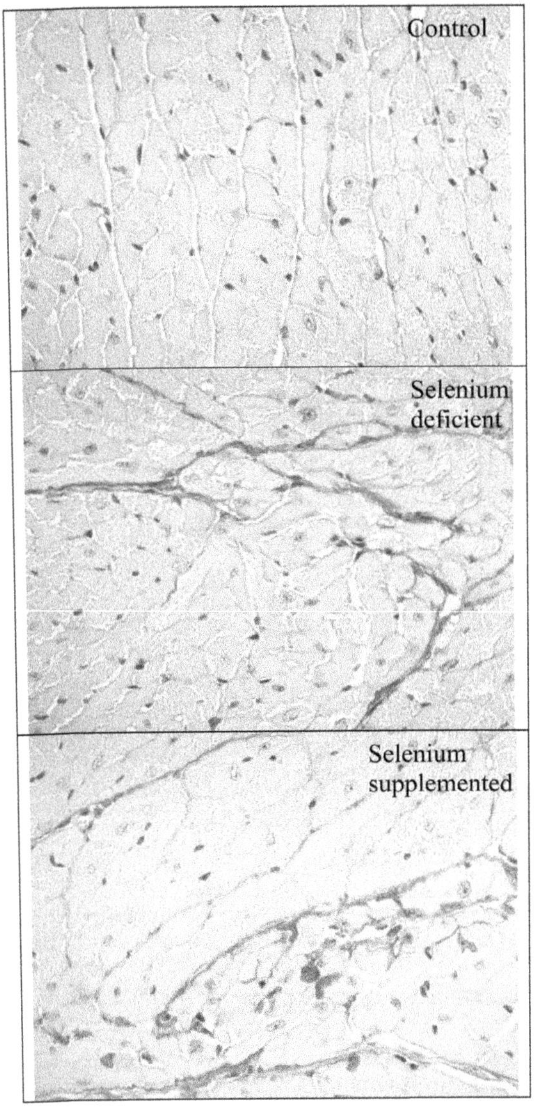

Figure 30.2 Sections of the myocardium stained with the special dye Picrosirius red (200× magnification). Collagen is stained pink, and is shown to be increased in the selenium deficient and selenium supplemented groups compared to mice fed control diet (reproduced with permission from Metes-Kosik *et al.*, 2012).

homocysteine, a major inhibitor of methylation reactions, was increased by selenium supplementation. Activity of DNA methyltransferase in the myocardium was decreased in the selenium supplemented group. Analysis of a highly methylated noncoding DNA sequence also demonstrated a time-dependent decline in DNA methylation with selenium supplementation.

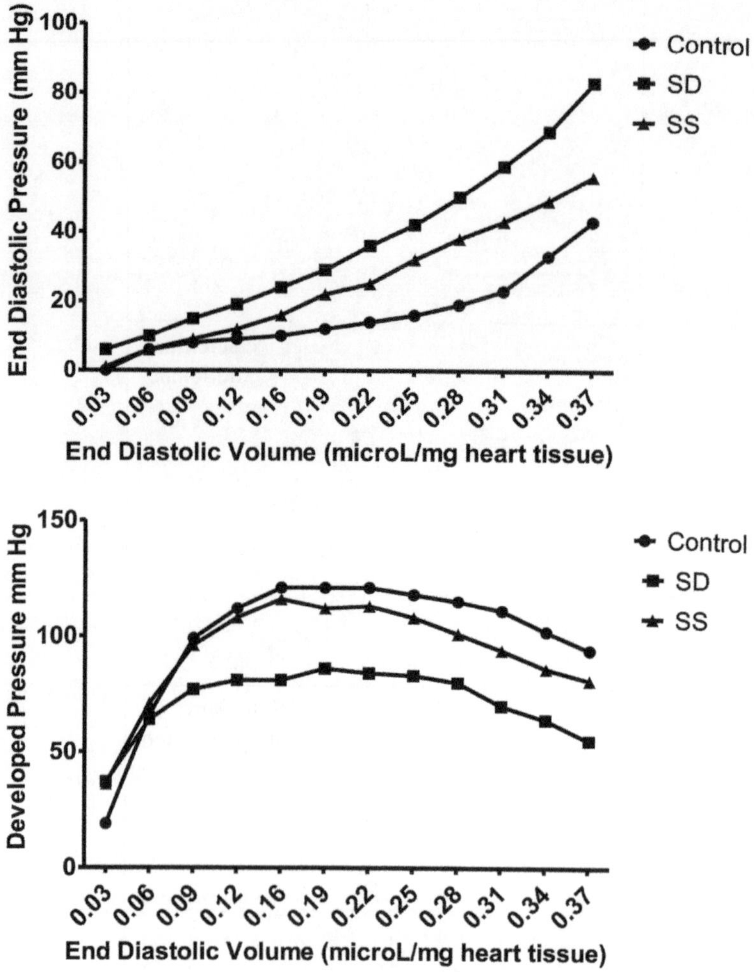

Figure 30.3 Effect of selenium on diastolic function (upper panel) and systolic function (lower panel) of the heart (reproduced with permission from Metes-Kosik *et al.*, 2012). Mice hearts were perfused outside the body and pressures inside the heart were measured after filling the heart to different volumes using a balloon. The upper panel shows that with the same heart volume, there was more pressure developed in the heart in the selenium-deficient (SD) and selenium-supplemented (SS) mice. This upward shift of the pressure–volume curve of the heart indicates diastolic dysfunction. The lower panel shows the amount of pressure developed in the mice hearts during contraction of the heart. Compared to control and selenium-supplemented groups, the selenium-deficient group had less pressure developed in response to the same heart volumes. This downward shift of the curve indicates contractile or systolic dysfunction in the selenium-deficient mice.

Table 30.1 Effects of dietary selenium on redox status (reproduced with permission from Metes-Kosik *et al.*, 2012). GPX – glutathione peroxidase; HNE – 4-hydroxynonenal (marker of oxidant stress).

	Dietary groups (n = 7–8 per group)		
Parameter	Selenium deficient	Control	Selenium supplemented
Plasma selenium (μg L^{-1})	61.8 ± 6.4[a]	370.7 ± 38.8	403.9 ± 1.2
Myocardial GPX activity (mU per mg protein)	16.7 ± 4.2	38.2 ± 1.4[b]	57.3 ± 9.7[c]
Myocardial HNE staining[e]	2.37 ± 0.2[d]	2.04 ± 0.2	1.54 ± 0.2
Plasma isoprostane level (pg mL^{-1})	320 ± 58[a]	83 ± 15	102 ± 12

[a] $p < 0.001$ *vs.* other groups.
[b] $p < 0.05$ *vs.* selenium-deficient group.
[c] $p < 0.001$ *vs.* selenium-deficient group.
[d] $p < 0.05$ *vs.* selenium-supplemented group.
[e] Arbitrary scale of 0–4.

Data from this study demonstrated that selenium had concomitant effects on redox balance and methylation. While selenium deficiency was associated with oxidant stress and normal methylation, selenium supplementation was associated with decreased methylation. It is possible that the similar effects seen on myocardial matrix expression and fibrosis may have been the result of oxidant stress (selenium deficiency) and decreased DNA methylation (selenium supplementation) producing similar effects on gene expression. It is also possible that reductive stress (see below) played a role in the adverse phenotype induced by selenium supplementation in the heart.

30.5.2.3 Selenium and Redox–Methylation Balance: A Potential Mechanism for In Vivo Effects of Selenium

The above data suggest that redox balance and methylation reactions are intricately linked, with selenium status being an important regulator of this redox–methylation balance (Joseph, 2013). As shown in Figure 30.4, both redox balance and methylation can be viewed as continuums. In the case of redox balance, normal levels of reactive species are required for normal cell signaling, while an excess produces oxidant stress and marked reduction of reactive oxygen species leads to reductive stress. While a plethora of reports link oxidant stress to cardiovascular pathology, reductive stress also has been shown to cause myocardial dysfunction (Rajasekaran *et al.*, 2007). Similarly, normal DNA methylation (of CpG dinucleotides) is necessary for epigenetic control of gene expression. An increase or decrease in methylation can lead to improper gene expression; with hypomethylation generally increasing, and hypermethylation decreasing, gene expression. As shown in Figure 30.1 and data from our study presented above, selenium can be a crucial influence

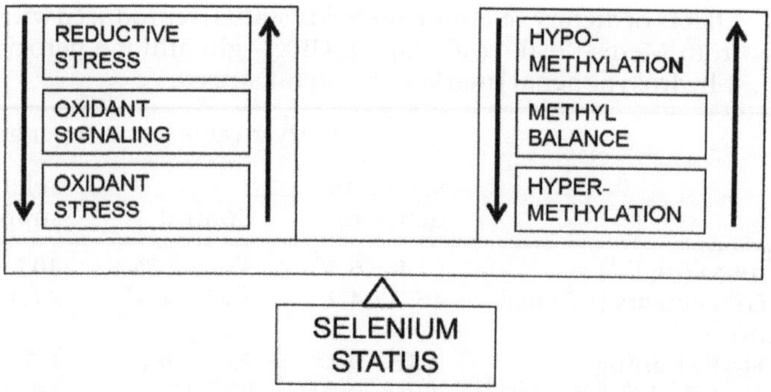

Figure 30.4 Selenium integrates redox methylation balance (reproduced with permission from Joseph and Loscalzo, 2013). A continuum exists in redox balance from normal oxidant signaling in cells to the extremes of oxidant stress and reductive stress. Similarly, methylation reactions also exist in a continuum between hypomethylation to hypermethylation with normal methylation reactions being crucial to normal cellular function and gene expression. Due to the close interaction between selenium metabolism and methionine–homocysteine cycle described in Figure 30.1, changes in selenium lead to changes in redox–methylation balance.

on redox–methylation balance. Hence, the effects on integrating redox–methylation balance offers a novel explanation for the similar effects of selenium-deficient and supplemental states, and for the narrow beneficial range of selenium intake.

30.6 Selenistasis – Selenium in Health and Disease

The above-described schema (Figure 30.4) relates to the healthy state. Since pathologic states could be associated with changes in redox and methylation balance, the effects of selenium is expected to be modulated differently in healthy and disease states. For example, Lymbury and colleagues showed that selenium supplementation induced beneficial effects in a rat model of hypertension-induced myocardial dysfunction, but not in normal rats (Lymbury *et al.*, 2010). Figure 30.5 demonstrates the concept of selenistasis, a schema to integrate redox–methylation balance with the continuum of health and disease (Joseph, 2013). In Figure 30.5, the green oval denotes normal oxidant signaling and methylation in the context of a healthy organism. Note that the plasma selenium level is not part of this schema. The red square and rectangle represent two possible combinations of disturbance of selenistasis leading to disease: oxidant stress and hypomethylation in the case of the red square, and hypermethylation and reductive stress in the case of the red rectangle. In terms of the therapeutic use of selenium, the intent of modifying selenium intake would be to move the integrated redox–methylation state towards selenistasis.

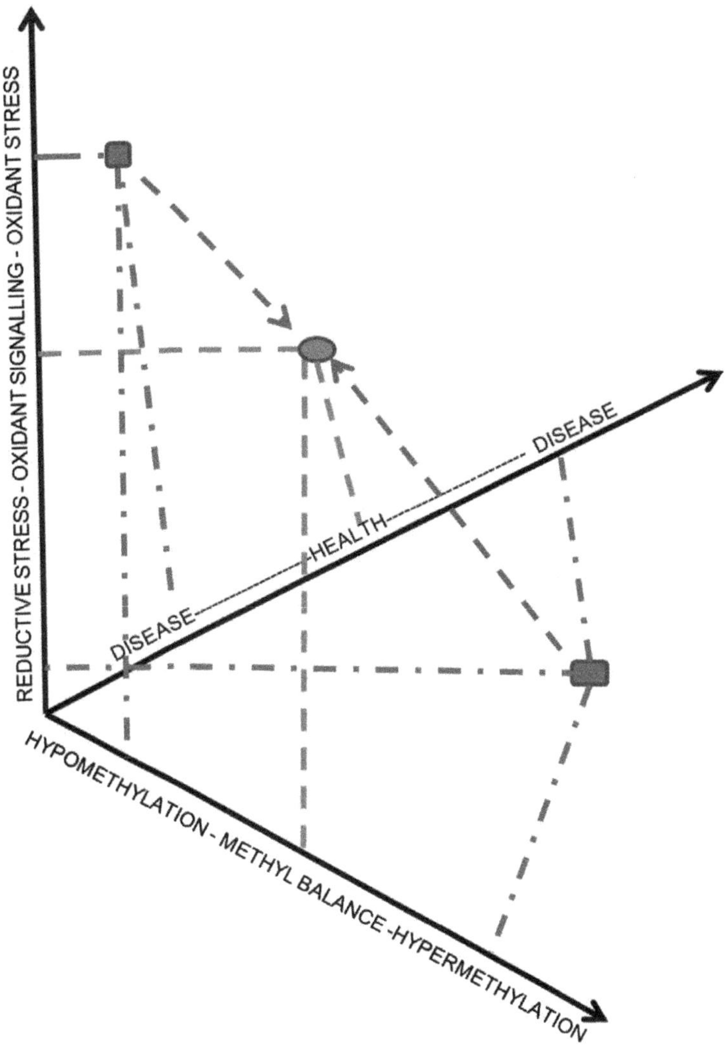

Figure 30.5 The novel concept of *selenistasis* (reproduced with permission from Joseph and Loscalzo, 2013). The 3 axes represent the continuums of redox balance, methylation balance, and the health–disease continuum. The green oval represents the ideal situation of redox–methylation balance in health or selenistasis, while the red square and red rectangle represent deviations for selenistasis. Note that this construct does not depend on the actual selenium level but on redox–methylation balance as a measure of selenium status.

30.7 Conclusions

Selenium deficiency is associated with cardiovascular disease; however, the exact mechanisms of selenium's actions on the heart and how to optimize myocardial selenium status to prevent and treat disease still needs to elucidated by preclinical research. A novel approach to understanding the effects of selenium on the heart is the concept of redox–methylation balance as an influence on cell signaling and gene expression. The concept of selenistasis offers a new model to conduct experiments in healthy and disease states and thereby understand how to utilize selenium to maintain health and treat disease in humans.

Summary Points

- This chapter focuses on preclinical studies of the effects of changing dietary selenium on the heart.
- Dietary selenium content modulates cardiovascular health; the exact mechanisms involved are not well elucidated.
- Studies have mostly focused on the effects of selenium on countering oxidant stress in the myocardium.
- Novel data suggest that the selenium content of the diet affects myocardial matrix deposition and thereby heart function.
- The effects of selenium on the heart are mediated *via* concomitant effects on redox balance and methylation.
- Selenium deficiency promotes oxidant stress, while modest selenium supplementation decreases methylation of DNA.
- These divergent effects on the redox–methylation balance promote similar effects on heart by selenium deficiency and selenium supplementation.
- Further preclinical research is needed to understand how selenium modulates this redox–methylation balance and myocardial gene expression to enable the use of selenium in preventing and treating heart disease.

Key Facts

Key facts regarding the normal structure and function of the heart and the changes that lead to heart failure

1. The primary function of the heart is to maintain blood flow to all organs.
2. The heart muscle, or myocardium, consists of the heart muscle cells or myocytes, and the nonmyocyte compartment.
3. The nonmyocyte compartment of the myocardium consists of a scaffolding of proteins (cardinal among which are the collagens), blood vessels supplying oxygen and nutrients, nerves, as well as a large number of cells and molecules.
4. The myocytes provide the contractile or squeezing function of the heart that leads to blood flow to other organs.
5. The nonmyocyte compartment is also crucial to heart function; changes in its components, especially collagen, can promote heart dysfunction.

6. Failure of the heart can occur due to a problem with filling or with emptying.
7. Systolic dysfunction and systolic heart failure occur from a problem with the contractile function of the heart; this is primarily a problem with the myocytes.
8. Diastolic dysfunction and diastolic heart failure occur due to changes in myocytes and nonmyocyte compartments; an increased deposition of collagen in the heart is a major cause of diastolic dysfunction.

Definitions of Words and Terms

Cardiomyopathy. Disorder of heart muscle that leads to heart failure.

Myocarditis. Inflammation of the heart muscle, which can lead to permanent heart damage.

Myocardial infarction. Death of heart muscle caused by interruption of blood flow.

Ischemia–reperfusion. Interruption of blood flow for a brief period of time followed by restoration of blood flow.

Gap junction. A specialized connection that allows direct communication and passage of molecules between two cells.

Diastolic dysfunction. Abnormal filling of the heart due to either an increased stiffness of the myocardium or impaired ability of the heart muscle to relax. This abnormality accounts for approximately half of human heart failure.

Systolic dysfunction. Abnormal contractility of the heart muscle leading to ineffective perfusion of organs. Systolic dysfunction underlies approximately half of all cases of human heart failure.

Redox–methylation balance. A novel concept of close interaction between redox reactions and methylation reactions that may be integrated by selenium.

Methylation potential. The ratio of levels of the methyl donor *S*-adenosyl methionine and the methylation inhibitor *S*-adenosyl homocysteine. This ratio is an important determinant of methylation reactions.

Epigenetics. Changes in genes that are not the result of alteration of DNA sequence, examples are DNA methylation and modification of histone proteins. These changes are heritable and alter gene expression.

List of Abbreviations

KD	Keshan disease
TRR	Thioredoxin reductase
GPX	Glutathione peroxidase
CVD	Cardiovascular disease
NFkB	Nuclear factor kappa light chain enhancer of activated B cells

I–R	Ischemia–reperfusion
DNA	Deoxyribonucleic acid
RNA	Ribonucleic acid
CpG	Cytosine guanine
TGF-β	Transforming growth factor-beta
TNF-α	Tumor necrosis factor-alpha

References

Baljinnyam, E., Hasebe, N., Morihira, M., Sumitomo, K., Matsusaka, T., Fujino, T., Fukuzawa, J., Ushikubi, F. and Kikuchi, K., 2006. Oral pretreatment with ebselen enhances heat shock protein 72 expression and reduces myocardial infarct size. *Hypertension Research*. 29: 905–913.

Beck, M. A., Esworthy, R. S., Ho, Y. S. and Chu, F. F., 1998. Glutathione peroxidase protects mice from viral-induced myocarditis. *FASEB Journal*. 12: 1143–1149.

Beck, M. A., Kolbeck, P. C., Rohr, L. H., Shi, Q., Morris, V. C. and Levander, O. A., 1994a. Benign human enterovirus becomes virulent in selenium-deficient mice. *Journal of Medical Virology*. 43: 166–170.

Beck, M. A., Kolbeck, P. C., Shi, Q., Rohr, L. H., Morris, V. C. and Levander, O. A., 1994b. Increased virulence of a human enterovirus (coxsackievirus B3) in selenium-deficient mice. *Journal of Infectious Diseases*. 170: 351–357.

Beck, M. A., Shi, Q., Morris, V. C. and Levander, O. A., 1995. Rapid genomic evolution of a non-virulent coxsackievirus B3 in selenium-deficient mice results in selection of identical virulent isolates. *Nature Medicine*. 1: 433–436.

Cermelli, C., Vinceti, M., Scaltriti, E., Bazzani, E., Beretti, F., Vivoli, G. and Portolani, M., 2002. Selenite inhibition of Coxsackie virus B5 replication: implications on the etiology of Keshan disease. *Journal of Trace Elements in Medicine and Biology*. 16: 41–46.

Cheng, Y. Y. and Qian, P. C., 1990. The effect of selenium-fortified table salt in the prevention of Keshan disease on a population of 1.05 million. *Biomedical and Environmental Sciences*. 3: 422–428.

Conrad, M., Jakupoglu, C., Moreno, S. G., Lippl, S., Banjac, A., Schneider, M., Beck, H., Hatzopoulos, A. K., Just, U., Sinowatz, F., Schmahl, W., Chien, K. R., Wurst, W., Bornkamm, G. W. and Brielmeier, M., 2004. Essential role for mitochondrial thioredoxin reductase in hematopoiesis, heart development, and heart function. *Molecular and Cellular Biology*. 24: 9414–9423.

Coudray, C., Hida, H., Boucher, F., De Leiris, J. and Favier, A., 1995. Modulation by selenium supplementation of lipid peroxidation induced by chronic administration of adriamycin in rats. *Nutrition*. 11(Suppl. 5): 512–516.

Fischer, J. G., Tackett, R. L., Howerth, E. W. and Johnson, M. A., 1992. Copper and selenium deficiencies do not enhance the cardiotoxicity in rats due to chronic doxorubicin treatment. *Journal of Nutrition*. 122: 2128–2137.

Ge, K., Xue, A., Bai, J. and Wang, S., 1983. Keshan disease – an endemic cardiomyopathy in China. *Virchows Archiv A: Pathological Anatomy and Histopathology*. 401: 1–15.

Jia, C., Chen, X., Li, X., Li, M., Miao, C., Sun, B., Fan, Z. and Ren, L., 2011. The effect of DHEA treatment on the oxidative stress and myocardial fibrosis induced by Keshan disease pathogenic factors. *Journal of Trace Elements in Medicine and Biology*. 25: 154–159.

Joseph, J. and Loscalzo, J., 2013. Selenistasis: epistatic effects of selenium on cardiovascular phenotype. *Nutrients*. 5: 340–358.

Joseph, J., 2013. Selenium and cardiometabolic health: inconclusive yet intriguing evidence. *American Journal of the Medical Sciences*. 346: 216–220.

Jotti, A., Maiorino, M., Paracchini, L., Piccinini, F. and Ursini, F., 1994. Protective effect of dietary selenium supplementation on delayed cardiotoxicity of adriamycin in rat: is PHGPX but not GPX involved?. *Free Radical Biology and Medicine*. 16: 283–288.

Labunskyy, V. M., Lee, B. C., Handy, D. E., Loscalzo, J., Hatfield, D. L. and Gladyshev, V. N., 2011. Both maximal expression of selenoproteins and selenoprotein deficiency can promote development of type 2 diabetes-like phenotype in mice. *Antioxidants and Redox Signaling*. 14: 2327–2336.

Liu, Z. W., Niu, X. L., Chen, K. L., Xing, Y. J., Wang, X., Qiu, C. and Gao, D. F., 2013. Selenium attenuates adriamycin-induced cardiac dysfunction *via* restoring expression of ATP-sensitive potassium channels in rats. *Biological Trace Element Research*. 153: 220–228.

Lymbury, R. S., Marino, M. J. and Perkins, A. V., 2010. Effect of dietary selenium on the progression of heart failure in the ageing spontaneously hypertensive rat. *Molecular Nutrition & Food Research*. 54: 1436–1444.

Metes-Kosik, N., Luptak, I., Dibello, P. M., Handy, D. E., Tang, S. S., Zhi, H., Qin, F., Jacobsen, D. W., Loscalzo, J. and Joseph, J., 2012. Both selenium deficiency and modest selenium supplementation lead to myocardial fibrosis in mice *via* effects on redox-methylation balance. *Molecular Nutrition & Food Research*. 56: 1812–1824.

Novoselov, S. V., Calvisi, D. F., Labunskyy, V. M., Factor, V. M., Carlson, B. A., Fomenko, D. E., Moustafa, M. E., Hatfield, D. L. and Gladyshev, V. N., 2005. Selenoprotein deficiency and high levels of selenium compounds can effectively inhibit hepatocarcinogenesis in transgenic mice. *Oncogene*. 24: 8003–8011.

Panicker, S., Swathy, S. S., John, F. and Indira, M., 2012. Impact of selenium on the leukotriene B4 synthesis pathway during isoproterenol-induced myocardial infarction in experimental rats. *Inflammation*. 35: 74–80.

Panicker, S., Swathy, S. S., John, F. and Madambath, I., 2010. Impact of selenium on NFκB translocation in isoproterenol-induced myocardial infarction in rats. *Biological Trace Element Research*. 138: 202–211.

Pei, J., Fu, W., Yang, L., Zhang, Z. and Liu, Y., 2013. Oxidative stress is involved in the pathogenesis of Keshan disease (an endemic dilated cardiomyopathy) in China. *Oxidative Medicine and Cellular Longevity*. 2013: 474203.

Rajasekaran, N. S., Connell, P., Christians, E. S., Yan, L. J., Taylor, R. P., Orosz, A., Zhang, X. Q., Stevenson, T. J., Peshock, R. M., Leopold, J. A., Barry, W. H., Loscalzo, J., Odelberg, S. J. and Benjamin, I. J., 2007. Human alpha B-crystallin mutation causes oxido-reductive stress and protein aggregation cardiomyopathy in mice. *Cell*. 130: 427–439.

Rakotovao, A., Tanguy, S., Toufektsian, M. C., Berthonneche, C., Ducros, V., Tosaki, A., de Leiris, J. and Boucher, F., 2005. Selenium status as determinant of connexin-43 dephosphorylation in ex vivo ischemic/reperfused rat myocardium. *Journal of Trace Elements in Medicine and Biology*. 19: 43–47.

Tanguy, S., Boucher, F., Besse, S., Ducros, V., Favier, A. and de Leiris, J., 1998. Trace elements and cardioprotection: increasing endogenous glutathione peroxidase activity by oral selenium supplementation in rats limits reperfusion-induced arrhythmias. *Journal of Trace Elements in Medicine and Biology*. 12: 28–38.

Tanguy, S., Morel, S., Berthonneche, C., Toufektsian, M. C., de Lorgeril, M., Ducros, V., Tosaki, A., de Leiris, J. and Boucher, F., 2004. Preischemic selenium status as a major determinant of myocardial infarct size *in vivo* in rats. *Antioxidants and Redox Signaling*. 6: 792–796.

Tanguy, S., Rakotovao, A., Jouan, M. G., Ghezzi, C., de Leiris, J. and Boucher, F., 2011. Dietary selenium intake influences Cx43 dephosphorylation, TNF-alpha expression and cardiac remodeling after reperfused infarction. *Molecular Nutrition & Food Research*. 55: 522–529.

Toufektsian, M. C., Boucher, F., Pucheu, S., Tanguy, S., Ribuot, C., Sanou, D., Tresallet, N. and de Leiris, J., 2000. Effects of selenium deficiency on the response of cardiac tissue to ischemia and reperfusion. *Toxicology*. 148: 125–132.

Turan, B., Saini, H. K., Zhang, M., Prajapati, D., Elimban, V. and Dhalla, N. S., 2005. Selenium improves cardiac function by attenuating the activation of NF-kappaB due to ischemia-reperfusion injury. *Antioxidants and Redox Signaling*. 7: 1388–1397.

Uthus, E. O. and Ross, S. A., 2007. Dietary selenium affects homocysteine metabolism differently in Fisher-344 rats and CD-1 mice. *Journal of Nutrition*. 137: 1132–1136.

Uthus, E. O., Yokoi, K. and Davis, C. D., 2002. Selenium deficiency in Fisher-344 rats decreases plasma and tissue homocysteine concentrations and alters plasma homocysteine and cysteine redox status. *Journal of Nutrition*. 132: 1122–1128.

Van Vleet, J. F. and Ferrans, V. J., 1980. Evaluation of vitamin E and selenium protection against chronic adriamycin toxicity in rabbits. *Cancer Treatment Reports*. 64: 315–317.

Van Vleet, J. F., Ferrans, V. J. and Weirich, W. E., 1980. Cardiac disease induced by chronic adriamycin administration in dogs and an evaluation

of vitamin E and selenium as cardioprotectants. *American Journal of Pathology*. 99: 13–42.
Venardos, K., Ashton, K., Headrick, J. and Perkins, A., 2005. Effects of dietary selenium on post-ischemic expression of antioxidant mRNA. *Molecular and Cellular Biochemistry*. 270: 131–138.
Venardos, K., Harrison, G., Headrick, J. and Perkins, A., 2004a. Effects of dietary selenium on glutathione peroxidase and thioredoxin reductase activity and recovery from cardiac ischemia-reperfusion. *Journal of Trace Elements in Medicine and Biology*. 18: 81–88.
Venardos, K., Harrison, G., Headrick, J. and Perkins, A., 2004b. Selenium supplementation and ischemia-reperfusion injury in rats. *Redox Report*. 9: 317–320.

CHAPTER 31

Selenium and Kashin–Beck Disease

XIONG GUO*[a], YU-JIE NING[a], AND XI WANG[a]

[a]Institute of Endemic Diseases, School of Public Health, Xi'an Jiaotong University Health Science Centre, No. 76 Yan Ta West Road, Shaanxi 710061, China
*E-mail: guox@mail.xjtu.edu.cn

31.1 The History of Kashin–Beck Disease

KBD is a chronic osteochondropathy endemic in some parts of China. The history of this disease might go back as far as the 16th century, but it was first mentioned by S. G. Filatov as deformed polyarthritis observed in eastern Zabaikale in 1830 and by I. M. Yurensky (or Jurensky) in 1849 (Voshchenko and Ivanov, 1990). It was named after doctors Nikolai Kashin studied from 1859 to 1868 and Eugene Beck from 1899 to 1903 in appreciation of their outstanding works in this field (Allander, 1994). In 1908, the disease was described in Chang Bai Mountain, an endemic area of the northern-east of China (Zhai, 1990). From 1919 to 1945, Japanese scientists also studied this disease (Egashira and Staff, 1990; Table 31.1). For the last 65 years, many epidemiological, clinical, radiological, medical geography and biochemical studies have investigated the etiology and pathogenesis of this disease.

Food and Nutritional Components in Focus No. 9
Selenium: Chemistry, Analysis, Function and Effects
Edited by Victor R. Preedy

Published by the Royal Society of Chemistry, www.rsc.org

Table 31.1 Summary of the Kashin–Beck disease classification.[a]

The diseases name suggested for Kashin–Beck disease	Authors	Year
A complication of endemic goiter	Kashin N. I.	1859
Osteoarthritis deformans endemica	Beck E. B.	1906
Chronic progressive polyosteoathrititis	Okano T. A	1919
Endemic osteochondritis	Dobrovolsky L. O.	1925
Panarthritis deformans endemica	Hiyeda K.	1936
Osteoarthropathia deformans endemica	Kubo H.	1936
Food fusariotoxicosis	Sergiycvski, Rubinstein	1943
Dysostosis enchondrosis meta-epiphysis	Takamori Tokio	1943
Degenerative arthropathy	Sella E. J.	1973
Osteodysplasia congenital	Poznanski A. K.	1974
Other specific arthropathies (Kashin–Beck disease)	WHO	1975
Endemic chromic osteoarthrosis	Peipu Yin	1983
Endemic osteoarthrophy with unknown cause	Nutrition, USA	1987
Spondyloepiphyseal dysostosis	Beighton P.	1988
Epidemic osteoarthrosis	Rheumatological association, Russian	1988
M12 other specific arthropathies (Kashin–Beck disease)	International classification of diseases, revision 10	1989
Acquired chondronecrosis	Sokoloff Leon	1990
Exopathic bone metabolism impair or dystrophy	Zhizhong Qian	1991
Osteochondrosis endemica humanis	Dongxu Mo	1994
Endemic deformed osteoarthrosis	Ministry of Health, PR China	1994
Osteoarthritis deformans endemica	Classification for bone	2004

[a]Source: Guo, X., 2009. Combined deficiency-toxicity disorders. In: Wang, W. (ed.) *A Profile of Genetic and Environmental Diseases in China*. Science Press, Beijing, pp. 254–255.

31.1.1 Epidemiological Distribution

In China, KBD distributes in thirteen provinces, municipalities, and autonomous regions. KBD-endemic areas spread obliquely across the northeast of Heilongjiang Province to the northwest of the Tibetan autonomous region. It mainly spread across in the remote mountainous and poor rural areas, where the soil is relatively infertile and lacks several chemical elements, especially selenium (Se). It suggested that Se deficiency is a risk factor for the KBD. The disease is mostly in local agricultural households. The analysis of association between geographical distribution of Se and KBD shows four types of affected areas: (1) low-Se areas with KBD only, such as the Yulin area, north part of Shaanxi; (2) low-Se areas with both KBD and Keshan disease (KD, an endemic myocardosis that is characterized by multifocal myocardial necrosis and fibrosis), such as the middle and south part of Shaanxi and Sichuan Province; (3) low-Se areas with KD only, such as the Yunnan Province; and (4) low-Se areas with no KBD or KD, such as the Nanbu County in Sichuan Province.

Table 31.2 Prevalence of KBD in China from 1995 to 2011.

Year	Number of endemic counties	Population in endemic regions (10 000)	Current patients (10 000)
1995	335	8995.71	104.17
1996	335	9514.97	108.99
1997	351	9560.00	108.45
2000	317	9895.61	102.52
2001	352	10 734.02	80.87
2002	353	10 119.40	81.97
2003	354	10 162.90	81.18
2004	358	10 225.80	79.52
2005	354	10 255.78	80.87
2006	354	10 342.79	74.05
2007	363	10 484.30	72.25
2008	366	10 528.34	71.48
2009	366	10 583.80	69.50
2010	366	10 508.30	66.09
2011	377	10 613.20	64.21

The prevalence of KBD in China from 1995 to 2011 is presented in Table 31.2. By 2012, there were 106 million residents at risk and approximately 0.64 million patients, of which 22 979 were under 13 years old (Ministry of Health, P. R. C., 2012). Children aged 3 to 12 are in the KBD high-risk group; the disease has been known to attack babies as young as 3 months. So far, no association has been observed between gender and the risk of the disease. The clinical changes are often not obvious enough to diagnose KBD at an early stage. A child suffering from KBD at the age of 5 might not show clinical symptoms until the age of 15, missing the period at which the treatment would be most effective.

31.1.2 Clinical and Radiographic Changes

KBD is characterized by multifocal and symmetrical enlarged joints and/or shortened metacarpal-phalanges and limbs according to the national diagnosis criteria (WS/T207-2010). Pathological changes happen before clinical manifestation of the disease; therefore, a method as cheap and convenient as X-ray imaging naturally becomes a conventional choice of noninvasive diagnosis tool.

The severity of the disease varies substantially among affected individuals. Based on clinical symptoms and X-ray images, KBD was divided into four stages—the early stage, the first, second, and third degrees. So far, there are no biochemical markers that could be used for early diagnosis. At first, patients commonly present multifocal and symmetrical arthritic pain, enlarged phalanges and limbs, limited motion, and mild muscle atrophy. The symptoms worsen as the disease progresses; brachydactylia and even dwarfism is possible (see Figure 31.1).

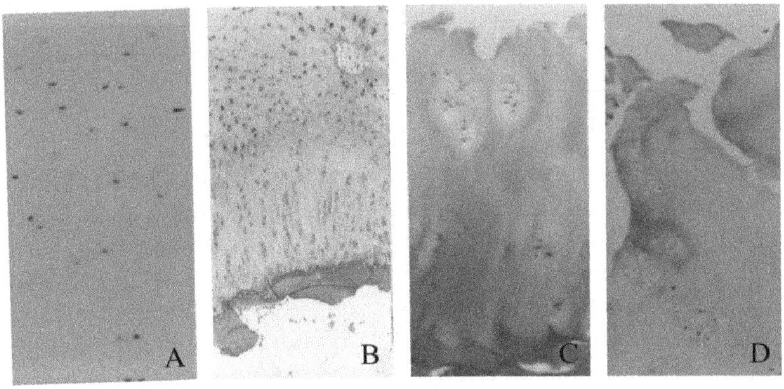

Figure 31.1 Pathological changes in epiphyseal cartilage and articular cartilage. Typically, in the growth plate cartilage of KBD, large chondronecrotic areas, lack of normal cells in the deep zone, and significantly reduced chondrocyte sizes are observed (B), but cell membranes are intact. In contrast with control (A and C), in the upper and middle zone there are chondrocyte clusters, chondronecrosis foci with red-staining plasma, compressed nuclei, and incomplete cell membranes (D). Source: the same as Table 31.1, p. 272.

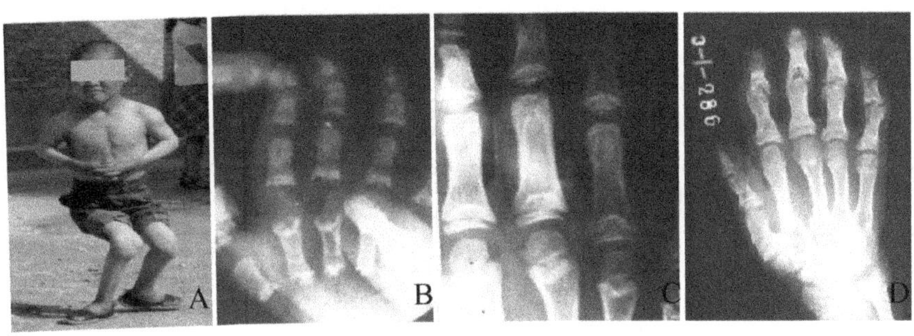

Figure 31.2 Clinical KBD manifestation in a 7 year-old boy (A); radiographic images of right hands of children with KBD, aged 1 (B), 7 (C) and 13 (D). Source: (A): Ning Y. *et al.*, 2013. Biological Trace Element Research. **153**(1–3): 58–68. (B–D) The same as Table 31.1, p. 258.

The most commonly affected sites of KBD are the growth plate cartilage and articular cartilage (see Figure 31.1). With focal chondronecrosis in the deep zone and impaired endochondral ossification, the disease can lead to a secondary chronic osteoarthrosis and impaired skeletal development. It mainly damages joints in the limbs, such as finger joints and wrists, and then the spine.

Radiography (see Figure 31.2) shows typical signs of defects or blurred, interrupted, or disappearing calcification line in the metaphysis; irregular and sclerous of the subarticular surface in the distal ends of the phalanges; premature closure of the epiphysial line; and substantially enlarged and deformed phalanges and shortened calcaneus. These signs could appear separately or in various combinations. KBD can be also categorized as active, silent, or relatively silent type, based on clinical manifestations and radiographic signs in the metaphysis and/or the distal ends of the phalanges.

31.1.3 Hypothesis of Se Deficiency Causing KBD

In the 1970s, trace-element investigations in 26 provinces, autonomous regions, and municipalities in China showed that KBD mainly spread in low-Se areas, where the Se content in water, soil, grain and hair, serum, and urine of children were significantly lower than in non-KBD areas. The results of administration of Se supplements such as sodium selenite tablets and Se-enriched salt demonstrated a significant effect of Se on the incidence and the repair of the pathological changes in the metaphysis and the distal ends of children's phalanges caused by KBD. It was proposed that environmental Se deficiency is the main factor causing low-Se nutritional status of residents in KBD-endemic areas. Studies of the etiological relationship between Se deficiency and KBD and its clinical trials have made substantial progress in China.

31.2 The Etiological Relationship between Se Deficiency and Kashin–Beck Disease

In China, KBD mainly distributes in areas with environmental Se deficiency, but not all such regions are KBD-endemic areas (see Figure 31.3). However, the available data indicated that Se deficiency plays an important role in the pathogenesis of KBD (Tan *et al.*, 1990).

A reasonable explanation for the origin of this disease would be that the residents get inadequate Se *via* the environment-food-body route (Figure 31.4). During the period of 1975 to 1979, Se contents in hair and urine of 400 KBD children in Shaanxi province (hair: 59 ± 25 to 98 ± 24 ng g^{-1}, urine: 2.09 ± 0.81 to 2.88 ± 1.20 μg per 24 h) were much lower than those of healthy children in non-KBD areas (hair: 195 ± 54 to 343 ± 124 ng g^{-1}, urine: 5.08 ± 1.13 to 7.22 ± 1.93 μg per 24 h). This Se deficiency continued for decades in Shaanxi, Sichuan, Tibet, and other endemic provinces. The low internal Se level (*e.g.*, in hair and urine) was consistent with the low Se content in the soil (1.04–4.30 ng g^{-1}), wheat (27–31 ng g^{-1}), corn (20–23 ng g^{-1}) and drinking water (0.12–0.32 ng g^{-1}) in the locality (Mo *et al.*, 1997).

Se-balance tests verified that children from KBD areas had a slightly positive balance of Se, which was substantially lower than for children from non-KBD areas (Lv and Bai, 1992). The children from KBD areas had only 1/8 of the Se intake and half of the Se surface absorptivity but much higher Se excretion levels in urine and feces in comparison with the later period of 1989–1992.

31.2.1 Cross-Sectional Study

Se content in hair, urine, and serum of KBD-affected children were lower than in KBD-free children. Low hair/urine Se content was associated with high X-ray detection rates of metaphysis lesions.

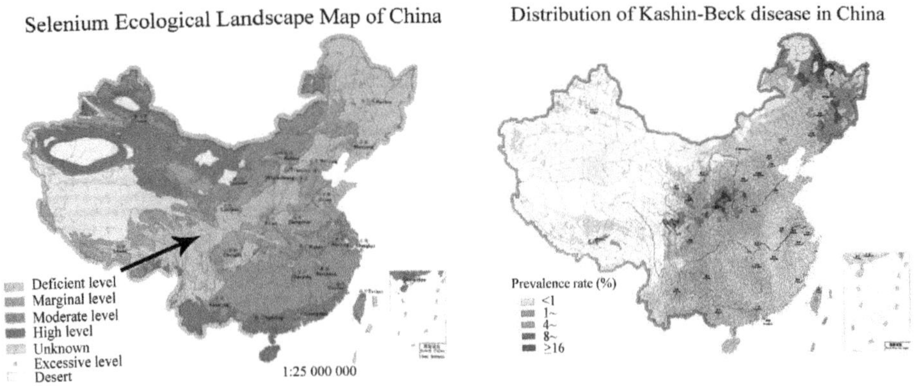

Figure 31.3 (Left) Distribution of low-Se areas in China. (Right) Distribution of Kashin–Beck disease in China. Source: Tan, J., 1989. *The Atlas of Endemic Diseases and Their Environments in the People's Republic of China*. Science Press, Beijing, p. 39 and 92.

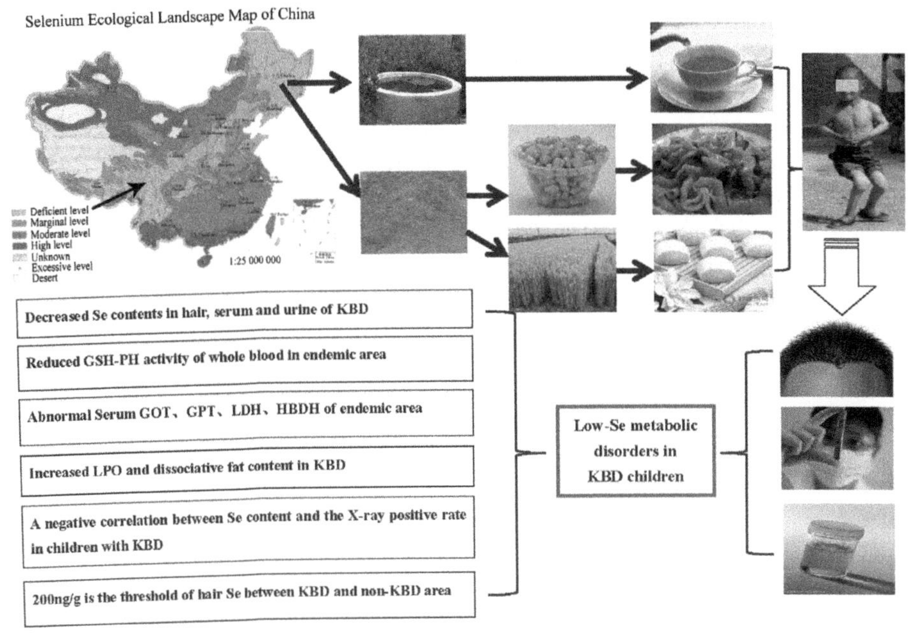

Figure 31.4 Schematic diagrams of low-Se nutritional metabolic disorders in children with KBD caused by low-Se environment in KBD-endemic areas.

Generally, the environmental and internal Se levels have remained very low in the Tibetan KBD areas since 1977. A positive correlation has been found between Se content in crops/staple food and hair/serum; the latter reflects the internal Se level and is related to the severity of KBD (Zhaxi *et al.*, 2003; Li *et al.*, 2008). However, the internal Se level and selenoprotein glutathione peroxidase (GPX) activity have increased in the inhabitants of Se-supplemented KBD areas (Li *et al.*, 2008).

In 1998, an investigation of the relationship between 39 chemical elements and KBD was conducted among 150 male children aged 6–14 using ICP-MS. The levels of Se, B, and I in the hair of children from KBD areas were significantly lower than in similar samples collected in non-KBD areas. The data showed that, apart from Se, the levels of several nutrients and toxic elements were affected. However, only low Se content was significantly correlated with KBD prevalence, with a consistent geographic distribution. Low environmental levels of iodine and Se have been observed in KBD areas in Tibet and Gansu Province since 1998 (Moreno-Reyes *et al.*, 1998; Xu *et al.*, 1999). A large proportion of high KBD prevalence also observed in the regions where iodine deficiency was very severe (Moreno-Reyes *et al.*, 1998). In Lhasa Prefecture, the KBD prevalence was 49%, accompanied by 46% of goiter and 1% of cretinism. In this region, in 66% of cases, urinary iodine concentration was less than 2 μg dL^{-1} and in 38% of cases, serum Se concentration was less than 5 ng mL^{-1}. However, no clear relationship has been found between the exposure and metabolism of iodine and KBD (Xu *et al.*, 1999; Li *et al.*, 2013).

31.2.2 Prospective Studies

In prospective studies performed in Shaanxi province from 1993 to 1995, the incidence of KBD was high in children with low Se levels (hair Se contents <110 ng g^{-1}), but no dose–response relationship was observed. During 1987–1988, with incidence of 14.58%, 21 new cases with metaphyseal changes were detected among 144 children whose hair Se content was 45 ± 9.2 to 99 ± 8.6 ng g^{-1}. In the later period of 1993–1995, a study involving 597 healthy children selected from low-, middle-, and high-prevalence KBD areas was conducted. A hair Se content of <200 ng g^{-1} was established as a critical level; the children were divided into low-Se exposure and non low-Se exposure groups. Six new KBD cases (incidence was 0.57% per person year) were diagnosed; all of these were in the low-Se exposure group. Standard computed mortality ratios indicated that the number of real-life cases was significantly lower than the number predicted for low-Se exposure group. However, hair Se content increased, and the X-ray detection rate of KBD decreased in some endemic areas without supplemental Se treatment (Mo *et al.*, 1997). Therefore, Se deficiency cannot be considered the only environmental causative factor of KBD.

31.2.3 The Natural Decrease in KBD Prevalence

The high KBD prevalence in many areas where Se supplements had not been used has spontaneously decreased with time. For example, in the Qiaoting area of Ankang City, the south of Shaanxi province, the X-ray detection rate was 40% in 1970 and fell to 1.0% in 1993, when the hair Se content rose from 73.60 ± 2.96 to 104.80 ± 3.75 ng g^{-1}. During that period, living conditions and diet in these areas improved considerably. People had moved from damp cave-dwellings to dry and clean houses. Wheat and rice replaced cornmeal as the staple food. Eating habits became more regular, especially for children; the proportion of vegetables, meat, eggs, fruit, and snacks gradually increased. Recently, some schools have introduced the "milk-egg project" to improve the nutrition of students.

31.3 Biological Changes Related to Low Se in KBD Patients

In the wake of some studies indicating a low rate of Se transfer from the environment to the human body, biological changes in KBD children and adult patients have been studied since 1970s.

31.3.1 Metabolic Disorders in Kashin–Beck Disease Caused by Se Deficiency

31.3.1.1 Metabolic Disorders Caused by Se Deficiency

Se-loading tests were performed among children from KBD-endemic areas with confirmed low-Se status. The results showed decreased urinary Se excretion and the delayed excretion peak after Se loading (Lv and Bai, 1992). There are several metabolic disorders caused by Se deficiency. In individuals from KBD areas, the concentration of Se in whole blood, hair, urine, and erythrocyte membranes and GPX activity in blood were significantly lower than the normal levels, while the concentration of lipid peroxide (LOP) and free fatty acid (FFA) notably increased, which indicated that the antioxidant capacity of KBD patients was substantially reduced (Han *et al.*, 1983).

31.3.1.2 Metabolic Changes Associated with Cartilage Lesions

Some biological indicators associated with cartilage metabolism were substantially affected in KBD patients compared to normal individuals. The activity of serum alkaline phosphatase increases, while the levels of non-sulfated glycosaminoglycans decrease. Total sulfhydryl and protein-binding sulfhydryl levels in plasma also significantly decrease, so does the activity of serum sulfating factor. All these changes might affect the cartilage growth. There is also a significantly positive correlation between the levels of urinary hydroxyproline, aminohexose, hexuronic acid, and sulfate in children with low-Se status in KBD areas.

31.3.1.3 Biological Markers of Muscle Damages

KBD patients, especially of the second and third degrees, show secondary muscle disuse atrophy. In 1975, it observed that the 24 h urine creatinine excretion of children in KBD areas was significantly lower than that in non-KBD areas. The creatinine levels in serum and urine samples from KBD areas decreased, while the activity of creatine phosphokinase increased. Moreover, there was a positive correlation between the urine Se content and the creatinine index.

31.3.1.4 Morphological Changes and Dysfunction of Erythrocytes

Some morphological changes and enzyme dysfunction were observed in erythrocytes of KBD patients (Tongshu Yang, 1984). In comparison with erythrocytes of healthy subjects, the levels of erythrocyte phosphatidyl choline (PC) decreased, while the sphingomyelin (SM) increased. The SM/PC value was thus raised, suggesting erythrocyte aging. Moreover, the erythrocytes become more fragile, but the membrane fluidity and the Na^+–K^+–ATPase activity decreased; these changes might reflect biological membrane defects of erythrocytes in KBD.

The glutamic oxaloacetic transaminase of plasma and lactate dehydrogenase activity significantly increased in KBD. The substantial decrease in cyclic guanosine monophosphate (cGMP) levels leads to elevated cAMP/cGMP ratio. Increased 17-hydroxyl corticosteroid levels combined with a decrease in 17-ketosteroid content might partly explain the dysfunction of the adrenal cortex in KBD patients; the raised level of T_3 and T_4 (especially T_4) hormones might be associated with a mild hyperthyroidism accompanying KBD.

31.3.2 Regulatory Factors and Gene Expression in KBD

Agilent microarray analysis revealed 26 genes related to Se deficiency in KBD, such as *POSTN*, *CLEC3B*, *BCL-2*, and *BAX*. *POSTN* is involved in the recruitment of cells to the chondrocytes lineage, initiating the cartilage loss. *CLEC3B*, which is associated with osteogenesis and bone mineralization (Valdes, 2011), was downexpressed (0.18-fold). The expression levels of apoptosis-related factors BCL-2, BAX, Fas, and iNos are higher in the surface and middle zone of articular cartilage in KBD patients than in normal cartilage (Zou *et al.*, 2009). Se deficiency might induce the excessive apoptosis of chondrocytes and impair the growth of bone and cartilage.

Apoptotic chondrocytes in KBD have activated caspases (−3, −8, −9), upregulated APAF1, reduced ATP levels, and an increased proportion of cells with de-energized mitochondria, and suggests that dysfunction of mitochondria plays an important role in the induction of apoptosis and, at the same time, in antiapoptotic reactions of articular chondrocytes in KBD.

The expression of *HS6ST2* is significantly downregulated in KBD cartilage at both mRNA and protein levels. This can inhibit cell viability through the FGF signal pathway, and it might affect aggrecan (major proteoglycan in the articular cartilage) metabolism *via* related regulatory genes. Downregulation of *CSGALNAC-1* and *PIM2* in KBD may lead to decreased aggrecan levels in chondroitin sulfate chains, weakening the water-binding function; cell death and apoptosis observed in KBD patients might also be accelerated.

Recently, it was confirmed that sodium selenite is effective in protecting chondrocyte damages from KBD patients by increasing the expression of genes *CSGALNAC-1* and *HAPLN-1*. Moreover, because the cooperating genes *LEF1*, *APCDD1*, and *PTN* are strongly upregulated in KBD cartilage, the inhibition of Wnt/β-catenin pathway was considered a potential novel target in KBD therapy. *CAV2* gene, highly expressed in KBD cartilage, could partly regulate the integrin-mediated extracellular matrix interactions in human cartilage and the process of premature senescence induced by stress factors (*e.g.*, oxygen free radicals) in chondrocytes.

The expression of enchondral ossification-related factors, such as PTHrP, TGF-β, VEGF, COMP, Col II, and Col X, substantially altered in synovial fluid and blood of adult patients, indicating degenerative changes in KBD cartilage, which eventually lead to chondronecrosis. The decrease in the content of hyaluronic acid and Se is negatively correlated with the levels of TGF-β, TNF-α, NO, VEGF, and IL-1β in KBD serum; this phenomenon might be associated with the repair of cartilage lesions.

The proteins that have been associated with KBD are mainly involved in cellular redox homeostasis and stress response (MnSOD, Hsp27, Peroxiredoxin-1, and Cofilin-1), glycolysis (PGK-1, PGM-1, α-enolase), cell motility, and cytoskeletal organization (Actin, Calponin-2, and Keratin). Disturbances in these processes might be partly responsible for the abnormalities of cytoskeletal remodeling, glycometabolism, and oxidative stress in KBD articular cartilage (Ma *et al.*, 2011). GPX activity and Se levels in KBD serum are lower than in the serum of healthy individuals in non-KBD areas (Xiong *et al.*, 2010). Polymorphisms in *GPX* gene and its decreased mRNA levels suggest that *GPX4* might be a candidate susceptibility gene for KBD (Du *et al.*, 2012). Decreased activity of GPX1 and increased expression of GPX7 observed in KBD chondrocytes from different individuals. As these two selenoproteins are involved in the protection of chondrocytes against free-radical damage, the data might reflect some damage in the protection system in advanced KBD cases.

31.4 Short Tandem Repeats (STRs) and Single Nucleotide Polymorphisms (SNPs) in KBD Patients

STR loci on chromosomes 2, 11, and 12 and SNPs of genes for GPX1, GPX4, TNF-α, IL-1β, and HLA–DRB1 have been shown to be correlated with KBD with low-Se nutritional states. STRs D2S165, D2S2333, D2S338, and D2S305 significantly differed between the KBD group, the healthy controls in the KBD area, and the healthy controls in the non-KBD areas. The 143-bp fragment of D2S165 and 89-bp and 105-bp fragments of D2S2333 loci were only in the KBD group. The frequency of 159-bp fragment of D2S165 and 79- and 81-bp fragments of D2S2333 in the KBD group were higher than in the healthy groups from KBD and non-KBD areas. The percentages of 290-bp fragment in locus D2S338 (OR < 1) in the KBD group were significantly lower than in the control groups from KBD and non-KBD areas; the opposite was true of 320-bp fragment in locus D2S305 (OR > 1).

Variants of the chromosomal STRs D11S4094, D11S4149, D11S917, and D11S912 are associated with KBD. In comparison with control groups, KBD patients have a lower percentage of 194-bp fragments in locus D11S4094 (OR < 1) but a higher percentage of 221-bp fragment in locus D11S4149 (OR > 1). However, the percentage of the 217-bp allele in D11S4149 in the KBD group was significantly lower than in the healthy group from KBD areas. The percentage of the 106-bp allele in locus D11S912 in healthy controls from non-KBD areas was significantly higher than in KBD group and the control group from KBD areas. With relatively high LOD values, the variants of the chromosomal STR D12S1725 were associated with KBD.

Linkage disequilibrium of SNP rs713041 and rs4807542, significantly decreased mRNA level and lower haplotype A-T were found for *GPX4* in KBD subjects (Du *et al.*, 2012). SNP rs1050450, rs1800668, and rs3811699 in gene *GPX1* showed a significant association with KBD (Huang *et al.*, 2013). A significantly increased risk of KBD was found for individuals with SNP Pro/Leu or Leu/Leu (odds ratio OR = 1.78) in comparison with SNP Pro/Pro. Moreover, the activity of GPX1 in whole blood was significantly decreased in a subgroup of individuals representing Pro/Leu and Leu/Leu in comparison with Pro/Pro SNP group (Fairweather-Tait *et al.*, 2011). With these odds ratios, the *GPX1* gene might play a protective role, decreasing individual susceptibility to KBD. SNPs rs6457617, rs6457620, rs9275295, and rs7745040 in the *HLA–DRB1* gene locus and SNP rs16944 of gene *IL-1β* were also partly associated with susceptibility to KBD. A protective haplotype GTCC (OR = 1.40) and a risk haplotype ACGT (OR = 0.77) in gene HLA–DRB1 observed by Shi *et al.* (2011). Frequencies of genotype GA and allele A in TNF-α gene of KBD patients were higher than in the healthy controls, suggesting that SNP TNF-α–308G/A is associated with the susceptibility of KBD.

31.5 Animal KBD Model with Low-Se Diet

To test the role of Se deficiency in KBD, many studies with animals such as rats, chicken, dogs, pigs, and rhesus monkeys kept on low-Se diet (either produced in KBD areas or artificially made) was conducted since 1950s. Between 1972 and 1973, a study using rats fed with water and grain from KBD areas reported a high level of liver necrosis (39.71%). The study demonstrated low Se content in the diet. In the 1980s, the pathological features in growth plate cartilage and articular cartilage in some livestock were observed, which led to diagnosis of osteochondrosis. The pathological changes and development patterns in growth plate cartilage and articular cartilage in domestic pigs with osteochondrosis were some similarities with KBD. Puppy and baby rhesus fed with water and grain from serious KBD areas for 7–17 months (Zhang *et al.*, 1987) appeared multifocal and focal chondronecrosis in the deep zone of articular and growth plate cartilage, which is similar to the cartilage damages in KBD.

Some pathological features of KBD were observed in mice with deletion of *Trsp* gene (the gene encodes selenocysteine $tRNA^{[Ser]Sec}$ and is necessary for the synthesis of selenoproteins), such as chondronecrosis in articular cartilage, growth retardation (Downey *et al.*, 2009).

In rats fed with low-Se diet, GPX activity in erythrocytes and plasma Se concentration were strongly reduced in comparison with pair-fed Se-adequate rats. The low-Se diet was associated with a 68% reduction of pituitary growth hormone levels and a 50% reduction of plasma IGF-I. In plasma, PTH and $1,25(OH)_2D_3$ levels increased two-fold, while osteocalcin levels fell by 25%. Urinary deoxypyridoline was down to 57% of the normal level; a 23% reduction of bone mineral density in the femur and 21% in tibia were observed. The trabecular bone volume in the femoral metaphys is reduced by 43%. The expression of *BCL-2* and *BAX* was higher than in controls. The positive rate of apoptotic chondrocytes in the upper and middle zone of articular cartilage was significantly higher in the Se-deficient group, the iodine-deficient group or the Se- and iodine-deficient group than in the control group. However, Se or iodine deficiency alone was not sufficient for inducing KBD-like focal chondronecrosis. Several protective effects of Se were reported, besides increasing the level of type II collagen (Chen *et al.*, 2011), it is also observed to enhance antioxidative effect, maintenance of thyroid function, and promotion of trabecular bone formation. It has been shown that the combined Se and iodine deficiency significantly affects the development of bone and cartilage; Se-adequate and iodine-adequate supplementation protects chondrocytes in growth plate cartilage and promotes the formation of trabecular bone. These results show that animal models fed with low-Se diet could suggest some developmental and pathological features of KBD in humans, and iodine deficiency might also partly associate with those processes.

However, some experiments showed that cartilage lesions in rats and minipigs fed with low-Se diet did not appear the chondronecrosis in metaphysis like human KBD. The effects of Se and/or iodine deficiency on cartilage development and chondrocyte differentiation in two generations of rats showed Se deficiency or iodine deficiency did not induce KBD-like focal chondronecrosis in the deep zone of cartilage. A combined Se and iodine deficiency significantly impaired the development of bone and cartilage. The changes in the expression of Col X and PTHrP in chondrocytes induced by combined Se and iodine deficiency were similar to those caused by human KBD. Meanwhile, Chinese minipigs fed with low-Se diet from Se-deficient non-KBD areas for 103 days showed no KBD-like pathological cartilage damages. Therefore, Se deficiency can cause some damages in cartilage similar to KBD, but no animal model could completely reproduce the KBD cartilage pathology.

31.6 The Preventive and Therapeutical Effect of Se Supplement

As Se deficiency is considered a main risk factor for the occurrence and development of KBD, various forms of Se supplements have been applied in most endemic areas since the 1970s. The most common types of Se supplements in China are oral sodium selenite tablets (Se tablet), Se-enriched salt, Se-enriched egg, Se-enriched tea, Se-enriched yeast, and combination therapies (*e.g.*, vitamin E, iodine or zinc gluconate, and Se). Crop dusting with selenium, Se-enriched fertilizer, and some more comprehensive measures like improvements in water source and food quality and improvements in general hygiene conditions have also been implemented (Li *et al.*, 1988).

31.6.1 The Effect of Se Supplements on Articular Chondrocytes

The effect of Se on the proliferation and apoptosis of KBD chondrocytes used different doses of sodium selenite (at 1.0 mg L^{-1}, 0.50 mg L^{-1}, 0.25 mg L^{-1}, 0.10 mg L^{-1}, 0.05 mg L^{-1}, 0.025 mg L^{-1}, and 0.0125 mg L^{-1}). The concentrations of sodium selenite between 0.05 mg L^{-1} and 0.10 mg L^{-1} impaired the growth rate of normal chondrocytes; in contrast, the growth of KBD chondrocytes accelerated at those selenite concentrations. Se concentrations between 0.025 mg L^{-1} and 0.10 mg L^{-1} seemed to protect KBD chondrocytes from damage and reduce the apoptosis rate, G2/M rate, and the expression of Fas and caspase-3. However, at selenite concentrations above 0.5 mg L^{-1}, the KBD chondrocytes were damaged. These results demonstrate that a suitable concentration (0.25–0.1 mg L^{-1}) of sodium selenite can improve the growth of KBD chondrocytes and reduce the apoptosis rate, but excessively high concentration (Se > 0.5 mg L^{-1}) can have a damaging effect.

31.6.2 Sodium Selenite Tablets

In the late 1970s, oral Se tablets were adopted to treat KBD. Dosage was determined according to the patient's age. Usually, children under five were given 0.5 mg Se per week, five- to nine-year-olds took 1 mg per week and children older than nine took 2 mg per week. A long-term observation experiment, conducted among children aged three to ten years, observed that the X-ray detection rate of metaphysis in KBD significantly decreased from 42.55% in 1976 to 4.09% in 1983. Among the children who had started the treatment before three years of age, none contracted KBD during this period.

A series of trials has proved that this method is effective in improving the membrane stability of chondrocytes, ameliorating the metaphyseal lesions, improving the Se nutritional status, promoting the repairing action of the lesions both in metaphyses and the distal end of phalanges, and preventing its deterioration in the hands of KBD children. Before the intervention, the mean hair Se content was lower than 100 ng g^{-1}; two years after accepting this treatment, it rose to over 200 ng g^{-1} or even reached the level found in samples from the non-KBD area. A trial involving 420 children showed that no one deteriorated in the intervention group, with 81.9% of X-ray improvement rate, while in the control group, only 39.6% of subjects improved and for 13.19% of patients in this group, the disease symptoms aggravated.

31.6.3 Se-Enriched Salt

Using cooking salt containing sodium selenite (16.7 mg kg^{-1}) introduced the prevention and treatment of KBD into the daily life of residents in KBD areas; the method achieved good results. Now, more than 1.03 million residents in Chinese KBD areas with low-Se nutrition are supplied with Se-enriched salt.

In 1986, a comprehensive case-control study presented the results of a Se-enriched salt supplementation trial (Liang *et al.*, 1986). At the baseline, average X-ray detection rates of metaphyseal lesions in hands of KBD children were 57.1% in the Se supplement group (Se+ group, supplemented with Se-enriched salt later) and 61.9% in the control group (Se− group, without Se supplement). Two and a half years later, average X-ray detection rates of metaphyseal lesions in the Se+ group decreased to 36.3% and it was 45.4% in the control group. Moreover, no new cases were detected in the Se+ group, while 0.8% incidence was observed in the control group.

Eight months later, in the Se+ group, the total rates of metaphyseal changes decreased from 45.65% to 9.10%, with no new cases reported. The severity of KBD decreased, and the high prevalence in KBD areas dropped to mild levels or even levels close to those in non-KBD areas. However, in the Se− group, the incidence was 8.56%.

Se-rich salt supplementation also substantially increased the Se nutritional status *in vivo*; the hair Se content of children from KBD area in Tibet rose from 74 ± 45 ng g^{-1} in 2001 to 256 ± 133 ng g^{-1} in 2003.

31.6.4 Se-Fortified Wheat

The soil-food-body route is the major pathway from the environment to the human body. Se-fortified fertilizer and Se-fortified wheat (sodium selenite spraying on the leaf at 1 g per 0.17 acre during wheat flowering stage) have been widely used in KBD areas. The Se content in Se-sprayed wheat increased to 75 ± 0.7 ng g^{-1} (17 ± 0.6 ng g^{-1} in the controls), but with no differences of Fe, Mn, Mg, Ca, Mo, Sr, Cu and P contents between the two groups. After consumption of Se-fortified wheat for 6 months, the Se content of children's hair and urine reached the levels in samples from the non-KBD area (hair Se: 200 ng g^{-1}, urine Se: 5 μg per 24 h).

In 1983–1988, a five-year trial verified the efficacy of Se-fortified wheat with Se-containing preparations. A significant negative correlation discovered between Se content in whole blood/hair/urine and the X-ray detection rate of metaphyseal lesions of the hands in children. At the base line, no significant difference was observed in the Se contents of whole blood between the Se-supplemented group and the control group. But only half a year later, the Se content of whole blood in supplemented group quickly reached 133 ± 27 ng g^{-1}, the X-ray detection rate of metaphyses in phalanges of children declined from 58.76% to 46.15% and the improvement rate of metaphyseal lesions reached 80.0%. Meanwhile, the Se content of whole blood in the supplemented group was only 35 ± 15 ng g^{-1}, the X-ray detection rate of metaphyses in phalanges of children increased from 42.67% to 48.32% and the improvement rate of metaphyseal lesions only 43.1%. Metabolically, in comparison with the levels before the Se-spray intervention, the level of urine creatinine significantly increased but urine creatine substantially decreased. The urinary hydroxyproline, ester sulfur, and inorganic sulfur showed an upward trend, while the urinary glycosaminoglycan levels changed with the season. The contents of urinary creatinine, urinary glycosaminoglycan, and the Se content of hair/urine were negatively correlated with the X-ray detection rate of metaphyseal lesions in hands of children.

31.6.5 Se Supplement Combined with Other Elements

After decades of practical experience, it has become clear that the comprehensive, combined methods are more effective than Se tablets or Se-enriched salt alone.

The most common practice was using Se-tablet combined with vitamin E (both having antioxidant properties) (Wang *et al.*, 1983). The results showed that this method accelerated the increase of hair/serum Se content and increased GPX activity, taking only two months to be effective. This treatment also shortened the repair process of metaphyseal lesions; after four months, the differences between the Se/vitamin E group and Se-enriched salt or Se-tablet group became obvious. Se-enriched salt or Se tablet was better at enhancing Se nutritional status and preventing deterioration than using either Se tablet combined with vitamin C, vitamin C only, or iodine-enriched salt alone. Some KBD-endemic areas are also iodine deficient. Therefore, iodine supplement was valued as much as Se and certainly improved the control of KBD. One year of Se supplementation proved that Se might prevent KBD occurrence among unaffected children in KBD-effected areas and hold back new lesions in affected children; new cases were only detected in the Se-deficient but iodine-adequate group. Growth retardation diminished significantly in both the Se and iodine-adequate group and Se the deficient but iodine the adequate group but not in the nonsupplemented group, which was probably due to the correction of iodine deficiency.

A Se-zinc therapeutic method (Se tablet combined with zinc gluconate) was also used. After a year and a half of treatment, the net repair rate of metaphyseal lesions in the Se-zinc group rose from 33.53% to 81.87% and was notably higher than in the Se-tablet group (1.57–24.84%) in the same period.

In addition to the purely oral supplements, some more comprehensive approaches were used; whole grain foods, improvements in water source and hygiene conditions, and some other factors were combined. After more than three decades, detection of a new case is scarce. For most children from the KBD areas, the hair Se content has reached the levels found in nonendemic areas.

31.6.6 Nano-Se-Chondroitin Sulfate

Nano-Se-chondroitin sulfate, a novel nanoscale Se preparation with 30–200 nm particles in distilled water, has less toxicity than sodium selenite and can effectively inhibit the apoptosis caused by T-2 toxin.

31.7 Conclusions

Low Se content in soil, water, and cereals in KBD-endemic areas can lead to low Se levels in the human body. There is a negative correlation between Se content in hair/urine of children and the KBD prevalence. Se supplements can diminish the pathological changes in the metaphysis and distal ends of phalanges, and it can decrease, but not completely prevent the KBD incidence of children. Low-Se diets in animal models do not reproduce the human KBD pathology completely. Se deficiency is clearly an important environmental factor; selenium supplementation can be used to good effect in the prevention of KBD in children.

Summary Points

- This chapter focuses on selenium and Kashin–Beck disease (KBD).
- KBD is an endemic osteoarthropathy and mainly distributed in the low-selenium areas in China.
- Selenium deficiency is one of the main etiological hypotheses of KBD through decades of studies focused on epidemiology, pathophysiology, genetics and effects of selenium-supplement control measures.
- There are several selenium-related genes and proteins expressed abnormally in KBD patients.
- The incidence of KBD in most endemic areas decreased with the increasing level of selenium nutrition after different forms of selenium supplement but some of them decreased naturally without any selenium supplement.
- Therefore, low selenium is an important environmental risk factor and has a good effect to prevent KBD.

Key Facts of Kashin–Beck Disease (KBD)

This table lists the key facts of KBD including the basic concept, the general risk factors and effective control measures.

1. Kashin–Beck disease (KBD) has been a big medical mystery to China, especially the western areas, the disease was firstly discovered in Siberia, Russia.
2. KBD is a chronic, deformed osteoarthropathy, and the severe cases are short stature, disability and even loss of life skills.

3. KBD mainly distributes in the low-selenium areas, but the low selenium level *in vivo* correlated with the incidence of KBD more intensely.
4. In the endemic areas, KBD is more than one patient detected in one family, however, it is not a familial hereditary disease.
5. Selenium deficiency can lead to biochemical metabolic disorders in KBD patient, such as the lipid peroxidation disturbance and ROS disorders.
6. Selenium supplements can effectively prevent the incidence of KBD in children, promote the repair of metaphyseal lesions in phalanges and alleviate the damages of chondrocyte cultivated *in vitro* treated with T-2 toxin.
7. Except for the selenium supplement, intake of diversified food that comes from the non-KBD areas with normal selenium levels can also improve the selenium nutrition of residents living in the KBD areas.

Definitions of Words and Terms

Kashin–Beck disease (KBD). KBD is an endemic, multifocal and deformed osteoarthropathy. It mainly damages the articular hyaline cartilage of children during the development stage that leads to degeneration, apoptosis and secondary osteoarthritis, or can even be as serious as dwarfism and permanent disability.

X-ray detection rate of KBD. This means the positive frequency or intensity of KBD detected by X-ray examination. It is the most commonly used indicator in field investigations of KBD.

Cartilage damage of KBD. The growth plate cartilage and articular cartilage are the most commonly affected sites of KBD. Focal chondronecrosis in the deep zone and impaired endochondral ossification in these cartilage results in a secondary chronic deforming osteoarthrosis and impaired skeletal development.

Selenium deficiency or low-selenium hypothesis. This holds the idea that KBD might be caused through the chain: endemic environmental selenium deficiency – scanty intake of selenium from local food – low selenium nutrition *in vivo* – selenium metabolic disorders – KBD.

Selenium level. This means the content of selenium. Selenium level in soil, selenium level in food and selenium level in human body are the commonly used indicators in KBD investigations.

Selenium level in soil. This means the content of selenium in soil, which is lower than 0.15 mg kg^{-1} in most KBD-endemic areas.

Selenium level in food. This means the content of selenium in all kinds of food eaten daily by residents. Generally, it is lower than 0.02 mg kg^{-1} in KBD-endemic areas, especially the selenium level of wheat and corn significantly lower than in nonendemic areas.

Selenium level in blood. This means the selenium content in human blood, which is remarkably lower in KBD endemic areas compared with the population in nonendemic areas.

Selenium level in hair. This means the selenium content in human hair, which is notably lower in KBD endemic areas compared with the population in nonendemic areas. 200 ng g^{-1} is considered as the critical value for distinguishing between endemic and nonendemic areas.

Comprehensive prevention measures of KBD. Under the situation of unclear etiology of KBD, performing the comprehensive measures including supplement selenium in different ways, change food, change the clear drinking water, and improve the hygienic and living conditions, which turned out to be effective control of KBD.

Selenium supplement. Sodium selenite tablet, salt-enriched Se, water enriched with Se and other food enriched with selenium, *etc.* are the main effective approaches used to improve the low selenium level of children and decrease the incidence of KBD, promote the repair of metaphyseal damages and prevent deterioration in phalanges of children. There is a negative correlation between low selenium level and X-ray detection rate of KBD according to the epidemiological investigation.

Sodium selenite tablet. This contains 1 mg selenium per tablet. The method is 1 mg per week for children under seven years old and 2 mg per week for children over eight years old.

Salt-enriched Selenium. The proportion of sodium selenite and edible salt is 1 to 60 000, which is supplied as cooking salt for the residents living in KBD-endemic areas.

Selenium-fortified wheat. Use 15 g of sodium selenite dissolved in 150 L of water and then spray on the leaf of wheat during flowering phase; it is enough for a hectare.

List of Abbreviations

KBD	Kashin–Beck disease
Se	Selenium
KD	Keshan disease
STRs	Short tandem repeats
SNPs	Single nucleotide polymorphisms
OR	Odds ratio
GPX	Glutathione peroxidase

References

Allander, E., 1994. Kashin–Beck disease. An analysis of research and public health activities based on a bibliography 1849–1992. *Scandinavian Journal of Rheumatology*. 99: 1–36.

Chen, J., Chu, Y., Cao, J., Wang, W., Liu, J. and Wang, J., 2011. Effects of T-2 toxin and selenium on chondrocyte expression of matrix metalloproteinases (MMP-1, MMP-13), alpha2-macroglobulin (alpha2M) and TIMPs. *Toxicology In Vitro*. 25: 492–499.

Downey, C. M., Horton, C. R., Carlson, B. A., Parsons, T. E., Hatfield, D. L., Hallgrimsson, B. and Jirik, F. R., 2009. Osteo-chondroprogenitor-specific deletion of the selenocysteine tRNA gene, Trsp, leads to chondronecrosis and abnormal skeletal development: a putative model for Kashin–Beck disease. *PLoS Genetics*. 5: e1000616.

Du, X. H., Dai, X. X., Xia Song, R., Zou, X. Z., Yan Sun, W., Mo, X. Y., Lu Bai, G. and Xiong, Y. M., 2012. SNP and mRNA expression for glutathione peroxidase 4 in Kashin–Beck disease. *British Journal of Nutrition*. 107: 164–169.

Egashira, Y. and Staff, H., 1990. *An outline of history of research on Kashin–Beck disease in Japan*. In: *Proceedings of International Workshop on Kashin–Beck Disease and Non-Communicable Diseases*. Chinese Academy of Preventive Medicine (CAPM). World Health Organization, Beijing, pp. 149–151.

Fairweather-Tait, S. J., Bao, Y., Broadley, M. R., Collings, R., Ford, D., Hesketh, J. E. and Hurst, R., 2011. Selenium in human health and disease. *Antioxidants and Redox Signaling*. 14: 1337–1383.

Han, Z., Li, G. Y., Cao, S. G., Yang, J. G., Wei, Y. Z. and Ren, Y. Y., 1983. Determination of blood glutathione peroxidase, vitamine E, lipid peroxide and free fatty acid of children in the Kashin–Beck disease effected areas. *Chinese Journal of Endemiology*. 2: 65–68.

Huang, L., Shi, Y., Lu, F., Zheng, H., Liu, X., Gong, B., Yang, J., Lin, Y., Cheng, J., Ma, S., Lin, H. and Yang, Z., 2013. Association study of polymorphisms in selenoprotein genes and Kashin–Beck disease and serum selenium/iodine concentration in a Tibetan population. *PLoS One*. 8: e71411.

Li, C. Z., Huang, J. R., Wang, W. L., Zhang, Y. X., Xu, J. M. and Wang, L., 1988. The experimental use of Se-diacetic acid in the prophylaxis and treatment of Kashin–Beck disease. *Endemic Diseases Bulletin*. 3: 81–83.

Li, L., Wang, Q., Tian, J. H., Yang, K. H., Tian, H. L., Shi, C. H., Jiang, J., Wu, T. X. and Liao, Y. J., 2013. The Relationship between the exposure and metabolism of iodine and Kashin–Beck disease: a systematic review. *Chinese Journal of Evidence Based Pediatrics*. 8: 176–180.

Li, S. J., Li, W., Hu, X., Yang, L. S. and Xirao, R. D., 2008. Distribution of Kashin–Beck disease (KBD) and its relation to selenium content in soil-plant-animal (human being) ecosystem in Tibet. *Chinese Journal of Ecology*. 27: 2167–2170.

Liang, S. T., Mu, X. Z. and Zhang, F. J., 1986. The observation on the effects of Kashin–Beck disease prevention by selenium salt. *Journal of Practical Endemiology*. 1: 21–23.

Lv, S. M. and Bai, C., 1992. Selenium load tests in the children from Kashin–Beck disease affected areas. *Journal of Xi'an Medical University*. 13: 333–342.

Ma, W. J., Guo, X., Liu, J. T., Liu, R. Y., Hu, J. W., Sun, A. G., Yu, Y. X. and Lammi, M. J., 2011. Proteomic changes in articular cartilage of human endemic osteoarthritis in China. *Proteomics*. 11: 2881–2890.

Ministry of Health, P. R. C., 2012. The prevention and control status of Kashin–Beck disease in 2011. http://www.nhfpc.gov.cn/htmlfiles/zwgkzt/ptjnj/year2012/index2012.html.

Mo, D. X., Lv, S. M. and Bai, C., 1997. Progress of the relationship between low selenium and Kashin–Beck disease. *Chinese Journal of Control of Endemic Diseases*. 9: 199–121.

Moreno-Reyes, R., Suetens, C., Mathieu, F., Begaux, F., Zhu, D., Rivera, M. T., Boelaert, M., Neve, J., Perlmutter, N. and Vanderpas, J., 1998. Kashin–Beck osteoarthropathy in rural Tibet in relation to selenium and iodine status. *New England Journal of Medicine*. 339: 1112–1120.

Shi, Y., Lu, F., Liu, X., Wang, Y., Huang, L. L., Liu, X. Q., Long, W. B., Lv, B., Zhang, K., Ma, S., Lin, H., Cheng, J., Zhou, B., Hu, M., Deng, J. Y., Zhu, J. X., Hao, P., Yang, X., Zeng, M. C., Wang, X. Q., Shen, S. K. and Yang, Z. L., 2011. Genetic variants in the HLA-DRB1 gene are associated with Kashin-Beck disease in the Tibetan population. *Arthritis and Rheumatism*. 63: 3408–3416.

Tan, J. A., Zhu, W. Y., Li, R. B., Zheng, D. X., Hou, S. F., Zhu, Z. Y. and Wang, W. Y., 1990. *Geographic distribution of Kashin–Beck disease in China and the relation of ecological chemico-geography to its occurrence*. In: *Proceedings of International Workshop on Kashin–Beck Disease and Non-Communicable Diseases*. Chinese Academy of Preventive Medicine (CAPM). World Health Organization, Beijing, pp. 12–15.

Tokio, T., 1968. *Kashin–Beck Disease. Dysostosis Enchondralis Endemica, Japan*. vol. 1, p. 188.

Valdes, A. M., 2011. Involvement of the CLEC3B gene in osteoarthritis. *Osteoarthritis Cartilage*. 19: 249.

Voshchenko, A. V. and Ivanov, V. N., 1990. Kashin–Beck disease in the USSR. In: *Proceedings of International Workshop on Kashin–Beck Disease and Non-Communicable Diseases*. Chinese Academy of Preventive Medicine (CAPM). World Health Organization, Beijing, pp. 152–196.

Wang, D. S., Zhu, Z. Y. and Luo, J. F., 1983. The observation on the treatment of Kashin–Beck disease with sodium selenite and vitamin E. *Chinese Journal of Endemiology*. 2: 204–207.

Xiong, Y. M., Mo, X. Y., Zou, X. Z., Song, R. X., Sun, W. Y., Lu, W., Chen, Q., Yu, Y. X. and Zang, W. J., 2010. Association study between polymorphisms in selenoprotein genes and susceptibility to Kashin–Beck disease. *Osteoarthritis Cartilage*. 18: 817–824.

Xu, J. M., Wu, M. L., Cheng, J. H., Zhou, X. G., Wang, G. D., Wang, J. and Wang, W. L., 1999. Study on the relations between Kashin–Beck disease and iodine. *Endemic Diseases Bulletin*. 14: 36–38.

Yang, T. S., 1984. *The progress of biochemistry in Kashin-Beck disease*. In: *Anthology of scientific investigation of Kashin-Beck disease in Yongshou County*. People's Medical Publishing House, Beijing, pp. 275–286.

Zhai, S. S., 1990. *Investigation on the relationship between Kashin-Beck disease and drinking water*. In: *Proceedings of International Workshop on Kashin-Beck Disease and Non-Communicable Diseases*. Chinese academy of preventive Medicine (CAPM). World health organization, Beijing, pp. 96–101.

Zhang, G. Q., Liu, J. X., Wang, D. Z., Zhang, H. X., Li, C. F., Sun, K. and Li, L. Q., 1987. The replication of pathological model on animal Kashin–Beck disease once more. *Chinese Journal of Control of Endemic Diseases*. 2: 8–10.

Zhaxi, S. Z., Gong, H. Q., Gesang, D. J. and Baima, C. W., 2003. Analysis of the selenium content of Kashin–Beck disease endemic area in Tibet. *Chinese Journal of Control of Endemic Diseases*. 18: 177.

Zou, K., Liu, G., Wu, T. and Du, L., 2009. Selenium for preventing Kashin–Beck osteoarthropathy in children: a meta-analysis. *Osteoarthritis Cartilage*. 17: 144–151.

CHAPTER 32

Selenium, the Placenta and Trophoblast Mitochondrial Oxidative Stress

ANTHONY V. PERKINS*[a]

[a]School of Medical Science, Griffith Health Institute, Griffith Health Centre, Griffith University Gold Coast Campus, Parklands Drive, Southport, QLD, 9726, Australia
*E-mail: a.perkins@griffith.edu.au

32.1 The Importance of the Placenta to Lifelong Health

The placenta is arguably the most important organ you will have in your lifetime. Although the life span of the placenta and its role as a life support system for foetal development is only 40 weeks, the work of David Barker in the early 1990s demonstrated that a healthy start to life was essential for a future of lifelong health (Barker, 1992). Placental health has repeatedly been associated with the development of chronic diseases later in life and the developmental origins of human adult disease is now a major focus in reproductive biology research. The placenta requires an adequate supply of macronutrients and oxygen to develop and sustain the growing foetus. Cellular differentiation must be tightly controlled through the exquisite regulation of gene expression and the role of placental epigenetics is currently the subject of intense

Food and Nutritional Components in Focus No. 9
Selenium: Chemistry, Analysis, Function and Effects
Edited by Victor R. Preedy

Published by the Royal Society of Chemistry, www.rsc.org

scrutiny in many laboratories around the globe. But what of the importance of placental micronutrients? How essential are the so-called essential vitamins and minerals to placental development, placental function and their role in the development of a healthy baby? In this chapter the role of selenium will be examined in relation to placental mitochondrial function and the alleviation of mitochondrial oxidative stress in the placenta during pregnancy.

32.2 Importance of Micronutrition to Placental Health

The human placenta requires an adequate and appropriate supply of macronutrients, balanced by an adequate and appropriate supply of micronutrients. Many studies in mammalian species have demonstrated the detrimental effects of poor nutrition on placental and subsequently foetal development. Calorific restriction causes poor placental development and intrauterine growth restriction (Brenseke *et al.*, 2013). Equally an oversupply of macronutrients has been associated with complications of pregnancy. Hyperglycaemia, caused by underlying and/or pregnancy-induced diabetes mellitus results in foetal macrosomia and higher incidences of pregnancy complications such as pre-eclampsia and preterm labor (Dyer and Rosenfeld, 2011). Hyperlipidemia has also been associated with complications of pregnancy and women with BMI above 30, who are clinically obese, have a higher risk of developing complications of pregnancy such as pre-eclampsia (Gallos *et al.*, 2013). Adequate and balanced levels of macro- and micronutrients are essential for placental function and foetal development, resulting in a successful pregnancy.

Oxygenation is also critical to placental health and poor placental perfusion due to shallow trophoblast invasion and inadequate placentation is equally detrimental to placental function (Hung *et al.*, 2002). The metabolic junction between macronutrient oxidation and oxygenation occurs in the mitochondria, and mitochondrial adequacy is essential for placental function. An excess of electrons due to macronutrient oversupply or a lack of available oxygen to accept electrons harnessed by oxidative pathways can lead to a state of oxidative stress. First described by Sies in 1985, oxidative stress is an imbalance between reactive oxygen and nitrogen species (RONS) and the complex systems of antioxidants that have evolved to make life in an aerobic atmosphere possible (Sies, 1997). Reactive oxygen species include partially reduced forms of oxygen such as superoxide and hydrogen peroxide and damaging nitrogen radicals such as peroxynitrite. Whilst a balanced diet will contain many essential antioxidant compounds such a vitamins C and E, lycopenes and resveratrol, the reductive power of these exogenous antioxidants is limited and may in fact lead to the generation of further radical species. This is the case with vitamin E that can form a peroxyl radical as part of its mechanism of action. These partially oxidized or reduced exogenous antioxidants can be scavenged by the enzymatic antioxidants that exist in the cytoplasm and most importantly the mitochondria.

32.3 Placental Oxidative Stress

Reactive oxygen and nitrogen species (RONS) are formed through the partial reduction of molecular oxygen and can be lethal to cells causing extensive macromolecular oxidation and tissue damage (Halliwell and Gutteridge, 2007). Addition of a single electron to molecular oxygen results in the formation of superoxide, which can be negated by the action of either Mn-superoxide dismutase (SOD) in the mitochondria or the equivalent Cu/Zn-SOD isoform of the enzyme that exists in the cytoplasm or extracellularly. The resulting formation of hydrogen peroxide (H_2O_2) is equally threatening to cell homeostasis and needs to be eliminated by a variety of enzymes with peroxidase activity including catalase, glutathione peroxidase (GPx) and thioredoxin peroxidase (peroxiredoxin, ThioPx). Oxidised GPx and ThioPx are recycled through electron shuffling from glutathione and thioredoxin respectively, which are then returned to their active state by glutathione reductase and thioredoxin reductase (ThioRed). In each case the ultimate electron donor for these important reductive systems is NADPH. Many of these antioxidant proteins require trace elements to be active, for example Mn-SOD or Cu-SOD, and without an adequate intake of essential trace elements they are catalytically inert. This is the case for GPx and ThioRed that are selenoproteins and are acutely sensitive to an adequate intake of selenium.

The human placenta is a hemochorial structure that has a vast surface area due to the extensive branching of the chorionic villi. This surface area is covered by a continuous syncytium that is constantly renewed by fusion of the underlying mononuclear cytotrophoblast into the overlaying multinuclear syncytium (Huppertz and Borges, 2008). This fusion is the initial step in trophoblast turnover that ultimately results in shedding of syncytial debris into the maternal circulation. Syncytial debris is extruded through normal apoptotic processes and is shed directly into the maternal circulation and this can be a significant physiological and immunological challenge to the mother (Redman and Sargent, 2000). Each day approximately 2–3 g of placental debris is shed into the maternal circulation in the form of macroparticles (1–100 μm), microparticles (100 nm–1 μm) and nanoparticles including exosomes (20–100 nm). There is also an abundance of free macromolecules including placental lipids, proteins and nucleic acids as well as the plethora of biologically active hormones and mediators produced by the placenta. These all originate from an organ that is 50% genetically distinct from the maternal host as the placenta is derived from embryonic tissue and so contains the paternal genome. The removal of this debris is potentially dangerous as the maternal immune system can react adversely and mount a hostile response directed at the foetal–placental unit if the shedding of placental debris is excessive and/or inappropriately packaged. This is the case in many complications of pregnancy including severe conditions such as pre-eclampsia that affects 3–5% of all human pregnancies and still causes approximately 60 000 maternal deaths globally each year (Figure 32.1).

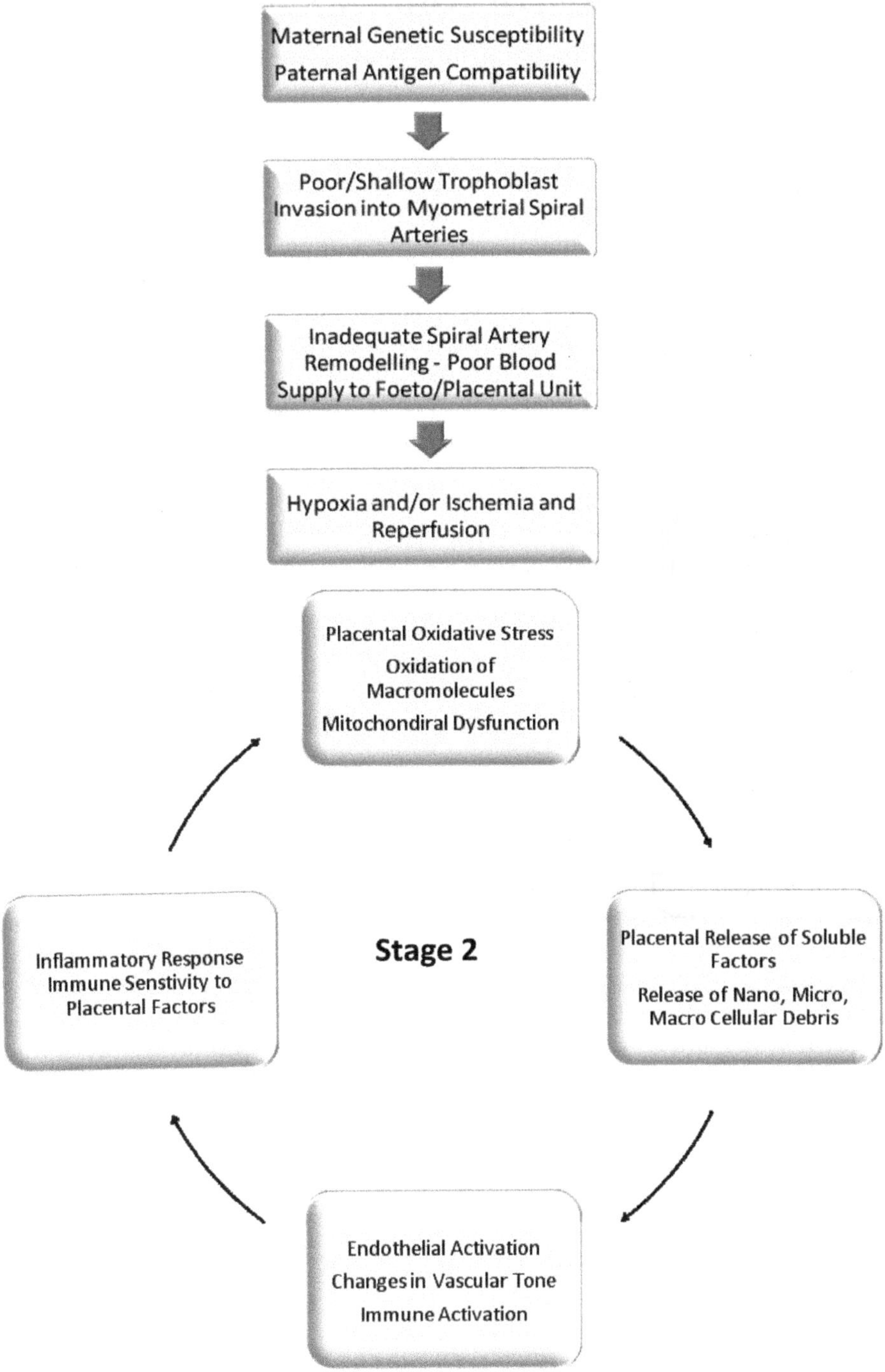

Figure 32.1 Schematic representation of the pathogenesis of pre-eclampsia.

32.4 Mitochondrial Oxidative Stress

The role of placental mitochondria in regulating placental function and viability is paramount in placental biology. Mitochondria are required to deliver ATP through oxidative phosphorylation but they are also critical in regulating cellular turnover, especially when they are subjected to oxidative stress. Mitochondria that exist in placental cytotrophoblasts have a relatively normal appearance and exist in a mononuclear environment surrounded by the normal cellular structures and processes that we associate with cell homeostasis and metabolism. But what of the mitochondria that exist in the syncytium, the enormous, continuous, multinuclear structure that covers the entire surface of the placenta? These mitochondria are constantly bathed in oxygenated maternal blood and susceptible to over- or undersupply of maternal nutrients and so may be particularly vulnerable to oxidative stress. They rely heavily on antioxidant defence mechanisms to counter the generation of reactive oxygen and nitrogen species. When not present or inadequately expressed, increased free-radical generation and corresponding mitochondrial oxidative stress in the syncytium results in placental apoptosis and necrosis and subsequent shedding of placental debris that may lead to adverse reactions by the maternal endothelium. This sequence of events underpins the pathophysiology of placental diseases of pregnancy such as pre-eclampsia (Burton *et al.*, 2009).

Placental mitochondria are slightly different from mitochondria found in other tissues. They contain significantly fewer respiratory cytochromes, suggesting that they may have a reduced capacity for oxidation of substrates than other tissues (Martinez *et al.*, 2012). This enables the placenta to act as a transport organ, more than a metabolic organ, and fulfil its role of maximizing nutrient supply to the developing foetus. In fact, the placenta uses mainly glucose as its preferred fuel and oxidises this through anaerobic glycolysis, thereby minimizing its own utilization of oxygen, fatty acids and amino acids that are all in high demand by the developing foetus (Desoye and Shafrir, 1994). There is also accumulating evidence of both morphological and biochemical differences in mitochondria from the cytotrophoblast and those present in the overlaying syncytium. Mitochondria from mononuclear cytotrophoblast are large with lamellar cristae and are similar in form to mitochondria from the liver and other tissues. However, mitochondria isolated from syncytiotrophoblast are much smaller with a condensed matrix and vesicle-like cristae. They are also biochemically distinct in that they express large quantities of the enzyme P450scc that is required as an essential step in the conversion of cholesterol to progesterone, linking mitochondrial function with the essential endocrine changes required to sustain pregnancy (De Los Rios Castillo *et al.*, 2011). The ability of syncytial mitochondria to produce large amounts of progesterone is an absolute requirement for a healthy pregnancy and inhibition of progesterone production results in pregnancy loss. Furthermore, Martinez *et al.* were able to demonstrate that these morphological and biochemical changes were a result of the differentiation process as they observed similar changes in BeWo cells, a mononuclear trophoblastic cell line, which can be induced to fuse and form a syncytium (Martinez *et al.*, 1997).

Ultrastructural studies have also shown that mitochondria from pre-eclamptic pregnancies are further modified and display degenerative changes suggestive of apoptosis including lysis of cristae and matrix vacuolization (Castejon, 2011). There are also significant biochemical changes in mitochondria from pre-eclamptic placenta and a recent study using proteomic comparisons has shown the downregulation of 22 key proteins involved in regulating apoptosis, fatty acid oxidation, generation of RONS and oxidative stress (Shi *et al.*, 2013). Many of these changes have been attributed to hypoxia that can alter the expression and assembly of critical iron–sulfur centers at the heart of complexes I and III in the electron-transport chain, making pre-eclamptic placental mitochondria more susceptible to oxidative stress and more likely to initiate apoptosis through cytochrome c leakage (Colleoni *et al.*, 2013). Understanding mitochondrial function in terminally differentiated trophoblast is central to understanding the pathogenesis of pregnancy complications such as pre-eclampsia and foetal growth restriction.

Our work and that of others has examined the expression and activity of key antioxidant proteins in the syncytium (Vanderlelie *et al.*, 2005). There is a paucity of both Cu/Zn-SOD and Mn-SOD in syncytiotrophoblast when examined by immunohistochemistry (Myatt *et al.*, 1997). Similarly, there appears to be very little expression of Glutathione peroxidase 1 and 4 (Mistry *et al.*, 2010). Our work has shown very little thioredoxin staining in the syncytial layer but strong expression in the underlying mononuclear villous cytotrophoblast and invading extravillous cytotrophoblast in first trimester placental tissues (Figure 32.2B). The expression of thioredoxin appears to be related to differentiation from cytotrophoblast to syncytiotrophoblast as the expression is maintained in undifferentiated cells in term membranes (Figure 32.2D) whereas it disappears in differentiated syncytium in the same-term placenta (Figure 32.2C).

Other studies have shown that the associated enzyme Thioredoxin reductase is also poorly expressed in the syncytium (Ejima *et al.*, 1999) and the expression seems to be related to cell fusion and differentiation from the mononuclear form to the multinuclear syncytiotrophoblast. In an elegant series of experiments, Frendo *et al.* treated mononuclear trophoblast cell line BeWo with Forskolin to induce differentiation and noted a concomitant decrease in SOD expression (Frendo *et al.*, 2000). From these observations it can be concluded that syncytiotrophoblast is poorly equipped with antioxidant enzymes and is particularly vulnerable to oxidative stress due to poor antioxidant protection and relatively high oxygenation due to the direct contact with maternal blood.

Measurements in pre-eclamptic tissues have supported these conclusions and indicate that the situation may be exacerbated in pathological tissues. Placental protein extracts produced from pre-eclamptic tissues contained less activity of SOD, GPx, thioredoxin and thioredoxin reductase when compared to

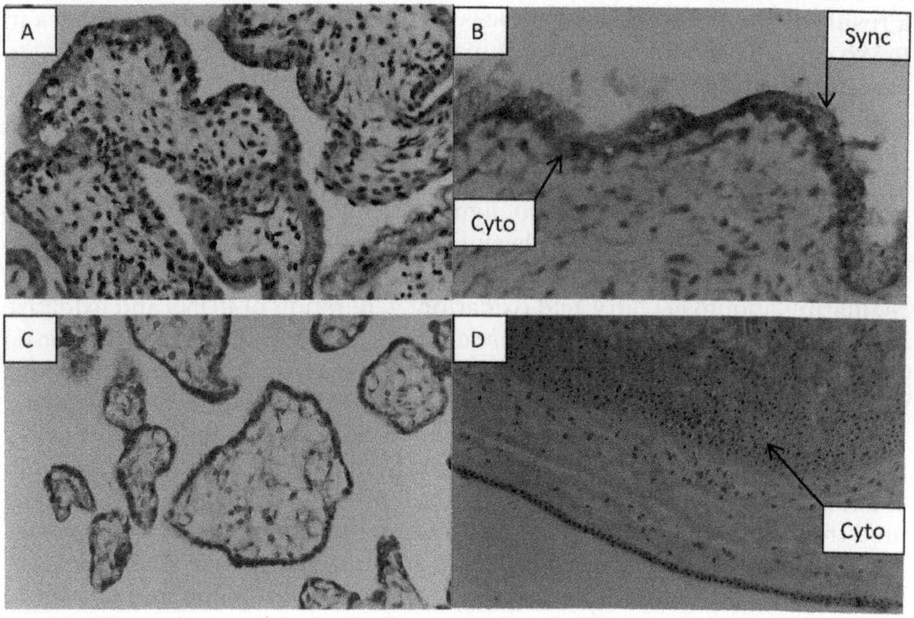

Figure 32.2 Thioredoxin immunocytochemistry in placental tissues: (A) positive control with cytotrophoblast and syncyiotrophoblast (Sync) immunostained for cytokeratin 18 in 1st trimester placental villi. (B) Cytotrophoblasts (Cyto) are immunopositive for thioredoxin, whilst syncytiotrophoblast is not in 1st trimester placental villi. (C) Absence of thioredoxin immunostaining in term placental villi. (D) Presence of thioredoxin staining in undifferentiated cytotrophoblast in chorionic membranes. (Experimental details in Perkins *et al.* (1995).)

gestational age matched tissues from healthy placentae. This was accompanied by evidence of increased oxidative stress with elevated levels of oxidized lipids, proteins and DNA (Vanderlelie *et al.*, 2005). These observations raise several interesting questions about the pathophysiology of pre-eclampsia and whether increased placental expression of these enzymes could have a protective effect and prevent oxidative insult in trophoblast cells. As mentioned earlier, GPx and thioredoxin reductase are selenoproteins and their expression and activity is acutely sensitive to selenium supplementation. In a series of experiments, we have used selenium supplementation as an effective *in vitro* method of upregulating the expression of GPx and ThRed in trophoblastic cell lines JEG-3, BeWo and Swann-71. Concentrations of 100 nM NaSe and 500 nM selenomethionine significantly increased the expression of both GPx and ThRed. Oxidative stress was then induced by either exogenous addition of various peroxides (Watson *et al.*, 2012) or endogenous ROS production by selectively inhibiting the electron-transport chain in trophoblast mitochondria (Khera *et al.*, 2013). In each case the addition of selenium as either the inorganic salt sodium selenite (NaSe) or organic selenomethionine resulted in mitochondrial protection from oxidative stress and promoted cell viability. More recently, our attention has turned

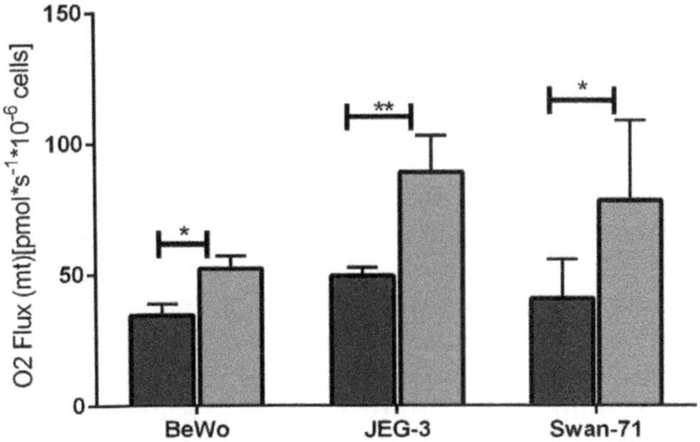

Figure 32.3 Mitochondrial respiration in trophoblast cells lines BeWO, Jeg-3 and Swan-71 after 24 h incubation in 100 nm Na selenite. Mean ± SD across 5 repeat experiments. $*p < 0.05$, $**p < 0.01$.

to the beneficial effects of selenium on mitochondrial respiration. Treatment of trophoblast cells with 100 nm Na selenite for 24 h significantly increased mitochondrial respiration (Figure 32.3) confirming similar observations in other cell types (Mekta *et al.*, 2012; Mendelev *et al.*, 2012). These observations are highly significant as they indicate that a simple treatment such a selenium supplementation can have substantial protective actions on placental mitochondria subjected to oxidative insult.

32.5 The Role of Selenium in Human Health and Reproduction

Elemental selenium was discovered in 1817 by Jons Jacob Berzelius, a Swedish chemist. For 140 years selenium was considered a toxic element and it was not until the late 1950s that selenium was identified as an essential component of many important biological processes. This was reinforced during the 1960s when it was discovered that the Keshan region of China had particularly low levels of selenium and this appeared to be associated with cardiomyopathy, a condition endemic in this region. The disease occurred in areas where daily selenium intakes were estimated to be between 10 and 15 µg per day and resulted in considerable morbidly and mortality in teenage children. Large-scale selenium supplementation was shown to alleviate this condition and many subsequent studies have shown the importance of selenium in cardiovascular health (Li, 2007). Selenium deficiency has been associated with various disease states, such as neurodegenerative and cardiovascular diseases, cancer, Kashin–Beck disease (a deforming arthritis) and dysfunction of immune systems, and reproductive health.

Adequate selenium is required for spermatogenesis in males, an intricate process requiring expression of GPx, ThRed as well as selenoprotein P. It has been hypothesized that these redox active proteins are important in defending against oxidative stress but also in stabilizing intramolecular disulfide bonds that are required in protein structure and formation of sperm architecture. Testes contain the highest expression of GPx-4 in the human body and it constitutes a significant component of the proteins in the mitochondria of spermatozoon (Maiorino *et al.*, 1999). Presumably this is to help alleviate the oxidative stress that could be generated by actively swimming sperm in a limited oxygen environment. Certainly it has been demonstrated that selenium deprivation results in poor sperm quality and impaired motility. Decreased expression of GPx was significantly lower in 75 infertile men when compared to matched controls and this was strongly correlated with poorer morphological integrity, viability and motility. There have been some limited reports of increased sperm quality and improved fertility in patients postselenium supplementation with 100 μg per day (Scott *et al.*, 1998).

Female fertility is equally dependent upon an adequate supply of this trace element. Investigations in healthy women indicate a tight correlation between oestrogen content, selenium levels and GPx activity. Serum concentrations and circulating GPx levels are lowest in the early follicular phase and increase rapidly to peak around ovulation and then decrease during the luteal phase (Ha and Smith, 2009). Pazkowski and colleagues have reported lower selenium concentrations in follicular fluid from women with unexplained infertility and animal studies have reinforced this, as selenium stimulates the proliferation of granular cells and follicular development in bovine ovaries (Pazkowski *et al.*, 1995). In other ruminants such as sheep and alpacas, selenium sufficiency is essential for successful fertility and sustaining a pregnancy.

32.6 Importance of Selenium in Human Pregnancy

During human pregnancy there is an increased demand for selenium, which results in a decrease in circulating Se and GPx concentrations as gestation proceeds. Plasma selenium in pregnant women is approximately 65 μg L^{-1} in the 1st trimester, falling to 50 μg L^{-1} in the third trimester and this has resulted in dietary recommendations of an increase in Se intake of an additional 10 μg per day for pregnant women (Zachara *et al.*, 1993). This increased requirement during pregnancy has been attributed to the increased need to power the antioxidant enzymes required to protect placental tissues from oxidative stress. Healthy pregnant women have a lower selenium level than nonpregnant women and further decreases are associated with recurrent miscarriage and an inability to sustain a successful pregnancy. This observation is supported by hair and toenail selenium analysis in women with recurrent miscarriage. Hair and toenail Se analysis is able to predict Se status preonset of disease and was significantly lower in women who experienced recurrent miscarriage, suggesting that selenium plays an important role in the peri-implantation phase (Al-Kunani *et al.*, 2001).

Selenium deficiency has also been associated with the development of hypertensive diseases of pregnancy such as pregnancy-induced hypertension and pre-eclampsia (PET). Epidemiological data indicates that countries that report lower selenium intakes also report higher incidences of pre-eclampsia (Vanderlelie and Perkins, 2011). The rate of pre-eclampsia in countries with levels of Se less that 95 $\mu g\ L^{-1}$ had an average pre-eclampsia rate of 4.5% whereas those countries with a Se above 95 $\mu g\ L^{-1}$ reported rates of 2.5%. Interestingly, New Zealand and Finland took action to increase selenium intakes over the period 1984–2004 and in that time the reported incidence of this disease almost halved.

In 2003, an important study investigated the association between Se toenail status and pre-eclampsia in 53 pre-eclamptic and 53 matched control pregnancies (Rayman *et al.*, 2003). The authors concluded that median toenail Se concentration of pre-eclamptics was significantly lower than their matched counterparts, with low Se status in the pre-eclamptic group significantly associated with the most severe form of the disease. Women with the lowest level of selenium were 4.4 times more likely to develop pre-eclampsia. The significance of this study is that (i) toenail Se status may be utilized as a clinical estimation of maternal Se status pre onset of disease, and (ii) increasing periconceptional Se status may reduce the risk of pre-eclampsia. More recently, Maleki *et al.* (2011) reported significantly decreased plasma Se levels in pre-eclamptics compared to healthy pregnant women ($p = 0.001$), highlighting the maternal serum Se–PET relationship. These observations have been repeated in numerous studies and the consensus is that low selenium status significantly increases the risk of pre-eclampsia.

A more direct relationship has been suggested by animal models where removal of selenium from the diets of pregnant rats resulted in hypertension, proteinuria and decreased pup weight (Vanderlelie *et al.*, 2004), all essential hallmarks of pre-eclampsia. As discussed earlier selenium is essential in protecting placental tissue from oxidative injury and in this rodent model of pre-eclampsia, selenium depletion resulted in enhanced placental production of oxidized lipids and proteins with a concomitant decrease is the cellular expression of GPx and ThRed.

With this combination of evidence in both animal models and human studies we have recently witnessed controlled clinical studies on the use of selenium supplementation in pre-eclampsia. These studies have indicated that a daily supplement from the start of the second trimester of pregnancy may lower the incidence of this disease. In 2010, Tara *et al.* conducted a randomized double-blind, placebo-controlled trial of 166 pregnant women with administration of selenium in the first trimester. This study concluded that supplementation of 100 μg Se alone reduced the frequency of PET in the supplementation group, however this did not reach significance ($p > 0.05$). Reasons why this trial did not reach significance could be attributed to the (i) small sample size ($n = 166$) of the cohort; (ii) late supplementation during the first trimester rather than periconception; and (iii) that Iranian women included in this study on average have comparatively high Se intakes and supplementing micronutrient sufficient women will have little effect. The outcomes of these studies suggest that sustained periconceptional Se supplementation in women at high PET risk, or low Se status, may be associated with a lower incidence, delayed onset or reduced outcome severity of PET.

These preliminary studies need to be confirmed in larger intervention trials. They have indicated that selenium may play a protective role at this dose, however, the administration may need to be earlier, indeed prior to conception, to be of maximum benefit. The origins of pre-eclampsia occur immediately postimplantation when the extra villous trophoblast cells invade and remodel the maternal spiral arteries and this is a time of substantial changes in the oxygenation of placental tissues. Prior to week 10–12 of gestation the invading extra villous trophoblast essentially plug the spiral arteries, occluding the vascular supply of oxygen and resulting in a relatively hypoxic environment. At around week 12 the plugs are eroded and the spiral arteries are transformed from constricted low-volume vessels to flaccid high-volume conduits. There is a substantial increase in the supply of oxygenated blood from the maternal circulation resulting in a reperfusion like challenge. If placental expression of antioxidant systems is not maximized during this critical period then oxidative stress can arise and trophoblastic debris enters the maternal circulation in increased quantities and in forms that are potentially immunogenic.

Preterm labor and foetal growth restriction are additional complications of human pregnancy that have been associated with poor selenium status and increased placental oxidation stress. In a series of studies, low Se status was observed in both the maternal serum and cord blood from babies born before 37 weeks when compared to those born at term, post 37 weeks (Dodbrynski *et al.*, 1998). Similarly, babies born in the lower quartiles of birth weight had lower levels of selenium, as did their mothers. Collectively these studies indicate the importance of an adequate selenium supply during conception, the establishment of pregnancy and throughout gestation. Selenium status should be an important indicator of a woman's preparedness for child bearing and if plasma selenium levels are lower than 50 $\mu g\ L^{-1}$ a selenium supplement should be recommended.

Increasingly, pregnant women and women preparing for pregnancy are using broad-spectrum multinutrient preparations. This has gained popularity following the important studies 20 years ago of the essential role of folate in preventing neural tube defects. Many of these multinutrient supplements contain selenium in the form of selenomethionine from yeast extracts, which is a safe readily absorbed form of selenium. Most supplements contain approximately 60 μg of selenium per tablet and women are encouraged to take one capsule per day. Several studies have been conducted to examine the benefit of this approach on pregnancy outcome and the message appears to be generally positive. In a study of 18 551 women from the Danish Birth Cohort, 63% reported use of micronutrients during pregnancy and this was associated with 20% decrease in the risk of developing pre-eclampsia in this group (Catov *et al.*, 2009). The micronutrient commonly used by this cohort contained 50 μg of selenium. Similar findings have been reported in a controlled trial using a broad spectrum antioxidant supplement containing selenium in women with low antioxidant status where the incidence of pre-eclampsia was significantly reduced (Rumaris *et al.*, 2006). In a recently published study our group has made similar findings and have observed the significant benefit of multinutrient supplementation in women with BMI >30 who have a higher risk of adverse pregnancy outcome, including the development of pre-eclampsia.

Over the past 5–10 years evidence from epidemiological studies, direct measurement in humans and animals and *in vitro* cellular models have all indicated the importance of an adequate selenium supply during pregnancy. This appears to be important in maximizing and maintaining antioxidant capacity in placental trophoblast cells and protecting trophoblast mitochondria from oxidative stress. Selenium status should be a routine consideration when providing advice to mothers who are planning a pregnancy or find themselves pregnant. Ensuring that expectant mothers have a well-balanced micronutrient supply, including adequate selenium intake, will maximize placental health and ensure a healthy start to life for the next generation.

Summary Points

- A healthy placenta is essential to a healthy start to life.
- Placental cell function is compromised by oxidative stress.
- Oxidative stress causes release of placental debris into the maternal circulation.
- Selenium is a key component of key antioxidant enzymes.
- Selenium is essential to placental cell function and homeostasis.
- The demand for selenium increases during gestation and poor selenium status has been associated with complications of pregnancy.
- Selenium supplementation may be important in maintaining placental health and guaranteeing a healthy start to life.

Key Facts

1. Placental cell debris is released into the maternal circulation during pregnancy as the surface of the placenta renews itself. This is a normal process and part of normal cell turnover. However, in some cases this cell turnover becomes unregulated and changes so that the maternal immune system sees this debris as foreign and mounts a hostile immune response. This is very dangerous to the mother and the baby and results in poor placental function. Placental cells are derived from fetal tissue and hence contain the paternal genome and may be immunogenic.
2. Recent research has shown that increased placental cell shedding is associated with complications of pregnancy such as preterm labor and pre-eclampsia and that this was linked to a state of oxidative stress. Oxidative stress occurs when oxygen, essential for normal metabolism

is reduced to form free radicals and these must be removed by antioxidants. Some of the important antioxidant proteins that work in the human placenta contain selenium, an essential micronutrient. This chapter summarizes what is known about placental oxidative stress and the importance of selenium in maintaining placental health and ensuring a healthy start to life.

Definitions of Words and Terms

Antioxidants. Compounds that are able to negate the detrimental effects of oxidants.
Chorionic villi. Branched structures that make up the ultrastructure of the human placenta.
Cytotrophoblast. Trophoblast cells that line the villous structures of the human placenta.
Oxidative stress. An imbalance between antioxidants and reactive oxygen and nitrogen species, favoring the build-up of the latter.
Pre-eclampsia. A severe hypertensive disorder of pregnancy.
Preterm labor. Labor that occurs before 37 completed weeks of gestation.
Periconception. Occurring immediately before or immediately after conception.
Peri-implantation. Occurring immediately before or immediately after implantation.
Proteinuria. The presence of protein in the urine.
Reactive oxygen and nitrogen species. Partially reduced forms of oxygen and nitrogen that result in free-radical generation.
Selenium. An essential trace element that occurs in nature and is required in the human diet.
Selenoproteins. Proteins that have a Se atom as part of their functional unit.
Syncytium. Layer of fused trophoblast cells that line the chorionic villi.
Syncytial debris. Small fragments of syncytium that are shed from the surface of the placenta.
Trophoblast. Cells that form the human placenta.

List of Abbreviations

ATP	Adenosine triphosphate
BMI	Body mass index
Cyto	Cytotrophoblast
DNA	Deoxyribonucleic acid
GPx	Glutathione peroxidase
H_2O_2	Hydrogen peroxide
NADPH	Nicotinamide adenine dinucleotide phosphate
NaSe	Sodium selenite
nM	Nanomolar

PET	Pre-eclampsia
RONS	Reactive oxygen and nitrogen species
SOD	Superoxide dismutase
Sync	Syncytiotrophoblast
ThioPx	Thioredoxin peroxidase
ThioRed	Thioredoxin reductase
µg	Micrograms
µM	Micromolar

References

Al-Kunani, A. S., Knight, R., Haswell, S. J., Thompson, J. W. and Lindow, S. W., 2001. The selenium status of women with a history of recurrent miscarriage. *British Journal of Obstetrics and Gynaecology*. 108(10): 1094–1097.

Barker, D. J. P. (ed.), 1992. *Fetal and Infant Origins of Adult Disease*. British Medical Journal, London.

Brenseke, B., Prater, M. R., Bahamonde, J. and Gutierrez, J. C., 2013. Current thoughts on maternal nutrition and fetal programming of the metabolic syndrome. *Journal of Pregnancy*. 13(368461): 1–13.

Burton, G. J., Yung, H. W., Cindrova-Davies, T. and Charnock-Jones, D. S., 2009. Placental endoplasmic reticulum stress and oxidative stress in the pathophysiology of unexplained intrauterine growth restriction and early onset preeclampsia. *Placenta*. 30: 43–48.

Castejon, O. C., 2011. Mitochondrial dysfunction and apoptosis in trophoblast cells during preeclampsia: an ultrastructural study. *Electronic Journal of Biomedicine*. 2011(2): 30–38.

Catov, J. M., Nohr, E. A., Bodnar, L. M., Knudson, V. K., Olsen, S. F. and Olsen, J., 2009. Association of periconceptional multivitamin use with reduced risk of preeclampsia among normal-weight women in the Danish National Birth Cohort. *American Journal of Epidemiology*. 169(11): 1304–1311.

Colleoni, F., Padmanabhan, N., Yung, H. W., Watson, E. D., Cetin, I., Tissot van Patot, M. C., Burton, G. J. and Murray, A. J., 2013. Suppression of mitochondrial electron transport chain function in the hypoxic human placenta: a role for miRNA-210 and protein synthesis inhibition. *PLoS One*. 8(1): e55194. DOI: 10.1371/journal.pone.0055194. Epub 2013 Jan 30.

De Los Rios Castillo, D., Zarco-Zavala, M., Olvera-Sanchez, S., Pardo, J. P., Juarez, O., Martinez, F., Mendoza-Hernandez, G., García-Trejo, J. J. and Flores-Herrera, O., 2011. A typical cristae morphology of human syncytiotrophoblast mitochondria. Role for complex v. *Journal of Biological Chemistry*. 286(27): 23911–23919, ISSN 0098–2997.

Desoye, G. and Shafrir, E., 1994. Placental metabolisms and its regulation in health and diabetes. *Molecular Aspects of Medicine*. 15(6): 505–682.

Dobrzynski, W., Trafikowska, U., Trafikowska, A., Pilecki, A., Szymanski, W. and Zachara, B. A., 1998. Decreased selenium concentration in maternal and cord blood in preterm compared with term delivery. *Analyst*. 123(1): 93–97.

Dyer, J. S. and Rosenfeld, C. R., 2011. Metabolic imprinting by prenatal, perinatal, and postnatal overnutrition: a review. *Seminars in Reproductive Medicine*. 29(3): 266–276.

Ejima, K., Nanri, H., Toki, N., Kashimura, M. and Ikeda, M., 1999. Localization of thioredoxin reductase and thioredoxin in normal human placenta and their protective effect against oxidative stress. *Placenta*. 20(1): 95–101.

Frendo, J. L., Thérond, P., Guibourdenche, J., Bidart, J. M., Vidaud, M. and Evain-Brion, D., 2000. Modulation of copper/zinc superoxide dismutase expression and activity with *in vitro* differentiation of human villous cytotrophoblasts. *Placenta*. 21(8): 773–781.

Gallos, I. D., Sivakumar, K., Kilby, M. D., Coomarasamy, A., Thangaratinam, S. and Vatish, M., 2013. Pre-eclampsia is associated with, and preceded by, hypertriglyceridaemia: a meta-analysis. *British Journal of Obstetrics and Gynaecology*. 120(11): 1321–1332.

Ha, E. J. and Smith, A. M., 2009. Selenium-dependent glutathione peroxidase activity is increased in healthy post-menopausal women. *Biology and Trace Element Research*. 131(1): 90–95.

Halliwell, B. and Gutteridge, J., 2007. *Free Radicals in Biology and Medicine*. 4th edn. University of Oxford Press.

Hung, T. H., Skepper, J. N., Charnock-Jones, D. S. and Burton, G. J., 2002. Hypoxia-reoxygenation a potent inducer of apoptotic changes in the human placenta and possible etiological factor in preeclampsia. *Circulation Research*. 90(12): 1274–1281.

Huppertz, B. and Borges, M., 2008. Placenta trophoblast fusion. In: E. Chen (ed.) *Cell Fusion*. Humana Press, pp. 135–147.

Khera, A., Vanderlelie, J. J. and Perkins, A. V., 2013. Selenium supplementation protects trophoblast cells from mitochondrial oxidative stress. *Placenta*. 34(7): 594–598.

Li, C., 2007. Selenium deficiency and endemic heart failure in China: a case study of biogeochemistry for human health. *Ambio*. 36(1): 90–93.

Maiorino, M., Flohé, L., Roveri, A., Steinert, P., Wissing, J. B. and Ursini, F., 1999. Selenium and reproduction. *Biofactors*. 10(2–3): 251–256.

Maleki, A., Fard, M. K., Zadeh, D. H., Mamegani, M. A., Abasaizadeh, S. and Mazloomzadeh, S., 2011. The relationship between plasma level of Se and preeclampsia. *Hypertension in Pregnancy*. 30(2): 180–187.

Martinez, F., Milan, R., Flores-Herrera, O., Olvera-Sanchez, S., Gomez-Chang, E. and Espinosa-Garcia, M. T., 2012. The role of mitochondria in syncytiotrophoblast cells: bioenergetics and steroidogenesis. In: Zheng, J. (ed.) *Recent Advances in Research on the Human Placenta*. InTech, ISBN 978-953-51-0194-9 Publisher.

Martinez, F., Kiriakidou, M. and Strauss III, J. F., 1997. Structural and functional changes in mitochondria associated with trophoblast differentiation: methods to isolate enriched preparations of syncytiotrophoblast mitochondria. *Endocrinology*. 138(5): 2172–2183.

Mehta, S. L., Kumari, S., Mendelev, N. and Li, P. A., 2012. Selenium preserves mitochondrial function, stimulates mitochondrial biogenesis, and reduces infarct volume after focal cerebral ischemia. *BMC Neuroscience*. 9(13): 79. DOI: 10.1186/1471-2202-13-79.

Mendelev, N., Mehta, S. L., Idris, H., Kumari, S. and Li, P. A., 2012. Selenite stimulates mitochondrial biogenesis signaling and enhances mitochondrial functional performance in murine hippocampal neuronal cells. *PLoS One*. 7(10): e47910. DOI: 10.1371.

Mistry, H. D., Kurlak, L. O., Williams, P. J., Ramsay, M. M., Symonds, M. E. and Broughton Pipkin, F., 2010. Differential expression and distribution of placental glutathione peroxidases 1, 3 and 4 in normal and preeclamptic pregnancy. *Placenta*. 31(5): 401–408.

Myatt, L., Eis, A. L., Brockman, D. E., Kossenjans, W., Greer, I. A. and Lyall, F., 1997. Differential localization of superoxide dismutase isoforms in placental villous tissue of normotensive, pre-eclamptic, and intrauterine growth-restricted pregnancies. *Journal of Histochemistry and Cytochemistry*. 45(10): 1433–1438.

Paszkowski, T., Traub, A. I., Robinson, S. Y. and McMaster, D., 1995. Selenium dependent glutathione peroxidase activity in human follicular fluid. *Clinica Chimica Acta*. 236: 173–180.

Perkins, A. V., Di Trapani, G., McKay, M. S. and Clarke, F. M., 1995. Immunocytochemical localization of thioredoxin in human trophoblast and decidua. *Placenta*. 16(7): 635–642.

Rayman, M. P., Bode, P. and Redman, C. W., 2003. Low selenium status is associated with the occurrence of the pregnancy disease preeclampsia in women from the United Kingdom. *American Journal of Obstetrics and Gynecology*. 189(5): 1343–1349.

Redman, C. W. G. and Sargent, I. L., 2000. Placental debris, oxidative stress and pre-eclampsia. *Placenta*. 21(7): 597–602.

Rumiris, D., Purwosunu, Y., Wibowo, N., Farina, A. and Sekizawa, A., 2006. Lower rate of preeclampsia after antioxidant supplementation in pregnant women with low antioxidant status. *Hypertension in Pregnancy*. 25: 241–253.

Scott, R., MacPherson, A., Yates, R. W., Hussain, B. and Dixon, J., 1998. The effect of oral selenium supplementation on human sperm motility. *British Journal of Urology*. 82(1): 76–80.

Shi, Z., Long, W., Zhao, C., Guo, X., Shen, R. and Ding, H., 2013. Comparative proteomics analysis suggests that placental mitochondria are involved in the development of pre-eclampsia. *PLoS One*. 8(5): e64351.

Sies, H., 1997. Oxidative stress: oxidants and antioxidants. *Experimental Physiology*. 82(2): 291–295.

Tara, F., Maamouri, G., Rayman, M. P., Ghayour-Mobarhan, M., Sahebkar, A., Yazarlu, O. and Ferns, G., 2010. Selenium supplementation and the incidence of preeclampsia in pregnant Iranian women: a randomized, double-blind, placebo-controlled pilot trial. *Taiwanese Journal of Obstetrics and Gynecology*. 49(2): 181–187.

Vanderlelie, J., Venardos, K. and Perkins, A. V., 2004. Selenium deficiency as a model of experimental pre-eclampsia in rats. *Reproduction*. 128(5): 635–641.

Vanderlelie, J., Venardos, K., Clifton, V. L., Gude, N. M., Clarke, F. M. and Perkins, A. V., 2005. Increased biological oxidation and reduced antioxidant enzyme activity in pre-eclamptic placentae. *Placenta*. 26(1): 53–58.
Vanderlelie, J. and Perkins, A. V., 2011. Selenium and preeclampsia: a global perspective. *Pregnancy Hypertension: An International Journal of Women's Cardiovascular Health*. 1(3): 213–224.
Watson, M., van Leer, L., Vanderlelie, J. and Perkins, A. V., 2012. Selenium supplementation protects trophoblast cells from oxidative stress. *Placenta*. 33(12): 1012–1019.
Zachara, B. A., Wardak, C., Didkowski, W., Maciag, A. and Marchaluk, E., 1993. Changes in blood selenium and glutathione concentrations and glutathione peroxidase activity in human pregnancy. *Gynaecological and Obstetric Investigations*. 35(1): 12–17.

CHAPTER 33

Selenium and Antioxidant Microgels

YUSHI BAI[a], TIEZHENG PAN[a], QUAN LUO*[a], AND JUNQIU LIU[a]

[a]State Key Laboratory of Supramolecular Structure and Materials, Institute of Theoretical Chemistry, Jilin University, 2699 Qianjin Street, Changchun, China
*E-mail: luoquan@jlu.edu.cn

33.1 Introduction

Selenium (Se) is a trace element and essential nutrient mainly found in food sources such as nuts, cereals, meats and fish. A voluminous body of research on selenium has accumulated for nearly two centuries: Se was discovered in 1817 by Swedish chemist Berzelius and defined as a micronutrient in diet for disease prevention in the late 1950s (Schwarz and Foltz, 1957). However, the biochemical role of Se was really established after it was identified as an important functional component of antioxidant selenoenzyme glutathione peroxidase (GPx) in 1973 (Flohé *et al.*, 1973). Se was suggested to exert its biochemical function in the form of an unusual amino acid selenocysteine (Sec), which is encoded in a special way by a stop UGA codon and directly involved in the catalytic activity by positioning in the active site of selenoenzymes (Allmang *et al.*, 2009).

The major dietary source of selenium intake for humans is food, especially plant foods. In general, selenium supplements are recommended to

Food and Nutritional Components in Focus No. 9
Selenium: Chemistry, Analysis, Function and Effects
Edited by Victor R. Preedy

Published by the Royal Society of Chemistry, www.rsc.org

be a daily dose of approximately 30–85 μg per day (Rayman, 2008). People dependent on food grown from selenium-deficient soil are usually associated with low nutritional selenium status and increased risk of a series of diseases such as cataract (Taylor, 1993), Keshan disease (Chen, 2012), cancer (Loft and Poulsen, 1996), heart disease (Salonen and Huttunen, 1986), and the acquired immunodeficiency syndrome (Favier *et al.*, 1994). *In vitro* and *in vivo* studies have shown that a normal selenoenzyme activity level is believed to be essential for human health: the major group of selenoeznymes in the human body evolves as redox regulators to protect cells against oxidative stress by removing excess endogenous or exogenous reactive oxygen species (ROS) that damage a variety of intracellular macromolecules (*e.g.* nucleic acid, proteins, and lipids) and may contribute to the development of the diseases mentioned above.

So far, at least 25 genes for selenoenzymes have been identified in the human genome and the number has grown substantially in recent years (Lu and Holmgren, 2009). An increased understanding of the structure and mechanism of these selenoenzymes would provide insight into the way in which selenium exerts its physiological effects. GPx is one of the well-characterized selenoenzymes. It adopts a thioredoxin-like α/β fold domain and Sec is located on a relatively exposed surface of the protein N-terminus, proximal to the binding site of substrate glutathione (GSH), thus allowing to catalyze GSH-dependent degradation of various hydroperoxides. In addition, some specific residues such as arginine and glutamine are involved in stabilizing the GSH and GPx interactions. The detailed redox cycle is the oxidation of selenol (ESeH) to selenenic acid (ESeOH), which reacts with GSH to form the key intermediate selenenyl sulfide (E-SeSG), and then a second GSH attacks the —Se—S— bond to produce a GSH disulfide (GSSG). Although these studies have unraveled the mystery of selenoenzyme catalysis, its practical applications are still limited due to the intrinsic disadvantages of natural proteins such as: instability, short half-life, and difficult preparation.

Over the past decade, considerable effort has been devoted towards the development of novel GPx mimics to replace natural selenoenzymes for biomedical applications. New design concepts that combine recognition and catalysis were proposed and various strategies including semisynthesis, genetic engineering, molecular imprinting, immune induction and supramolecular assembly were employed to construct artificial selenoenzymes by introducing a GPx-like active site into chemically synthetic or naturally occurring scaffolds, ranging from the small molecules to macrocyclic compounds, polymers, biomacromolecules and even supramolecular assemblies. In the initial stages, many organoselenium compounds with less toxicity (*e.g.* ebselen, selenenamides, diselenides, a-phenylselenoketones, selenium-containing cyclodextrins) were synthesized for biochemical and pharmacological studies and some of them have shown great potential in the treatment of cancer. Inspired by these studies, the design of advanced antioxidants that rival the specificity and efficiency of natural selenoenzymes has prompted attempts to develop more sophisticated materials. Polymers like dendrimers and hyperbranched polymers were prepared in a controllable manner to obtain a favorable substrate microenvironment and selenium was incorporated into the core of them, leading to remarkably enhanced activities as compared to small-molecule-based GPx mimics. Moreover, some biomacromolecules such as enzymes, antibodies, and other proteins were also exploited to comprehensively simulate the active center of GPx by site-specific engineering with Sec and a GSH-binding site. Because of the intramolecular synergistic effect, their activities were often found to achieve the level of natural GPx. Very recently, the nano/microsized composites formed by crosslinking or supramolecular assembly have attracted much attention. These materials with three-dimensional (3D) architectures and structural dynamic characteristics will provide an opportunity for the development of efficient, smart, synergistic antioxidants.

Microgels are typical nano/microsized colloidal particles that consist of crosslinked polymer networks and can swell or deswell in response to external stimuli by conformational changes of the subchains between two neighboring crosslinking points inside the gel network. It is worth noting that microgel particles exhibit unique properties of both colloidal and polymer-like characteristics and the particle/polymer duality enable them to be extensively exploited in the biomedical field. Utilizing the biocompatible surface and surface-coating property of microgel particles, biomolecules such as antibody, protein, deoxyribonucleic acid (DNA), and enzyme can be immobilized for immunosensing, bioassays, and bioseparation technology. Furthermore, they possess a highly stable 3D network structure capable of offering reaction space for chemical synthesis as a microreactor (Antonietti *et al.*, 1997) and are easily engineered to be a tissue-mimetic environment that can serve as ideal matrices for the encapsulation of bioactive substances for tissue regeneration (Jia *et al.*, 2006). More importantly, polymer materials for building microgels can be modified to respond precisely to environmental signals, such as temperature (Wang *et al.*, 2007), pH (Hoare and Pelton, 2004), ionic strength (Neyret and Vincent, 1997), and electromagnetic field (Fernández-Nieves and Márquez, 2005). Stimuli-responsive microgels undergo reversible phase–volume transition to release the desired molecules in a controlled manner, which is the major requirement for an advanced drug-delivery system. Owing to the tunable structures and properties, microgels have recently applied as a new class of platform to design artificial antioxidants. This application for microgels thereby needs the novel design strategies that can combine their inherent advantages such as water solubility, biocompatibility, multicomponents, and environmental responsiveness to further promote the efficiencies and properties of the antioxidant system.

In this chapter, we will give a brief review on the recent works of antioxidant microgels and focus on the biomimetic design strategies, the synthesis and characterization of different microgels as biochemical scaffolds, and the new properties of the antioxidant system endowed by microgels.

33.2 Antioxidant Microgels

Both the construction of artificial selenoenzymes and the fabrication of microgels for different applications have achieved considerable progress as two independent research branches. However, a combination of these two research fields to construct the antioxidant microgels is still a young field and most reported work is only to use microgel particles as supports for the immobilization of the commercially available antioxidant agents, for example horseradish peroxidase (Xu *et al.*, 2007). Herein, we will describe the recent synthetic efforts toward the construction of advanced antioxidant microgels for intelligent or synergistic catalysis by taking advantage of both biomimetic design strategies and versatile microgel properties. Two selected examples of these antioxidant microgels will be discussed in detail.

33.2.1 Microgels as "Smart" Materials to Build an Artificial Antioxidant

ROS are products of cell metabolism, which act as signal molecules and do not cause cell damage and diseases unless they are overproduced. This dual property of ROS depends on a fine regulation of their concentration. Therefore, an ideal artificial antioxidant should keep the ROS level in a balance rather than eliminate the ROS thoroughly. If the designed antioxidant could respond to external environmental stimuli and adaptively change its antioxidant activity, it will be possible to regulate the ROS level in a smart way. Recently, a proof-of-concept example of such regulation has succeeded in the temperature-responsive microgel system described as follows (Huang *et al.*, 2009).

33.2.1.1 Design of the Glutathione Peroxidase-Like Catalytic Center

Since selenium is known to be an essential part of the active site of natural GPx enzymes, scientists have synthesized a series of Se-containing small molecular GPx mimics with considerable activities for antioxidant studies (Engman *et al.*, 1992). In contrast to Se, tellurium (Te), a member of the same group VI in the periodic table, exhibits higher active redox activity due to its intrinsically lower redox potential and thus allows for the design and synthesis of various organotellurium compounds to enhance the antioxidant ability. If these synthetic Te-containing compounds have the ability to crosslink with each other to form microgels, the antioxidant efficiency should be further enhanced by generating a favorable substrate microenvironment and highly integrated multiple active centers. Meanwhile, the stimuli-sensitive functional groups can be incorporated by a copolymerization method for activity regulation. Figure 33.1 shows the functional components of this type of antioxidant microgels.

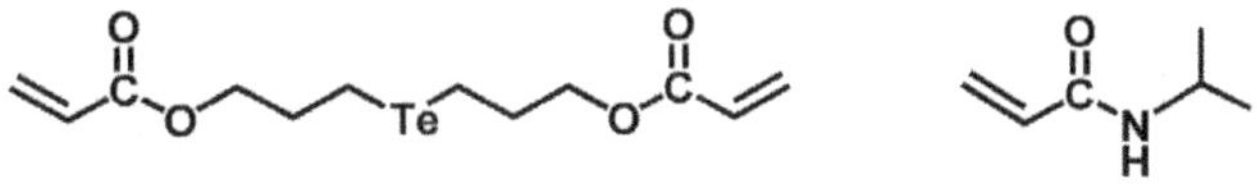

(1) (2)

Figure 33.1 Chemical structures of the functional components of the antioxidant microgels. Chemical structures of the designed tellurium-containing small-molecule (1) and *N*-isopropylacrylamide monomer (2). Reproduced from Huang *et al.* (2009).

In this system, compound 1 exerts a dual function: the construction of the catalytic center and function as a linker in the formation of microgel (see Section 33.1.2), and compound 2 is an acrylamide group monomer used to produce poly(*N*-isopropylacrylamide) (PNIPAM) thermosensitive polymer.

33.2.1.2 *Design and Construction of "Smart" Antioxidant Microgels*

PNIPAM is a hydrophilic polymer that is extensively used to form hydrogels. The PNIPAM-based microgel system has received considerable attention due to its good chemical stability, high water solubility, and low toxicity, especially its temperature responsiveness. PNIPAM microgels can undergo thermally induced phase–volume transitions in a reversible way: below the lower critical solution temperature (LCST) of 32 °C (McPhee *et al.*, 1993), it is swollen and highly solvated due to the formation of hydrogen bonds between the amide groups of the polymer chains and surrounding water molecules. However, when the temperature increases above the LCST, the hydrogen bonds are increasingly broken and the microgels tend to shrink and collapse. Because the LCST of PNIPAM is close to the human body temperature, it is especially suitable for applications in biomedical areas.

Considering this attractive temperature responsiveness, PNIPAM microgels were utilized as a platform to design "smart" antioxidants by introducing GPx-like active centers into the microgel system. In traditional PNIPAM microgels, *N,N*-methylene(bis)acrylamide (MBA) is a commonly used crosslinker for polymer polymerization. As for this microgel system, however, the Te-containing small molecule with two double bonds (Figure 33.1) was designed to react with *N*-isopropylacrylamide *via* free radical copolymerization and thus the catalytic GPx centers can be incorporated spontaneously to form multiple active centers in the microgel networks. In addition, the microgel system exhibited reversible swelling responses to temperature stimulus due to the thermosensitive property provided by PNIPAM, leading to a controlled antioxidant activity. The regulatory mechanism is illustrated in Figure 33.2.

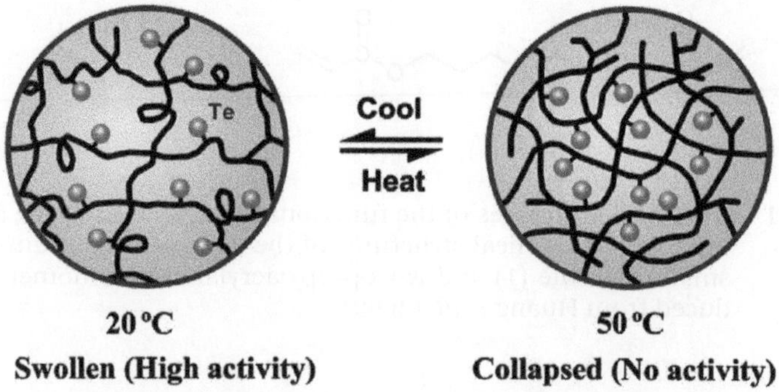

Figure 33.2 The activity regulation mechanism of the temperature-responsive antioxidant microgels. The antioxidant activity regulation depends on the phase-volume transition of poly(*N*-isopropylacrylamide) microgels. Below the lower critical solution temperature (20 °C), they are highly swollen *via* the formation of hydrates by hydrogen bonding, allowing substrate access to their "expanded" hydrophobic pockets for efficient catalysis; Above the lower critical solution temperature (50 °C), hydrogen bonds break and poly(*N*-isopropylacrylamide) microgels shrink to prevent substrate from entering the condensed microgel core, leading to very low activity. Reproduced from Huang *et al.* (2009).

33.2.1.3 The Catalytic Behavior of "Smart" Antioxidant Microgels

With the 3D micropore nanostructure, the PNIPAM-based antioxidant microgels can provide a good microenvironment for enzymatic catalysis in the reduction of cumene hydroperoxide (CUOOH) by 3-carboxyl-4-nitrobenzenethiol (TNB). Thus, they exhibit high GPx catalytic activity, which is about 339 000-fold more efficient than that of PhSeSePh, a typical GPx mimic. Compared to the Te-containing small-molecular GPx mimic, the catalytic activity of this microgel system is enhanced about 20-fold, suggesting that the porous microgel structure indeed plays an important role in the GPx-like catalytic process.

In the TNB assay system with CUOOH as a substrate, the temperature-responsive property of the antioxidant microgels was evaluated at different temperatures from 10 °C to 60 °C. The results showed that the catalytic activity is gradually increasing until the temperature reaches 32 °C. However, at a higher temperature, the catalytic activity goes down quickly and it almost loses the catalytic activity above 50 °C (Figure 33.3a).

This temperature-responsive catalytic behavior is closely related to the hydrophobic and structural changes of PNIPAM microgels.

The antioxidant activity regulation mainly depends on the phase–volume transition of PNIPAM microgels. Below its LCST at 32 °C, PNIPAM microgels are highly swollen through the intermolecular crosslinking by hydrogen bonds. When the temperature increases above 32 °C, the hydrogen bonds are broken, resulting in the shrinkage of the microgels to reduce the size of the porous microgel structure. Therefore, it is less probable for the substrate

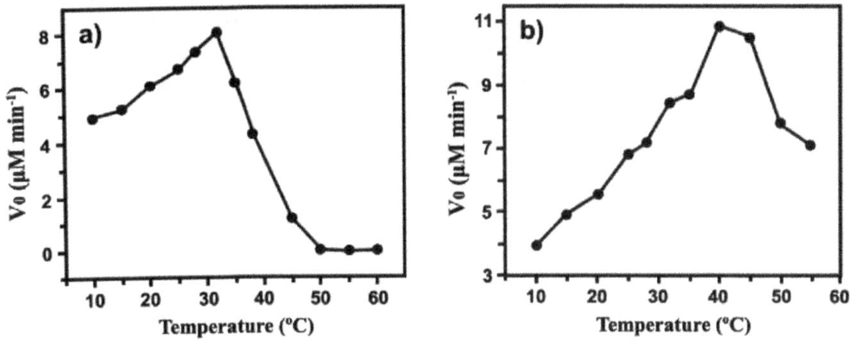

Figure 33.3 The catalytic rates of the temperature-responsive antioxidant microgels with different substrates. Plot of the catalytic rates of microgel catalyst *vs.* temperature during the catalytic reduction of cumene hydroperoxide (0.25 mM) with 1.00 mM catalytic center by (a) 3-carboxyl-4-nitrobenzenethiol (0.15 mM) and (b) *p*-nitrobenzenethiol (0.15 mM). Reproduced from Huang *et al.* (2009).

to enter the condensed microgel core to achieve the catalytic process. This is also confirmed by an alternative assay system, in which the microgel catalyst still exhibit high catalytic activity even at 55 °C using a smaller-sized *p*-nitrobenzenethiol (NBT) substrate (Figure 33.3b). Instead, no activity was observed in the same condition by utilizing TNB as a substrate, indicating that the structural change of PNIPAM microgels plays a pivotal role in the antioxidant activity regulation.

The other factor that contributes to the antioxidant activity regulation is the hydrophobic transition of the microgel system. In PNIPAM-based microgels, the hydrophilic core will change to hydrophobic above the LCST (32 °C), which is attributed to the cleavage of hydrogen bonds. This property is preserved in the organotellurium small molecule linked PNIPAM microgels and has been confirmed by optical transition from transparent to turbid of the microgel solution with the increment of temperature and fluorescence spectra measurements. The resulting hydrophobic core will hinder the relatively hydrophilic TNB to enter the core to suppress the catalytic activity. On the other hand, the antioxidant activity changed slightly in the NBT assay system when temperature altered from 32 °C to 50 °C using hydrophobic peroxide CUOOH as a substrate; however, a 30-fold decreased activity was observed from 32 °C to 50 °C by substituting CUOOH with a hydrophilic H_2O_2. This phenomenon also confirmed that a hydrophilic to hydrophobic transition will affect the antioxidant activity of the microgel system.

In conclusion, both structural and hydrophobic transitions of the "smart" antioxidant microgel system function together in the regulation of its catalytic behavior. The designed PNIPAM polymer offers the temperature-responsive characteristic and enables to switch on–off the antioxidant activity by the heating–cooling cycles around the physiological temperature. This strategy is attractive in the preparation of stimulus-responsive antioxidants for biomedical applications. Other polymers with a similar responsiveness such as pH, light, magnetic or electric field, ionic, *etc.* will open up a wide range of new designs for "smart" antioxidant microgels.

33.2.2 Microgels as Platform to Construct a Synergistic Antioxidant Enzymatic System

GPx and superoxide dismutase (SOD) are major members of the antioxidant defense system. They work together in a synergistic way to protect cellular machines against the oxidative stress of excessive ROS. SOD catalyzes the dismutation of superoxide radical anion (O_2^-) into molecular oxygen (O_2) and less reactive hydrogen peroxide (H_2O_2), which can be further converted to water (H_2O) by GSH-dependent GPx catalysis. Therefore, the design and construction of a synergistic GPx/SOD enzyme system that directly transfers the reactants to consecutive active sites for sequential reactions can support more efficient antioxidant defense. As uniformly dispersed colloidal particles, microgels offer a number of potential advantages such as water solubility, stability, biocompatibility, and multicomponents to construct this dual enzymatic system. Typically, a bifunctional antioxidant microgel model was reported by a combination of the engineered ferritin with selenium (Se-Fn) and manganese (Mn) porphyrin derivative as the GPx and SOD mimics respectively (Gao *et al.*, 2013). By crosslinking the benzaldehyde group of Mn-porphyrin derivative and the lysine residue of Se-Fn, GPx and SOD mimics are spatially organized in the microgel network to increase its synergistic antioxidant property (Figure 33.4).

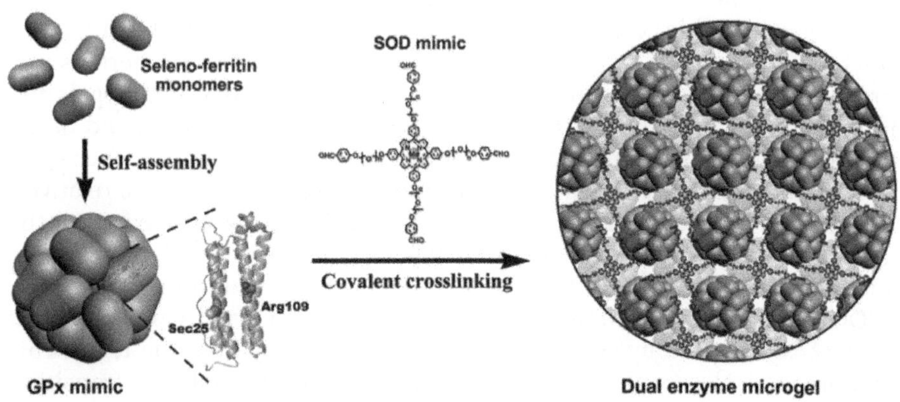

Figure 33.4 Schematic depictions of the design of dual enzyme microgels. The glutathione peroxidase-like selenium containing ferritin and the superoxide dismutase-like manganese porpyrin derivative were crosslinked by covalent bonding between the aldehyde group and the lysine residue to construct the synergistic antioxidant microgels. Data are from Gao *et al.* (2013), with permission from Wiley-VCH Verlag GmbH & Co. KGaA, Weinheim.

33.2.2.1 The Design Strategy for the Construction of Glutathione Peroxidase Mimic on Natural Protein Scaffolds

Proteins are naturally evolved amino acid polymers that share similar folding structures with enzymes and thereby allow the design and modification of them with new activity for various biological functions. During the past decade, significant progress has been made in protein-based artificial antioxidant by grafting GPx-like machinery into protein scaffolds (Liu *et al.*, 2009). These studies have revealed the basic principles of antioxidant enzyme design: the synergistic effect between the substrate recognition and catalysis should be taken into consideration.

The "compute and build" strategy based on the structure and catalytic mechanism of natural GPx enzymes was utilized to site-specifically incorporate the catalytic residue Sec and a GSH-binding site into protein scaffolds. Computational design procedure was carried out to investigate the detail mechanism of enzyme–substrate recognition and identify the critical residues for protein engineering. Then, these selected sites were mutated to the target residues to create the active center. In this design process, a proper orientation of Sec and high binding affinity of substrate are two critical factors for the fabrication of a GPx-like biocatalyst with high catalytic efficiency and selectivity.

It is well known that the active site Sec residue is essential for the catalytic activity of natural GPx and is difficult to be inserted into protein using the traditional recombinant DNA approach. New technological innovations such as auxotrophic expression technology make it possible to replace the selected site by Sec. Recently, several unnatural amino acids have been successfully incorporated into proteins with high fidelity and efficiency using the auxotrophic expression system. This is based on the hypothesis that an efficient charging of transfer ribonucleic acid with amino acid analog occurs when the amino acid is omitted and thus allowing the efficient biosynthetic substitution of amino acid by amino acid analog (Budisa, 2004).

Typically, many antioxidative selenoenzymes have been produced by incorporation of Sec into naturally occurring proteins using a cysteine auxotrophic expression system (Huang *et al.*, 2011). For this synergistic antioxidant microgel system, human heavy-chain ferritin (Fn) was chosen as a scaffold to construct the GPx mimic using the same method.

Using molecular docking, the recognition mechanism between Fn and GSH has been determined for the computational design of the active site. In the Fn-GSH complex model, GSH was located in a binding cleft of Fn surface with steric shape complementarity (Figure 33.5) and the side chain of Asn 25 was observed to make a close contact with the sulfhydryl group of GSH, suggesting that it's a perfect position to introduce the GPx catalytic moiety (Sec) for nucleophilic attack of GSH. Furthermore, Asn 109 can be mutated to Arg that forms salt bridges with GSH to enhance the substrate recognition. Fn is a superfamily of multisubunit proteins that can self-assemble into nanoscale cages. After the overexpression of Fn mutant, the designed Fn will aggregate together to generate a large, stable and highly symmetric protein architecture with multi-GPx active centers. Finally, an effective GPx mimic has been successfully designed and constructed.

33.2.2.2 The Construction of Synergistic Protein Microgels with Superoxide Dismutase and Glutathione Peroxidase Activities

SOD is a synergistic partner of GPx in antioxidant defense against oxidative stress. It is a well-studied metalloenzyme and has three major families distinguished by different metal cofactors: copper/zinc type, ferrum (Fe) and manganese types, and nickel type. In the active site of Mn-SOD, the manganese ions are coordinated by three histidines, an aspartic acid, and a water molecule or hydroxyl ligand. Based on the structural feature

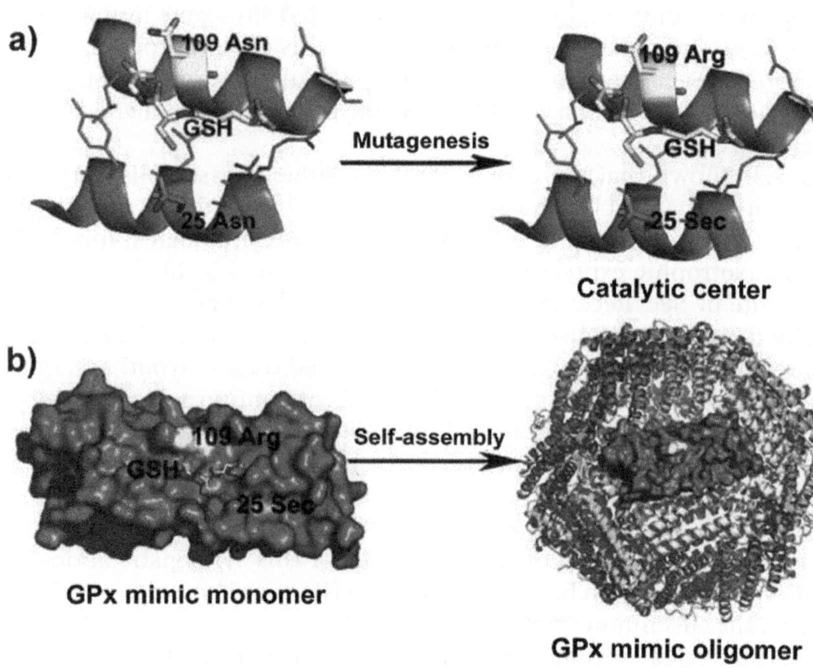

Figure 33.5 The design strategy to engineer ferritin with the glutathione peroxidase-like active center. (a) Computer simulation of ferritin to guide site-directed mutagenesis for the construction of the glutathione peroxidase mimic. Selenocysteine 25 is the catalytic residue for nucleophilic attack of glutathione, while arginine 109 is the recognition site that can stabilize the reactive intermediates by forming salt bridges with glutathione. (b) Selenium-containing ferritin monomers self-assemble into a nanostructure with multisubunits. Data are from Gao *et al.* (2013), with permission from Wiley-VCH Verlag GmbH & Co. KGaA, Weinheim.

of natural SOD enzymes, many Mn-porphyrins and their derivatives have been reported as SOD mimics with high stability and chemical versatility (Batinić-Haberle *et al.*, 2006).

For the purpose of combining GPx and SOD mimics in a microgel system, a manganese-tetrakis-4-hydroxyphenylporphyrin-poly(ethylene glycol)$_{2000}$-benzaldehyde (Mn-THPP-(PEG$_{2000}$-BA)$_4$) molecule was designed and synthesized. Besides the central Mn-porphyrin part to mimic the SOD enzyme, the molecule has also modified with four reactive functional groups (benzaldehyde) which can react with the lysine residues of Se-Fn to construct the protein-based antioxidant microgel system. The synthetic route of this SOD mimic is shown in Figure 33.6.

With the stoichiometric ratio between the aldehyde group of the porphyrin derivative and the lysine residue of Se-Fn was determined to be 1 : 1, the GPx-like Se-Fn and the SOD-like Mn-porpyrin derivative were crosslinked in aqueous solution under ultrasonication to form the synergistic antioxidant microgels, whose average diameters were determined to be ~270 nm by

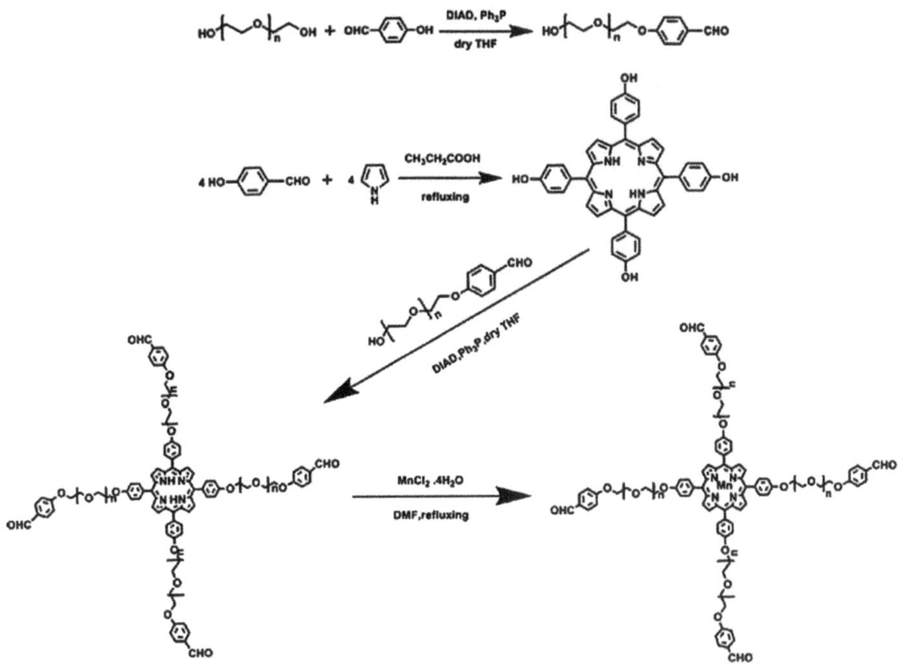

Figure 33.6 Synthetic route of the superoxide dismutase-like catalytic center. Manganese-tetrakis-4-hydroxyphenylporphyrin-poly(ethylene glycol)$_{2000}$-benzaldehyde was prepared by four steps: (I) synthesis of poly(ethylene glycol)$_{2000}$-benzaldehyde; (II) synthesis of tetrakis-4-hydroxyphenylporphyrin; (III) synthesis of tetrakis-4-hydroxyphenylporphyrin-poly(ethylene glycol)$_{2000}$-benzaldehyde; (IV) metallation of tetrakis-4-hydroxyphenylporphyrin-poly(ethylene glycol)$_{2000}$-benzaldehyde. Data are from Gao *et al.* (2013), with permission from Wiley-VCH Verlag GmbH & Co. KGaA, Weinheim.

scanning electron microscopy (SEM), atomic force microscopy (AFM), and dynamic light scattering (DLS) (Figure 33.7).

33.2.2.3 *The Catalytic Behavior of the Synergistic Antioxidant Microgels*

The antioxidant properties of individual components of the microgels and their possible synergistic associations were evaluated, respectively. In the GPx analyzing system, the catalytic reaction rates of the antioxidant microgel catalyst were measured to be 257.6 μmol min^{-1} μmol^{-1} using natural substrates GSH and H_2O_2. Under the same condition, however, the well-known GPx mimic ebselen (Sies, 1993) exhibited the activity of 1.02 μmol min^{-1} μmol^{-1}, indicating that the designed microgels exhibit high GPx activity.

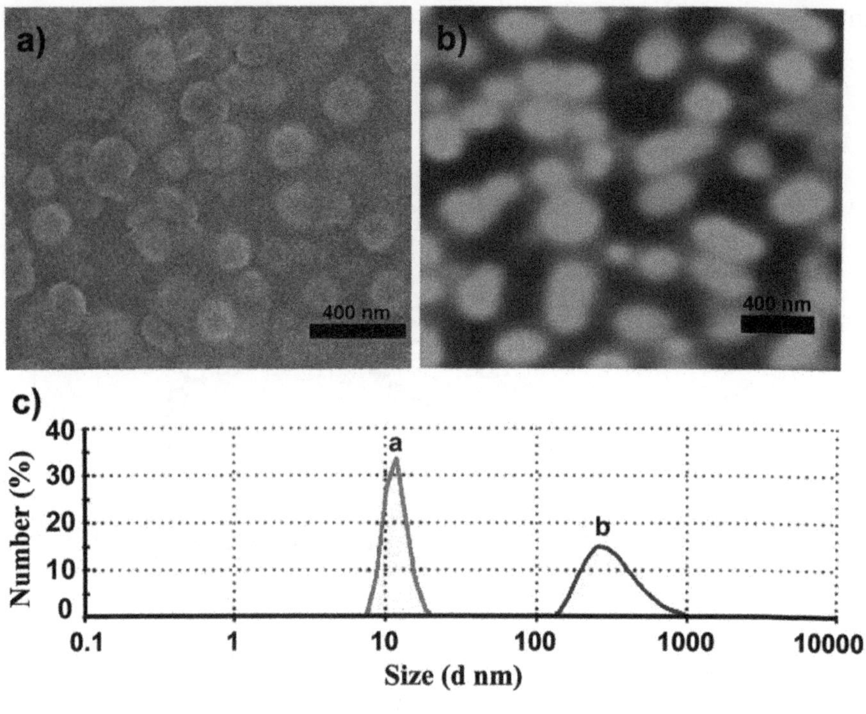

Figure 33.7 Scanning electron microscopy and atomic force microscopy characterization of the synergistic antioxidant microgels. (a) Scanning electron microscopy and (b) atomic force microscopy images show that the selenium containing ferritin cages have been crosslinked into microgels by Mn-porpyrin derivatives. (c) The average size distribution of (a) selenium-containing ferritin and (b) selenium-containing ferritin/manganese-tetrakis-4-hydroxyphenylporphyrin-poly(ethyleneglycol)$_{2000}$-benzaldehyde composites are 12 nm and 270 nm, respectively. Data are from Gao *et al.* (2013), with permission from Wiley-VCH Verlag GmbH & Co. KGaA, Weinheim.

The SOD activity was determined by spectrophotometric analysis of the reduction of nitro blue tetrazelium to blue formazane in a standard xanthine/xanthine oxidase (XOD) assay system (McCord and Fridovich, 1969). The reduction rate is linearly related to the XOD activity and is inhibited by SOD. Therefore, the SOD activity can be defined as the SOD concentration that is required for 50% inhibition (IC_{50}) of the reaction. In the case of the antioxidant microgel system, its IC_{50} value was found to be 0.217 mM, which approaches 1.37% of the native SOD activity and achieves the same level of the previously reported SOD mimics (Yu *et al.*, 2010). This means that the designed SOD active sites are not affected by the formation of microgels.

In order to evaluate the synergistic antioxidant effect of the designed antioxidant microgels, mitochondria swelling and mitochondrial lipid peroxidation were selected as standards. The analytical system includes a free-radical damage system (xanthine/XOD) that contains a mixture of superoxide anions, H_2O_2 and OH and bovine heart mitochondria isolated from fresh bovine heart. The swelling level of mitochondria was characterized by UV absorption at 520 nm. The experimental results indicated that no single mimic is effective for all stages to provide continuous protection of mitochondria. In the synergistic antioxidant microgels, however, GPx mimics and SOD mimics can reinforce each other by cooperative effects for long-term, effective protection of the cellular components against oxidative damage.

The lipid peroxidation was measured by assessing the level of malondialdehyde (MDA), the end product of the oxidative process. Compared to the standard damaged system that contains xanthine/XOD/Fe^{2+}, the MDA level has decreased by 3.9%, 19.7%, 37.1%, and 46.2% after adding single SOD mimic, ebselen, GPx mimic and the synergistic antioxidant microgels, respectively. This result suggests that the synergistic antioxidant microgel is more effective than single GPx or SOD mimic in the protection of mitochondria lipid peroxidation.

In conclusion, microgels offer a good platform to construct different mimic components with multifunctional antioxidant activities and support a synergistic effect between them to scavenge most kinds of ROS for enhanced antioxidant ability.

33.3 Summary and Perspective

In this chapter, we describe the biochemical role of Se and the development of Se-related antioxidant system based on various scaffolds, especially microgel scaffolds. Two typical studies that use the biomimetic design strategies in combination with versatile microgel properties to construct advanced antioxidant microgels for intelligent or synergistic catalysis were highlighted. However, the growing need for some special applications will require higher demands on the efficiencies and properties of antioxidant microgel system. How to further push forward this field is still a challenge. The recent emergence of various sophisticated materials may provide the possibility to construct antioxidant microgels with new properties, while more flexible and innovative design strategies are urgently needed to maximize the potential of microgels. It is anticipated that the designed microgels could act as potent antioxidant in the treatment of human diseases and have practical applications in biomedical field.

Summary Points

- This chapter focuses on selenium and antioxidant microgels.
- Selenium is a trace element and an essential nutrient for human health.
- Selenium exerts its antioxidant effect *in vivo* mainly through selenoenzymes.
- However, the practical application of selenoenzymes is limited due to their intrinsic disadvantages.
- Therefore, artificial antioxidants were developed as a new source to replace natural selenoenzymes.
- In recent years, a variety of molecular scaffolds were exploited to incorporate selenium.
- Microgels offer a lot of advantages to construct artificial antioxidants compared with other molecular scaffolds.
- The biomimetic strategies that combine recognition and catalysis were utilized in the design of antioxidant microgels.
- The versatile properties of microgels endow artificial antioxidants with particular abilities for biomedical applications.
- Two selected antioxidant microgel systems for intelligent or synergistic catalysis were highlighted.

Key Facts

Key Facts of Antioxidants

1. Oxidative stress occurs when free radicals are overproduced and are unable to be adequately neutralized by the body's antioxidant defense system.
2. Free radicals are chemically active atoms or molecular fragments with unpaired electrons.
3. Reactive oxygen species are the most biologically significant free radicals that contain the element oxygen.
4. Reactive oxygen species include singlet oxygen, superoxide radical anion, hydrogen peroxide, hydroxyl radical, lipid peroxide and nitric oxide.
5. Reactive oxygen species react with important cellular components such as deoxyribonucleic acid, proteins, and cell membrane to cause oxidative damage and cell death.
6. Antioxidants are nutrients that inhibit the oxidative damage to human tissues.
7. Antioxidants are involved in the prevention of the pathway of aging-related diseases.

Key Facts of Enzyme Assay

1. Enzyme assays are experimental protocols for measuring enzymatic activity.
2. Enzyme activity is measured as a function of substrate concentration to obtain the kinetic constants.
3. In the glutathione peroxidase-like enzyme assay system, the initial rates of the reduction of cumene hydroperoxide by 3-carboxy-4-nitrobenzenethiol in the presence of the glutathione peroxidase mimics were determined by monitoring the ultraviolet absorption of 3-carboxy-4-nitrobenzenethiol at 410 nm.
4. To illustrate the substrate selectivity for glutathione peroxidase mimics, *p*-nitrobenzenethiol was also utilized as substrate for crosscomparison.
5. In the synergistic antioxidant system, the glutathione peroxidase-like enzymatic activity is measured indirectly by monitoring the ultraviolet absorption of nicotinamide adenine dinucleotide phosphate at 340 nm, in which glutathione and hydrogen peroxide were utilized as substrates and the produced glutathione disulfide was recycled back to glutathione by nicotinamide adenine dinucleotide phosphate catalyzed by glutathione reductase.

Definitions of Words and Terms

Reactive oxygen species (ROS). ROS are highly active oxygen free radicals generated as byproducts of cellular metabolism. They are essential for cell signaling and homeostasis. However, when ROS are overproduced and exceed the capacity of antioxidant defense system, they will cause oxidative stress and cell death, resulting in the progression of a series of human diseases such as cancer, heart disease, and the acquired immunodeficiency syndrome *etc.* This dual property of ROS depends on a fine regulation of its concentration.

Selenocysteine (Sec). Sec is a cysteine analog with a selenol group instead of the thiol group of cysteine, known as the 21st amino acid. It is encoded by a stop codon and undergoes a very complicated biosynthetic pathway to be inserted into selenoproteins that are involved in antioxidant activity to protect cell from oxidative stress.

Glutathione (GSH). GSH is a natural tripeptide that is composed of three amino acids: glutamic acid, cysteine, and glycine. It has the antioxidant activity regulated by the reduction of cysteine thiol and acts as an important antioxidant to protect cellular machines from oxidative stress.

Glutathione peroxidase (GPx). GPx is a selenoenzyme family with peroxidase activity. It catalyzes the reduction of lipid hydroperoxides by glutathione. Selenocysteine is an important constituent and is responsible for the biological function of GPx.

Tellurium (Te). Te is a chemical element from Group VI of the periodic table. It has a lower redox potential than Se, and thus is often utilized in the fabrication of glutathione peroxidase mimics for high catalytic efficiency.

Superoxide dismutase (SOD). SOD is an important member of the antioxidant defense system. It catalyzes the dismutation of superoxide radical anion into molecular oxygen and hydrogen peroxide and functions in a synergistic way with glutathione peroxidase to scavenge reactive oxygen species.

Microgels. Microgels are nano/microsized colloidal particles that consist of crosslinked polymer networks. They exhibit the properties of both colloidal and polymer-like characteristics and are extensively exploited in biomedical field due to their versatile properties and unique three-dimensional structures.

Poly(*N*-isopropylacrylamide) (PNIPAM). PNIPAM is a thermosensitive polymer that undergoes a reversible phase transition from hydrophilic to hydrophobic at physiological temperature and is widely used to construct "smart" microgels.

Lower critical solution temperature (LCST). LCST is the temperature below which all the components in the system are totally miscible. Poly(*N*-isopropylacrylamide) exhibits a typical reversible phase-volume transition at the LCST of 32 °C.

Auxotrophic expression system. The auxotrophic expression system is an efficient protocol utilized to incorporate unnatural amino acid into proteins, based on the hypothesis that biosynthetic substitution of an amino acid by an amino acid analog occurs when the amino acid is omitted.

List of Abbreviations

Se	Selenium
GPx	Glutathione peroxidase
Sec	Selenocysteine
ROS	Reactive oxygen species
GSH	Glutathione
ESeH	Selenol
ESeOH	Selenenic acid
E-SeSG	Selenenyl sulfide
GSSG	Glutathione disulfide
DNA	Deoxyribonucleic acid
3D	Three-dimensional
Te	Tellurium
PNIPAM	Poly(*N*-isopropylacrylamide)
LCST	Lower critical solution temperature
MBA	Methylene(bis)acrylamide
CUOOH	Cumene hydroperoxide
TNB	3-Carboxyl-4-nitrobenzenethiol
NBT	*p*-Nitrobenzenethiol
SOD	Superoxide dismutase
Mn	Manganese

Fe	Ferrum
O_2^-	Superoxide radical anion
O_2	Molecular oxygen
H_2O_2	Hydrogen peroxide
H_2O	Water
Fn	Ferritin
Se-Fn	Engineered ferritin with selenium
SEM	Scanning electron microscopy
AFM	Atomic force microscopy
DLS	Dynamic light scattering
XOD	Xanthine oxidase
IC_{50}	50% inhibition concentration
MDA	Malondialdehyde
Mn-THPP-(PEG_{2000}-BA)$_4$	Manganese-tetrakis-4-hydroxyphenylporphyrin-poly(ethylene glycol)$_{2000}$-benzaldehyde

References

Antonietti, M., Gröhn, F., Hartmann, J. and Bronstein, L., 1997. Nonclassical shapes of noble-metal colloids by synthesis in microgel nanoreactors. *Angewandte Chemie International Edition*. 36: 2080–2083.

Allmang, C., Wurth, L. and Krol, A., 2009. The selenium to selenoprotein pathway in eukaryotes: more molecular partners than anticipated. *Biochimica et Biophysica Acta*. 1790: 1415–1423.

Budisa, N., 2004. Prolegomena to future experimental efforts on genetic code engineering by expanding its amino acid repertoire. *Angewandte Chemie International Edition*. 43: 6426–6463.

Batinić-Haberle, I., Spasojević, I., Stevens, R. D., Bondurant, B., Okado-Matsumoto, A., Fridovich, I., Vujasković, Z. M. and Dewhirst, W., 2006. New PEG-ylated Mn(III) porphyrins approaching catalytic activity of SOD enzyme. *Dalton Transactions*. 4: 617–624.

Chen, J., 2012. An original discovery: selenium deficiency and Keshan disease (an endemic heart disease). *Asia Pacific Journal of Clinical Nutrition*. 21: 320–326.

Engman, L., Stern, D., Cotgreave, I. A. and Andersson, C. M., 1992. Thiol peroxidase activity of diaryl ditellurides as determined by a ^1H NMR method. *Journal of the American Chemical Society*. 114: 9737–9743.

Flohé, L., Gunzler, W. A. and Schock, H. H., 1973. Glutathione peroxidase: a selenoenzyme. *FEBS Letters*. 32: 132–134.

Favier, A., Sappey, C., Leclerc, P., Faure, P. and Micoud, M., 1994. Antioxidant status and lipid peroxidation in patients infected with HIV. *Chemico-Biological Interactions*. 91: 165–180.

Fernández-Nieves, A. and Márquez, M., 2005. Electrophoresis of ionic microgel particles: from charged hard spheres to polyelectrolyte-like behavior. *Journal of Chemical Physics*. 122: 84702–84702.

Gao, Y. Z., Hou, C. X., Zhou, L. P., Zhang, D. M., Zhang, C. Q., Miao, L., Wang, L., Dong, Z. Y., Luo, Q. and Liu, J. Q., 2013. A dual enzyme microgel with high antioxidant ability based on engineered seleno-ferritin and artificial superoxide dismutase. *Macromolecular Bioscience*. 13: 808–816.

Hoare, T. and Pelton, R., 2004. Highly pH and temperature-responsive microgels functionalized with vinylacetic acid. *Macromolecules*. 37: 2544–2550.

Huang, X., Yin, Y. Z., Tang, Y., Bai, X. L., Zhang, Z. M., Xu, J. Y., Liu, J. Q. and Shen, J. C., 2009. Smart microgel catalyst with modulatory glutathione peroxidase activity. *Soft Matter*. 5: 1905–1911.

Huang, X., Liu, X. M., Luo, Q., Liu, J. Q. and Shen, J. C., 2011. Artificial selenoenzymes: designed and redesigned. *Chemical Society Reviews*. 40: 1171–1184.

Jia, X., Yeo, Y., Clifton, R. J., Jiao, T., Kohane, D. S., Kobler, J. B., Zeitels, S. M. and Langer, R., 2006. Hyaluronic acid-based microgels and microgel networks for vocal fold regeneration. *Biomacromolecules*. 7: 3336–3344.

Loft, S. and Poulsen, H. E., 1996. Cancer risk and oxidative DNA damage in man. *Journal of Molecular Medicine (Berlin)*. 74: 297–312.

Lu, J. and Holmgren, A., 2009. Selenoproteins. *Journal of Biological Chemistry*. 284: 723–727.

Liu, X. M., Silks, L. A., Liu, C. P., Ollivault-Shiflett, M., Huang, X., Li, J., Luo, G. M., Hou, Y. M., Liu, J. Q. and Shen, J. C., 2009. Incorporation of tellurocysteine into glutathione transferase installs high glutathione peroxidase efficiency. *Angewandte Chemie International Edition*. 48: 2020–2023.

McCord, J. M. and Fridovich, I., 1969. Superoxide dismutase: an enzymic function for erythrocuprein (hemocuprein). *Journal of Biological Chemistry*. 244: 6049–6055.

McPhee, W. K., Tam, C. and Pelton, R., 1993. Poly(*N*-isopropylacrylamide) latices prepared with sodium dodecyl sulfate. *Journal of Colloid and Interface Science*. 156: 24–30.

Neyret, S. and Vincent, B., 1997. The properties of polyampholyte microgel particles prepared by microemulsion polymerization. *Polymer*. 38: 6129–6134.

Rayman, M. P., 2008. Food-chain selenium and human health: emphasis on intake. *British Journal of Nutrition*. 100: 254–268.

Schwarz, K. and Foltz, C. M., 1957. Selenium as an integral part of Factor 3 against dietary necrotic liver degeneration. *Journal of the American Chemical Society*. 79: 3292–3293.

Salonen, J. T. and Huttunen, J. K., 1986. Selenium in cardiovascular diseases. *Annals of Clinical Research*. 18: 30–35.

Sies, H., 1993. Ebselen, a selenoorganic compound as glutathione peroxidase mimic. *Free Radical Biology and Medicine*. 14: 313–323.

Taylor, A., 1993. Cataract: relationship between nutrition and oxidation. *Journal of the American College of Nutrition*. 12: 138–146.

Wang, Q., Zhao, Y. B., Yang, Y. J., Xu, H. B. and Yang, X. L., 2007. Thermosensitive phase behavior and drug release of in situ gelable poly(*N*-isopropylacrylamide-co-acrylamide) microgels. *Colloid and Polymer Science*. 285: 515–521.

Xu, J., Zeng, F., Wu, S., Liu, X., Hou, C. and Tong, Z., 2007. Gold nanoparticles bound on microgel particles and their application as an enzyme support. *Nanotechnology*. 18: 265704–265712.

Yu, S. J., Yin, Y. Z., Zhu, J. Y., Huang, X., Luo, Q., Xu, J. Y., Shen, J. C. and Liu, J. Q., 2010. A modulatory bifunctional artificial enzyme with both SOD and GPx activities based on a smart star-shaped pseudo-block copolymer. *Soft Matter*. 6: 5342–5350.

Subject Index

Illustrations and figures are in **bold**. Tables are in *italics*.